			13	14	15	16	17	18
								2 He 4.00
			5 B 10.81	6 C 12.01	7 N 14.01	8 O 16.00	9 F 19.00	10 Ne 20.18
10	11	12	13 Al 26.98	14 Si 28.09	15 P 30.97	16 S 32.06	17 Cl 35.45	18 Ar 39.95
28 Ni 58.71	29 Cu 63.54	30 Zn 65.37	31 Ga 69.72	32 Ge 72.59	33 As 74.92	34 Se 78.96	35 Br 79.91	36 Kr 83.80
46 Pd 106.4	47 Ag 107.87	48 Cd 112.40	49 In 114.82	50 Sn 118.69	51 Sb 121.75	52 Te 127.60	53 I 126.90	54 Xe 131.30
78 Pt 195.09	79 Au 196.97	80 Hg 200.59	81 Tl 204.37	82 Pb 207.19	83 Bi 208.98	84 Po 210	85 At 210	86 Rn 222
110 Uun	111 Uuu	112 Uub						

←— Metals | Nonmetals —→

Semi-metals

| 62
Sm
150.35 | 63
Eu
151.96 | 64
Gd
157.25 | 65
Tb
158.92 | 66
Dy
162.50 | 67
Ho
164.93 | 68
Er
167.26 | 69
Tm
168.93 | 70
Yb
173.04 | Lanthanoids |
| 94
Pu
239.05 | 95
Am
241.06 | 96
Cm
247.07 | 97
Bk
249.08 | 98
Cf
251.08 | 99
Es
254.09 | 100
Fm
257.10 | 101
Md
258.10 | 102
No
255 | Actinoids |

Descriptive
Inorganic Chemistry

THIRD EDITION

Geoff Rayner-Canham

Sir Wilfred Grenfell College
Memorial University

Tina Overton

University of Hull

W. H. FREEMAN AND COMPANY
New York

An Inorganic Chemist's Periodic Table
© Geoff Rayner-Canham, 2002

The major links in the Periodic Table, as shown on the cover, are those of the Groups and Periods. In addition, there are other patterns:
The (n) and ($n + 10$) groups linkages (grey)
The diagonal relationships (green)
The "knights move" relationships (tan)
The aluminum–iron link (red)
The lanthanoid and actinoid relationships (grey)
The "combo" elements (violet)
The "pseudo" elements (blue)
See Chapter 9 for details.

Acquisitions Editor: Jessica Fiorillo
Developmental and New Media and Supplements Editor: Guy Copes
Marketing Manager: Mark Santee
Text Design: Cambraia Fernandes
Production Coordinator: Susan Wein
Composition: TechBooks
Manufacturing: Hamilton Printing Company

Library of Congress Cataloging-in-Publication Data

Rayner-Canham, Geoffrey.
   Descriptive inorganic chemistry / Geoff Rayner-Canham, Tina Overton.– 3rd ed.
     p. cm.
   Includes index.
   ISBN 0-7167-4620-4
   1. Chemistry, Inorganic. I. Overton, Tina. II. Title.
QD151 .5 .R39 2002
546—dc21   2002072594

Printed in the United States of America

First printing 2002

W. H. Freeman and Company
41 Madison Avenue, New York, NY 10010
Houndmills, Basingstoke RG21 6XS, England

Overview

Contents

Chapter 15   The Group 15 Elements   323

Chapter 21   The Group 12 Elements   523

Chapter 22   Organometallic Chemistry   534

Color Insert

Applications of Inorganic Chemistry
Memory Metal: The Shape of Things to Come
Chemosynthesis: Redox Chemistry on the Sea Floor
Concrete: An Old Material with a New Future
New Minerals: Going Beyond the Limitations of Geochemistry
Cosmochemistry: Io, the Sulfur-Rich Moon
Technetium: A Rare Element with an Important Medical Use
Glass: Ancient Bottles and Modern Lenses

What Is Descriptive Inorganic Chemistry?

Descriptive inorganic chemistry was traditionally concerned with the properties of the elements and their compounds. Now, in the renaissance of the subject, the properties are being linked with explanations for the formulas and structures of compounds together with an understanding of the chemical reactions they undergo. In addition, we are no longer looking at inorganic chemistry as an isolated subject but as a part of essential scientific knowledge with applications throughout science and our lives. And it is because of a need for greater contextualization that we have added more features and more applications.

In many colleges and universities, descriptive inorganic chemistry is offered as a sophomore or junior course. In this way, students come to know something of the fundamental properties of important and interesting elements and their compounds. Such knowledge is important for careers not only in pure or applied chemistry but also in pharmacy, medicine, geology, environmental science, and so on. This course can then be followed by a junior or senior course that focuses on the theoretical principles and the use of spectroscopy to a greater depth than is covered in a descriptive text. In fact, the theoretical course builds nicely on the descriptive background. Without the descriptive grounding, however, the theory becomes sterile, uninteresting, and irrelevant.

Education has often been a case of the "swinging pendulum" and this has been very true of inorganic chemistry. Up until the 1960s, it was very much pure descriptive, requiring exclusively memorization. In the 1970s and 1980s, upper-level texts focused exclusively on the theoretical principles. Now it is apparent that descriptive is very important—but not the traditional memorization of facts, but the linking of facts, where possible, to underlying principles. Students need to have modern descriptive inorganic chemistry as part of their education. Thus, we must ensure that chemists are aware of the "new descriptive inorganic chemistry."

Preface

Inorganic chemistry goes beyond academic interest; it is an important part of our lives.

Inorganic chemistry is interesting—more than that—it is exciting! So much of our 21st century science and technology relies on natural and synthetic materials–often inorganic compounds. Inorganic chemistry is ubiquitous in our daily lives: household products; some pharmaceuticals; our transportation—both the vehicles themselves and the synthesis of the fuels; battery technology; and medical treatments. There is the industrial aspect, the production of all the chemicals that are required to drive our economy, everything from steel to sulfuric acid to glass and cement. Environment chemistry is largely a question of the inorganic chemistry of the atmosphere, water, and soil. Finally, there are the profound issues of the inorganic chemistry of our planet, the solar system, and of the universe.

This text is designed to focus on the properties of selected interesting, important, and unusual elements and compounds. However, to understand inorganic chemistry, it is crucial to tie this knowledge to the underlying chemical principles and hence provide explanations for the existence and behavior of compounds. For this reason, almost half the chapters survey the relevant concepts of atomic theory, bonding, intermolecular forces, thermodynamics, acid–base behavior, and reduction-oxidation properties as a prelude to and preparation for the descriptive material.

For this third edition, the major improvements are as follows:

A Chapter on Periodic Patterns. To provide a basis for the subsequent chapters of each group, there is now an overview chapter that systematically reviews trends in groups and then bonding and acid–base trends in periods. We then introduce other relationships: the links between the (n) and ($n + 10$) groups, isoelectronic similarities, the diagonal relationship, isomorphism among ionic compounds, and other patterns.

A Chapter on Organometallic Chemistry. In response to requests, the coverage of this important branch of inorganic chemistry has been expanded to a whole chapter. The chapter follows exactly the same style as the rest of the text, with a logical development and appropriate contextual material. Thus, the topic can be either treated as an independent unit or blended with the coverage of each Group in the earlier chapters.

Expanded Coverage of Bioinorganic Chemistry. The end of chapter sections on the biological roles of elements has been expanded, particularly those of arsenic, bismuth, boron, bromine, chlorine, chromium, copper, fluorine, iodine, iron, lead, phosphorus, selenium, silicon, and sulfur.

Expanded Coverage of Environmental Chemistry. Sections or features on the following topics have been added: *Crystal Structures and Nuclear Waste Disposal; Ionic Liquids; Water: The New Wonder Solvent; The Hydrogen Economy; Cyanide and Tropical Fish; Green Chemistry; Sequestration of Carbon Dioxide;* and *Enriched and Depleted Uranium.*

Expanded Coverage of Materials Chemistry. Sections or features on the following topics have been added: *Lithium Batteries; Biomineralization: A New Interdisciplinary "Frontier"; Borides; Carbon Nanotubes; Moissanite: The Diamond Substitute; Inorganic Polymers; the Pentanitrogen Cation;* and *New Pigments through Perovskites.*

Expanded Coverage of Medicinal Chemistry. Features on the following topics have been added: *Medicinal Inorganic Chemistry: An Introduction; Antacids; Boron Neutron Capture Therapy; The Fluoridation of Water;* and *Technetium: The Most Important Radiopharmaceutical.*

Expanded Coverage of Geochemistry and Cosmochemistry. Features on the following topics have been added: *Searching the Depths of Space for the Trihydrogen Ion; Is There Life Elsewhere in Our Solar System;* and *The Earth and Crystal Structures.*

Increased Number of End-of-Chapter Problems. More problems have been added, bringing the total up to close to 1000.

Revision of Element Reaction Flowcharts. Several of the element flowcharts have been redrawn to endeavor, where possible, to follow a similar format for elements in a group.

Web-Based Material. Resources will be increasingly web-based, and supplementary material to this text will be found at: www.whfreeman.com/raynercanham. To reduce the length of the text, the *Additional Readings, Chapter 23,* and some of the *Appendix* material will be found at this site. A set of *Test-Tube Experiments* keyed to the text can be found at this address. *Answers to Alternate Text Questions* and other supporting material will also be posted to this site.

Solutions Manual. Accompanying the text will be a *Student Solutions Manual* (0716793849), containing solutions to all the odd-numbered problems, and an *Instructor's Solutions Manual* (0716793830) containing solutions to all of the even-numbered problems.

Finally, we are delighted to add an outstanding and gifted co-author to this (and future) editions of the text: Dr. Tina Overton, Senior Lecturer in inorganic chemistry at the University of Hull, England, who brings with her a strong background in chemical education. She has exactly the right mix of experience to ensure that this text remains the leader of its type in the world and a trend setter for the 21st century.

This book was written to pass on to another generation our fascination with descriptive inorganic chemistry. Thus, the comments of the readers, both students and instructors, will be sincerely appreciated. Any suggestions for added or updated additional readings would also be welcome. Our current e-mail addresses are: grcanham@swgc.mun.ca and T.L.Overton@hull.ac.uk.

Acknowledgments

Many thanks must go to the team at W.H. Freeman who have contributed their talents to the three editions of this book. We offer our sincere gratitude to our editors of the 3rd edition, Jessica Fiorillo and Guy Copes; of the 2nd edition, Michelle Julet and Mary Louise Byrd; and a special thanks to Deborah Allen, who bravely commissioned the 1st edition of the text. Each one of our fabulous editors has been a source of encouragement, support, and helpfulness.

We wish to acknowledge the following reviewers of this edition, whose criticisms and comments were much appreciated: François Caron at Laurentian University; Thomas D. Getman at Northern Michigan University; Janet R. Morrow at the State University of New York at Buffalo; Robert D. Pike at the College of William and Mary; Michael B. Wells at Cambell University; and particularly Joe Takats of the University of Alberta for his comprehensive critique of the 2nd edition. The contributions of the reviewers of the 2nd edition are gratefully acknowledged: F.C. Hentz at North Carolina State University; Richard B. Kaner at the University of California, Los Angeles; Michael D. Johnson at New Mexico State University; Richard H. Langley at Stephen F. Austin State University; James M. Mayer at the University of Washington; Jon Melton at Messiah College; Joseph S. Merola at Virginia Technical Institute; David Phillips at Wabash College; John R. Pladziewicz at the University of Wisconsin, Eau Claire; Daniel Rabinovich at the University of North Carolina at Charlotte; David F. Reich at Salisbury State University; Todd K. Trout at Mercyhurst College; Steve Watton at the Virginia Commonwealth University; and John S. Wood at the University of Massachusetts, Amherst. Likewise, we thank the reviewers of the 1st edition: E. Joseph Billo at Boston College; David Finster at Wittenberg University; Stephen J. Hawkes at Oregon State University; Martin Hocking at the University of Victoria; Vake Marganian at Bridgewater State College; Edward Mottel at the Rose-Hulman Institute of Technology; and Alex Whitla at Mount Allison University.

As a personal acknowledgment, Geoff Rayner-Canham wishes to especially thank three teachers and mentors who played a major influence in his career: Briant Bourne, Harvey Grammar School; Margaret Goodgame, Imperial College, London University; and Derek Sutton, Simon Fraser

University. And he expresses his eternal gratitude to his spouse, Marelene, for her support and encouragement.

    Tina Overton would like to thank her colleague Phil King for his invaluable suggestions for improvements and his assistance with the illustrations. Thanks must also go to her family, Dave, John, and Lucy, for their patience during the months when this project filled all her waking hours.

Dedication

Chemistry is a human endeavor. New discoveries are the result of the work of enthusiastic individuals and groups of individuals who want to explore the molecular world. We hope that you, the reader, will come to share our own fascination with inorganic chemistry. We have chosen to dedicate this book to two persons who, for very different reasons, never did receive the ultimate accolade of a Nobel Prize.

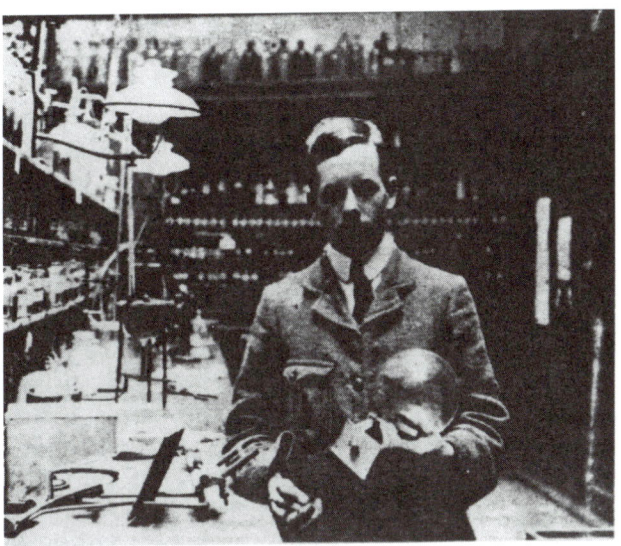

Henry Moseley (1887–1915)

Though Mendeléev is identified as the discoverer of the Periodic Table, his version was based on an increase in atomic mass. In some cases, the order of elements had to be reversed to match properties with location. It was a British scientist, Henry Moseley, who put the Periodic Table on a much firmer footing by discovering that, upon bombardment with electrons, each element emitted X-rays of characteristic wavelengths. The wavelengths fitted a formula related by an integer number unique to each element. We now know that number to be the number of protons. With the establishment of the atomic number of an element, chemists at last knew the fundamental organization of the Table. Sadly, Moseley was killed at the battle of Gallipoli in the First World War. Thus, one of the brightest scientific talents of the 20th century died at the age of 27. The famous American scientist, Robert Milliken, commented: "Had the European War had no other result than the snuffing out of this young life, that alone would make it one of the most hideous and most irreparable crimes in history." Unfortunately, Nobel Prizes are only awarded to living scientists. In 1924, there was the claim of the discovery of element 43, and it was named moseleyum, however, the claim was disproved by the very method that Moseley had pioneered.

Lise Meitner (1878–1968)

In the 1930s, scientists were bombarding atoms of heavy elements such as uranium with subatomic particles to try to make new elements and extend the Periodic Table. Austrian scientist, Lise Meitner, had shared leadership with Otto Hahn of the German research team working on the synthesis of new elements. They thought they had discovered nine new elements. Shortly after the claimed discovery, Meitner was forced to flee Germany because of her Jewish ancestory, and she settled in Sweden. Hahn reported to her that one of the new elements behaved chemically just like barium. During a famous "walk in the snow" with her nephew, the physicist Otto Frisch, Meitner realized that an atomic nucleus could break in two just like a drop of water. No wonder the element formed behaved like barium: it was barium! Thus was born the concept of nuclear fission. She informed Hahn of her proposal. When Hahn wrote the research paper on the work, he barely mentioned the vital contribution of Meitner and Frisch. As a result, Hahn and his colleague, Fritz Strassmann, received the Nobel Prize. Meitner's flash of genius was ignored. Only recently has Meitner received the acclaim she deserved by naming an element after her, element 109, meitnerium.

Additional reading

Heilbron, J.L., *H.G.J. Moseley*, University of California Press, Berkeley, CA, 1974.

Rayner-Canham, M.F., and Rayner-Canham, G.W. *Women in Chemistry: Their Changing Roles from Alchemical Times to the Mid-Twentieth Century*, Chemical Heritage Foundation, Philadelphia, PA, 1998.

Sime, R.L., *Lise Meitner: A Life in Physics*, University of California Press, Berkeley, CA, 1996.

Weeks M.E., and Leicester, H.M. *Discovery of the Elements*, Journal of Chemical Education, Easton, PA, 7th edition, 1968.

The Electronic Structure of the Atom: A Review

Chapter 1

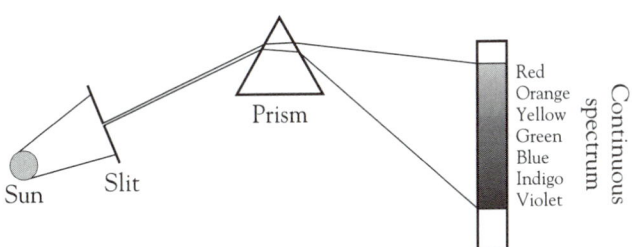

p_x p_y p_z

To understand the behavior of inorganic compounds, we need to study the nature of chemical bonding. Bonding, in turn, relates to the behavior of electrons in the constituent atoms. Our study of inorganic chemistry, therefore, starts with a review of the probability model of the atom and a survey of the model's applications to the electron configurations of atoms and ions.

It is amazing that Isaac Newton discovered anything at all, for he was the original model for the absent-minded professor. Supposedly, he always timed the boiled egg he ate at breakfast; one morning, his maid found him standing by the pot of boiling water, holding an egg in his hand and gazing intently at the watch in the bottom of the pot! Nevertheless, he initiated the study of the electronic structure of the atom in about 1700, when he noticed that the passage of sunlight through a prism produced a continuous visible spectrum (Figure 1.1). Much later, in 1860, Robert Bunsen (of burner fame) investigated the light emissions from flames and gases. Bunsen observed that the emission spectra, rather than being continuous, were series of colored lines (line spectra). He noted that each chemical element produced a unique and characteristic spectrum (Figure 1.2). Other investigators subsequently showed that there were, in fact, several sets of spectral lines for the hydrogen atom: one set in the ultraviolet region, one in the visible region, and a number of sets in the infrared part of the electromagnetic spectrum (Figure 1.3).

Figure 1.1 A prism splits white light into the wavelengths of the visible spectrum.

Sun — Slit — Prism

Red
Orange
Yellow
Green
Blue
Indigo
Violet

Continuous spectrum

1

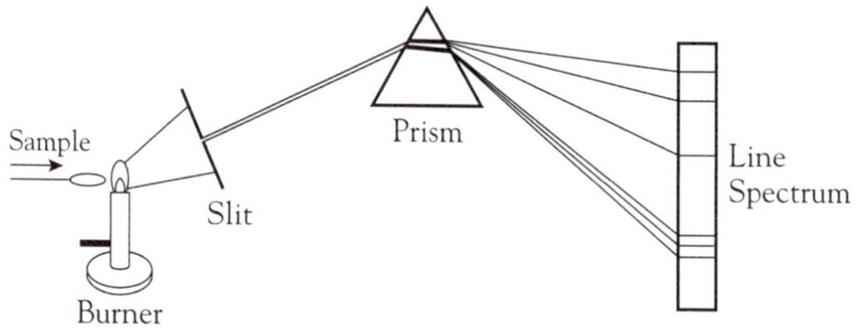

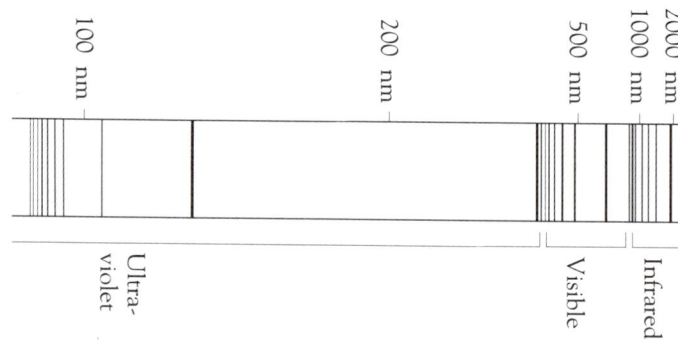

Figure 1.2 A line spectrum is produced when an element is heated in a flame.

Figure 1.3 Emission spectrum of hydrogen.

The explanation of the spectral lines was one of the triumphs of the *Bohr model* of the atom. In 1913, Niels Bohr proposed that the electrons could occupy only certain energy levels that corresponded to various circular orbits around the nucleus. He identified each of these levels with an integer, which he called a *quantum number*. The value of this parameter could range from 1 to ∞. He argued that, as energy was absorbed by an atom from a flame or electrical discharge, electrons moved from one quantum level to one or more higher energy levels. The electrons sooner or later returned to lower quantum levels, closer to the nucleus; and when they did, light was emitted. The wavelength of the emitted light directly corresponded to the energy of separation of the initial and final quantum levels. When the electrons occupied the lowest possible energy level, they were said to be in the *ground state*. If one or more electrons absorbed enough energy to move away from the nucleus, then they were said to be in an *excited state*.

The energy of an electron in each level could be found from the relationship:

$$E = -\mathrm{R_H}\left(\frac{1}{n^2}\right)$$

where E is the electron energy, n is the quantum number, and $\mathrm{R_H}$ is the Rydberg constant for hydrogen. The energy of the light emitted could be calculated from the difference in energies of the initial and final energy levels. The wavelength of the light could then be found from the relationships $E = h\nu$ and $c = \lambda\nu$ where h is Planck's constant, ν is the frequency, and λ is the wavelength of the emitted light.

However, the Bohr model had a number of flaws. For example, the spectra of multielectron atoms had far more lines than the simple Bohr model predicted. Nor could the Bohr model explain the splitting of the spectral

Atomic Absorption Spectroscopy

A glowing body, such as the Sun, is expected to emit a continuous spectrum of electromagnetic radiation. However, in the early nineteenth century, a German scientist, Josef von Fraunhofer, noticed that the visible spectrum from the Sun actually contained a number of dark bands. Later investigators realized that the bands were the result of the absorption of particular wavelengths by cooler atoms in the "atmosphere" above the surface of the Sun. The electrons of these atoms were in the ground state, and they were absorbing radiation at wavelengths corresponding to the energies needed to excite them to higher energy states. A study of these "negative" spectra led to the discovery of helium. Such spectral studies are still of great importance in cosmochemistry—the study of the chemical composition of stars.

In 1955, two groups of scientists, one in Australia and the other in Holland, finally realized that the absorption method could be used to detect the presence of elements at very low concentrations. Each element has a particular absorption spectrum corresponding to the various separations of (differences between) the energy levels in its atoms. When light from a powerful source is passed through a vaporized sample of an element, the particular wavelengths corresponding to the various energy separations will be absorbed. The higher the concentration of the atoms, the greater the proportion of the light that will be absorbed. This linear relationship between light absorption and concentration is known as *Beer's Law*. The sensitivity of this method is extremely high, and concentrations of parts per million are easy to determine; some elements can be detected at the parts per billion level. Atomic absorption spectroscopy has now become a routine analytical tool in chemistry, metallurgy, geology, medicine, forensic science, and many other fields of science—and it simply requires the movement of electrons from one energy level to another.

lines in a magnetic field (a phenomenon known as the *Zeeman effect*). Within a short time, a radically different model, the quantum mechanical model, was proposed to account for these observations.

1.1   *The Schrödinger Wave Equation and Its Significance*

The more sophisticated quantum mechanical model of atomic structure was derived from the work of Louis de Broglie. de Broglie showed that, just as electromagnetic waves could be treated as streams of particles (photons), moving particles could exhibit wavelike properties. Thus, it was equally valid to picture electrons either as particles or as waves. Using this wave–particle duality, Erwin Schrödinger developed a partial differential equation to represent the behavior of an electron around an atomic nucleus. This equation, given here for a one-electron atom, shows the relationship between the wave function of the electron, Ψ, and E and V, the total and potential energies of the system, respectively. The second differential terms represent the wave function along each of the Cartesian coordinates x, y, and z; m is the mass of an electron, and h is the Planck constant.

$$\frac{\partial \Psi}{\partial x^2} + \frac{\partial \Psi}{\partial y^2} + \frac{\partial \Psi}{\partial z^2} + \frac{8\pi^2 m}{h^2}\,(E - V)\Psi = 0$$

The derivation of this equation and the method of solving it are in the realm of physics and physical chemistry, but the solution itself is of great importance to inorganic chemists. We should always keep in mind, however, that the wave equation is simply a mathematical formula. We attach meanings to the solution simply because most people need concrete images to think about subatomic phenomena. The images that we create corresponding to our macroscopic world can only vaguely resemble the subatomic reality.

Schrödinger argued that the real meaning of the equation could be found from the square of the wave function, Ψ^2, which represents the probability of finding the electron at any point in the region surrounding the nucleus. There are a number of solutions to a wave equation. Each solution describes a different orbital and, hence, a different probability distribution for an electron in that orbital. Each of these orbitals is uniquely defined by a set of three integers: n, l, and m_l. Like the integers in the Bohr model, these integers are also called quantum numbers.

In addition to the three quantum numbers derived from the original theory, a fourth quantum number had to be defined to explain the results of a later experiment. In this experiment, it was found that passing a beam of hydrogen atoms through a magnetic field caused about half the atoms to be deflected in one direction and the other half in the opposite direction. Other investigators proposed that the observation was the result of two different electronic spin orientations. The atoms possessing an electron with one spin were deflected one way, and the atoms whose electron had the opposite spin were deflected in the opposite direction. This spin quantum number was assigned the symbol m_s.

The possible values of the quantum numbers are defined as follows:

> n, the *principal quantum number*, can have all positive integer values from 1 to ∞.
>
> l, the *angular momentum quantum number*, can have all integer values from $n - 1$ to 0.
>
> m_l, the *magnetic quantum number*, can have all integer values from +1 through 0 to −1.
>
> m_s, the *spin quantum number*, can have values of +½ and −½.

When the value of the principal quantum number is 1, there is only one possible set of quantum numbers n, l, and m_l (1, 0, 0); whereas for a principal quantum number of 2, there are four sets of quantum numbers (2, 0, 0; 2, 1, −1; 2, 1, 0; 2, 1, +1). This situation is shown diagrammatically in Figure 1.4. To identify the electron orbital that corresponds to each set of quantum numbers, we use the value of the principal quantum number n, followed by a letter for the angular momentum quantum number l. Thus, when $n = 1$, there is only the 1s orbital.

When $n = 2$, there is one 2s orbital and three 2p orbitals (corresponding to the m_l values of +1, 0, and −1). The letters s, p, d, and f are derived from

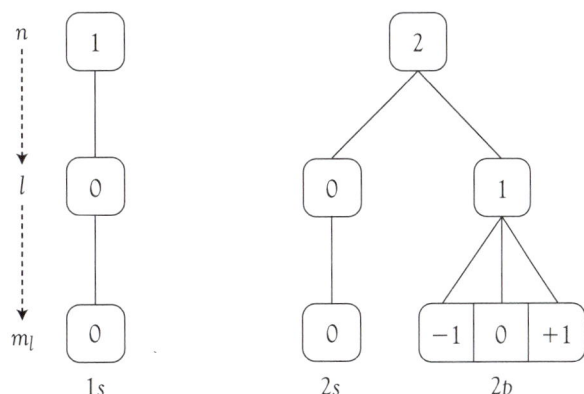

Figure 1.4 The possible sets of quantum numbers for $n = 1$ and $n = 2$.

Table 1.1 Correspondence between angular momentum number l and orbital designation

l Value	Orbital designation
0	s
1	p
2	d
3	f

Table 1.2 Correspondence between angular momentum number l and number of orbitals

l Value	Number of orbitals
0	1
1	3
2	5
3	7

categories of the spectral lines: sharp, principal, diffuse, and fundamental. The correspondences are shown in Table 1.1.

When the principal quantum number $n = 3$, there are nine sets of quantum numbers (Figure 1.5). These sets correspond to one $3s$, three $3p$, and five $3d$ orbitals. A similar diagram for the principal quantum number $n = 4$ would show 16 sets of quantum numbers, corresponding to one $4s$, three $4p$, five $4d$, and seven $4f$ orbitals (Table 1.2). Theoretically, we can go on and on, but as we shall see, the f orbitals represent the limit of orbital types among the elements of the periodic table for atoms in their electronic ground states.

The Schrödinger wave equation is usually presented as the definitive representation of the electrons of an atom, but it is not. As we discuss in Chapter 2, Section 2.5, the equation fails to take into account the fact that some of the electrons in the more massive elements are traveling at extremely high velocities. As a result, the electron masses are influenced by relativistic effects. Although the Schrödinger equation can be modified to account for the problem, in 1928, the English physicist P.A.M. Dirac developed a better wave equation that integrates relativity factors. The Dirac equation provides four quantum numbers directly, although only the principal quantum number, n, has the same significance in both Schrödinger and Dirac equations. Even the shapes of the orbitals derived from the Dirac equation are different from those derived from the Schrödinger equation. However, because this is a descriptive chemistry text, we emphasize the features of the simpler and more commonly used Schrödinger-derived orbitals.

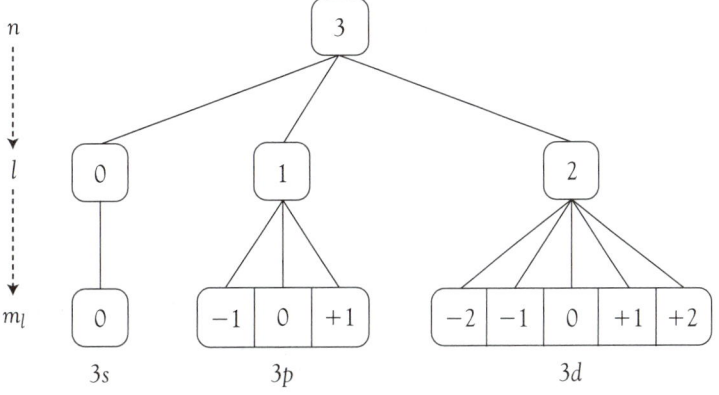

Figure 1.5 The possible sets of quantum numbers for $n = 3$.

1.2   Shapes of the Atomic Orbitals

Representing the solutions to a wave equation on paper is not an easy task. In fact, we would need four-dimensional graph paper (if it existed) to display the complete solution for each orbital. As a realistic alternative, we break the wave equation into two parts: a radial part and an angular part.

Each of the three quantum numbers derived from the wave equation represents a different aspect of the orbital:

> The principal quantum number n indicates the size of the orbital.
>
> The angular momentum quantum number l represents the shape of the orbital.
>
> The magnetic quantum number m_l represents the spatial direction of the orbital.
>
> The spin quantum number m_s has little physical meaning; it merely allows two electrons to occupy the same orbital.

The value of the principal quantum number and, to a lesser extent, that of the angular momentum quantum number, determines the energy of the electron. Although the electron may not literally be spinning, it behaves as if it were spinning and has the magnetic properties expected for a spinning particle.

An orbital diagram is used to indicate the probability of finding an electron at any instant at any location. An alternative viewpoint is to consider the locations of an electron over a lengthy period of time. We define a location where an electron seems to spend most of its time as an area of high electron density. Conversely, locations rarely visited by an electron are called areas of low electron density.

The s Orbitals

The s orbitals are spherically symmetric about the atomic nucleus. As the principal quantum number increases, the electron tends to be found farther from the nucleus. To express this idea in a different way, we say that, as the principal quantum number increases, the orbital becomes more diffuse. A unique feature of electron behavior in an s orbital is that there is a finite probability of finding the electron close to and even within the nucleus. This penetration by s orbital electrons plays a role in atomic radii (see Chapter 2) and as a means of studying atomic structure. In fact, the technique of Mössbauer spectroscopy involves the study of the effect of changes in s orbital density on nuclear energies.

Same-scale representations of the shapes (angular functions) of the 1s and 2s orbitals of an atom are compared in Figure 1.6. The volume of a 2s orbital is about four times greater than that of a 1s orbital. In both cases, the tiny nucleus is located at the center of the spheres. These spheres represent the region in which there is a 99 percent probability of finding an electron. The total probability cannot be represented, for the probability of finding an electron drops to zero only at an infinite distance from the nucleus.

The probability of finding the electron within an orbital will always be positive (as the probability is derived

Figure 1.6   Representations of the shapes of the 1s and 2s orbitals.
(Adapted from D.A. McQuarrie and P.A. Rock, General Chemistry, 2nd ed. [New York: W.H. Freeman, 1991], p. 322.)

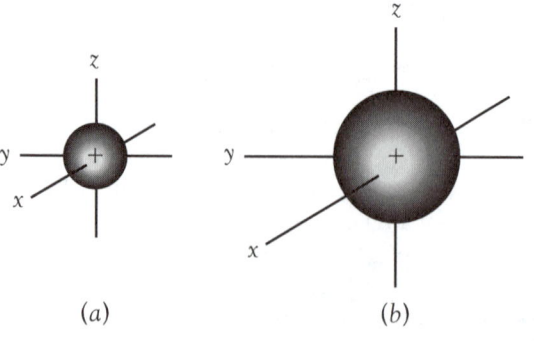

(a)   (b)

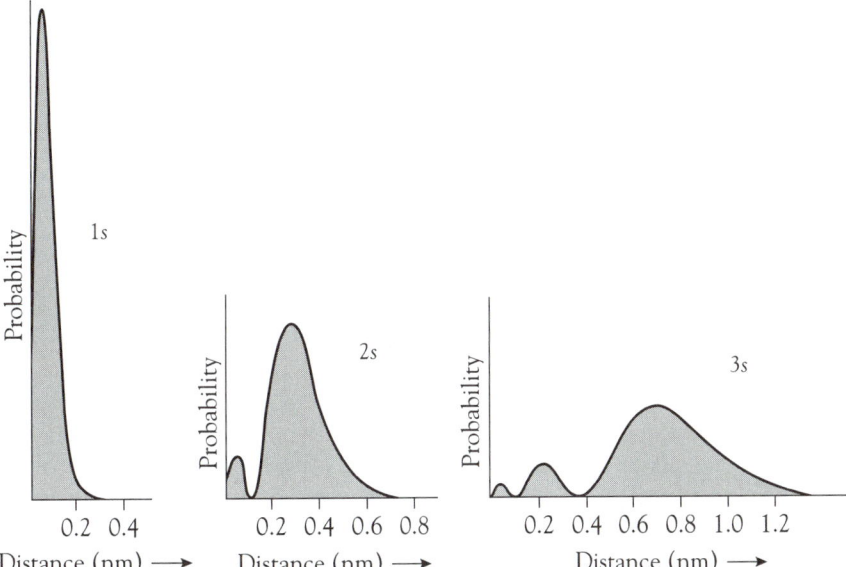

Figure 1.7 The variation of the radial density distribution function with distance from the nucleus for electrons in the 1s, 2s, and 3s orbitals of a hydrogen atom.

from the square of the wave function and squaring a negative makes a positive). However, when we discuss the bonding of atoms, we find that the sign related to the original wave function has importance. For this reason, it is conventional to superimpose the sign of the wave function on the representation of each atomic orbital. For an *s* orbital, the sign is positive.

In addition to the enormous difference in size between the 1*s* and the 2*s* orbitals, the 2*s* orbital has, at a certain distance from the nucleus, a spherical surface on which the electron density is zero. A surface on which the probability of finding an electron is zero is called a *nodal surface*. When the principal quantum number increases by 1, the number of nodal surfaces also increases by 1. We can visualize nodal surfaces more clearly by plotting a graph of the radial density distribution function as a function of distance from the nucleus for any direction. Figure 1.7 shows plots for the 1*s*, 2*s*, and 3*s* orbitals. These plots show that the electron tends to be farther from the nucleus as the principal quantum number increases. The areas under all three curves are the same.

Electrons in an *s* orbital are different from those in *p*, *d*, or *f* orbitals in two significant ways. First, only the *s* orbital has an electron density that varies in the same way in every direction out from the atomic nucleus. Second, there is a finite probability that an electron in an *s* orbital is at the nucleus of the atom. Every other orbital has a node at the nucleus.

The *p* Orbitals

Unlike the *s* orbitals, the *p* orbitals are not spherically symmetric. In fact, the *p* orbitals consist of two separate volumes of space (lobes), with the nucleus located between the two lobes. Because there are three *p* orbitals, we assign each orbital a direction according to Cartesian coordinates: we have p_x, p_y, and p_z. Figure 1.8 shows representations of the three 2*p* orbitals. At right angles to the axis of higher probability, there is a nodal plane through the nucleus. For example, the $2p_z$ orbital has a nodal surface in the *xy* plane. In terms of wave function sign, one lobe is positive and the other negative.

Figure 1.8   Representations of the shapes of the $2p_x$, $2p_y$, and $2p_z$ orbitals.
(Adapted from D.F. Shriver, P. Atkins, and C.H. Langford, Inorganic Chemistry, 2nd ed. [New York: W.H. Freeman, 1994], p. 24.)

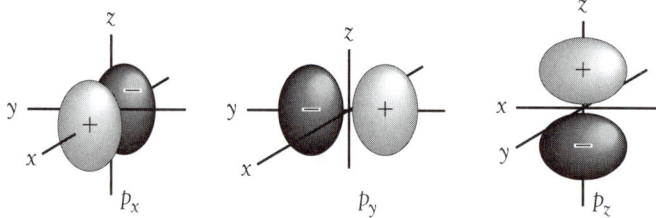

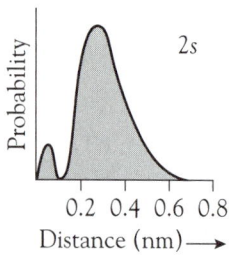 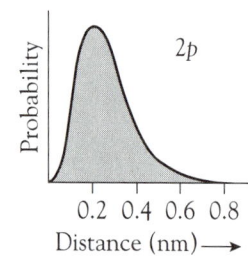

Figure 1.9   The variation of the radial density distribution function with distance from the nucleus for electrons in the 2s and 2p orbitals of a hydrogen atom.

If we compare graphs of electron density as a function of atomic radius for the $2s$ orbital and a $2p$ orbital (the latter plotted along the axis of higher probability), we find that the $2s$ orbital has a much greater electron density close to the nucleus than does the $2p$ orbital (Figure 1.9). Conversely, the second maximum of the $2s$ orbital is farther out than the single maximum of the $2p$ orbital. However, the mean distance of maximum probability is the same for both orbitals.

Like the s orbitals, the p orbitals develop additional nodal surfaces within the orbital structure as the principal quantum number increases. Thus, a $3p$ orbital does not look exactly like a $2p$ orbital as it has an additional nodal surface. However, the detailed differences in orbital shapes for a particular angular momentum quantum number are of little relevance in the context of basic inorganic chemistry.

The d Orbitals

The five d orbitals have more complex shapes. Three of them are located between the Cartesian axes, and the other two are oriented along the axes. In all cases, the nucleus is located at the intersection of the axes. Three orbitals each have four lobes that are located between pairs of axes (Figure 1.10). These orbitals are identified as d_{xy}, d_{xz}, and d_{yz}. The other two d or-

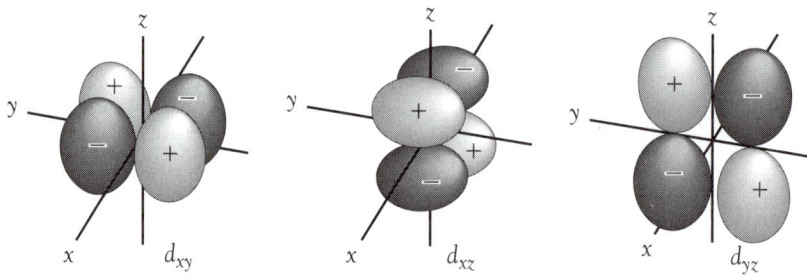

Figure 1.10   Representations of the shapes of the $3d_{xy}$, $3d_{xz}$, and $3d_{yz}$ orbitals.
(Adapted from D.F. Shriver, P. Atkins, and C.H. Langford, Inorganic Chemistry, 2nd ed. [New York: W.H. Freeman, 1994], p. 25.)

bitals, d_{z^2} and $d_{x^2-y^2}$, are shown in Figure 1.11. The d_{z^2} orbital looks somewhat similar to a p_z orbital (see Figure 1.8), except that it has an additional donut-shaped ring of high electron density in the xy plane. The $d_{x^2-y^2}$ orbital is identical to the d_{xy} orbital but has been rotated through 45°.

The f Orbitals

The f orbitals are even more complex than the d orbitals. There are seven f orbitals, four of which have eight lobes. The other three look like the d_{z^2} orbital, but have two donut-shaped rings instead of one. These orbitals are rarely involved in bonding, so we do not need to consider them in any detail.

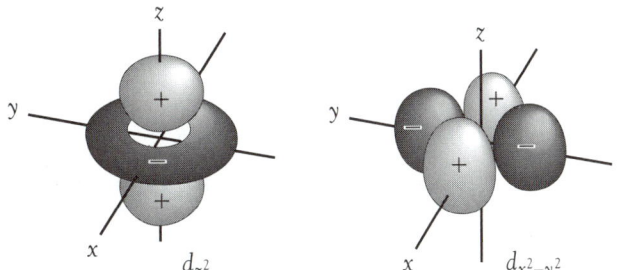

Figure 1.11   Representations of the shapes of the $3d_{z^2}$ and $3d_{x^2-y^2}$ orbitals.
(Adapted from D.F. Shriver, P. Atkins, and C.H. Langford, Inorganic Chemistry, 2nd ed. [New York: W.H. Freeman, 1994], p. 25.)

1.3   The Polyelectronic Atom

In our model of the polyelectronic atom, the electrons are distributed among the orbitals of the atom according to the *Aufbau* (building-up) *principle*. This simple idea proposes that, when the electrons of an atom are all in the ground state, they occupy the orbitals of lowest energy, thereby minimizing the atom's total electronic energy. Thus, the configuration of an atom can be described simply by adding electrons one by one until the total number required for the element has been reached.

Before starting to construct electron configurations, there is a second rule to be considered: the *Pauli exclusion principle*. According to this rule, no two electrons in an atom may possess identical sets of the four quantum numbers. Thus, there can be only one orbital of each three-quantum-number set per atom and each orbital can hold only two electrons, one with $m_s = +\frac{1}{2}$ and the other with $m_s = -\frac{1}{2}$.

Filling the *s* Orbitals

The simplest configuration is that of the hydrogen atom. According to the Aufbau principle, the single electron will be located in the $1s$ orbital. This configuration is the ground state of the hydrogen atom. Adding energy would raise the electron to one of the many higher energy states. These configurations are referred to as excited states. In the diagram of the ground state of the hydrogen atom (Figure 1.12), a half-headed arrow is used to indicate the direction of electron spin. The electron configuration is written as $1s^1$, with the superscript 1 indicating the number of electrons in that orbital.

With a two-electron atom (helium), there is a choice: The second electron could go in the $1s$ orbital (Figure 1.13a) or the next higher energy orbital, the $2s$ orbital (Figure 1.13b). Although it might seem obvious that the second electron would enter the $1s$ orbital, it is not so simple. If the second electron entered the $1s$ orbital, it would be occupying the same volume of space as the electron already in that orbital. The very strong electrostatic repulsions would discourage the occupancy of the same orbital. For helium, the *pairing energy*, the energy needed to overcome the interelectronic repulsive forces, is about 3 MJ·mol^{-1}. However, by occupying an orbital with a high probability closer to the nucleus, the second electron will experience a much greater nuclear attraction. The nuclear attraction is greater than the interelectron repulsion. Hence, the actual configuration will be $1s^2$, although it must be emphasized that electrons pair up in the same orbital only when pairing is the lower energy option.

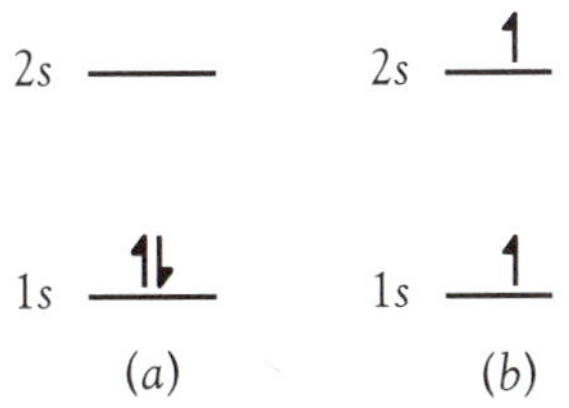

In the lithium atom the $1s$ orbital is filled by two electrons, and the third electron must be in the next higher energy orbital, the $2s$ orbital. Thus,

$1s$ ⟍↑⟋

Figure 1.12   Electron configuration of a hydrogen atom.

Figure 1.13   Two possible electron configurations for helium.

$2s$ ——— $2s$ ⟍↑⟋

$1s$ ⟍↑↓⟋ $1s$ ⟍↑⟋

(a) (b)

lithium has the configuration of $1s^2 2s^1$. Because the energy separation of an s and its corresponding p orbitals is always greater than the pairing energy in a polyelectronic atom, the electron configuration of beryllium will be $1s^2 2s^2$ rather than $1s^2 2s^1 2p^1$.

Filling the p Orbitals

Boron marks the beginning of the filling of the $2p$ orbitals. A boron atom has an electron configuration of $1s^2 2s^2 2p^1$. As the p orbitals are degenerate (that is, they all have the same energy), it is impossible to decide which one of the three orbitals contains the electron.

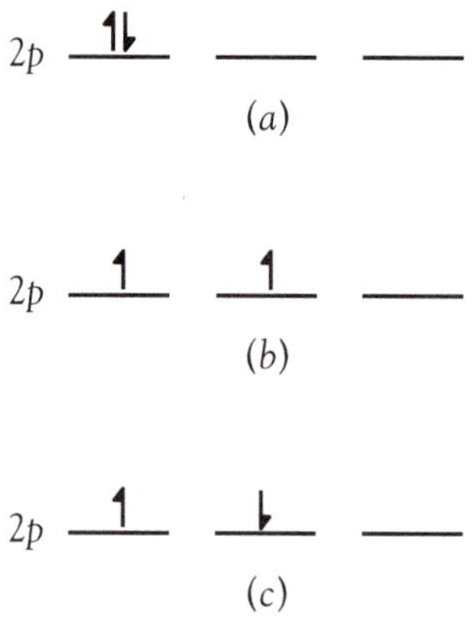

Carbon is the second ground-state atom with electrons in the p orbitals. Its electron configuration provides another challenge. There are three possible arrangements of the two $2p$ electrons (Figure 1.14): (a) both electrons in one orbital; (b) two electrons with parallel spins in different orbitals; and (c) two electrons with opposed spins in different orbitals. On the basis of electron repulsions, the first possibility (a) can be rejected immediately. The decision between the other two possibilities is less obvious and requires a deeper knowledge of quantum theory. In fact, if the two electrons have parallel spins, there is a zero probability of them occupying the same space. However, if the spins are opposed, there is a finite possibility that the two electrons will occupy the same region in space, thereby resulting in some repulsion and a higher energy state. Hence, the parallel spin situation (b) will have the lowest energy. This preference for unpaired electrons with parallel spins has been formalized in *Hund's rule*: When filling a set of degenerate orbitals, the number of unpaired electrons will be maximized and these electrons will have parallel spins.

Figure 1.14 showing three possible $2p$ electron configurations:

$2p$ (a) one orbital with paired electrons ↑↓

$2p$ (b) two orbitals each with one electron, parallel spins ↑ ↑

$2p$ (c) two orbitals each with one electron, opposed spins ↑ ↓

Figure 1.14 Possible 2p electron configurations for carbon.

After the completion of the $2p$ electron set at neon ($1s^2 2s^2 2p^6$), the $3s$ and $3p$ orbitals start to fill. Rather than write the full electron configurations, a shortened form can be used. In this notation, the inner electrons are represented by the noble-gas symbol having that configuration. Thus, magnesium, whose full electron configuration would be written as $1s^2 2s^2 2p^6 3s^2$, can be represented as having a neon noble-gas core, and its configuration is written as $[\text{Ne}]3s^2$. An advantage of the noble-gas core representation is that it emphasizes the outermost (valence) electrons, and it is these electrons that are involved in chemical bonding. At this point, we have finished our analysis of the electron configuration of the two short *periods* (rows) of the periodic table (Figure 1.15): Period 2, lithium to neon, and Period 3, sodium to argon.

Filling the d Orbitals

Once the $3p$ orbitals are filled (argon), the $3d$ and $4s$ orbitals start to fill. It is here that the simple orbital energy level concept breaks down, because the energy levels of the $4s$ and $3d$ orbitals are very close. What becomes most important is not the minimum energy for a single electron but the configuration that results in the least number of interelectron repulsions for all the electrons. For potassium, this is $[\text{Ar}]4s^1$; for calcium, $[\text{Ar}]4s^2$. To illustrate how this delicate balance changes with increasing numbers of protons and electrons, the outer electrons in each of the Group 3 to Group 12 elements are listed here. These configurations are not important in themselves, but they do show how close in energy the ns and $(n-1)d$ electrons are in energy.

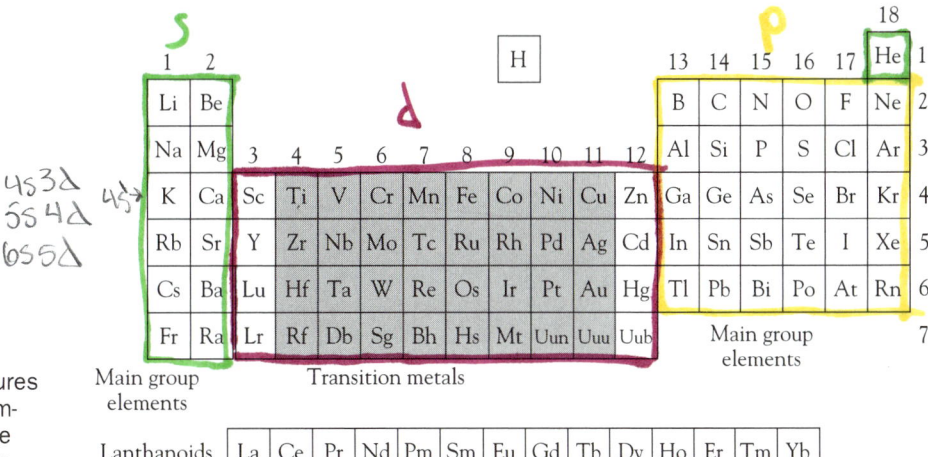

Figure 1.15 Essential features of the periodic table. The numbers across the top designate groups (columns) of elements. The numbers down the right-hand side designate periods (rows).

Atom	Configuration	Atom	Configuration	Atom	Configuration
Sc	$4s^2 3d^1$	Y	$5s^2 4d^1$	Lu	$6s^2 5d^1$
Ti	$4s^2 3d^2$	Zr	$5s^2 4d^2$	Hf	$6s^2 5d^2$
V	$4s^2 3d^3$	Nb	$5s^1 4d^4$	Ta	$6s^2 5d^3$
Cr	$4s^1 3d^5$	Mo	$5s^1 4d^5$	W	$6s^2 5d^4$
Mn	$4s^2 3d^5$	Tc	$5s^2 4d^5$	Re	$6s^2 5d^5$
Fe	$4s^2 3d^6$	Ru	$5s^1 4d^7$	Os	$6s^2 5d^6$
Co	$4s^2 3d^7$	Rh	$5s^1 4d^8$	Ir	$6s^2 5d^7$
Ni	$4s^2 3d^8$	Pd	$5s^0 4d^{10}$	Pt	$6s^1 5d^9$
Cu	$4s^1 3d^{10}$	Ag	$5s^1 4d^{10}$	Au	$6s^1 5d^{10}$
Zn	$4s^2 3d^{10}$	Cd	$5s^2 4d^{10}$	Hg	$6s^2 5d^{10}$

In general, the lowest overall energy for each transition metal is obtained by filling the s orbitals first; the remaining electrons then occupy the d orbitals. However, for certain elements, the lowest energy is obtained by shifting one or both of the s electrons to d orbitals. Looking at the first series in isolation would lead to the conclusion that there is some preference for a half-full or full set of d orbitals by chromium and copper. Although palladium and silver in the second transition series both favor the $4d^{10}$ configuration, it is more accurate to say that the interelectron repulsion between the two s electrons is sufficient in several cases to result in an s^1 configuration.

For the elements from lanthanum (La) to ytterbium (Yb), the situation is even more fluid because the $6s$, $5d$, and $4f$ orbitals all have similar energies. For example, lanthanum has a configuration of $[Xe]6s^2 5d^1$, whereas the next element, cerium, has a configuration of $[Xe]6s^2 4f^2$. The most interesting electron configuration in this row is that of gadolinium, $[Xe]6s^2 5d^1 4f^7$, rather than the predicted $[Xe]6s^2 4f^8$. This configuration provides more evidence of the importance of interelectron repulsion in the determination of electron configuration when adjacent orbitals have similar energies. Similar complexities occur among the elements from actinium

Though on Earth, essentially all atoms are in their ground state, it is not true in interstellar space. Electron transitions for the hydrogen atom from as high as the n = 253 to n = 252 level have been observed from this region of the universe. The Bohr radius for such a hydrogen atom would be about 0.34 mm, making the atom so large it could (theoretically) be seen.

(Ac) to nobelium (No), in which the 7*s*, 6*d*, and 5*f* orbitals have similar energies.

Although there are minor fluctuations in configurations throughout the *d*-block and *f*-block elements, the order of filling is quite consistent:

$$1s\ 2s\ 2p\ 3s\ 3p\ 4s\ 3d\ 4p\ 5s\ 4d\ 5p\ 6s\ 4f\ 5d\ 6p\ 7s\ 5f\ 6d\ 7p$$

This order is also shown in Figure 1.16. The orbitals fill in this order because the energy differences between the *s*, *p*, *d*, and *f* orbitals of the same principal quantum number become so great beyond *n* = 2 that they overlap with the orbitals of the following principal quantum numbers. It is important to note that Figure 1.16 shows the filling order, not the order for any particular element. For example, for elements beyond zinc, electrons in the 3*d* orbitals are far lower in energy than those in the 4*s* orbitals. Thus, at this point, the 3*d* orbitals have become "inner" orbitals and have no role in chemical bonding. Hence, their precise ordering is unimportant.

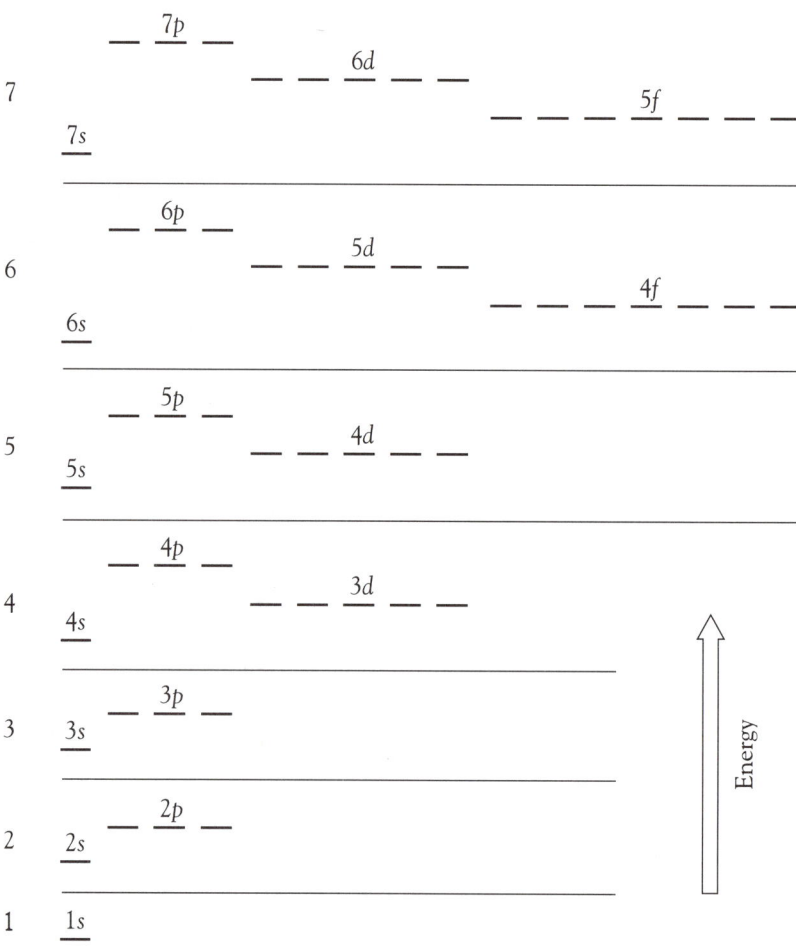

Figure 1.16 Representation of the comparative energies of the atomic orbitals for filling order purposes.

1.4 *Ion Electron Configurations*

For the early main group elements, the common ion electron configurations can be predicted quite readily. Thus, metals tend to lose all the electrons in the outer orbital set. This situation is illustrated for the *isoelectronic* series (same electron configuration) of sodium, magnesium, and aluminum cations:

Atom	Electron configuration	Ion	Electron configuration
Na	$[Ne]3s^1$	Na^+	$[Ne]$
Mg	$[Ne]3s^2$	Mg^{2+}	$[Ne]$
Al	$[Ne]3s^23p^1$	Al^{3+}	$[Ne]$

Nonmetals gain electrons to complete the outer orbital set. This situation is shown for nitrogen, oxygen, and fluorine anions:

Atom	Electron configuration	Ion	Electron configuration
N	$[He]2s^22p^3$	N^{3-}	$[Ne]$
O	$[He]2s^22p^4$	O^{2-}	$[Ne]$
F	$[He]2s^22p^5$	F^-	$[Ne]$

Some of the later main group metals form two ions with different charges. For example, lead forms Pb^{2+} and (rarely) Pb^{4+}. The 2+ charge can be explained by the loss of the $6p$ electrons only, whereas the 4+ charge results from loss of both $6s$ and $6p$ electrons:

Atom	Electron configuration	Ion	Electron configuration
Pb	$[Xe]6s^24f^{14}5d^{10}6p^2$	Pb^{2+}	$[Xe]6s^24f^{14}5d^{10}$
		Pb^{4+}	$[Xe]4f^{14}5d^{10}$

Notice that the electrons of the higher principal quantum number are lost first. This rule is found to be true for all the elements. For the transition metals, the *s* electrons are always lost first when a metal cation is formed. In other words, for the transition metal cations, the $3d$ orbitals are always lower in energy than the $4s$ orbitals; and a charge of 2+, representing the loss of the two *s* electrons, is common for the transition metals and the Group 12 metals. For example, zinc always forms an ion of 2+ charge:

Atom	Electron configuration	Ion	Electron configuration
Zn	$[Ar]4s^23d^{10}$	Zn^{2+}	$[Ar]3d^{10}$

Iron forms ions with charges of 2+ and 3+ and, as shown here, it is tempting to ascribe the formation of the 3+ ion to a process in which interelectron repulsion "forces out" the only paired *d* electron:

Atom	Electron configuration	Ion	Electron configuration
Fe	$[Ar]4s^23d^6$	Fe^{2+}	$[Ar]3d^6$
		Fe^{3+}	$[Ar]3d^5$

It is dangerous, however, to read too much into the electron configurations of atoms as a means of predicting the ion charges. The series of nickel, palladium, and platinum illustrate this point: They have different configurations as atoms, yet their common ionic charges and corresponding ion electron configurations are similar:

Atom	Electron configuration	Ion	Electron configuration
Ni	$[Ar]4s^2 3d^8$	Ni^{2+}	$[Ar]3d^8$
Pd	$[Kr]5s^0 4d^{10}$	Pd^{2+}, Pd^{4+}	$[Kr]4d^8$, $[Kr]4d^6$
Pt	$[Xe]6s^1 5d^9$	Pt^{2+}, Pt^{4+}	$[Xe]5d^8$, $[Xe]5d^6$

1.5   Magnetic Properties of Atoms

In the discussions of electron configuration, we saw that some atoms possess unpaired electrons. The presence of unpaired electrons in the atoms of an element can be determined easily from the element's magnetic properties. If atoms containing only spin-paired electrons are placed in a magnetic field, they are weakly repelled by the field. This phenomenon is called *diamagnetism*. Conversely, atoms containing one or more unpaired electrons are attracted by the magnetic field. This behavior of unpaired electrons is named *paramagnetism*. The attraction of each unpaired electron is many times stronger than the repulsion of all the spin-paired electrons in that atom.

To explain paramagnetism in simple terms, we can visualize the electron as a particle spinning on its axis and generating a magnetic moment, just as an electric current flowing through a wire does. This permanent magnetic moment results in an attraction into the stronger part of the field. When electrons have their spins paired, the magnetic moments cancel each other. As a result, the paired electrons are weakly repelled by the lines of force of the magnetic field. We will encounter this phenomenon again in our discussions of covalent bonding and the bonding in transition metal compounds.

There is a third relatively common form of magnetic behavior—*ferromagnetism*. In paramagnetic materials, application of a magnetic field aligns some of the normally randomly oriented electron spins with the applied magnetic field (Figure 1.17*a* and *b*). It is this alignment that results in the attraction of the material into the magnetic field. In ferromagnetic materials, the unpaired electrons are aligned with their neighbors even in the absence of a magnetic field. These groups of mutually aligned spins are known as *magnetic domains*. Application of a magnetic field causes all these domains

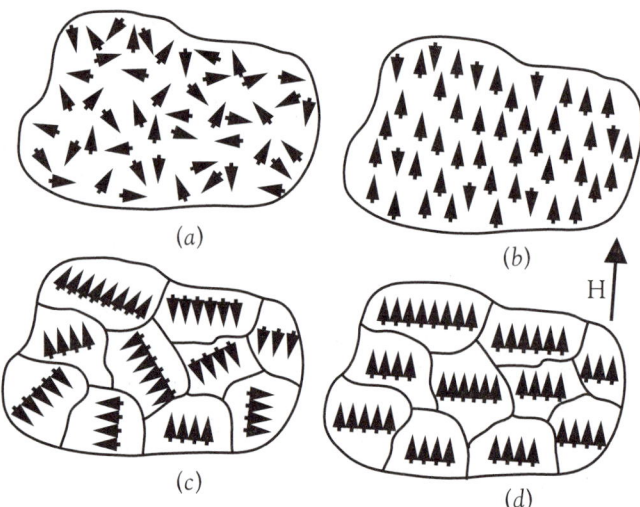

Figure 1.17   The behavior of paramagnetic materials without (*a*) and with (*b*) an applied magnetic field and ferromagnetic materials without (*c*) and with (*d*) an applied magnetic field.

to align with the magnetic field (Figure 1.17c and d). This alignment is much stronger than that of paramagnetism, and it can be permanent. We use the ferromagnetism of γ-iron(III) oxide and chromium(IV) oxide as a recording medium in audio- and videotape surfaces.

EXERCISES

1.1 Define the following terms: (a) nodal surface; (b) Pauli exclusion principle; (c) paramagnetic.

1.2 Define the following terms: (a) orbital; (b) degenerate; (c) Hund's rule.

1.3 Construct a quantum number tree for the principal quantum number $n = 4$ similar to that depicted for $n = 3$ in Figure 1.5.

1.4 Determine the lowest value of n for which m_l can (theoretically) have a value of $+4$.

1.5 Identify the orbital that has $n = 5$ and $l = 1$.

1.6 Identify the orbital that has $n = 6$ and $l = 0$.

1.7 How does the quantum number n relate to the properties of an orbital?

1.8 How does the quantum number l relate to the properties of an orbital?

1.9 Explain concisely why carbon has two electrons in different p orbitals with parallel spins rather than the other possible arrangements.

1.10 Explain concisely why beryllium has a ground-state electron configuration of $1s^2 2s^2$ rather than $1s^2 2s^1 2p^1$.

1.11 Write noble-gas core ground-state electron configurations for atoms of (a) sodium; (b) nickel; (c) copper.

1.12 Write noble-gas core ground-state electron configurations for atoms of (a) calcium; (b) chromium; (c) lead.

1.13 Write noble-gas core ground-state electron configurations for ions of (a) potassium; (b) scandium 3+; (c) copper 2+.

1.14 Write noble-gas core ground-state electron configurations for ions of (a) chlorine; (b) cobalt 2+; (c) manganese 4+.

1.15 Predict the common charges of the ions of thallium. Explain your reasoning in terms of electron configurations.

1.16 Predict the common charges of the ions of tin. Explain your reasoning in terms of electron configurations.

1.17 Predict the common charge of the silver ion. Explain your reasoning in terms of electron configurations.

1.18 Predict the highest possible charge of a zirconium ion. Explain your reasoning in terms of electron configurations.

1.19 Use diagrams similar to Figure 1.14 to determine the number of unpaired electrons in atoms of (a) oxygen; (b) magnesium; (c) chromium.

1.20 Use diagrams similar to Figure 1.14 to determine the number of unpaired electrons in atoms of (a) nitrogen; (b) silicon; (c) iron.

1.21 Write the electron configuration expected for element 113 and the configurations for the two cations that it is most likely to form.

1.22 Which of the following species are hydrogen-like? (a) He^+; (b) He^-; (c) Li^+; (d) Li^{2+}.

BEYOND THE BASICS

1.23 The next set of orbitals after the f orbitals are the g orbitals. How many g orbitals would there be? What would be the lowest principal quantum number n that would possess g orbitals? Deduce the atomic number of the first element at which g orbitals would begin to be filled on the basis of the patterns of the d and f orbitals.

1.24 Use physical chemistry texts for additional information on the Dirac wave equation and contrast it with the Schrödinger wave equation.

1.25 Use an advanced inorganic chemistry text as a source of information on the f orbitals. What are their common features? How do they differ among themselves?

1.26 In Section 1.3, gadolinium is mentioned as having an electron configuration that deviates from the lanthanoid pattern. Which element in the actinoids should show a similar deviation? What would be its electron configuration?

1.27 A philosphical question: Does an orbital exist even if it does not contain an electron? Discuss.

An Overview of the Periodic Table

Chapter 2

Main						
2s					1s	Main 1s
3s						2p
4s				Transition		3p
5s				3d		4p
6s	4f	Lanthanoids		4d		5p
7s	5f	Actinoids		5d		6p
				6d		

The periodic table is the framework upon which much of our understanding of inorganic chemistry is based. In this chapter, we provide the essential information that you will need for the more detailed discussions of the individual groups in later chapters.

The search for patterns among the chemical elements really started with the work of the German chemist Johann Döbereiner in 1817. He noticed that there were similarities in properties among various groups of three elements, such as calcium, strontium, and barium. He named these groups "triads." Almost 50 years later, John Newlands, a British sugar refiner, realized that, when the elements were placed in order of increasing atomic weights, a cycle of properties repeated with every eight elements. Newlands called this pattern the law of octaves. At the time, scientists had started to look for a unity of the physical laws that would explain everything, so to correlate element organization with the musical scale seemed natural. Unfortunately, his proposal was laughed at by most chemists of the time. A few years later, the Russian chemist Dmitri Mendeleev (pronounced Mendelé-ev) independently devised the same concept (without linking it to music) and made the crucial advance of using the law as a predictive tool. It attracted little attention until Lothar Meyer, a German chemist, published his own report on the periodic relationship. Meyer did acknowledge that Mendeleev had had the same idea first.

Mendeleev and Meyer would hardly recognize the contemporary periodic table. In Mendeleev's proposal, the elements known at the time were organized in an eight-column format in order of increasing atomic mass.

I		II		III		IV		V		VI		VII		VIII
	H													
Li		Be		B		C		N		O		F		
	Na		Mg		Al		Si		P		S		Cl	
K		Ca				Ti		V		Cr		Mn		Fe Ni Co Cu
	Cu?		Zn						As		Se		Br	
Rb		Sr		Yt		Zr		Nb		Mo				Ru Pd Rh Ag
	Ag?		Cd		In		Sn		Sb		Te		I	
Cs		Ba				Ce								
				Er		La		Ta		W				Os Pt Ir Au
	Au?		Hg		Tl		Pb		Bi					
						Th				U				

Figure 2.1 The organization of one of Mendeleev's designs for the periodic table.

He claimed that each eighth element had similar properties. Groups I to VII each contained two subgroups, and Group VIII contained four subgroups. The organization of one of his designs is shown in Figure 2.1. To ensure that the patterns in the properties of elements fitted the table, it was necessary to leave spaces. Mendeleev assumed that these spaces corresponded to unknown elements. He argued that the properties of the missing elements could be predicted on the basis of the chemistry of its neighbors in the same group. For example, the missing element between silicon and tin, called eka-silicon (Es) by Mendeleev, should have properties intermediate between those of silicon and tin. Table 2.1 compares Mendeleev's predictions with the properties of germanium, discovered 15 years later.

However, the Mendeleev periodic table had three major problems:

1. If the order of increasing atomic mass was consistently followed, elements did not always fit in the group that had the matching properties. Thus, the order of nickel and cobalt had to be reversed, as did that of iodine and tellurium.

2. Elements were being discovered, such as holmium and samarium, for which no space could be found. This difficulty was a particular embarrassment.

3. Elements in the same group were sometimes quite different in their chemical reactivity. This discrepancy was particularly true of the first group, which contained the very reactive alkali metals and the very unreactive coinage metals (copper, silver, and gold).

Table 2.1 Comparison of Mendeleev's predictions for eka-silicon and the actual properties of germanium

Element	Atomic weight	Density ($g \cdot cm^{-3}$)	Oxide formula	Chloride formula
Eka-silicon	72	5.5	EsO_2	$EsCl_4$
Germanium	72.3	5.47	GeO_2	$GeCl_4$

As we now know, there was another flaw: To establish a group of elements, at least one element has to be known already. Because none of the noble gases was known at that time, no space was left for them. Conversely, some spaces in Mendeleev's table were completely erroneous. This was because he tried to fit the elements into repeating rows (periods) of eight. Now, of course, we know that the periods are not consistently eight members long but, instead, increase regularly; successive rows have 2, 8, 8, 18, 18, 32, and 32 elements.

The crucial transition to our modern ideas was provided by Henry Moseley, a British physicist, as we discuss in the Dedication of this text. Placing the elements in order of the atomic number that he derived from spectroscopic measurements removed the irregularities of the table that was based on atomic masses, and it defined exactly the spaces in the table where elements still needed to be found.

2.1   Organization of the Modern Periodic Table

In the modern periodic table, the elements are placed in order of increasing atomic number (the number of protons). There have been numerous designs of the table over the years, but the two most common are the long form and the short form. The long form (Figure 2.2) shows all the elements in numerical order.

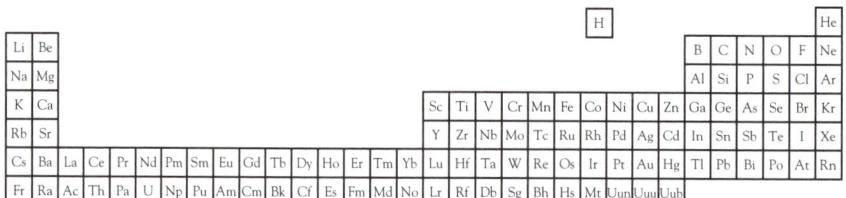

Figure 2.2 The long form of the periodic table.

The start of a new period always corresponds to the introduction of the first electron into the *s* orbital of a new principal quantum number. The number of elements in each period corresponds to the number of electrons required to fill those orbitals (Figure 2.3). In a particular period, the principal quantum number of the *p* orbitals is the same as that of the *s* orbitals, whereas the *d* orbitals are one less and the *f* orbitals are two less.

Each group contains elements of similar electron configuration. For example, all Group 1 elements have an outer electron that is ns^1, where n is the principal quantum number. Although elements in a group have similar

Main						Transition				Main		1s
2s											2p	
3s											3p	
4s							3d				4p	
5s							4d				5p	
6s		4f	Lanthanoids				5d				6p	
7s		5f	Actinoids				6d					

Figure 2.3 Electron orbital filling sequence in the periodic table.

1	2																18
												H					He
Li	Be											13	14	15	16	17	
												B	C	N	O	F	Ne
Na	Mg	3	4	5	6	7	8	9	10	11	12	Al	Si	P	S	Cl	Ar
K	Ca	Sc	Ti	V	Cr	Mn	Fe	Co	Ni	Cu	Zn	Ga	Ge	As	Se	Br	Kr
Rb	Sr	Y	Zr	Nb	Mo	Tc	Ru	Rh	Pd	Ag	Cd	In	Sn	Sb	Te	I	Xe
Cs	Ba	*	Lu	Hf	Ta	W	Re	Os	Ir	Pt	Au	Hg	Tl	Pb	Bi	Po	At	Rn
Fr	Ra	**	Lr	Rf	Db	Sg	Bh	Hs	Mt	Uun	Uuu	Uub					

Lanthanoids *	La	Ce	Pr	Nd	Pm	Sm	Eu	Gd	Tb	Dy	Ho	Er	Tm	Yb
Actinoids **	Ac	Th	Pa	U	Np	Pu	Am	Cm	Bk	Cf	Es	Fm	Md	No

Figure 2.4 Short form of the periodic table displaying the group numbers. Main group elements are shaded.

properties, it is important to realize that every element is unique. Thus, although nitrogen and phosphorus are sequential elements in the same group, nitrogen gas is very unreactive and phosphorus is so reactive that it spontaneously reacts with the oxygen in the air.

Because the long form of the periodic table is a very elongated diagram and because the elements from lanthanum to ytterbium and from actinium to nobelium show similar chemical behavior, the short form displays these two sets of elements in rows beneath the remainder of the table and the resulting space is closed up. Figure 2.4 shows this more compact, short form of the table.

According to the recommendations of the International Union of Pure and Applied Chemistry (IUPAC), the main and transition groups of elements are numbered from 1 to 18. This system replaces the old system of using a mixture of Roman numerals and letters, a notation that caused confusion because of differences in numbering between North America and the rest of the world. For example, in North America, IIIB referred to the group containing scandium, whereas in the rest of the world this designation was used for the group starting with boron. Numerical designations are not used for the series of elements from lanthanum (La) to ytterbium (Yb) and from actinium (Ac) to nobelium (No), because there is much more resemblance in properties within each of those rows of elements than vertically in groups.

Groups 1 and 2 and 13 through 18 represent the *main group elements* and these groups correspond to the filling of the s and p orbitals. Groups 4 through 11, corresponding to the filling of the d orbitals, are classified as the *transition metals*. The discussion of the main groups will take up the majority of space in this text because it is these elements that cover the widest range of chemical and physical properties. The elements of Group 12, although sometimes included among the transition metals, have a very different chemistry from that series; hence, Group 12 will be considered separately. Several of the main groups have been given specific names: *alkali metals* (Group 1), *alkaline earth metals* (Group 2), *chalcogens* (a little-used term for Group 16), *halogens* (Group 17), and *noble gases* (Group 18). The elements in Group 11 are sometimes called the *coinage metals*.

The elements corresponding to the filling of the $4f$ orbitals are called the *lanthanoids*, and those corresponding to the filling of the $5f$ orbitals are called the *actinoids*. They used to be named the lanthanides and actinides, but the *-ide* ending more correctly means a negative ion, such as oxide or chloride. For a few years, IUPAC was suggesting the names of lanthanons and actinons, but because the ending *-on* is preferred for nonmetals (and the lanthanoids and actinoids are all metallic elements), the *-oid* ending is now recommended. The chemistry of the elements of Group 3, scandium (Sc), yttrium (Y), and lutetium (Lu), more closely resembles that of the lanthanoids than that of the transition metals. For this reason, these three elements are usually discussed together with those of the lanthanoid elements, lanthanum to ytterbium (see Chapter 23). There is, in fact, a collective name for the Group 3 and lanthanoid elements, the *rare earth elements.*

Although the elements in the periodic table are arranged according to electron structure, we make an exception for helium ($1s^2$). Rather than placing it with the other ns^2 configuration elements, the alkaline earth metals, it is placed with the other noble gases (of configuration ns^2np^6) because of chemical similarities (see Figure 2.3). Hydrogen is even more of a problem. Although some versions of the periodic table show it as a member of Group 1 or Group 17, or both, its chemistry is unlike either the alkali metals or the halogens. For this reason, it is placed on its own in the tables in this text to indicate its uniqueness.

Four lanthanoid elements were named after the small Swedish town of Ytterby, where the elements were first discovered. These elements are yttrium, ytterbium, terbium, and erbium.

2.2   Existence of the Elements

To understand why there are so many elements and to explain the pattern of the abundances of the elements, we must look at the most widely accepted theory of the origin of the universe. This is the big bang theory, which assumes that the universe started from a single point. About one second after the universe came into existence, the temperature had dropped to about 10^{10} K, at which point protons and neutrons could exist. During the next three minutes, hydrogen-1, hydrogen-2, helium-3, helium-4, beryllium-7, and lithium-7 nuclei formed. (The number following the hyphen represents the *mass number,* the sum of protons and neutrons, of that isotope.) After these first few minutes, the universe had expanded and cooled to the point where nuclear fusion reactions could no longer occur. At this point, as is still true today, most of the universe consisted of hydrogen-1 and some helium-4.

Through gravitational effects, the atoms became concentrated in small volumes of space; indeed, the compression was great enough to cause exothermic nuclear reactions. These volumes of space we call stars. In the stars, hydrogen nuclei are fused to give more helium-4 nuclei. About 10 percent of the helium in the present universe has come from hydrogen fusion within stars. As the larger stars become older, buildup of helium-4 and additional gravitational collapse cause the helium nuclei to combine to form beryllium-8, carbon-12, and oxygen-16. At the same time, the fragile helium-3, beryllium-7, and lithium-7 are destroyed. For most stars, oxygen-16 and traces of neon-20 are the largest (highest atomic number) elements produced. However, the temperature of the very massive stars increases to a maximum as high as 10^9 K, and their density increases to about 10^6 g·cm^{-3}. Under these conditions, the tremendous repulsion between the high positive charges of carbon and oxygen nuclei can be overcome, a condition leading to the formation of all the elements up to iron. However, iron is the limit because, beyond iron, synthesis (fusion) is endothermic rather than exothermic.

When the more massive elements have accumulated in the core of the star and the energy from nuclear syntheses is no longer balancing the enormous gravitational forces, a catastrophic collapse occurs. This can happen in as short a time as a few seconds. It is during the brief time of this explosion, what we see as a supernova, that there is sufficient free energy to cause the formation of large atomic nuclei (greater than 26 protons) in endothermic nuclear reactions. All the elements from the supernovas that happened early in the history of the universe have spread throughout the universe. It is these elements that make up our solar system and, indeed, ourselves. So it is really true when songwriters and poets say that we are "stardust."

2.3 Stability of the Elements and Their Isotopes

In the universe, there are only 81 stable elements (Figure 2.5). For these elements, one or more isotopes do not undergo spontaneous radioactive decay. No stable isotope exists for any element above bismuth, and 2 elements in the earlier part of the table, technetium and promethium, exist only as radioactive isotopes. Two elements, uranium and thorium, for which only radioactive isotopes exist, are found quite abundantly on Earth because the half-lives of some of their isotopes—10^8 to 10^9 years—are almost as great as the age of Earth itself.

The fact that the number of stable elements is limited can be explained by recalling that the nucleus contains positively charged protons. Repulsive forces exist between the protons, just like the repulsive forces between electrons discussed in Chapter 1. We can visualize the neutrons simply as material separating the positive charges. Table 2.2 shows that, as the number of protons increases, the number of neutrons in the most common isotope of each element increases at a faster rate. Beyond bismuth, the number of positive charges in the nucleus becomes too large to maintain nuclear stability, and the repulsive forces prevail.

The longest half-life of a radioactive isotope is 9×10^{15} years for cadmium-113. This compares with the age of the universe being about 1.3×10^{10} years.

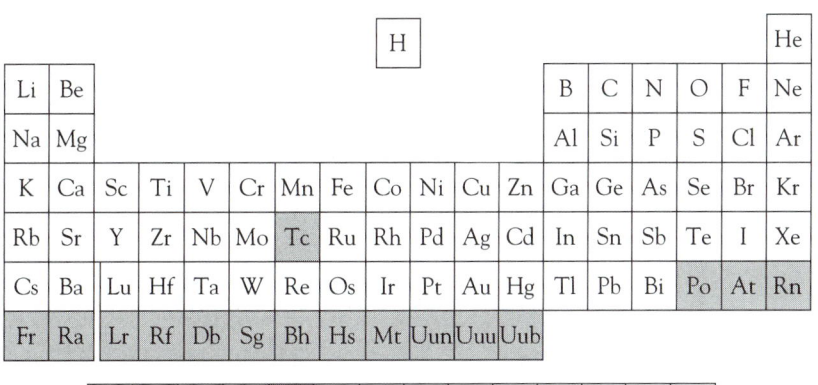

Figure 2.5 Elements that have only radioactive isotopes (shaded).

Table 2.2   Neutron–proton ratios for common isotopes

Element	No. of protons	No. of neutrons	Neutron–proton ratio
Hydrogen	1	0	0.0
Helium	2	2	1.0
Carbon	6	6	1.0
Iron	26	30	1.2
Iodine	53	74	1.4
Lead	82	126	1.5
Bismuth	83	126	1.5
Uranium	92	146	1.6

To gain a better understanding of the nucleus, we can devise a quantum (or shell) model of the nucleus. Just as Bohr visualized electrons as existing in quantum levels, so we can visualize layers of protons and neutrons (together called the *nucleons*). Thus, within the nucleus, protons and neutrons will independently fill energy levels corresponding to the principal quantum number n. However, the angular momentum quantum number l is not limited as it is for electrons. In fact, for nucleons, the filling order starts with $1s$, $1p$, $2s$, $1d$. . . . Each nuclear energy level is controlled by the same magnetic quantum number rules as electrons, so there are one s level, three p levels, and five d levels. Both nucleons have spin quantum numbers that can be $+\frac{1}{2}$ or $-\frac{1}{2}$.

Using these rules, we find that, for nuclei, completed quantum levels contain 2, 8, 20, 28, 50, 82, and 126 nucleons of one kind (compared with 2, 10, 18, 36, 54, and 86 for electrons). Thus, the first completed quantum level corresponds to the $1s^2$ configuration, the next with the $1s^21p^6$ configuration, and the following one with $1s^21p^62s^21d^{10}$. These levels are filled independently for protons and for neutrons. We find that, just like the quantum levels of electrons, completed nucleon levels confer a particular stability on a nucleus. For example, the decay of all naturally occurring radioactive elements beyond lead results in the formation of lead isotopes, all of which have 82 protons.

The influence of the filled energy levels is apparent in the patterns among stable isotopes. Thus, tin, with 50 protons, has the largest number of stable isotopes (10). Similarly, there are seven different elements with isotopes containing 82 neutrons (isotones) and six different elements with isotopes containing 50 neutrons.

If the possession of a completed quantum level of one nucleon confers a stability to the nucleus, then we might expect that nuclei with filled levels for both nucleons—so-called "doubly-magic" nuclei—would be even more favored. This is indeed the case. In particular, helium-4 with $1s^2$ configurations of both protons and neutrons is the second most common isotope in the universe, and the helium-4 nucleus (the α-particle) is ejected in many nuclear reactions. Similarly, we find that it is the next doubly completed nucleus, oxygen-16, that makes up 99.8 percent of oxygen on this planet. Calcium follows the trend with 97 percent of the element being calcium-40. As we saw in Table 2.2, the number of neutrons increases more

The Origin of the Shell Model of the Nucleus

The proposal that the nucleus might have a structure came much later than Bohr's work on electron energy levels. Of the contributors to the discovery, probably the most crucial work was accomplished by Maria Goeppert Mayer. In 1946, Mayer was studying the abundances of the different elements in the universe, and she noticed that certain nuclei were far more abundant than those of their neighbors. The higher abundances had to reflect a greater stability of those particular nuclei. She realized that the stability could be explained by considering that the protons and neutrons were not just a solid core but were themselves organized in energy levels just like the electrons.

Mayer published her ideas, but the picture was not complete. She could not understand why the numbers of nucleons to complete each energy level were 2, 8, 20, 28, 50, 82, and 126. After working on the problem for 3 years, the flash of inspiration came one evening and she was able to derive theoretically the quantum levels and sublevels. Another physicist, Hans Jensen, read her ideas on the shell model of the nucleus and, in the same year as Mayer, independently came up with the same theoretical results. Mayer and Jensen met and collaborated to write the definitive book on the nuclear structure of the atom. Becoming good friends, Mayer and Jensen shared the 1963 Nobel Prize in physics for their discovery of the structure of the nucleus.

rapidly than that of protons. Thus, the next doubly stable isotope is lead-208 (82 protons and 126 neutrons). This is the most massive stable isotope of lead, and the most common in nature.

Nuclear physicists have also had a fascination with doubly-magic nuclei. Five of these are well-established stable nuclei: helium-4 (2p, 2n), oxygen-16 (8p, 8n), calcium-40 (20p, 20n), calcium-48 (20p, 28n), and lead-208 (82p, 126n). In addition, radioactive nickel-56 (28p, 28n) and tin-132 (50p, 82n) have been long known. Atoms of neutron-rich helium-10 (2p, 8n) were synthesized in 1993, while the synthesis of proton-rich tin-100 (50p, 50n) was announced the following year. The neutron–proton ratio of 1.0 of tin-100 is much less than the average of 1.4 for the other tin isotopes, yet the half-life of tin-100 is in the seconds range, much greater than the millisecond range that would otherwise be expected. The synthesis of neutron-rich nickel-78 (28p, 50n) in 1995 again showed the stability conferred by the completion of nuclear shells.

The latest addition to the doubly-magic nuclei was the synthesis of nickel-48 (28p, 20n) in 2000. The existence of this isotope (and its significant half-life of at least 0.5 μs), with the lowest neutron–proton ratio found to date, can only be explained in terms of the Shell Model of the nucleus. Of wider significance, the synthesis of this isotope completes the list of all feasible doubly-magic isotopes. In particular, we now have the three doubly-magic isotopes of nickel, the only element for which this is possible: nickel-48 (28p, 20n), nickel-56 (28p, 28n), and nickel-78 (28p, 50n). The synthesis of nickel-48 also provides us with the only known example of

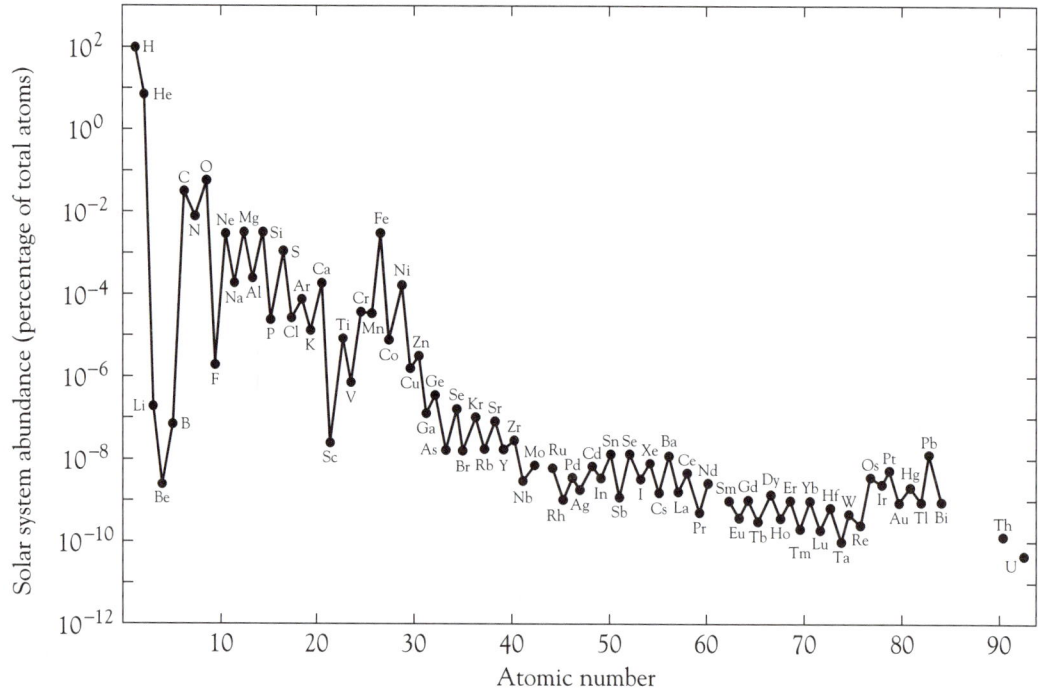

Figure 2.6   Solar system abundances of the elements as percentages on a logarithmic scale. (*Adapted from P.A. Cox, The Elements [Oxford: Oxford University Press, 1989], p. 17.*)

doubly-magic mirror pairs: calcium-48 (20p, 28n) and nickel-48 (28p, 20n). This will enable nuclear scientists to compare nucleon energy levels in the two related species.

Different from electron behavior, spin pairing is an important factor for nucleons. In fact, of the 273 stable nuclei, only 4 have odd numbers of both protons and neutrons. Elements with even numbers of protons tend to have large numbers of stable isotopes, whereas those with odd numbers of protons tend to have 1 or, at the most, 2 stable isotopes. For example, cesium (55 protons) has just 1 stable isotope, whereas barium (56 protons) has 7 stable isotopes. Technetium and promethium, the only elements before bismuth to exist only as radioactive isotopes, both have odd numbers of protons.

The greater stability of even numbers of protons in nuclei can be related to the abundance of elements on Earth. As well as the decrease of abundance with increasing atomic number, we see that elements with odd numbers of protons have an abundance about one-tenth that of their even-numbered neighbors (Figure 2.6).

2.4   Classifications of the Elements

There are numerous ways in which the elements can be classified. The most obvious is by phase at *standard ambient temperature* (25°C) *and pressure* (100 kPa), conditions that are referred to as SATP (not to be confused with the old standard, STP, of 0°C and 101 kPa pressure). Among all the elements, there are only 2 liquids and 11 gases (Figure 2.7) at SATP. It is important to define the temperature precisely, because two metals have melting

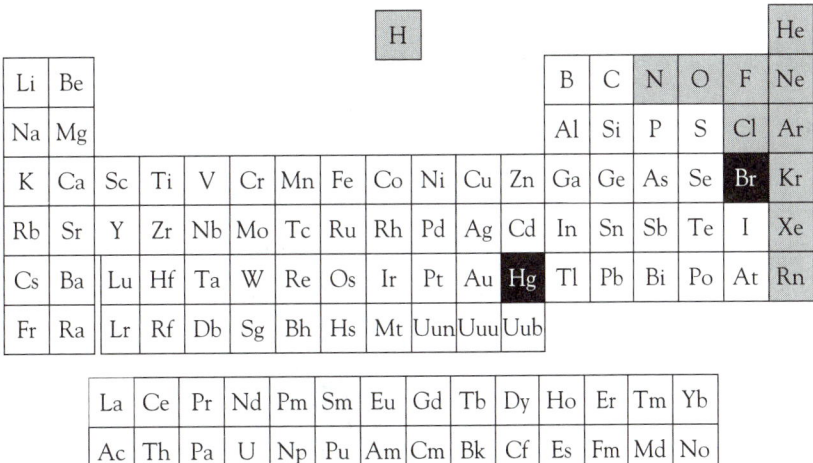

Figure 2.7 Classification of the elements into gases (shaded), liquids (black), and solids (white).

Metals show a similar range of thermal conductivities, silver being the highest and copper the second highest. It is because of its high thermal conductivity that copper is the preferred metal for cooking pots. The thermal conductivity of manganese is only 2 percent of that of silver.

points (m.p.) just slightly above the standard temperatures: cesium, m.p. 29°C, and gallium, m.p. 30°C. Also, highly radioactive francium may be a liquid at room temperature, but the numbers of atoms studied have been so small that its bulk properties could not be ascertained.

Another very common classification scheme has two categories: metals and nonmetals. But what is meant by a metal? A lustrous surface is not a good criterion, for several elements that are regarded as nonmetals—silicon and iodine are two examples—have very shiny surfaces. Even a few compounds, such as the mineral pyrite, FeS_2 (also known as fool's gold), look metallic. Density is not a good guide either, because lithium has a density ½ that of water, and osmium has a density 40 times that of lithium. Hardness is an equally poor guide, because the alkali metals are very soft. The ability of the element to be flattened into sheets (*malleability*) or to be pulled into wires (*ductility*) is sometimes cited as a common property of metals, but some of the transition metals are quite brittle. High thermal conductivity is common among the elements we call metals, but diamond, a nonmetal, has one of the highest thermal conductivities of any element. So that classification is not valid, either.

High three-dimensional electrical conductivity is the best criterion of a metal. We have to stipulate three rather than two dimensions because graphite, an allotrope of carbon, has high electrical conductivity in two dimensions. There is a difference of 10^2 in conductivity between the best electrical conducting metal (silver) and the worst (plutonium). But even plutonium has an electrical conductivity about 10^5 times better than the best conducting nonmetallic element. But, to be precise, the SATP conditions of 100 kPa pressure and 25°C have to be stipulated, because below 18°C, the stable allotrope of tin is nonelectrically conducting. Furthermore, under readily obtainable pressures, iodine becomes electrically conducting. A more specific physical criterion is the temperature dependence of the electrical conductivity, for the conductivity of metals decreases with increasing temperature, whereas that of nonmetals increases.

For chemists, however, the most important feature of an element is its pattern of chemical behavior, in particular, its tendency toward covalent bond formation or its preference for cation formation. But no matter which

Medicinal Inorganic Chemistry: An Introduction

Inorganic chemistry affects our lives directly in two ways. First, as we discuss at the end of this and later chapters, many chemical elements are required for the functioning of living organisms. Second, inorganic elements and compounds have been used as medicines since earliest times. Periodically in this text, we give examples of the use of inorganic compounds as medicinal substances, but it is useful to provide an overview.

There have been many inorganic compounds used as medicines through the ages. A fashionable habit in European countries was to "drink the waters" at Spa cities. In some cases, the springs were mineral-rich, for example, the water in Vichy, France (now available bottled), is rich in magnesium ion, which acts as a potent laxative. That water, therefore, should only be drunk in small quantities. The solid salt, magnesium sulfate heptahydrate, $MgSO_4 \cdot 7H_2O$, has the same effect. It was named Epsom salts after the town in England where it was first discovered. During the nineteenth century, one British hospital was using 2.5 tonnes per year on its patients!

Some cultures practice geophagy, the eating of soil—usually clay. Clays are a complex class of minerals as we discuss in Chapter 14. One form of clay is kaolin—a substance that is known for its absorptive abilities. Several types of tablets to combat stomach upsets employ kaolin, which, it is believed, can surface-absorb toxins produced by ingested harmful bacteria. Other clays and soils can supply trace nutrients. However, persistent clay-eating is not advised as the clay can line the stomach and prevent nutrient adsorption. Also, many natural clays contain high concentrations of harmful elements, such as lead.

Inorganic medicinal chemistry can appear in the most unusual context. For example, religious statues made from the mineral, realgar, diarsenic disulfide (As_2S_2), were popular among devotees of the Chinese Taoist religion. Handling the statues was believed to restore health. In this particular case, chemistry rather than faith might have contributed, for many people in tropical areas suffer from internal parasites and handling the statues would result in arsenic absorption through the skin, enough to kill the parasites but not enough to kill the devotee.

In this text, we mention a few of the many modern medicinal applications of inorganic compounds:

The antacids (Chapter 7)

Tellurium as a radiopharmaceutical (Chapter 8)

Lithium in the treatment of bipolar disorder (Chapter 9)

Boron neutron capture therapy (Chapter 13)

Bismuth as an antibacterial agent (Chapter 15)

Platinum complexes as anticancer agents (Chapter 19)

Gold in the treatment of rheumatoid arthritis (Chapter 20)

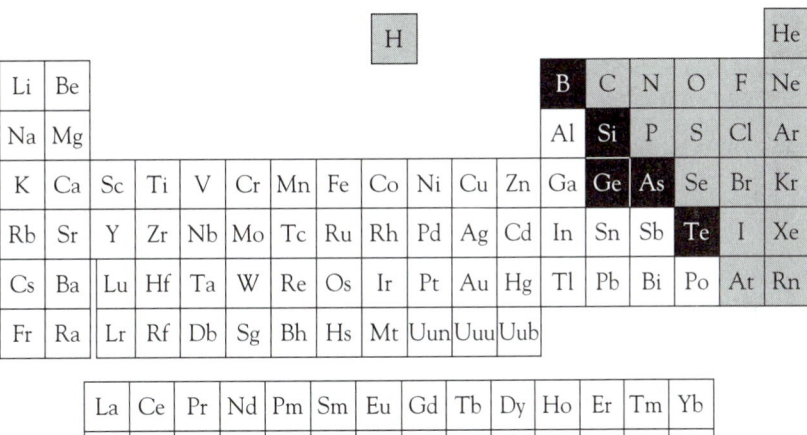

Figure 2.8 Classification of the elements into metals (white), semimetals (black), and non-metals (shaded).

criteria are used, some elements always fall in the border region of the metal–nonmetal divide. As a result, most inorganic chemists agree that boron, silicon, germanium, arsenic, and tellurium can be assigned an ambiguous status as *semimetals* (Figure 2.8), formerly called metalloids.

Even then, the division of elements into three categories is a simplification. There is a subgroup of the metals, the ones closest to the borderline, that exhibit some chemical behavior that is more typical of the semimetals, particularly formation of anionic species. These nine *weak metals* are beryllium, aluminum, zinc, gallium, tin, lead, antimony, bismuth, and polonium. As an example of one of the anionic species, we can choose aluminum. In very basic solution, as we see in Chapter 13, aluminum forms aluminates, $Al(OH)_4^-(aq)$ [sometimes written as $AlO_2^-(aq)$]. The other "weak metals" similarly form beryllates, zincates, gallates, stannates, plumbates, antimonates, bismuthates, and polonates.

2.5 *Periodic Properties: Atomic Radius*

One of the most systematic periodic properties is atomic radius. What is the meaning of atomic size? Because the electrons can be defined only in terms of probability, there is no real boundary to an atom. Nevertheless, there are two common ways in which we can define atomic radius. The *covalent radius, r_{cov},* is defined as the half-distance between the nuclei of two atoms of the same element joined in a single covalent bond. The *van der Waals radius, r_{vdw},* is defined as the half-distance between the nuclei of two atoms of neighboring molecules (Figure 2.9). Furthermore, for the metallic elements, it is possible to measure a *metallic radius:* the half-distance between the nuclei of two neighboring atoms in the solid metal.

Covalent radii are experimental values and there are significant variations in the values derived from different sets of measurements. It is important to realize that some values are obtained only from extrapolation of trends. For example, many general chemistry texts list covalent radii for all the noble gases. However, as isolable compounds of helium, neon, and argon have never been synthesized, it is obvious these values are, at best, rough estimates or theoretical values. Likewise, covalent radii for metallic elements are of limited validity and the values cited are often those of the metallic

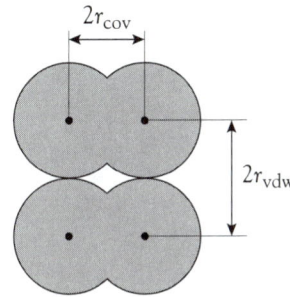

Figure 2.9 Comparison of the covalent radius, r_{cov}, and the van der Waals radius, r_{vdw}.

Be	B	C	N	O	F
106	88	77	70	66	64

					Cl
					99

					Br
					114

					I
					133

Figure 2.10   Covalent radii (pm) of a typical group and short period.

The largest atom in the periodic table is francium.

Figure 2.11   The variation of electron probability with distance from the nucleus for electrons in 1s and 2s orbitals.

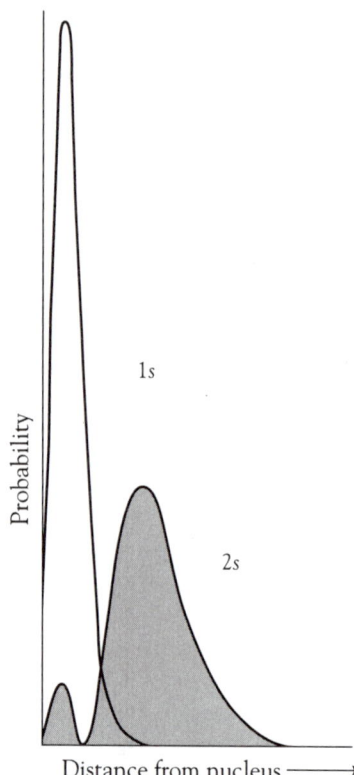

radius. Figure 2.10 provides some of the more reliable values of covalent radii (in picometers, 10^{-12} m) for the Group 17 elements and for the nonmetals, semimetal, and "weak" metal of Period 2. Though the numerical values may differ between sources, the trend is always the same: with few exceptions, radii decrease from left to right across a period and increase descending a group.

To explain these trends, we must examine the model of the atom. Let us start with lithium. A lithium atom contains three protons, and its electron configuration is $1s^2 2s^1$. The apparent size of the atom is determined by the size of the outermost orbital containing electrons, in this case, the $2s$ orbital. The electron in the $2s$ orbital is shielded from the full attraction of the protons by the electrons in the $1s$ orbital (Figure 2.11). Hence, the *effective nuclear charge*, Z_{eff}, felt by the $2s$ electron will be much less than 3 and closer to 1. The electrons in the inner orbitals will not completely shield the $2s$ electron; however, because the volumes of the $2s$ and $1s$ orbitals overlap, so the Z_{eff} will be slightly greater than 1. In fact, its value can be estimated as 1.3 charge units.

A beryllium nucleus has four protons, and the electron configuration of beryllium is $1s^2 2s^2$. There will be two factors to consider: the increased Z_{eff} as a result of the increased number of protons and the repulsions between the two negative electrons themselves. Each $2s$ electron will experience the higher Z_{eff} as each $2s$ electron, having the same average distance from the nucleus, will offer little shielding to the other. It is found that the nuclear attraction is more significant than the interelectron repulsion; hence, there will be a contraction of the $2s$ orbitals. Proceeding across the period, the contraction continues and we can explain this trend in terms of the effect of Z_{eff}. The Z_{eff} increases as a result of the increasing nuclear charge and has a greater and greater effect on the electrons that are being added to orbitals (s and p of the same principal quantum number) that overlap substantially. In other words, we can consider the value of Z_{eff} for the outer electrons to determine the apparent outer orbital size and hence the radii of the atoms across a period.

Descending a group, the atoms become larger (see Figure 2.10). This trend is also explainable in terms of the increasing size of the orbitals and the influence of the shielding effect. Let us compare a lithium atom (3 protons) with the larger sodium atom (11 protons). Because of the greater number of protons, one might expect the greater nuclear charge to cause sodium to have the smaller atomic radius. However, sodium has 10 "inner" electrons, $1s^2 2s^2 2p^6$, shielding the electron in the $3s^1$ orbital. As a result, the $3s$ electron will feel a much reduced nuclear attraction. Thus, the outermost orbital of sodium will be quite large (radially diffuse), hence accounting for the larger measured covalent radius of sodium than that of lithium.

There are a few minor variations in the smooth trend. For example, gallium has the same covalent radius (126 pm) as aluminum, the element above it. If we compare the electron configurations—aluminum is $[Ne]3s^2 3p^1$ and gallium is $[Ar]4s^2 3d^{10} 4p^1$—we see that gallium has 10 additional protons in its nucleus; these protons correspond to the electrons in the $3d$ orbitals. However, the $3d$ orbitals do not shield outer orbitals very well. Thus, the $4p$ electrons are exposed to a higher Z_{eff} than expected. As a result, the radius is reduced to a value similar to that of the preceding member of the group.

Slater's Rules   Up to now, we have used Z_{eff} in very vague terms. In 1930, J.C. Slater proposed a set of empirical rules to semiquantify the concept of effective nuclear charge. He proposed a formula that related Z_{eff} to the actual nuclear charge, Z.

$$Z_{eff} = Z - \sigma$$

where σ is called *Slater's screening constant*. Slater derived a series of empirical rules for the calculation of σ. To use this series of rules, we must order the orbitals by principal quantum number, that is, $1s$, $2s$, $2p$, $3s$, $3p$, $3d$, $4s$, $4p$, $4d$, $4f$, and so on. To find the screening constant for a particular electron, the rules are as follows:

1. All electrons in orbitals of greater principal quantum number contribute zero.

2. Each electron in the same principal quantum number contributes 0.35, except when the electron studied is in a d or f orbital, then those in the s and p orbitals count 1.00 each.

3. Electrons in the $(n - 1)$ principal quantum level contribute 0.85 each, except when the electron studied is in a d or f orbital, then they count 1.00 each.

4. All electrons in the lesser principal quantum levels count 1.00 each.

For example, to calculate the effective nuclear charge on one of the $2p$ electrons in the oxygen atom ($1s^2 2s^2 2p^4$), we first find the screening constant:

$$\sigma = (2 \times 0.85) + (5 \times 0.35) = 3.45$$

Hence, $Z_{eff} = Z - \sigma = 8 - 3.45 = 4.55$. Thus, a $2p$ electron in oxygen does not experience the full attraction of the eight protons in the nucleus, nor is there total shielding by the inner electrons. A net charge of about 4.55 is still a very strong nuclear attraction.

Although the results of calculations using Slater's rules provide a more quantitative feel for the concept of effective nuclear charge, their simplicity makes them less than perfect. For example, the rules assume that both s and p electrons in the same principal quantum number experience the same nuclear charge. From the orbital diagrams that we discussed in Chapter 1, this is obviously not the case. Using calculations based on the atomic wave functions, Clementi and Raimondi derived more precise values for the values of effective nuclear charge, some of which are shown in Table 2.3. The Clementi and Raimondi values do indeed show a small but

Table 2.3   Values of effective nuclear charge for electrons in second period elements according to Clementi and Raimondi

Element	Li	Be	B	C	N	O	F	Ne
Z	3	4	5	6	7	8	9	10
$1s$	2.69	3.68	4.68	5.67	6.66	7.66	8.65	9.64
$2s$	1.28	1.91	2.58	3.22	3.85	4.49	5.13	5.76
$2p$			2.42	3.14	3.83	4.45	5.10	5.76

significant difference in effective nuclear charge for the more penetrating s electrons compared with the p electrons in the same principal quantum number. Note that the increasing Z_{eff} on the outermost electrons along the period correlates well with the trend of decreasing atomic radii.

Relativistic Effects

There is a contraction in radius for the elements in the sixth period and beyond, compared with that predicted from classical calculations. We can explain this in terms of a shortcoming of the Schrödinger wave equation: that it fails to take into account the effects of relativity on the electrons. This simplification is acceptable for the lower mass elements, but toward the bottom of the periodic table, relativistic effects cannot be ignored. For example, the $1s$ electrons of mercury have been estimated to travel at over half the velocity of light. Such a speed results in about a 20 percent increase in mass and hence an approximately 20 percent decrease in orbital size for these electrons. This reduction in size is particularly apparent for s orbitals, as electrons in these orbitals have high probabilities close to the nucleus. The p orbitals experience a similar but lesser contraction. The d and f orbitals do not penetrate the core and, with the contraction of the s and p orbitals, they are more strongly shielded from the nucleus. As a result, the d and f orbitals expand. However, the outermost orbitals that determine atomic radius are usually s and p; hence, the net effect is a shrinkage in radius among the later-period elements. As we see in later chapters, the relativistic effect can also be used to explain anomalies in the chemistry of the more massive elements.

2.6 Periodic Properties: Ionization Energy

One trend that relates very closely to electron configuration is that of ionization energy. Usually we are interested in the *first ionization energy,* that is, the energy needed to remove one electron from the outermost occupied orbital of a free atom X:

$$X(g) \rightarrow X^+(g) + e^-$$

Whereas the values of covalent radii depend on which molecules are studied and the errors in measurement, ionization energies can be measured with great precision. Figure 2.12 shows the first ionization energies for the first two periods.

The explanation for the substantial increase from hydrogen to helium (1.3 MJ·mol^{-1} to 2.4 MJ·mol^{-1}) involves the second proton in the nucleus. Each electron in the $1s$ orbital of helium is only slightly shielded by the other. Thus, the nuclear attraction, Z_{eff}, on each electron is almost twice

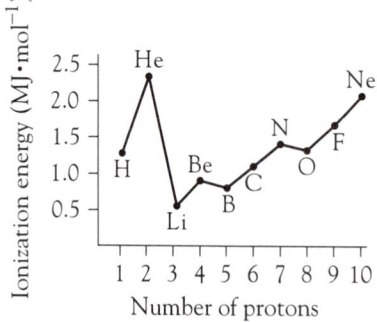

Figure 2.12   First ionization energies for the elements in Periods 1 and 2.

that for the electron of the hydrogen atom. In fact, there is a good correlation between the measured first ionization energy of an atom and the calculated value of Z_{eff} for the outermost electron.

In the lithium atom, the ionizing $2s$ electron is shielded from the nuclear attraction by the two electrons in the $1s$ orbital. With a weaker attraction to overcome, the energy needed should be much less, and that is what we find experimentally. The first ionization energy of beryllium is higher than that of lithium, and again we use the concept of little shielding between electrons in the same orbital set—in this case, the $2s$ orbital—to explain this result.

The slight drop in ionization energy for boron shows a phenomenon that is not apparent from a comparison of covalent radii; it is an indication that the s orbitals do partially shield the corresponding p orbitals. This effect is not unexpected; we show in Chapter 1 that the s orbitals penetrate closer to the nucleus than do the matching p orbitals. Following boron, the trend of increasing ionization energy resumes as the Z_{eff} increases and the additional electrons are placed in the same p orbital set.

The final deviation from the trend comes with oxygen. The drop in first ionization energy here can only be explained in terms of interelectron repulsions. That is, the one paired electron can be lost more readily than would otherwise be the case, leaving the oxygen ion with an electron configuration of $1s^2 2s^2 2p^3$. Beyond oxygen, the steady rise in first ionization energy continues to the completion of Period 2. Again, this pattern is expected as a result of the increase in Z_{eff}.

Proceeding down a group, the first ionization energy generally decreases. Using the same argument as that for the increase in atomic radius, we conclude that the inner orbitals shield the electrons in the outer orbitals and the successive outer orbitals themselves are larger. For example, compare lithium and sodium again. Although the number of protons has increased from 3 in lithium to 11 in sodium, sodium has 10 shielding inner electrons. Thus, the Z_{eff} for the outermost electron of each atom will be essentially the same. At the same time, the volume occupied by the electron in the $3s$ orbital of sodium will be significantly larger (thus, on average, the electron will be further from the nucleus) than that occupied by the electron in the $2s$ orbital of lithium. Hence, the $3s$ electron of sodium will require less energy to ionize than the $2s$ electron of lithium. A similar trend is apparent among the halogen atoms, although the values themselves are much higher than those of the alkali metals (Figure 2.13).

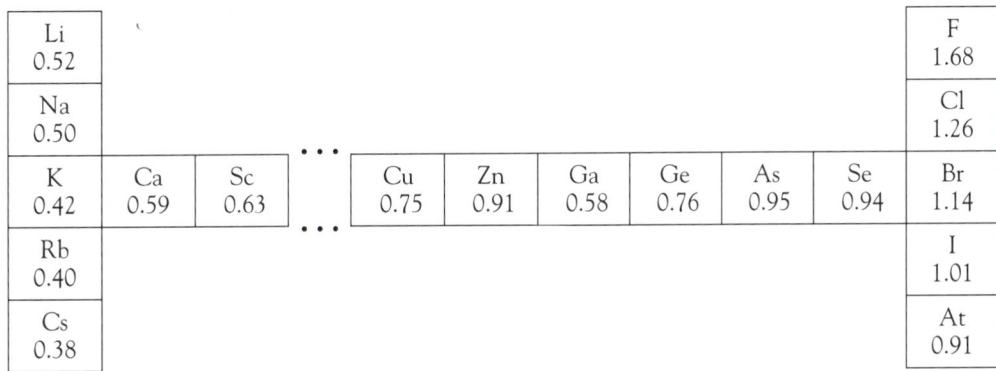

Figure 2.13 Ionization energies ($MJ \cdot mol^{-1}$) of two typical groups and part of a long period.

Looking at a long period, such as that from potassium to bromine, we find that the ionization energy rises slowly across the transition metals. Once the filling of the $4p$ orbitals commences, we find that the ionization energy for gallium is quite low, in fact, similar to that of calcium. The sudden drop can be explained by the electrons in the $3d$ orbitals becoming part of the core and hence more effective at shielding.

We can also gain information from looking at successive ionizations of an element. For example, the *second ionization energy* corresponds to the process

$$X^+(g) \rightarrow X^{2+}(g) + e^-$$

Lithium provides a simple example of such trends, with a first ionization energy of 0.52 MJ·mol^{-1}, a second ionization energy of 7.4 MJ·mol^{-1}, and a third ionization energy of 11.8 MJ·mol^{-1}. The second electron, being one of the $1s$ electrons, requires greater than 10 times more energy to remove than it takes to remove the $2s$ electron. To remove the third and last electron takes even more energy. The lesser value for removing the second electron compared to the third can be accounted for by two factors: First, there are always electron–electron repulsions when two electrons occupy the same orbital; second, even within the same orbital, one electron does partially shield the other electron.

2.7   Periodic Properties: Electron Affinity

Just as ionization energy represents the loss of an electron by an atom, so electron affinity represents the gain of an electron. *Electron affinity* is defined as the energy change when an electron is added to the lowest energy unoccupied orbital of a free atom:

$$X(g) + e^- \rightarrow X^-(g)$$

Note that addition of an electron to an alkali metal is an exothermic process. Because losing an electron by ionization is endothermic (requires energy) and gaining an electron is exothermic (releases energy), for the alkali metals, forming a negative ion is energetically preferred to forming a positive ion! This statement contradicts the dogma often taught in introductory chemistry. (The point is relevant to the discussion of ionic bonding in Chapter 5.) However, we must not forget that ion formation involves competition between a pair of elements for the electrons. Because the formation of an anion by a nonmetal is more exothermic (releases more energy) than that for a metal, it is the nonmetals that gain an electron rather than the metals.

There are conflicting sets of values for experimental electron affinities, but the trends are always consistent, and it is the trends that are important to inorganic chemists. One source of confusion is that electron affinity is sometimes defined as the energy released when an electron is added to an atom. This definition yields signs that are the opposite of those on the values discussed here. To identify which sign convention is being used, recall that halogen atoms become halide ions exothermically (that is, the electron affinities for this group should have a negative sign). A typical data set is shown in Figure 2.14.

To explain the weakly positive electron affinity for beryllium, we have to assume that the electrons in the $2s$ orbital shield any electron added to the $2p$ orbital. Thus, the attraction of a $2p$ electron to the nucleus is close

Li −60	Be +ve	B −26	C −154	N −7	O −141	F −328	Ne +ve
Na −53						Cl −398	
K −48						Br −321	
Rb −47						I −295	
Cs −46							

Figure 2.14 Electron affinities ($kJ \cdot mol^{-1}$) of a typical group and two periods.

to zero. The highly negative electron affinity of carbon indicates that addition of an electron to give the $1s^2 2s^2 2p^3$ half-filled p orbital set of the C^- ion does provide some energy advantage. The near-zero value for nitrogen suggests that the added interelectron repulsion when a $2p^3$ configuration is changed to that of $2p^4$ is a very significant factor. The high values for oxygen and fluorine, however, suggest that the high Z_{eff} for $2p$ electrons for these two atoms outweighs the interelectron repulsion factor.

Finally, just as there are sequential ionization energies, there are sequential electron affinities. These values, too, have their anomalies. Let us

Alkali Metal Anions

It is so easy to become locked into preconceptions. Everyone "knows" that the alkali metals "want" to lose an electron and form cations. In fact, this is not true, 502 kJ is required to remove the $3s$ electron from a mole of free sodium atoms. However, as we have just seen, 53 kJ of energy is released when an additional $3s$ electron is added to a mole of sodium atoms. In other words, sodium would actually prefer a filled $3s$ set of electrons than an empty $3s$ set. The reason why this does not happen normally is that the sodium is usually combined with an element with a higher electron affinity; hence, it "loses out" in the competition for the valence electrons.

It was a Michigan State chemist, James Dye, who realized that, by finding the right conditions, it might just be possible to stabilize the alkali metal anion. After a number of attempts, he found a complex organic compound of formula $C_{20}H_{36}O_6$ that could just contain a sodium cation within its structure. He was hoping that, by adding this compound to a sample of sodium metal, some of the sodium atoms would pass their s electrons to neighboring sodium atoms to produce sodium anions. This happened, as predicted:

$$2\,Na(s) + C_{20}H_{36}O_6 \rightarrow [Na(C_{20}H_{36}O_6)]^+ \cdot Na^-$$

The metallic-looking crystals were shown to contain the sodium anion, but the compound was found to be very reactive with almost everything. Although there are few uses for the compound, its existence does remind us to question even the most commonly held beliefs.

look at the first and second electron affinities for oxygen:

$$O_{(g)} + e^- \rightarrow O^-_{(g)} \qquad -141 \text{ kJ·mol}^{-1}$$

$$O^-_{(g)} + e^- \rightarrow O^{2-}_{(g)} \qquad +744 \text{ kJ·mol}^{-1}$$

Thus, the addition of a second electron is an endothermic process. This energetically unfavorable process is not surprising from the point of view of adding an electron to a species that is already negatively charged, but then one has to explain how the oxide ion exists in chemical compounds. In fact, as we see in Chapter 5, the oxide ion can only exist where there is some other driving force, such as the formation of a crystal lattice.

2.8   *Biochemistry of the Elements*

Bioinorganic chemistry, the study of the elements in the context of living organisms, is one of the most rapidly growing areas of chemistry. The details of the role of the elements in biological systems is discussed in the framework of each group, but an overview of the essential elements for life is provided here. An element is considered essential when a lack of that element produces an impairment of function and addition of the element restores the organism to a healthy state. Fourteen chemical elements are required in considerable quantities (Figure 2.15).

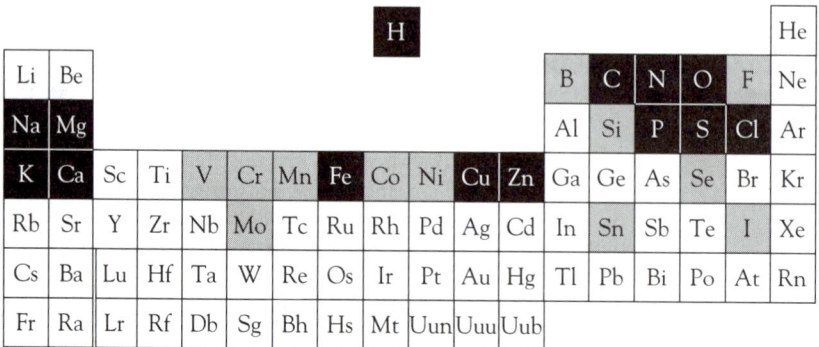

Figure 2.15 The elements necessary for life in large quantities (black) and in ultratrace amounts (shaded).

The essentiality of these bulk-requirement elements is easy to determine, but it is very challenging to identify elements that organisms need in tiny quantities—the ultratrace elements. Because we need so little of them, it is almost impossible to eliminate them from a normal diet to examine the effects of any deficiency. Up to now, 12 additional elements are confirmed as being needed for a healthy life. It is amazing that our bodies require over one-fourth of the stable elements for healthy functioning. The precise functions of some of these ultratrace elements are still unknown. As biochemical techniques become more sophisticated, it is possible that more elements will be added to the list of those required.

For almost all the essential elements, there is a range of intake that is optimum, whereas below and above that range some harmful effects are experienced. This principle is known as *Bertrand's rule* (Figure 2.16). Many people are aware of the Bertrand rule in the context of iron intake. Too little iron can cause anemia, yet children also have died after consuming too large quantities of iron supplement pills. The range of optimum intake varies

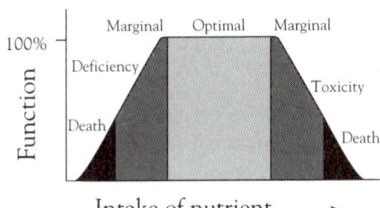

Figure 2.16 Variation of response with intake dose, Bertrand's rule.

tremendously from element to element. One of the narrow ranges is that of selenium, for which the optimum intake is between 50 $\mu g \cdot day^{-1}$ and 200 $\mu \cdot day^{-1}$. Less than 10 $\mu g \cdot day^{-1}$ will cause severe health problems, whereas death ensues from intake levels above 1 $mg \cdot day^{-1}$. Fortunately, most people, through their normal food intake, ingest levels of selenium in the required range.

EXERCISES

2.1 Define the following terms: (a) rare earth metals; (b) van der Waals radius; (c) effective nuclear charge.

2.2 Define the following terms: (a) second ionization energy; (b) electron affinity; (c) Bertrand's rule.

2.3 Explain the two reasons why the discovery of argon posed problems for the original Mendeleev periodic table.

2.4 Explain why the atomic mass of cobalt is greater than that of nickel even though the atomic number of cobalt is less than that of nickel.

2.5 Give one advantage and one disadvantage of the long form of the periodic table.

2.6 Suggest why the Group 11 elements are sometimes called the coinage metals.

2.7 Suggest why it would be more logical to call element 2 "helon" rather than helium. Why is the *-ium* ending inappropriate?

2.8 Why were the names lanthanides and actinides inappropriate for those series of elements?

2.9 Why is iron the highest atomic number element formed in stellar processes?

2.10 Why must the heavy elements on this planet have been formed from the very early supernovas that exploded?

2.11 Identify
(a) the highest atomic number element for which stable isotopes exist;
(b) the only transition metal for which no stable isotopes are known;
(c) the only liquid nonmetal at SATP.

2.12 Identify the only two radioactive elements to exist in significant quantities on Earth. Explain why they are still present.

2.13 Which element—sodium or magnesium—is likely to have only one stable isotope? Explain your reasoning.

2.14 Suggest the number of neutrons in the most common isotope of calcium.

2.15 Yttrium, element 39, exists in nature as only one isotope. Without consulting tables, deduce the number of neutrons in this isotope.

2.16 Suggest why polonium-210 and astatine-211 are the isotopes of those elements with the longest half-lives.

2.17 In the classification of elements into metals and nonmetals,
(a) why is a metallic luster a poor guide?
(b) why can't thermal conductivity be used?
(c) why is it important to define electrical conductivity in three dimensions as the best criteria for metallic behavior?

2.18 On what basis are elements classified as semimetals?

2.19 Which atom should have the larger covalent radius, potassium or calcium? Give your reasoning.

2.20 Which atom should have the larger covalent radius, fluorine or chlorine? Give your reasoning.

2.21 Suggest a reason why the covalent radius of germanium (122 pm) is almost the same as that of silicon (117 pm), even though germanium has 18 more electrons than silicon.

2.22 Suggest a reason why the covalent radius of hafnium (144 pm) is less than that of zirconium (145 pm), the element above it in the periodic table.

2.23 In Table 2.3, we show the values of effective nuclear charge for the second period elements calculated by the sophisticated method of Clementi and Raimondi. For each of those elements, calculate the effective nuclear charge on each of the $1s$, $2s$, and $2p$ orbitals according to Slater's rules. Compare them to the Clementi and Raimondi values and discuss whether the differences are really significant.

2.24 Using Slater's rules, calculate the effective nuclear charge on an electron in each of the orbitals in an atom of potassium.

2.25 Using Slater's rules, calculate the relative effective nuclear charge on one of the $3d$ electrons compared to that on one of the $4s$ electrons for an atom of manganese.

2.26 Using Slater's rules, calculate the effective nuclear charge on a $3p$ electron in (a) aluminum and (b) chlorine. Explain how your results relate to:
 (i) the relative atomic radii of the two atoms,
 (ii) the relative first ionization energies of the two atoms.

2.27 Which element should have the higher ionization energy, silicon or phosphorus? Give your reasoning.

2.28 Which element should have the higher ionization energy, arsenic or phosphorus? Give your reasoning.

2.29 An element has the following first through fourth ionization energies in $MJ \cdot mol^{-1}$: 0.7, 1.5, 7.7, 10.5. Deduce to which group in the periodic table it probably belongs. Give your reasoning.

2.30 Which one, in each pair of elements—boron and carbon, and carbon and nitrogen—will have the higher second ionization energy? Give your reasoning in each case.

2.31 For the elements sodium and magnesium, which has the higher first ionization energy? Second ionization energy? Third ionization energy?

2.32 Which element, sodium or magnesium, should have an electron affinity closer to zero? Give your reasoning.

2.33 Would you expect the electron affinity of helium to be positive or negative in sign? Explain your reasoning.

2.34 What part of the periodic table contains the elements that we need in large quantities? How does this correspond to the element abundances?

2.35 Without consulting data tables or a periodic table, write the mass number for: (a) the most common isotope of lead (element 82); (b) the only stable isotope of bismuth (element 83); (c) the longest lived isotope of polonium (element 84).

BEYOND THE BASICS

2.36 Contrary to the general trend, the first ionization energy of lead ($715 \, kJ \cdot mol^{-1}$) is higher than that of tin ($708 \, kJ \cdot mol^{-1}$). Suggest a reason for this.

2.37 Why are elemental hydrogen and helium not present in any significant amounts in the Earth's atmosphere even though they are the two most abundant elements in the universe?

2.38 Why is it wise for your food to come from a number of different geographic locations?

2.39 When element 117 is synthesized, what would you expect qualitatively in terms of its physical and chemical properties?

2.40 Use an advanced inorganic chemistry or bioinorganic chemistry text to identify a role in human nutrition for as many of the ultratrace elements shown in Figure 2.15 as possible.

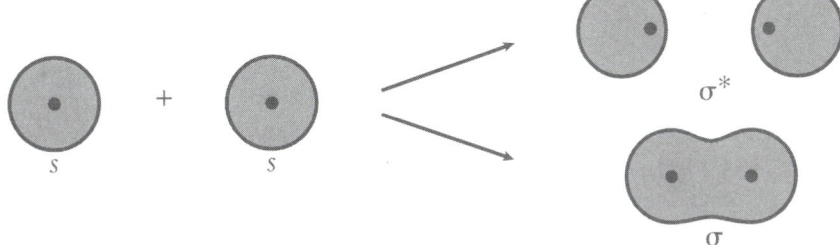

Chapter 3 Covalent Bonding

The covalent bond, one of the most crucial concepts in chemistry, is explained best by molecular orbital theory. For complex molecules, more simplistic theories can be used to predict their shapes. To account for the values of the melting points and boiling points of small covalently bonded molecules, we invoke the existence of intermolecular forces between neighboring molecules.

One of the tantalizing questions raised at the beginning of the twentieth century was: How do atoms combine to form molecules? A great pioneer in the study of bonding was Gilbert N. Lewis, who was raised on a small farm in Nebraska. In 1916, he suggested that the outer (valence) electrons could be visualized as sitting at the corners of an imaginary cube around the nucleus. An atom that was deficient in the number of electrons needed to fill eight corners of the cube could share edges with another atom to complete its octet (Figure 3.1).

As with most revolutionary ideas, many of the chemists of the time rejected the proposal. The well-known chemist Kasimir Fajans commented:

> Saying that each of two atoms can attain closed electron shells by sharing a pair of electrons is equivalent to a husband and wife, by having a total of two dollars in a joint bank account and each having six dollars in individual bank accounts, have got eight dollars apiece.

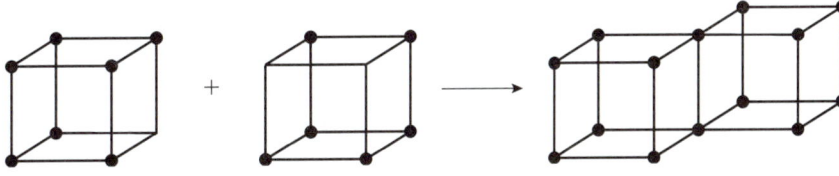

Figure 3.1 The Lewis cube model of the bonding of two halogen atoms.

3.15 Dipole–Dipole Forces

3.16 Hydrogen Bonding

3.17 Covalent Bonding and the Periodic Table

Despite the initial criticism, the Lewis concept of shared electron pairs was generally accepted, although the cube diagrams lost favor.

The classical view of bonding was soon overtaken by the rise of quantum mechanics. In 1937, Linus Pauling devised a model that involved the overlapping of atomic orbitals. Pauling was awarded the Nobel Prize in chemistry in 1954 for his work on the nature of the chemical bond.

3.1 Introduction to Covalent Bonding

In Chapter 1, we see that the quantum mechanical model of the atom provides the best means of understanding the properties and trends among the elements. For example, the underlying structure of the periodic table could be explained by means of the occupancy of the s, p, d, and f orbitals. The trends in such properties as ionization energy also become understandable using the probability model and such concepts as shielding.

Just as the properties of atoms can best be interpreted in terms of atomic orbitals, so the properties of covalent compounds can best be explained in terms of molecular orbitals. An electron in a molecular orbital is the property of the whole molecule, not of an individual atom. Molecular orbital theory enables us to explain aspects of chemical bonding that are difficult to comprehend in terms of the simple Lewis electron-dot representations that are taught in general chemistry.

Interestingly, it is one of the simplest molecules, dioxygen, O_2, that provides one of the greatest shortcomings of electron-dot diagrams. In 1845, Michael Faraday showed that oxygen gas was the only common gas to be attracted into a magnetic field; that is, dioxygen is paramagnetic, so it must possess unpaired electrons (two, as we now know). Later, bond strength studies showed that the dioxygen molecule has a double bond. Thus, any acceptable electron-dot diagram should possess these two properties. In fact, we cannot devise a reasonable electron-dot diagram that will combine both attributes. We can draw a diagram with a double bond (Figure 3.2)—but that has no unpaired electrons. Alternatively, we can draw a diagram with two unpaired electrons (Figure 3.3)—but that has a single bond. However, the ground-state molecular orbital diagram for dioxygen does, as we will see, correspond to a double bond and two unpaired electrons.

As the molecular orbital model flows naturally from the atomic orbital model of Chapter 1, it is with molecular orbitals that we start the chapter. However, molecular orbital theory is complex and is only simply applied to diatomic molecules (our focus here). Thus, when we look at the properties and shapes of more complex molecules and ions, we will revert to the electron-dot diagrams and the valence shell electron pair repulsion theory (VSEPR) that enables us to deduce molecular shape.

There is an intermediate theory that can be applied to explain the shapes of molecules—atomic orbital hybridization theory. In this theory, electrons are still the property of individual atoms, but the theory proposes that the individual atomic orbitals of an atom mix (hybridize) to give optimal bonding directions. This theory is described later in the chapter.

$$\ddot{O}::\ddot{O}$$

Figure 3.2 Electron-dot diagram for dioxygen with a double bond.

$$\cdot\ddot{O}:\ddot{O}\cdot$$

Figure 3.3 Electron-dot diagram for dioxygen with two unpaired electrons.

3.2 *Introduction to Molecular Orbital Theory*

When two atoms approach each other, according to molecular orbital theory, their atomic orbitals overlap. The electrons no longer belong to one atom but to the molecule as a whole. To represent this process, we can combine the two atomic wave functions to give two molecular orbitals. This realistic representation of the bonding in covalent compounds involves the linear combination of atomic orbitals and is called *LCAO theory*.

If it is *s* orbitals that mix, then the molecular orbitals formed are given the representation of σ and σ* (pronounced sigma and sigma-star). Figure 3.4 shows simplified electron density plots for the atomic orbitals and the resulting molecular orbitals.

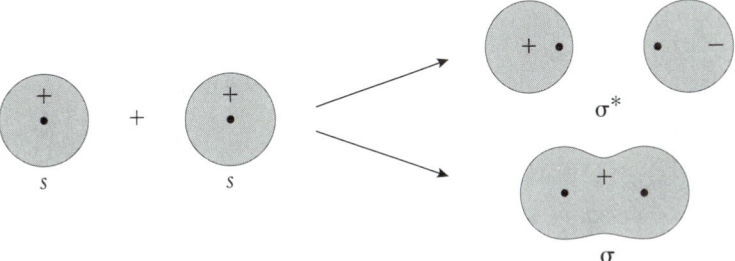

Figure 3.4 The combination of two s atomic orbitals to form σ and σ* molecular orbitals.

For the σ orbital, the electron density between the two nuclei is increased relative to that between two independent atoms. There is an electrostatic attraction between the positive nuclei and this area of higher electron density, and the orbital is called a *bonding orbital*. Conversely, for the σ* orbital, the electron density between the nuclei is decreased, and the partially exposed nuclei cause an electrostatic repulsion between the two atoms. Thus, the σ* orbital is an *antibonding orbital*. Figure 3.5 illustrates the variation in the energies of these two molecular orbitals as the atoms are brought together.

When the atoms are an infinite distance apart, there is no attraction or repulsion; thus, under those conditions they can be considered as having a zero energy state. Bringing together two atoms with electrons in bonding orbitals results in a decrease in energy as a result of electron–proton electrostatic attraction. Figure 3.5 shows that the energy of the bonding orbital reaches a minimum at a certain internuclear separation. This point represents the normal bond length in the molecule. At that separation, the attractive force between the electron of one atom and the nuclear protons of the other atom is just balanced by the repulsions between the two nuclei. When the atoms are brought closer together, the repulsive force between the nuclei becomes greater, and the energy of the bonding orbital starts to rise. For electrons in the antibonding orbital, there is no energy minimum. Electrostatic repulsion increases continuously as the partially exposed nuclei come closer and closer.

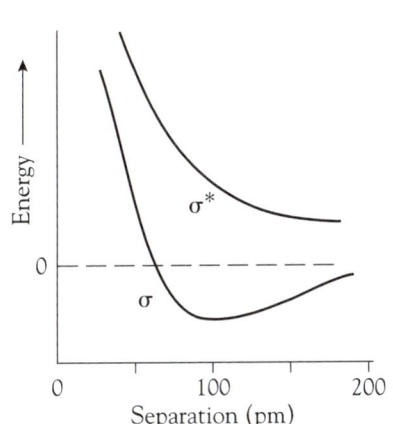

Figure 3.5 Molecular orbital energies as a function of atom separation for two hydrogen-like atoms.

Another way to picture the two types of molecular orbitals is to consider them as wave combinations. The overlap of electron wave functions of the constituent atoms in constructive interference corresponds to a bonding orbital. Destructive interference, however, corresponds to an antibonding orbital.

Several general statements can be made about molecular orbitals:

1. For orbitals to overlap, the signs on the overlapping lobes must be the same.

2. Whenever two atomic orbitals mix, two molecular orbitals are formed, one of which is bonding and the other antibonding. The bonding orbital is always lower in energy than the antibonding orbital.

3. For significant mixing to occur, the atomic orbitals must be of similar energy.

4. Each molecular orbital can hold a maximum of two electrons, one with spin $+\frac{1}{2}$, the other $-\frac{1}{2}$.

5. The electron configuration of a molecule can be constructed by using the Aufbau principle by filling the lowest energy molecular orbitals in sequence.

6. When electrons are placed in different molecular orbitals of equal energy, the parallel arrangement (Hund rule) will have the lowest energy.

7. The bond order in a diatomic molecule is defined as the number of bonding electron pairs minus the number of antibonding pairs.

In the next section, we see how the molecular orbital theory can be used to explain the properties of diatomic molecules of Period 1 and then, in the following section, look at the slightly more complex cases of Period 2 elements.

3.3 Molecular Orbitals for Period 1 Diatomic Molecules

The simplest diatomic species is that formed between a hydrogen atom and a hydrogen ion, the H_2^+ molecular ion. Figure 3.6 is an energy level diagram that depicts the occupancy of the atomic orbitals and the resulting molecular orbitals. Subscripts are used to indicate from which atomic orbitals the molecular orbitals are derived. Hence, the σ orbital arising from the mixing of two $1s$ atomic orbitals is labeled as σ_{1s}. Notice that the energy of the electron is lower in the σ_{1s} molecular orbital than it is in the $1s$ atomic

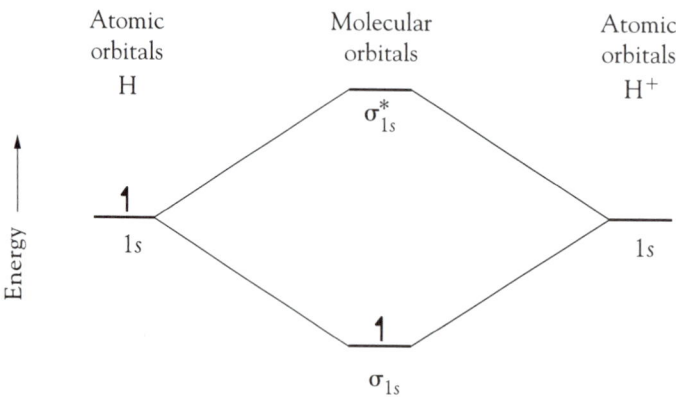

Figure 3.6 Molecular orbital diagram for the H_2^+ molecular ion.

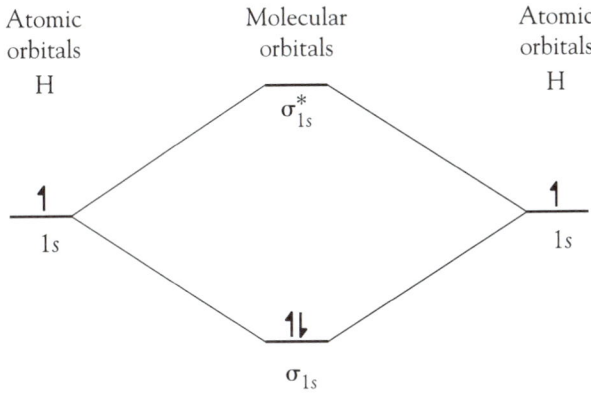

Figure 3.7 Molecular orbital diagram for the H_2 molecule.

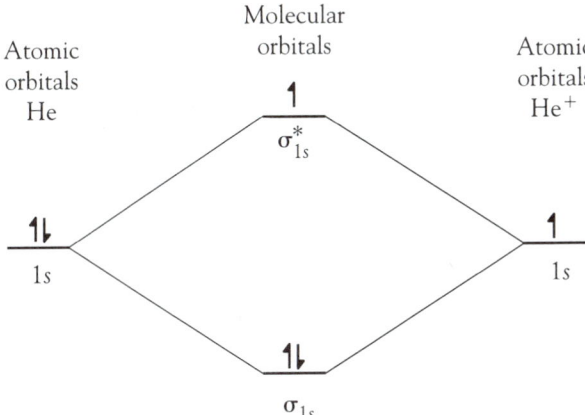

Figure 3.8 Molecular orbital diagram for the He_2^+ ion.

orbital. This is a result of the simultaneous attraction of the electron to two hydrogen nuclei. It is the net reduction in total electron energy that is the driving force in covalent bond formation.

The electron configuration of the dihydrogen cation is written as $(\sigma_{1s})^1$. A "normal" covalent bond consists of one pair of electrons. Because there is only one electron in the dihydrogen ion bonding orbital, the bond order is ½. Experimental studies of this ion show that it has a bond length of 106 pm and a bond strength of 255 kJ·mol^{-1}.

The energy level diagram for the hydrogen molecule, H_2, is shown in Figure 3.7. With a second bonding electron, the bond order is 1. The greater the bond order, the greater the strength of the bond and the shorter the bond length. This correlation matches our experimental findings of a shorter bond length (74 pm) and a much stronger bond (436 kJ·mol^{-1}) than that in the dihydrogen cation. The electron configuration is written as $(\sigma_{1s})^2$.

It is possible under extreme conditions to combine a helium atom and a helium ion to give the He_2^+ molecular ion. In this species, the third electron will have to occupy the σ^* orbital (Figure 3.8). The molecular ion has an electron configuration of $(\sigma_{1s})^2(\sigma_{1s}^*)^1$ and the bond order is $(1 - ½)$ or ½. The existence of a weaker bond is confirmed by the bond length (108 pm) and bond energy (251 kJ·mol^{-1})—values about the same as those of the dihydrogen ion.

We can make up a molecular orbital diagram for the He_2 molecule (Figure 3.9). Two electrons decrease in energy upon formation of the molecular orbitals while two electrons increase in energy by the same quantity. Thus, there is no net decrease in energy by bond formation. An alternative way of expressing the same point is that the net bond order will be zero. Thus, no covalent bonding would be expected to occur, and, indeed, helium is a monatomic gas.

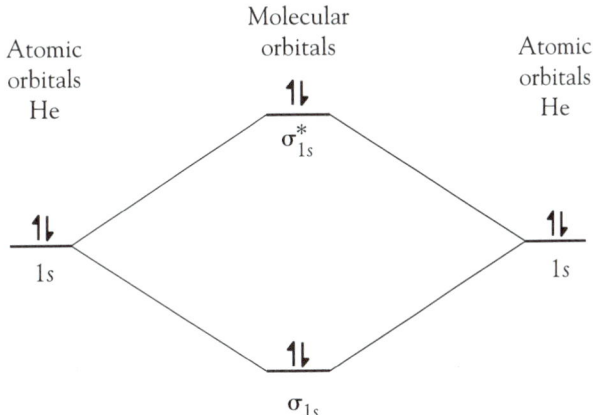

Figure 3.9 Molecular orbital diagram for the 1s atomic orbitals of the (theoretical) He_2 molecule.

3.4 Molecular Orbitals for Period 2 Diatomic Molecules

With greater effective nuclear charge—in all of the Period 2 elements (and those of later periods)—the $1s$ atomic orbitals contract closer to the nucleus. As a result, they are not involved in the bonding process. Hence, for the first two elements of Period 2, we need only construct a molecular orbital energy diagram corresponding to the $2s$ atomic orbitals. These outermost occupied orbitals, at the "edge" of the molecule, are often called the *frontier orbitals*, and they are always the crucial orbitals for bonding.

Lithium is the simplest of the Period 2 elements. In both solid and liquid phases, the bonding is metallic, a topic that we discuss in Chapter 4. In the gas phase, however, there is evidence for the existence of diatomic molecules. The two electrons from the $2s$ atomic orbitals occupy the σ_{2s} molecular orbital, thereby producing a bond order of 1 (Figure 3.10). Both the measured bond length and the bond energy are consistent with this value for the bond order. The occupancy of the frontier (valence) molecular orbitals is represented as $(\sigma_{2s})^2$.

Before we consider the heavier Period 2 elements, we must examine the formation of molecular orbitals from $2p$ atomic orbitals. These orbitals can mix in two ways. First, they can mix end to end. When this orientation occurs, a pair of bonding and antibonding orbitals is formed and resembles those of the σ_{1s} orbitals. These orbitals are designated the σ_{2p} and σ_{2p}^* molecular orbitals (Figure 3.11). In fact, a σ bond is defined as one formed by atomic orbital overlap along the axis joining the two nuclear centers. As we mentioned earlier, orbitals can only overlap if the signs of the lobes are the same—in this case, we show positive to positive.

Alternatively, the $2p$ atomic orbitals can mix side to side. To satisfy the sign requirement, there has to be overlap of positive with positive and negative with negative. The bonding and antibonding molecular orbitals formed in this way are designated π orbitals (Figure 3.12). For π orbitals, the increased electron density in the bonding orbital is not between the two nuclei, but above and below a plane containing the nuclei. Thus, by contrast to a σ bond, a π bond is formed by overlap of orbitals at right angles to the axis joining the two nuclear centers. Because every atom has three $2p$ atomic orbitals, when two such atoms combine there will be three bonding

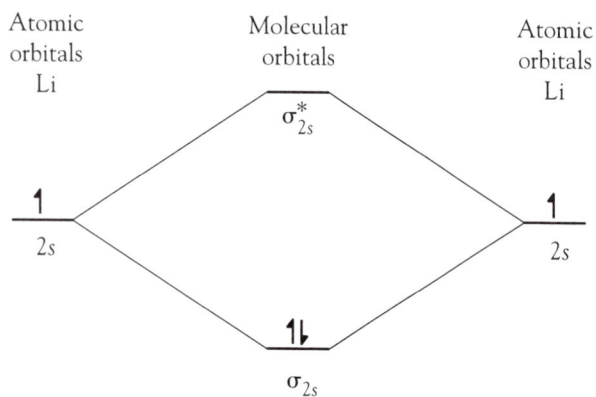

Figure 3.10 Molecular orbital diagram for the 2s atomic orbitals of the Li₂ (gas-phase) molecule.

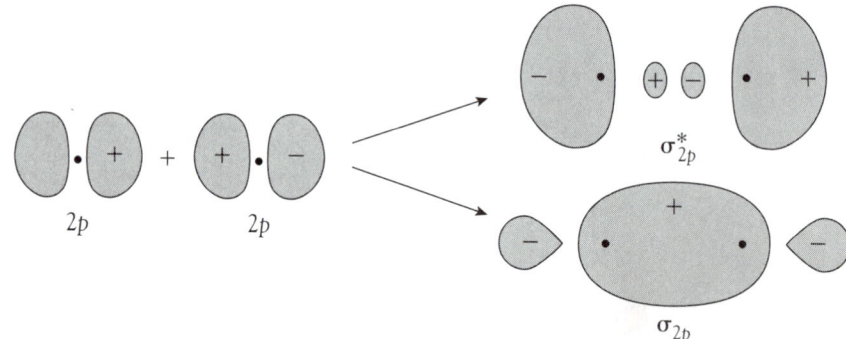

Figure 3.11 The combination of two 2p atomic orbitals end to end to form σ_{2p} and σ_{2p}^* molecular orbitals.

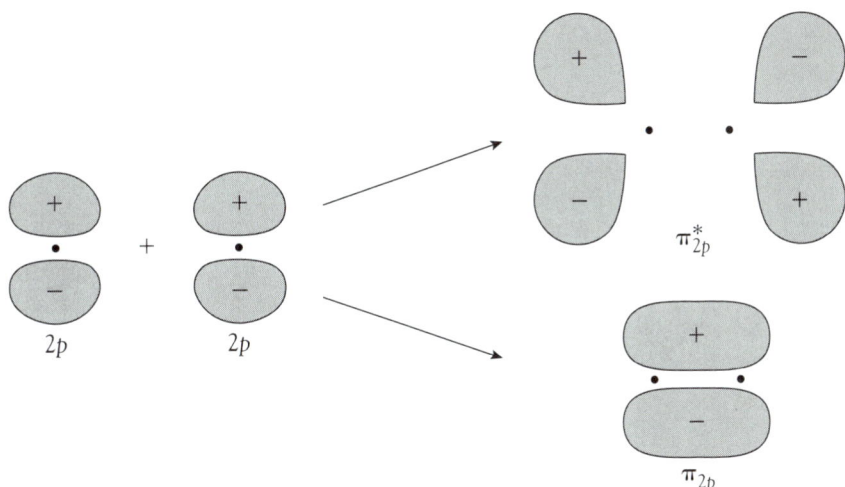

Figure 3.12 The combination of two $2p$ atomic orbitals side to side to form π_{2p} and π_{2p}^{*} molecular orbitals.

and three antibonding molecular orbitals produced in total from the $2p$ orbital set. If we assume the bonding direction to be along the z-axis, the orbitals formed in that direction will be σ_{2p_z} and $\sigma_{2p_z}^{*}$. At right angles, the other two $2p$ atomic orbitals form π_{2p_x}, $\pi_{2p_x}^{*}$, π_{2p_y}, and $\pi_{2p_y}^{*}$ molecular orbitals.

It must be emphasized that bonding models are developed to explain experimental observations. The shorter the bond length and the higher the bond energy, the stronger the bond. For the Period 2 elements, bonds with energies of 200–300 kJ·mol^{-1} are typical for single bonds; those with energies of 500–600 kJ·mol^{-1} are defined as double bonds; and those with energies of 900–1000 kJ·mol^{-1} are defined as triple bonds. Thus, for dinitrogen, dioxygen, and difluorine, the molecular orbital model must conform to the bond orders deduced from the measured bond information shown in Table 3.1

For all three of these diatomic molecules, N_2, O_2, and F_2, the bonding and antibonding orbitals formed from both $1s$ and $2s$ atomic orbitals are filled; thus, there will be no net bonding contribution from these orbitals. Hence, we need only consider the filling of the molecular orbitals derived from the $2p$ atomic orbitals.

For the Period 2 elements beyond dinitrogen, the σ_{2p} orbital is the lowest in energy, followed in order of increasing energy by π_{2p}, π_{2p}^{*}, and σ_{2p}^{*}. When the molecular orbital diagram is completed for the dioxygen molecule (Figure 3.13), we see that, according to Hund's rule, there are indeed two unpaired electrons; this diagram conforms with experimental measurements. Furthermore, the bond order of two $[3 - (2 \times \frac{1}{2})]$ is consistent with bond length and bond energy measurements. Thus, the molecular orbital model explains our experimental observations perfectly.

Table 3.1 Bond order information for the heavier Period 2 elements

Molecule	Bond length (pm)	Bond energy (kJ·mol^{-1})	Assigned bond order
N_2	110	942	3
O_2	121	494	2
F_2	142	155	1

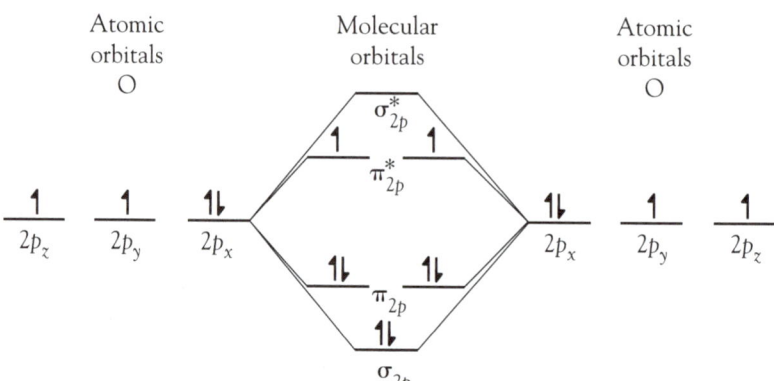

Figure 3.13 Molecular orbital diagram for the 2p atomic orbitals of the O_2 molecule.

In difluorine, two more electrons are placed in the antibonding orbital (Figure 3.14). Hence, the bond order of 1 represents the net bonding arising from three filled bonding orbitals and two filled antibonding orbitals. The valence electron configuration is represented as $(\sigma_{2s})^2(\sigma_{2s}^\star)^2(\sigma_{2p})^2(\pi_{2p})^4(\pi_{2p}^\star)^4$.

Neon is the last (heaviest) element in Period 2. If a molecular orbital diagram is constructed for the theoretical Ne_2 molecule, all the bonding and antibonding orbitals derived from the $2p$ atomic orbitals are filled; as a result, the net bond order is 0. This prediction is consistent with the observation that neon exists as a monatomic gas.

Up to now, we have avoided discussion of the elements lying in the middle part of Period 2, particularly dinitrogen. The reason concerns the relative energies of the $2s$ and $2p$ orbitals. For fluorine, with a high Z_{eff}, the $2s$ atomic energy level is about 2.5 MJ·mol^{-1} lower in energy than that of the $2p$ level. This difference results from the penetration of the s orbital close to the nucleus (as discussed in Chapter 2, Section 2.5), hence an electron in an s orbital is more strongly influenced by the increasing nuclear charge.

However, at the beginning of the period, the levels differ in energy by only about 0.2 MJ·mol^{-1}. In these circumstances, the wave functions for the $2s$ and $2p$ orbitals become mixed. One result of the mixing is an increase in energy of the σ_{2p} molecular orbital to the point where it has greater energy than the π_{2p} orbital. This ordering of orbitals applies to dinitrogen and the preceding elements in Period 2, the σ-π crossover occurring between dinitrogen and dioxygen. When we use this modified molecular orbital

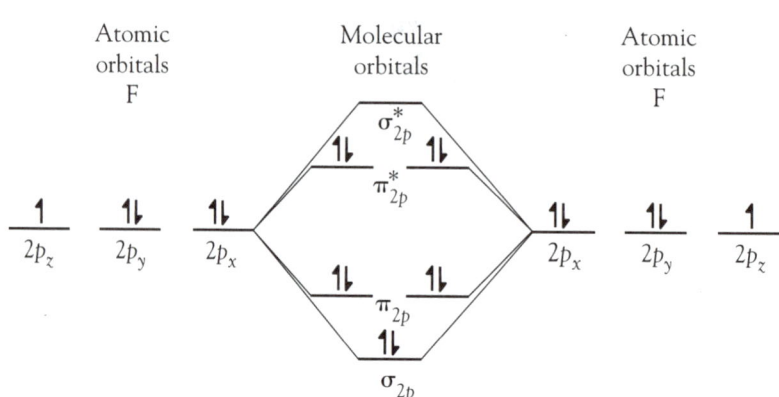

Figure 3.14 Molecular orbital diagram for the 2p atomic orbitals of the F_2 molecule.

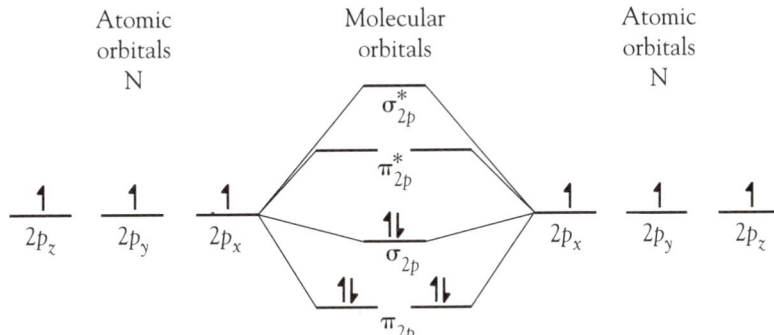

Figure 3.15   Molecular orbital diagram for the 2p atomic orbitals of the N_2 molecule.

diagram to fill the molecular orbitals from the 2p atomic orbitals for the dinitrogen molecule, we get a configuration with a bonding order of 3 (Figure 3.15). This calculation corresponds with the strong bond known to exist in the molecule. The valence electron configuration of dinitrogen is $(\sigma_{2s})^2(\sigma_{2s}^*)^2(\pi_{2p})^4(\sigma_{2p})^2$.

How are we so sure that the molecular orbital energies are actually what we predict? The orbital energies can be measured by a technique known as ultraviolet photoelectron spectroscopy (UV-PES). High-frequency ultraviolet radiation is directed at the molecule, causing an electron from one of the outer orbitals to be ejected. When an electron is lost from a dinitrogen molecule, a dinitrogen cation remains:

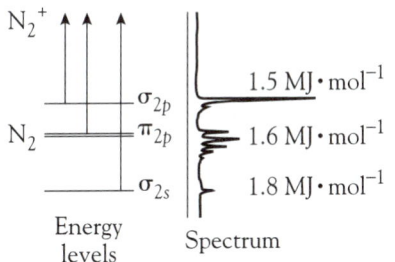

Figure 3.16   Correspondence between the three highest occupied molecular orbitals of the dinitrogen molecule and its photoelectron spectrum.

$$N_2(g) \xrightarrow{\text{UV}} N_2^+(g) + e^-$$

The various electrons removed have specific energies, and we can correlate those energies with the different molecular orbitals. This technique is illustrated for dinitrogen in Figure 3.16, where the three highest occupied molecular orbitals are matched in energy to the observed UV-PES spectrum. (The several lines for the electrons ejected from π_{2p} orbitals result from molecular vibrations.)

3.5   *Molecular Orbitals for Heteronuclear Diatomic Molecules*

When we combine atomic orbitals from different elements, we have to consider that the atomic orbitals will have different energies. For elements from the same period, the higher the atomic number, the higher the Z_{eff}, and hence the lower the orbital energies. We can use molecular orbital theory to visualize the bonding of carbon monoxide. A simplified diagram of the molecular orbitals derived from the 2s and 2p atomic orbitals is shown in Figure 3.17. The oxygen atomic orbitals are lower in energy than those of carbon as a result of the greater Z_{eff}, but they are close enough in energy that we can construct a molecular orbital diagram similar to that of the homonuclear diatomic molecules.

A major difference between homonuclear and heteronuclear diatomic molecules is that the molecular orbitals derived primarily from the 2s atomic orbitals of one element overlap significantly in energy with those derived from the 2p atomic orbitals of the other element. Thus, we must consider molecular orbitals derived from both these atomic orbitals in our diagram. Furthermore, because of the asymmetry of the orbital energies, the bonding

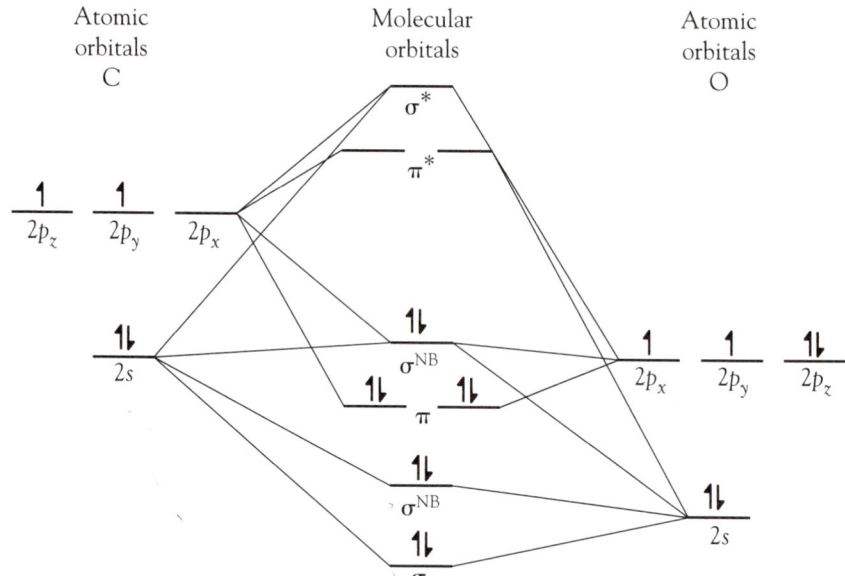

Figure 3.17 Simplified molecular orbital diagram for the 2s and 2p atomic orbitals of the CO molecule.

molecular orbitals are derived mostly from the lower energy oxygen atomic orbitals, whereas the antibonding molecular orbitals are derived mostly from the higher energy carbon atomic orbitals. Finally, there are two molecular orbitals whose energies are between those of the contributing atomic orbitals. These orbitals, σ_{NB}, are defined as *nonbonding molecular orbitals*, that is, they do not contribute significantly to the bonding.

To determine the bond order of carbon monoxide, the number of antibonding pairs (0) is subtracted from the number of bonding pairs (3), a calculation leading to the prediction of a triple bond. The very high bond energy of 1072 $kJ \cdot mol^{-1}$ supports this prediction. However, the molecular orbital diagram is much more meaningful by providing us with a grasp of electron energies. The molecular orbital diagram also indicates that the triple bond is not made up by three equivalent bonds as the electron-dot diagram suggests but by a combination of one σ and two π bonds.

The molecular orbital approach can be applied to diatomic molecules containing atoms of different periods. However, it is then necessary to identify the orbitals of similar energies on the two atoms, a task well beyond the scope of this descriptive inorganic chemistry course. It is instructive, however, to do one example. Consider the hydrogen chloride molecule (Figure 3.18). Calculations show that the 3p orbitals of chlorine have a slightly lower energy than that of the 1s orbital of hydrogen. The 1s orbital can only form a σ bond and this has to be with the 3p orbital that is aligned along the bonding axis (traditionally chosen as the p_z orbital). Hence, we conclude that a σ bonding and σ antibonding pair of orbitals will be formed between the 1s (H) and 3p (Cl) orbitals. With each atom contributing one electron, the bonding molecular orbital will be filled. This configuration yields a single bond. The two other 3p orbitals are oriented in such a way that no net overlap (hence, no mixing) with the 1s orbital of hydrogen can occur. As a result, the electron pairs in these orbitals are considered to be nonbonding. That is, they have the same energy in the molecule as they did in the independent chlorine atom.

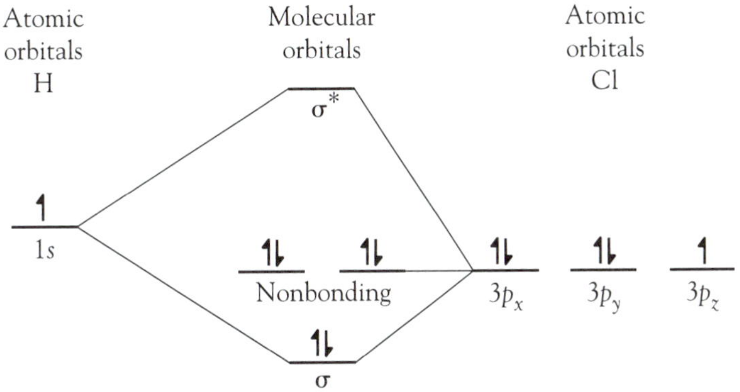

Figure 3.18 Molecular orbital diagram for the 1s atomic orbital of hydrogen and the 3p atomic orbitals of chlorine in the HCl molecule.

Molecular orbital theory can also be used to develop bonding schemes for molecules containing more than two atoms. However, the energy diagrams and the orbital shapes become more and more complex. In later chapters, we look specifically at the π molecular orbital energy diagrams of some triatomic molecules in order to explain the bond orders that we find from experiment. Nevertheless, for most of the polyatomic molecules, we are more interested in the prediction of the shapes of molecules rather than in the finer points of orbital energy levels.

3.6 A Brief Review of the Lewis Theory

Molecular orbital theory is a very powerful theory, as it can be used to provide a conceptual basis for our observations of bond length and bond energy in covalent molecules. The theory categorizes bonds as σ or π and the theory can cope with the concept of fractional electron bonds. The simplistic bonding approach of G.N. Lewis—that atoms combine by sharing electron pairs—tells us nothing in detail about the bonds themselves. Nevertheless, for complex molecules, constructing simple electron-dot representations is very useful. In particular, we can use such representations to deduce molecular shape.

The Lewis, or electron-dot, approach to covalent bond formation is covered extensively in high school and freshman chemistry; hence, only a brief review is provided here. The Lewis theory explains the driving force of bond formation as being the desire of each atom in the molecule to attain an octet of electrons in its outer (valence) energy level (except hydrogen, where a duet is required). Completion of the octet is accomplished by a sharing of electron pairs between bonded atoms.

$$H : \ddot{\underset{..}{Cl}} :$$

Figure 3.19 The electron-dot diagram for hydrogen chloride.

We can illustrate using the two examples from the previous section, hydrogen chloride and carbon monoxide. In hydrogen chloride, the sharing of an electron pair results in a single bond (Figure 3.19)—the σ bond of molecular orbital theory. The other electron pairs are lone pairs—the equivalent of being in nonbonding molecular orbitals (see Figure 3.18).

To make up the octet around both carbon and oxygen in carbon monoxide, three bonding pairs are needed (Figure 3.20). This is equivalent to the one σ and two π bonds of molecular orbital theory (see Figure 3.17). The lone pair on each atom corresponds in the molecular orbital representation to the two electron pairs in σ_{NB} orbitals.

$$: C ::: O :$$

Figure 3.20 The electron-dot diagram for carbon monoxide.

Constructing Electron-Dot Diagrams

The most commonly used method for constructing electron-dot diagrams is the following set of procedures:

1. Identify the central atom, usually the atom of lower electronegativity (though hydrogen is never central). Write the symbol of the central atom and place the symbols of the other atoms around the central atom.

2. Count the total number of valence electrons. If it is a charged ion and not a neutral molecule, add on the number of negative charges or subtract the number of positive charges.

3. Place an electron pair (single covalent bond) between the central atom and each of the surrounding atoms. Add lone pairs to the surrounding atoms. Then any excess electrons are added to the central atom.

4. If the number of electrons on the central atom is less than eight, and there are "left-over" electrons, add lone pairs to the central atom. If the number of electrons on the central atom is less than eight, and there are no more electrons, construct double and triple bonds using lone pairs from surrounding atoms.

Figure 3.21 Electron-dot diagram for nitrogen trifluoride.

As an example, we can use nitrogen trifluoride. The lower electronegativity nitrogen atom will be surrounded by the three fluorine atoms. The total number of valence electrons is $[5 + (3 \times 7)] = 26$. Six will be used to form single covalent bonds. Eighteen electrons are needed to provide lone pairs to the fluorine atoms. The remaining electron pair will provide a lone pair on the nitrogen atom (Figure 3.21).

Exceeding the Octet

There are a few anomalous molecules in which the central atom has fewer than 8 electrons. There also are a substantial number of molecules in which the central atom has shares in more than 8 bonding electrons. Lewis did not realize that the maximum of 8 was only generally applicable to the Period 2 elements, for which the sum of the *s* and *p* electrons could not exceed 8. For elements in Period 3 and higher periods, *d* orbitals can be used in bonding to give a theoretical maximum of 18 bonding electrons. In reality, compounds of some of the higher period elements frequently have central atoms with 8, 10, or 12 bonding electrons. For example, phosphorus pentafluoride "pairs up" its 5 outer electrons with 1 electron of each of the fluorine atoms to attain an outer set of 10 electrons (Figure 3.22).

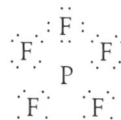

Figure 3.22 Electron-dot diagram for phosphorus pentafluoride.

3.7 Partial Bond Order

In some cases, the only structure that can be drawn does not correlate with our measured bond information. The nitrate ion illustrates this situation. A conventional electron-dot diagram for the nitrate ion is shown in Figure 3.23. In this structure, one nitrogen–oxygen bond is a double bond, whereas the other two nitrogen–oxygen bonds are single bonds.

However, it has been shown that the nitrogen–oxygen bond lengths are all the same at 122 pm. This length is significantly less than the "true" (theoretical) nitrogen–oxygen single bond length of 141 pm. We explain this discrepancy by arguing that the double bond is shared between the three

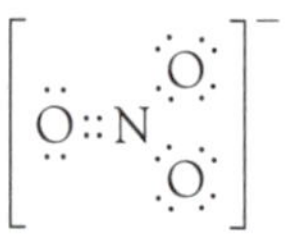

Figure 3.23 Electron-dot diagram for the nitrate ion.

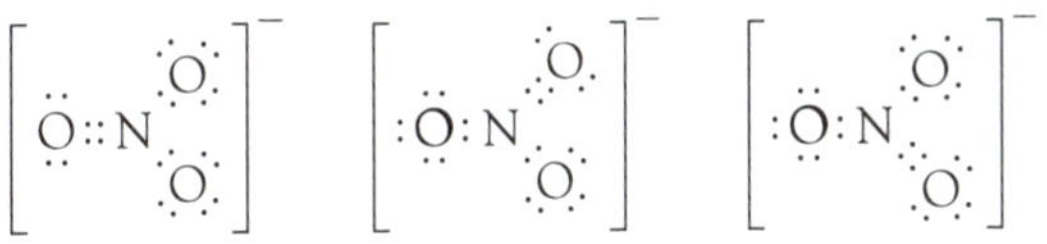

$$\left[\ddot{O} :: N \begin{matrix} \ddot{O} \\ \vdots \\ \ddot{O} \end{matrix} \right]^{-} \quad \left[:\ddot{O} : N \begin{matrix} \ddot{O} \\ \vdots \\ \ddot{O} \end{matrix} \right]^{-} \quad \left[:\ddot{O} : N \begin{matrix} \ddot{O} \\ \vdots \\ \ddot{O} \end{matrix} \right]^{-}$$

Figure 3.24 The three resonance structures of the nitrate ion.

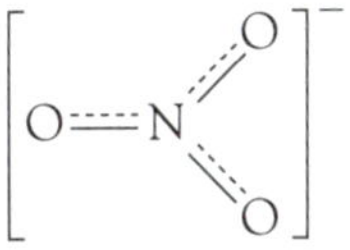

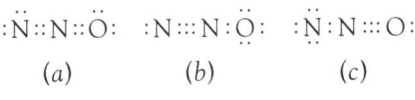

Figure 3.25 Representation of the partial multiple bond character of the nitrate ion.

nitrogen–oxygen bond locations—a concept called *resonance*. (This is equivalent to saying in molecular orbital theory that the electron pair is in a molecular orbital derived from $2p$ atomic orbitals on all four constituent atoms.) The three alternatives could be represented by three electron-dot diagrams for the nitrate ion, each with the double bond in a different location (Figure 3.24).

A third approach is to use a structural formula with broken lines to represent a fractional bond order (Figure 3.25). In this case, because the double bond character is shared by three bonds, the average bond order would be $1\frac{1}{3}$. This representation is, in most ways, the best of the three. It indicates the equivalency of the three bonds and that they each have a bond order between 1 and 2. The partial bond structure is also the closest to the molecular orbital representation, where we would depict each pair of atoms being joined by a σ bond while a two-electron π bond is shared over the whole ion (see Chapter 15, Section 15.16).

3.8 Formal Charge

In some cases, we can draw more than one feasible electron-dot diagram, one such example being dinitrogen oxide. It is known that N_2O is an asymmetric linear molecule with a central nitrogen atom, but there are a number of possible electron-dot diagrams, three of which are shown in Figure 3.26.

$$:\ddot{N}::N::\ddot{O}: \quad :N:::N:\ddot{O}: \quad :\ddot{N}:N:::O:$$
$$\quad (a) \qquad\qquad (b) \qquad\qquad (c)$$

Figure 3.26 Three possible electron-dot diagrams for the dinitrogen oxide molecule.

To help decide which possibilities are unrealistic, we can use the concept of *formal charge*. To find the formal charge, we divide the bonding electrons equally among the constituent atoms and compare the number of assigned electrons for each atom with its original number of valence electrons. Any difference is identified by a charge sign (Figure 3.27). For example, in structure (*a*), the left-hand nitrogen atom is assigned six electrons; the free atom has five. Hence, its formal charge is $(5 - 6)$, or -1. The central nitrogen atom has four assigned electrons and a formal charge of $(5 - 4)$, or $+1$; the oxygen atom has the same number of electrons as a free atom $(6 - 6)$, so its formal charge is 0.

Figure 3.27 Assignment of formal charges to three electron-dot diagrams for the dinitrogen oxide molecule.

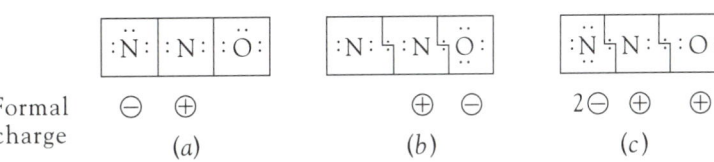

Formal charge

$$\begin{array}{ccc} \ominus \quad \oplus & \oplus \quad \ominus & 2\ominus \quad \oplus \quad \oplus \\ (a) & (b) & (c) \end{array}$$

$$N \equiv\!\equiv N \equiv\!\equiv O$$

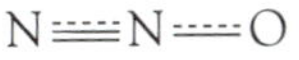

Figure 3.28 Representation of the partial bond order in the dinitrogen oxide molecule.

According to the concept of formal charge, the lowest energy structure will be the one with the smallest formal charges on the atoms. In the case of dinitrogen oxide, structure (*c*) is eliminated, but both (*a*) and (*b*) have equal but different formal charge arrangements. The optimum representation, then, is likely to be a resonance mixture of these two possibilities. This resonance form can best be represented by partial bonds, as shown in Figure 3.28. If these two resonance forms contributed equally, there would be an N–N bond order of $2\frac{1}{2}$ and an N–O bond order of $1\frac{1}{2}$, which is close to that estimated from measurement of bond lengths.

3.9   Valence Shell Electron Pair Repulsion Theory

Electron-dot diagrams can be used to derive the probable molecular shape. To accomplish this, we can use a very simplistic concept that tells us nothing about the bonding but is surprisingly effective at predicting molecular shapes—the valence shell electron pair repulsion theory, known as the *VSEPR* theory. According to the VSEPR approach, it is assumed that repulsions between electron pairs in the outermost occupied energy levels on a central atom will cause those electron pairs to be located as far from one another as is geometrically possible. For the purposes of this theory, we must ignore the differences between the energies of the *s*, *p*, and *d* orbitals and simply regard them as degenerate. It is these outer electrons that are traditionally called the valence electrons. Although conceptually there are many flaws with the VSEPR concept, it is an effective tool for deducing molecular shape.

The VSEPR theory is concerned with electron groupings around the central atom. An electron grouping can be an electron pair of a single bond, the two electron pairs in a double bond, the three electron pairs in a triple bond, a lone pair of electrons, or the rare case of a single electron. For simplicity, lone pairs are shown only for the central atom in diagrams of molecular geometry. In the following sections, we look at each of the common configurations in turn.

Linear Geometry

All diatomic molecules and ions are, by definition, linear. However, our main interest is the few common examples of this simplest geometry with triatomic molecules and ions. The most often used example is beryllium chloride. This compound has a complex structure in the room-temperature solid phase, but when it is heated above its boiling point of 820°C, it forms simple triatomic molecules. According to the Lewis theory, the two outer electrons of beryllium pair with one electron of each chlorine atom and form two electron pairs around the central beryllium atom. Because there are only two electron groupings around the central atom, the bonds will be farthest apart when the angle between them is 180°. Hence, the molecule should be linear, and that is indeed what we find (Figure 3.29).

Another example of a molecule with two electron groupings is carbon dioxide. Although both carbon–oxygen bonds involve two electron pairs, each double bond represents only one electron grouping (Figure 3.30). Hence, the carbon dioxide molecule is linear.

Trigonal Planar Geometry

Boron trifluoride is the common example of trigonal planar geometry. The three outer electrons of the boron atom pair with one electron of each of the fluorine atoms to produce three electron pairs. The maximum separation of three electron pairs requires an angle of 120° between each pair, as shown in Figure 3.31.

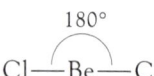

180°

Cl—Be—Cl

Figure 3.29 Predicted and actual geometry for the gaseous beryllium chloride molecule.

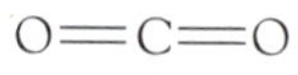

O=C=O

Figure 3.30 Predicted and actual geometry for the carbon dioxide molecule.

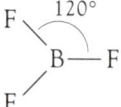

Figure 3.31 Predicted and actual geometry for the boron trifluoride molecule.

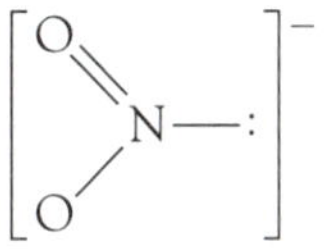

Figure 3.32 Predicted and actual geometry for the nitrite ion.

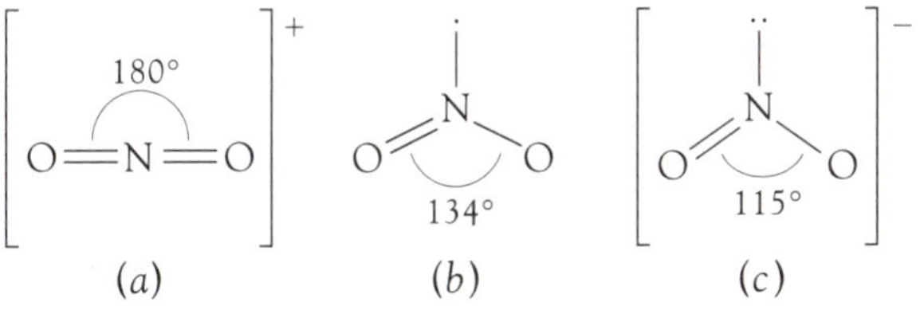

Figure 3.33 Actual geometries for the nitryl ion (NO_2^+), the nitrogen dioxide molecule (NO_2), and the nitrite ion (NO_2^-).

The nitrite ion is a good example of a species containing a lone pair on the central atom. The electron pair arrangement around the nitrogen atom is trigonal planar (Figure 3.32). However, we cannot detect lone pairs experimentally. The molecular shape we actually observe is V-shaped (also called angular or bent). According to VSEPR theory, the lone pair must occupy that third site; otherwise the molecule would be linear.

The nitrite ion illustrates a phenomenon that is generally true for all molecules and ions containing lone pairs. The molecular angles deviate from those of the theoretical geometric figure. For example, the O–N–O bond angle is "squashed" down to 115° from the anticipated 120° value. One suggested explanation is that the lone pairs of electrons occupy a greater volume of space than do the bonding pairs. We can use a series of ions and molecules to illustrate this concept. The nitronium ion, NO_2^+, with only two electron groupings, is linear; the neutral nitrogen dioxide molecule, NO_2, with three electron groupings (one an "odd" electron), has an O–N–O bond angle of 134°; the nitrite ion, which has a lone pair rather than a single electron, has an observed bond angle of 115° (Figure 3.33). Thus, even though we cannot experimentally "see" lone pairs, they must play a major role in determining molecular shape. The names of the shapes for central atoms that have three electron groupings are given in Table 3.2.

Table 3.2 Molecules and ions with trigonal planar geometry

Bonding pairs	Lone pairs	Shape
3	0	Trigonal planar
2	1	V

Tetrahedral Geometry

The tetrahedral geometry about the carbon atom was first proposed in 1874 by the 22-year-old Dutch chemist van't Hoff. The idea was denounced by the established chemist, Kolbe, as "phantasmagorical puffery," "fantastic foolishness," and "shallow speculations."

The most common of all molecular geometries is that of the tetrahedron. To place four electron pairs as far apart as possible, molecules adopt this particular three-dimensional geometry in which the bond angles are 109½°. The simplest example is the organic compound methane, CH_4, shown in Figure 3.34. To represent the three-dimensional shape on two-dimensional paper, it is conventional to use a solid wedge to indicate a bond directed above the plane of the paper and a broken line to denote a bond angled below the plane of the paper.

Ammonia provides the simplest example of a molecule where one of the four electron pairs on the central atom is a lone pair. The resulting molecular shape is trigonal pyramidal (Figure 3.35). Like the earlier example of the nitrite ion, the H–N–H bond angle of 107° is slightly less than the expected 109½°. The most familiar molecule with two lone pairs is water (Figure 3.36).

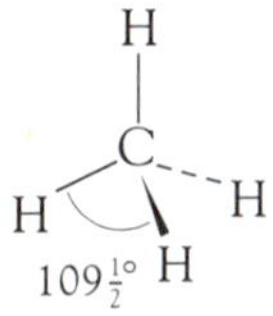

Figure 3.34 Predicted and actual geometry for the methane molecule.

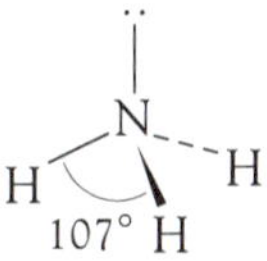

Figure 3.35 Actual geometry for the ammonia molecule.

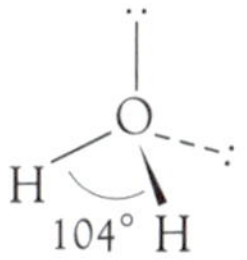

Figure 3.36 Actual geometry for the water molecule.

Table 3.3 Molecules and ions with tetrahedral geometry

Bonding pairs	Lone pairs	Shape
4	0	Tetrahedral
3	1	Trigonal pyramidal
2	2	V

The H–O–H bond angle in this V-shaped molecule is reduced from the expected 109½° to 104½°. The names of the shapes for central atoms that have four electron groupings are given in Table 3.3.

Trigonal Bipyramidal Geometry

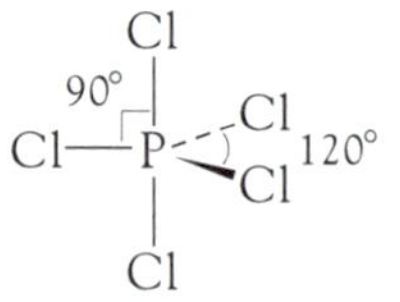

As mentioned earlier, atoms beyond Period 2 can possess more than four electron pairs when occupying the central position in a molecule. An example of five electron pairs around the central atom is provided by phosphorus pentafluoride in the gas phase (Figure 3.37). This is the only common molecular geometry in which the angles are not equal. Thus, three (equatorial) bonds lie in a single plane and are separated by angles of 120°; the other two (axial) bonds extend above and below the plane and make an angle of 90° with it.

Sulfur tetrafluoride provides an example of a molecule that has a trigonal bipyramidal electron pair arrangement with one lone pair. There are two possible locations for the lone pair: one of the two axial positions (Figure 3.38*a*) or one of the three equatorial locations (Figure 3.38*b*). In fact, we find that lone pairs are located so that, first, they are as far from one another as possible and, second, they are as far from the bonding pairs as possible. Sulfur tetrafluoride possesses one lone pair, so only the second guideline is applicable. If the lone pair were in an axial position, there would be three bonding pairs at 90° and one at 180°. However, if the lone pair were in an equatorial position, there would be only two bonding pairs at an angle of 90° and the other two at 120°. It is the second possibility, in which the atoms are arranged in the seesaw shape, that provides the optimum situation. This arrangement has been confirmed by bond angle measurements. In the measured angles, the axial fluorine atoms are bent away from the lone pair by 93½° rather than by 90°. Much more striking is the compression of the F–S–F equatorial angle from 120° to 103°, presumably as a result of the influence of the lone pair (Figure 3.39).

The bromine trifluoride molecule provides an example of trigonal bipyramidal electron pair arrangement with two lone pairs (Figure 3.40). The minimum electron repulsions occur with both lone pairs in the equatorial plane. Hence, the molecule is essentially T-shaped, but the axial

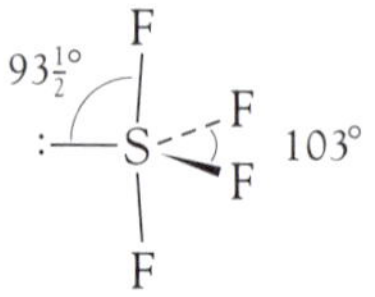

Figure 3.37 Predicted and actual geometry for the gaseous phosphorus pentafluoride molecule.

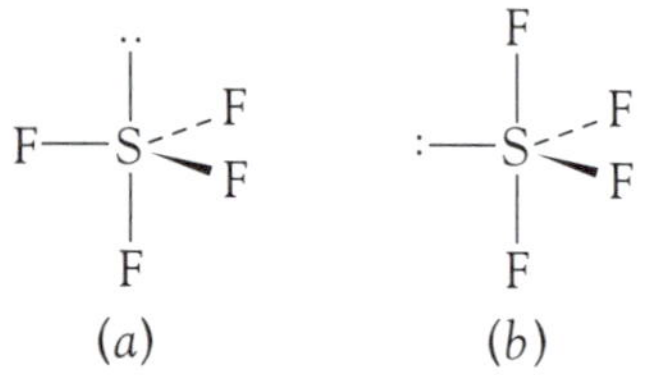

Figure 3.38 Possible geometries for the sulfur tetrafluoride molecule (a) with the lone pair in the axial position and (b) with the lone pair in the equatorial position.

Figure 3.39 Actual geometry for the sulfur tetrafluoride molecule.

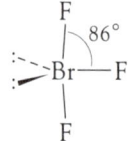

Figure 3.40 Actual geometry for the bromine trifluoride molecule.

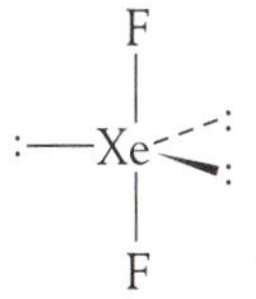

Figure 3.41 Predicted and actual geometry for the xenon difluoride molecule.

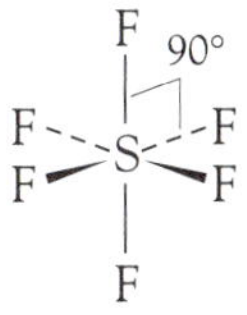

Figure 3.42 Predicted and actual geometry for the sulfur hexafluoride molecule.

Table 3.4 Molecules and ions with trigonal bipyramidal geometry

Bonding pairs	Lone pairs	Shape
5	0	Trigonal bipyramidal
4	1	Seesaw
3	2	T
2	3	Linear

fluorine atoms are bent away from the vertical to form a F_{axial}–Br–$F_{equatorial}$ angle of only 86°.

There are a number of examples of molecules with trigonal bipyramidal electron pair arrangement having three lone pairs. One of these is the xenon difluoride molecule (Figure 3.41). The third lone pair occupies the equatorial position as well. Hence, the observed molecular shape is linear. The names for the shapes of molecules and ions that have a trigonal bipyramidal geometry are given in Table 3.4.

Octahedral Geometry

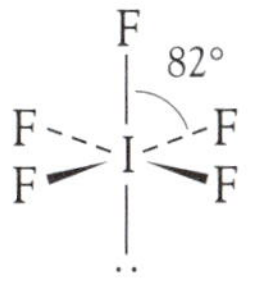

Figure 3.43 Actual geometry for the iodine pentafluoride molecule.

The common example of a molecule with six electron groupings is sulfur hexafluoride. The most widely spaced possibility arises from bonds at equal angles of 90°, the octahedral arrangement (Figure 3.42).

Iodine pentafluoride provides an example of a molecule with five bonding electron pairs and one lone pair around the central atom. Because theoretically all the angles are equal, the lone pair can occupy any site (Figure 3.43), thus producing an apparent square-based pyramidal shape. However, experimental measurements show that the four equatorial fluorine atoms are slightly above the horizontal plane, thus giving a F_{axial}–I–$F_{equatorial}$ angle of only 82°. Once again, this result indicates that the lone pair occupies a greater volume than do the bonding pairs.

Finally, xenon tetrafluoride proves to be an example of a molecule that has four bonding pairs and two lone pairs around the central xenon atom. The lone pairs occupy opposite sides of the molecule, thereby producing a square planar arrangement of fluorine atoms (Figure 3.44). The names for the shapes of molecules and ions that have an octahedral geometry are given in Table 3.5.

Greater Than Six Bonding Directions

There are a few examples of molecules and ions in which the central atom is bonded to more than six neighbors. To accommodate seven or eight atoms around a central atom, the central atom itself has to be quite large and the surrounding atoms and ions quite small. Thus, heavier elements from the

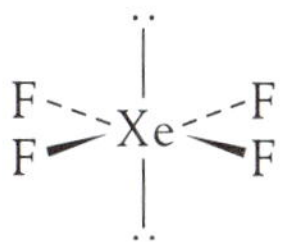

Figure 3.44 Predicted and actual geometry for the xenon tetrafluoride molecule.

Table 3.5 Molecules and ions with octahedral geometry

Bonding pairs	Lone pairs	Shape
6	0	Octahedral
5	1	Square-based pyramidal
4	2	Square planar

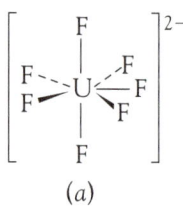

(a)

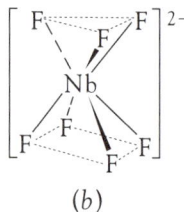

(b)

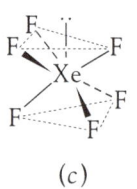

(c)

Figure 3.45   (a) The pentagonal bipyramidal structure of uranium(V) fluoride; (b) the capped trigonal prismatic structure of niobium(V) fluoride; (c) the probable capped octahedral structure of xenon hexafluoride.

lower portion of the periodic table combined with the small fluoride ion provide examples of these structures. The MX_7 species are particularly interesting because they can assume three possible geometries: pentagonal bipyramid, capped trigonal prism, and capped octahedron. The pentagonal bipyramid resembles the trigonal bipyramid and octahedron, except it has five rather than three and four bonds, respectively, in the equatorial plane. The capped trigonal prism has three atoms in a triangular arrangement above the central atom and four atoms in a square plane below the central atom. The capped octahedron is simply an octahedral arrangement in which three of the bonds are opened up from the 90° angle and a seventh bond inserted between.

These three structures must be almost equally favored in terms of relative energy and atom spacing because all are found: The uranium(V) fluoride ion, UF_7^{2-}, adopts the pentagonal bipyramidal arrangement, whereas the niobium(V) fluoride ion, NbF_7^{2-}, adopts the capped trigonal prismatic structure, and it is believed that xenon hexafluoride, XeF_6, adopts the capped octahedral structure in the gas phase (Figure 3.45).

3.10   Valence-Bond Theory

The valence-bond theory builds on the Lewis concept that bonding results from electron pairing between neighboring atoms. The Lewis approach was put into a quantum-mechanical context and then the results (valence-bond theory) were refined by Linus Pauling. The theory is used much less now than it used to be, but the concepts are still employed by some chemists, particularly those in organic chemistry. We see in Chapter 19, Section 19.6 that valence-bond theory can also be applied to the bonding in transition metal compounds.

The principles of valence-bond theory can be summarized in a series of statements:

1. A covalent bond results from the pairing of unpaired electrons in neighboring atoms.

2. The spins of the paired electrons must be antiparallel (one up and one down).

3. To provide enough unpaired electrons in each atom for the maximum bond formation, it is considered that electrons can be excited to fill empty orbitals during bond formation.

4. The shape of the molecule results from the directions in which the orbitals of the central atom point.

Orbital Hybridization

The simplistic valence-bond theory becomes untenable as soon as we look at bonding angles in common molecules. A good example is the ammonia molecule, NH_3. Assuming the three unpaired $2p$ electrons on the central

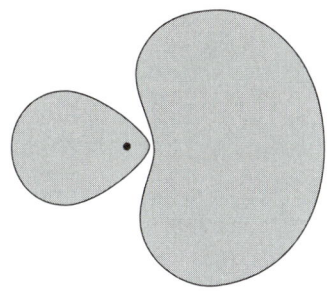

Figure 3.46 The 90 percent probability surface of a hybrid orbital involving combinations of s and p orbitals. The black dot in the smaller lobe identifies the location of the nucleus.

nitrogen atom are used, statement 4 above tells us that the bonds to the hydrogen atoms should follow the axes of the bonding orbitals, $2p_x$, $2p_y$, and $2p_z$. That is, the hydrogen atoms should be 90° apart. We now know from actual measurements that the bond angles in ammonia are 107°. To account for the substantial difference in theoretical versus actual bond angles for this and other covalent compounds, we invoke the modification known as *orbital hybridization*.

The orbital hybridization concept asserts that the wave functions of electrons in atomic orbitals of an atom (usually the central atom of a molecule) can mix together during bond formation to occupy hybrid atomic orbitals. According to this approach, electrons in these hybrid orbitals are still the property of the donor atom. If the wave functions of an *s* orbital and one or more *p* orbitals are combined, the possible hybrid orbitals produced are all similar to that shown in Figure 3.46. Such hybrid orbitals are given the symbols sp, sp^2, and sp^3 depending on whether the wave functions of one, two, or three *p* orbitals are "mixed in" with the *s* orbital wave function. These hybrid orbitals are oriented in a particular direction and should overlap more with the orbitals of another atom than do those of a spherical *s* orbital or of a two-lobed *p* orbital. A greater overlap means that the wave functions of the two atoms will mix better and form a stronger covalent bond.

The number of hybrid orbitals formed will equal the sum of the number of atomic orbitals that are involved in the mixing of wave functions. Like *s* and *p* orbitals, *d* orbitals can also be mixed in, though theoretical chemists now contend that *d* orbitals play a minimal role in covalent bonding. Nevertheless, for our simplistic bonding approach, it is often useful to propose *d* orbital involvement to account for the shapes of molecules where the central atom has more than four neighbors. The number of atomic orbitals used, the symbol for the hybrid orbital, and the geometry of the resulting molecule are all listed in Table 3.6.

We can illustrate the concept of hybridization using boron trifluoride. Prior to compound formation, the boron atom has an electron configuration of $[He]2s^2 2p^1$ (Figure 3.47a). Suppose that one of the $2s$ electrons moves to a $2p$ orbital (Figure 3.47b). The wave functions of the three orbitals each containing a single electron then mix to provide three equivalent sp^2 orbitals (Figure 3.47c). These orbitals, oriented at 120° to one another, overlap with the singly occupied $2p$ orbital on each fluorine atom

Table 3.6 Numbers of hybrid orbitals and type of hybridization for various molecular geometries

s	p	d	Type of hybridization	Number of hybrid orbitals	Resulting molecular geometry
1	1	0	sp	2	Linear
1	2	0	sp^2	3	Trigonal planar
1	3	0	sp^3	4	Tetrahedral
1	3	1	$sp^3 d$	5	Trigonal bipyramidal
1	3	2	$sp^3 d^2$	6	Octahedral

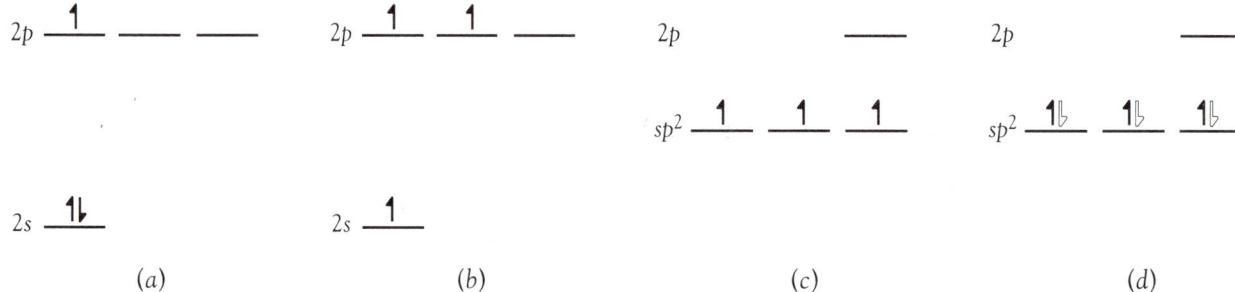

Figure 3.47   The concept of hybrid orbital formation applied to boron trifluoride. (*a*) The electron configuration of the free atom. (*b*) The shift of an electron from the 2s orbital to the 2p orbital. (*c*) The formation of three sp^2 hybrid orbitals. (*d*) Pairing of boron electrons with three electrons (open half-arrows) of the fluorine atoms.

to give three σ covalent bonds (Figure 3.47*d*). This explanation matches our experimental findings of equivalent boron–fluorine bonds, each forming 120° angles with the other two—the trigonal planar geometry.

Carbon dioxide provides an example of a molecule in which we assume that not all of the occupied orbitals are hybridized. We assume that the $[He]2s^2 2p^2$ configuration of the carbon atom (Figure 3.48*a*) is altered to $[He]2s^1 2p^3$ (Figure 3.48*b*). The *s* orbital and one of the *p* orbitals hybridize (Figure 3.48*c*). The resulting *sp* hybrid orbitals are 180° apart, and they overlap with one 2*p* orbital on each oxygen atom to provide a single σ bond and a linear structure. This leaves single electrons in the other two 2*p* orbitals of the carbon atom. Each *p* orbital overlaps side to side with a singly occupied 2*p* orbital on an oxygen atom to form a π bond with each of the two oxygen atoms (Figure 3.48*d*). Thus, the concept of hybridization can be used to explain the linear nature of the carbon dioxide molecule and the presence of two carbon–oxygen double bonds.

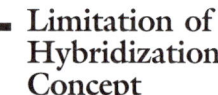

Limitation of Hybridization Concept

To review, the formation of hybrid orbitals can be used successfully to account for a particular molecular shape. However, hybridization is simply a mathematical manipulation of wave functions, and we have no evidence that it actually happens. Furthermore, the hybridization concept is not a

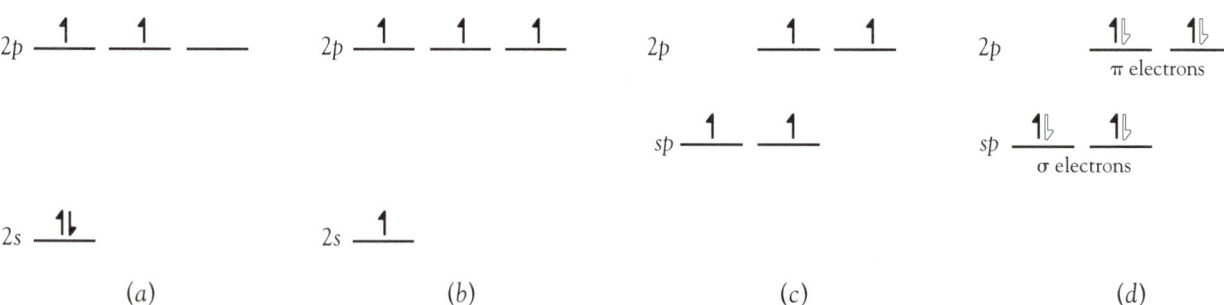

Figure 3.48   The concept of hybrid orbital formation applied to carbon dioxide. (*a*) The electron configuration of the free carbon atom. (*b*) The shift of an electron from the 2s orbital to the 2p orbital. (*c*) The formation of two *sp* hybrid orbitals. (*d*) Pairing of the carbon electrons with four oxygen electrons (open half-arrows).

predictive tool; we can only use it when the molecular structure has actually been established. Molecular orbital theory, in which the electrons are considered the property of the molecule as a whole, is predictive. The problem, of course, with molecular orbital theory is the complexity of the calculations that are needed to deduce molecular shape.

3.11 Network Covalent Substances

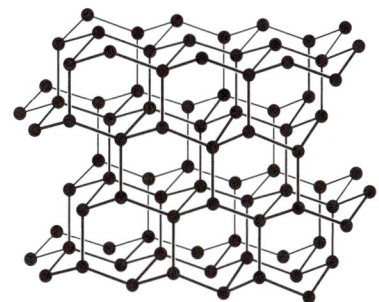

Figure 3.49 Arrangement of carbon atoms in a diamond.

Up to now, we have been discussing elements and compounds that exist as small, individual molecules. However, there are some structures, such as diamond or quartz, in which all the atoms are held together by covalent bonds. Such linking by covalent bonds throughout a substance is known as *network covalent bonding*. The whole crystal is one giant molecule. Diamond is a form of carbon in which each and every carbon atom is bonded in a tetrahedral arrangement to all its neighbors (Figure 3.49). The second common example of network covalent bonding is quartz, the common crystalline form of silicon dioxide, SiO_2. In this compound, each silicon atom is surrounded by a tetrahedron of oxygen atoms, and each oxygen atom is bonded to two silicon atoms.

To melt a substance that contains network covalent bonds, one must break the covalent bonds. But covalent bonds have energies in the range of hundreds of kilojoules per mole, so very high temperatures are needed to accomplish this cleavage. Thus diamond sublimes at about 4000°C, and silicon dioxide melts at 2000°C. For the same reason, network covalent substances are extremely hard; a diamond is the hardest naturally occurring substance known. Furthermore, such substances are insoluble in all solvents.

3.12 Intermolecular Forces

Network covalent molecules are rare. Almost all covalently-bonded substances consist of independent molecular units. If there were only intramolecular forces (the covalent bonds), there would be no attractions between neighboring molecules, and, consequently, all covalently bonded substances would be gases at all temperatures. We know this is not the case. Thus, there must be forces between molecules, or intermolecular forces. Indeed, there is one intermolecular force that operates between all molecules: induced dipole attractions, also called dispersion forces or London forces (after the scientist Fritz London, not the British capital). The other types of forces—dipole–dipole, ion–dipole, and hydrogen bonding—only occur in specific circumstances, which we discuss later in this chapter.

3.13 Dispersion (London) Forces

In the orbital representation of atoms and molecules, the probability distribution of the electrons (electron density) is a time-averaged value. It is the oscillations from this time-averaged value that lead to the attractions between

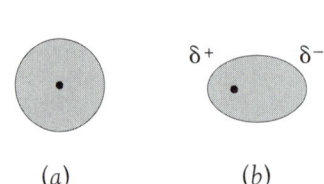

(a)   (b)

Figure 3.50 (a) Average electron density for an atom. (b) Instantaneous electron density producing a temporary dipole.

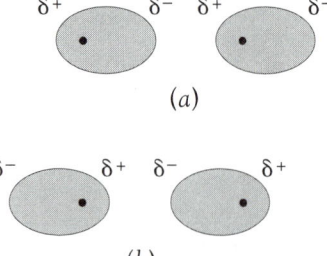

(a)

(b)

Figure 3.51 (a) The instantaneous attraction between neighboring molecules. (b) The reversal of polarity in the next instant.

neighboring molecules. The noble-gas atoms provide the simplest example. On average, the electron density should be spherically symmetric around the atomic nucleus (Figure 3.50a). However, most of the time, the electrons are asymmetrically distributed; consequently, one part of the atom has a higher electron density and another part has a lower electron density (Figure 3.50b). The end at which the nucleus is partially exposed will be slightly more positive ($\delta+$), and the end to which the electron density had shifted will be partially negative ($\delta-$). This separation of charge is called a *temporary dipole*. The partially exposed nucleus of one atom will attract electron density from a neighboring atom (Figure 3.51a), and it is this induced dipole between molecules that represents the dispersion force between atoms and molecules. However, an instant later, the electron density will have shifted, and the partial charges involved in the attraction will be reversed (Figure 3.51b).

The strength of the *dispersion force* depends on a number of factors and their discussion is more appropriate to an advanced physical chemistry course. However, a qualitative and predictive approach is to consider that the dispersion force relates to the number of electrons in the atom or molecule. On this basis, it is the number of electrons that determines how easily the electron density can be polarized, and the greater the polarization, the stronger the dispersion forces. In turn, the stronger the intermolecular forces, the higher will be both the melting and the boiling points. This relationship is illustrated by the chart in Figure 3.52, which shows the dependence of the boiling points of the Group 14 hydrides on the number of electrons in the molecule.

Molecular shape is a secondary factor affecting the strength of dispersion forces. A compact molecule will allow only a small separation of charge, whereas an elongated molecule can allow a much greater charge separation. A good comparison is provided by sulfur hexafluoride, SF_6, and decane, $CH_3CH_2CH_2CH_2CH_2CH_2CH_2CH_2CH_2CH_3$. The former has 70 electrons and a melting point of $-51°C$, whereas the latter has 72 electrons and a melting point of $-30°C$. Hence, the dispersion forces are greater between the long decane molecules than between the near-spherical sulfur hexafluoride molecules.

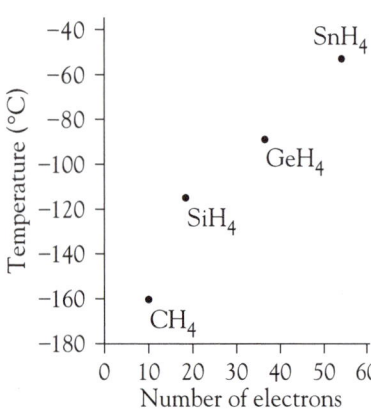

Figure 3.52 Dependence of the boiling points of the Group 14 hydrides on the number of electrons.

3.14 Electronegativity

There is a very simple experiment that shows the existence of two types of molecules. In this experiment, a positively charged rod is held near a stream of liquid. Many liquids (for example, carbon tetrachloride) are unaffected by the charged rod, whereas others (for example, water) are attracted by the rod. If the positively charged rod is replaced by a negatively charged rod, those unaffected by the positive charge are also unaffected by the negative charge, whereas those attracted by the positive charge are also attracted by the negative charge. To explain these observations, we infer that the deflected liquids consist of molecules in which there is a permanent charge separation (a permanent dipole). Thus, the partially negative ends of the molecules are attracted toward the positively charged rod, and the partially positive ends are attracted toward the negatively charged rod. But why should some molecules have a permanent charge separation? For an explanation, we need to look at another concept of Linus Pauling's—*electronegativity*.

Pauling defined electronegativity as the power of an atom in a molecule to attract shared electrons to itself. This relative attraction for bonding electron pairs really reflects the comparative Z_{eff} of the two atoms on the shared electrons. Thus, the values increase from left to right across a period and decrease down a group in the same way as ionization energies do. Electronegativity is a relative concept, not a measurable function. The Pauling electronegativity scale is an arbitrary one, with the value for fluorine defined as 4.0. Some useful electronegativity values are shown in Figure 3.53.

			H 2.2			
Li 1.0	Be 1.6	B 2.0	C 2.5	N 3.0	O 3.4	F 4.0
			Si 1.9	P 2.2	S 2.6	Cl 3.2
						Br 3.0
						I 2.7

Figure 3.53 Pauling electronegativity values of various main group elements.

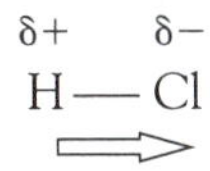

$$\overset{\delta+}{H}\overset{\delta-}{—Cl}$$

Figure 3.54 Permanent dipole of the hydrogen chloride molecule.

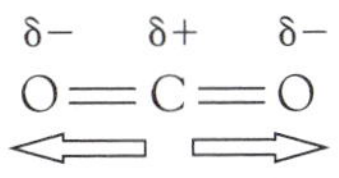

$$\overset{\delta-}{O}=\overset{\delta+}{C}=\overset{\delta-}{O}$$

Figure 3.55 Because it has opposing bond dipoles, the carbon dioxide molecule is nonpolar.

Thus, in a molecule such as hydrogen chloride, the bonding electrons will not be shared equally between the two atoms. Instead, the higher Z_{eff} of chlorine will cause the bonding pair to be more closely associated with the chlorine atom than with the hydrogen atom. As a result, there will be a permanent dipole in the molecule. This dipole is depicted in Figure 3.54, using the σ sign to indicate a partial charge and an arrow to indicate the dipole direction.

Individual bond dipoles can act to "cancel" each other. A simple example is provided by carbon dioxide, where the bond dipoles are acting in opposite directions. Hence, the molecule does not possess a net dipole; in other words, the molecule is nonpolar (Figure 3.55).

The Origins of the Electronegativity Concept

Electronegativity is probably the most widely used concept in chemistry, yet its roots seem to have become forgotten. As a result, Pauling electronegativity values are sometimes imbued with a greater significance than was originally intended.

In his book, *The Nature of the Chemical Bond*, Pauling made it clear that his development of the concept of electronegativity in the 1930s arose from studies of bond energies, for which he used the symbol D. He considered two elements A and B and argued that, for a purely covalent bond, the A–B bond energy should be the geometric mean of the A–A and B–B bond energies. However, he found this was often not the case. He defined this difference as Δ', where

$$\Delta' = D(A\text{–}B) - \{D(A\text{–}A)(B\text{–}B)\}^{1/2}$$

For example, the Cl–Cl bond has an energy of 242 kJ·mol^{-1} and that of the H–H bond is 432 kJ·mol^{-1}. The geometric mean is 323 kJ·mol^{-1}, but the experimental value for the H–Cl bond energy is 428 kJ·mol^{-1}. Thus, Δ'(H–Cl) is 105 kJ·mol^{-1}. Pauling ascribed this difference to an ionic contribution to the bonding, making the heterogenous bond stronger than the mean of the two homogenous bonds.

Pauling produced a table of data for combinations of 14 main group elements that expressed the "excess ionic energy" of heteronuclear covalent bonds; for example, that of C–H was 0.4, while that of H–F was 1.5. To provide a better fit, he adjusted some of the numbers; for example, he upped that of the H–F bond to 1.9.

Taking the electronegativity of hydrogen as zero, Pauling assigned the balance of the ionic energy difference to the other element. Then, he added 2.05 to all values to produce a simple numerical sequence across the second period elements. In 1960, Allred recalculated and published values of Pauling electronegativities based on more modern and reliable bond energy data. The Allred's first period values are compared with Pauling's original values in Table 3.7.

Since Pauling published his first work on a scale of electronegativity, others have derived scales using different parameters, such as the Allred–Rochow electronegativity scale that uses the concept of effective nuclear charge to derive electronegativity values. The Allred–Rochow scale is widely used among inorganic chemists. Mulliken suggested that the average of ionization energy and electron affinity could be used to derive information on the most important resonance structures in

Table 3.7 Original values of Pauling electronegativities compared with Allred's corrected values for the first period elements

Element	Li	Be	B	C	N	O	F
Pauling	1.0	1.5	2.0	2.5	3.0	3.5	4.0
Allred	0.98	1.57	2.04	2.55	3.04	3.44	3.98

heteronuclear covalent bonds. The approach was further developed by Jaffé, and the values obtained by this method are known as the Mulliken–Jaffé electronegativities.

3.15 Dipole–Dipole Forces

A permanent dipole results in an enhancement of the intermolecular forces. For example, carbon monoxide has higher melting and boiling points ($-205°C$ and $-191°C$, respectively) than does dinitrogen ($-210°C$ and $-196°C$), even though the two compounds are isoelectronic.

It is important to realize that *dipole–dipole attractions* are often a secondary effect in addition to the induced dipole effect. This point is illustrated by comparing hydrogen chloride with hydrogen bromide. The electronegativity difference between hydrogen and chlorine is 1.0 for hydrogen chloride and is 0.8 between the atoms in hydrogen bromide, so the dipole–dipole attractions between neighboring hydrogen chloride molecules will be stronger than those between neighboring hydrogen bromide molecules. Yet the boiling point of hydrogen bromide ($-67°C$) is higher than that of hydrogen chloride ($-85°C$). Therefore, induced dipole (dispersion) forces, which will be higher for the hydrogen bromide (36 electrons in HBr and 18 in HCl), must be the predominant factor. In fact, complex calculations show that dispersion forces account for 83 percent of the attraction between neighboring hydrogen chloride molecules and 96 percent of the attraction between neighboring hydrogen bromide molecules.

3.16 Hydrogen Bonding

If we look at the trend in boiling points of the Group 17 hydrides (Figure 3.56), we see that hydrogen fluoride has an anomalously high value. Similar plots for the Groups 15 and 16 hydrides show that the boiling points of ammonia and water also are anomalous. The elements involved have high electronegativities; thus, it is argued that the much stronger intermolecular forces are a result of exceptionally strong dipole–dipole forces. These forces are given the special name of *hydrogen bonds*. In chemistry, the hydrogen bonding of the water molecule is particularly important as we see in Chapter 10.

Hydrogen bonding, then, is by far the strongest intermolecular force; indeed, it can represent 5 to 20 percent of the strength of a covalent bond. The strength of the hydrogen bond between molecules does depend on the identity of the nonhydrogen element. Thus, hydrogen bond strength decreases in the order H–F > H–O > H–N, and this order parallels the decrease in electronegativity differences. However, this factor cannot be the whole answer, because the H–Cl bond is more polar than the H–N bond, yet hydrogen chloride molecules do not exhibit very strong intermolecular attractions.

Because the distances between two molecules sharing a hydrogen bond are significantly less than the sum of the van der Waals radii, it is argued that electron density is shared across a hydrogen bond. In this approach, a hydrogen bond is less of an intermolecular force and more of a weak covalent bond.

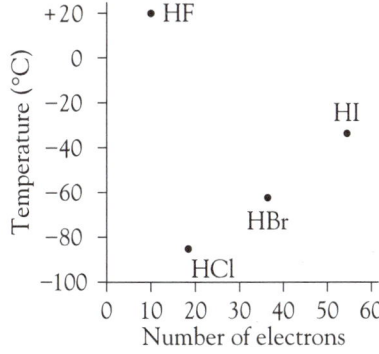

Figure 3.56 Boiling points of the Group 17 hydrides.

3.17   Covalent Bonding and the Periodic Table

In this chapter, we have emphasized the existence of covalent bonds in compounds containing nonmetals and semimetals. The chemistry of nonmetals is dominated by the covalent bond. Yet as we will see throughout this text, covalent bonding is also very important for compounds containing metals. Metal-containing polyatomic ions provide examples of these. The permanganate ion, MnO_4^-, contains manganese covalently bonded to four oxygen atoms. We see in Chapter 5 that the bonding of many metallic compounds is more easily described as covalent rather than ionic. However, it is better to discuss this topic in the context of the chapter on ionic bonds, because we need to discuss the nature of the ionic bond before we can appreciate why a compound adopts one or the other of the bonding types.

EXERCISES

3.1   Define the following terms: (a) LCAO theory; (b) σ orbital; (c) VSEPR theory; (d) hybridization.

3.2   Define the following terms: (a) network covalent molecules; (b) intramolecular forces; (c) electronegativity; (d) hydrogen bonding.

3.3   Use a molecular orbital diagram to determine the bond order of the H_2^- ion. Would the ion be diamagnetic or paramagnetic?

3.4   Would you expect Be_2 to exist? Use a molecular orbital energy diagram to explain your reasoning.

3.5   Use a molecular orbital diagram to determine the bond order in the N_2^+ ion. Write a valence electron configuration $[(\sigma_{2s})^2, \ldots]$ for this ion.

3.6   Use a molecular orbital diagram to determine the bond order in the O_2^+ ion. Write a valence electron configuration $[(\sigma_{2s})^2, \ldots]$ for this ion.

3.7   Assuming that it has similar molecular orbital energies to those of carbon monoxide, deduce the bond order of the NO^+ ion.

3.8   Assuming that it has similar molecular orbital energies to those of carbon monoxide, deduce the bond order of the NO^- ion.

3.9   Construct a molecular orbital diagram for diboron, B_2. What would you predict for the bond order? Construct a similar diagram for diboron using the ordering for the heavier Period 2 elements and compare the two results. What experimental property could be used to confirm this different ordering?

3.10   Construct a molecular orbital diagram and write the valence electron configuration of the dicarbon anion and cation, C_2^- and C_2^+. Determine the bond order in each of these ions.

3.11   Construct electron-dot diagrams for: (a) oxygen difluoride; (b) phosphorus trichloride; (c) xenon difluoride; (d) the tetrachloroiodate ion, ICl_4^-.

3.12   Construct electron-dot diagrams for: (a) the ammonium ion; (b) carbon tetrachloride; (c) the hexafluorosilicate ion, SiF_6^{2-}; (d) the pentafluorosulfate ion, SF_5^-.

3.13   Construct an electron-dot diagram for the nitrite ion. Draw the structural formulas of the two resonance possibilities for the ion and estimate the average nitrogen–oxygen bond order. Draw a partial bond representation of the ion.

3.14   Construct an electron-dot diagram for the carbonate ion. Draw the structural formulas of the three resonance possibilities for the ion and estimate the average carbon–oxygen bond order. Draw a partial bond representation of the ion.

3.15   The thiocyanate ion, NCS^-, is linear, with a central carbon atom. Construct all feasible electron-dot diagrams for this ion; then use the concept of formal charge to identify the most probable contributing structures. Display the result using a partial bond representation.

3.16   The boron trifluoride molecule is depicted as having three single bonds and an electron-deficient central boron atom. Use the concept of formal charge to suggest why a structure involving a double bond to one fluorine, which would provide an octet to the boron, is not favored.

3.17   For each of the molecules and polyatomic ions in Exercise 3.11, determine the electron pair arrangement and the molecular shape according to the VSEPR theory.

3.18 For each of the molecules and polyatomic ions in Exercise 3.12, determine the electron pair arrangement and the molecular shape according to the VSEPR theory.

3.19 Which of the following triatomic molecules would you expect to be linear and which would you expect to be V-shaped? For those V-shaped, suggest approximate bond angles. (a) carbon disulfide, CS_2; (b) chlorine dioxide, ClO_2; (c) gaseous tin(II) chloride, $SnCl_2$; (d) nitrosyl chloride, $NOCl$ (nitrogen is the central atom); (e) xenon difluoride, XeF_2.

3.20 Which of the following triatomic ions would you expect to be linear and which would you expect to be V-shaped? For those V-shaped, suggest approximate bond angles. (a) BrF_2^+; (b) BrF_2^-; (c) CN_2^{2-}.

3.21 For each of the molecules and polyatomic ions in Exercise 3.11, identify in which cases distortion from the regular geometric angles will occur due to the presence of one or more lone pairs.

3.22 For each of the molecules and polyatomic ions in Exercise 3.12, identify in which cases distortion from the regular geometric angles will occur due to the presence of one or more lone pairs.

3.23 For each of the electron pair arrangements determined in Exercise 3.11, identify the hybridization that would correspond to the shape.

3.24 For each of the electron pair arrangements determined in Exercise 3.12, identify the hybridization that would correspond to the shape.

3.25 Using Figure 3.45 as a model, show how the concept of hybrid orbitals can be used to explain the shape of the gaseous beryllium chloride molecule.

3.26 Using Figure 3.45 as a model, show how the concept of hybrid orbitals can be used to explain the shape of the methane molecule.

3.27 Which would you expect to have the higher boiling point, hydrogen sulfide, H_2S, or hydrogen selenide, H_2Se? Explain your reasoning clearly.

3.28 Which would you expect to have the higher melting point, dibromine, Br_2, or iodine monochloride, ICl? Explain your reasoning clearly.

3.29 For each of the molecules and polyatomic ions in Exercise 3.11, determine whether they are polar or nonpolar.

3.30 For each of the molecules and polyatomic ions in Exercise 3.12, determine whether they are polar or nonpolar.

3.31 Which would you expect to have the higher boiling point, ammonia, NH_3, or phosphine, PH_3? Explain your reasoning clearly.

3.32 Which would you expect to have the higher boiling point, phosphine, PH_3, or arsine, AsH_3? Explain your reasoning clearly.

3.33 For each of the following covalent compounds, deduce their molecular shape and the possible hybridization of the central atom: (a) indium(I) iodide, InI; (b) tin(II) bromide, $SnBr_2$; (c) antimony tribromide, $SbBr_3$; (d) tellurium tetrachloride, $TeCl_4$; (e) iodine pentafluoride, IF_5.

3.34 Arsenic trifluoride and arsenic trichloride have bond angles of 96.2° and 98.5° respectively. Suggest reasons for the difference in angles.

BEYOND THE BASICS

3.35 Research the molecular orbitals formed by overlap of d orbitals with s and p orbitals. Draw diagrams to show how the d atomic orbitals overlap with the s and p orbitals to form σ and π molecular orbitals.

3.36 The CO_2^- ion has been synthesized. What shape and approximate bond angles would you expect?

3.37 The dinitrogen oxide molecule has the atomic order NNO rather than the symmetrical NON arrangement. Suggest a possible reason.

3.38 The cyanate ion, OCN^-, forms many stable salts, while the salts of the isocyanate ion, CNO^-, are often explosive. Suggest a possible reason.

3.39 Following from Exercise 3.38, a third possible arrangement would be CON^-. Explain why this order is unlikely to be a stable ion.

3.40 Carbon forms similar free radical species with hydrogen and fluorine: $CH_3\cdot$ and $CF_3\cdot$ yet one is planar whereas the other is pyramidal in shape. Which one adopts which geometry? Suggest an explanation.

3.41 Predict which of the following gas-phase reactions is the more favored and give your reasoning.

$$NO + CN \rightarrow NO^+ + CN^-$$

$$NO + CN \rightarrow NO^- + CN^+$$

3.42 It is possible to synthesize the ion $C(CN)_3^-$. Draw an electron-dot structure and deduce its most likely geometry. In fact, the ion is planar. Draw one of the resonance structures that would be compatible with this finding.

3.43 Phosphorus pentafluoride has a higher boiling point ($-84°C$) than phosphorus trifluoride ($-101°C$), while antimony pentachloride, $SbCl_5$, has a lower boiling point ($140°C$) than antimony trichloride, $SbCl_3$ ($283°C$). Suggest why the two patterns are different.

3.44 The pentafluoroxenate(IV) anion, XeF_5^- has been prepared. The actual shape matches that predicted by the VSEPR theory. What is that shape?

Chapter 4 — Metallic Bonding

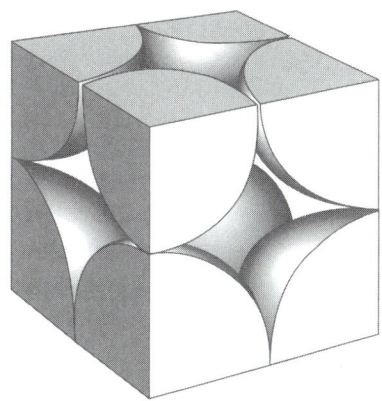

The bonding in metals is explained best by the molecular orbital theory that we have already discussed in the context of covalent bonding. The arrangement of atoms in a metal crystal can be interpreted in terms of the packing of hard spheres. These packing arrangements are common to both metals and ionic compounds. Thus, a study of metallic bonding provides a link between covalent and ionic bonding.

The extraction of metals from their ores coincided with the rise of civilization. Bronze, an alloy of copper and tin, was the first metallic material to be widely used. As smelting techniques became more sophisticated, iron became the preferred metal because it is a harder material and more suitable than bronze for swords and ploughs. For decorative use, gold and silver were easy to use because they are very malleable metals (that is, they can be deformed easily).

Over the ensuing centuries, the number of metals known climbed to its present large number, the great majority of the elements in the periodic table. Yet in the contemporary world, it is still a small number of metals that dominate our lives, particularly iron, copper, aluminum, and zinc. The metals that we choose must suit the purpose for which we need them, yet the availability of an ore and the cost of extraction are often the main reasons why one metal is chosen over another.

4.1   Metallic Bonding

In our discussions on the classification of elements (Chapter 2), we noted that high three-dimensional electrical conductivity at SATP was the one key characteristic of metallic bonding. We can relate this property to the bonding in metals. Unlike nonmetals, where electron sharing is almost always within discrete molecular units, metal atoms share outer (valence) electrons with all nearest neighboring atoms. It is the free movement of electrons throughout the metal structure that can be used to explain the high electrical and thermal conductivity of metals together with their high reflectivity.

The lack of directional bonding can be used to account for the high malleability and ductility of most metals in that metal atoms can readily slide over one another to form new metallic bonds. The ease of formation of metal bonds accounts for our ability to sinter the harder metals; that is, we can produce solid metal shapes by filling a mold with metal powder and placing the powder under conditions of high temperature and pressure. In those circumstances, metal–metal bonds are formed across the powder grain boundaries without the metal actually bulk-melting.

Whereas simple covalent molecules generally have low melting points and ionic compounds have high melting points, metals have melting points ranging from $-39°C$ for mercury to $+3410°C$ for tungsten. Metals continue to conduct heat and electricity in their molten state. (In fact, molten alkali metals are often used as heat transfer agents in nuclear power units.) This is evidence that metallic bonding is maintained in the liquid phase. It is the boiling point that correlates most closely with the strength of the metallic bond. Thus, mercury has a boiling point of $357°C$ and an enthalpy of atomization of 61 kJ·mol^{-1}, while the boiling point of tungsten is $5660°C$ and its enthalpy of atomization is 837 kJ·mol^{-1}. Thus, the metallic bond in mercury is as weak as some intermolecular forces, whereas that in tungsten is comparable in strength to a multiple covalent bond. In the gas phase, however, metallic elements like lithium exist as pairs, Li_2, or like beryllium as individual atoms and hence lose their bulk metallic properties. Metals in the gas phase do not even look metallic; for example, in the gas phase, potassium has a green color.

4.2   Bonding Models

Any theory of metallic bonding must account for the key properties of metals, the most important feature of which is the high electrical conductivity (see Chapter 2). Furthermore, any model should account for the high thermal conductivity and the high reflectivity (metallic luster) of metals.

The simplest metallic bonding model is the electron-sea (or electron-gas) model. In this model, the valence electrons are free to move through the bulk metal structure (hence the term *electron sea*) and even leave the metal, thereby producing positive ions. It is valence electrons, then, that convey electric current, and it is the motion of the valence electrons that transfers heat through a metal. However, this model is more qualitative than quantitative.

Molecular orbital theory provides a more comprehensive model of metallic bonding. This extension of molecular orbital theory is sometimes called *band theory*, which we will illustrate by looking at the orbitals of lithium. In Chapter 3, we saw that two lithium atoms combined in the gas phase to form the dilithium molecule. The molecular orbital diagram showing the mixing of two 2s atomic orbitals is given in Figure 4.1. (Both sets of atomic orbitals

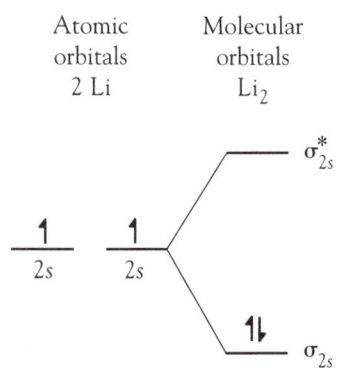

Figure 4.1 Molecular orbital diagram for the dilithium (gas phase) molecule.

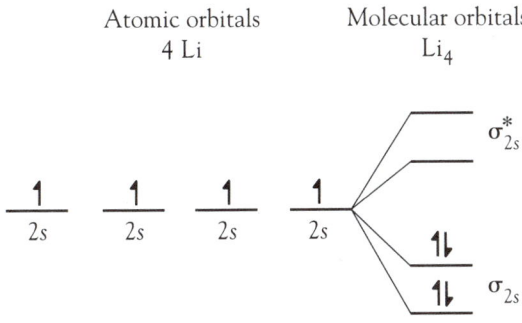

Atomic orbitals
4 Li

Molecular orbitals
Li$_4$

Figure 4.2 Molecular orbital diagram for the combination of four lithium atoms.

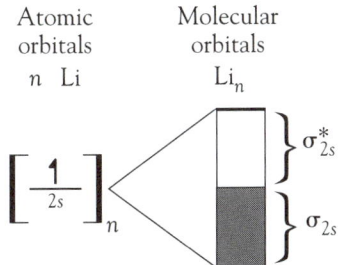

Atomic orbitals
n Li

Molecular orbitals
Li$_n$

Figure 4.3 Band derived from the 2s atomic orbitals by the combination of n lithium atoms.

Figure 4.4 Bands derived from the frontier orbitals (2s and 2p) of beryllium.

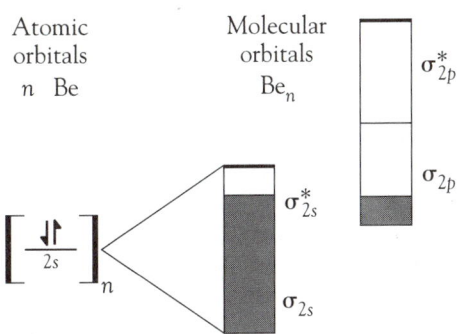

Atomic orbitals
n Be

Molecular orbitals
Be$_n$

are shown on the left.) Now suppose that the atomic orbitals of four lithium atoms are mixed. Again, there must be the same number of σ_{2s} molecular orbitals as 2s atomic orbitals, half of which are bonding and the other half antibonding. To avoid violating the quantum rules, the energies of the orbitals cannot be degenerate. That is, one σ_{2s} cannot have exactly the same energy as the other σ_{2s} orbital. Figure 4.2 shows the resulting orbital arrangement.

In a large metal crystal, the orbitals of n atoms, where n is some enormous number, are mixed. These orbitals interact throughout the three dimensions of the metal crystal, yet the same principles of bonding apply. There will be ½n σ_{2s} (bonding) molecular orbitals and ½n σ_{2s}^* (antibonding) molecular orbitals. With such a large number of energy levels, the spacing of levels becomes so close that they essentially constitute a continuum. This continuum is referred to as a band. For lithium, the band derived from the 2s atomic orbitals will be half-filled. That is, the σ_{2s} part of the band will be filled and the σ_{2s}^* part will be empty (Figure 4.3).

We can visualize electrical conductivity simplistically as the gain by an electron of the infinitesimally small quantity of energy needed to raise it into the empty antibonding orbitals. It can then move freely through the metal structure, as electric current. Similarly, the high thermal conductivity of metals can be visualized as "free" electrons transporting translational energy throughout the metal structure. It is important to remember, however, that the "real" explanation of these phenomena requires a more thorough study of band theory.

In Chapter 1, we saw that light is absorbed and emitted when electrons move from one energy level to another. The light emissions are observed as a line spectrum. With the multitudinous energy levels in a metal, there is an almost infinite number of possible energy level transitions. As a result, the atoms on a metal surface can absorb any wavelength and then re-emit light at that same wavelength as the electrons release that same energy when returning to the ground state. Hence, band theory accounts for the reflectivity of metals.

Beryllium also fits the band model. With an atomic configuration of [He]2s^2, both the σ_{2s} and σ_{2s}^* molecular orbitals will be fully occupied. That is, the band derived from overlap of the 2s atomic orbitals will be completely filled. At first, the conclusion would be that beryllium could not exhibit metallic properties because there is no space in the band in which the electrons can "wander." However, the empty 2p band overlaps with the 2s band, thus enabling the electrons to "roam" through the metal structure (Figure 4.4).

We can use band theory to explain why some substances are electrical conductors, some are not, and some are semiconductors. In the metals, bands overlap and allow a free movement of electrons. In nonmetals, bands are widely separated, so no electron movement can occur (Figure 4.5a). These elements are called insulators. In a few elements, the bands are close enough to allow only a small amount of electron excitation in an upper unoccupied band (Figure 4.5b). These elements are known as intrinsic *semiconductors*. Our modern technology depends on the use of semiconducting materials,

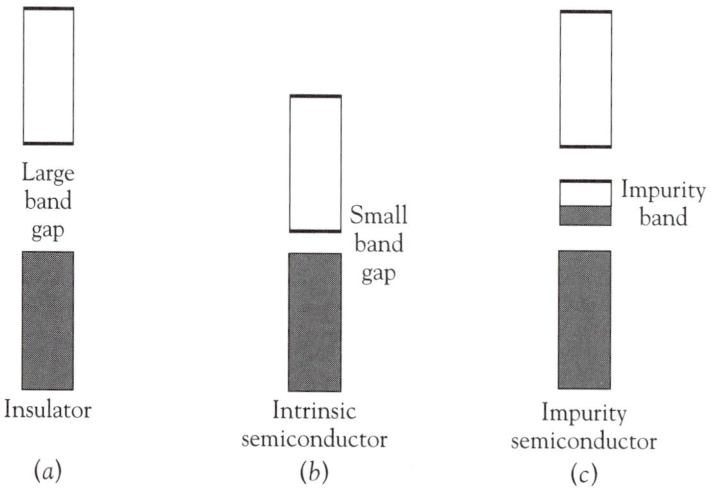

Figure 4.5 Schematic of the band structure of (*a*) a nonmetal, (*b*) an intrinsic semiconductor, and (*c*) an impurity semiconductor.

Large band gap

Insulator

(*a*)

Small band gap

Intrinsic semiconductor

(*b*)

Impurity band

Impurity semiconductor

(*c*)

and it has become necessary to synthesize semiconductors with very specific properties. This can be done by taking an element with a wide band gap and "doping" it with some other element, that is, adding a trace impurity. The added element has an energy level between that of the filled and empty energy levels of the main component (Figure 4.5*c*). This impurity band can be accessed by the electrons in the filled band, enabling some conductivity to occur. By this means, the electrical properties of semiconductors can be adjusted to meet any requirements.

4.3   *Structure of Metals*

Figure 4.6 Simple cubic packing. Successive layers are superimposed over the first array.

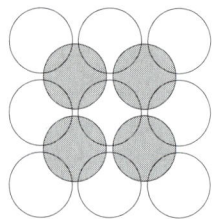

Figure 4.7 Body-centered cubic packing. The second layer (shaded) is placed over holes in the first layer and the third over the holes in the second.

The way in which metal atoms pack together in a crystal is interesting in itself, but, of equal importance to an inorganic chemist, it also provides a basis from which to discuss the ion packing in an ionic compound (which we do in Chapter 5, Section 5.5). The concept of crystal packing assumes that the atoms are hard spheres. In a metal crystal, the atoms are arranged in a repeating array that is called a *crystal lattice*. The packing of metal atoms is really a problem in geometry. That is, we are concerned with the different ways in which spheres of equal size can be arranged.

It is easiest to picture the atomic arrays by arranging one layer and then placing successive layers over it. The simplest possible arrangement is that in which the atoms in the base are packed side by side. The successive layers of atoms are then placed directly over the layer below. This is known as *simple cubic packing* (*sc*) (Figure 4.6), or primitive cubic. Each atom is touched by four other atoms in its own plane plus one atom above and one below, a total of six neighboring atoms. As a result, it is said that each atom has a *coordination number* of six.

The simple cubic arrangement is not very compact and is known only for polonium in metal structures, although as we will see in Chapter 5, it is found in some ionic compounds. An alternative cubic packing arrangement is to place the second layer of atoms over the holes in the first layer. The third layer then fits over the holes in the second layer—which happens to be exactly over the first layer. This more compact arrangement is called *body-centered cubic* (*bcc*) (Figure 4.7). Each atom is touched by four atoms above and four atoms below its plane. Thus body-centered cubic results in a coordination number of eight.

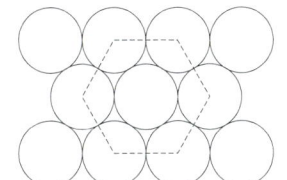

Figure 4.8 The first layer of the hexagonal arrangement.

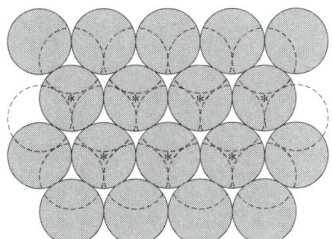

Figure 4.9 In hexagonal packing, the second layer (shaded) fits over alternate holes in the first layer. The hexagonal close-packed arrangement involves placing the third layer (locations marked with asterisks) over the top of the first layer.

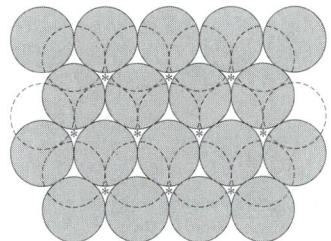

Figure 4.10 The cubic close-packed arrangement has the third layer (locations marked with asterisks) placed over voids in both the first and second layers.

The other two possibilities are based on a hexagon arrangement for each layer; that is, each atom is surrounded by six neighbors in the plane. Notice that in the hexagonal arrangement the holes between the atoms are much closer together than in the cubic arrangement (Figure 4.8). When the second hexagonal layer is placed over the first, it is physically impossible to place atoms over all the holes in the first layer. In fact, only half the holes can be covered. If the third layer is placed over holes in the second layer so that it is superimposed over the first layer, then a *hexagonal close-packed* (*hcp*) arrangement is obtained. The fourth layer is then superimposed over the second layer. Hence, it is also known as an *abab* packing arrangement (Figure 4.9).

An alternative hexagonal packing arrangement involves placement of the third layer over the holes in the first and second layers. It is the fourth layer, then, that repeats the alignment of the first layer. This *abcabc* packing arrangement is known as *cubic close-packed* (*ccp*) or *face-centered cubic* (*fcc*) (Figure 4.10). Both packings based on the hexagonal arrangement are 12-coordinate.

The types of packing, the coordination numbers, and the percentage of occupancy (filling) of the total volume are shown in Table 4.1. An occupancy of 60 percent means that 60 percent of the crystal volume is occupied by atoms and the spaces between atoms accounts for 40 percent. Hence, the higher the percentage of occupancy, the more closely packed the atoms.

Most metals adopt one of the three more compact arrangements (*bcc*, *hcp*, *fcc*); polonium is the sole metal to adopt simple cubic packing. Some metals, particularly those to the right-hand side of the periodic table, adopt distorted or nonstandard packing arrangements. The hard sphere model of packing does not enable us to predict which arrangement a particular metal will adopt. However, there seems to be a general rule that as the number

Table 4.1 Properties of the different packing types

Packing type	Coordination number	Occupancy (%)
Simple cubic (*sc*)	6	52
Body-centered cubic (*bcc*)	8	68
Hexagonal close-packed (*hcp*)	12	74
Cubic close-packed (*ccp/fcc*)	12	74

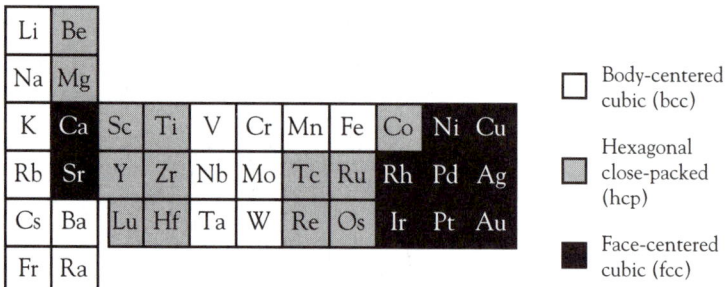

Figure 4.11 Part of the periodic table, showing the common stable packing arrangements of the metals at SATP.

□ Body-centered cubic (bcc)

▨ Hexagonal close-packed (hcp)

■ Face-centered cubic (fcc)

of outer electrons increases, the preferred packing arrangement changes from *bcc* to *hcp* and finally to *fcc*. Figure 4.11 shows the common packing arrangements for the metals. The packing arrangement for several metals is temperature dependent. For example, iron adopts a *bcc* structure (α-iron) at room temperature, converting to a *fcc* structure (γ-iron) above 910°C, and back to a *bcc* structure (δ-iron) about 1390°C. As we will see in Chapter 14, Section 14.23, the temperature dependence of packing is of particular importance in the case of tin.

4.4   Unit Cells

The simplest arrangement of spheres, which, when repeated, will reproduce the whole crystal structure is called a *unit cell*. The cell is easiest to see in the simple cubic case (Figure 4.12). In the unit cell, we have cut a cube from the center of eight atoms. Inside the unit cell itself, there are eight pieces, each one-eighth of an atom. Because $8(\frac{1}{8}) = 1$, we can say that each unit cell contains one atom.

To obtain a unit cell for the body-centered cubic, we must take a larger cluster, one that shows the repeating three-layer structure. Cutting out a cube provides one central atom with eight one-eighth atoms at the corners. Hence, the unit cell contains $1 + [8(\frac{1}{8})]$ or two atoms (Figure 4.13).

At first inspection, the cubic close-packed arrangement does not provide a simple unit cell. However, if a slice is taken through the corner of the face-centered cubic array, we can construct a face-centered cube in which there is an atom at each corner and an atom in the middle of each face (Figure 4.14). When we do this, the cube contains $6(\frac{1}{2}) + 8(\frac{1}{8})$ or four atoms.

Once we know the crystal packing arrangement and the density of a metal, we can calculate the metallic radius of the element.

Figure 4.12 (a) Eight atoms in the simple cubic array. (b) The resulting unit cell.
(From G. Rayner-Canham et al., Chemistry: A Second Course [Don Mills, ON: Addison–Wesley, 1989], p. 64.)

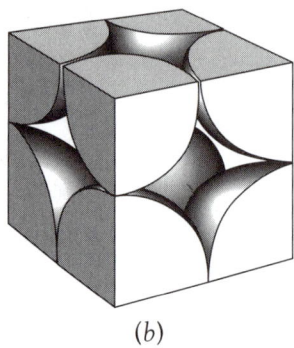

(a)   (b)

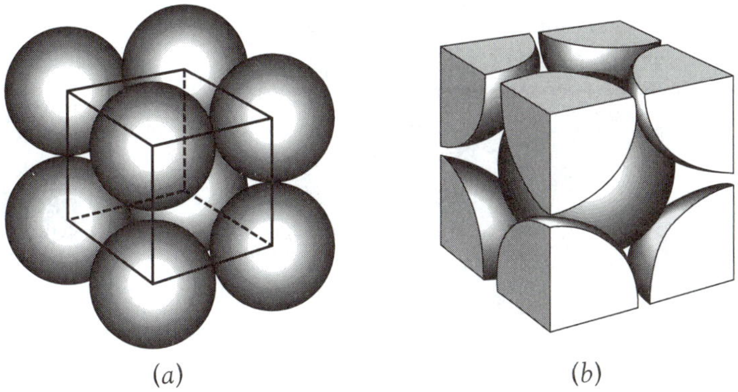

Figure 4.13 (a) Nine atoms in the body-centered cubic array. (b) The resulting unit cell. *(From G. Rayner-Canham et al., Chemistry: A Second Course [Don Mills, ON: Addison–Wesley, 1989], p. 64.)*

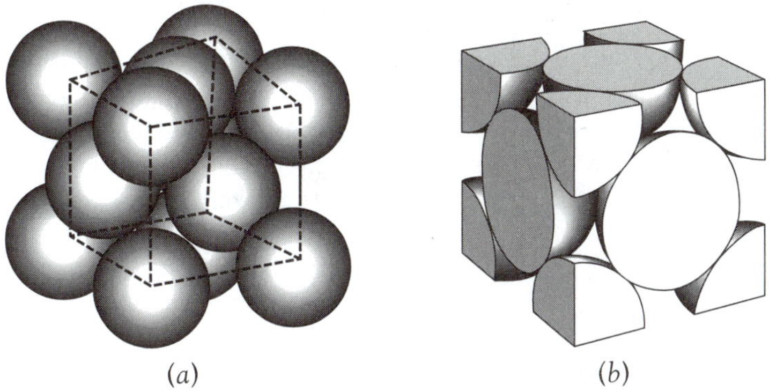

(a)

(b)

Figure 4.14 (a) Fourteen atoms in the face-centered cubic array. (b) The resulting unit cell. *(From G. Rayner-Canham et al., Chemistry: A Second Course [Don Mills, ON: Addison–Wesley, 1989], p. 65.)*

(a)

(b)

4.5 Alloys

A combination of two or more solid metals is called an *alloy*. The number of possible alloys is enormous. Alloys play a vital role in our lives, yet chemists rarely mention them. The atoms in alloys are held together by metallic bonds just like the component metallic elements. This bonding parallels the covalent bonding of nonmetals. Covalent bonds hold together molecules formed from pairs of different nonmetallic elements as well as from pairs of identical nonmetallic elements. Similarly, the metallic bonds of an alloy hold together atoms of different metallic elements.

There are two types of alloys: the solid solutions and the alloy compounds. In the former, the molten metals blend to form a homogeneous mixture. To form a solid solution, the atoms of the two metals have to be about the same size, and the two metallic crystals must have the same structure. In addition, the metals must have similar chemical properties. Gold and copper, for example, form a single phase all the way from 100 percent gold to 100 percent copper. These two metals have similar metallic radii (144 pm for gold and 128 pm for copper), and they both adopt cubic close-packed structures. Lead and tin have very similar metallic radii (175 pm and 162 pm, respectively), but lead forms a face-centered cubic structure and tin has a complex packing arrangement. Very little of one metal will crystallize with the other. Plumber's solder contains 30 percent tin and 70 percent lead, and it is this very immiscibility that enables it to function. During the cooling of the liquid, there is a range of about 100°C over which crystallization occurs. A solid solution cannot exist with more than 20 percent

Mercury Amalgam in Teeth

Many of us have mercury in our mouths in the form of dental fillings. The fillings consist of an amalgam—homogeneous mixture of a liquid metal (mercury) and a number of solid metals. Typically, the dental amalgam has compositions in the following range: mercury (50–55 percent), silver (23–35 percent), tin (1–15 percent), zinc (1–20 percent), and copper (5–20 percent). The soft mixture is placed in the excavated tooth cavity while it is still a suspension of particles of the solid metals in mercury. In the cavity, the mercury atoms infiltrate the metal structure to give a solid amalgam (the equivalent of an alloy). As reaction occurs, there is a slight expansion that holds the filling in place.

Mercury is a very toxic element. However, its amalgamation with solid metals decreases its vapor pressure, so it does not present the same degree of hazard as pure liquid mercury. It is the verdict of the American Dental Association that mercury fillings are quite safe, but there are some who argue that even at very low levels the mercury released from fillings presents a hazard. It is hard to quantify the degree of hazard. There have been claims of people having their mercury fillings removed and rapidly recovering from a chronic disease. Such sudden recoveries are unlikely to be related to the removal of the mercury, as it would take a long while for the mercury to be excreted from the body. In fact, the process of removal of mercury fillings will result in a short-term increase in mercury exposure.

The real problem is that we currently have no substitute that has the low thermal expansion and the high strength of the amalgam. Researchers are currently trying to synthesize a material that will chemically bond to the tooth surface and be strong enough to withstand the immense pressures that we place on our back (grinding) teeth. A significant portion of mercury pollution comes from dentist's offices. For a dentist still using mercury amalgam for fillings of back teeth, over one-half a kilogram of mercury waste is produced per year, most of which is flushed down the drain into the sewer system. Depending on location, sewage sludge can be incinerated, spread on farmland, or dumped. Particularly through incineration, the mercury can escape into the environment. To prevent this, in some jurisdictions, laws require dentists to install separators on their wastewater lines. The mercury waste can then be collected periodically from the trap and sent back to a mercury recycling center.

As cremation becomes more common, particularly in heavily populated countries, we have to recognize the potential for mercury pollution from this source. During the incineration of the bodies, the mercury amalgam decomposes, releasing mercury vapor into the atmosphere. Thus, environmental controls of crematoria emissions are a new concern.

tin. As a result, the crystals are richer in the higher melting lead and the remaining solution has a lower solidification temperature. This "slushy" condition enables plumbers to work with the solder.

Table 4.2 Common alloys

Name	Composition (%)	Properties	Uses
Brass	Cu 70–85, Zn 15–30	Harder than pure Cu	Plumbing
Gold, 18-carat	Au 75, Ag 10–20, Cu 5–15	Harder than pure (24-carat) gold	Jewelry
Stainless steel	Fe 65–85, Cr 12–20, Ni 2–15, Mn 1–2, C 0.1–1, Si 0.5–1	Corrosion resistance	Tools, chemical equipment

In some cases in which the crystal structures of the components are different, mixing molten metals results in the formation of precise stoichiometric phases. For example, copper and zinc form three "compounds": $CuZn$, Cu_5Zn_8, and $CuZn_3$. Table 4.2 lists some of the common alloys, their compositions, and their uses.

EXERCISES

4.1 Define the following terms: (a) electron-sea model of bonding; (b) unit cell; (c) alloy.

4.2 Explain the meanings of the following terms: (a) crystal lattice; (b) coordination number; (c) amalgam.

4.3 What are the three major characteristics of a metal?

4.4 What are the four most widely used metals?

4.5 Using a band diagram, explain how magnesium can exhibit metallic behavior when its $3s$ band is completely full.

4.6 Construct a band diagram for aluminum.

4.7 Explain why metallic behavior does not occur in the gas phase.

4.8 What is the likely formula of potassium in the gas phase? Draw a molecular orbital diagram to show your reasoning.

4.9 What are the two types of layer arrangements in metals? Which has the closer packing?

4.10 What is the difference in layer structure between cubic close-packed and hexagonal close-packed arrangements?

4.11 Draw the simple cubic unit cell and show how the number of atoms per unit cell is derived.

4.12 Draw the body-centered cubic unit cell and show how the number of atoms per unit cell is derived.

4.13 What conditions are necessary for the formation of a solid solution alloy?

4.14 Suggest two reasons why zinc and potassium are unlikely to form a solid solution alloy.

BEYOND THE BASICS

4.15 Use geometry to show that about 48 percent of a simple cubic lattice is empty space.

4.16 Use geometry to show that about 32 percent of a body-centered cubic lattice is empty space.

4.17 In a face-centered cubic unit cell, the atoms usually touch across the diagonal of the face. If the atomic radius is r, calculate the length of each side of the unit cell.

4.18 In a body-centered cubic unit cell, the atoms usually touch along the diagonal from one corner through the center of the cell to the opposite corner. If the atomic radius is r, calculate the length of each side of the unit cell.

4.19 Chromium forms a body-centered cubic lattice in which the edge length of the unit cell is 288 pm. Calculate (a) the metallic radius of a chromium atom and (b) the density of chromium metal.

4.20   The atoms in barium metal are arranged in a body-centered cubic unit cell. Calculate the radius of a barium atom if the density of barium is 3.50 $g \cdot cm^{-3}$ [hint: use your answer to Exercise 4.18].

4.21   The atoms in silver metal are arranged in a face-centered cubic unit cell. Calculate the radius of a silver atom if the density of silver is 10.50 $g \cdot cm^{-3}$ [hint: use your answer to Exercise 4.17].

4.22   Suggest what might be a problem with placing two clean metal surfaces in contact in outer space.

4.23   Undertake a critical study of the controversy on the use of mercury amalgam in teeth fillings.

Chapter 5 Ionic Bonding

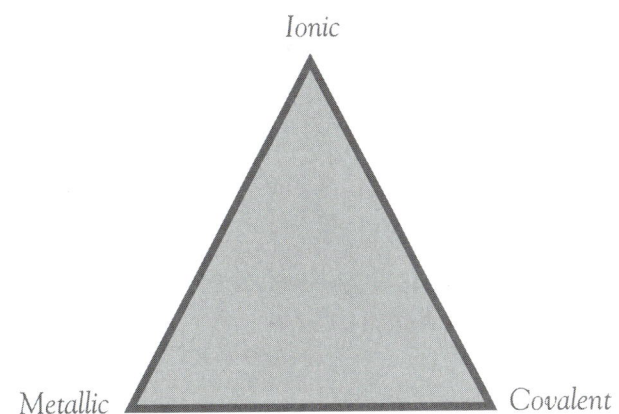

*C*hemical bonding can also occur through transfer of electrons and the subsequent electrostatic attraction between the charged particles. Elementary chemistry courses usually imply that there is a rigid division between ionic and covalent bonding. In fact, there are few cases of "pure" ionic compounds, and chemical bonding is best regarded as a continuum between the extremes of the two models.

One of the simplest chemical experiments is to take a beaker of deionized water and insert a light bulb conductivity tester. The bulb does not light. Then some salt is stirred into the water and the bulb lights. This experiment was crucial in the history of chemistry. In 1884, Svante Arrhenius proposed the modern explanation of this experiment. At the time, hardly anyone accepted his theory of electrolytic dissociation. In fact, his doctoral dissertation on the subject was given a low grade in view of the unacceptability of his conclusions. Not until 1891 was there general support for his proposal that the particles in salt solutions dissociated into ions. In 1903, when the significance of his work was finally realized, Arrhenius's name was put forward to share in both the chemistry and the physics Nobel Prizes. The physicists balked at the proposal and, as a result, Arrhenius was the recipient of the 1903 Nobel Prize in chemistry for this work.

Although we ridicule those who opposed Arrhenius, at the time the opposition was quite understandable. The scientific community was divided between those who believed in atoms (the atomists) and those who did not.

The atomists were convinced of the indivisibility of atoms. Enter Arrhenius, who argued against both sides: He asserted that sodium chloride broke down into sodium *ions* and chloride *ions* in solution but that these ions were not the same as sodium *atoms* and chlorine *atoms*. That is, the sodium was no longer reactive and metallic, nor was the chlorine green and toxic. No wonder his ideas were rejected until the era of J.J. Thomson and the discovery of the electron.

5.1 Characteristics of Ionic Compounds

Whereas covalent substances at room temperature can be solids, liquids, or gases, all conventional ionic compounds are solids and have the following properties.

1. Crystals of ionic compounds are hard and brittle.

2. Ionic compounds have high melting points.

3. When heated to the molten state (if they do not decompose), ionic compounds conduct electricity.

4. Many ionic compounds dissolve in high-polarity solvents (such as water), and, when they do, the solutions are electrically conducting.

5.2 The Ionic Model and the Size of Ions

According to the Pauling concept of electronegativity, as the electronegativity difference between two covalently bonded atoms increases, the bond becomes increasingly polar. Eventually, the difference becomes so large that any "sharing" of electrons is negligible, and we define the bond as ionic. An ionic bond is simply the electrostatic attraction between a positive ion (cation) and a negative ion (anion).

Figure 5.1 shows a relationship devised by Pauling between electronegativity difference and the ionic character. Notice that the relationship is a continuum and that there is no actual dividing line between covalent and ionic behaviors. In fact, there are many compounds for which the bonding seems to fall into the intermediate range, where it could equally well be considered as very polar covalent bonding or partially covalent ionic bonding.

Because metals have low electronegativities and nonmetals have high electronegativities, the combinations from these two classes generally fit the category of the ionic bond. According to the "pure" ionic model, some of the outermost electrons have been completely transferred from the element of lower electronegativity to the element of higher electronegativity. This model is surprisingly useful, even though Pauling's plot indicates that there is always some small degree of covalency, even when the electronegativity difference is very large. As we study the chemistry of the different groups, we will see many examples of covalent character in supposedly ionic compounds.

In Chapter 2, we saw that the size of atoms decreases from left to right in a period as a result of an increase in Z_{eff}. However, the conversion of many atoms to ions results in a significant change in size. The most noticeable examples are the main group metals where cation formation

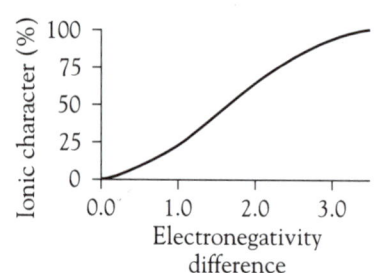

Figure 5.1 Relationship between electronegativity and ionic character.

Table 5.1 Selected isoelectronic period third-cation radii

Ion	Radius (pm)
Na^+	116
Mg^{2+}	86
Al^{3+}	68

Table 5.2 Selected isoelectronic period 2 anion radii

Ion	Radius (pm)
N^{3-}	132
O^{2-}	124
F^-	117

usually involves the removal of all the outer (valence) electrons. The cation that remains possesses only the core electrons. Thus, the cation will be very much smaller than the parent atom. For example, the metallic radius of sodium is 186 pm, whereas its ionic radius is only 116 pm. In fact, the decrease in size is really more dramatic. The volume of a sphere is given by the formula $V = (\frac{4}{3})\pi r^3$. Hence, the reduction of the radius of sodium upon ionization actually means that the ion is one-fourth the size of the atom!

It should be noted that ionic radii cannot be measured directly and are thus subject to error. For example, we can measure precisely the distance between the centers of a pair of sodium and chloride ions in a salt crystal, but this gives the sum of the two radii. The choice of how to apportion the distance between the two ions relies on an empirical formula rather than on a definitive measurement. In this text we will use what are known as the Shannon–Prewitt values of ionic radii for consistency.

Trends in Ionic Radii

The cation radii become even smaller if the ions have a multiple charge. We can see this from the set of isoelectronic ions in Table 5.1. Each of the ions has a total of 10 electrons ($1s^2 2s^2 2p^6$). The only difference is the number of protons in the nucleus; the larger the proton number, the higher the effective nuclear charge, Z_{eff}, and hence the stronger the attraction between electrons and nucleus and the smaller the ion.

For anions, the reverse situation is true: The negative ion is larger than the corresponding atom. For example, the covalent radius of the oxygen atom is 74 pm, whereas the radius of the oxide ion is 124 pm, resulting in a fivefold increase in volume. It can be argued that, with added electrons, the Z_{eff} on each individual outer electron will be less, resulting in a weaker nuclear attraction. As well, there will be additional interelectron repulsions between the added electron and those present in the atom. Hence, the anion will be larger than the atom. Table 5.2 shows that, for an isoelectronic series, the smaller the nuclear charge, the larger the anion. These anions are isoelectronic with the cations in Table 5.1 and illustrate how much larger anions are than cations. It is *generally* true, then, that the metal cations are smaller than the nonmetal anions.

Proceeding down a group, as the atoms become larger, so do the ions, both anions and cations. Values for the Group 17 anions are given in Table 5.3.

Table 5.3 Radii of the Group 17 anions

Ion	Radius (pm)
F^-	119
Cl^-	167
Br^-	182
I^-	206

Trends in Melting Points

The ionic bond is a result of the attraction of one ion by the ions of opposite charge that surround it in the crystal lattice. The melting process involves partially overcoming the strong ionic attractions and allowing the free movement of the ions in the liquid phase. The smaller the ion, the shorter

Table 5.4 Melting points of the potassium halides

Compound	Melting point (°C)
KF	857
KCl	772
KBr	735
KI	685

the interionic distance, hence the stronger the electrostatic attraction, and the higher the melting point. As shown in Table 5.3, the anion radii increase down the halogen group. This increase in radii corresponds to a decrease in melting points of the potassium halides (Table 5.4).

A second and usually more crucial factor in determining the value of melting points is ion charge, the higher the charge, the higher the melting point. Thus magnesium oxide ($Mg^{2+}O^{2-}$) has a melting point of 2800°C, whereas that of isoelectronic sodium fluoride (Na^+F^-) is only 993°C.

5.3   Polarization and Covalency

Even though most combinations of metals and nonmetals have the characteristics of ionic compounds, there are a variety of exceptions. These exceptions arise when the outermost electrons of the anion are so strongly attracted to the cation that a significant degree of covalency is generated in the bond; that is, the electron density of the anion is distorted toward the cation. This distortion from the spherical shape of the ideal anion is referred to as *polarization*.

The chemist Kasimir Fajans developed the following rules summarizing the factors favoring polarization of ions and hence the increase in covalency.

1. A cation will be more polarizing if it is small and highly positively charged.

2. An anion will be more easily polarized if it is large and highly negatively charged.

3. Polarization is favored by cations that do not have a noble-gas configuration.

A measure of the polarizing power of a cation is its *charge density*. The charge density is the ion charge (number of charge units times the proton charge in coulombs) divided by the ion volume. For example, the sodium ion has a charge of 1+ and an ionic radius of 116 pm (we use radii in millimeters to give an exponent-free charge density value). Hence,

$$\text{charge density} = \frac{1 \times (1.60 \times 10^{-19}\ C)}{(\frac{4}{3}) \times \pi \times (1.16 \times 10^{-7}\ mm)^3} = 24\ C \cdot mm^{-3}$$

Similarly, the charge density of the aluminum ion can be calculated as 364 $C \cdot mm^{-3}$. With a much greater charge density, the aluminum ion is much more polarizing than the sodium ion and hence more likely to favor covalency in its bonding.

One of the most obvious ways of distinguishing ionic behavior from covalent behavior is by observing melting points (m.p.): Those of ionic compounds (and network covalent compounds) tend to be high; those of small-molecule covalent compounds, low. To illustrate the effects of anion size, we can compare aluminum fluoride (m.p. 1290°C) and aluminum iodide (m.p. 190°C). The fluoride ion, with an ionic radius of 117 pm, is much smaller than the iodide ion of radius 206 pm. In fact, the iodide ion has a volume more than five times greater than that of the fluoride ion. The fluoride ion cannot be polarized significantly by the aluminum ion. Hence, the bonding is essentially ionic. The electron density of the iodide ion, however, is distorted toward the high-charge-density aluminum ion to such an extent that covalently bonded aluminum iodide molecules are formed.

Because the ionic radius is itself dependent on ion charge, we find that the value of the cation charge is often a good guide in determining the degree of covalency in a simple metal compound. With a cation charge of 1+ or 2+, ionic behavior will usually predominate. With a cation charge of 3+, only compounds with poorly polarizable anions, such as fluoride, are likely to be ionic. Cations that theoretically have charges of 4+ or above do not actually exist as ions, and their compounds can always be considered to have a predominantly covalent character. This principle is illustrated by a comparison of two of the manganese oxides. Manganese(II) oxide, MnO, has a melting point of 1785°C, whereas manganese(VII) oxide, Mn_2O_7, is a liquid at room temperature. Studies have confirmed that manganese(II) oxide forms an ionic crystal lattice, whereas manganese(VII) oxide consists of covalently bonded Mn_2O_7 molecules. Comparing charge densities, we find that the manganese(II) ion has a charge density of 84 C·mm^{-3}, whereas that of the manganese(VII) ion, if it existed, would be 1240 C·mm^{-3}. The latter figure is so high that the (theoretical) manganese(VII) ion is capable of polarizing all anions and exclusively forming covalent bonds.

The third Fajans rule relates to cations that do not have a noble-gas electron configuration. Most common cations, such as calcium, have an electron configuration that is the same as that of the preceding noble gas (for calcium, [Ar]). However, some do not. The silver ion (Ag^+), with an electron configuration of $[Kr]4d^{10}$, is a good example (among the others are Cu^+, Sn^{2+}, and Pb^{2+}). As the size of the silver ion is between that of sodium and potassium, for the purely ionic model we might expect the melting points of silver salts to be between those of sodium and potassium. But they are not. For example, the melting points of sodium chloride and potassium chloride are 801°C and 770°C, respectively, while that of silver chloride is only 455°C.

We explain the comparatively low melting point of silver chloride as follows. In the solid phase, the silver ions and halide ions are arranged in a crystal lattice, like any other "ionic" compound. However, the overlap of electron density between each anion and cation is sufficiently high, it is argued, that we can consider the melting process to involve the formation of actual silver halide molecules. Apparently, the energy needed to change from a partial ionic solid to covalently bonded molecules is less than that needed for the normal melting process of an ionic compound.

Another indication of a difference in the bonding behaviors of the potassium ion and the silver ion is their different aqueous solubilities. All of the potassium halides are highly water soluble, whereas the silver chloride, bromide, and iodide are essentially insoluble in water. The solution process, as we will see later, involves the interaction of polar water molecules with the charged ions. If the ionic charge is decreased by partial electron sharing (covalent bonding) between the anion and the cation, then the ion–water interaction will be weaker and the tendency to dissolve will be less. Unlike the other silver halides, silver fluoride is soluble in water. This observation is consistent with the Fajans rules that predict that silver fluoride should have the weakest polarization and the most ionic bonding of all the silver halides.

An Alternative View of Bonding in Boron Trifluoride

It is commonly accepted that boron trifluoride is covalently bonded. However, Ron Gillespie, a Canadian chemist, has argued that the bonding is essentially ionic. He notes that, across the Period 2 fluorides (from LiF through BeF_2 and BF_3 to CF_4), it is assumed that there is an abrupt change in bonding from ionic to covalent between beryllium and boron. He contends that the boron–fluorine bond distance is, in fact, that expected for three close-packed fluoride ions around a boron 3+ ion. The greater bond distance in the tetrahedral tetrafluoroborate ion, BF_4^-, in his view, confirms this hypothesis; four fluoride ions cannot approach the tiny boron ion as closely as three can. In fact, the ratio of B–F bond lengths between BF_3 and BF_4^- is that predicted by the close-packed model. To those who would argue that ionic compounds should not be gases at room temperature, he contends that ionic substances can indeed be gases, provided they are fully coordinated. Thus, aluminum fluoride, AlF_3, is a solid because six fluorides can surround the aluminum ion, resulting in a three-dimensional (high-melting) lattice. Boron, by virtue of its small size, can only accommodate the three fluoride ions. So which is the "correct" model? In chemistry, we often forget that explanations are used to account for observed facts, not vice versa. Sometimes we use simplistic models that have good predictive value even when they have limited experimental validity. It would be nice to have the "correct" model, but chemistry is rarely as definitive as it is sometimes portrayed in general chemistry courses.

Often in chemistry there is more than one way to explain an observed phenomenon. This is certainly true for the properties of ionic compounds. To illustrate this point, we can compare the oxides and sulfides of sodium and of copper(I). Both these cations have about the same radius, yet sodium oxide and sodium sulfide behave as typical ionic compounds, reacting with water, whereas copper(I) oxide and copper(I) sulfide are almost completely insoluble in water. We can explain this in terms of the third rule of Fajans, that is, the non-noble gas configuration cation has a greater tendency toward covalency. Alternatively, we can use the Pauling concept of electronegativity and say that the electronegativity difference for, say, sodium oxide of 2.5 would indicate predominantly ionic bonding, whereas that of copper(I) oxide (1.5) would indicate a major covalent character to the bonding.

5.4 *Hydration of Ions*

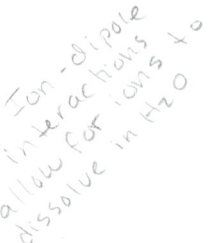
Ion–dipole interactions allow for ions to dissolve in H₂O

If the electrostatic attractions between the ions provide the ionic bond, what is the driving force that allows many ionic compounds to dissolve in water? It is the formation of ion–dipole interactions with the water molecules. The water molecules are polar as a result of the electronegativity difference between the constituent oxygen and hydrogen atoms (as we discussed in Section 3.14). In the dissolving process, we picture the $\delta-$ oxygen ends of water molecules surrounding the cations and the $\delta+$ hydrogen ends of other water molecules surrounding the anions. If the ion–dipole interactions are

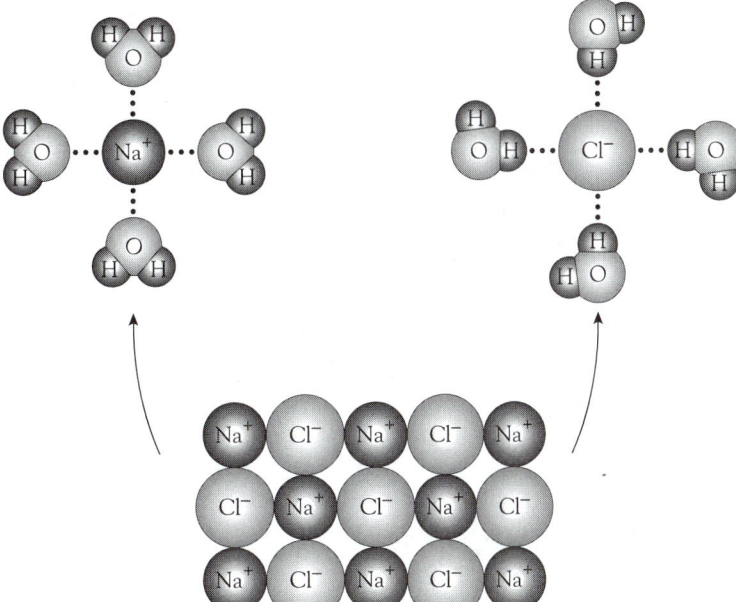

Figure 5.2 Representation of the dissolving process for sodium chloride in water.
(From G. Rayner-Canham et al., Chemistry: A Second Course [Don Mills, ON: Addison–Wesley, 1989], p. 350.)

stronger than the sum of the ionic attractions and the intermolecular forces of the water molecules, then solution will occur. The dissolution process is illustrated in Figure 5.2.

A more quantitative discussion of the process is given in Section 6.4, but it is important to realize that ionic compounds will only dissolve when the ion–dipole interactions formed with the solvent are very strong compared with the ionic bond. Thus, the solvent itself must be very polar. Among the common liquids, only water is polar enough to dissolve ionic compounds.

When an ionic compound crystallizes from an aqueous solution, water molecules often become incorporated in the solid crystal. These water-containing ionic compounds are known as hydrates. In some hydrates, the water molecules are simply sitting in holes in the crystal lattice, but in the majority of hydrates, the water molecules are associated closely with either the anion or the cation, usually the cation. For example, aluminum chloride crystallizes as aluminum chloride hexahydrate, $AlCl_3 \cdot 6H_2O$. In fact, the six water molecules are organized in an octahedral arrangement around the aluminum ion, with the oxygen atoms oriented toward the aluminum ion. Thus the solid compound is more accurately represented as $[Al(OH_2)_6]^{3+} \cdot 3Cl^-$, hexaaquaaluminum chloride (the water molecule is written reversed to indicate that it is the $\delta-$ oxygen that forms the ion–dipole interaction with the positive aluminum ion). Thus in the crystal of hydrated aluminum chloride, there are alternating hexaaquaaluminum cations and chloride anions.

The extent of hydration of ions in the solid phase usually correlates with the ion charge and size, in other words, the charge density. We can therefore account for the anhydrous nature of the simple binary alkali metal salts, such as sodium chloride, because both ions have low charge densities. Crystallization of an ion with a 3+ charge from aqueous solution always results in a hexahydrated ion in the crystal lattice. That is, the small, highly charged cation causes the ion–dipole interaction to be particularly strong. The extent of hydration of anions also depends on charge density. Thus, the more highly charged oxyanions are almost always hydrated, although not to the extent of the cations. For example, zinc sulfate, $ZnSO_4$, is a heptahydrate with six of the water molecules associated with the zinc ion and the seventh with the sulfate ion. Hence, the compound is more accurately represented as $[Zn(OH_2)_6]^{2+}[SO_4(H_2O)]^{2-}$. Many other dipositive metal sulfates form heptahydrates with the same structure as that of the zinc compound.

5.5   The Ionic Lattice

In Section 4.3 we showed four different packing arrangements for metal atoms. The same packing arrangements are common among ionic compounds as well. Generally, the anions are much larger than the cations; thus, it is the anions that form the array and the smaller cations fit in holes (called *interstices*) between the anions. Before discussing the particular types of packing, however, we should consider general principles that apply to ionic lattices.

1. Ions are assumed to be charged, incompressible, nonpolarizable spheres. We have seen that there is usually some degree of covalency in all ionic compounds, yet the hard sphere model seems to work quite well for most of the compounds that we classify as ionic.

2. Ions try to surround themselves with as many ions of the opposite charge as possible and as closely as possible. This principle is of particular importance for the cation. Usually, in the packing arrangement adopted, the cation is just large enough to allow the anions to surround it without touching one another.

3. The cation to anion ratio must reflect the chemical composition of the compound. For example, the crystal structure of calcium chloride, $CaCl_2$, must consist of an array of chloride anions with only half that number of calcium cations fitting in the interstices in the crystal lattice.

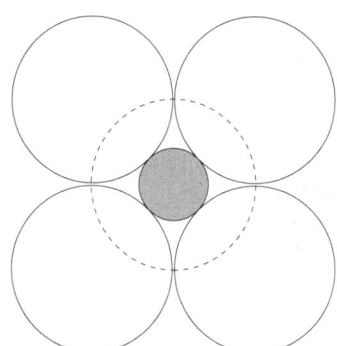

Figure 5.3 Representation of six anions surrounding a cation (shaded).

As mentioned in point 2, the packing arrangement adopted by an ionic compound is usually determined by the comparative sizes of the ions. Figure 5.3 shows four solid circles representing the anions of part of a body-centered cubic anion array and a dashed circle representing the anions below and above the plane. To fit exactly in the space between these six anions, the cation has to be the size shown by the shaded circle. By using the Pythagorean theorem, we can calculate that the optimum ratio of cation radius to anion radius is 0.414. The numerical value, r_+/r_-, is called the *radius ratio*.

If the cation is larger than one giving the optimum 0.414 ratio, then the anions will be forced apart. In fact, this happens in most cases, and the increased anion–anion distance decreases the anion–anion electrostatic repulsion. However, when the radius ratio reaches 0.732, it becomes possible for eight anions to fit around the cation. Conversely, if the radius ratio is less than 0.414, the anions will be in close contact and the cations will be "rattling around" in the central cavity. Rather than allowing this to happen, the anions rearrange to give smaller cavities surrounded by only four anions. A summary of the radius ratios and packing arrangements is given in Table 5.5.

The Cubic Case

The best way to picture the ionic lattice is to consider the anion arrangement first and then look at the coordination number of the interstices in

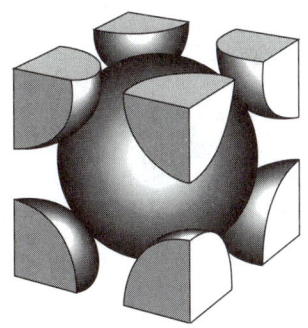

Figure 5.4 Unit cell of cesium chloride.
(Adapted from G. Rayner-Canham et al., Chemistry: A Second Course [Don Mills, ON: Addison–Wesley, 1989], p. 72.)

Table 5.5 The range of radius ratios corresponding to different ion arrangements

r_+/r_- values	Coordination number preferred	Name
0.732 to 0.999	8	Cubic
0.414 to 0.732	6	Octahedral
0.225 to 0.414	4	Tetrahedral

the anion arrays. The packing arrangement that can accept the largest cation is the simple cubic (see Figure 4.6). In this arrangement, there are eight anions surrounding each cation. The classic example is *cesium chloride*, and this compound gives its name to the lattice arrangement. The chloride anions adopt a simple cubic packing arrangement, with each cation sitting at the center of a cube. In cesium chloride, the radius ratio of 0.934 indicates that the cations are sufficiently large to prevent the anions from contacting one another. The unit cell of this crystal is shown in Figure 5.4; it contains one cesium ion and $8(\frac{1}{8})$ chloride ions. Hence, each unit cell contains, in total, one formula unit. The cesium cation separates the chloride anions, so the ions only make contact along a diagonal line that runs from one corner through the center of the unit cell to the opposite corner. This diagonal has a length equal to the sum of two anion radii and two cation radii.

To enhance our visualization of the various ion arrangements, we will display most of the ionic structures as *ionic lattice diagrams*. In these diagrams, the ionic spheres have been shrunk in size and solid lines have been inserted to represent points of ionic contact. Ionic lattice diagrams are better at showing the coordination numbers of the ions more clearly than the space-filling representation. On the other hand, they give a false impression that the lattice is mostly empty space when, in reality, it consists of closely packed ions of very different ionic radii. The ionic lattice diagram for cesium chloride is shown in Figure 5.5.

If the stoichiometry of cation to anion is not 1:1, then the less common ion occupies a certain proportion of the spaces. A good example is calcium fluoride, CaF_2, in which the cation to anion ratio is 1:2. This is called the *fluorite* structure, after the mineral name of calcium fluoride. Each calcium ion is surrounded by eight fluoride ions, similar to the cesium chloride structure. However, each alternate cation location (that is, every other space) in the lattice is empty, thus preserving the 1:2 cation to anion ratio (Figure 5.6).

It is also possible to have cation to anion ratios of 2:1, as found for lithium oxide. The structure is again based on the cesium chloride lattice, but this time every alternate anion site is empty. Because the unoccupied lattice spaces in the lithium oxide structure are the opposite of those left unoccupied in the calcium fluoride (fluorite) structure, the name given to this arrangement is the *antifluorite* structure.

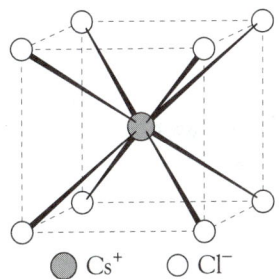

Figure 5.5 Ionic lattice diagram of cesium chloride.
(Adapted from A.F. Wells, Structural Inorganic Chemistry, 5th ed. [New York: Oxford University Press, 1984], p. 246.)

○ Cs^+ ○ Cl^-

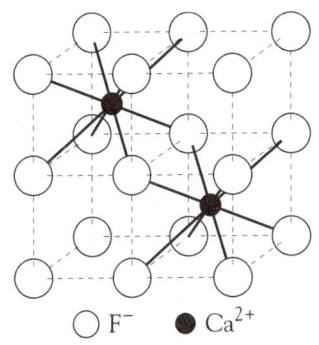

○ F^- ● Ca^{2+}

Figure 5.6 Partial ionic lattice diagram of calcium fluoride.
(Adapted from A.F. Wells, Structural Inorganic Chemistry, 5th ed. [New York: Oxford University Press, 1984], p. 256.)

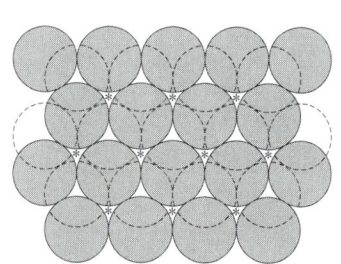

Figure 5.7 The first two layers of the cubic close-packed anion array, showing the octahedral holes (*) in which cations can fit.

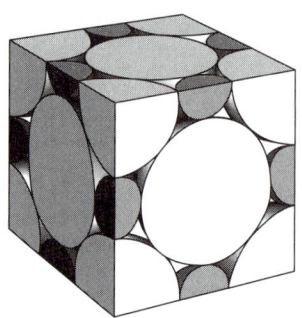

Figure 5.8 Unit cell of sodium chloride.
(From G. Rayner-Canham et al., Chemistry: A Second Course [Don Mills, ON: Addison–Wesley, 1989], p. 71.)

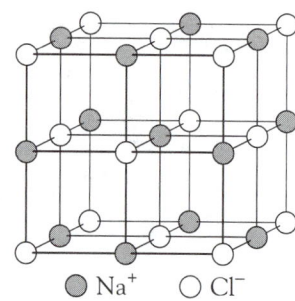

$\bigcirc$ Na$^+$ $\bigcirc$ Cl$^-$

Figure 5.9 Ionic lattice diagram of sodium chloride.
(Adapted from A.F. Wells, Structural Inorganic Chemistry, 5th ed. [New York: Oxford University Press, 1984], p. 239.)

The Octahedral Case

When the radius ratio falls below 0.732, the anions in the cesium chloride structure are no longer held apart by the cations. The potential repulsions between the anions cause the octahedral geometry to become the preferred arrangement. For this smaller radius ratio, six anions can fit around a cation without touching one another (see Figure 5.3). The actual anion arrangement is based on the cubic close-packed array in which there are octahedral holes and tetrahedral holes. Figure 5.7 shows the array with an asterisk (*) marking the location of the octahedral holes in which the cations can fit.

In the octahedral packing, all the octahedral holes are filled with cations and all of the tetrahedral holes are empty. *Sodium chloride* adopts this particular packing arrangement, and it gives its name to the structure. In the unit cell—the smallest repeating unit of the structure—the chloride anions form a face-centered cubic arrangement. Between each pair of anions, a cation is located. Because the cations are acting as separators of the anions, alternating cations and anions touch along the edge of the cube (Figure 5.8). The unit cell contains one central sodium ion plus 12($\frac{1}{4}$) sodium ions along the edges. The centers of the faces hold 6($\frac{1}{2}$) chloride ions, and the corners of the cube hold 8($\frac{1}{8}$) more. As a result, the sodium chloride unit cell contains four formula units. The length of the side of the cube is the sum of two anion radii and two cation radii. The ionic lattice diagram shows that each sodium ion has six nearest-neighbor chloride ions and each chloride anion is surrounded by six sodium ions (Figure 5.9).

It is also possible to have octahedral packing for compounds with stoichiometries other than 1 : 1. The classic example is that of titanium(IV) oxide, TiO_2 (mineral name, *rutile*). For the crystal, it is easiest to picture the titanium(IV) ions as forming a distorted body-centered array (even though they are much smaller than the oxide anions), with the oxide ions fitting in between (Figure 5.10).

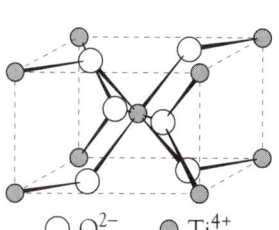

$\bigcirc$ O^{2-} $\bigcirc$ Ti^{4+}

Figure 5.10 Ionic lattice diagram of titanium(IV) oxide.
(Adapted from A.F. Wells, Structural Inorganic Chemistry, 5th ed. [New York: Oxford University Press, 1984], p. 247.)

The Tetrahedral Case

Ionic compounds in which the cations are very much smaller than the anions can be visualized as close-packed arrays of anions, with the cations fitting into the tetrahedral holes (the octahedral holes are always empty).

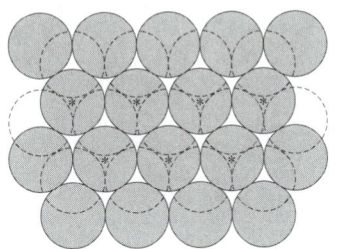

Figure 5.11   The first two layers of the cubic close-packed anion array, showing the tetrahedral holes (*) in which cations can fit.

Both hexagonal close-packed (*hcp*) and cubic close-packed (*ccp*) arrangements are possible, and usually a compound will adopt one or the other, although the reasons for particular preferences are not well understood. Figure 5.11 shows the cubic close-packed array; an asterisk (*) marks the location of the tetrahedral holes.

The prototype of this class is zinc sulfide, ZnS, which exists in nature in two crystal forms: the common mineral *sphalerite* (formerly called zinc blende), in which the sulfide ions form a cubic close-packed array; and *wurtzite*, in which the anion array is hexagonal close packed (see Chapter 4). Both structures have twice as many tetrahedral holes as cations, so only alternate cation sites are filled (Figure 5.12).

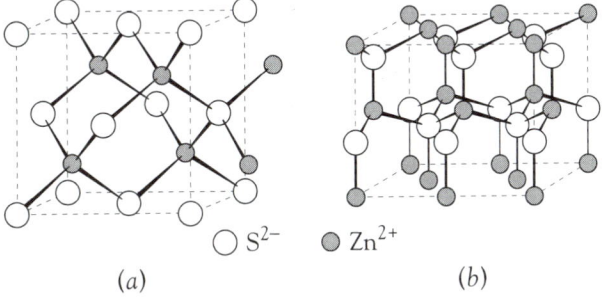

Figure 5.12   Two ionic lattice diagrams of zinc sulfide:
(a) sphalerite; (b) wurtzite.
(Adapted from A.F. Wells, Structural Inorganic Chemistry, 5th ed. [New York: Oxford University Press, 1984], p. 121.)

$\bigcirc$ S^{2-} $\bullet$ Zn^{2+}

(a)   (b)

Exceptions to the Packing Rules

Up to now, we have discussed the different packing arrangements and their relationship to the radius ratio. However, the radius ratio is only a guide, and although many ionic compounds do adopt the predicted packing arrangement, there are many exceptions. In fact, the packing rules appear to predict the correct arrangement in about two-thirds of the cases. Table 5.6 shows three of the exceptions.

Chemistry is not a simplistic subject, and to reduce the reasons for a particular packing arrangement to one particular criterion, the radius ratio, is to disregard many factors. In particular, we discussed earlier that there is an appreciable degree of covalent bonding in most ionic compounds. Thus, the hard sphere model of ions is not considered valid for many compounds. For example, mercury(II) sulfide is likely to have such a high degree of covalency in its bonding that the compound might equally well be regarded as a network covalent substance, like diamond or silicon dioxide (see Chapter 3). A high degree of covalency would specifically explain the preference of mercury(II) sulfide for the tetrahedral coordination of the ZnS structure, because in its covalent compounds, mercury(II) often forms four covalent bonds arranged at the tetrahedral angles.

Table 5.6   Selected examples of exceptions to the packing predicted by the radius ratio rule

Compound	r_+/r_-	Expected packing	Actual packing
HgS	0.55	NaCl	ZnS
LiI	0.35	ZnS	NaCl
RbCl	0.84	CsCl	NaCl

Partial covalent behavior is also observed in lithium iodide. On the basis of standard values for ionic radii, its adoption of the octahedral coordination of the sodium chloride lattice makes no sense. The iodide anions would be in contact with one another, and the tiny lithium ions would "rattle around" in the octahedral holes. However, the bonding in this compound is believed to be about 30 percent covalent, and crystal structure studies show that the electron density of the lithium is not spherical but stretched out toward each of the six surrounding anions. Thus lithium iodide, too, cannot be considered a "true" ionic compound.

Furthermore, there is evidence that the energy differences between the different packing arrangements are often quite small. For example, rubidium chloride normally adopts the unexpected sodium chloride structure (see Table 5.6), but crystallization under pressure results in the cesium chloride structure. Thus, the energy difference in this case between the two packing arrangements must be very small.

Finally, we must keep in mind that the values of the ionic radii are not constant from one environment to another. For example, the cesium ion has a radius of 181 pm only when it is surrounded by six anion neighbors. With eight neighbors, such as we find in the cesium chloride lattice, it has a Shannon–Prewitt radius of 188 pm. This is not a major factor in most of our calculations, but with small ions there is a very significant difference. For lithium, the four-coordinated ion has a radius of 73 pm, whereas that of the crowded six-coordinated ion is 90 pm. For consistency in this text, all ionic radii quoted are for six-coordination, except for the Period 2 elements, for which four-coordination is much more common and realistic.

Crystal Structures Involving Polyatomic Ions

Up to this point, only binary ionic compounds have been discussed, but ionic compounds containing polyatomic ions also crystallize to give specific structures. In these crystals, the polyatomic ion occupies the same site as a monatomic ion. For example, calcium carbonate forms a distorted sodium chloride structure, with the carbonate ions occupying the anion sites and the calcium ions, the cation sites.

In some cases, properties of a compound can be explained in terms of a mismatch between the anions and the cations, usually a large anion with a very small cation. One possible way of coping with this problem is for the compound to absorb moisture and form a hydrate. In the hydration process, the water molecules usually surround the tiny cation. The hydrated cation is then closer in size to the anion. Magnesium perchlorate is a good example of this arrangement. The anhydrous compound absorbs water so readily that it is used as a drying agent. In the crystal of the hydrate, the hexaaquamagnesium ion, $Mg(OH_2)_6^{2+}$, occupies the cation sites and the perchlorate ion occupies the anion sites.

Ion mismatch may cause compounds to be thermally unstable. The alkali metal carbonates are stable to heating up to very high temperatures with one exception—lithium carbonate. Lithium carbonate, with a smallish, low-charge cation and a larger, high-charge anion, decomposes on heating to give lithium oxide:

$$Li_2CO_{3(s)} \xrightarrow{\Delta} Li_2O_{(s)} + CO_{2(g)}$$

Some ions are so mismatched in size that they cannot form compounds under any circumstances. The effects of ion mismatch constrain large, low-charge anions to form stable compounds only with large, single-charge cations. Thus the hydrogen carbonate ion, HCO_3^-, forms solid stable compounds only with the alkali metals and the ammonium ion.

5.6 The Bond Triangle

In these last three chapters, we have discussed the three types of bonding: covalent, metallic, and ionic. Covalent bonding involves orbital overlap between specific pairs of atoms in an element or compound. If the orbital overlap is not between specific atoms but is delocalized throughout the entire crystal, then the bonding is regarded as metallic. The third alternative, ionic bonding, involves the electrostatic interaction between individual ions.

Although we categorize bonding in a substance as being of one specific type, in fact, all three types are related, and the bonding in many species is a combination of two or even all three types. Figure 5.13 shows the bond triangle (or more correctly, the Van Arkel–Ketelaar triangle) for a few elements and compounds of Period 3. This figure is not to scale, but it does illustrate the continuum of bonding that seems to exist.

Looking first at the covalent–ionic side of the triangle, we start with chlorine, a nonpolar molecule. The bonding in this molecule, involving two identical, strongly electronegative atoms, is essentially covalent. As we progress up the right side of the triangle, the electronegativity difference between the constituent atoms increases, causing the bonds to become increasingly polar. A point is reached, in this case at magnesium chloride, at which the electronegativity difference is so large that there is very little overlap of orbitals and the species can be regarded as independent ions with ionic bonds. However, as we will see in Chapter 12, Section 12.5, there are some aspects of magnesium chemistry that can best be interpreted in terms of covalent bond contributions. Finally, sodium chloride exhibits almost "pure" ionic bonding.

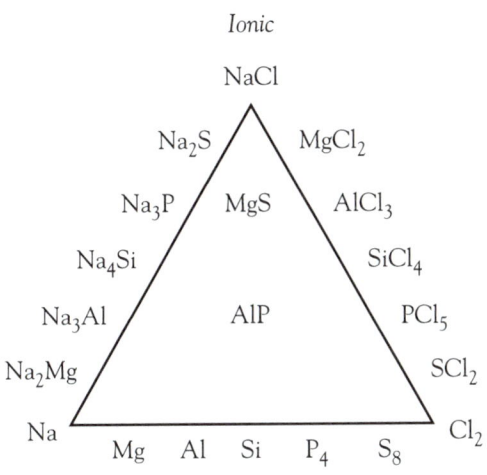

Figure 5.13 The bond triangle.

The covalent–metallic axis corresponds to the change in bonding from the overlap of orbitals in particular directions (covalent) to the delocalized bonding in the metallic situation. Proceeding from right to left across the base of the triangle, the first three elements—dichlorine, octasulfur, S_8, and tetraphosphorus, P_4—are small-molecule covalent. However, at phosphorus, the trend toward less localized bonding is already apparent, for there are other allotropes in which the covalent bonding is directed toward several neighbors. As a result, the bonding in these allotropes is network covalent rather than small-molecule covalent: The orbitals are no longer localized between pairs of atoms but overlap in all directions. Silicon more commonly exists as a network covalent structure, but it can be synthesized as a metallic allotrope. The transition from network covalent to metallic corresponds to the decreasing separation of the molecular orbitals from the $3s$ and $3p$ atomic orbitals to the point where they overlap, thereby allowing electrons to move freely through the entire crystal lattice. The metallic, delocalized bonding is the normal state for the low-electronegativity elements: aluminum, magnesium, and sodium.

Finally, the left side of the bond triangle represents the metallic–ionic transition. From the metallic corner, we start with alloys of low-electronegativity elements. The delocalization of electrons is high; thus, the alloys exhibit metallic bonding. Proceeding along the side of the triangle toward the apex, the electronegativity difference between the element pairs becomes greater and greater and the orbital overlap becomes less and less, with the electron density becoming centered on the more electronegative atom. The final situation is the ionic bond.

As we mentioned earlier, it is possible for compounds to have characteristics of all three bonding types, examples being magnesium sulfide and aluminum phosphide. Fortunately, as we discuss in later chapters, most elements and compounds appear to have properties that can be explained in terms of one bonding type or, at the most, a combination of two bonding types.

EXERCISES

5.1 Define the following terms: (a) polarization; (b) interstices; (c) bond triangle.

5.2 Define the following terms: (a) ion–dipole interactions; (b) radius ratio; (c) cubic arrangement.

5.3 What properties of a compound would lead you to expect that it contains ionic bonds?

5.4 Which would you expect to contain ionic bonds, $MgCl_2$ or SCl_2? Explain your reasoning.

5.5 Which one of each of the following pairs will be smaller? Explain your reasoning in each case. (a) K or K^+; (b) K^+ or Ca^{2+}; (c) Br^- or Rb^+.

5.6 Which one of each of the following pairs will be smaller? Explain your reasoning in each case. (a) Se^{2-} or Br^-; (b) O^{2-} or S^{2-}.

5.7 Which one, NaCl or NaI, would be expected to have the higher melting point? Explain your reasoning.

5.8 Which one, NaCl or KCl, would be expected to have the higher melting point? Explain your reasoning.

5.9 Compare the charge density values of the three silver ions: Ag^+, Ag^{2+}, and Ag^{3+} (on-line Appendix). Which is most likely to form compounds exhibiting ionic bonding?

5.10 Which would be more polarizable, the fluoride ion or the iodide ion. Give your reason.

5.11 Explain the difference between the 227°C melting point of tin(II) chloride, $SnCl_2$, and the −33°C melting point of tin(IV) chloride, $SnCl_4$.

5.12 Magnesium ion and copper(II) ion have almost the same ionic radius. Which would you expect to have the lower melting point, magnesium chloride, $MgCl_2$, or copper(II) chloride, $CuCl_2$? Explain your reasoning.

5.13 Would you expect sodium chloride to dissolve in carbon tetrachloride, CCl_4? Explain your reasoning.

5.14 Suggest a reason why calcium carbonate, $CaCO_3$, is insoluble in water.

5.15 Which of sodium chloride and magnesium chloride is more likely to be hydrated in the solid phase? Explain your reasoning.

5.16 Nickel(II) sulfate commonly exists as a hydrate. Predict the formula of the hydrate and explain your reasoning.

5.17 Of lithium nitrate and sodium nitrate, which is more likely to exist as a hydrate in the solid phase? Explain your reasoning.

5.18 What are the key assumptions in the ionic lattice concept?

5.19 Explain the factor affecting the ion coordination number in an ionic compound.

5.20 Why, in the study of an ionic lattice, is the anion packing considered to be the frame into which the cations fit?

5.21 Though calcium fluoride adopts the fluorite structure, magnesium fluoride adopts the rutile structure. Suggest an explanation.

5.22 Suggest the probable crystal structure of (a) barium fluoride; (b) potassium bromide; (c) magnesium sulfide. You can use comparisons or obtain ionic radii from data tables.

5.23 Use Figure 5.6 as a model to draw a partial ionic lattice diagram for the antifluorite structure of lithium oxide.

5.24 Would the hydrogen sulfate ion be more likely to form a stable solid compound with sodium ion or magnesium ion? Explain your reasoning.

5.25 Using the bond triangle concept, state the probable combinations of bonding in (a) $CoZn_3$; (b) BF_3.

5.26 Using the bond triangle concept, state the probable combinations of bonding in (a) As; (b) K_3As; (c) AsF_3.

5.27 The element gallium melts in your hand (m.p. 30°C). What would be your first suggestion as to the bond type in this element? What one test would you suggest would be most effective in confirming the bond type?

BEYOND THE BASICS

5.28 It has been said that a long liquid range is characteristic of metals. Is this true? Give some examples using data tables. Suggest why a long liquid range cannot be used as the sole criterion for metallic bonding.

5.29 The internuclear distance between sodium and chloride ions in the sodium chloride lattice, $NaCl(s)$, is 281 pm, while the bond distance in $NaCl(g)$ from the vaporized lattice is 236 pm. Suggest why the gas-phase distance is much shorter.

5.30 In melting point, the value for sodium fluoride is higher than that for sodium chloride while that for carbon tetrafluoride is less than that for carbon tetrachloride. Explain the reason for the difference in trends.

5.31 Which member of the following pairs has the higher melting point? Give your reasoning in each case: (a) copper(I) chloride, CuCl, or copper(II) chloride, $CuCl_2$; (b) lead(II) chloride, $PbCl_2$, or lead(IV) chloride, $PbCl_4$.

5.32 In a sodium chloride lattice, the ions usually touch along the edge of the unit cell. If the ionic radii are r_+ and r_-, calculate the length of each side of the unit cell.

5.33 In a cesium chloride lattice, the atoms usually touch along the diagonal from one corner through the center of the cell to the opposite corner. If the ionic radii are r_+ and r_-, calculate the length of each side of the unit cell.

5.34 Calculate the radius of a cesium ion in cesium chloride if the density of cesium chloride is 3.97 $g \cdot cm^{-3}$, and it is assumed that the ions touch through the diagonal of the unit cell.

5.35 Rubidium chloride adopts the sodium chloride structure. Calculate the radius of a rubidium ion if the density of rubidium chloride is 2.76 $g \cdot cm^{-3}$, and it is assumed that the ions touch along the edges of the unit cell.

5.36 Ammonium chloride crystallizes in the cesium chloride lattice. The cations and anions are in contact across the body diagonal of the unit cell and the edge length is 386 pm. Determine a value for the radius of the ammonium ion.

5.37 Sodium hexafluoroantimonate(V), $NaSbF_6$, density 4.37 $g \cdot cm^{-3}$, crystallizes in the sodium chloride lattice. The sodium cations and hexafluoroantimonate(V) anions are in contact along the edge of the unit cell. Determine a value for the radius of the hexafluoroantimonate(V) ion.

5.38 The unit cell of a particular solid has tungsten atoms at the corners, oxygen atoms in the centers of each cube edge, and a sodium atom in the cube center. What is the empirical formula of the compound?

5.39 Astatine, the lowest (and radioactive) member of the halogen series can form the astatinide ion, At^-, which has an approximate ionic radius of 225 pm. What lattice type would be expected for each of the alkali metal compounds of the astatinide ion?

Inorganic Thermodynamics

Chapter **6**

$$\Delta G° = \Delta H° - T\Delta S°$$

Descriptive inorganic chemistry is not simply a study of the chemical elements and the myriad compounds that they form. It also involves the need to explain why some compounds form and others do not. The explanation usually relates to the energy factors involved in the formation of compounds. This topic is a branch of thermodynamics, and in this chapter we present a simplified introduction to inorganic thermodynamics.

Although much of the development of chemistry took place in Britain, France, and Germany, two Americans played important roles in the development of thermodynamics. The first of these was Benjamin Thompson, whose life would make a good movie script. Born in 1753 in Woburn, Massachusetts, he became a major in the (British) Second Colonial Regiment and subsequently spied for the British, using his chemical knowledge to send messages in invisible ink. When Boston fell to the Revolutionary forces, he fled to England and then to Bavaria, now part of Germany. For his scientific contributions to Bavaria's military forces, he was made a count and he chose the title of Count Rumford. At the time, heat was thought to be a type of fluid, and among many significant discoveries, Rumford showed conclusively that heat is a physical property of matter, not a material substance. In fact, it can be argued that he was the first thermodynamicist.

A little less than 100 years later, J. Willard Gibbs was born in New Haven, Connecticut. At Yale University in 1863, he gained one of the first Ph.D. degrees awarded in the United States. It was Gibbs who first derived the mathematical equations that are the basis of modern thermodynamics. About that time, the concept of entropy had been proposed in Europe, and the German physicist Rudolf Clausius had summarized the laws of thermodynamics in the statement "The energy of the universe is a constant, the entropy of the universe tends to a maximum." Yet Clausius and other physicists did not appreciate the importance of the entropy concept. It was Gibbs who showed that everything from miscibility of gases to positions of chemical equilibria depend on entropy factors. In recognition of his role, the thermodynamic function free energy was assigned the symbol G and given the full name of Gibbs free energy.

6.1 Thermodynamics of the Formation of Compounds

Compounds are produced from elements by chemical reactions. For example, our table salt, sodium chloride, can be formed by the combination of the reactive metal sodium with a toxic green gas, chlorine:

$$2 \text{ Na}(s) + \text{Cl}_2(g) \rightarrow 2 \text{ NaCl}(s)$$

Because this reaction occurs without need for external "help," it is said to be a *spontaneous reaction*. (Although being spontaneous does not give any indication of how fast or slow the reaction may be.) The reverse reaction, the decomposition of sodium chloride, is a nonspontaneous process, which is just as well; we certainly would not want the salt on the dining table to start releasing clouds of poisonous chlorine gas! One way to obtain sodium metal and chlorine gas back again is to pass an electric current (an external energy source) through molten sodium chloride:

$$2 \text{ NaCl}(l) \xrightarrow{\text{energy}} 2 \text{ Na}(l) + \text{Cl}_2(g)$$

The study of the causes of chemical reactions is a branch of thermodynamics. This chapter provides a simplified coverage of the topic as it relates to the formation of inorganic compounds. In this section, we will show that the feasibility of chemical reactions depends on two factors: enthalpy and entropy.

Enthalpy

Enthalpy is often defined as the heat content of a substance. When the products of a chemical reaction have a lower enthalpy than the reactants, the reaction releases heat to the surroundings; that is, the process is exothermic. If the products have a higher enthalpy than the reactants, then heat energy is acquired from the surroundings and the reaction is said to be endothermic. The difference between the enthalpy of the products and the enthalpy of the reactants is called the enthalpy change, ΔH.

Entropy

Entropy is often related to the degree of disorder of the substance (although the concept of entropy is really more complex). Thus, the solid phase has a lower entropy than the liquid phase, whereas the gas phase, in which there is random motion, has a very high entropy. The entropy change is indicated by the symbol ΔS.

The Driving Force of a Reaction

For a spontaneous reaction, there must be an increase in entropy overall (that is, the entropy change of the universe must be positive). The universe, to a physical chemist, consists of the reaction (system) that we are studying and its surroundings. It is comparatively easy to measure entropy changes of the reaction, but those of the surroundings are more difficult to determine directly. Fortunately, the change in entropy of the surroundings usually results from the heat released to, or absorbed from, the reaction. Heat released to the surroundings (an exothermic reaction) will increase the entropy of the surroundings while absorption of heat (an endothermic reaction) will lead to a decrease in entropy of the surroundings. Thus we can determine whether a reaction is spontaneous from the entropy and enthalpy changes of the reaction itself.

The changes that favor spontaneity, then, are an increase in entropy and a decrease in enthalpy for the reaction. If both factors occur in a particular chemical reaction, then it is certain to be spontaneous under all conditions. If the reaction would lead to an increase in enthalpy and a decrease in entropy, then it would be nonspontaneous under all conditions. Many reactions fit into the other two categories: a decrease in both enthalpy and entropy or an increase in both factors. For these cases, reaction spontaneity depends upon temperature. This dependency is apparent in the function that combines the enthalpy and entropy factors, the Gibbs free energy, G. The relationship is $\Delta G = \Delta H - T\Delta S$, where T is the Kelvin temperature.

For a spontaneous chemical reaction, there must be a decrease in free energy for the system (that is, ΔG must be negative). Both enthalpy and entropy values are temperature dependent, but it is the entropy factor that is directly multiplied by the Kelvin temperature. Thus, a reaction with positive enthalpy and entropy changes will always become spontaneous above a certain temperature. An increase in entropy, then, is the driving force that results in the decomposition of compounds upon heating. The decomposition process results in more moles of products than of reactants, and some of the products are usually in the gas phase. For example, heating solid mercury(II) oxide gives liquid mercury and gaseous oxygen:

$$2\ HgO(s) \xrightarrow{\Delta} 2\ Hg(l) + O_2(g)$$

The process is spontaneous at high temperatures even though the process is endothermic, for the gas and liquid products will have a higher entropy than the reactants. The possible sign combinations for the thermodynamic functions are summarized in Table 6.1.

Table 6.1   Factors affecting the spontaneity of a reaction

ΔH	ΔS	ΔG	Result
Negative	Positive	Always negative	Spontaneous
Positive	Negative	Always positive	Nonspontaneous
Positive	Positive	Negative at high T	Spontaneous at high T
Negative	Negative	Negative at low T	Spontaneous at low T

In this course, we are rarely concerned with exact numerical calculations. It is just as well, for many of the data values are only known approximately. The reason for performing calculations in this text is to try to understand why some compounds form and others do not. Thus, it is often the sign and the exponent of the number that are meaningful rather than the precise numerical value itself.

To illustrate this point, we can study the formation of ammonia from its elements:

$$\tfrac{1}{2}\ N_2(g) + \tfrac{3}{2}\ H_2(g) \rightarrow NH_3(g)$$

For this reaction, the enthalpy change ΔH has a value of 246 kJ·mol^{-1}, whereas the entropy change ΔS has a value of -0.099 kJ·mol^{-1}·K^{-1}. The negative enthalpy and entropy terms correspond to the last category in Table 6.1. Hence, the reaction should be spontaneous at low temperature but not at high temperature. Inserting the values in the formula $\Delta G = \Delta H - T\Delta S$ at 298 K, the free energy change will be

$$(-46 \text{ kJ·mol}^{-1}) - (298 \text{ K})(-0.099 \text{ kJ·mol}^{-1}\text{·K}^{-1}) = -16 \text{ kJ·mol}^{-1}$$

Thus, the reaction is thermodynamically favored at room temperature. However, at 600 K, the calculation has a different result:

$$(-46 \text{ kJ·mol}^{-1}) - (600 \text{ K})(-0.099 \text{ kJ·mol}^{-1}\text{·K}^{-1}) = +13 \text{ kJ·mol}^{-1}$$

At this temperature, the reaction will not be thermodynamically favored. In fact, the reverse reaction—the decomposition of ammonia—will occur.

Enthalpy of Formation

The enthalpy of a compound is usually listed in data tables as the enthalpy of formation value. The enthalpy of formation is defined as the change in heat content when 1 mol of a compound is formed from its elements in their standard phases at 298 K and 100 kPa. By definition, the enthalpy of formation of an element is zero. This is an arbitrary standard, just as our geographical measure of altitude is taken as height above mean sea level rather than from the center of the Earth. The symbol for the enthalpy of formation under standard conditions is ΔH_f°. Thus we find in data tables values such as $\Delta H_f^\circ(CO_{2(g)}) = -394 \text{ kJ·mol}^{-1}$. This datum indicates that 394 kJ of energy is released when 1 mol of carbon (graphite) reacts with 1 mol of oxygen gas at 298 K and a pressure of 100 kPa to give 1 mol of carbon dioxide:

$$C_{(s)} + O_{2(g)} \rightarrow CO_{2(g)} \qquad\qquad \Delta H_f^\circ = -394 \text{ kJ·mol}^{-1}$$

Enthalpies of formation can be combined to calculate the enthalpy change in other chemical reactions. For example, we can determine the enthalpy change when carbon monoxide burns in air to give carbon dioxide:

$$CO_{(g)} + \tfrac{1}{2} O_{2(g)} \rightarrow CO_{2(g)}$$

First, we collect the necessary data from tables: $\Delta H_f^\circ(CO_{2(g)}) = -394 \text{ kJ·mol}^{-1}$ and $\Delta H_f^\circ(CO_{(g)}) = -111 \text{ kJ·mol}^{-1}$; by definition, $\Delta H_f^\circ(O_{2(g)})$ is zero. The enthalpy change for the reaction can be obtained from the expression

$$\Delta H^\circ(\textit{reaction}) = \Sigma \Delta H^\circ(\textit{products}) - \Sigma \Delta H^\circ(\textit{reactants})$$

Hence,

$$\begin{aligned} \Delta H^\circ(\textit{reaction}) &= (-394 \text{ kJ·mol}^{-1}) - (-111 \text{ kJ·mol}^{-1}) \\ &= -283 \text{ kJ·mol}^{-1} \end{aligned}$$

Thus, the reaction is exothermic, as are almost all combustion reactions.

Bond Energies (Enthalpies)

Enthalpies of formation values are very convenient for the calculation of enthalpy changes in reactions. However, inorganic chemists are often interested in the strength of the attraction between atoms in covalent molecules, the bond enthalpy. The term is commonly called *bond energy*, although there are small differences between the two in terms of their definition and numerical values. We have already mentioned bond energies in the context of the strength of the covalent bond in simple diatomic molecules (see Chapter 2). Bond energy is defined as the energy needed to break 1 mol of

Table 6.2 Bond energies of the diatomic molecules of the halogens

Molecule	Bond energy (kJ·mol^{-1})
F—F	158
Cl—Cl	242
Br—Br	193
I—I	151

Table 6.3 Average bond energies for various carbon–nitrogen bonds

Bond	Bond energy (kJ·mol^{-1})
C—N	305
C=N	615
C≡N	887

the particular covalent bond. Energy is released when bonds are formed, and energy must be supplied when bonds are broken.

We can measure the exact bond energy for a particular pair of atoms joined by a covalent bond. For example, Table 6.2 lists the bond energies in the diatomic molecules of the halogen series. If we look at elements within a group, we see that the bond energies usually decrease as one goes down the group as a result of the increase in atomic size and decrease in orbital overlap of electron density. We will see in this and later chapters that the anomalously low F—F bond energy has a major effect on fluorine chemistry.

The bond energy depends on the other atoms that are present in the molecule. For example, the value of the O—H bond energy is 492 kJ·mol^{-1} in water (HO—H) but 435 kJ·mol^{-1} in methanol, CH_3O—H. Because of this variability, data tables provide average bond energies for a particular covalent bond.

The energy of a specific bond, and thus the bond strength, increases substantially as the bond order increases. Table 6.3 shows this trend for the series of carbon–nitrogen average bond energies.

Lattice Energies (Enthalpies)

The *lattice energy* is the energy change for the formation of 1 mol of an ionic solid from its constituent gaseous ions (we are really considering lattice enthalpy here, but the difference is negligible). We can illustrate the process with sodium chloride. The lattice energy of sodium chloride corresponds to the energy change for

$$Na^+(g) + Cl^-(g) \rightarrow Na^+Cl^-(s)$$

The lattice energy is really a measure of the electrostatic attractions and repulsions of the ions in the crystal lattice. This series of interactions can be illustrated by the sodium chloride crystal lattice (Figure 6.1).

Surrounding the central cation, there are six anions at a distance of r, where r is the distance between centers of nearest neighbors. This is the major attractive force holding the lattice together. However, at a distance of $(2)^{\frac{1}{2}}r$, there are 12 cations. These will provide a repulsion factor. Adding layers of ions beyond the unit cell, we find that there are 8 more anions at a distance of $(3)^{\frac{1}{2}}r$, then 6 more cations at $2r$. Hence, the true balance of charge is represented by an infinite series of alternating attraction and repulsion terms, although the size of the contributions drops off rapidly with increasing distance. The summation of all of the attraction and repulsion terms, then, becomes a converging series. Each type of lattice has a different arrangement of cations and anions and hence its

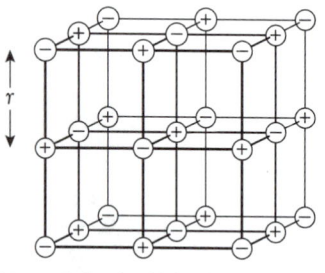

Figure 6.1 Ionic lattice diagram of the sodium chloride structure, showing the ion charge.
(Adapted from A.F. Wells, Structural Inorganic Chemistry, 5th ed. [New York: Oxford University Press, 1984], p. 239.)

Table 6.4   Madelung constants for common lattice types

Type of lattice	Madelung constant, A
Sphalerite, ZnS	1.638
Wurtzite, ZnS	1.641
Sodium chloride, NaCl	1.748
Cesium chloride, CsCl	1.763
Rutile, TiO_2	2.408
Fluorite, CaF_2	2.519

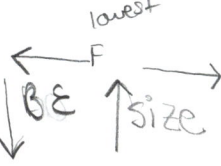

own converging series. The numerical values for these series are known as *Madelung constants, A*. Examples of common lattice types are shown in Table 6.4.

In addition to a dependence upon lattice type, lattice energy varies considerably with ion charge. Doubling the charge from +1 to +2 (or −1 to −2) approximately triples the lattice energy. For example, the lattice energy of potassium chloride is 701 $kJ \cdot mol^{-1}$, whereas that of calcium chloride is 2237 $kJ \cdot mol^{-1}$. In fact, for the series MX, MX_2, MX_3, MX_4, the lattice energies are related in the ratios of 1:3:6:10. The lattice energy is also greater if the ions are smaller, a factor that results in closer ion centers.

It is possible to determine the lattice energy of an ionic crystal from experimental measurements as we will see shortly, but we often need an estimate of lattice energy when the experimental data are unavailable. To determine a theoretical value, we can use the *Born–Landé equation*. This equation, shown in the following, derives remarkably accurate values of the lattice energy, U, from combinations of simple functions:

$$U = -\frac{NAz^+z^-e^2}{4\pi\varepsilon_0 r_0}\left(1 - \frac{1}{n}\right)$$

where N = Avogadro's number (6.023×10^{23} mol^{-1}); A = the Madelung constant; z^+ = the relative cation charge; z^- = the relative anion charge; e = the actual electron charge (1.602×10^{-19} C); $\pi = 3.142$; ε_0 = the permittivity of free space (8.854×10^{-12} $C^2 \cdot J^{-1} \cdot m^{-1}$); r_0 = the sum of the ionic radii; and n = the average Born exponent (discussed in the following).

The Madelung constant accounts for the attractive forces between ions. But when ions approach one another closely, their filled electron orbitals begin to add a repulsion factor that increases rapidly as the two nuclei approach one another. To account for this repulsive term, the *Born exponent, n*, is included in the Born–Landé equation. In fact, the sum of the ionic radii represent the equilibrium position between the ion attractions and the interelectron repulsions. Some values of the Born exponent are given in Table 6.5.

To illustrate how a lattice energy can be calculated from the Born–Landé equation, we will use the example of sodium chloride. The charges for this ionic compound are +1 (z^+) and −1 (z^-). The ionic radii are 116 pm and 167 pm, respectively, giving an r_0 value of 283 pm or 2.83×10^{-10} m. The

Table 6.5 Values for the Born exponent

Electronic configuration of ion	Born exponent, n	Examples
He	5	Li^+
Ne	7	Na^+, Mg^{2+}, O^{2-}, F^-
Ar	9	K^+, Ca^{2+}, S^{2-}, Cl^-, Cu^+
Kr	10	Rb^+, Br^-, Ag^+
Xe	12	Cs^+, I^-, Au^+

value for the Born constant will be the average for that of the sodium and chloride ions: $(7 + 9)/2 = 8$. Substituting in the equation gives

$$U = -\frac{(6.023 \times 10^{23}\ mol^{-1}) \times 1.748 \times 1 \times 1 \times (1.602 \times 10^{-19}\ C)^2}{4 \times 3.142 \times (8.854 \times 10^{-12}\ C^2 \cdot J^{-1} \cdot m^{-1})(2.83 \times 10^{-10}\ m)}\left(1 - \frac{1}{8}\right)$$
$$= 2751\ kJ \cdot mol^{-1}$$

which is close to the best experimental value of 2770 kJ·mol^{-1} (an error of only 2.5 percent). When there is a significant difference between the experimental value of lattice energy and that obtained from the Born–Landé equation, it is usually an indication that the interaction between the ions is not purely ionic but also contains significant covalent character.

The Born–Landé equation can only be used to find the lattice energy if the crystal structure of the compound is known. If it is not, the *Kapustinskii equation* can be used, where v is the number of ions in the empirical formula (for example, for calcium fluoride, $v = 3$) and the other symbols have the same meaning and values as the Born–Landé equation except r_0 is in units of pm.

$$U = -\frac{1.202 \times 10^5\ v z^+ z^-}{r_0}\left(1 - \frac{34.5}{r_0}\right) kJ \cdot mol^{-1}$$

Finally, it should be mentioned that all crystals have a lattice energy. For simple covalent compounds, lattice energy is attributable to intermolecular attractions; for network covalent substances, lattice energy is the energy of the covalent bonds; for metals, it is the attractions that create the metallic bond. However, the energy term for simple covalent molecules is commonly called the enthalpy of sublimation rather than the lattice energy.

Enthalpies of Atomization

Another useful measurement is that of *energy of atomization*. This is defined as the energy needed to produce 1 mol of gaseous atoms of that element from the element in its normal phase at room temperature. This energy term can be used to represent the breaking of the metallic bond in metals or the overcoming of the covalent bonds and intermolecular forces in nonmetals. For example, both of the following equations show atomization processes:

$$Cu(s) \rightarrow Cu(g)$$

$$I_2(s) \rightarrow 2\ I(g)$$

Ionic Liquids

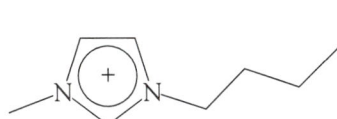

[bmim]$^+$

Figure 6.2   The 1-butyl-3-methylimidazolium ion, commonly abbreviated to [bmim]$^+$.

Students learn that a characteristic of ionic compounds is their high melting points. For conventional ionic compounds, this is true. But like most chemistry, to every rule, there are exceptions. The first ionic liquid was synthesized by accident in the late 1940s. Two chemists at the Rice Institute, Texas, mixed two solids, an N-alkylpyridinium salt with aluminum chloride, and on warming, obtained a liquid. This liquid consisted of free-moving N-alkylpyridinium ions and tetrachloroaluminate, $AlCl_4^-$, ions. There was very little interest in these compounds until recently when chemists started to realize that ionic liquids may be part of the green solution to minimize environmental problems in industrial chemical processes.

These ionic compounds exist as liquids at low temperatures because of their extremely low lattice energy. Two criteria are required: The cation must be an unsymmetrical organic cation together with a low-charge inorganic halo-anion. Most of the anions are nitrogen derivatives, such as the alkylammonium ion $(NR_xH_{4-x})^+$, or ions based on the five-member (three-carbon two-nitrogen) imidazolium ring, or the six-member (five-carbon one-nitrogen) pyridinium ring. Figure 6.2 shows the structure of the 1-butyl-3-methylimidazolium ion, commonly abbreviated to [bmim]$^+$. Examples of the inorganic ions are: the tetrachloroaluminate ion, $AlCl_4^-$; the tetrafluoroborate ion, BF_4^-; and the hexafluorophosphate ion, PF_6^-. Depending upon the cation–anion combination, melting points are often below room temperature, the lowest to date being $-96°C$.

Up until the discovery of ionic liquids, all solution-phase chemical reactions were performed in covalent solvents. As a result, our ideas of what will occur and what will not are biased by our use of the one class of solvent. Ionic solvents have opened the doors to the possibility of entirely new reactions. But it is the use of ionic liquids for conventional industrial syntheses where the interest is most intense. Ionic liquids have the following advantages:

1. They have near-zero vapor pressures, thus avoiding the environmental problems of escaping vapors of most organic solvents. They can also be used in vacuum systems.

2. By combining different anions and cations, ionic liquids with specific solvent properties can be synthesized.

3. They are good solvents for a wide range of inorganic and organic compounds, enabling unusual combinations of reagents to be brought into the same phase.

4. As ionic liquids are immiscible with many organic solvents, they provide the possibility of nonaqueous two-phase reaction systems.

Two of the current problems are: Ionic liquids are, at present, comparatively expensive; and methods of separation and recycling are in their infancy.

Entropy Changes

Unlike enthalpy, which is usually tabulated as relative values (such as enthalpy of formation), entropy is discussed on an absolute basis. Thus, even elements have a listed value of entropy. The zero point is taken to be that of a perfect crystal of a substance at the absolute zero of temperature. We can calculate the standard entropy change for a reaction in the same way as that of the enthalpy change:

$$\Delta S^\circ(reaction) = \sum S^\circ(products) - \sum S^\circ(reactants)$$

For example, we can calculate the standard entropy change for the formation of sodium chloride from sodium metal and chlorine gas:

$$Na(s) + \tfrac{1}{2} Cl_2(g) \rightarrow NaCl(s)$$

Hence,

$$\begin{aligned}
\Delta S^\circ &= [S^\circ(NaCl(s))] - [S^\circ(Na(s))] - \tfrac{1}{2}[S^\circ(Cl_2(g))] \\
&= (+72 \text{ J·mol}^{-1}\text{·K}^{-1}) - (+51 \text{ J·mol}^{-1}\text{·K}^{-1}) - \tfrac{1}{2}(+223 \text{ J·mol}^{-1}\text{·K}^{-1}) \\
&= -90 \text{ J·mol}^{-1}\text{·K}^{-1}
\end{aligned}$$

We would expect an entropy decrease for the system in this process, because it involves the net loss of ½ mol of gas. The reaction is spontaneous at ambient temperatures, so it must be enthalpy driven. (In fact, the recorded value of the enthalpy of formation ΔH_f° of sodium chloride is -411 kJ·mol^{-1}.)

6.2   Formation of Ionic Compounds

When an ionic compound is formed from its elements, there is usually a decrease in entropy, for the ordered, solid crystalline compound has a very low entropy and often the nonmetal reactant, such as oxygen or chlorine, is a high-entropy gas. For example, in the previous section we determined that the entropy change for the formation of sodium chloride was negative. Thus, for the formation of a thermodynamically stable compound from its constituent elements, a negative enthalpy change of the reaction is usually necessary. It is the exothermicity that becomes the driving force of the reaction.

To attempt to understand why particular compounds form and others do not, we will break the formation process of an ionic compound into a series of theoretical steps: first breaking the reactant bonds, then forming those of the products. In this way, we can identify which enthalpy factors are crucial to the spontaneity of reaction. We consider again the formation of sodium chloride:

$$Na(s) + \tfrac{1}{2} Cl_2(g) \rightarrow NaCl(s) \qquad \Delta H_f^\circ = -411 \text{ kJ·mol}^{-1}$$

1. The solid sodium is converted to free (gaseous) sodium atoms. This process requires the enthalpy of atomization:

$$Na(s) \rightarrow Na(g) \qquad \Delta H^\circ = +108 \text{ kJ·mol}^{-1}$$

2. The gaseous chlorine molecules must be dissociated into atoms. This transformation requires one-half of the bond energy of chlorine molecules:

$$\tfrac{1}{2} Cl_2(g) \rightarrow Cl(g) \qquad \Delta H^\circ = +121 \text{ kJ·mol}^{-1}$$

3. The sodium atoms must then be ionized. This process requires the first ionization energy. (If we had a metal that formed a divalent cation, such as calcium, then we would have to add both the first and the second ionization energies.)

$$Na(g) \rightarrow Na^+(g) + e^- \qquad\qquad \Delta H^\circ = +502 \text{ kJ·mol}^{-1}$$

4. The chlorine atoms must gain electrons. This value is the electron affinity of chlorine atoms.

$$Cl(g) + e^- \rightarrow Cl^-(g) \qquad\qquad \Delta H^\circ = -349 \text{ kJ·mol}^{-1}$$

The value for the addition of the first electron is usually exothermic, whereas addition of a second electron is usually endothermic (such as the formation of O^{2-} from O^-).

5. The free ions then associate to form a solid ionic compound. This bringing together of ions is a highly exothermic process—the lattice energy. The lattice energy can be looked on as the major driving force for the formation of an ionic compound:

$$Na^+(g) + Cl^-(g) \rightarrow NaCl(s) \qquad\qquad \Delta H^\circ = -793 \text{ kJ·mol}^{-1}$$

Appendix 2 contains enthalpy, entropy, and free energy data for many common inorganic elements and compounds. Some data, such as ionization energies, are based on actual measurements; however, lattice energies cannot be found from actual measurements. They can be determined only from Born–Haber cycles or theoretical calculations. Furthermore, some measurements are not known to a high degree of precision. As a result, many calculations based on thermodynamic data provide only approximate answers, but such semi-quantitative values can often provide useful insight into chemical stability and reactivity.

6.3 The Born–Haber Cycle

It is usually easier to comprehend information if it is displayed graphically. This representation can be done for the theoretical components of the formation of an ionic compound from its elements. The "up" direction is used to indicate endothermic steps in the process, and the "down" direction corresponds to exothermic steps. The resulting diagram is called a *Born–Haber cycle*. Figure 6.3 shows such a cycle for the formation of sodium chloride.

These enthalpy diagrams can be used in two ways: (1) to gain a visual image of the key enthalpy terms in the formation of the compound and (2) to determine any one unknown enthalpy value in the thermodynamic cycle, for we know that the sum of the component terms should equal the overall enthalpy change for the formation process.

The major endothermic step results from the ionization of the metal atom, whereas the most exothermic step derives from the formation of the ionic crystal lattice. This balance, with the lattice energy exceeding the ionization energies, is common among stable ionic compounds. Magnesium fluoride, MgF_2, can be used to illustrate this point. The sum of the first and second ionization energies of the magnesium ion is $+2190$ kJ·mol^{-1}, much higher than the single ionization energy of the monopositive sodium ion.

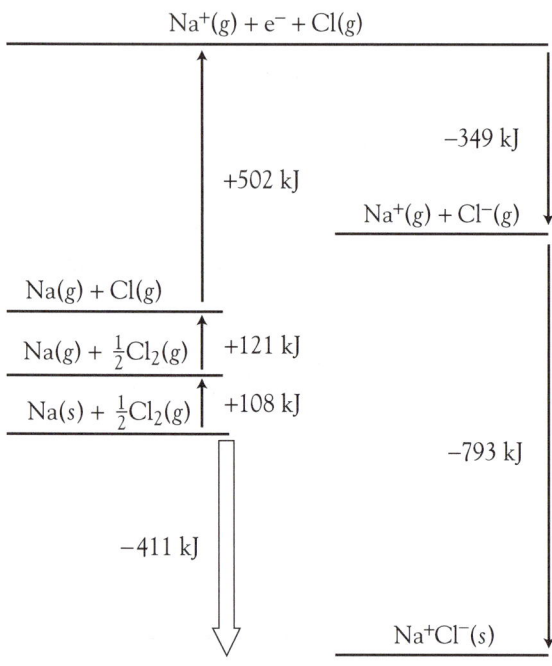

$Na^+(g) + e^- + Cl(g)$

$+502\ kJ$

$-349\ kJ$

$Na^+(g) + Cl^-(g)$

$Na(g) + Cl(g)$

$Na(g) + \frac{1}{2}Cl_2(g)$ $+121\ kJ$

$Na(s) + \frac{1}{2}Cl_2(g)$ $+108\ kJ$

$+121\ kJ$

$-793\ kJ$

$-411\ kJ$

$Na^+Cl^-(s)$

Figure 6.3   Born–Haber cycle for the formation of sodium chloride.

However, with the small, highly charged magnesium ion, the lattice energy is also much higher at $-2880\ kJ\cdot mol^{-1}$. Incorporating the other terms from the enthalpy cycle results in a relatively large negative enthalpy of formation for magnesium fluoride: $-1100\ kJ\cdot mol^{-1}$.

If the lattice energy increases so much with greater cation charge, why do magnesium and fluorine form MgF_2 and not MgF_3? If we estimate the enthalpy of formation of MgF_3, the lattice energy will be even greater than that of MgF_2 because of the greater electrostatic attraction by an Mg^{3+} ion. However, for magnesium, the electrons that must be ionized to give the 3+ ion are core electrons, and the third ionization energy is enormous ($7740\ kJ\cdot mol^{-1}$)—far larger than the gain from the lattice energy. Combined with a negative entropy term, there is no possibility that the compound will exist.

Conversely, why do magnesium and fluorine form MgF_2 and not MgF? As we have seen previously, the two largest energy factors are the ionization energies of magnesium (endothermic) and the lattice energy (exothermic). Far less energy is required to ionize only one electron to form a monopositive magnesium ion than a dipositive ion. However, lattice energy is highly charge dependent; thus, a monopositive cation will result in a much smaller lattice energy. Table 6.6 compares the numerical values for the thermodynamic components of the Born–Haber cycles for the formation of MgF, MgF_2, and MgF_3. These show that the formation of MgF_2 is the most favored in terms of enthalpy factors. The three Born–Haber cycles can be compared graphically, where the importance of the balance between ionization energy and lattice energy becomes apparent (Figure 6.4).

If MgF_2 is preferred over MgF, then we might expect the compound of sodium and fluorine to be NaF_2 and not NaF. In the case of sodium, however, the second electron to be ionized is a core electron, an extremely endothermic process. The enormous ionization energy would not be balanced by the increased lattice energy. There are also two minor factors that make the enthalpy of formation of NaF more negative than what we would expect for MgF. With the lower Z_{eff}, we find that the first ionization energy of sodium is about $200\ kJ\cdot mol^{-1}$ less than that of magnesium.

Table 6.6   Thermodynamic factors in the formation of three possible magnesium fluorides

Enthalpy factors ($kJ\cdot mol^{-1}$)	MgF	MgF$_2$	MgF$_3$
Mg atomization	$+150$	$+150$	$+150$
F—F bond energy	$+80$	$+160$	$+240$
Mg ionization (total)	$+740$	$+2190$	$+9930$
F electron affinity	-330	2660	-990
Lattice energy	≈ -900	-2880	≈ -5900
ΔH_f° (estimated)	-260	-1040	$+3430$

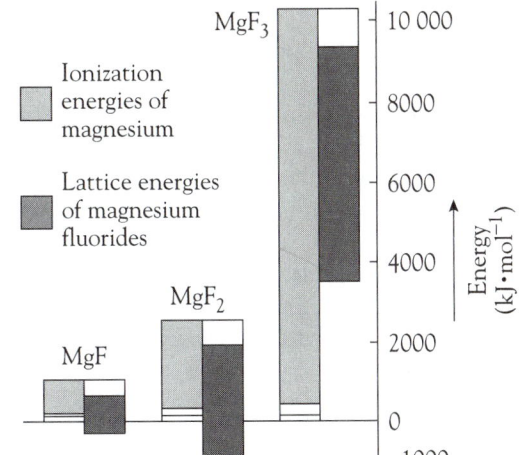

Figure 6.4 Graphical comparison of the Born–Haber cycles for three possible magnesium fluorides.

Furthermore, the Mg^+ ion, with its one $3s$ electron remaining, is a larger cation than the Na^+ ion. Hence, the lattice energy of NaF would be higher than that of MgF. The combination of lower ionization energy and higher lattice energy results in an enthalpy of formation for sodium fluoride of -574 kJ·mol^{-1} compared to the estimated value of -260 kJ·mol^{-1} for MgF.

6.4 Thermodynamics of the Solution Process for Ionic Compounds

Just as the formation of a compound from its constituent elements can be considered as a series of theoretical steps, so can the solution process be broken down into several steps. For this analysis, we visualize first that the ions in the crystal lattice are dispersed into the gas phase and then, in a separate step, that water molecules surround the gaseous ions to give the hydrated ions. Thus ion–ion interactions (ionic bonds) are broken and ion–dipole interactions are formed. The degree of solubility, then, depends on the balance of these two factors, each of which has both enthalpy and entropy components.

There is one key difference between the two analyses. In the formation of a compound, we use the thermodynamic factors simply to determine whether or not a compound will form spontaneously. With the thermodynamics of the solution process, we are concerned with the degree of solubility—that is, where a compound fits on the continuum from very soluble through soluble, slightly soluble, and insoluble to very insoluble. Even for a very insoluble compound, there will be a measurable proportion of aqueous ions present in equilibrium with the solid compound.

Lattice Energy

To break the ions free from the lattice—overcoming the ionic bond—requires a large energy input. The value of the lattice energy depends on the strength of the ionic bond, and this, in turn, is related to ion size and charge. That is, magnesium oxide, with dipositive ions, will have a much higher lattice energy than sodium fluoride with its monopositive ions (3933 kJ·mol^{-1} and 915 kJ·mol^{-1}, respectively). At the same time, the entropy factor will always be highly favorable as the system changes from the highly ordered solid crystal to the disordered gas phase. Consequently, both ΔS and ΔH for lattice dissociation are always positive.

Energy of Hydration

In aqueous solution, the ions are surrounded by polar water molecules. A primary hydration sphere of water molecules (usually six) surrounds the cations, with the partially negative oxygen atoms oriented toward the cation. Similarly, the anion is surrounded by water molecules, with the partially positive hydrogen atoms oriented toward the anion. Beyond the first shell of water molecules, we find additional layers of oriented water molecules (Figure 6.5). The total number of water molecules that effectively surround an ion is called the *hydration number*.

The smaller and more highly charged ions will have a larger number of water molecules in hydration spheres than do larger, less highly charged ions. As a result, the effective size of a hydrated ion in solution can be very different from that in the solid phase. This size difference is illustrated in Table 6.7. It is the smaller size of the hydrated potassium ion that enables it to pass through biological membranes more readily than the larger hydrated sodium ions.

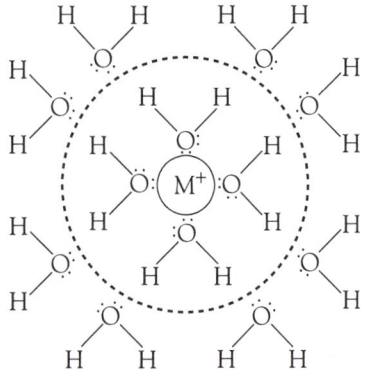

Figure 6.5 Primary and secondary hydration spheres of a metal cation.
(Adapted from G. Wulfsberg, Principles of Descriptive Inorganic Chemistry [New York: University Science Books, 1990], p. 66.)

Table 6.7 Hydration effects on the size of sodium and potassium ions

Ion	Radius (pm)	Hydrated ion	Hydrated radius (pm)
Na^+	116	$Na(OH_2)_{13}{}^+$	276
K^+	152	$K(OH_2)_7{}^+$	232

The formation of the ion–dipole interactions of hydrated ions is highly exothermic. The value of the enthalpy of hydration is also dependent on ion charge and ion size, that is, the charge density. Table 6.8 shows the strong correlation between enthalpy of hydration and charge density for an isoelectronic series of cations.

The entropy of hydration is also negative, mainly because the water molecules surrounding the ions are in a more ordered state than they would be as free water molecules. With the small, more highly charged cations, such as magnesium and aluminum, the hydration spheres are larger than that of sodium; hence, there is a strong ordering of the water molecules around the two higher-charge cations. For these cations, there is a very large decrease in entropy for the hydration process.

Table 6.8 Enthalpy of hydration and charge density for three isoelectronic cations

Ion	Hydration enthalpy ($kJ \cdot mol^{-1}$)	Charge density ($C \cdot mm^{-3}$)
Na^+	−406	24
Mg^{2+}	−1920	120
Al^{3+}	−4610	364

Energy Change of the Solution Process

We can use the solution process for sodium chloride to illustrate an enthalpy of solution cycle. First, the lattice must be vaporized (dissociate into gaseous ions):

$$NaCl(s) \rightarrow Na^+(g) + Cl^-(g) \qquad \Delta H° = +788 \text{ kJ·mol}^{-1}$$

Then the ions are hydrated:

$$Na^+(g) \rightarrow Na^+(aq) \qquad \Delta H° = -406 \text{ kJ·mol}^{-1}$$

$$Cl^-(g) \rightarrow Cl^-(aq) \qquad \Delta H° = -378 \text{ kJ·mol}^{-1}$$

Thus, the enthalpy change $\Delta H°$ for the solution process is

$$(+788) + (-406) + (-378) = +4 \text{ kJ·mol}^{-1}$$

The process can be displayed as a diagram (Figure 6.6).

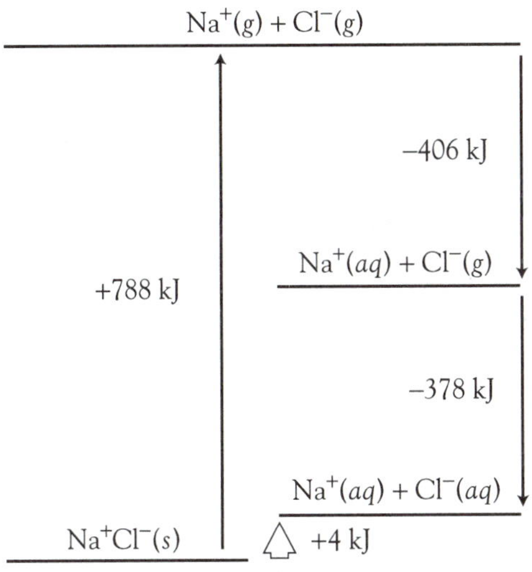

The enthalpy changes are usually far larger than entropy changes at normal temperatures. However, in this case, the very large enthalpy changes essentially "cancel" each other, making the small entropy change a major factor in determining the solubility of sodium chloride. Thus, we now need to make a similar calculation for the entropy factors. So that we can compare the results with the enthalpy values, we will use $T\Delta S°$ data at 298 K. First, the lattice must be vaporized:

$$NaCl(s) \rightarrow Na^+(g) + Cl^-(g) \qquad T\Delta S° = +68 \text{ kJ·mol}^{-1}$$

Then the ions are hydrated:

$$Na^+(g) \rightarrow Na^+(aq) \qquad T\Delta S° = -27 \text{ kJ·mol}^{-1}$$

$$Cl^-(g) \rightarrow Cl^-(aq) \qquad T\Delta S° = -28 \text{ kJ·mol}^{-1}$$

Thus entropy change (as $T\Delta S°$) for the solution process is

$$(+68) + (-27) + (-28) = +13 \text{ kJ·mol}^{-1}$$

The process can be displayed as a diagram (Figure 6.7).

After calculating the free energy change for the solution process, we see that it is the net entropy change that favors solution, whereas the net enthalpy change does not; it is the former that is greater than the latter. Hence, as we know from experience, sodium chloride is quite soluble in water at 298 K.

$$\begin{aligned} \Delta G° &= \Delta H° - T\Delta S° \\ &= (+4 \text{ kJ·mol}^{-1}) - (+13 \text{ kJ·mol}^{-1}) \\ &= -9 \text{ kJ·mol}^{-1} \end{aligned}$$

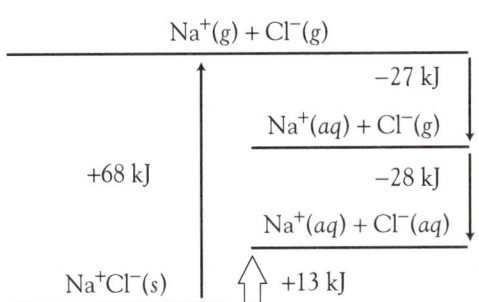

Figure 6.6 Theoretical enthalpy cycle for the solution process for sodium chloride.

Figure 6.7 Theoretical entropy (as $T\Delta S°$) cycle for the solution process for sodium chloride.

6.5 *Formation of Covalent Compounds*

To study the thermodynamics of covalent compound formation, it is possible to construct a cycle similar to the Born–Haber cycle that we used for ionic compounds. However, there is a major difference. The cycle does not involve ion formation; instead, we are concerned with covalent bond energies. The process can be illustrated for the formation of nitrogen trifluoride.

The Hydrogen Economy

We can write a similar energy cycle for the formation of (gaseous) water from hydrogen and oxygen.

$$H_2(g) + \tfrac{1}{2} O_2(g) \rightarrow H_2O(g)$$

Focussing on the enthalpy change, we can write:

$$\Delta H_f^\circ = (\Delta H(H\!-\!H) + \tfrac{1}{2}\Delta H(O\!=\!O)) - (2\,\Delta H(O\!-\!H))$$

$$= [(432 + \tfrac{1}{2} \times 494) - (2 \times 459)]\ \text{kJ·mol}^{-1} = -239\ \text{kJ·mol}^{-1}$$

Thus, the formation of water is a significantly exothermic reaction.

In the view of many scientists, this exothermic reaction holds the long-term key to energy consumption. Hydrocarbon energy sources will run out—it is only the time frame that is in question. But of increasing importance, many scientists believe that we should wean ourselves off the hydrocarbon economy sooner rather than later. The major reason is the contribution of fossil fuels to global climate change. The sooner we stabilize carbon dioxide emissions, the less the effect on the global environment. Also, the liquid hydrocarbons are a valuable source of complex molecules that are of more value as feedstock for the chemical industry for the synthesis of polymers, pharmaceuticals, and other complex organic compounds. The hydrogen–oxygen reaction has the additional advantage that the reaction product is nonpolluting.

The U.S. Department of Energy has now teamed up with the auto industry to develop hydrogen-based fuel cells to power vehicles of the future, a program called Freedom Car. However, before the hydrogen economy becomes a reality, there are three questions that need to be answered regarding synthesis, storage, and usage. The oxygen can be obtained from the air—in fact, for most purposes the atmospheric nitrogen/oxygen mix is quite suitable. How can we generate hydrogen? At present, the simplest method is the electrolysis of a dilute solution of sodium or potassium hydroxide:

$$\text{Cathode} \quad 2\,H_2O(l) + 2\,e^- \rightarrow 2\,OH^-(aq) + H_2(g)$$

$$\text{Anode} \quad 2\,OH^-(aq) \rightarrow H_2O(l) + \tfrac{1}{2}\,O_2(g) + 2\,e^-$$

giving an overall reaction of

$$H_2O(l) \rightarrow H_2(g) + \tfrac{1}{2}\,O_2(g)$$

As the alkali metal hydroxide is not consumed, addition of fresh water will provide a continuously generating cell. Any electrical energy source can be used. For example, hydroelectric dams, wind- and wave-generating systems all produce under-utilized power in off-peak periods. There is now the exciting alternative of biological synthesis. Photosynthetic bacteria have been selectively bred that produce hydrogen gas in the presence of light and nutrients. A different photosynthetic pathway is the use of light and transition metal catalysts to provide the decomposition energy.

Storage is probably the greatest problem. Hydrogen, having the lowest molar mass of all gases, has the lowest density. Thus, a large volume of gas is required compared to the same energy from a liquid fuel. One route is the development of rupture-resistant high-pressure tanks. These have now been developed. Better still is the use of liquid hydrogen, but with a boiling point of $-252°C$, much of the hydrogen is lost by vaporization unless considerable insulation is used. A novel storage method is the use of metal hydrides (see Chapter 10, Section 10.4). Certain metals act as hydrogen sponges with heating releasing the hydrogen. One possible new route is the use of carbon nanotubes (Chapter 14, Section 14.2). Hydrogen molecules can be absorbed in these tubes, though at the time of writing this text, the reversibility of the process has yet to be demonstrated.

The hydrogen could be burned in an internal combustion engine, but a more efficient route is a fuel cell. This functions like a continuously fueled battery employing the reverse electrochemical reactions to that of electrolysis. The city of Chicago, a pioneering city in low-emission public transport, has completed 3 years of testing with hydrogen fuel buses. The experiences learned have lead to a new generation of hydrogen-fueled buses that will be tested in 10 European cities. Unfortunately, at present, the cost of these experimental buses is far higher than that of the most energy-efficient diesel buses.

To transform our society to a nonpolluting hydrogen economy has to be seen as a long-term, not a short-term, solution to the environmental challenges. A tremendous investment will have to be made in hydrogen synthesis and storage plants. And, of course, hydrogen has to become an economically viable fuel. There is a long way to go, but the first steps are already in place with the International Association for Hydrogen Energy coordinating advances worldwide in hydrogen technology.

Once again, the calculation will focus on the enthalpy terms, and the entropy factor will be considered later.

$$\tfrac{1}{2}\,N_2(g) + \tfrac{3}{2}\,F_2(g) \rightarrow NF_3(g) \qquad\qquad \Delta H_f^\circ = -125 \text{ kJ·mol}^{-1}$$

1. The dinitrogen triple bond is broken. This cleavage requires one-half of the $N{\equiv}N$ bond energy:

$$\tfrac{1}{2}\,N_2(g) \rightarrow N(g) \qquad\qquad \Delta H^\circ = +471 \text{ kJ·mol}^{-1}$$

2. The difluorine single bond is broken. For the stoichiometry, three-halves of the F—F bond energy is required:

$$\tfrac{3}{2}\,F_2(g) \rightarrow 3\,F(g) \qquad\qquad \Delta H^\circ = +232 \text{ kJ·mol}^{-1}$$

3. The nitrogen–fluorine bonds are formed. This process releases three times the N—F bond energy as three moles of bonds are being formed:

$$N(g) + 3\,F(g) \rightarrow NF_3(g) \qquad\qquad \Delta H^\circ = -828 \text{ kJ·mol}^{-1}$$

The enthalpy diagram for the formation of nitrogen trifluoride is shown in Figure 6.8.

Turning to the entropy factor, in the formation of nitrogen trifluoride from its elements, there is a net decrease of 1 mol of gas. Thus, a decrease

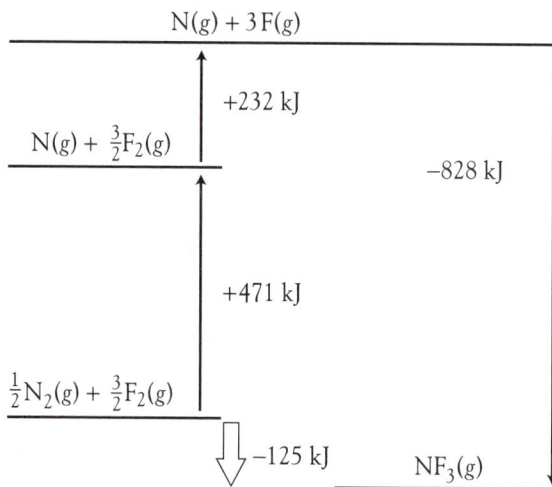

Figure 6.8 Theoretical enthalpy cycle for the formation of nitrogen trifluoride.

in entropy would be expected. In fact, this is the case, and the overall entropy change is -140 J·mol^{-1}·K^{-1}. The resulting free energy change is

$$\Delta G_f^\circ = \Delta H_f^\circ - T\Delta S_f^\circ$$

$$\Delta G_f^\circ = (-125 \text{ kJ·mol}^{-1}) - (298 \text{ K})(-0.140 \text{ kJ·mol}^{-1}\cdot\text{K}^{-1})$$

$$\Delta G_f^\circ = -83 \text{ kJ·mol}^{-1}$$

a value indicating that the compound is quite stable thermodynamically.

6.6 Thermodynamic versus Kinetic Factors

Thermodynamics is concerned with the feasibility of reaction, the position of equilibrium, and the stability of a compound. There is no information about the rate of reaction—the field of kinetics. The rate of a reaction is, to a large extent, determined by the activation energy for the reaction; that is, the energy barrier involved in the pathway for compound formation. This concept is illustrated in Figure 6.9.

A very simple example of the effect of activation energy is provided by the two more common allotropes of carbon, graphite, and diamond. Diamond is thermodynamically unstable with respect to graphite:

$$\text{C}(diamond) \rightarrow \text{C}(graphite) \qquad \Delta G^\circ = -3 \text{ kJ·mol}^{-1}$$

Yet, of course, diamonds in diamond rings do not crumble to a black powder on a daily basis. They do not because an extremely high activation energy is required to rearrange the covalent bonds from the tetrahedral arrangement in diamond to the planar arrangement in graphite. Furthermore, all forms of carbon are thermodynamically unstable with respect to oxidation to carbon dioxide in the presence of dioxygen. Once again, it is the high activation energy that prevents diamonds in rings and the graphite ("lead") in pencils from bursting into flame:

$$\text{C}(s) + \text{O}_2(g) \rightarrow \text{CO}_2(g) \qquad \Delta G^\circ = -390 \text{ kJ·mol}^{-1}$$

Figure 6.9 Kinetic and thermodynamic energy factors in a chemical reaction.

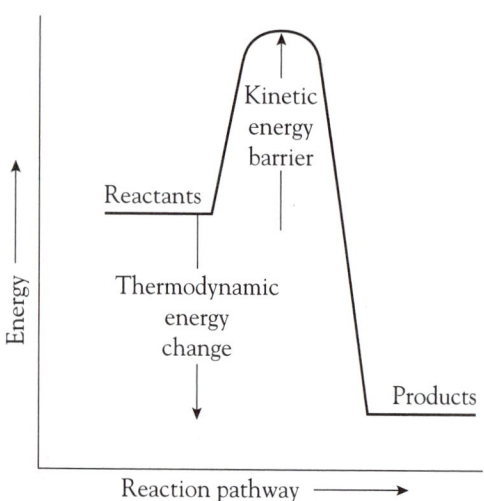

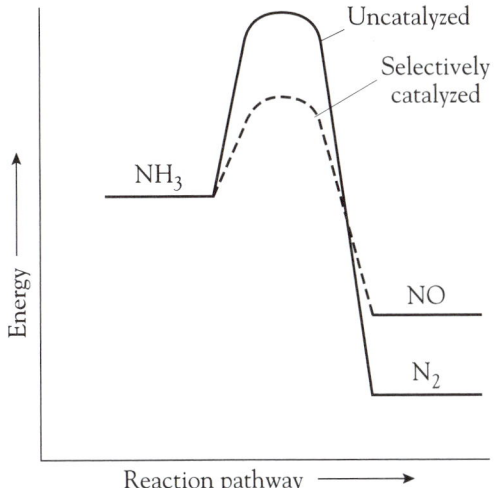

Figure 6.10 Diagram (not to scale) of the kinetic and thermodynamic energy factors in the two pathways for the combustion of ammonia.

We can actually make use of kinetics to alter the product of a chemical reaction. A particularly important example is the combustion of ammonia. Ammonia burns in air to form dinitrogen and water vapor:

$$4\ NH_3(g) + 3\ O_2(g) \rightarrow 2\ N_2(g) + 6\ H_2O(g)$$

This is the thermodynamically favored path, with a free energy change of $-1306\ kJ \cdot mol^{-1}$. When the combustion is performed in the presence of a catalyst, the activation energy of a competing reaction, that to produce nitrogen monoxide, is, in fact, lower than that of the reaction producing dinitrogen gas:

$$4\ NH_3(g) + 5\ O_2(g) \rightarrow 4\ NO(g) + 6\ H_2O(g)$$

The latter reaction, which is a key step in the industrial preparation of nitric acid, occurs even though the free energy change for the reaction is only $-958\ kJ \cdot mol^{-1}$. Thus, we are using kinetics to control the products of reaction and overriding the thermodynamically preferred path (Figure 6.10).

It is also possible to synthesize compounds that have a positive free energy of formation. For example, trioxygen (ozone) and all the oxides of nitrogen have positive free energies of formation. The synthesis of such substances is feasible if there is a pathway involving a net decrease in free energy and if the decomposition of the compound is kinetically slow. Alternatively, there must be a pathway allowing for the input of energy, such as light in the case of photosynthesis and electrical energy in the case of electrolysis.

An interesting example is provided by nitrogen trichloride. We saw in the previous section that nitrogen trifluoride is thermodynamically stable. In contrast, nitrogen trichloride is thermodynamically unstable; yet it exists:

Figure 6.11 Theoretical enthalpy cycle for the formation of nitrogen trichloride.

$$\tfrac{1}{2}\ N_2(g) + \tfrac{3}{2}\ F_2(g) \rightarrow NF_3(g) \qquad \Delta G^\circ = -84\ kJ \cdot mol^{-1}$$
$$\tfrac{1}{2}\ N_2(g) + \tfrac{3}{2}\ Cl_2(g) \rightarrow NCl_3(l) \qquad \Delta G^\circ = +240\ kJ \cdot mol^{-1}$$

To understand this difference, we need to compare the key terms in each energy cycle. First of all, the reduction in the number of moles of gas from reactants to product means that, in both cases, the entropy term will be negative. Hence, for a spontaneous process, the enthalpy change must be negative.

In the synthesis of nitrogen trifluoride, the fluorine–fluorine bond to be broken is very weak ($158\ kJ \cdot mol^{-1}$), whereas the nitrogen–fluorine bond to be formed is very strong ($276\ kJ \cdot mol^{-1}$). As a result, the enthalpy of formation of nitrogen trifluoride is quite negative. The chlorine–chlorine bond ($242\ kJ \cdot mol^{-1}$) is stronger than that of the fluorine–fluorine bond, and the nitrogen–chlorine bond ($188\ kJ \cdot mol^{-1}$) in nitrogen trichloride is weaker than the nitrogen–fluorine bond in nitrogen trifluoride. As a result, the enthalpy change for the formation of nitrogen trichloride is positive (Figure 6.11) and, with a negative entropy change giving a positive $-T\Delta S$ term, the free energy change will be positive.

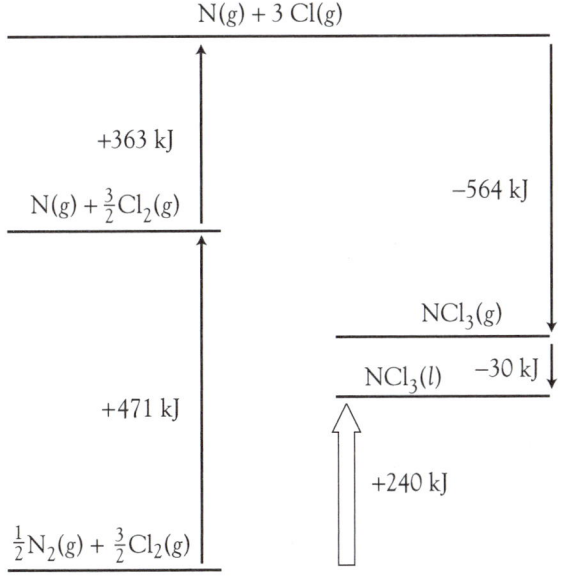

How is it possible to prepare such a compound? The reaction between ammonia and dichlorine to give nitrogen trichloride and hydrogen chloride has a slightly negative free energy change as a result of the formation of strong hydrogen–chlorine bonds:

$$NH_3(g) + 3\ Cl_2(g) \rightarrow NCl_3(l) + 3\ HCl(g)$$

The thermodynamically unstable nitrogen trichloride decomposes violently when warmed:

$$2\ NCl_3(l) \rightarrow N_2(g) + 3\ Cl_2(g)$$

Thermodynamics, then, is a useful tool for understanding chemistry. At the same time, we should always be aware that kinetic factors can cause the product to be other than the most thermodynamically stable one (as in the case of the oxidation of ammonia). In addition, it is sometimes possible to synthesize compounds that have a positive free energy of formation, provided the synthetic route involves a net decrease in free energy and the decomposition of the compound is kinetically slow (as we saw for the existence of nitrogen trichloride).

EXERCISES

6.1 Define the following terms: (a) spontaneous process; (b) entropy; (c) standard enthalpy of formation.

6.2 Define the following terms: (a) enthalpy; (b) average bond energy; (c) enthalpy of hydration.

6.3 For the formation of solid calcium oxide from solid calcium and gaseous oxygen, what is the probable sign of the entropy change? What, then, must be the sign of the enthalpy change if the formation of the product occurs spontaneously? Do not consult data tables.

6.4 At very high temperatures, water will decompose to hydrogen and oxygen gas. Explain why this is to be expected in terms of the formula relating free energy with the other two common thermodynamic functions. Do not consult data tables.

6.5 Using enthalpy of formation and absolute entropy values from the data tables in Appendix 2, determine the enthalpy, entropy, and free energy of reaction for the following reaction. Use this information to identify whether the reaction is spontaneous at standard temperature and pressure.

$$H_2(g) + \tfrac{1}{2}\ O_2(g) \rightarrow H_2O(l)$$

6.6 Using enthalpy of formation and absolute entropy values from the data tables in the Appendices, determine the enthalpy, entropy, and free energy of reaction for the following reaction. Use this information to identify whether the reaction is spontaneous at standard temperature and pressure.

$$\tfrac{1}{2}\ N_2(g) + O_2(g) \rightarrow NO_2(g)$$

6.7 Deduce whether the synthesis of sulfuryl chloride, SO_2Cl_2 ($\Delta G_f^\circ(SO_2Cl_2(g)) = -314$ kJ·mol^{-1}), is thermodynamically feasible from phosphorus pentachloride and sulfur dioxide. The other product is phosphoryl chloride, $POCl_3$. Use free energy of formation values from the data tables in Appendix 2.

6.8 Use free energy of formation values from Appendix 2 to determine whether the following reaction is thermodynamically feasible:

$$PCl_5(g) + 4\ H_2O(l) \rightarrow H_3PO_4(s) + 4\ HCl(g)$$

6.9 Which one of the N—N or N=N bonds will be stronger? Do not look at data tables. Explain your reasoning.

6.10 The molecules of dinitrogen and carbon monoxide are isoelectronic. Yet the C≡O bond energy (1072 kJ·mol^{-1}) is stronger than that of the N≡N bond (942 kJ·mol^{-1}). Suggest an explanation.

6.11 Use bond energy data from Appendix 4 to calculate an approximate value for the enthalpy of reaction for

$$4\ H_2S_2(g) \rightarrow S_8(g) + 4\ H_2(g)$$

6.12 Use bond energy data from Appendix 4 to decide whether the following reaction is thermodynamically possible.

$$CH_4(g) + 4\ F_2(g) \rightarrow CF_4(g) + 4\ HF(g)$$

6.13 Place the following compounds in order of increasing lattice energy: magnesium oxide, lithium fluoride, and sodium chloride. Give the reasoning for this order.

6.14 Calculate the first three terms of the series for the Madelung constant for the sodium chloride lattice. How does this compare with the limiting value?

6.15 Calculate the first two terms of the series for the Madelung constant for the cesium chloride lattice. How does this compare with the limiting value?

6.16 Using the Born–Landé equation, calculate the lattice energy of cesium chloride.

6.17 Using the Born–Landé equation, calculate the lattice energy of calcium fluoride.

6.18 Construct a Born–Haber cycle for the formation of aluminum fluoride. Do not perform any calculation.

6.19 Construct a Born–Haber cycle for the formation of magnesium sulfide. Do not perform any calculation.

6.20 Calculate the enthalpy of formation of copper(I) fluoride. This compound adopts a sphalerite structure.

6.21 The lattice energy of sodium hydride is 2782 $kJ \cdot mol^{-1}$. Using additional data from the Appendices, calculate a value for the electron affinity of atomic hydrogen.

6.22 In the discussion of thermodynamic and kinetic factors, we compared the two reactions:

$$4\,NH_3(g) + 3\,O_2(g) \rightarrow 2\,N_2(g) + 6\,H_2O(g)$$

$$4\,NH_3(g) + 5\,O_2(g) \rightarrow 4\,NO(g) + 6\,H_2O(g)$$

Without consulting data tables:
(a) Is there any major difference in entropy factors between the two reactions? Explain.
(b) Considering your answer to (a) and the fact that the free energy for the second reaction is less negative than for the first, deduce the sign of the enthalpy of formation of nitrogen monoxide, NO.

6.23 The electron affinities of the oxygen atom are:

$$O(g) + e^- \rightarrow O^-(g) \qquad -141\ kJ \cdot mol^{-1}$$

$$O^-(g) + e^- \rightarrow O^{2-}(g) \qquad +744\ kJ \cdot mol^{-1}$$

If the second electron affinity is so endothermic, why are ionic oxides so prevalent?

6.24 For an ionic compound, MX, the lattice energy is 1205 $kJ \cdot mol^{-1}$ and the enthalpy of solution is $-90\ kJ \cdot mol^{-1}$. If the enthalpy of hydration of the cation is 1.5 times that of the anion, what are the enthalpies of hydration of the ions?

BEYOND THE BASICS

6.25 The Born–Landé equation utilizes a term called "the permittivity of free space." Use a physics text to explain the significance of this term in physical science.

6.26 Calculate the enthalpy of formation of calcium oxide using a Born–Haber cycle. Obtain all necessary information from the data tables in the Appendices. Compare the value that you obtain with the actual measured value of $\Delta H_f^\circ(CaO(s))$. Then calculate a similar cycle assuming that calcium oxide is Ca^+O^- rather than $Ca^{2+}O^{2-}$. Take the lattice energy of Ca^+O^- to be $-800\ kJ \cdot mol^{-1}$. Discuss why the second scenario is less favored in enthalpy terms.

6.27 Construct Born–Haber cycles for the theoretical compounds, $NaCl_2$ and $NaCl_3$. Calculate the enthalpy of formation for both of these two compounds using information from data tables plus the following values: theoretical lattice energy, $NaCl_2 = -2500\ kJ \cdot mol^{-1}$, theoretical lattice energy, $NaCl_3 = -5400\ kJ \cdot mol^{-1}$, second ionization energy (IE) Na $= 4569\ kJ \cdot mol^{-1}$, third

IE Na $= 6919\ kJ \cdot mol^{-1}$. Compare the cycles and suggest why $NaCl_2$ and $NaCl_3$ are not the preferred products.

6.28 The lattice energy of sodium tetrahydridoborate(III), $NaBH_4$, is 2703 $kJ \cdot mol^{-1}$. Using additional data from the Appendices, calculate the enthalpy of formation of the tetrahydridoborate ion.

6.29 Magnesium chloride is very soluble in water, whereas magnesium oxide is very insoluble in water. Offer an explanation for this difference in terms of the theoretical steps of the solution process. Do not use data tables.

6.30 Use lattice energy and enthalpy of hydration values from data tables to determine the enthalpy of solution of (a) lithium chloride; (b) magnesium chloride. Explain the major difference in the two values in terms of the theoretical steps.

6.31 Construct an energy diagram, similar to a Born–Haber cycle, for the formation of carbon tetrafluoride. Then calculate the enthalpy of formation from the steps, using numerical values from the data tables in the Appendices. Finally, compare your value with the tabulated value of $\Delta H_f^{\circ}(\text{CF}_4(g))$.

6.32 Construct an energy diagram, similar to a Born–Haber cycle, for the formation of sulfur hexafluoride. Then calculate the enthalpy of formation from the steps, using numerical values from data tables. Finally, compare your value with the tabulated value of $\Delta H_f^{\circ}(\text{SF}_6(g))$.

6.33 Calculate the chlorine–fluorine bond energy in chlorine monofluoride, ClF, using an energy diagram.

6.34 Using enthalpy of formation and absolute entropy values from data tables, determine the free energy of formation for the following reactions:

$$S(s) + O_2(g) \rightarrow SO_2(g)$$
$$S(s) + \tfrac{3}{2}\, O_2(g) \rightarrow SO_3(g)$$

(a) Account for the sign of the entropy change in the formation of sulfur trioxide.
(b) Which combustion leads to the greatest decrease in free energy (that is, which reaction is thermodynamically preferred)?
(c) Which of the oxides of sulfur is most commonly discussed?
(d) Suggest an explanation to account for the conflict between your answers to parts (b) and (c).

6.35 Although the hydration energy of the calcium ion, Ca^{2+}, is much greater than that of the potassium ion, K^+, the molar solubility of calcium chloride is much less than that of potassium chloride. Suggest an explanation.

6.36 The enthalpy of solution for sodium chloride is $+4\ \text{kJ·mol}^{-1}$ while that of silver chloride is $+65\ \text{kJ·mol}^{-1}$.
(a) What would you suspect about the comparative solubilities of the two compounds? In drawing this conclusion, what assumption do you have to make?
(b) Using enthalpy of hydration data, calculate values for the two lattice energies. (Both adopt the sodium chloride structure.)
(c) Calculate values for the lattice energies using the Born–Landé equation and compare to the values calculated in part (b). Suggest a reason for the significant discrepancy in the case of one of the compounds.

6.37 Using the following data, plus any other necessary data from the Appendices, calculate three values

for the hydration enthalpy of the sulfate ion. Are the values consistent?

Compound	ΔH° solution (kJ·mol^{-1})	Lattice energy (kJ·mol^{-1})
CaSO$_4$	-17.8	-2653
SrSO$_4$	-8.7	-2603
BaSO$_4$	$+19.4$	-2423

6.38 It is possible to calculate standard enthalpy/entropy of solution values from the difference between the standard enthalpy/entropy of formation of the solid compound and the standard enthalpy/entropy of formation of the constituent aqueous ions. Thus, determine the free energy of solution of calcium phosphate at 20°C. Suggest what your value means in terms of the solubility of calcium phosphate.

6.39 Magnesium and lead have similar first and second ionization energies, yet their reactivity with acid:

$$M(s) + 2\, H^+(aq) \rightarrow M^{2+}(aq) + H_2(g)$$
(where M = Mg, Pb)

is very different. Construct a suitable cycle and obtain the appropriate data from the Appendices (hint: as the reduction of hydrogen ion is common to both cycles, you need only consider the formation of the aqueous metal cations), and then deduce the factor that causes the reactivity difference. Suggest a fundamental reason for this difference.

6.40 For the main group elements, the most thermodynamically stable compound is that when all of the valence electrons have been lost. Yet for all of the lanthanoids, it is the 3+ state which provides the most stable state. Explain in terms of ionization energies and their role in Born–Haber cycles.

6.41 Use the Kapustinskii equation to calculate a lattice energy for cesium chloride and compare it to the experimental value and to that obtained from the Born–Landé equation (Exercise 6.14).

6.42 Thallium has the two oxidation numbers of +1 and +3. Use the Kapustinskii equation to calculate values for the lattice energies of TlF and TlF$_3$. The radii of Tl$^+$ and Tl^{3+} are 164 pm and 102 pm, respectively.

6.43 Under very high pressure, rubidium chloride will adopt a cesium chloride structure. Calculate the enthalpy change for the transition from its sodium chloride lattice structure to that of the cesium chloride lattice structure.

6.44 One of the most spectacular chemical demonstrations is the thermite reaction.

$$2\ Al(s) + Fe_2O_3(s) \rightarrow Al_2O_3(s) + 2\ Fe(l)$$

This reaction is so exothermic that molten iron is produced and it was formerly used as a means of welding railroad track. Using the data tables in the Appendices, account for the enormous exothermicity of this reaction.

6.45 Calculate the proton affinity of ammonia:

$$NH_3(g) + H^+(g) \rightarrow NH_4^+(g)$$

given that ammonium fluoride, NH_4F, crystallizes in a wurtzite structure, with distance of 256 pm between ammonium ion and fluoride ion centers, and that the Born exponent for the crystal lattice is 8. The ionization energy for the hydrogen atom is $+1537$ kJ·mol^{-1}. Hint: Among the additional data that you will need from the Appendices are the enthalpy of formation of ammonia, hydrogen fluoride, and of ammonium fluoride.

6.46 The following reaction sequence has been proposed as a thermochemical method for the production of dihydrogen and dioxygen from water. Calculate the free energy changes for each step at 298 K and those for the overall process.

$$CaBr_2(s) + H_2O(g) \xrightarrow{\Delta} CaO(s) + 2\ HBr(g)$$

$$Hg(l) + 2\ HBr(g) \xrightarrow{\Delta} HgBr_2(s) + H_2(g)$$

$$HgBr_2(s) + CaO(s) \rightarrow HgO(s) + CaBr_2(s)$$

$$HgO(s) \xrightarrow{\Delta} Hg(l) + \tfrac{1}{2}\ O_2(g)$$

Suggest reasons why the process has little possibility of commercial adoption.

6.47 The ionic bond is often described as being formed as a result of metals "wanting to lose electrons" and nonmetals "wanting to gain electrons." Critique this statement using appropriate thermodynamic values.

Chapter 7 — Acids and Bases

$$pK_a + pK_b = pK_w$$

The study of inorganic acids and bases overlaps the fields of inorganic, analytical, and physical chemistry. In this chapter, we focus on the structural and theoretical aspects of acid–base behavior. For most purposes, the Brønsted–Lowry interpretation of acid–base properties is quite adequate, although Lewis concepts are also discussed. The latter part of the chapter focusses on the hard–soft acid–base concept, a particularly useful way of explaining some of the properties of inorganic compounds.

When we study theories, we often forget that new theories are devised when the former theories no longer explain all the known facts. The theories of acids and bases are an excellent example of this progression of knowledge. A simple concept of acids and bases was first devised by Svante Arrhenius in 1884. According to the Arrhenius theory, acids contain hydrogen ions and bases contain hydroxide ions. However, there were two major flaws in this theory: the solvent problem and the salt problem.

The Arrhenius theory assumes that the solvent has no influence on acid–base properties. If hydrogen chloride is dissolved in water (to give hydrochloric acid), the solution conducts electricity, but if it is dissolved in a nonpolar solvent such as benzene, C_6H_6, the solution does not conduct an electric current. The difference in properties of hydrogen chloride in the two solvents means that the solvent does affect the behavior of a solute. The second problem of the Arrhenius theory concerns the behavior of salts. Salts should be neutral compounds, yet there are many salts that contradict this rule. For example, solutions of phosphate ion and carbonate ion are basic, whereas those of ammonium ion are slightly acidic and those of aluminum ion are very acidic. To add to the confusion, a solution of sodium dihydrogen phosphate is acidic, but that of disodium hydrogen phosphate is basic.

To provide a more realistic model of acid–base behavior, Thomas Lowry in England and Johannes Brønsted in Denmark independently devised a theory that involved the solvent in the acid–base phenomenon. Even though there have been newer and more sophisticated theories of acid–base behavior, the Brønsted–Lowry theory still provides the most convenient framework for understanding acids and bases.

7.1 Ionization and Dissociation

Though the terms ionization and dissociation are often used interchangeably in discussion of solution processes, their definitions are different. The term *dissociation* means separation, and is used when a solvent such as water separates the ions that are present in ionic compounds. As an example, we can use sodium hydroxide. The crystal lattice of this white solid contains alternating sodium ions and hydroxide ions. The solution process would be represented as

$$Na^+OH^-(s) \xrightarrow{\text{dissolve in water}} Na^+(aq) + OH^-(aq)$$

However, most acids contain covalent bonds. The hydrogen halides and hydrogen cyanide are colorless gases, while pure nitric acid, perchloric acid, and sulfuric acid are very strongly oxidizing, oily liquids. Solution in water results in *ionization*, the breaking of the covalent bonds resulting in ion formation. We can illustrate with hydrochloric acid:

$$HCl(g) + H_2O(l) \xrightarrow{\text{dissolve in water}} H_3O^+(aq) + OH^-(aq)$$

7.2 Brønsted–Lowry Theory

According to the Brønsted–Lowry theory, acids are proton donors and bases are proton acceptors. The language is somewhat misleading, for it is more accurately a competition for the proton between two chemical substances (with the base winning). The acid does not "donate" the proton any more willingly than you would "donate" your wallet or purse to a mugger. However, the central feature of the theory is the importance of the solvent, which self-ionizes by its own acid–base reaction. Thus water undergoes a slight self-ionization, or *autoionization,* to give the hydronium ion and the hydroxide ion:

$$H_2O(l) + H_2O(l) \rightleftharpoons H_3O^+(aq) + OH^-(aq)$$

Like all equilibria, the autoionization of water is temperature dependent. Thus the 1.0×10^{-14} value of the ion product, $[H_3O^+][OH^-]$, used in many acid–base calculations is only true at 25°C. In fact, the constant has a value of 1.2×10^{-15} at 0°C and 4.8×10^{-13} at 100°C. Furthermore, the value of the ion product is different when there are other solutes in the water; thus, at 25°C, it is about 1.5×10^{-14} for blood.

During autoionization, the water molecule that donates the hydrogen ion is an acid, and the water molecule that accepts the hydrogen ion is a base. When we consider the reverse process, we see that the hydronium ion acts as a hydrogen ion donor (an acid) and that the hydroxide ion is a hydrogen ion acceptor (a base). Two species that differ in formula by a hydrogen ion are called a *conjugate acid–base pair*. In this case, water is both the conjugate base of the hydronium ion and the conjugate acid of the hydroxide ion. A substance that can act either as an acid or as a base is said to be *amphiprotic*.

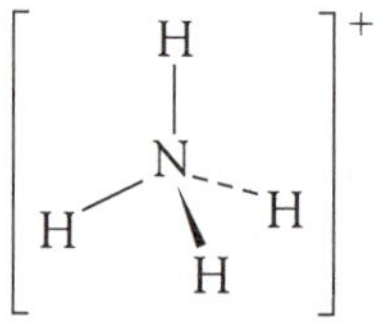

The Brønsted–Lowry theory obviously depends on the actual existence of the hydronium ion. Brønsted and Lowry proposed their theory in 1923, and the first evidence for the existence of the hydronium ion came a year later when it was shown that crystals of perchloric acid monohydrate, $HClO_4 \cdot H_2O$, had the same appearance as crystals of ammonium perchlorate, $NH_4^+ \cdot ClO_4^-$. This similarity in appearance suggested a similarity in formula. The obvious conclusion was that the solid perchloric acid contained a hydronium ion that was analogous to the ammonium ion and that the actual structure of solid perchloric acid was $OH_3^+ \cdot ClO_4^-$ (or more conventionally, $H_3O^+ \cdot ClO_4^-$). Decades later, when crystal structures of acid hydrates were obtained, the proposal was shown to be correct. The shapes of the ammonium ion and hydronium ion are shown in Figure 7.1. Later still, evidence was obtained that the hydronium ion was present in solution. In fact, the hydronium ion is now known to be hydrogen bonded to three neighboring water molecules; thus, it is more correctly written as $H_9O_4^+$. However, for simplicity, we usually ignore the three molecules of hydration.

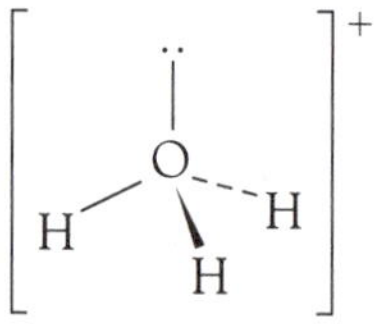

Acid or base behavior, then, usually depends on a chemical reaction with the solvent—in this case, water. This behavior can be illustrated for hydrofluoric acid:

$$HF(aq) + H_2O(l) \rightleftharpoons H_3O^+(aq) + F^-(aq)$$

In this reaction, water functions as a base, and the fluoride ion is the conjugate base of hydrofluoric acid. Similarly, ammonia can be used as an example of a typical base, reacting with water (which, in this case, acts like an acid) to give the conjugate acid, the ammonium ion:

$$NH_3(aq) + H_2O(l) \rightleftharpoons NH_4^+(aq) + OH^-(aq)$$

The reaction between a strong acid and a strong base can be represented simply as the reaction between the hydronium ion of the acid and the hydroxide of the base. This reaction is the reverse of the autoionization of water. Because water autoionizes to such a small extent, the reaction between hydronium ion and hydroxide ion goes essentially to completion:

$$H_3O^+(aq) + OH^-(aq) \rightarrow 2\,H_2O(l)$$

In aqueous solution, the strongest possible acid is the hydronium ion and the strongest possible base is the hydroxide ion. Thus, if a stronger base, such as the oxide ion, O^{2-}, is placed in water, it immediately reacts to give hydroxide ion:

$$O^{2-}(aq) + H_2O(l) \rightarrow 2\,OH^-(aq)$$

A stronger acid, such as perchloric acid, $HClO_4$, will similarly ionize to give hydronium ion:

$$HClO_4(aq) + H_2O(l) \rightarrow H_3O^+(aq) + ClO_4^-(aq)$$

Figure 7.1 (*a*) The ammonium ion and (*b*) the hydronium ion.

Antacids

One of the major categories of over-the-counter medications are antacids. In fact, the treatment of upset stomachs is a billion-dollar business. Antacids are the most common of the types of inorganic pharmaceuticals. The stomach contains acid—hydrochloric acid—as the hydronium ion is an excellent catalyst for the breakdown of complex proteins (hydrolysis) into the simpler peptide units that can be absorbed through the stomach wall. Unfortunately, some people have stomachs that can overproduce acid. To ameliorate the unpleasant effects of excess acid, a base is required. But the choice of bases is not as simple as in a chemistry lab. For example, ingestion of sodium hydroxide would cause severe and possibly life-threatening throat damage.

One commonly used remedy is baking soda, sodium hydrogen carbonate. The hydrogen carbonate ion reacts with hydrogen ion as follows:

$$HCO_3^-(aq) + H^+(aq) \rightarrow H_2O(l) + CO_2(g)$$

This compound has one obvious and one less obvious disadvantage. The compound might increase stomach pH, but it will also lead to the production of gas (so-called flatulence). In addition, the sodium intake is unwise for those with high blood pressure.

Some proprietary antacids contain calcium carbonate. This, too, produces carbon dioxide.

$$CaCO_3(s) + 2\ H^+(aq) \rightarrow Ca^{2+}(aq) + H_2O(l) + CO_2(g)$$

Though the beneficial aspects of increasing one's calcium intake are mentioned by companies selling such antacid compositions, they rarely mention that calcium ion acts as a constipative.

Another popular antacid compound is magnesium hydroxide. This is available in tablet formulations, but it is also marketed as a finely ground solid mixed with colored water to form a slurry called "milk of magnesia." The low solubility of the magnesium hydroxide means that there is a negligible concentration of free hydroxide ion in the suspension. In the stomach, the insoluble base reacts with acid to give a solution of magnesium ion.

$$Mg(OH)_2(s) + 2\ H^+(aq) \rightarrow Mg^{2+}(aq) + 2\ H_2O(l)$$

While calcium ion is a constipative, magnesium ion is a laxative. For this reason, some formulations contain a mixture of calcium carbonate and magnesium hydroxide, balancing the effects of the two ions.

Aluminum hydroxide is the active ingredient in a few antacid formulations. This base is also water insoluble; thus, the hydroxide ions are not released until the tablet reaches the stomach.

$$Al(OH)_3(s) + 3\ H^+(aq) \rightarrow Al^{3+}(aq) + 3\ H_2O(l)$$

As we will discuss in Chapter 13, Section 13.12, aluminum ion is toxic. There is no evidence that the occasional intake from an aluminum-containing antacid tablet will cause long-term health effects, but regular users of antacids might consider using calcium and/or magnesium formulations.

Brønsted–Lowry acid–base chemistry need not be performed in water—any solvent system containing ionizable hydrogen will suffice. For example, liquid ammonia undergoes an autoionization to produce ammonium ion (the acid) and amide ion, NH_2^- (the base):

$$NH_3(l) + NH_3(l) \rightleftharpoons NH_4^+(NH_3) + NH_2^-(NH_3)$$

Acid–base reactions can be performed in this solvent, although with a boiling point of $-33°C$, it is only used for specialized purposes. An example of an acid–base reaction in ammonia is that between ammonium chloride and sodium amide, $NaNH_2$, to give sodium chloride and ammonia.

$$NH_4^+Cl^-(NH_3) + Na^+NH_2^-(NH_3) \rightarrow Na^+Cl^-(s) + 2\,NH_3(l)$$

This reaction parallels the acid–base reaction in water between hydrochloric acid and sodium hydroxide. The similarity can be seen more clearly if we write hydrochloric acid as "hydronium chloride," $H_3O^+ \cdot Cl^-$:

$$H_3O^+Cl^-(aq) + Na^+OH^-(aq) \rightarrow Na^+Cl^-(aq) + 2\,H_2O(l)$$

7.3   Acid–Base Equilibrium Constants

The strength of an acid is a measure of how easily a hydrogen ion can be removed from the substance. The usual index of this is the position of equilibrium with water, in other words, the value of the equilibrium constant. In the case of an acid, the constant is identified as the *acid ionization constant*, K_a. For a general acid, HA, the equilibrium can be written as

$$HA(aq) + H_2O(l) \rightleftharpoons H_3O^+(aq) + A^-(aq)$$

The corresponding acid ionization expression would be

$$K_a = \frac{[H_3O^+][A^-]}{[HA]}$$

Because the values of acid ionization constants can involve very large or very small exponents, the most useful quantitative measure of acid strength is the pK_a, where

$$pK_a = -(\log_{10} K_a)$$

The stronger the acid, the more negative the pK_a. Typical values are shown in Table 7.1. Throughout this text, the equilibrium constants used are those measured at the standard temperature of 25°C.

Table 7.1   Acid ionization constants of various inorganic acids

Acid	HA	A$^-$	K_a (at 25°C)	pK_a
Perchloric acid	$HClO_4$	ClO_4^-	10^{10}	-10
Hydrochloric acid	HCl	Cl^-	10^2	-2
Hydrofluoric acid	HF	F^-	3.5×10^{-4}	3.45
Ammonium ion	NH_4^+	NH_3	5.5×10^{-10}	9.26

Cyanide and Tropical Fish

Cyanide ion is the conjugate base of the weak acid, hydrocyanic acid. Thus, a solution of sodium cyanide is not only toxic from the presence of the base and its conjugate acid but is also very basic:

$$CN^-(aq) + H_2O(l) \rightleftharpoons HCN(aq) + OH^-(aq)$$

Cyanide is used as a complexing agent in the extraction of precious metals (see Chapter 20, Section 20.9). The accidental release of cyanide-containing solutions from mining operations can cause major localized damage to aquatic organisms. However, the greatest environmental catastrophe involving cyanide has resulted from the tropical fish trade.

Saltwater tropical fish are prized for their brilliant colors. As they are almost impossible to breed in captivity, collectors rely on the harvesting of fish from tropical coral reefs. It is estimated that about 35 million tropical fish are collected each year for the aquarium trade. In the United States alone, about 700,000 households and businesses keep marine aquariums.

The trade started in 1957 in the Philippines. Cyanide poisoning was the simplest method of collecting the fish, and it is estimated that over the past 40 years, in Philippine waters alone, over 1 million kg of sodium cyanide has been squirted onto tropical reefs. Divers crush one or two tablets of sodium cyanide and mix the powder with water, then squirt the solution over a portion of the reef. The hydrogen cyanide is absorbed through the mouth or gills, immediately disabling enzymes such as cytochrome oxidase, resulting in diminished oxygen uptake. The agile fish become asphyxiated, making it easy to capture them before they can flee into crevices in the coral. It is estimated that about half of the fish are killed immediately. Of the remainder, the long-term effects of liver-absorbed cyanide will kill about 40 percent, leaving only about 10 percent to make it to the collector's tank.

In addition to the fish kill, cyanide has a major effect on the reef organisms themselves. Concentrations of cyanide as low as 50 mg·L^{-1} are enough to cause the death of corals. About 30 percent of coral reefs are in southeast Asian waters, and they have the greatest diversity of marine life anywhere on the planet. It is believed that the effects of cyanide on the reefs in the Philippines and Indonesia, the sources of 85 percent of tropical fish, have contributed to the destruction of vast areas of reef. In fact only 4 percent of Philippine reefs and 7 percent of Indonesian reefs are in excellent condition. There are suspicions of cyanide use in Vietnam and Kiribati.

Cyanide is still the easiest way to collect the fish. Fishing practices are difficult to change when livelihoods are at stake. To discourage the trade, it is the fish purchaser who must be targeted. The Marine Aquarium Council (MAC) has been founded to develop methods of cyanide-free collection of marine fish. MAC officials are hoping that collectors of marine fish will insist on purchasing from their pet stores only fish that have been MAC-certified. It is hoped that insistence on certified fish will effectively destroy the market for cyanide-caught fish and require the harvesters to adopt less ecologically damaging methods.

Table 7.2   Base ionization constants of various inorganic bases

Base	A$^-$	HA	K_b (at 25°C)	pK_b
Phosphate ion	PO$_4^{3-}$	HPO$_4^{2-}$	4.7×10^{-2}	1.33
Cyanide ion	CN$^-$	HCN	1.6×10^{-5}	4.79
Ammonia	NH$_3$	NH$_4^+$	1.8×10^{-5}	4.74
Hydrazine	N$_2$H$_4$	N$_2$H$_5^+$	8.5×10^{-7}	6.07

The relevant constant for bases is identified as the *base ionization constant, K_b*. For a general base, A$^-$, the equilibrium can be written as

$$A^-(aq) + H_2O(l) \rightleftharpoons HA(aq) + OH^-(aq)$$

The corresponding base ionization expression would be

$$K_b = \frac{[HA][OH^-]}{[A^-]}$$

Similarly, pK_b is defined as

$$pK_b = -(\log_{10} K_b)$$

Typical values are shown in Table 7.2.

There is a mathematical relationship between the acid ionization constant K_a of an acid and the base ionization constant K_b of its conjugate base. The product of the two terms equals the ion product constant for water, K_w:

$$K_w = [K_a] \times [K_b]$$

This can be expressed more conveniently in logarithmic form as

$$pK_w = pK_a + pK_b \quad \text{where } pK_w = 14.00 \text{ at } 25°C.$$

Thus, the stronger the base, the weaker the conjugate acid. Conversely, a strong acid will have a weak conjugate base.

7.4 Brønsted–Lowry Acids

For an inorganic chemist, it is trends in the strengths of acids that are interesting. The common acids with $K_a > 1$ (negative pK_a values), such as hydrochloric acid, nitric acid, sulfuric acid, and perchloric acid, are all regarded as strong acids. Those with $K_a < 1$ (positive pK_a values), such as nitrous acid, hydrofluoric acid, and most of the other inorganic acids, are weak acids; that is, there are appreciable proportions of the molecular acid present in solution.

In water, all the strong acids seem equally strong, undergoing close to 100 percent ionization; that is, water acts as a leveling solvent. (As we saw earlier in the chapter, the hydronium ion is the strongest possible acid in aqueous solution.) To qualitatively identify the stronger acid, we dissolve the acids in a base weaker than water. A weaker base—often a pure weak acid—will function as a differentiating solvent for acids. This test can be illustrated by the equilibrium for perchloric acid in hydrofluoric acid:

$$HClO_4(HF) + HF(l) \rightleftharpoons H_2F^+(HF) + ClO_4^-(HF)$$

Table 7.3 Correlation between the acid strengths of the hydrohalic acids and the energies of the hydrogen–halogen bonds

Acid	pK_a	Bond energy (kJ·mol^{-1})
HF(aq)	+3	565
HCl(aq)	−7	428
HBr(aq)	−9	362
HI(aq)	−10	295

The weaker acid, hydrofluoric acid, functions in this case as a proton acceptor (base) for the stronger perchloric acid. However, because hydrofluoric acid is a weaker base than water, the equilibrium does not lie completely to the right, as does that for the reaction of perchloric acid with water.

The experiment can be repeated with the other strong acids, and the strongest acid is the acid that causes the equilibrium to lie furthest to the right. Of the common acids, this is perchloric acid.

 **Binary Acids**

The most common binary acids are the hydrohalic acids, whose pK_a values are shown in Table 7.3. With a positive pK_a, hydrofluoric acid is clearly a much weaker acid than the other three. The others are all strong acids, and they ionize almost completely. Hydroiodic acid is the strongest; hence, it has the most negative pK_a.

If we write the equation for the ionization process, using HX to represent any hydrohalic acid, we see that the main changes are the breaking of the H—X bond and the formation of an additional O—H bond as the water molecule becomes the hydronium ion:

$$HX(aq) + H_2O(l) \rightleftharpoons H_3O^+(aq) + X^-(aq)$$

The values of the various H—X bond energies are also given in Table 7.3, and the O—H bond energy is 459 kJ·mol^{-1}. Because the tendency of any reaction is toward the formation of the stronger bond, it is apparent that ionization is not energetically favored for hydrofluoric acid but is favored for the other hydrohalic acids. In fact, the bond energy differences correlate remarkably well with the trends in acid strength. However, it should be kept in mind that this is a simplistic explanation; the comparative enthalpies and entropies of hydration of the ions and molecules must be included in any quantitative calculations of acid behavior.

Oxyacids

Oxyacids are ternary acids containing oxygen. For all the common inorganic acids, the ionizable hydrogen atoms are covalently bonded to oxygen atoms. For example, nitric acid, HNO_3, is more appropriately written as $HONO_2$.

In a series of oxyacids of one element, there is a correlation between acid strength and the number of oxygen atoms. Thus nitric acid is a strong acid ($pK_a = -1.4$), whereas nitrous acid, HONO, is a weak acid ($pK_a = +3.3$). Electronegativity arguments can be used to provide an explanation. Oxyacids are like the hydrohalic acids in that their acid strength depends on the weakness of the covalent bond between the ionizable hydrogen atom and its neighbor. For the oxyacids, the greater the number of highly electronegative oxygen atoms in the molecule, the more the electron density is pulled away from the hydrogen atom and the weaker the hydrogen–oxygen bond. As a result, an acid with numerous oxygen atoms is more easily ionized and hence stronger. This tendency is illustrated in Figure 7.2.

The considerable dependence of acid strength on the number of oxygen atoms can actually be used in a semiquantitative fashion. If the formula of an oxyacid is written as $(HO)_nXO_m$, then when $m = 0$, the value of pK_a for the first ionization is about 8; for $m = 1$, it is about 2; for $m = 2$, it is about -1; and for $m = 3$, it is about -8.

One of the strongest simple acids is fluorosulfonic acid, HSO_3F.

Polyprotic Acids

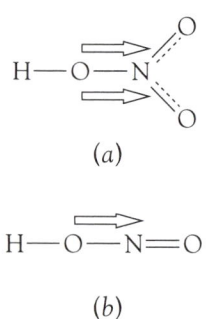

(a)

(b)

Figure 7.2 Nitric acid (a) is a stronger acid than nitrous acid (b) because the electron flow away from the H—O bond is greater in nitric acid.

There are several acids, including sulfuric acid and phosphoric acid, that have more than one ionizable hydrogen atom. The successive ionizations always proceed to a lesser and lesser extent. This trend can be illustrated by the two ionization steps for sulfuric acid:

$$H_2SO_4(aq) + H_2O(l) \rightarrow H_3O^+(aq) + HSO_4^-(aq) \quad pK_a = -2$$
$$HSO_4^-(aq) + H_2O(l) \rightleftharpoons H_3O^+(aq) + SO_4^{2-}(aq) \quad pK_a = +1.9$$

The first step proceeds essentially to completion; hence, sulfuric acid is identified as a strong acid. The equilibrium for the second step lies slightly to the left at common acid concentrations. Thus, in an aqueous solution of sulfuric acid, the hydrogen sulfate ion, HSO_4^-, is one of the major species. We can explain the decreasing values of successive ionizations in terms of the increasing negative charge of the resulting anion, making the loss of an additional hydronium ion more difficult.

Yet when we add a dipositive (or tripositive) metal ion, it is the metal sulfate, not the hydrogen sulfate, that crystallizes. The reason for this lies in the comparative lattice energies. As we discussed in Chapter 6; Section 6.1, the lattice energy depends to a significant extent on the ionic charges, that is, the electrostatic attraction between the ions. Hence, the lattice energy of a crystal containing a 2+ cation and a 2− anion is greater than that of a crystal containing a 2+ cation and two 1− anions. For example, the lattice energy of magnesium fluoride, MgF_2, is 2.9 MJ·mol^{-1}, whereas that of magnesium oxide, MgO, is 3.9 MJ·mol^{-1}. Thus, the formation of the solid metal sulfate will be about 1 MJ·mol^{-1} more exothermic than that of the solid metal hydrogen sulfate. Reaction is favored by a decrease in enthalpy; thus, it is the sulfate rather than the hydrogen sulfate that is formed by dipositive and tripositive metal ions. As the sulfate ion is removed by precipitation, more is generated from the supply of hydrogen sulfate ions, in accord with the Le Châtelier principle:

$$HSO_4^-(aq) + H_2O(l) \rightarrow H_3O^+(aq) + SO_4^{2-}(aq)$$

It is only the monopositive ions, such as the alkali metal ions, that form stable crystalline compounds with the hydrogen sulfate ion. In these cases, the large, low-charge cations can form a stable lattice with the low-charge polyatomic anion.

Generally, the salts of acid anions behave like acids themselves. In fact, sodium hydrogen sulfate is used as a household cleaner because its solution is acidic and the reagent itself is an easily handled powder. Dissolving the compound in water yields the free hydrogen sulfate ion and, hence, hydronium ion:

$$HSO_4^-(aq) + H_2O(l) \rightleftharpoons H_3O^+(aq) + SO_4^{2-}(aq)$$

7.5   *Brønsted–Lowry Bases*

After the hydroxide ion itself, ammonia is the next most important Brønsted–Lowry base. This compound reacts with water to produce the hydroxide ion, and it is the production of the hydroxide ion that makes ammonia solutions a useful glass cleaner (the hydroxide ion reacts with fat molecules to form water-soluble salts):

One of the strongest bases used in the laboratory is n-butyllithium, LiC$_4$H$_9$.

$$NH_3(aq) + H_2O(l) \rightleftharpoons NH_4^+(aq) + OH^-(aq)$$

However, there are many other common bases—the conjugate bases of weak acids. It is these anions that are present in many metal salts and yield basic solutions when the salts are dissolved in water. These conjugate bases include the phosphate ion and the sulfide ion, both quite strong bases; and the fluoride ion, a weak base:

$$PO_4^{3-}(aq) + H_2O(l) \rightleftharpoons HPO_4^{2-}(aq) + OH^-(aq) \qquad pK_b = 1.35$$

$$S^{2-}(aq) + H_2O(l) \rightleftharpoons HS^-(aq) + OH^-(aq) \qquad pK_b = 2.04$$

$$F^-(aq) + H_2O(l) \rightleftharpoons HF(aq) + OH^-(aq) \qquad pK_b = 10.55$$

A second equilibrium step occurs to a significant extent for the sulfide ion, and this reaction is the cause of the hydrogen sulfide smell that can always be detected above sulfide ion solutions:

$$HS^-(aq) + H_2O(l) \rightleftharpoons H_2S(aq) + OH^-(aq) \qquad pK_b = 6.96$$

The equilibrium for the phosphate ion also lies far to the right, so here, too, the second equilibrium step occurs to an appreciable extent:

$$HPO_4^{2-}(aq) + H_2O(l) \rightleftharpoons H_2PO_4^-(aq) + OH^-(aq) \qquad pK_b = 6.79$$

Thus, the hydrogen phosphate ion acts as a base, not as an acid as might be expected. The third equilibrium reaction for the phosphate ion does not proceed to the right to any meaningful degree:

$$H_2PO_4^-(aq) + H_2O(l) \rightleftharpoons H_3PO_4(aq) + OH^-(aq) \qquad pK_b = 11.88$$

Sodium phosphate is effective as a household cleaner partly because of the very high pH of the solution from the first and second steps of the reaction with water and the resulting reaction of hydroxide ion with fat molecules.

It is important to note that the conjugate bases of strong acids do not react to any significant extent with water; that is, they are very weak bases (or in other words, the base ionization equilibrium lies far to the left). Therefore, solutions of the nitrate and halide ions (except fluoride) are essentially pH neutral, and those of the sulfate ion are very close to neutral.

7.6   *Trends in Acid–Base Behavior*

The Acidity of Metal Ions

Dissolving sodium chloride in water provides an essentially neutral solution, while dissolving aluminum chloride in water gives a strongly acidic solution. As we have discussed in Chapter 5, Section 5.4, and Chapter 6, Section 6.4, ions are hydrated in solution. Most commonly, the first solvation sphere around a metal ion consists of six water molecules. We account for the ions forming a neutral solution by considering those water molecules to be comparatively weakly held, such as in the case of the sodium ion. As the charge

Table 7.4   Acidity of some metal ions

Li^+	Be^{2+}		
Slightly acidic	Weakly acidic		
Na^+	Mg^{2+}	Al^{3+}	
Neutral	Weakly acidic	Acidic	
K^+	Ca^{2+}	Sc^{3+}	Ti^{4+}
Neutral	Slightly acidic	Acidic	Very acidic

density increases, that is, with smaller ions and more highly charged ions, the lone pairs on the oxygen become strongly attracted to the metal ion, essentially forming a covalent bond. As a result the hydrogen atoms become increasingly positively charged to the point where they act as proton donors to a neighboring water molecule. The process can be illustrated using the iron(III) ion:

$$[Fe(OH_2)_6]^{3+}(aq) + H_2O(l) \rightleftharpoons H_3O^+(aq) + [Fe(OH_2)_5(OH)]^{2+}(aq)$$

The higher the charge density of the metal ion, then, the more the equilibrium shifts toward the production of hydronium ion, leaving the metal ion as a hydroxy-species. As a result of this equilibrium, metal ions with a $3+$ or higher positive charge will only exist as the true hydrated ion at very low pH. This process of reaction with water is sometimes called *hydrolysis*. Table 7.4 shows the trends among some common main group metals, from the neutral to the slightly acidic to the weakly acidic to the acidic to the very acidic cations.

There will be as many pK_a values as the ion has charge. For example, for magnesium there are the two equilibria:

$$[Mg(OH_2)_6]^{2+}(aq) + H_2O(l) \rightleftharpoons H_3O^+(aq) + [Mg(OH_2)_5(OH)]^+(aq)$$

$$[Mg(OH_2)_5(OH)]^+(aq) + H_2O(l) \rightleftharpoons H_3O^+(aq)$$
$$+ Mg(OH)_2(s) + 4\,H_2O(l)$$

Thus, for an acidic metal ion, as the pH increases, the series of equilibria shift to the right, resulting in precipitation of the metal hydroxide. This behavior is illustrated for the first three metal ions of Period 3 (Figure 7.3). For the aluminum ion, at even higher pH values, the metal hydroxide

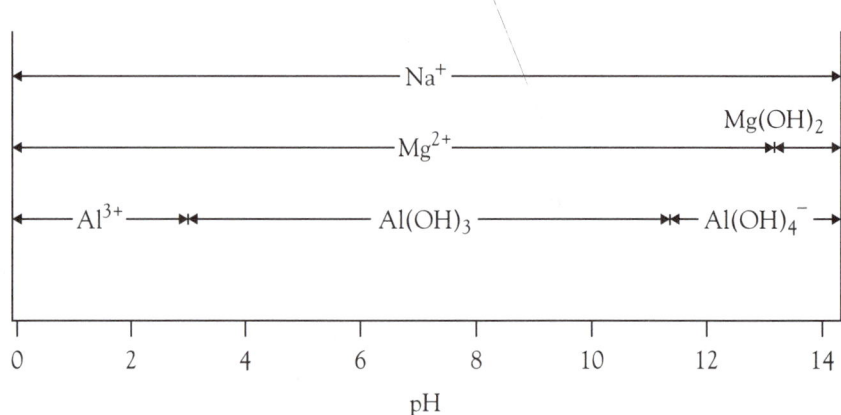

Figure 7.3   The dependence of predominant species upon pH for the cations of sodium, magnesium, and aluminum.

Table 7.5 Basicity of some nonmetal ions

N^{3-}	O^{2-}	F^-
Very basic	Very basic	Weakly basic
P^{3-}	S^{2-}	Cl^-
Very basic	Basic	Neutral
As^{3-}	Se^{2-}	Br^-
Very basic	Basic	Neutral
	Te^{2-}	I^-
	Weakly basic	Neutral

redissolves to form an oxyanion:

$$Al(OH)_3(s) + OH^-(aq) \rightleftharpoons [Al(OH)_4]^-(aq)$$

The Basicity of Nonmetal Anions

Just as there are patterns in the acidity of cations, so are there patterns in the basicity of simple anions. In fact, the pattern is almost the mirror image of the cations. Thus, the halide ions (with the exception of fluoride) are neutral, just as the alkali metals (except lithium) are neutral. Again, the factor affecting basicity is charge density. Table 7.5 categorizes the basicity among the common anions, from neutral to slightly basic to weakly basic to basic to very basic anions.

Thus, in solution, the fluoride ion partially hydrolyzes to hydrofluoric acid:

$$F^-(aq) + H_2O(l) \rightleftharpoons HF(aq) + OH^-(aq)$$

While the hydrolysis of the sulfide ion is almost totally complete, with few sulfide ions being present [followed by the second hydrolysis step]:

$$S^{2-}(aq) + H_2O(l) \rightleftharpoons HS^-(aq) + OH^-(aq)$$
$$[\text{plus } HS^-(aq) + H_2O(l) \rightleftharpoons H_2S(aq) + OH^-(aq)]$$

And the oxide ion is immediately and totally hydrolyzed in aqueous solution:

$$O^{2-}(aq) + H_2O(l) \rightarrow 2\,OH^-(aq)$$

Likewise, none of the 3− ions can exist in solution, thus all soluble nitrides, phosphides, and arsenides react rapidly with water.

The Basicity of Oxyanions

Just as the acidity of oxyacids depends upon the number of oxygen atoms and on the number of hydrogen atoms, so the basicity of the corresponding anion depends upon the number of oxygen atoms and on the ion charge. Again the ions can be categorized according to neutral, slightly basic, moderately basic, and strongly basic.

First, looking at the series XO_n^-, we see that as n decreases, so the basicity increases. This is the converse of the series for oxyacids, where as n decreases for HXO_n, the acidity decreases. The cause of these two trends has the same origin. That is, the greater the number of oxygen atoms around the element X, the weaker any O–H bond, and the less prone the anion will be toward hydrolysis (Table 7.6).

For the series of oxyacids having the common formula XO_4^{n-}, as the charge increases, so the basicity increases (Table 7.7). This pattern resembles that of the monatomic anions where the higher the charge, the more basic the ion.

Table 7.6   Basicity of some common XO_n^- oxyanions

Classification	Type	Examples
Neutral	XO_4^-	ClO_4^-, MnO_4^-
	XO_3^-	NO_3^-, ClO_3^-
Weakly basic	XO_2^-	NO_2^-, ClO_2^-
Moderately basic	XO^-	ClO^-

Table 7.7   Basicity of some common XO_4^{n-} oxyanions

Classification	Type	Examples
Neutral	XO_4^-	ClO_4^-, MnO_4^-
Weakly basic	XO_4^{2-}	SO_4^{2-}, CrO_4^{2-}, MoO_4^{2-}
Moderately basic	XO_4^{3-}	PO_4^{3-}, VO_4^{3-}
Strongly basic	XO_4^{4-}	SiO_4^{4-}

This trend is illustrated in Figure 7.4 by the pH dependence of the predominant species of the Period 3 isoelectronic oxyanions, PO_4^{3-}, SO_4^{2-}, ClO_4^-. The silicate ion has been excluded from the series, as its chemistry is very complex. At very high pH values, the orthosilicate ion, SiO_4^{4-} predominates; at a very low pH, hydrated silicon dioxide; however, over the middle of the pH range there are numerous polymeric ions whose proportions depend upon the solution concentration as well as the pH.

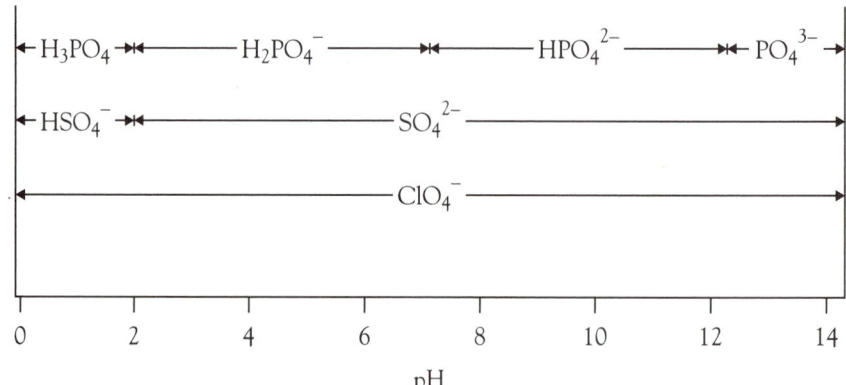

Figure 7.4   The dependence of predominant species upon pH for the isoelectronic XO_4^{n-} oxyanions of phosphorus, sulfur, and chlorine.

The pattern among XO_4^{n-} ions can also be seen in the XO_3^{n-} ions, with the basicity increasing as the oxyanion charge increases (Table 7.8).

Table 7.8   Basicity of some common XO_3^{n-} oxyanions

Classification	Type	Examples
Neutral	XO_3^-	NO_3^-, ClO_3^-
Moderately basic	XO_3^{2-}	CO_3^{2-}, SO_3^{2-}

7.7 *Acid–Base Reactions of Oxides*

In this section, we will focus specifically on the acid–base reactions involving oxides. Oxides can be classified as basic (most metal oxides), acidic (generally nonmetal oxides), amphoteric (the "weak" metal oxides with both acidic and basic properties), and neutral (a few nonmetal and metal oxides). In Chapter 9, Section 9.3, we will look at periodic trends in the acid–base behavior of oxides; then in Chapter 16, Section 16.5, we will see how the acid–base behavior relates to the oxidation state of the non-oxygen atom.

The most typical of the oxide reactions is the reaction of an acidic oxide with a basic oxide to form a salt. For example, sulfur dioxide (an acidic oxide) is a major waste product from metal smelters and other industrial processes. Traditionally, it was released into the atmosphere, but now a number of acid–base reactions have been devised to remove the acidic gas from the waste emissions. The simplest of these reacts with basic calcium oxide to give solid calcium sulfite:

$$CaO(s) + SO_2(g) \rightarrow CaSO_3(s)$$

The calcium sulfite produced in the neutralization reaction then oxidizes in the presence of air to calcium sulfate:

$$2\ CaSO_3(s) + O_2(g) \rightarrow 2\ CaSO_4(s)$$

In the industrial process, cheap powdered limestone (calcium carbonate) is mixed with the very hot sulfur dioxide-containing gas. Thus, the calcium oxide is formed just prior to reaction with the sulfur dioxide:

$$CaCO_3(s) \xrightarrow{\Delta} CaO(s) + CO_2(g)$$

Acidic oxides often react with bases. For example, carbon dioxide reacts with sodium hydroxide solution to produce sodium carbonate:

$$CO_2(g) + 2\ NaOH(aq) \rightarrow Na_2CO_3(aq) + H_2O(l)$$

Conversely, many basic oxides react with acids. For example, magnesium oxide reacts with nitric acid to form magnesium nitrate:

$$MgO(s) + 2\ HNO_3(aq) \rightarrow Mg(NO_3)_2(aq) + H_2O(l)$$

To determine an order of acidity among acidic oxides, the free energy of reaction of the various acidic oxides with the same base can be compared. The larger the free energy change, the stronger the acidic oxide. Let us use calcium oxide as the common basic oxide:

$$CaO(s) + CO_2(g) \rightarrow CaCO_3(s) \qquad \Delta G° = -134\ kJ\cdot mol^{-1}$$
$$CaO(s) + SO_3(g) \rightarrow CaSO_4(s) \qquad \Delta G° = -347\ kJ\cdot mol^{-1}$$

Thus, sulfur trioxide is the more acidic of these two acidic oxides. We can do an analogous test to find the most basic of several basic oxides, by comparing the free energy of reaction of various basic oxides with an acid (in this case, water):

$$Na_2O(s) + H_2O(l) \rightarrow 2\ NaOH(s) \qquad \Delta G° = -142\ kJ\cdot mol^{-1}$$
$$CaO(s) + H_2O(l) \rightarrow Ca(OH)_2(s) \qquad \Delta G° = -59\ kJ\cdot mol^{-1}$$
$$Al_2O_3(s) + 3\ H_2O(l) \rightarrow 2\ Al(OH)_3(s) \qquad \Delta G° = -2\ kJ\cdot mol^{-1}$$

These calculations suggest that sodium oxide is the most basic of the three basic oxides and aluminum oxide, the least. Again it is important to recall that the thermodynamic values predict the energetic feasibility rather than the rate of reaction (the kinetic feasibility).

Traditionally, geochemists classified silicate rocks on an acid–base scale. Such rocks contain metal ions, silicon, and oxygen, and we can think of them as a combination of metal oxides (basic) and silicon dioxide (acidic). We consider a rock such as granite, with more than 66 percent silicon dioxide, to be acidic; those with 52–66 percent SiO_2, intermediate; those with 45–52 percent SiO_2, such as basalt, basic; and those with less than 45 percent, ultrabasic. For example, the mineral olivine, a common component of ultrabasic rocks, has a chemical composition $MgFeSiO_4$, which can be thought of as a combination of oxides, $MgO \cdot FeO \cdot SiO_2$ (though the proportions of magnesium and iron(II) ions can vary). The percentage by mass of silicon dioxide is 35 percent; therefore, this mineral is classified as ultrabasic. In general, the acidic silicate rocks tend to be light in color (granite is pale gray), whereas the basic rocks are dark (basalt is black).

7.8 Lewis Theory

The Brønsted–Lowry theory worked well for the study of reactions in hydrogen-containing solvents, and it is still used in such circumstances. However, it soon became apparent that acid–base concepts could be applied to reactions where hydrogen transfer did not occur. For these reactions, the Lewis definition is much more appropriate. G.N. Lewis defined an acid as an electron pair acceptor and a base as an electron pair donor. The classic example of a Lewis acid–base reaction is that between boron trifluoride and ammonia. Using electron-dot diagrams, we can see that boron trifluoride, with its empty p orbital, is the Lewis acid; ammonia, with its available lone pair, is the Lewis base (Figure 7.5).

There are other Lewis acid–base reactions that more closely parallel Brønsted–Lowry reactions, and some of these apply to nonaqueous solvents. Any pure liquid that has a measurable electrical conductivity must contain ions, one such liquid being bromine trifluoride, BrF_3. It has been shown that the conductivity of this compound results from an autoionization to give the BrF_2^+ cation and the BrF_4^- anion:

$$2 \, BrF_3(l) \rightleftharpoons BrF_2^+(BrF_3) + BrF_4^-(BrF_3)$$

Compounds containing either the cation or the anion are known, for example, $(BrF_2)(SbF_6)$ and $Ag(BrF_4)$, the former being an acid and the latter a base in the BrF_3 solvent system. With bromine trifluoride as the solvent, they can be reacted together in a Lewis neutralization reaction:

$$(BrF_2)(SbF_6)(BrF_3) + Ag(BrF_4)(BrF_3) \rightarrow Ag(SbF_6)(s) + 2 \, BrF_3(l)$$

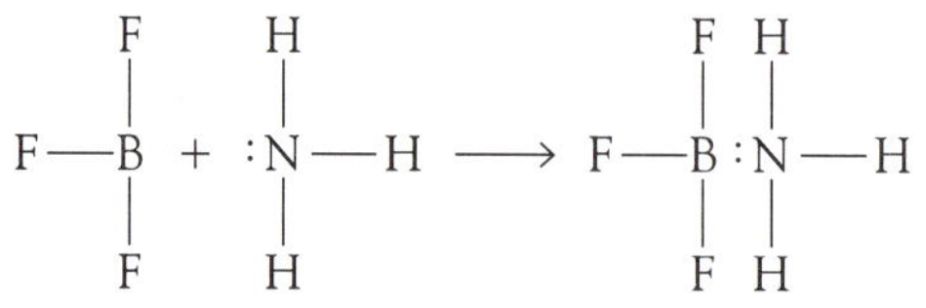

Figure 7.5 The reaction of boron trifluoride with ammonia.

Superacids

A superacid can be defined as an acid that is stronger than 100 percent sulfuric acid. In fact, chemists have synthesized superacids that are from 10^7 to 10^{19} times stronger than sulfuric acid. There are four categories of superacids: Brønsted, Lewis, conjugate Brønsted–Lewis, and solid superacids. A common Brønsted superacid is perchloric acid. When perchloric acid is mixed with pure sulfuric acid, the sulfuric acid actually acts like a base:

$$HClO_4(H_2SO_4) + H_2SO_4(l) \rightleftharpoons H_3SO_4^+(H_2SO_4) + ClO_4^-(H_2SO_4)$$

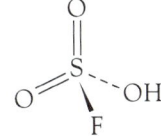

Fluorosulfuric acid, HSO_3F, is the strongest Brønsted superacid; it is more than 1000 times more acidic than sulfuric acid. This superacid, which is an ideal solvent because it is liquid from $-89°C$ to $+164°C$, has the structure shown.

A Brønsted–Lewis superacid is a mixture of a powerful Lewis acid and a strong Brønsted–Lowry acid. The most potent combination is a 10 percent solution of antimony pentafluoride, SbF_5, in fluorosulfuric acid. The addition of antimony pentafluoride increases the acidity of the fluorosulfuric acid several thousand times. The reaction between the two acids is very complex, but the super-hydrogen-ion donor present in the mixture is the $H_2SO_3F^+$ ion. This acid mixture will react with many substances, such as hydrocarbons, that do not react with normal acids. For example, propene, C_3H_6, reacts with this ion to give the propyl cation:

$$C_3H_6(HSO_3F) + H_2SO_3F^+(HSO_3F) \rightarrow C_3H_7^+(HSO_3F) + HSO_3F(l)$$

The solution of antimony pentafluoride in fluorosulfuric acid is commonly called "Magic Acid." The name originated in the Case Western Reserve University laboratory of George Olah, a pioneer in the field of superacids (and recipient of the Nobel Prize for Chemistry in 1994). A researcher working with Olah put a small piece of Christmas candle left over from a lab party into the acid and found that it dissolved rapidly. He studied the resulting solution and found that the long-chain hydrocarbon molecules of the paraffin wax had added hydrogen ions, and the resulting cations had rearranged themselves to form branched-chain molecules. This unexpected finding suggested the name "Magic Acid," and it is now a registered trade name for the compound. This family of superacids is used in the petroleum industry for the conversion of the less important straight-chain hydrocarbons to the more valuable branched-chain molecules, which are needed to produce high-octane gasoline.

7.9 Pearson Hard–Soft Acid–Base Concepts

In Chapter 6, we saw that thermodynamics can be used to predict the feasibility of chemical reactions. However, we need complete thermodynamic data with which to perform the calculations, and these are not always available. Chemists have therefore tried to find a more qualitative empirical approach to reaction prediction. For example, will sodium iodide react with silver nitrate to give silver iodide and sodium nitrate, or will silver iodide react with sodium nitrate to give sodium iodide and silver nitrate? To make such predictions of reaction, a very effective method was devised by R.G. Pearson and it is known as the *Hard–Soft Acid–Base (HSAB) Concept.*

Pearson proposed that Lewis acids and bases could be categorized as either "hard" or "soft." Using these categories, he showed that a reaction generally proceeded in the direction that would pair the softer acid with the softer base and the harder acid with the harder base. The elements were divided as follows.

Hard Acids   The *hard acids,* also known as *class a metal ions,* consist of most of the metal ions in the periodic table. They are characterized by low electronegativities and, often, high charge densities. The charge density is sometimes the better guide to hardness, for we categorize the theoretical H^+, B^{3+}, and C^{4+} ions as hard acids, and these have extremely high charge densities.

Soft Acids   The *soft acids,* also known as *class b metal ions,* are the group of metal ions that are in the lower right portion of the metallic elements (Figure 7.6). They have low charge densities and tend to have among the highest electronegativities of the metallic elements. With low charge densities, these cations will be easily polarized; hence, they tend towards covalent bond formation. The softest of all acids is gold(I).

Borderline Acids   As in so many categorizations, there are the borderline cases. The borderline acids form a border between the soft and hard acids and they have intermediate values as charge densities. Oxidation state becomes a crucial factor in determining hardness. For example, copper(I) with a charge density of 51 $C \cdot mm^{-3}$ is categorized as soft, while copper(II), charge density 116 $C \cdot mm^{-3}$ is considered borderline. Likewise, iron(III) and cobalt(III) ions, both with charge densities over 200 $C \cdot mm^{-3}$, are assigned to the hard category, while the iron(II) and cobalt(II) ions (charge densities of about 100 $C \cdot mm^{-3}$) are designated as borderline.

Hard Bases   The *hard bases* or *class a ligands* are fluorine and oxygen-bonded species, including oxide, hydroxide, nitrate, phosphate, carbonate,

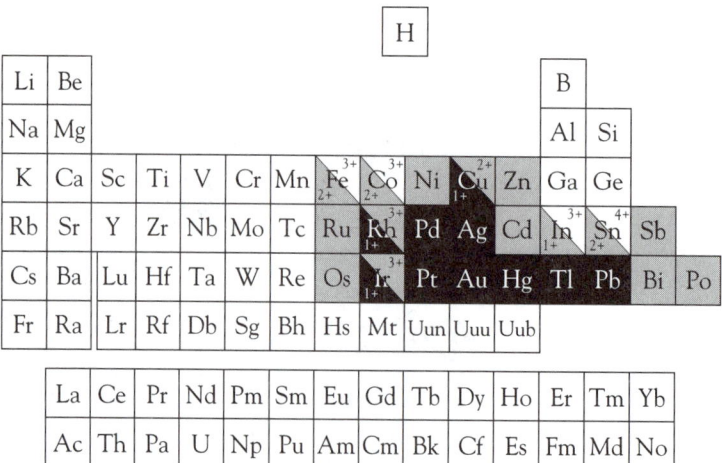

Figure 7.6 The classification of the HSAB acid ions into hard (white), borderline (shaded), and soft (black).

Table 7.9 Common hard, borderline, and soft bases

Hard	Borderline	Soft
F^-, O^{2-}, OH^-, H_2O, CO_3^{2-}, NH_3, NO_3^-, SO_4^{2-}, ClO_4^-, PO_4^{3-}, (Cl^-)	Br^-, N_3^-, NCS^-	I^-, S^{2-}, P^{3-}, H^-, CN^-, CO, SCN^-, $S_2O_3^{2-}$

sulfate, and perchlorate. The monatomic ions have comparatively high charge densities. Chloride is considered borderline hard.

Soft Bases The *soft bases* or *class b ligands* are the less electronegative nonmetals including carbon, sulfur, phosphorus, and iodine. These large, polarizable ions (low charge density) tend to favor covalent bond formation.

Borderline Bases Just as we have borderline acids, so are there borderline bases. It should be realized that the categories are not rigidly divided. For example, the halide ions form a series from the very hard fluoride ion, through the hard-borderline chloride ion to the borderline bromide ion and the soft iodide ion.

In a few cases, an anion will fit more than one category of bases. These anions are capable of covalently bonding to a metal ion through two different atoms. One common example is the thiocyanate ion, NCS^-. This ion can bond through the nitrogen atom ($-NCS$), in which case it behaves as a borderline base. While bonding through the sulfur atom ($-SCN$), however, it behaves as a soft base (Table 7.9). Ions that can bond through different elements are called *ambidentate ligands*.

7.10 Applications of the HSAB Concept

In this section, we will review some of the applications of the HSAB concept to simple inorganic chemistry. In Chapter 19, Section 19.12, we will return to the concept in the context of transition metal complexes.

The most important application of the HSAB concept is in the prediction of chemical reactions. For example, we can predict the gas-phase reaction of mercury(II) fluoride with beryllium iodide because the soft-acid mercury(II) ion is paired with the hard-base fluoride ion while the hard-acid beryllium ion is paired with soft-base iodide ion. According to the HSAB concept, the ions would prefer to be partnered with their own type. Hence, the following reaction would be expected and it does, in fact, occur:

$$HgF_{2(g)} + BeI_{2(g)} \rightarrow BeF_{2(g)} + HgI_{2(g)}$$
soft–hard   hard–soft   hard–hard   soft–soft

The HSAB concept can be used even when less than half of the species are hard. That is, one can say that softer acids prefer softer bases. Among the nonmetallic elements, softness increases from the upper right in the periodic table to the lower left. For example, iodine is the softest of the halogens. Thus, we expect iodide ion to react with silver bromide as the soft-acid silver ion will prefer the soft-base iodide ion, over the borderline-base bromide ion.

$$AgBr_{(s)} + I^-_{(aq)} \rightarrow AgI_{(s)} + Br^-_{(aq)}$$

Table 7.10   Solubilities of sodium and silver halides ($mol \cdot L^{-1}$)

	Fluoride	Chloride	Bromide	Iodide
Sodium	1.0	6.1	11.3	12.3
Silver	14.3	1.3×10^{-5}	7.2×10^{-7}	9.1×10^{-9}

Another example is the reaction between cadmium selenide with mercury(II) sulfide, where the soft-acid mercury(II) ion prefers the softer-base selenide ion, while the borderline-acid, cadmium ion prefers the less-soft sulfide ion:

$$CdSe(s) + HgS(s) \rightarrow CdS(s) + HgSe(s)$$

We can also use HSAB concept in the interpretation of solubility patterns. Table 7.10 shows that the solubility trend for sodium halides is completely reversed to that for silver halides. The difference can be explained as the hard-acid sodium preferring the harder bases while the soft-acid silver prefers softer bases. This is one approach to the discussion of solubility patterns. In Chapter 11, Section 11.2, we examine the trend in solubilities of the sodium halides in terms of thermodynamic contributions.

Application of the HSAB Concept to Qualitative Analysis

We can use the HSAB concept to account for the common system of cation analysis. The groups into which we classify cations have no direct connection with the periodic table; instead the cations are categorized according to their solubilities with different anions. To distinguish the two uses of "groups" we will use the traditional Roman numerals to denote the analysis groups. Group I comprises those cations that form insoluble chlorides; Group II, those cations having soluble chlorides and very insoluble sulfides; and Group III, those cations having soluble chlorides and insoluble sulfides. To distinguish the Group II and III categories, we control the sulfide ion concentration by means of pH. According to the equilibria,

$$H_2S(aq) + H_2O(l) \rightleftharpoons H_3O^+(aq) + HS^-(aq)$$

$$HS^-(aq) + H_2O(l) \rightleftharpoons H_3O^+(aq) + S^{2-}(aq)$$

at low pH, the concentration of sulfide ion will be very low; hence, only those metal sulfides with very small solubility product values will precipitate. If we then increase the pH, the equilibrium concentration of sulfide ion will increase, and those metal sulfides that are not quite as insoluble will precipitate. Under such basic conditions, those metal ions that form soluble sulfides, but very insoluble metal hydroxides, will also precipitate in Group IV. The Group IV metal ions correspond to those cations that have soluble chlorides and sulfides. Group V contains the ions that form few, if any, insoluble salts. The precipitating species are shown in Table 7.11.

It is not obvious how the HSAB concept can be used to understand the cation group analysis scheme. However, if we recall from Chapter 6 that aqueous ions consist of the ions surrounded by a sphere of water molecules held by ion–dipole attraction, we can rewrite precipitation reactions, such as that of silver ion with chloride ion, as:

$$[Ag(OH_2)_n]^+(aq) + [Cl(H_2O)_m]^-(aq) \rightarrow AgCl(s) + (n + m) H_2O(l)$$

Table 7.11   Common scheme of cation analysis

Group I	Group II	Group III	Group IV	Group V
$AgCl$	HgS	MnS	$CaCO_3$	Na^+
$PbCl_2$	CdS	FeS	$SrCO_3$	K^+
Hg_2Cl_2	CuS	CoS	$BaCO_3$	NH_4^+
	SnS_2	NiS		Mg^{2+}
	As_2S_3	ZnS		
	Sb_2S_3	$Al(OH)_3$		
	Bi_2S_3	$Cr(OH)_3$		

Writing the equation this way (that is, according to HSAB concept) the driving force of the reaction is the preference of the soft-acid silver ion for the borderline-hard base, chloride, rather than for the hard-base oxygen of the water molecule.

The metal ions in Group II are those that are soft acids and borderline acids. Thus, they readily combine with the soft-base sulfide ion. We can illustrate this category with cadmium ion:

$$[Cd(OH_2)_n]^{2+}(aq) + [S(H_2O)_m]^{2-}(aq) \rightarrow CdS(s) + (n + m)\,H_2O(l)$$

The members of analysis Group III are the hard-acid together with some borderline-acid cations. Thus, it is argued that a higher concentration of sulfide ion is necessary for the precipitation of these metal sulfides. Using HSAB arguments, the aluminum and chromium(III) ions are sufficiently hard that they prefer to react with the hard-base hydroxide ion rather than with the soft-base sulfide ion. The members of Group IV are very hard acids; they give precipitates only with hard-base carbonate ions in this analysis scheme.

The HSAB Concept in Geochemistry

In 1923, the geochemist, V.M. Goldschmidt, devised a classification of the chemical elements that would relate to the geological history of the Earth. As the Earth cooled, some elements separated into the metallic phase of the Earth's core (the *siderophiles*), some formed sulfides (the *chalcophiles*), some formed silicates (the *lithophiles*), while others escaped to provide the atmosphere (the *atmophiles*). The classification is still used today in various modified forms. Here we will consider the categories in terms of the element forms on the surface of the Earth. Using this modification of Goldschmidt's classification, we consider atmophiles as the unreactive nonmetals found solely in the atmosphere in their elemental forms (the noble gases and dinitrogen); lithophiles as the metals and nonmetals that occur mainly as oxides, silicates, sulfates, or carbonates; chalcophiles as the elements that are usually found as sulfides; and the siderophiles as the metals that, on the surface of the Earth, are usually found in elemental form. According to this assignment, the distribution of the lithophiles, chalcophiles, and siderophiles in the periodic table are shown in Figure 7.7.

If we compare the geochemical classification with the HSAB classification, we see that the metallic lithophiles are those that are hard acids. They would therefore be expected to prefer the hard-base oxygen, either as the oxide ion or

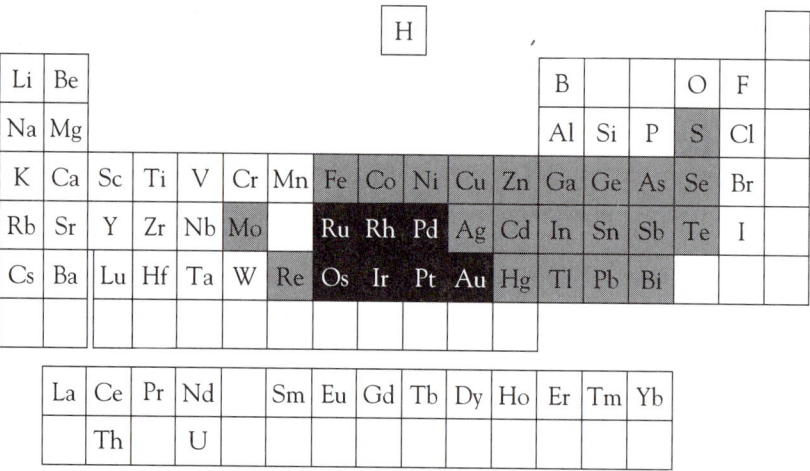

Figure 7.7 The modified geochemical classification of the chemical elements into lithophiles (white), chalcophiles (shaded), and siderophiles (black) according to their common surface combinations. The atmophiles and those elements that do not have stable isotopes have been deleted from the table.

as an oxygen-bonded oxyanion, such as silicate. For example, the common ore of aluminum is aluminum oxide, Al_2O_3 (bauxite), while the most common compound of calcium is calcium carbonate, $CaCO_3$ (limestone, chalk, marble), in both cases being hard-acid, hard-base combinations. The chalcophiles metals, on the other hand, are in the borderline- and soft-acid categories. These metals are found in combination with soft bases, in particular, sulfide ion. Thus, zinc is found mainly as zinc sulfide, ZnS (sphalerite, wurtzite), and mercury as mercury(II) sulfide, HgS (cinnabar). In the chalcophile category, we also find the soft-base nonmetals that combine with the other chalcophile elements, such as the common mineral of arsenic, di-arsenic trisulfide, As_2S_3 (orpiment).

There are some interesting comparisons that give credence to the application of the HSAB concept to mineralogy. First, we find the hard-acid iron(III) with hard-base oxide in iron(III) oxide (hematite), while the borderline iron(II) is found with soft-base sulfide in iron(II) disulfide, FeS_2 (pyrite). Second, among the Group 14 metals, tin is found as hard-acid tin(IV) in the compound tin(IV) oxide, SnO_2 (cassiterite), but lead is found primarily as the soft-acid lead(II) in the compound lead(II) sulfide, PbS (galena). However, it is always dangerous to place too much trust in general principles such as the HSAB concept. For example, soft-acid lead(II) is also found in a number of minerals in which it is combined with a hard base. One example is lead(II) sulfate, $PbSO_4$ (anglesite).

Interpretation of the HSAB Concept

The HSAB concept originated as a qualitative empirical approach that would enable chemists to predict whether or not a particular reaction is likely to occur. Since Pearson first proposed the concept, there have been attempts to understand why it works and to derive quantitative hardness parameters. Though the latter is best left to a more advanced course, it is useful to see how the HSAB concept fits into our other perspectives.

The Pearson approach can be related to the earlier discussions of ionic and covalent bonding (Chapters 3 and 5). The hard-acid–hard-base combination is really the pairing of a low-electronegativity cation with a high-electronegativity anion, properties that result in ionic behavior. Conversely, the soft acids are the metals that lie close to the nonmetal border and have comparatively high electronegativities. These metallic ions will form covalent bonds with the soft-base ions such as sulfide.

7.11 *Biological Aspects*

In Chapter 2, Section 2.8, we surveyed the elements that are essential to life. Here we will discuss those elements that are considered toxic. According to Bertrand's rule, each element is biochemically toxic above a certain level of intake characteristic of that element. It is the concentration at which toxicity commences that determines whether or not, for practical reasons, we would call an element "toxic." Here we will discuss only those elements for which the onset of toxicity is at very low concentrations. These elements are shown in Figure 7.8, and we can use the HSAB concept to help us understand why these elements are so toxic.

| | | | | | | | | | | | | | | | | | H | | | | | | | | | | | | | | | He |

(periodic table, Figure 7.8)

Figure 7.8 Elements that are regarded as particularly toxic (in black).

The thiol group (–SH) part of the amino acid, cysteine (Figure 7.9), is a common component of enzymes. Normally, zinc would bind to many of the thiol sites, but the toxic metals (except for beryllium) are softer than zinc and will bind preferentially. Beryllium, on the other hand, is a hard acid and, as such, it preferentially binds to sites occupied by another Group 2 element, hard-acid magnesium. The environmentally toxic nonmetals and semimetals, arsenic, selenium, and tellurium, are very soft bases. The biochemical form of their toxicity is not well understood, but it may arise from preferentially binding to borderline acids such as iron(II) and zinc, preventing the metals from performing their essential roles in enzymes. Selenium is a particularly interesting case for, as mentioned in Chapter 2, very small concentrations of selenium-containing enzymes are essential for our health. There are many diseases linked with low selenium levels in diets. Higher levels cause the disease of selenosis.

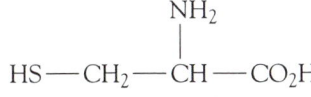

$$HS-CH_2-CH-CO_2H$$
with NH_2 above the CH

Figure 7.9 The structure of the thio-amino acid, cysteine.

EXERCISES

7.1 Write net ionic equations to correspond with each of the following molecular equations:

(a) $HCl(aq) + NaOH(aq) \rightarrow NaCl(aq) + H_2O(l)$

(b) $Na_2CO_3(aq) + CoCl_2(aq) \rightarrow 2\ NaCl(aq) + CoCO_3(s)$

(c) $NH_4OH(aq) + CH_3COOH(aq) \rightarrow CH_3COONH_4(aq) + H_2O(l)$

7.2 Write net ionic equations to correspond with each of the following molecular equations:

(a) $Na_2S(aq) + 2\ HCl(aq) \rightarrow 2\ NaCl(aq) + H_2S(g)$

(b) $HF(aq) + NaOH(aq) \rightarrow NaF(aq) + H_2O(l)$

(c) $Na_2HPO_4(aq) + H_2SO_4(aq) \rightarrow NaH_2PO_4(aq) + NaHSO_4(aq)$

7.3   Define the following terms: (a) conjugate acid–base pairs; (b) self-ionization; (c) amphiprotic.

7.4   Define the following terms: (a) acid ionization constant; (b) leveling solvent; (c) polyprotic.

7.5   Write a balanced net ionic equation for the reaction of each of the following compounds with water: (a) NH_4NO_3; (b) KCN; (c) $NaHSO_4$.

7.6   Write a balanced net ionic equation for the reaction of each of the following compounds with water: (a) Na_3PO_4; (b) $NaHSO_4$; (c) $(CH_3)_3NHCl$.

7.7   Write an equilibrium equation to represent the reaction of chloramine, $ClNH_2$, a base, with water.

7.8   Write an equilibrium equation to represent the reaction of fluorosulfonic acid, HSO_3F with water.

7.9   Pure sulfuric acid can be used as a solvent. Write an equilibrium to represent the self-ionization reaction.

7.10 The following species are amphiprotic. Write the formulas of the corresponding conjugate acids and bases: (a) HSe^-; (b) PH_3; (c) HPO_4^{2-}.

7.11 Using liquid ammonia as a solvent what is (a) the strongest acid? (b) the strongest base?

7.12 Hydrogen fluoride is a strong acid when dissolved in liquid ammonia. Write a chemical equation to represent the acid–base equilibrium.

7.13 Hydrogen fluoride behaves as a base when dissolved in pure sulfuric acid. Write a chemical equation to represent the acid–base equilibrium and identify the conjugate acid–base pairs.

7.14 Identify the conjugate acid–base pairs in the following equilibrium:

$$HSeO_4^-(aq) + H_2O(l) \rightleftharpoons H_3O^+(aq) + SeO_4^{2-}(aq)$$

7.15 Identify the conjugate acid–base pairs in the following equilibrium:

$$HSeO_4^-(aq) + H_2O(l) \rightleftharpoons OH^-(aq) + H_2SeO_4(aq)$$

7.16 Which will be the stronger acid, sulfurous acid, $H_2SO_3(aq)$, or sulfuric acid, $H_2SO_4(aq)$? Use electronegativity arguments to explain your reasoning.

7.17 Hydrogen selenide, H_2Se, is a stronger acid than hydrogen sulfide. Use bond strength arguments to explain your reasoning.

7.18 Addition of copper(II) ion to a hydrogen phosphate, HPO_4^{2-}, solution results in

precipitation of copper(II) phosphate. Use two chemical equations to suggest an explanation.

7.19 The hydrated zinc ion, $Zn(OH_2)_6^{2+}$, forms an acidic solution. Use a chemical equation to suggest an explanation.

7.20 A solution of the cyanide ion, CN^-, is a strong base. Write a chemical equation to illustrate this. What can you deduce about the properties of hydrocyanic acid, HCN?

7.21 The weak base, hydrazine, H_2NNH_2, can react with water to form a diprotic acid, $^+H_3NNH_3^+$. Write chemical equations to depict the two equilibrium steps. When hydrazine is dissolved in water, which of the three hydrazine species will be present in the lowest concentration?

7.22 When dissolved in water, which of the following salts will give neutral, acidic, or basic solutions: (a) potassium fluoride; (b) ammonium chloride? Explain your reasoning.

7.23 When dissolved in water, which of the following salts will give neutral, acidic, or basic solutions: (a) aluminum nitrate; (b) sodium iodide? Explain your reasoning.

7.24 When two sodium salts, NaX and NaY, are dissolved in water to give solutions of equal concentration, the pH values obtained are 7.3 and 10.9, respectively. Which is the stronger acid, HX or HY? Explain your reasoning.

7.25 The pK_b values of the bases A^- and B^- are 3.5 and 6.2, respectively. Which is the stronger acid, HA or HB? Explain your reasoning.

7.26 Pure liquid sulfuric acid can be dissolved in liquid acetic (ethanoic) acid, CH_3COOH. Write a balanced chemical equation for the equilibrium. Will the acetic acid act as a differentiating or leveling solvent? Explain your reasoning.

7.27 Write a net ionic equation for the equilibrium reaction between aqueous phosphoric acid and aqueous disodium hydrogen phosphate, Na_2HPO_4.

7.28 In a damp climate, sodium sulfide has a strong "rotten egg" smell, characteristic of hydrogen sulfide. Write two net ionic equilibria to indicate how the gas is produced.

7.29 Identify the oxides corresponding to the following acids: (a) nitric acid; (b) chromic acid, H_2CrO_4; (c) periodic acid, H_5IO_6.

7.30 Identify the oxides corresponding to the following bases: (a) potassium hydroxide; (b) chromium(III) hydroxide, $Cr(OH)_3$.

7.31 For each of the following nonaqueous reactions, identify the acid and the base.

(a) $SiO_2 + Na_2O \rightarrow Na_2SiO_3$

(b) $NOF + ClF_3 \rightarrow NO^+ + ClF_4^-$

(c) $Al_2Cl_6 + 2\,PF_3 \rightarrow 2\,AlCl_3{:}PF_3$

7.32 For each of the following nonaqueous reactions, identify the acid and the base.

(a) $PCl_5 + ICl \rightarrow PCl_4^+ + ICl_2^-$

(b) $POCl_3 + Cl^- \rightarrow POCl_4^-$

(c) $Li_3N + 2\,NH_3 \rightarrow 3\,Li^+ + 3\,NH_2^-$

7.33 What will be the effect on the pH of the water (if anything) when you add the following salt. Write a chemical equation where appropriate: (a) CsCl, (b) K_2Se, (c) $ScBr_3$, (d) KF. In the cases where the pH does change, would you expect a large or small change?

7.34 What will be the effect on the pH of the water (if anything) when you add the following salt. Write a chemical equation where appropriate: (a) Na_2O; (b) $Mg(NO_3)_2$; (c) K_2CO_3.

7.35 Identify the following oxyanions as neutral, weakly basic, moderately basic, or strongly basic: (a) WO_4^{2-}; (b) TcO_4^-; (c) AsO_4^{3-}; (d) GeO_4^{4-}.

7.36 Identify the following oxyanions as neutral, weakly basic, or moderately basic: (a) BrO_3^-; (b) BrO^-; (c) BrO_2^-.

7.37 The ion XeO_6^{4-} is moderately basic. What degree of basicity would you expect for the isoelectronic ions (a) IO_6^{5-}? (b) TeO_6^{6-}?

7.38 Each of the following reactions lies toward the product side. On this basis, arrange all of the Brønsted acids in order of decreasing strength.

$H_3PO_4(aq) + N_3^-(aq) \rightleftharpoons HN_3(aq) + H_2PO_4^-(aq)$

$HN_3(aq) + OH^-(aq) \rightleftharpoons H_2O(l) + N_3^-(aq)$

$H_3O^+(aq) + H_2PO_4^-(aq) \rightleftharpoons H_3PO_4(aq) + H_2O(l)$

$H_2O(l) + PH_2^-(aq) \rightleftharpoons PH_3(aq) + OH^-(aq)$

7.39 From free energy of formation values, determine the free energy of reaction of magnesium oxide with water to give magnesium hydroxide, and then deduce whether magnesium oxide is more or less basic than calcium oxide.

7.40 From free energy of formation values, determine the free energy of reaction of silicon dioxide with calcium oxide to give calcium silicate, $CaSiO_3$, and then deduce whether silicon dioxide is more or less acidic than carbon dioxide.

7.41 Nitrosyl chloride, NOCl, can be used as a nonaqueous solvent. It undergoes the following self-ionization:

$$NOCl(l) \rightleftharpoons NO^+(NOCl) + Cl^-(NOCl)$$

Identify which of the ions is a Lewis acid and which is a Lewis base. Also, write a balanced equation for the reaction between $(NO)^+(AlCl_4)^-$ and $[(CH_3)_4N]^+Cl^-$.

7.42 Liquid bromine trifluoride, BrF_3, undergoes self-ionization. Write a balanced equilibrium equation to represent this process.

7.43 In pure liquid ammonia, the self-ionization constant is 1×10^{-33}.
(a) Calculate the concentration of ammonium ion in liquid ammonia.
(b) Calculate the concentration of ammonium ion in a $1.0\ mol{\cdot}L^{-1}$ solution of sodium amide, $NaNH_2$.

7.44 Each of the following reactions in aqueous solution can also be performed using liquid hydrogen fluoride as a solvent and reactant in place of the water. Write the corresponding equations in hydrogen fluoride. Will the position of equilibrium be the same, further to the right, or further to the left than those for the aqueous reactions?

(a) $CN^-(aq) + H_2O(l) \rightleftharpoons HCN(aq) + OH^-(aq)$

(b) $HClO_4(aq) + H_2O(aq) \rightarrow H_3O^+(aq) + ClO_4^-(aq)$

7.45 Will either of the following high temperature gas-phase reactions be feasible? Give your reasoning in each case.

(a) $CuBr_2(g) + 2\,NaF(g) \rightarrow CuF_2(g) + 2\,NaBr(g)$

(b) $TiF_4(g) + 2\,TiI_2(g) \rightarrow TiI_4(g) + 2\,TiF_2(g)$

7.46 Will either of the following high temperature gas-phase reactions be feasible? Give your reasoning in each case.

(a) $CuI_2(g) + 2\,CuF(g) \rightarrow CuF_2(g) + 2\,CuI(g)$

(b) $CoF_2(g) + HgBr_2(g) \rightarrow CoBr_2(g) + HgF_2(g)$

7.47 In the following solution equilibria, suggest whether the equilibrium constant is likely to be greater or less than 1.

(a) $[AgCl_2]^-(aq) + 2\,CN^-(aq) \rightleftharpoons [Ag(CN)_2]^-(aq) + 2\,Cl^-(aq)$

(b) $CH_3HgI(aq) + HCl(aq) \rightleftharpoons CH_3HgCl(aq) + HI(aq)$

7.48 In the following solution equilibria, will the products or reactants be favored? Give your reasoning in each case.

(a) $AgF(aq) + LiI(aq) \rightleftharpoons AgI(s) + LiF(aq)$

(b) $2\ Fe(OCN)_3(aq) + 3\ Fe(SCN)_2(aq) \rightleftharpoons$
$2\ Fe(SCN)_3(aq) + 3\ Fe(OCN)_2(aq)$

7.49 Predict to which group in the common cation analysis each of the following ions will belong and write a formula for the probable precipitate (if any): (a) thallium(I), Tl^+; (b) rubidium, Rb^+; (c) radium, Ra^{2+}; (d) iron(III), Fe^{3+}.

7.50 Identify which compound is a common ore of the element listed:

(a) thorium: ThS_2 or ThO_2
(b) platinum: $PtAs_2$ or $PtSiO_4$
(c) fluorine: CaF_2 or PbF_2

7.51 Identify which compound is a common ore of the element listed and give your reasoning:
(a) magnesium: MgS or $MgSO_4$
(b) cobalt: CoS or $CoSO_4$

7.52 For the three most common antacids: magnesium hydroxide, calcium carbonate, and aluminum hydroxide, calculate which provides the greatest neutralization per gram of antacid. Would you use this factor as the sole reason for choosing an antacid?

BEYOND THE BASICS

7.53 Calculate the concentration of sulfide ion in a $0.010\ mol \cdot L^{-1}$ solution of hydrogen sulfide in $1.0\ mol \cdot L^{-1}$ strong acid. The values of the acid ionization constants, K_{a1} and K_{a2}, are 8.9×10^{-8} and 1.2×10^{-13}, respectively. If the solution contained cadmium and iron(II) ions at concentrations of $0.010\ mol \cdot L^{-1}$, would either of their sulfides precipitate? The solubility products of cadmium and iron(II) sulfides are 1.6×10^{-28} and 6.3×10^{-18}, respectively.

7.54 What pH would be required in a $0.010\ mol \cdot L^{-1}$ solution of hydrogen sulfide to just start to precipitate tin(II) sulfide from a $0.010\ mol \cdot L^{-1}$ tin(II) ion solution? The values of the acid ionization constants, K_{a1} and K_{a2}, are 8.9×10^{-8} and 1.2×10^{-13}, respectively. The solubility of tin(II) sulfide is 1.0×10^{-25}.

7.55 The only common ore of mercury is mercury(II) sulfide. Zinc, however, is found as the sulfide, carbonate, silicate, and oxide. Comment on this.

7.56 In Figure 7.3, we saw how acid–base properties changed across a period. Research the change in acid–base behavior of the highest oxidation states of the Group 15 elements as the group is descended.

7.57 Silicic acid, often written as $SiO_2 \cdot xH_2O(s)$, is a weaker acid than carbonic acid, $H_2CO_3(aq)$. Write a balanced equation for the reaction of carbonic acid with a simple silicate, such as $Mg_2SiO_4(s)$. Explain how this reaction is relevant to the reduction of atmospheric carbon dioxide levels over geological time.

7.58 Boric acid, $B(OH)_3(aq)$, acts as an acid in water. However, it does not do so as a proton donor, instead it acts as a Lewis acid toward the hydroxide ion. Depict this process in an equation for the reaction of boric acid with sodium hydroxide solution.

7.59 Copper(I) ion undergoes the following disproportionation reaction for which the equilibrium constant is about 10^6:

$$2\ Cu^+(aq) \rightleftharpoons Cu(s) + Cu^{2+}(aq)$$

If the copper(I) ion is dissolved in dimethylsulfoxide, $(CH_3)_2SO$, as a solvent, the equilibrium constant is only about 2. Suggest an explanation.

7.60 The molecule $(CH_3)_2N{-}PF_2$ has two atoms that can act as Lewis bases. With boron compounds, BH_3 attaches to phosphorus, while BF_3 attaches to nitrogen. Give your reasoning.

7.61 The molar solubility of calcium chloride is about four times greater than that of barium chloride. Suggest an explanation in terms of the HSAB concept and an explanation in terms of thermodynamic factors.

7.62 The reaction between calcium oxide and silicon dioxide to give calcium silicate, $CaSiO_3$, is an important reaction in a blast furnace for the production of iron, in that it removes silicate impurities as low-density slag that can be poured off. What would be the theoretical oxidation number of the silicon? Then explain the transfer of the oxide ion to the silicon dioxide in terms of HSAB theory.

Chapter 8

Oxidation and Reduction

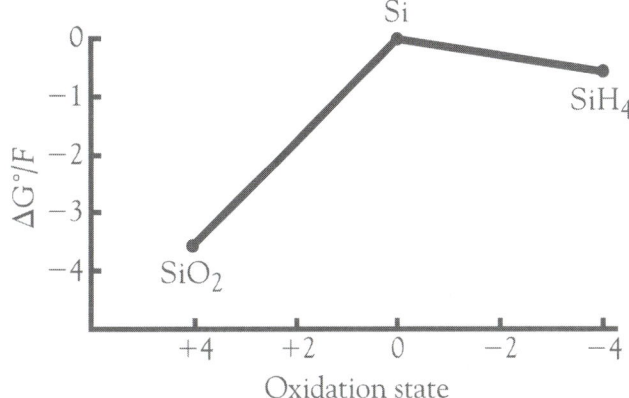

A large proportion of chemical reactions involves changes in oxidation state. In this chapter, we show how oxidation numbers can be determined. We then examine redox reactions. The redox properties of a particular oxidation state of an element can be displayed graphically, giving us information about the thermodynamic stability of that compound or ion.

In the history of chemistry, one of the most vehement disputes concerned the nature of oxidation. The story really begins in 1718. Georg Stahl, a German chemist, was studying the formation of metals from oxides by heating the oxide with charcoal (carbon). He proposed that the formation of the metal was caused by the absorption of a substance that he named "phlogiston." According to Stahl, the converse process of heating a metal in air to form its oxide caused the release of phlogiston to the atmosphere.

Fifty-four years later, the French chemist Louis-Bernard Guyton de Morveau performed careful experiments showing that, during combustion, metals increase in weight. However, the existence of phlogiston was so well established among chemists that he interpreted the results as meaning that phlogiston had a negative weight. It was his colleague, Antoine Lavoisier, who was willing to throw out the phlogiston concept and propose that combustion was due to the addition of oxygen to the metal (oxidation) and that the formation of a metal from an oxide corresponded to the loss of oxygen (reduction).

The editors of the French science journal were phlogistonists, so they would not publish this proposal. Thus, Lavoisier, with his new convert de

Table 8.1   Traditional definitions of oxidation and reduction

Oxidation	Reduction
Gain of oxygen atoms	Loss of oxygen atoms
Loss of hydrogen atoms	Gain of hydrogen atoms
Loss of electrons	Gain of electrons

Morveau, and others, had to establish their own journal to publish the "new chemistry." It was the overthrow of the phlogiston theory that caused chemists to realize that elements were the fundamental substances in chemistry—and modern chemistry was born.

8.1   Redox Terminology

Many inorganic reactions are redox reactions, and, like so many fields of chemistry, the study of oxidation and reduction has its own vocabulary and definitions. Traditionally, oxidation and reduction were each defined in three different ways, as shown in Table 8.1.

In modern chemistry, we use more general definitions of oxidation and reduction:

Oxidation: Increase in oxidation number

Reduction: Decrease in oxidation number

8.2   Oxidation Number Rules

Of course, now we have to define an oxidation number (also called oxidation state). Oxidation numbers are simply theoretical values used to simplify electron bookkeeping. We assign these values to the common elements on the basis of a simple set of rules:

1. The oxidation number, N_{ox}, of an atom as an element is zero.

2. The oxidation number of a monatomic ion is the same as its ion charge.

3. The algebraic sum of the oxidation numbers in a neutral polyatomic compound is zero; in a polyatomic ion, it is equal to the ion charge.

4. In combinations of elements, the more electronegative element has its characteristic negative oxidation number (for example, -3 for nitrogen, -2 for oxygen, -1 for chlorine), and the more electropositive element has a positive oxidation number.

5. Hydrogen usually has an oxidation number of $+1$ (except with more electropositive elements, when it is -1).

For example, to find the oxidation number of sulfur in sulfuric acid, H_2SO_4, we can use rule 3 to write

$$2[N_{ox}(H)] + [N_{ox}(S)] + 4[N_{ox}(O)] = 0$$

Because oxygen usually has an oxidation number of -2 (rule 4) and hydrogen, $+1$ (rule 5), we write

$$2(+1) + [N_{ox}(S)] + 4(-2) = 0$$

Hence, $[N_{ox}(S)] = +6$.

Now let us deduce the oxidation number of iodine in the ion ICl_4^-. For this, we can use rule 3 to write

$$[N_{ox}(I)] + 4[N_{ox}(Cl)] = -1$$

Chlorine is more electronegative than iodine, so chlorine will have the conventional negative oxidation number of -1 (rule 4). Thus,

$$[N_{ox}(I)] + 4(-1) = -1$$

Hence, $[N_{ox}(I)] = +3$.

8.3 *Determination of Oxidation Numbers from Electronegativities*

Memorizing rules does not necessarily enable us to understand the concept of oxidation number. Furthermore, there are numerous polyatomic ions and molecules for which there is no obvious way of applying the "rules." Rather than mechanically apply simplistic algebraic rules, we can always deduce an oxidation number from relative electronegativities. This method is particularly useful for cases in which there are two atoms of the same element in a molecule or ion that have different chemical environments. Using the electronegativity approach, we can identify the oxidation number of each atom in its own unique environment, whereas the algebraic method simply gives an average number.

To assign oxidation numbers to covalently bonded atoms, we draw the electron-dot formula of the molecule and refer to the electronegativity values of the elements involved (Figure 8.1). Although the electrons in a polar covalent bond are unequally shared, for the purpose of assigning oxidation numbers, we assume that they are completely "owned" by the more electronegative atom. Then we compare how many outer (valence) electrons an atom "possesses" in its molecule or ion with the number it has as a free monatomic element. The difference—number of valence electrons possessed by free atom minus number of valence electrons "possessed" by molecular or ionic atom—is the oxidation number.

The hydrogen chloride molecule will serve as an example. Figure 8.1 shows that chlorine has a higher electronegativity than hydrogen, so we assign the bonding electrons to chlorine. A chlorine atom in hydrogen chloride will "have" one more electron in its outer set of electrons than a neutral chlorine atom has. Hence, we assign it an oxidation number of $7 - 8$, or -1. The hydrogen atom has "lost" its one electron; thus, it has an oxidation number of $1 - 0$, or $+1$. This assignment is illustrated as

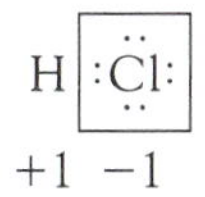

$$H\ \boxed{:\overset{..}{\underset{..}{Cl}}:}$$
$$+1\ \ -1$$

When we construct a similar electron-dot diagram for water, we see that each hydrogen atom has an oxidation number of $1 - 0$, or $+1$, and the oxygen atom, $6 - 8$, or -2, as rules 4 and 5 state.

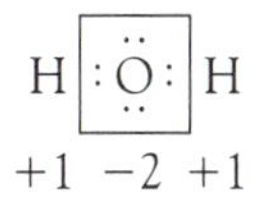

$$H\ \boxed{:\overset{..}{\underset{..}{O}}:}\ H$$
$$+1\ \ -2\ \ +1$$

Figure 8.1 Pauling electronegativity values of various nonmetals and semimetals.

H 2.2					He —
B 2.0	C 2.5	N 3.0	O 3.4	F 4.0	Ne —
	Si 1.9	P 2.2	S 2.6	Cl 3.2	Ar —
	Ge 2.0	As 2.2	Se 2.6	Br 3.0	Kr 3.0
			Te 2.1	I 2.7	Xe 2.6
				At 2.2	Rn —

In hydrogen peroxide, oxygen has an "abnormal" oxidation state. This is easy to comprehend if we realize that when pairs of atoms of the same electronegativity are bonded together, we must assume that the bonding electron pair is split between them. In this case, each oxygen atom has an oxidation number of $6 - 7$, or -1. The hydrogen atoms are still each $+1$.

$$H \; \overset{..}{\underset{..}{O}} \; \overset{..}{\underset{..}{O}} \; H$$
$$+1 \;\; -1 \;\; -1 \;\; +1$$

This method can be applied to molecules containing three (or more) different elements. Hydrogen cyanide, HCN, illustrates the process. Nitrogen is more electronegative than carbon, so it "possesses" the electrons participating in the C—N bond. And carbon is more electronegative than hydrogen, so it "possesses" the electrons of the H—C bond.

$$H \; :C \; :::N: $$
$$+1 \;\; +2 \;\;\;\; -3$$

Polyatomic ions can be treated in the same way as neutral molecules are. We can use the simple electron-dot structure of the sulfate ion to illustrate this. (However, in Chapter 16, Section 16.20, we will see that the bonding in the sulfate ion is actually more complex.) Following the same rules of assigning bonding electrons to the more electronegative atom, we assign an oxidation number of -2 for each oxygen. But sulfur has six outer electrons in the neutral atom and none in this structure. Hence, according to the rules, the sulfur atom is assigned an oxidation number of $6 - 0$, or $+6$. Note that this same oxidation number was assigned earlier to the sulfur atom in sulfuric acid.

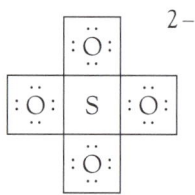

As already mentioned, two atoms of the same element can actually have different oxidation numbers within the same molecule. The classic example of this situation is the thiosulfate ion, $S_2O_3^{2-}$, which has sulfur atoms that are in different environments. Each oxygen atom has an oxidation number of -2. But, according to the rule stated earlier, the two equally electronegative sulfur atoms divide the two electrons participating in the S—S bond. Hence, the central sulfur atom has an oxidation number of $+5$ ($6 - 1$), and the other sulfur atom has an oxidation number of -1 ($6 - 7$). These assignments correlate with this ion's chemical reactions, in which the two sulfur atoms behave differently.

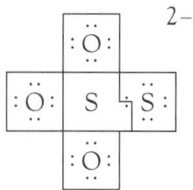

Rule 1 states that the oxidation number of any element is 0. How do we arrive at this value? In a molecule consisting of two identical atoms, such as difluorine, the electrons in the covalent bond will always be shared equally. Thus, we divide the shared electrons between the fluorine atoms. Each fluorine atom had seven valence electrons as a free atom and now it still has seven; hence, $7 - 7$ is 0.

$$:\ddot{F} | \ddot{F}:$$
$$0 \quad 0$$

8.4 The Difference Between Oxidation Number and Formal Charge

In Chapter 3, Section 3.8, we mentioned the concept of formal charge as a means of identifying feasible electron-dot structures for covalent molecules. To calculate formal charge, we divided the bonding electrons equally between the constituent atoms. The favored structures were generally those with the lowest formal charges. For example, the electron-dot diagram for carbon monoxide is shown in the following figure with the electrons allocated according to the rules for formal charge.

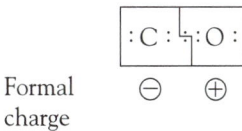

Formal charge $\quad \ominus \quad \oplus$

However, to determine oxidation numbers, which can quite often have large numerical values, we assign the bonding electrons to the atom with higher electronegativity. According to this method, the atoms in carbon monoxide are assigned electrons as shown in the following figure.

$$:C | :::O:$$

Oxidation number $\quad +2 \quad -2$

8.5 Periodic Variations of Oxidation Numbers

There are patterns to the oxidation numbers of the main group elements; in fact, they are one of the most systematic periodic trends. This pattern can be seen in Figure 8.2, which shows the oxidation numbers of the most common compounds of the first 25 main group elements (d-block elements have been omitted). The most obvious trend is the stepwise increase in the positive oxidation number as we progress from left to right across the periods. An atom's maximum positive oxidation number is equal to the number of electrons in its outer orbital set. For example, aluminum, with an electron configuration of $[Ne]3s^2 3p^1$ has an oxidation number of $+3$. Electrons in inner orbital sets do not enter into the calculation for main group elements. Hence the maximum oxidation number for bromine, which has an electron configuration of $[Ar]4s^2 3d^{10} 4p^5$, is $+7$, a value corresponding to the sum of the electrons in the $4s$ and $4p$ orbitals.

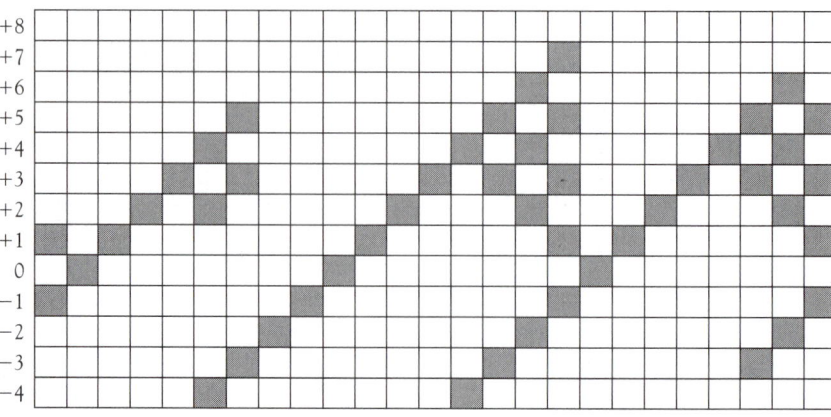

Figure 8.2 Common oxidation numbers in the compounds of the first 25 main group elements.

Many of the nonmetals and semimetals exhibit more than one oxidation number. For example, in its different compounds, nitrogen assumes every oxidation number between -3 and $+5$. The common oxidation states of nonmetals, however, tend to decrease in units of two. This pattern can be seen in the oxidation numbers chlorine has in the various oxyanions it forms (Table 8.2).

8.6 Redox Equations

In a redox reaction, one substance is oxidized and another is reduced. This process is sometimes easy to see; for example, when a rod of copper metal is placed in a silver nitrate solution, shiny crystals of silver metal are formed on the copper surface and the solution turns blue. In this case, the oxidation number of copper has increased from 0 to $+2$, and that of silver has decreased from $+1$ to 0:

$$Cu(s) + 2\,Ag^+(aq) \rightarrow Cu^{2+}(aq) + 2\,Ag(s)$$

We can think of the process as two separate half-reactions, the loss of electrons by the copper metal and the gain of electrons by the silver ions:

$$Cu(s) \rightarrow Cu^{2+}(aq) + 2\,e^-$$

$$2\,Ag^+(aq) + 2\,e^- \rightarrow 2\,Ag(s)$$

A more complex example is the reaction between hydrogen sulfide gas and a solution of iron(III) ions, which yields the products solid sulfur, iron(II) ions, and hydrogen (or hydronium) ions:

$$H_2S(g) + 2\,Fe^{3+}(aq) \rightarrow S(s) + 2\,Fe^{2+}(aq) + 2\,H^+(aq)$$

In this case, it is necessary to comb through the formulas and calculate the oxidation number of each element. It is easy to see that the iron has been reduced from $+3$ to $+2$, but it requires a quick calculation to see that the oxidation state of sulfur has changed from -2 to 0. Thus we can write the half-reactions as

$$H_2S(g) \rightarrow S(s) + 2\,H^+(aq) + 2\,e^-$$

$$2\,Fe^{3+}(aq) + 2\,e^- \rightarrow 2\,Fe^{2+}(aq)$$

Table 8.2 Oxidation number of chlorine in common oxyanions

Ion	Oxidation number
ClO^-	$+1$
ClO_2^-	$+3$
ClO_3^-	$+5$
ClO_4^-	$+7$

We have just seen that a redox equation can be divided into oxidation and reduction half-reactions. It is equally useful to perform the reverse process, that is, to construct balanced redox reactions from the oxidation and reduction components. For example, a solution of purple permanganate ion oxidizes an iron(II) ion solution to iron(III) ion in acid solution, itself being reduced to colorless manganese(II) ion. We can write the following skeletal (unbalanced) equation:

$$MnO_4^-(aq) + Fe^{2+}(aq) \rightarrow Mn^{2+}(aq) + Fe^{3+}(aq)$$

The first step is to identify the two half-reactions:

$$Fe^{2+}(aq) \rightarrow Fe^{3+}(aq)$$

$$MnO_4^-(aq) \rightarrow Mn^{2+}(aq)$$

We can balance the iron half-reaction first because it is very simple, requiring just one electron:

$$Fe^{2+}(aq) \rightarrow Fe^{3+}(aq) + e^-$$

But in the reduction equation, we have oxygen atoms on the left, but none on the right. We can remedy this by adding the appropriate number of water molecules to the side lacking oxygen atoms:

$$MnO_4^-(aq) \rightarrow Mn^{2+}(aq) + 4\,H_2O(l)$$

This addition has balanced the oxygen atoms, but it has introduced hydrogen atoms in the process. To balance these, we add hydrogen ions to the left-hand side:

$$MnO_4^-(aq) + 8\,H^+(aq) \rightarrow Mn^{2+}(aq) + 4\,H_2O(l)$$

Finally, we balance the charges by adding electrons as needed:

$$MnO_4^-(aq) + 8\,H^+(aq) + 5\,e^- \rightarrow Mn^{2+}(aq) + 4\,H_2O(l)$$

Before adding the two half-reactions, the number of electrons required for the reduction must match the number of electrons produced during the oxidation. In this case, we achieve this balance by multiplying the iron oxidation half-reaction by 5:

$$5\,Fe^{2+}(aq) \rightarrow 5\,Fe^{3+}(aq) + 5\,e^-$$

The final balanced reaction will be

$$5\,Fe^{2+}(aq) + MnO_4^-(aq) + 8\,H^+(aq) \rightarrow 5\,Fe^{3+}(aq) + Mn^{2+}(aq) + 4\,H_2O(l)$$

Now consider the disproportionation reaction of dichlorine to chloride ion and chlorate ion in basic solution:

$$Cl_2(aq) \rightarrow Cl^-(aq) + ClO_3^-(aq)$$

A disproportionation reaction occurs when some ions (or molecules) are oxidized, while others of the same species are reduced. In the case of dichlorine, some of the chlorine atoms are oxidized, changing their oxidation number from 0 to +5; the remainder are reduced, changing from 0 to −1. As before, we can construct the two half-reactions:

$$Cl_2(aq) \rightarrow Cl^-(aq)$$

$$Cl_2(aq) \rightarrow ClO_3^-(aq)$$

Choosing the reduction half-reaction first, the initial step is to balance the number of atoms of chlorine:

$$Cl_2(aq) \rightarrow 2\ Cl^-(aq)$$

Then we can balance for charge:

$$Cl_2(aq) + 2\ e^- \rightarrow 2\ Cl^-(aq)$$

For the oxidation half-reaction, the number of chlorine atoms also has to be balanced:

$$Cl_2(aq) \rightarrow 2\ ClO_3^-(aq)$$

Then, because the reaction takes place in basic solution, we add twice the number of moles of hydroxide ion on one side as there are oxygen atoms on the other:

$$Cl_2(aq) + 12\ OH^-(aq) \rightarrow 2\ ClO_3^-(aq)$$

We balance the added hydrogen with water molecules:

$$Cl_2(aq) + 12\ OH^-(aq) \rightarrow 2\ ClO_3^-(aq) + 6\ H_2O(l)$$

Next, we balance the equation with respect to charge by adding electrons:

$$Cl_2(aq) + 12\ OH^-(aq) \rightarrow 2\ ClO_3^-(aq) + 6\ H_2O(l) + 10\ e^-$$

The reduction half-reaction must be multiplied by 5 to balance the electrons:

$$5\ Cl_2(aq) + 10\ e^- \rightarrow 10\ Cl^-(aq)$$

The sum of the two half-reactions will be

$$6\ Cl_2(aq) + 12\ OH^-(aq) \rightarrow 10\ Cl^-(aq) + 2\ ClO_3^-(aq) + 6\ H_2O(l)$$

Finally, the coefficients of the equation can be divided through by 2:

$$3\ Cl_2(aq) + 6\ OH^-(aq) \rightarrow 5\ Cl^-(aq) + ClO_3^-(aq) + 3\ H_2O(l)$$

8.7 *Quantitative Aspects of Half-Reactions*

The relative oxidizing or reducing power of a half-reaction can be determined from the half-cell potential, which is the potential of the half-reaction relative to the potential of a half-reaction in which hydrogen ion ($1\ mol \cdot L^{-1}$) is reduced to hydrogen gas (100 kPa pressure on a black platinum surface). This reference half-reaction is assigned a standard potential, $E°$, of zero:

$$2\ H^+(aq) + 2\ e^- \rightarrow H_2(g) \qquad\qquad E° = 0.00\ V$$

For a redox reaction to be spontaneous, the sum of its half-cell reduction potentials must be positive. For example, consider the reaction of copper metal with silver ion, which we discussed earlier. The values of the half-cell reduction potentials are

$$Cu^{2+}(aq) + 2\ e^- \rightarrow Cu(s) \qquad\qquad E° = +0.34\ V$$

$$Ag^+(aq) + e^- \rightarrow Ag(s) \qquad\qquad E° = +0.80\ V$$

Adding the silver ion reduction potential to the copper metal oxidation potential

$$2\,Ag^+(aq) + 2\,e^- \rightarrow 2\,Ag(s) \qquad\qquad E° = +0.80\ V$$

$$Cu(s) \rightarrow Cu^{2+}(aq) + 2\,e^- \qquad\qquad E° = -0.34\ V$$

gives a positive cell potential:

$$2\,Ag^+(aq) + Cu(s) \rightarrow 2\,Ag(s) + Cu^{2+}(aq) \qquad E° = +0.46\ V$$

The more positive the half-cell reduction potential, the stronger the oxidizing power of the species. For example, difluorine is an extremely strong oxidizing agent (or electron acceptor):

$$\tfrac{1}{2}\,F_2(g) + e^- \rightarrow F^-(aq) \qquad\qquad E° = +2.80\ V$$

Conversely, the lithium ion has a very negative reduction potential:

$$Li^+(aq) + e^- \rightarrow Li(s) \qquad\qquad E° = -3.04\ V$$

For lithium, the reverse half-reaction results in a positive potential; hence lithium metal is a very strong reducing agent (or electron provider):

$$Li(s) \rightarrow Li^+(aq) + e^- \qquad\qquad E° = +3.04\ V$$

However, it must always be kept in mind that half-cell potentials are concentration dependent. Thus, it is possible for a reaction to be spontaneous under certain conditions but not under others. The variation of potential with concentration is given by the Nernst equation:

$$E = E° - \frac{RT}{nF}\ln\frac{[products]}{[reactants]}$$

where R is the ideal gas constant ($8.31\ V{\cdot}C{\cdot}mol^{-1}{\cdot}K^{-1}$), T is the temperature in kelvins, n is the moles of transferred electrons according to the redox equation, F is the Faraday constant ($9.65 \times 10^4\ C{\cdot}mol^{-1}$), and $E°$ is the potential under standard conditions of 1 $mol{\cdot}L^{-1}$ for species in solution and 100 kPa pressure for gases.

To see the effects of nonstandard conditions, consider the permanganate ion to manganese(II) ion half-cell. This half-cell is represented by the half-reaction we balanced earlier:

$$MnO_4^-(aq) + 8\,H^+(aq) + 5\,e^- \rightarrow Mn^{2+}(aq) + 4\,H_2O(l) \qquad E° = +1.70\ V$$

The corresponding Nernst equation will be

$$E = +1.70\ V - \frac{RT}{5F}\ln\frac{[Mn^{2+}]}{[MnO_4^-][H^+]^8}$$

Suppose the pH is increased to 4.00 (that is, $[H^+]$ is reduced to 1.0×10^{-4} $mol{\cdot}L^{-1}$), but the concentrations of permanganate ion and manganese(II) ion are kept at 1.0 $mol{\cdot}L^{-1}$. Under the new conditions (first, solving for $RT/5\,F$), the half-cell potential becomes

$$E = +1.70\ V - (5.13 \times 10^{-3}\ V)\ln\frac{(1.00)}{(1.00)(1.0 \times 10^{-4})^8}$$

$$E = +1.70\ V - (5.13 \times 10^{-3}\ V)\ln(1.0 \times 10^{32})$$

$$E = +1.70\ V - 0.38\ V = +1.32\ V$$

The strongest oxidizing agent is oxygen difluoride ($E° = +3.29$ V in acid solution). The strongest reducing agent is the azide ion, N_3^- ($E° = -3.33$ V in acid solution)

Thus, permanganate ion is a significantly weaker oxidizing agent in less acid solutions. Notice, however, that the effect is substantial only because in the Nernst equation the hydrogen ion concentration is raised to the eighth power; as a result, the potential is exceptionally sensitive to pH. Fortunately, for the qualitative comparisons of half-cell potentials used in this text, standard values will suffice.

8.8   Electrode Potentials as Thermodynamic Functions

As we have just seen in the equation for the silver ion–copper metal reaction, electrode potentials are not altered when coefficients of equations are changed. The potential is the force driving the reaction, and it is localized either at the surface of an electrode or at a point where two chemical species come in contact. Hence, the potential does not depend on stoichiometry. Potentials are simply a measure of the free energy of the process. The relationship between free energy and potential is

$$\Delta G^\circ = -nFE^\circ$$

where ΔG° is the standard free energy change, n is the moles of electrons, F is the Faraday constant, and E° is the standard electrode potential. The Faraday constant is usually expressed as 9.65×10^4 C·mol^{-1}, but for use in this particular formula, it is best written in units of joules: 9.65×10^4 J·V^{-1}·mol^{-1}. For the calculations in this section, however, it is even more convenient to express the free energy change as the product of moles of electrons and half-cell potentials. In this way, we do not need to evaluate the Faraday constant.

To illustrate this point, let us repeat the previous calculation of the copper–silver reaction, using free energies instead of just standard potentials:

$$2\,Ag^+(aq) + 2\,e^- \rightarrow 2\,Ag(s) \qquad \Delta G^\circ = -2(F)(+0.80) = -1.60F$$

$$Cu(s) \rightarrow Cu^{2+}(aq) + 2\,e^- \qquad \Delta G^\circ = -2(F)(-0.34) = +0.68F$$

The free energy change for the process, then, is $(-1.60F + 0.68F)$, or $-0.92F$. Converting this value back to a standard potential gives

$$E^\circ = -\Delta G^\circ/nF = -(-0.92F)/2F = +0.46 \text{ V}$$

or the same value obtained by simply adding the standard potentials.

But suppose we want to combine two half-cell potentials to derive the value for an unlisted half-cell potential; then the shortcut of using standard electrode potentials does not work. Note that we are adding half-reactions to get another half-reaction, not a balanced redox reaction. The number of electrons in the two reduction half-reactions will not balance. Consequently, we must work with free energies. As an example, we can determine the half-cell potential for the reduction of iron(III) ion to iron metal,

$$Fe^{3+}(aq) + 3\,e^- \rightarrow Fe(s)$$

given the values for the reduction of iron(III) ion to iron(II) ion and from iron(II) ion to iron metal:

$$Fe^{3+}(aq) + e^- \rightarrow Fe^{2+}(aq) \qquad\qquad E^\circ = +0.77 \text{ V}$$

$$Fe^{2+}(aq) + 2\,e^- \rightarrow Fe(s) \qquad\qquad E^\circ = -0.44 \text{ V}$$

First, we calculate the free energy change for each half-reaction:

$$Fe^{3+}(aq) + e^- \rightarrow Fe^{2+}(aq) \qquad \Delta G° = -1(F)(+0.77) = -0.77F$$

$$Fe^{2+}(aq) + 2\,e^- \rightarrow Fe(s) \qquad \Delta G° = -2(F)(-0.44) = +0.88F$$

Adding the two equations results in the "cancellation" of the Fe^{2+} species. Hence, the free energy change for

$$Fe^{3+}(aq) + 3\,e^- \rightarrow Fe(s)$$

will be $(-0.77F + 0.88F)$, or $+0.11F$. Converting this $\Delta G°$ value back to potential for the reduction of iron(III) to iron metal gives

$$E° = -\Delta G°/nF = -(+0.11F)/3F = -0.04 \text{ V}$$

8.9 *Latimer (Reduction Potential) Diagrams*

It is easier to interpret data when they are displayed in the form of a diagram. The standard reduction potentials for a related set of species can be displayed in a reduction potential diagram, or what is sometimes called a *Latimer diagram*. The various iron oxidation states in acid solution are shown here in such a diagram.

$$\overset{6+}{FeO_4{}^{2-}} \xrightarrow{+2.20 \text{ V}} \overset{3+}{Fe^{3+}} \xrightarrow{+0.77 \text{ V}} \overset{2+}{Fe^{2+}} \xrightarrow{-0.44 \text{ V}} \overset{0}{Fe}$$

$$Fe^{3+} \xrightarrow{-0.04 \text{ V}} Fe$$

The diagram includes the three common oxidation states of iron ($+3$, $+2$, 0) and the uncommon oxidation state of $+6$. The number between each pair of species is the standard reduction potential for the reduction half-reaction involving those species. Notice that although the species are indicated, to use the information we have to write the corresponding complete half-reaction. For the simple ions, writing the half-reaction is very easy. For example, for the reduction of iron(III) ion to iron(II) ion, we simply write

$$Fe^{3+}(aq) + e^- \rightarrow Fe^{2+}(aq) \qquad\qquad E° = +0.77 \text{ V}$$

However, for the reduction of ferrate ion, $FeO_4{}^{2-}$, we have to balance the oxygen with water, then the hydrogen in the added water with hydrogen ion, and finally the charge with electrons:

$$FeO_4{}^{2-}(aq) + 8\,H^+(aq) + 3\,e^- \rightarrow Fe^{3+}(aq) + 4\,H_2O(l) \qquad E° = +2.20 \text{ V}$$

Latimer diagrams display the redox information about a series of oxidation states in a very compact form. More than that, they enable us to predict the redox behavior of the species. For example, the high positive value between the ferrate ion and the iron(III) ion indicates that the ferrate ion is a strong oxidizing agent (that is, it is very easily reduced). A negative number indicates that the species to the right is a reducing agent. In fact, iron metal can be used as a reducing agent, itself being oxidized to the iron(II) ion.

Let us look at another example of a reduction potential diagram, that for oxygen in acid solution.

$$\overset{0}{O_2} \xrightarrow{+0.68 \text{ V}} \overset{-1}{H_2O_2} \xrightarrow{+1.78 \text{ V}} \overset{-2}{H_2O}$$

$$O_2 \xrightarrow{+1.23 \text{ V}} H_2O$$

With a reduction potential of $+1.78$ V, hydrogen peroxide is a strong oxidizing agent with respect to water. For example, hydrogen peroxide will oxidize iron(II) ion to iron(III) ion:

$$H_2O_2(aq) + 2\ H^+(aq) + 2\ e^- \rightarrow 2\ H_2O(l) \qquad\qquad E° = +1.78\ \text{V}$$

$$Fe^{2+}(aq) \rightarrow Fe^{3+}(aq) + e^- \qquad\qquad E° = -0.77\ \text{V}$$

The diagram tells us something else about hydrogen peroxide. The sum of the potentials for the reduction and oxidation of hydrogen peroxide is positive ($+1.78$ V $- 0.68$ V). This value indicates that hydrogen peroxide will disproportionate:

$$H_2O_2(aq) + 2\ H^+(aq) + 2\ e^- \rightarrow 2\ H_2O(l) \qquad\qquad E° = +1.78\ \text{V}$$

$$H_2O_2(aq) \rightarrow O_2(g) + 2\ H^+(aq) + 2\ e^- \qquad\qquad E° = -0.68\ \text{V}$$

Summing the two half-equations gives the overall equation:

$$2\ H_2O_2(aq) \rightarrow 2\ H_2O(l) + O_2(g) \qquad\qquad E° = +1.10\ \text{V}$$

Even though the disproportionation is thermodynamically spontaneous, it is kinetically very slow. The decomposition happens rapidly, however, in the presence of a catalyst such as iodide ion or many transition metal ions. Our bodies contain the enzyme, *catalase,* to catalyse this reaction, and thus destroy harmful hydrogen peroxide in our cells.

In all the examples used so far in this section, the reactions occur in acid solution. The values are sometimes quite different in basic solution, because of the presence of different chemical species at high pH. For example, as the diagram at the beginning of this section shows, iron metal is oxidized in acid solution to the soluble iron(II) cation:

$$Fe(s) \rightarrow Fe^{2+}(aq) + 2\ e^-$$

However, in basic solution, the iron(II) ion immediately reacts with the hydroxide ion present in high concentration to give insoluble iron(II) hydroxide:

$$Fe(s) + 2\ OH^-(aq) \rightarrow Fe(OH)_2(s) + 2\ e^-$$

Thus, the Latimer diagram for iron in basic solution (shown in the following) contains several different species from the diagram under acid conditions, and, as a result, the potentials are different.

$$\overset{6+}{FeO_4^{2-}} \xrightarrow{+0.9\ \text{V}} \overset{3+}{Fe(OH)_3} \xrightarrow{-0.56\ \text{V}} \overset{2+}{Fe(OH)_2} \xrightarrow{-0.89\ \text{V}} \overset{0}{Fe}$$

We see that in basic solution iron(II) hydroxide is easily oxidized to iron(III) hydroxide ($+0.56$ V), and the ferrate ion is now a very weak oxidizing agent ($+0.9$ V in basic solution, 2.20 V in acid solution).

Although Latimer diagrams are useful for identifying reduction potentials for specific redox steps, they can become very complex. For example, a diagram for the five species of manganese has 10 potentials relating the various pairs of the 5 species. It is tedious to sort out the information that is stored in such a complex diagram. For this reason, it is more useful to display the oxidation states and their comparative energies as a two-dimensional graph. This is the topic of the next section.

8.10 Frost (Oxidation State) Diagrams

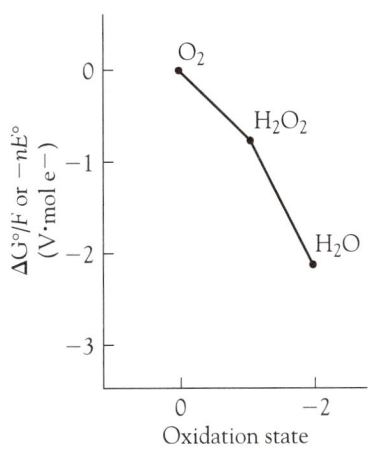

Figure 8.3 Frost diagram for oxygen in acid solution.

It is preferable to display the information about the numerous oxidation states of an element as an oxidation state diagram, or a *Frost diagram,* as it is sometimes called. Such a diagram enables us to extract information about the properties of different oxidation states visually, without the need for calculations. A Frost diagram shows the relative free energy (rather than potential) on the vertical axis and the oxidation state on the horizontal axis. Note that we denote energy as $-nE°$; thus, energy values are usually plotted in units of volts times moles of electrons for that redox step (V·mol e⁻). We obtain the same value by dividing the free energy by the Faraday constant, $\Delta G°/F$. For consistency, the element in oxidation state 0 is considered to have zero free energy. Lines connect species of adjacent oxidation states.

From the Latimer diagram for oxygen shown in the previous section, we can construct a Frost diagram for oxygen species in acid solution (Figure 8.3):

$$O_2 \xrightarrow{+0.68 \text{ V}} H_2O_2 \xrightarrow{+1.78 \text{ V}} H_2O$$
$$\text{(+1.23 V)}$$

The first point will simply be 0, 0 for dioxygen because its free energy is taken to be 0 when its oxidation state is 0. The point for hydrogen peroxide will then be -1, -0.68 because the oxidation state for oxygen in hydrogen peroxide is -1 and its free energy is -1 times the product of the moles of electrons (1) and the half-cell reduction potential (+0.68 V). Finally, the point for water will be at -2, -2.46 because the oxygen has an oxidation state of -2 and the free energy of the oxygen in water will be $-(1 \times 1.78)$ units below the hydrogen peroxide point. This diagram enables us to obtain a visual image of the redox chemistry of oxygen in acid solution. Water, at the lowest point on the plot, must be the most thermodynamically stable. Hydrogen peroxide, on a convex curve, will disproportionate.

All the features of a Frost diagram can be appreciated by studying the redox chemistry of manganese (Figure 8.4). From this diagram, we can draw the following conclusions:

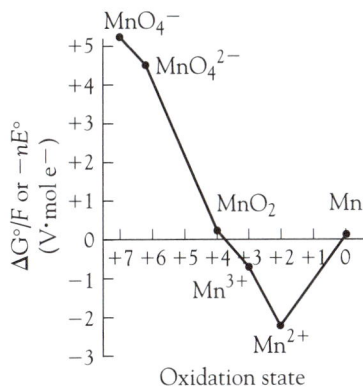

Figure 8.4 Frost diagram for manganese in acid solution.

1. More thermodynamically stable states will be found lower in the diagram. Thus, manganese(II) is the most stable (from a redox perspective) of all the manganese species.

2. A species on a convex curve [such as the manganate ion, MnO_4^{2-}, and the manganese(III) ion] will tend to disproportionate.

3. A species on a concave curve [such as manganese(IV) oxide, MnO_2] will not disproportionate.

4. A species that is high and on the left of the plot [such as the permanganate ion, MnO_4^-] will be strongly oxidizing.

5. A species that is high and on the right of the plot will be strongly reducing. Thus, manganese metal is moderately reducing.

However, interpretation of Frost diagrams has caveats. First, the diagram represents the comparative free energy for standard conditions, that is, a solution of concentration 1 mol·L^{-1} at pH 0 (a hydrogen ion concentration of 1 mol·L^{-1}). If the conditions are changed, then the energy will be different, and the relative stabilities might be different.

As pH changes, the potential of any half-reaction that includes the hydrogen ion also changes. But, even more important, often the actual species involved will change. For example, the aqueous manganese(II) ion does not exist at high pH values. Under these conditions, insoluble manganese(II) hydroxide, $Mn(OH)_2$, is formed. It is this compound, not Mn^{2+}, that appears on the Frost diagram for manganese(II) in basic solution.

Finally, we must emphasize that the Frost diagrams are thermodynamic functions and do not contain information about the rate of decomposition of a thermodynamically unstable species. Potassium permanganate, $KMnO_4$, is a good example. Even though the reduction of permanganate ion to a more stable lower oxidation state of manganese(II) ion is favored, it is kinetically slow (except in the presence of a catalyst). Thus, we can still work with permanganate ion solutions.

8.11   Pourbaix Diagrams

In the last section, we saw how a Frost diagram could be used to compare the thermodynamic stabilities of different oxidation states of an element. Frost diagrams can be constructed for both acid (pH = 0) and basic (pH = 14) conditions. It would be useful to be able to identify the thermodynamically stable species at any particular permutation of half-cell potential, E, and pH. A French chemist, M. Pourbaix, devised such a plot; hence, they are usually named *Pourbaix diagrams* after him, although they are also called $E° - pH$ diagrams and predominance-area diagrams.

Figure 8.5 shows a Pourbaix diagram for the manganese system. The more oxidized species, such as permanganate, are found towards the positive-potential upper part of the diagram, while the more reduced species, such as manganese metal, are found toward the negative-potential lower portion of the plot. Similarly, the more basic species are found to the right (high pH) and the more acidic species (low pH) to the left. A vertical divide, such as that between the manganese(II) ion and manganese(II) hydroxide indicates an equilibrium that is dependent solely on pH and not on a redox process:

$$Mn^{2+}(aq) + 2 \text{ OH}^-(aq) \rightleftharpoons Mn(OH)_2(s) \qquad K_{sp} = 2.0 \times 10^{-13}$$

Hence, when manganese(II) is in its standard concentration of 1 mol·L^{-1}, as $K_{sp} = [Mn^{2+}][OH^-]^2$, then $[OH^-] = \sqrt{(2.0 \times 10^{-13})} = 4.4 \times 10^{-7}$, and pH = 7.65. Thus at a pH greater than this, the hydroxide is the preferred form of manganese(II).

Conversely, a horizontal line will represent a pure redox transformation. An example of this is found between manganese metal and the manganese(II) ion:

$$Mn^{2+}(aq) + 2 \text{ e}^- \rightarrow Mn(s) \qquad E° = -1.18 \text{ V}$$

Most boundaries lie between these extremes as they are both pH and potential dependent. For example, the reduction of manganese(IV) oxide to manganese(II) ion is represented as

$$MnO_2(s) + 4 \text{ H}^+(aq) + 2 \text{ e}^- \rightarrow Mn^{2+}(aq) + 2 \text{ H}_2O(l) \quad E° = +1.23 \text{ V}$$

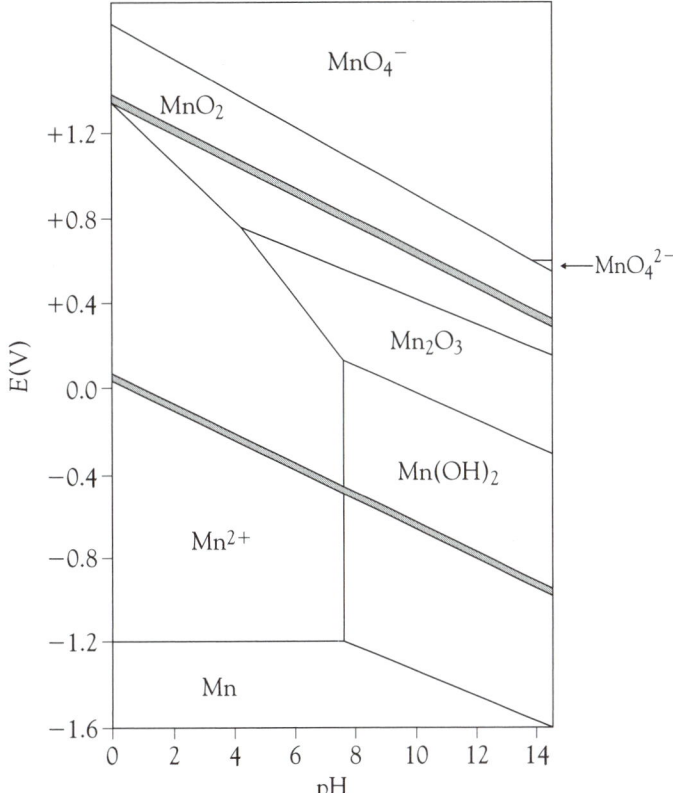

Figure 8.5 Pourbaix diagram showing the thermodynamically stable manganese species as a function of standard potential, $E°$, and pH.
(Adapted from J.E. Huheey, E.A. Keiter, and R.L. Keiter, Inorganic Chemistry, 4th ed. [New York: HarperCollins, 1993], p. 592.)

The Nernst expression can then be used to plot the boundary between the two states:

$$E = E° - \frac{RT}{nF} \ln \frac{[\text{Mn}^{2+}]}{[\text{H}^+]^4}$$

Inserting the values of $E°$, R, T, and F, setting $[\text{Mn}^{2+}]$ as 1 mol·L^{-1}, and converting ln to $\log_{10}$ (by multiplying by 2.303), gives

$$E = 1.23 \text{ V} - 0.118 \text{ pH}$$

Substituting different values of pH, we can calculate the corresponding value of E and construct the boundary line on the Pourbaix diagram.

The diagram also shows two shaded lines. The upper line represents the oxidation of water:

$$\tfrac{1}{2}\,\text{O}_2(g) + 2\,\text{H}^+(aq) + 2\,\text{e}^- \rightarrow \text{H}_2\text{O}(l) \qquad\qquad E° = +1.23 \text{ V}$$

while the lower dashed line represents the reduction of water to hydrogen gas:

$$\text{H}_2\text{O}(l) + \text{e}^- \rightarrow \tfrac{1}{2}\,\text{H}_2(g) + \text{OH}^-(aq) \qquad\qquad E° = -0.83 \text{ V}$$

which under conditions of 1 mol·L^{-1} H$^+$ is represented as

$$\text{H}^+(aq) + \text{e}^- \rightarrow \tfrac{1}{2}\,\text{H}_2(g) \qquad\qquad E° = 0.00 \text{ V}$$

These two shaded lines represent the boundaries within which reactions in aqueous solution are possible. A higher potential and the water will start oxidizing; a lower potential and the water will start reducing. So we can see

Technetium: The Most Important Radiopharmaceutical

There are several different ways of imaging the inside of a human body, for example, magnetic resonance imaging, MRI. However, it is particularly useful to be able to highlight the particular location of interest, such as a specific organ or a tumor. Metal compounds play a valuable role as they often concentrate in specific organs or in tumors. The reason why they concentrate in tumors is not well understood but it probably relates to the increased metabolism and altered chemistry of the tumor cells.

To be an effective isotope for diagnostic imaging, the nucleus must be a gamma emitter with a half-life long enough that the isotope can be produced and inserted into the patient's body before much has decayed. However, the half-life must be short enough that the emission intensity is measurable at very low concentrations. A short half-life also means that the patient is exposed to the radiation for only a brief period. A half-life of less than 8 days is preferable. Technetium-99 fits this role superbly and it is used for over 80 percent of all radiodiagnostics.

The technetium is obtained as the pertechnate ion, TcO_4^-, from the radioactive decay of the molybdate ion, MoO_4^{2-}. The synthesis is as follows. Nonradioactive molybdenum-98 is placed in a neutron source to give radioactive molybdenum-99:

$$^{98}_{42}Mo + ^{1}_{0}n \rightarrow ^{99}_{42}Mo$$

This decays with a half-life of 66 hr to technetium-99. The technetium-99 is in a nuclear excited state. That is, just as electrons can be in excited states and release visible, ultraviolet, or infrared electromagnetic radiation as the electron descends to the ground state, so the proton formed in the nuclear decay is in an excited state and emits a gamma ray as it drops to the nuclear ground state. It is the emission of the gamma ray that is recorded in the radiodiagnostic procedure.

$$^{99}_{42}Mo \rightarrow ^{99m}_{43}Tc + ^{0}_{-1}e \; (t_{1/2} = 66 \text{ hr})$$

$$^{99m}_{43}Tc \rightarrow ^{99}_{43}Tc + ^{0}_{0}\gamma \; (t_{1/2} = 6 \text{ hr})$$

The ground-state technetium-99 has such a long half-life that the radiation level is so low to represent a negligible hazard.

$$^{99}_{43}Tc \rightarrow ^{99}_{44}Ru + ^{0}_{-1}e \; (t_{1/2} = 2.1 \times 10^5 \text{ years})$$

Though the nuclear chemistry is of great importance, the aqueous chemistry is also crucial. The technetium must be in forms that are soluble in the body fluids. These fluids are typically near neutral and slightly oxidizing. As can be seen from Figure 8.6, the Pourbaix diagram shows that the pertechnate ion is stable over much of the physiologically relevant range. This is unlike the element above it in the Periodic Table, manganese (see Figure 8.5), where most of the physiological range is occupied by insoluble oxides and hydroxides and the permanganate ion is only stable under highly oxidizing conditions.

The pertechnate ion itself is about the same size (and charge density) as the iodide ion. Thus, the pertechnate ion has been used for the imaging of the iodine-rich thyroid gland (see Chapter 17). However, it is complex compounds of technetium that have been made to be specific for other organs and for tumors.

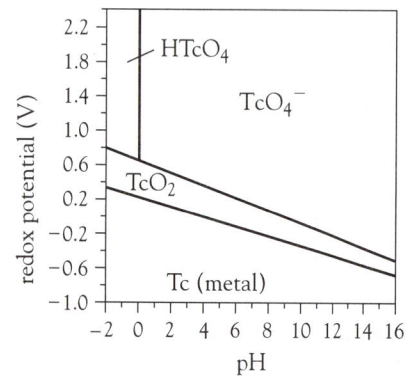

Figure 8.6 Pourbaix diagram for technetium.
(From W. Kaim and B. Schwederski, Bioinorganic Chemistry: Inorganic Elements in the Chemistry of Life, Wiley VCH, ch 18, p. 360.)

that the permanganate ion lies outside the limit of aqueous solutions. Yet permanganate ion can exist in aqueous solution. Even though permanganate ion is thermodynamically unstable in aqueous solution, there is a high activation energy barrier, providing kinetic stability. However, permanganate solutions are not stable over the long term, and they can decompose very rapidly in the presence of a catalyst species.

The manganate ion, MnO_4^{2-}, occupies a tiny niche at very high pH and outside of the water limits. Thus, to synthesize this ion, we resort to the oxidation of manganese(IV) oxide in molten potassium hydroxide.

$$MnO_2(s) + 4\ OH^-(KOH) \rightarrow MnO_4^{2-}(KOH) + 2\ H_2O(g) + 2\ e^-$$

It is easier to identify the major aqueous species under different pH and E conditions from a Pourbaix diagram, but the study of the relative stabilities of different oxidation states is best accomplished from a Frost diagram. It is important to realize that Pourbaix diagrams display only the common thermodynamically preferred species. Sometimes species are left off the diagram for simplicity. For example, Figure 8.5 does not include the mixed manganese(II) manganese(III) oxide, Mn_3O_4. Other species do not come in the range of the diagram. Thus, the aqueous manganese(III) ion only becomes the thermodynamically stable species when $[H^+]$ is about 10 mol·L^{-1} and the potential is about +1.5 V.

8.12   *Ellingham Diagrams and the Extraction of Metals*

Frost diagrams are very useful for the study of reactions in aqueous solution. However, one of the most important types of redox reactions is usually performed in the solid, liquid, and gaseous phases—the reduction of metal compounds to the pure metal.

For most metallic elements, the oxides are more thermodynamically stable than the metals themselves over the range of normal working temperatures. For example, zinc metal will spontaneously (although slowly) oxidize to zinc oxide at room temperature:

$$2\ Zn(s) + O_2(g) \rightarrow 2\ ZnO(s) \qquad \Delta G°(298\ K) = -636\ kJ·mol^{-1}$$

However, recalling the formula given in Chapter 6,

$$\Delta G° = \Delta H° - T\Delta S°$$

we can identify what factors actually lead to the spontaneity of the reaction. Because the number of moles of gas changes from 1 on the left to 0 on the right, the entropy change of this reaction must be negative. Hence, the driving force for this reaction has to be the enthalpy factor. And, in fact, when we check the thermodynamic tables, we see that the enthalpy of formation of zinc oxide is very negative. That is, the oxidation is highly exothermic.

The entropy term, $T\Delta S°$, involves the Kelvin temperature, so increasing the temperature will cause $\Delta G°$ to become less and less negative. (The actual values of both $\Delta H°$ and $\Delta S°$ change slightly with temperature as well, but we will ignore this complication here.) Finally, at a sufficiently high temperature, $\Delta G°$ will become 0; and above that temperature, it will have a positive value. In other words, the converse process, the reduction of zinc oxide to zinc metal, will become spontaneous. Figure 8.7, a plot of free energy per mole of dioxygen against Celsius temperature, displays this information.

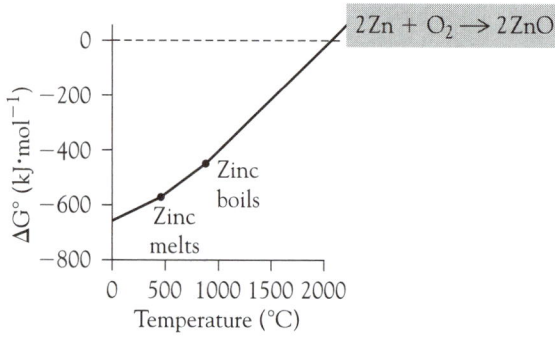

Figure 8.7 Free energy of oxidation of zinc as a function of temperature.

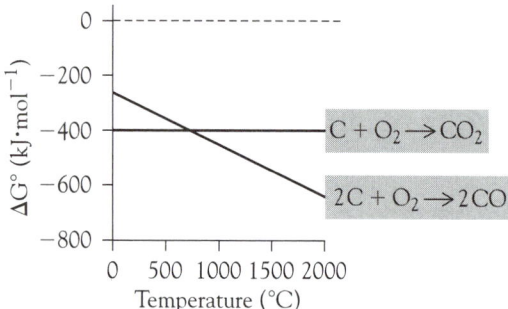

Figure 8.8 Free energy of oxidation of carbon as a function of temperature.

Figure 8.9 Superimposed plots of free energy of oxidation of carbon and zinc as a function of temperature.

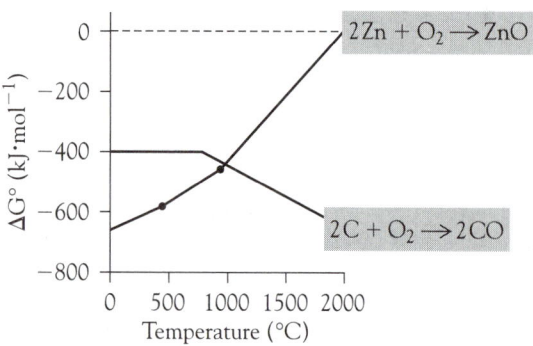

The slope of the curve steepens slightly above the melting point of zinc (lower dot on the line) and even more above the boiling point of zinc (upper dot on the line). Above the melting point of zinc, 2 mol of solid zinc oxide is being produced from 1 mol of gas and 2 mol of liquid; above the boiling point of zinc, 2 mol of solid zinc oxide form from 3 mol of gas (one of dioxygen and two of zinc). Hence, the entropy decrease at these temperatures and above will be greater. Even then, the temperature will have to be extremely high (about 2000°C) before the $T\Delta S°$ term exceeds the $\Delta H°$ term. In fact, the requisite temperature is so high that this reaction does not represent a realistic means of obtaining zinc metal from the oxide. Furthermore, the dioxygen and gaseous zinc would have to be separated before cooling, or the reverse reaction would occur, and we would end up with zinc oxide again.

There is a way around the problem. It is possible to couple the reduction reaction, which has a positive free energy, with an oxidation reaction that has a greater negative free energy to give a net negative free energy value for the combined reaction. The most useful oxidation reaction with a negative free energy is that of carbon. The coupled reaction has the right thermodynamic characteristics, as we will see, and carbon is a very inexpensive industrial reagent. The temperature dependence of this reaction is illustrated in Figure 8.8.

Up to 710°C, the oxidation of carbon to carbon dioxide is thermodynamically preferred:

$$C(s) + O_2(g) \rightarrow CO_2(g)$$

This slope of the line for the free energy change of the reaction is very close to zero because there is 1 mol of gas on each side of the equation. The line representing the free energy change during the oxidation to produce carbon monoxide, however, has a steep negative slope because the reaction produces 2 mol of gas for every mole consumed:

$$2\,C(s) + O_2(g) \rightarrow 2\,CO(g)$$

Thus, the production of carbon monoxide becomes thermodynamically preferred above 710°C, and, because both reactions are kinetically fast, the latter is the actual reaction observed above this temperature.

Figure 8.9 shows the two plots of Figures 8.7 and 8.8 superimposed. We can see that the lines cross at about 900°C. At this temperature, the oxidation of carbon becomes more negative than the reduction of zinc oxide is positive. Thus, the oxidation of carbon can cause the reduction of zinc oxide above this temperature:

$$ZnO(s) + C(s) \rightarrow Zn(g) + CO(g) \qquad T > 1000°C$$

Note that all the thermodynamic calculations relate to standard conditions of pressure. In an industrial smelter, conditions are far from this; consequently, the temperature that we have calculated is only an approximate guide to the actual minimum temperature for the reduction process.

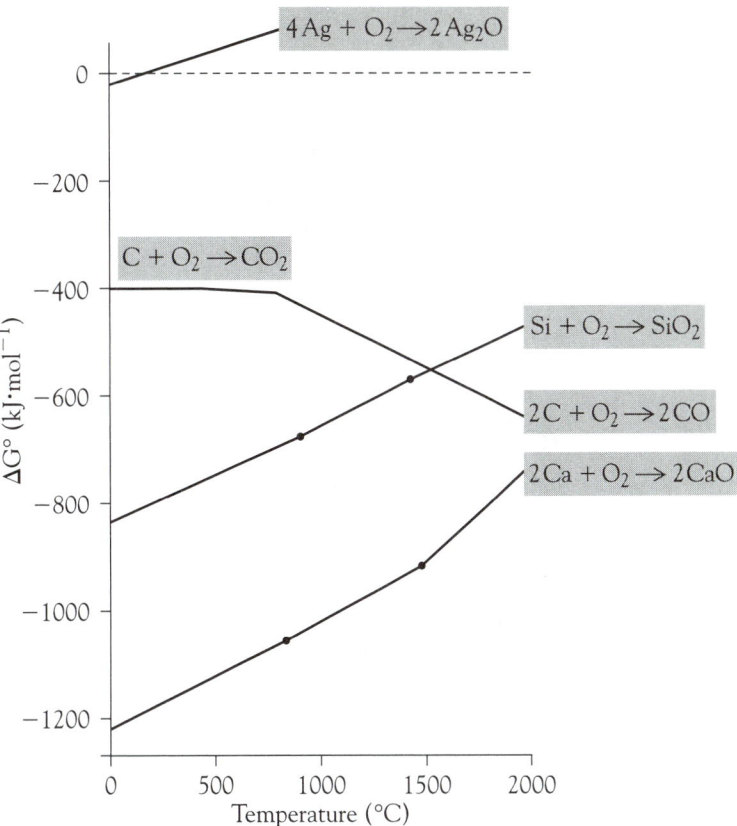

Figure 8.10 Ellingham diagram for silver, carbon, silicon, and calcium. Points indicate melting and boiling points of elements.

It was the chemist H.G.T. Ellingham who first recognized how useful plots of free energy as a function of temperature were for investigating the conditions under which redox reactions were feasible. As a result, these plots are usually referred to as *Ellingham diagrams*. Figure 8.10 shows an Ellingham diagram for oxides of calcium, carbon, silicon, and silver.

We can see that

1. After the silver plot crosses the $\Delta G° = 0$ line, the formation of silver(I) oxide is no longer spontaneous (about 300°C). Above this temperature, the reverse reaction, the decomposition of silver(I) oxide to silver metal is spontaneous ($\Delta G°$ negative).

2. The silicon plot crosses that of carbon at about 1500°C. Above this temperature, the free energy of formation of silicon dioxide is less than that of carbon monoxide. Hence, the sum of the free energy of decomposition of silicon dioxide coupled with the free energy of formation of carbon monoxide will result in a net negative value. In other words, silicon dioxide can be reduced to silicon by using carbon as a reductant above this temperature. (Below this temperature, the reverse reaction pair is spontaneous.)

3. The calcium plot does not cross the carbon plot at temperatures that are feasible in conventional smelters. Hence, thermochemical methods are not practical for extracting calcium metal. In fact, an electrolytic process is used to produce most calcium.

Green Chemistry

Traditionally, most metals, such as iron, copper, and nickel, have been produced by smelting. This is the high-temperature reduction of a metal ore, usually the oxide or sulfide, with carbon monoxide. There are many disadvantages of smelting. First, the carbon monoxide is produced from the partial oxidation of coke, a shiny brittle form of impure carbon. Coke is obtained by heating coal in the absence of air, breaking down the complex structure of the coal and releasing a mixture of aromatic hydrocarbons and nitrogen-containing heterocyclic compounds. This mixture condenses to a carcinogenic black liquid. Though there is a use for some components, traditionally much of the tarry waste was dumped. The classic example is the Tar Ponds of Sydney, Nova Scotia, Canada. The nearby steel works produced coke from coal over an 80-year period and the 700,000 tonnes of accumulated condensed organic compounds now represents North America's largest hazardous waste site. Traditional smelting also produces large volumes of gases, mostly carbon dioxide, but in the smelting of sulfides, sulfur dioxide is a major product.

Green chemistry is making a major difference. This involves the use of different branches of chemistry to make chemical industry eco-friendly and energy-efficient. Green chemists search for more environmentally benign methods of metal extraction, such as the hydrometallurgical process that we will discuss in the context of copper chemistry (Chapter 20, Section 20.9). In fact, green chemistry is actually a new industrial chemistry philosophy. The research of green chemists will enable chemical industry to:

1. Make existing chemical products in a way that minimizes the production of waste chemicals. This may be through the development of a completely new route of synthesis that has a higher yield or requires less extreme synthetic conditions. Alternatively, a new catalyst may be developed that favors the desired product, giving a higher yield. Most of these novel catalysts are newly discovered inorganic or organometallic compounds.

2. Minimize energy consumption. It would be most energy-efficient if the process could be performed at room temperature. Again, inorganic catalysts are one route, another being the use of enzyme or bacteriological processes.

3. Develop new chemical products, such as refrigerants (Chapter 14, Section 14.13) that serve the same purpose as existing ones, yet are less toxic and less harmful to the environment. These products should degrade into harmless compounds when they are released into the environment.

4. Use reagents that come from renewable resources rather than from nonrenewable ones.

5. Find uses for the by-products from existing processes.

In this text we will see several examples of green chemistry in action, from the development of novel solvents to the use of bacteria to extract metals from their ores.

There are Ellingham diagrams for the reduction of most oxides, sulfides, and chlorides. As a result, the feasibility of smelting processes can be identified simply by looking at the appropriate Ellingham plot rather than by laboratory testing; this alternative represents a significant savings in time and money.

8.13 Biological Aspects

Many biological processes, for example, photosynthesis and respiration, involve oxidation and reduction. Some plants, such as peas and beans, are able to use bacteria-filled nodules on their roots to convert the dinitrogen in the air to the ammonium ion that the plants require. This complex process, known as nitrogen fixation, involves the reduction of nitrogen from an oxidation state of 0 to a -3 oxidation state.

In all biological systems, we have to consider both the potential, E, and the acidity, pH, simultaneously when trying to decide what species of an element might be present (and we have to consider kinetic factors as well). Thus, Pourbaix diagrams have a particular importance for bioinorganic chemistry and inorganic geochemistry. Figure 8.11 shows the limits of pH and E that we find in natural waters. The upper dashed line, representing water in contact with the atmosphere, corresponds to a partial pressure of dioxygen of 20 kPa, the oxygen gas pressure at sea level. Rain tends to be slightly acidic as a result of the absorption of carbon dioxide from the atmosphere:

$$CO_2(g) + 2\ H_2O(l) \rightleftharpoons H_3O^+(aq) + HCO_3^-(aq)$$

Depending on the geology of an area, the water in streams tends to be closer to neutral; seawater is slightly basic. Open water is rarely more basic than pH 9 because of the carbonate–hydrogen carbonate buffer system that is present:

$$CO_3^{2-}(aq) + H_2O(l) \rightleftharpoons HCO_3^-(aq) + OH^-(aq)$$

All surface waters, however, are oxidizing as a result of the high partial pressure of dissolved oxygen.

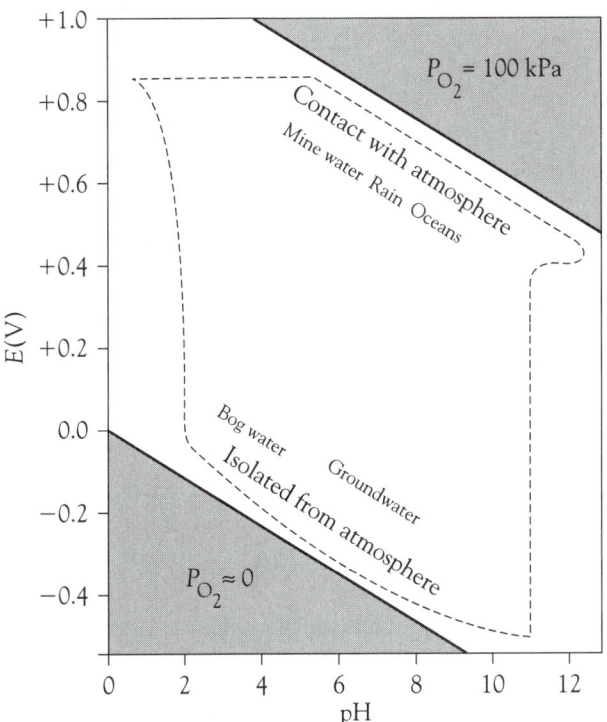

Figure 8.11 Pourbaix diagram showing the limits of E and pH conditions in natural waters (dashed line).
(Adapted from Gunter Faure, Principles and Applications of Inorganic Geochemistry [New York: Macmillan, 1991], p. 324.)

In a lake or river in which there is a high level of plant or algal growth, the level of oxygen is less. As a result, these waters have a lower potential. The lowest positive potentials occur in environments with a high biological activity and no atmospheric contact, typically bogs and stagnant lakes. Under these conditions, anaerobic bacteria flourish, the level of dissolved dioxygen will be close to 0, and the environment will be highly reducing. Bogs are also often highly acidic because of the decaying vegetation they contain.

Looking at the Pourbaix diagram of sulfur species within the limits of aqueous solution (Figure 8.12), it can be seen that the sulfate ion is the predominant species over most of the range of pH and $E°$. Because the hydrogen sulfate ion is the conjugate base of a fairly strong acid, only below about pH 2 is the HSO_4^- ion preferred. Such a situation can occur in mine

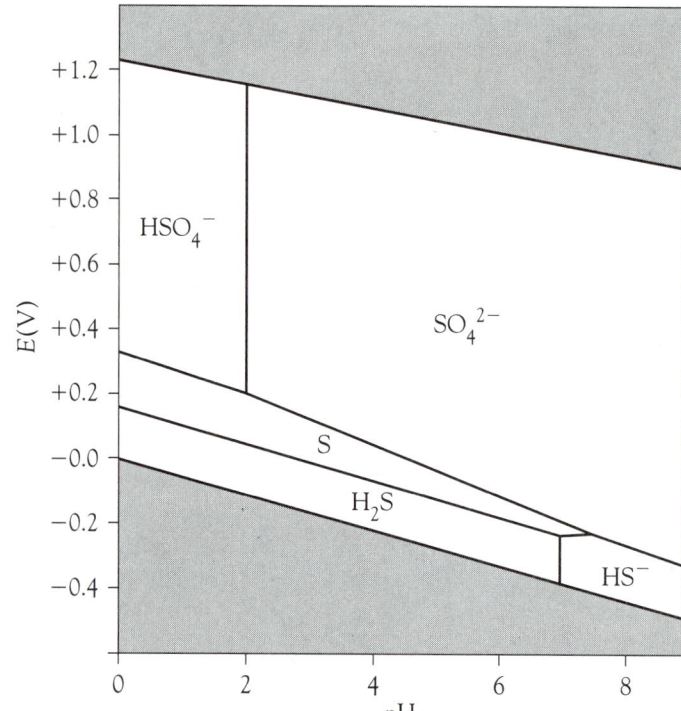

Figure 8.12 Pourbaix diagram showing the thermodynamically stable sulfur species in aqueous solution as a function of potential, E, and pH.
(Adapted from Gunter Faure, Principles and Applications of Inorganic Geochemistry [New York: Macmillan, 1991], p. 334.)

runoff, the acid conditions often being caused by the oxidation of iron(II) disulfide:

$$4\ FeS_2(s) + 15\ O_2(g) + 22\ H_2O(l) \rightarrow$$
$$4\ Fe(OH)_3(s) + 8\ H_3O^+(aq) + 8\ HSO_4^-(aq)$$

Over most of the pH range, a more reducing environment, such as in bogs, results in the conversion of sulfate ion to elemental sulfur:

$$SO_4^{2-}(aq) + 8\ H^+(aq) + 6\ e^- \rightarrow S(s) + 4\ H_2O(l)$$

In stronger reducing potentials, the sulfur is, in turn, reduced to hydrogen sulfide:

$$S(s) + 2\ H^+(aq) + 2\ e^- \rightarrow H_2S(aq)$$

It is this gas that can sometimes be smelled in boggy areas and in many waters of volcanic origin (such as those in Yellowstone National Park). Notice that aqueous hydrogen sulfide is the predominant reduced species. The reason relates to the weakness of this acid. Only under basic conditions will the hydrogen sulfide ion predominate:

$$H_2S(aq) + OH^-(aq) \rightleftharpoons HS^-(aq) + H_2O(l)$$

EXERCISES

8.1 Define the following terms: (a) oxidizing agent; (b) Ellingham diagrams.

8.2 Define the following terms: (a) Frost diagrams; (b) Pourbaix diagrams.

8.3 Using the oxidation state rules; determine the oxidation number of phosphorus in (a) P_4O_6; (b) H_3PO_4; (c) Na_3P; (d) PH_4^+; (e) $POCl_3$.

8.4 Using the oxidation state rules, determine the oxidation number of chlorine in (a) ClF_3; (b) Cl_2O; (c) Cl_2O_7; (d) HCl.

8.5 Using electron-dot diagrams, determine the oxidation number of sulfur in each of the following compounds: (a) H_2S; (b) SCl_2; (c) H_2S_2; (d) SF_6; (e) COS (structure O=C=S).

8.6 Using electron-dot diagrams; determine the formal charges and the oxidation numbers for each element in $SOCl_2$.

8.7 What are the likely oxidation states of iodine in its compounds?

8.8 What would you predict to be the highest oxidation state of xenon in its compounds? What other oxidation states are likely?

8.9 For each of the following compounds, deduce the oxidation number of the nonhalogen atom and identify the trend in oxidation numbers in the series: (a) indium(I) iodide, InI; (b) tin(II) chloride, $SnCl_2$; (c) antimony tribromide, $SbBr_3$; (d) tellurium tetrachloride, $TeCl_4$; (e) iodine pentafluoride, IF_5.

8.10 Identify the changes in oxidation states in the following equations:

(a) $Mg(s) + FeSO_4(aq) \rightarrow Fe(s) + MgSO_4(aq)$

(b) $2\ HNO_3(aq) + 3\ H_2S(aq) \rightarrow$
$\qquad 2\ NO(g) + 3\ S(s) + 4\ H_2O(l)$

8.11 Identify the changes in oxidation states in the following equations:

(a) $NiO(s) + C(s) \rightarrow Ni(s) + CO(g)$

(b) $2\ MnO_4^-(aq) + 5\ H_2SO_3(aq) + H^+(aq) \rightarrow$
$\qquad 2\ Mn^{2+}(aq) + 5\ HSO_4^-(aq) + 3\ H_2O(l)$

8.12 Write a half-reaction for the following reduction in acid solution:

$$H_2MoO_4(aq) \rightarrow Mo^{3+}(aq)$$

8.13 Write a half-reaction for the following oxidation in acid solution:

$$NH_4^+(aq) \rightarrow NO_3^-(aq)$$

8.14 Write a half-reaction for the following oxidation in basic solution:

$$S^{2-}(aq) \rightarrow SO_4^{2-}(aq)$$

8.15 Write a half-reaction for the following oxidation in basic solution:

$$N_2H_4(aq) \rightarrow N_2(g)$$

8.16 Balance the following redox reactions in acidic solution:

(a) $Fe^{3+}(aq) + I_2(aq) \rightarrow Fe^{2+}(aq) + I^-(aq)$

(b) $Ag(s) + Cr_2O_7^{2-}(aq) \rightarrow Ag^+(aq) + Cr^{3+}(aq)$

8.17 Balance the following redox reactions in acidic solution:

(a) $HBr(aq) + HBrO_3(aq) \rightarrow Br_2(aq)$

(b) $HNO_3(aq) + Cu(s) \rightarrow NO_2(g) + Cu^{2+}(aq)$

8.18 Balance the following redox equations in basic solution:

(a) $Ce^{4+}(aq) + I^-(aq) \rightarrow Ce^{3+}(aq) + IO_3^-(aq)$

(b) $Al(s) + MnO_4^-(aq) \rightarrow MnO_2(s) + Al(OH)_4^-(aq)$

8.19 Balance the following redox reactions in basic solution:

(a) $V(s) + ClO_3^-(aq) \rightarrow HV_2O_7^{3-}(aq) + Cl^-(aq)$

(b) $S_2O_4^{2-}(aq) + O_2(g) \rightarrow SO_4^{2-}(aq)$

8.20 Use standard reduction potentials from the on-line Appendices to determine which of the following reactions will be spontaneous under standard conditions.

(a) $SO_2(aq) + MnO_2(s) \rightarrow Mn^{2+}(aq) + SO_4^{2-}(aq)$

(b) $2\ H^+(aq) + 2\ Br^-(aq) \rightarrow H_2(g) + Br_2(aq)$

(c) $Ce^{4+}(aq) + Fe^{2+}(aq) \rightarrow Ce^{3+}(aq) + Fe^{3+}(aq)$

8.21 Use standard reduction potentials from the on-line Appendices to determine which of the following disproportionation reactions will be spontaneous under standard conditions:

(a) $2\ Cu^+(aq) \rightarrow Cu^{2+}(aq) + Cu(s)$

(b) $3\ Fe^{2+}(aq) \rightarrow 2\ Fe^{3+}(aq) + Fe(s)$

8.22 Use standard reduction potentials from the on-line Appendices to suggest a chemical reagent that could be used to oxidize hydrochloric acid to chlorine gas.

8.23 Use standard reduction potentials from the Appendices to suggest a chemical reagent that could be used to reduce chromium(III) ion to chromium(II) ion.

8.24 Silver can exist in two oxidation states, the more common silver(I) and the rarer silver(II):

$$Ag^+(aq) + e^- \rightarrow Ag(s) \qquad E° = +0.80\ V$$
$$Ag^{2+}(aq) + e^- \rightarrow Ag^+(aq) \qquad E° = +1.98\ V$$

(a) Is the silver(I) ion a good oxidizing agent or a good reducing agent?

(b) Which of the following is the most feasible reagent for oxidizing silver(I) ion to silver(II) ion: difluorine, fluoride ion, diiodine, iodide ion.

(c) You are thinking of preparing silver(I) hydride, for which:

$$\tfrac{1}{2}\,H_2(g) + e^- \rightarrow H^-(aq) \qquad E° = -2.25\text{ V}$$

Do you think the compound is likely to be thermodynamically stable? Explain your reasoning.

8.25 For the two half-reactions

$$Al^{3+}(aq) + 3\,e^- \rightarrow Al(s) \qquad E° = -1.67\text{ V}$$

$$Au^{3+}(aq) + 3\,e^- \rightarrow Au(s) \qquad E° = +1.46\text{ V}$$

(a) Identify the half-reaction that would provide the stronger oxidizing agent.

(b) Identify the half-reaction that would provide the stronger reducing agent.

8.26 Calculate the half-reaction potential for the reaction

$$Au^{3+}(aq) + 2\,e^- \rightarrow Au^+(aq)$$

given

$$Au^{3+}(aq) + 3\,e^- \rightarrow Au(s) \qquad E° = +1.46\text{ V}$$

$$Au^+(aq) + e^- \rightarrow Au(s) \qquad E° = +1.69\text{ V}$$

8.27 Calculate the half-cell potential for the reduction of lead(II) ion to lead metal in a saturated solution of lead(II) sulfate, concentration 1.5×10^{-5} mol·L^{-1}.

8.28 Calculate the half-cell potential for the reduction of aqueous permanganate ion to solid manganese(IV) oxide at a pH of 9.00 (all other ions being at standard concentration).

8.29 From the standard reduction potential value in the on-line Appendix, calculate the potential for the reduction of oxygen gas

$$O_2(g) + 4\,H^+(aq) + 4\,e^- \rightarrow 2\,H_2O(l)$$

at pH 7.00 and normal atmospheric partial pressure of dioxygen, 20 kPa. Note that functions in logarithms should be unitless. Hence, divide the pressure value by the standard pressure of 100 kPa before inserting into the Nernst function.

8.30 The following Latimer potential diagram shows bromine species under acidic conditions.

(a) Identify which species are unstable with respect to disproportionation.

(b) Determine the half-potential for the reduction of the bromate ion, $BrO_3^-(aq)$, to bromine.

$$\overset{+7}{BrO_4^-} \xrightarrow{+1.82\text{ V}} \overset{+5}{BrO_3^-} \xrightarrow{+1.49\text{ V}} \overset{+1}{HBrO} \xrightarrow{+1.59\text{ V}}$$

$$\overset{0}{Br_2} \xrightarrow{+1.07\text{ V}} \overset{-1}{Br^-}$$

8.31 The following Latimer potential diagram shows bromine species in basic conditions.

(a) Identify which species are unstable with respect to disproportionation.

(b) Determine the half-potential for the reduction of the bromate ion, $BrO_3^-(aq)$, to bromine.

(c) Explain why the bromine to bromide half-potential has the same value in both acidic and basic solutions.

$$\overset{+7}{BrO_4^-} \xrightarrow{+0.99\text{ V}} \overset{+5}{BrO_3^-} \xrightarrow{+0.54\text{ V}} \overset{+1}{BrO^-} \xrightarrow{+0.45\text{ V}}$$

$$\overset{0}{Br_2} \xrightarrow{+1.07\text{ V}} \overset{-1}{Br^-}$$

8.32 The Frost diagram in the following shows lead species (connected by a solid line) and silicon species (connected by a dashed line).

(a) Identify a strong oxidizing agent.

(b) Which is the most thermodynamically stable lead species?

(c) Which is the most thermodynamically stable silicon species?

(d) Which species could potentially disproportionate?

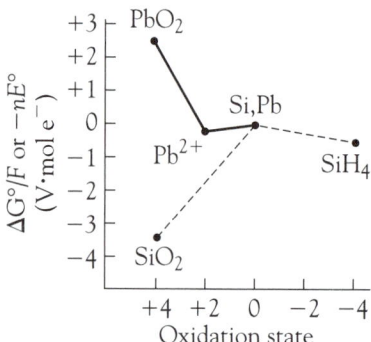

8.33 Construct a Frost diagram for cerium and discuss the relative stability of the oxidation states, given

$$Ce^{3+}(aq) + 3\,e^- \rightarrow Ce(s) \qquad E° = -2.33\text{ V}$$

$$Ce^{4+}(aq) + e^- \rightarrow Ce^{3+}(aq) \qquad E° = +1.70\text{ V}$$

8.34 From the Pourbaix diagram in Figure 8.11, identify the thermodynamically preferred sulfur species at a pH of 7.0 and an E of 0.0 V.

8.35 The Ellingham diagram shown below shows the free energy changes for oxides of mercury, carbon, hydrogen, magnesium, and aluminum.
  (a) Which metal oxide can be reduced to its metallic state simply by heating?
  (b) At a very high temperature, it is feasible to reduce one metal oxide with carbon. Which is it?
  (c) Which metal oxide cannot be reduced to its metal by means of carbon reduction?
  (d) Why is it that the slope of the magnesium–magnesium oxide line changes so dramatically beyond the second phase change?
  (e) Why is hydrogen gas rarely used for metal reduction?

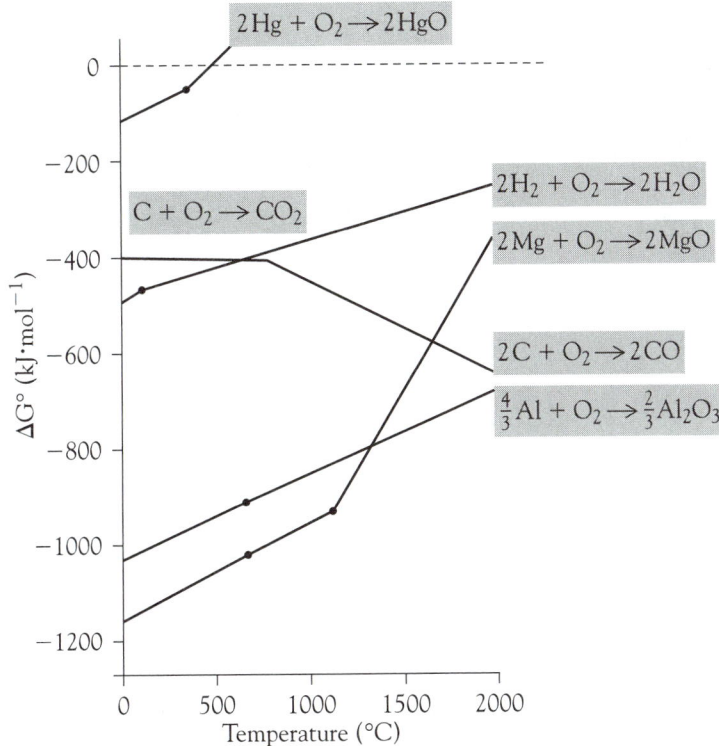

8.36 Using values of ΔH_f° and S° from the data tables in the Appendices, calculate approximate ΔG_f° values for the formation of uranium(IV) oxide at 0°C, 500°C, 1000°C, 1500°C, and 2000°C (both uranium and uranium oxide are solid throughout the temperature range). Plotting the values on the figure in Exercise 8.35, suggest how uranium might be produced from uranium(IV) oxide.

8.37 Explain briefly why the standard free energies of formation of metallic oxides decrease (become less negative) with increasing temperature while the free energy of formation of carbon monoxide increases (becomes more negative) with increasing temperature.

8.38 Thermodynamic data are often used to calculate standard potential values rather than using direct electrochemical methods. For the following half-reaction, calculate E° using values of ΔH_f° and S° from the Appendices for

$$Cr_2O_{3(s)} + 3\ H_2O(l) + 6\ e^- \rightarrow 2\ Cr(s) + 6\ OH^-(aq)$$

8.39 Is the perchlorate ion a stronger oxidizing agent at pH 0.00 or at pH 14.00? Give your reasoning.

BEYOND THE BASICS

8.40 Diarsenic trisulfide, As_2S_3, is oxidized to arsenate ion, AsO_4^{3-}, and sulfate ion by nitrate ion in acidic solution, the nitrate ion being oxidized to gaseous nitrogen monoxide. Write a balanced equation for the reaction and identify the changes in oxidation number.

8.41 Below about 710°C, the oxidation of carbon to carbon dioxide is preferred for metal reductions, but above that temperature, however, the oxidation of carbon to carbon monoxide is more effective. Discuss this statement qualitatively and then use a pair of $\Delta G°$ calculations to find an approximate value for the change-over temperature.

8.42 Construct a Pourbaix diagram for the nickel system. The only four species you will need to consider are nickel metal, nickel(II) ion, nickel(II) hydroxide ($K_{sp} = 6 \times 10^{-16}$), and nickel(IV) oxide.

8.43 Use the Pourbaix diagram in Figure 8.5 to suggest why, in well water containing manganese, a dark solid stain of an insoluble manganese compound forms in a toilet bowl even though the original well water is completely clear.

8.44 Use the standard electrode potential given in the Appendix for

$$MnO_4^-(aq) + 4\ H^+(aq) + 3\ e^- \rightarrow MnO_2(s) + 2\ H_2O(l)$$

and show that the value of $E°$ obtained for a pH of 14.00 is the same as that given in the Appendix for the half-reaction

$$MnO_4^-(aq) + 2\ H_2O(l) + 3\ e^- \rightarrow MnO_2(s) + 4\ OH^-(aq)$$

Suggest why it would be meaningless to calculate the $E°$ value for

$$MnO_4^-(aq) + 8\ H^+(aq) + 5\ e^- \rightarrow Mn^{2+}(aq) + 4\ H_2O(l)$$

at pH 14.00.

Chapter 9 Periodic Trends

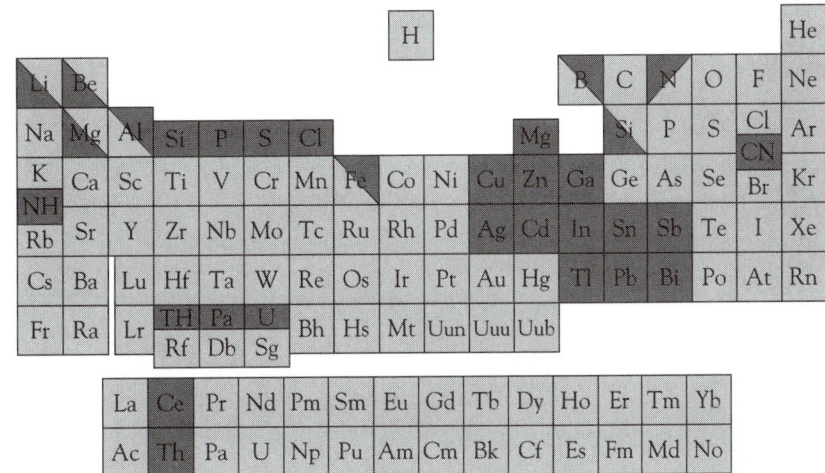

*I*norganic chemistry covers the chemistry of over 100 elements. Though each element has its individual characteristics, there are patterns and trends that provide a framework of order. It therefore makes sense to compile these global features in one chapter prior to an examination of each group. This overview chapter can then be used as a starting point for the discussion of the important and unique aspects of each element.

As chemists, we become used to the standard short form of the Periodic Table. This is indeed the most useful form, but it has its disadvantages. For example, the lanthanoids and actinoids are "orphaned" at the bottom of the table. Over the last 100 years, there have been many attempts to arrange the elements in the best possible format. In particular, the common formats fail to show the continuity of the elements. This can be remedied by the spiral arrangement shown in Figure 9.1. Such a spiral form avoids unnatural gaps between blocks and it shows the progression from the Group 3 elements into the lanthanoids and actinoids.

The reason for introducing alternative forms of the Periodic Table is to illustrate how the design of the table affects our perception of pattern and trends. We look at the short form of the table and search for patterns vertically and horizontally. Such trends do exist. But as we will see later in this chapter, beyond the groups and periods, there is a greater richness of trends and patterns among elements and the compounds and ions they form.

163

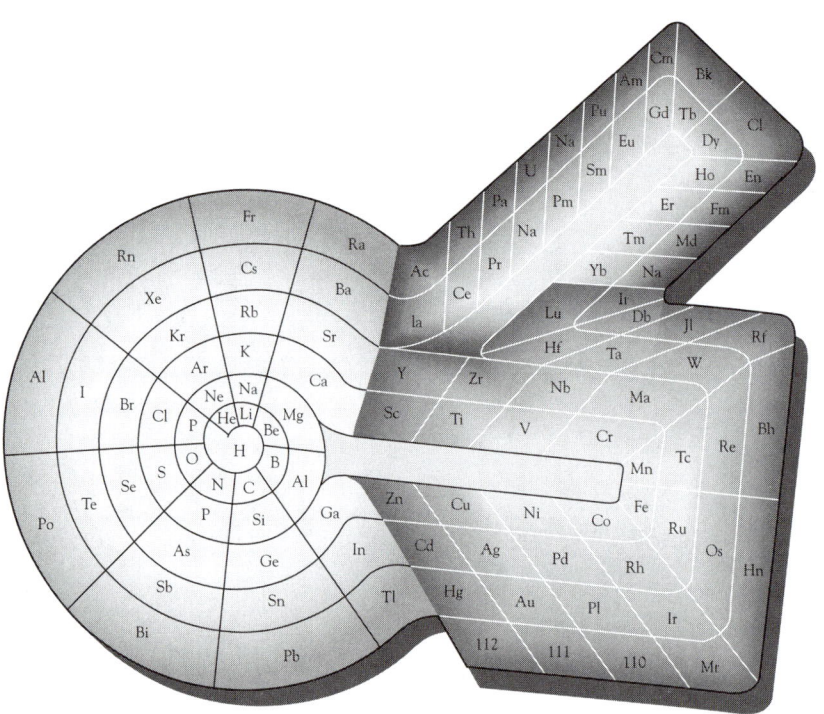

Figure 9.1 The spiral form of the Periodic Table (modified from a design by Ted Benfy).

9.1   *Group Trends*

The most significant and consistent trends in the Periodic Table are those within the groups. We can define the relationship as:

> Groups of elements tend to have characteristic properties. Descending a group in the periodic table, there are often smooth trends in these properties.

There are four main groups in which those trends are most clearly defined: Groups 1 and 2 and Groups 17 and 18. For Groups 13 through 16, some of the trends are much less systematic as a result of the shift down the group from metallic, through semimetallic, to nonmetallic properties. In this section, we will focus on the trends in three of the main groups, the alkali metals (Group 1), the halogens (Group 17), and Group 15.

The Alkali Metals

For the groups containing metals, such as the alkali metals, there is a decrease in melting and boiling point down the group (Table 9.1). This trend

Table 9.1   Melting and boiling points for the alkali metals

Element	Melting point (°C)	Boiling point (°C)
Lithium	180	1330
Sodium	98	892
Potassium	64	759
Rubidium	39	700
Cesium	29	690

Table 9.2 Some physical properties of the alkali metals

Element	Ionization energy ($kJ \cdot mol^{-1}$)	Hydration enthalpy ($kJ \cdot mol^{-1}$)	Electrode potential (V)
Lithium	526	-519	-3.04
Sodium	502	-406	-2.71
Potassium	425	-322	-2.94
Rubidium	403	-301	-2.92
Cesium	376	-276	-3.03

can be explained in terms of the weakening metallic bond as the atomic radius of the metal increases.

Though all of the alkali metals are very reactive, reactivity increases spectacularly down the group. The best illustration is the reaction with water to give the metal hydroxide and hydrogen gas:

$$2 \text{ M}(s) + 2 \text{ H}_2\text{O}(l) \rightarrow 2 \text{ MOH}(aq) + \text{H}_2(g)$$

Lithium bubbles quietly to produce the hydroxide and hydrogen gas. Sodium melts, skating around on the water surface as a silvery globule, and the hydrogen that is produced usually burns. For the heavier members of the group, the reaction is extremely violent; explosions often occur when small chunks of rubidium and cesium are dropped into water. The explosions are the result of the ignition of the dihydrogen gas.

Looking at the trends, the ionization energies and hydration enthalpies of all the alkali metals are low and they parallel one another (Table 9.2). Thus, the comparative chemical reactivity really reflects small differences in the component thermodynamic factors (Chapter 6). The overall standard electrode potential does not show any clear trends. It is commonly argued that all of the alkali metals are very reactive and that the observed trend in reactivity in fact reflects kinetic factors. That is, reactions with lithium are kinetically slow, those with sodium less so, and so on down the group.

The Halogens

For the nonmetal groups, such as the halogens, there is the reverse trend in melting and boiling points to that of Group 1, with the values increasing down the group (Table 9.3). For these elements, the explanation lies with the intermolecular forces between neighboring diatomic molecules. As we discussed in Chapter 3, Section 3.13, the dispersion forces increase in strength with the number of electrons, thus accounting for the trend.

Table 9.3 Melting and boiling points for the halogens

Element	Melting point (°C)	Boiling point (°C)
Fluorine	-219	-188
Chlorine	-101	-34
Bromine	-7	60
Iodine	114	185

Table 9.4   Some physical properties of the halogens

Element	Bond energy ($kJ \cdot mol^{-1}$)	Electron affinity ($kJ \cdot mol^{-1}$)	Electrode potential (V)
Fluorine	155	-328	$+3.05$
Chlorine	240	-349	$+1.36$
Bromine	190	-331	$+1.09$
Iodine	149	-301	$+0.54$

Just as the physical properties of the halogens are in reverse order to the alkali metals, so are the chemical reactivities. An illustration in this case is the reaction with hydrogen:

$$H_2(g) + X_2(g) \rightarrow 2\ HX(g)$$

A hydrogen and fluorine mixture is explosive. In fact, the H_2–F_2 reaction was once considered as a rocket propellant. The reaction with chlorine is violent but needs catalysis by light. Reaction with bromine is slow, and that with heated iodine vapor gives an equilibrium mixture of hydrogen iodide, hydrogen, and iodine.

The reactivity in this series depends largely upon the differences between the bond energies and the electron affinities (Table 9.4). With the abnormally low bond energy of fluorine, the net effect is to make fluorine by far the most oxidizing halogen as is shown from the standard electrode potentials in acidic solution.

The Group 15 Elements

Having seen the changes down a metallic group and then down a nonmetallic group, we will see the behavior of a group in which there is a transition from nonmetallic to metallic behavior. The melting and boiling points of the Group 15 elements (Table 9.5) are illustrative of this pattern found for Groups 13 through 16.

The colorless nonmetal nitrogen, N_2, has only weak dispersion forces between its molecules, accounting for its very low melting and boiling points. Phosphorus, too, is a nonmetal. The melting and boiling points listed here are for the white tetraphosphorus, P_4, allotrope. (The bonding in phosphorus allotropes is discussed in more detail in Chapter 15, Section 15.18). With a higher number of electrons per atom and a cluster of four atoms, higher melting and boiling points are to be expected.

Arsenic, antimony, and bismuth are all grey solids with the electrical and thermal conductivity increasing down the series. Arsenic has a layer structure

Table 9.5   Melting and boiling points for the Group 15 elements

Element	Melting point (°C)	Boiling point (°C)
Nitrogen	-210	-196
Phosphorus	44	281
Arsenic		615(sub)
Antimony	631	1387
Bismuth	271	1564

containing network covalent bonding. These moderately strong covalent bonds must be broken to escape the solid phase. In fact, arsenic sublimes directly into the gas phase when heated strongly, converting to As_4 molecular clusters analogous to those of phosphorus. Antimony and bismuth have similar solid-state structures, but there is much more interaction between layers, giving a predominantly metallic bonding type. Antimony and bismuth have the long liquid range characteristic of metals.

There is not a consistent trend in chemical reactivity descending this group. Nitrogen is unreactive, while white phosphorus is extremely reactive and the other Group 15 elements also being unreactive. This variation makes the point that trends in element reactivity are only apparent when the members of the group share a common bonding type.

In the chemistry of the compounds, we also see a transition in behavior from typical nonmetallic to metallic. The nonmetals readily form stable oxo-anions: nitrate, NO_3^-; phosphate, PO_4^{3-}; and arsenate, AsO_4^{3-}, respectively. The syntheses of oxo-anions of antimony and bismuth are more difficult, though they do exist, one example being the very strong oxidizing agent, sodium bismuthate, $NaBiO_3$. Exhibiting typical metallic behavior, both antimony and bismuth form salts, such as antimony(III) sulfate, $Sb_2(SO_4)_3$, and bismuth(III) nitrate, $Bi(NO_3)_3$. The formation of cations is consistent with their metallic nature. However, the fact that these elements form oxo-anions at all indicates they fit the category of weak metals.

9.2 *Periodic Trends in Bonding*

We will look first at the trends in the properties of elements across the second and third periods and relate the properties to the bonding type. We will see that trends in bonding types of elements follows the metallic–covalent side of the bond triangle (Chapter 5, Section 5.6). Then we will examine patterns in the fluorides, oxides, and hydrides of these two periods. There are usually systematic trends in the formulas of compounds across a period. However, there are rarely smooth trends in the physical and chemical properties of the compounds as the bonding type changes from ionic, through network covalent, to small molecule covalent (the second side of the bond triangle). Thus, our definition of trends across a period are best stated as:

> Crossing a period, we observe systematic patterns in chemical formulas of the compounds formed by the elements. In addition, there are partial trends in physical and chemical properties of the elements.

Bonding Trends in the Second Period Elements

The melting points of the Period 2 elements (Table 9.6) show a rapid increase in value, followed by an abrupt drop. However, the apparent trend in the first part of the period masks a significant change in bonding type. Lithium and beryllium are both metals, being shiny with high electrical conductivities, yet they differ profoundly in their properties. Lithium, with only one outer electron and a comparatively large size has weak metallic bonding resulting in a low melting point and a high chemical reactivity. Beryllium,

Table 9.6 Melting points of the Period 2 elements

Element	Li	Be	B	C	N_2	O_2	F_2	Ne
Melting point (°C)	180	1287	2180	4100(sub)	−210	−229	−219	−249

on the other hand, has two outer electrons for metallic bonding and a very much smaller radius. Hence, it has a strong metallic bond, resulting in a high melting point.

As boron follows the trend in increasing melting points, it is tempting to think that it, too, has metallic bonding. This cannot be the case as the pure element is dark red by transmitted light and a poor electrical conductor. Instead, boron is often classified as semimetallic. The element has a unique structure consisting of B_{12} units with the atoms joined by covalent bonds within the units and also between neighboring B_{12} units. To melt boron necessitates the breaking of these linking covalent bonds, thus accounting for the very high melting point. The other high-melting element is carbon. Graphite, the common form of carbon, sublimes at over 4000°C. This nonmetal consists of layers of multiple-bonded carbon atoms. Thus, as with boron, very strong network covalent bonds must be broken for the melting process.

The next three members of the period—nitrogen, oxygen, and fluorine—exhibit covalent bonding within the diatomic molecules. Between neighboring molecules there are very weak dispersion forces, accounting for the very low melting points of these elements. These second period nonmetals prefer multiple bonding when possible; thus, the dinitrogen molecule contains a triple bond and dioxygen, a double bond. Finally, the noble gas, neon, with its monatomic unit and weak dispersion forces between atoms, has the lowest melting point of the period.

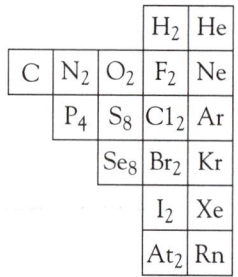

Figure 9.2 The common forms of some nonmetallic elements.

Bonding Trends in the Third Period Elements

In this period (Table 9.7), the first three elements have metallic bonding. Then, like the second period, the next member, shiny blue-grey silicon, an electrical semiconductor, has network covalent bonding.

The nonmetals of the third and subsequent periods do not exhibit multiple covalent bonds in their common forms. For example, the most common form of phosphorus is a waxy, white solid, containing groups of four atoms—P_4 units—bound together by single covalent bonds (Figure 9.2). Similarly, yellow sulfur has S_8 rings in which the constituent atoms are held together by single covalent bonds. Chlorine, like all the halogens, is found as simple, covalently bonded diatomic molecules. It is the weakness of the dispersion forces between the small molecules that results in the low melting points of these elements. Finally, argon is a monatomic gas, like all the other noble gases, accounting for its lowest melting point of the period.

Bonding Trends in the Highest Fluorides of the Second and Third Periods

Just as the progression of the elements across the second and third periods demonstrates the transition from metallic to covalent bonding, so there is a transition in the fluorides of these elements from ionic to covalent (Table 9.8), the network covalent region marking the borderline between the two bonding categories. We have stipulated the highest (oxidation state) fluorides as several of the nonmetals form more than one fluoride. For example, phosphorus also forms phosphorus trifluoride, PF_3, and sulfur also forms sulfur tetrafluoride, SF_4.

Table 9.7 Melting points of the Period 3 elements

Element	Na	Mg	Al	Si	P_4	S_8	Cl_2	Ar
Melting point (°C)	98	649	660	1420	44	119	−101	−189

Table 9.8   Formulas, bonding types, and phases at room temperature, of the highest fluorides of the Periods 2 and 3 elements

Compound	LiF	BeF_2	BF_3	CF_4	NF_3	OF_2	—
Bonding type (phase)	Ionic (solid)	Network covalent (solid)	Covalent (gas)	Covalent (gas)	Covalent (gas)	Covalent (gas)	—
Compound	NaF	MgF_2	AlF_3	SiF_4	PF_5	SF_6	ClF_5
Bonding type (phase)	Ionic (solid)	Ionic (solid)	Network covalent (solid)	Covalent (gas)	Covalent (gas)	Covalent (gas)	Covalent (gas)

The melting point of the ionic compounds is high as the melting process involves the breaking of ionic bonds in the crystal lattice. The melting points of network covalently bonded compounds also tends to be very high as covalent bonds must be broken in the process. On the other hand, the melting and boiling points of the (small molecule) covalent compounds tends to be very low, as the intermolecular forces such as dispersion and dipole–dipole are very weak.

The trends in the formulas themselves are interesting. Crossing the second period, the formulas rise to a maximum element–fluorine ratio at carbon, then decrease again. This trend is explicable in terms of the covalent bonding of the later second period being limited to the s and p orbitals, hence a maximum of eight electrons. For the third period elements, the oxidation number of the other element increases smoothly until sulfur, the additional bonds being possible evidence of the use of d orbitals on the central atom. On the basis of the other fluorides and the oxidation number trend, chlorine would be expected to form chlorine heptafluoride, ClF_7. The nonexistence of this compound is often attributed to the impossibility of fitting seven fluorine atoms around a central chlorine atom.

Bonding Trends in the Highest Oxides of the Second and Third Periods

The formulas of the highest (oxidation state) oxides, like those of the fluorides, correlate with the group number of the nonoxygen element, that is, +1 (Group 1), +2 (Group 2), +3 (Group 13), +4 (Group 14), +5 (Group 15), +6 (Group 16), and +7 (Group 17). The one exception is difluorine oxide, the only oxide in which the other element has a higher electronegativity than oxygen. As can be seen from Table 9.9, like the fluorides, there

Table 9.9   Formulas, bonding types, and phases at room temperature, of the highest oxides of the Periods 2 and 3 elements

Compound	Li_2O	BeO	B_2O_3	CO_2	N_2O_5	—	F_2O
Bonding type (phase)	Ionic (solid)	Ionic (solid)	Network covalent (solid)	Covalent (gas)	Covalent (gas)	—	Covalent (gas)
Compound	Na_2O	MgO	Al_2O_3	SiO_2	P_4O_{10}	$(SO_3)_3$	Cl_2O_7
Bonding type (phase)	Ionic (solid)	Ionic (solid)	Ionic (solid)	Network covalent (solid)	Covalent (solid)	Covalent (solid)	Covalent (liquid)

Table 9.10 Free energy of formation of the highest oxidation state covalent oxides of the Periods 2 and 3 elements

Compound	B_2O_3	CO_2	N_2O_5	—	F_2O
ΔG_f° (kJ·mol^{-1})	−1194	−386	+115	—	+42

Compound		SiO_2	P_4O_{10}	$(SO_3)_3$	Cl_2O_7
ΔG_f° (kJ·mol^{-1})		−856	−2700	−371	>+270

is a diagonal band of compounds adopting network covalent structures separating the ionic and covalently bonded regions.

The stability of the covalent oxides decreases to the right as can be seen from the free energies of formation (Table 9.10). Dinitrogen pentaoxide, difluorine oxide, and dichlorine heptaoxide are all very strong oxidizing agents, themselves being reduced. Dichlorine heptaoxide decomposes explosively (thus, its ΔG_f° is not known precisely), while dinitrogen pentaoxide is only stable below room temperature. Such behavior is expected, as the free energies of formation of these three compounds are all positive.

Bonding Trends in the Hydrides of the Second and Third Periods

The formulas of the hydrides correlate with the lowest common oxidation state of the nonhydrogen element, that is, +1 (Group 1), +2 (Group 2), +3 (Group 13), ±4 (Group 14), −3 (Group 15), −2 (Group 16), and −1 (Group 17). Table 9.11 shows the patterns in bonding for the hydrides of Periods 2 and 3. The nonmetal hydrides are small-molecule covalent with very low boiling points; in fact, all are gases at room temperature except for water and hydrogen fluoride. These two particular hydrides are liquids as a result of the strong (intermolecular) hydrogen bonds (see Chapter 3, Section 3.16). Again we see the bonding pattern of ionic–network covalent–covalent on crossing the periods.

It is in the chemical reactivity of the hydrides that we see the major differences. The ionic hydrides, such as sodium hydride, are stable in dry air but react rapidly with water:

$$NaH(s) + H_2O(l) \rightarrow NaOH(aq) + H_2(g)$$

The polymeric covalent hydrides are also dry-air stable and react with water.

The small-molecule covalent hydrides do not react with water, but in the trend from right to left, their reactivity toward oxygen increases. For example, diborane, B_2H_6, spontaneously burns in air to form diboron trioxide:

$$B_2H_6(g) + 3\ O_2(g) \rightarrow B_2O_3(s) + 3\ H_2O(l)$$

Table 9.11 Formulas, bonding types, and phases at room temperature, of the hydrides of the Periods 2 and 3 elements

Compound	LiH	$(BeH_2)_x$	B_2H_6	CH_4	NH_3	H_2O	HF
Bonding type (phase)	Ionic (solid)	Network covalent (solid)	Covalent (gas)	Covalent (gas)	Covalent (gas)	Covalent (liquid)	Covalent (liquid)

Compound	NaH	MgH_2	$(AlH_3)_x$	SiH_4	PH_3	H_2S	HCl
Bonding type (phase)	Ionic (solid)	Ionic (solid)	Network covalent (solid)	Covalent (gas)	Covalent (gas)	Covalent (gas)	Covalent (gas)

Table 9.12 Free energy of formation of the gaseous covalent hydrides of the Periods 2 and 3 elements

Compound	B_2H_6	CH_4	NH_3	H_2O	HF
ΔG_f° (kJ·mol^{-1})	+87	−51	−16	−237	−275
Compound	SiH_4	PH_3	H_2S		HCl
ΔG_f° (kJ·mol^{-1})	+57	+13	−34		−95

Silane, SiH_4, is also spontaneously flammable in air, while phosphine, PH_3, often ignites due to trace impurities. Methane, ammonia, and hydrogen sulfide need an ignition source before they will burn. The flammability of the nonmetal hydrides correlates with the pattern in the free energy of formation of the compounds (Table 9.12). Thus, the three most reactive hydrides are actually thermodynamically unstable. In Chapter 10, Section 10.5, we will see that the pattern in hydride reactivity can be interpreted in terms of bond polarity.

9.3 Isoelectronic Series in Covalent Compounds

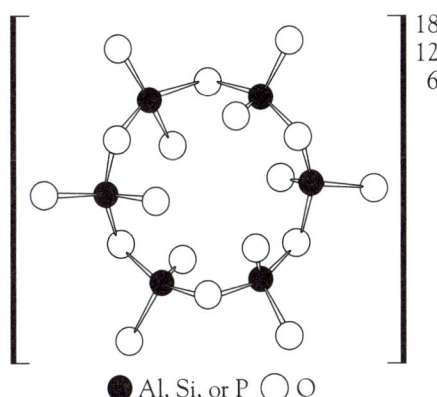

18–(Al)
12–(Si)
6–(P)

● Al, Si, or P ○ O

Figure 9.3 Structure of the isoelectronic $[Al_6O_{18}]^{18-}$, $[Si_6O_{18}]^{12-}$, and $[P_6O_{18}]^{6-}$ ions.

Among the elements that form covalent compounds, we often find patterns in formulas. One example is that of the highest oxyacids of the third period: SiO_4^{4-}, PO_4^{3-}, SO_4^{2-}, and ClO_4^{-}. As the number of electrons on the central atom increases by one, so the charge decreases by one unit. Thus, all of these four oxyanions are *isoelectronic*. In the strict definition of isoelectronic that we use in this text, *isoelectronic species share the same total number of valence electrons **and** the same total sum of electrons.*

There are numerous sets of isoelectronic species but a particularly interesting trio is that of the cyclic ions: $[Al_6O_{18}]^{18-}$, $[Si_6O_{18}]^{12-}$, and $[P_6O_{18}]^{6-}$ (Figure 9.3). These share a common structure, even though in the first ion, the nonoxygen element is a metal; the second, a semimetal; and the third, a nonmetal. Calcium aluminate, $Ca_9[Al_6O_{18}]$, is a major component of cement, while beryllium aluminum silicate, $Be_3Al_2[Si_6O_{18}]$, is the mineral and gemstone, beryl.

9.4 Trends in Acid–Base Properties

As we saw in Chapter 7, Section 7.6, the variations in acid–base behavior are an important dimension of inorganic chemistry. Here we will review the acid–base properties of the oxides and hydrides.

Acid–Base Patterns in the Highest Oxides of the Second and Third Periods

Progressing across a period, there is a transition from basic oxide for the metals to acidic oxide for the nonmetals (Table 9.13). For example, sodium oxide reacts with water to give sodium hydroxide, while sulfur trioxide reacts with water to give sulfuric acid.

$$Na_2O(s) + H_2O(l) \rightarrow 2\,NaOH(aq)$$

$$SO_3(s) + H_2O(l) \rightarrow H_2SO_4(l)$$

Table 9.13   Acid–base properties of the highest oxides of the Period 3 elements

Compound	Na_2O	MgO	Al_2O_3	SiO_2	P_4O_{10}	$(SO_3)_3$	Cl_2O_7
Acid–base behavior	Basic	Basic	Amphoteric	Acidic	Acidic	Acidic	Acidic

Not all oxides react with water, some being insoluble. For example, magnesium oxide is water-insoluble but it will react with acids.

$$MgO(s) + 2\ HCl(aq) \rightarrow MgCl_2(aq) + H_2O(l)$$

While acidic oxides, such as silicon dioxide, will react with bases:

$$SiO_2(s) + 2\ NaOH(l) \xrightarrow{\Delta} Na_2SiO_3(l) + H_2O(g)$$

Table 9.14   Acid–base properties of the highest oxides of the Group 15 elements

N_2O_5	Acidic
P_4O_{10}	Acidic
As_2O_3	Acidic
Sb_2O_3	Amphoteric
Bi_2O_3	Basic

Aluminum is one of the borderline or weak metals. These metals, other examples being zinc and tin, have chemical properties of both metals and nonmetals. Thus, their oxides react with acids (like metals) and bases (like nonmetals). For example, aluminum oxide reacts with acids to give the hexaaquaaluminum cation, $[Al(OH_2)_6]^{3+}$, which we can simply represent as $Al^{3+}(aq)$, and it reacts with bases to give the tetrahydroxoaluminate, $[Al(OH)_4]^-$, anion:

$$Al_2O_3(s) + 6\ H^+(aq) \rightarrow 2\ Al^{3+}(aq) + 3\ H_2O(l)$$

$$Al_2O_3(s) + 2\ OH^-(aq) + H_2O(l) \rightarrow [Al(OH)_4]^-(aq)$$

Acid–Base Patterns in the Highest Oxides of the Group 15 Elements

Just as there is a transition from basic oxide to acidic oxide across a period, so there is a similar pattern down a group in the middle of the Periodic Table. Here we have chosen the Group 15 elements to illustrate the pattern (Table 9.14).

Acid–Base Patterns in the Covalent Hydrides of the Second and Third Periods

The nonmetal hydrides show a very different pattern to the oxides (Table 9.15). Those of the low electronegativity elements, such as methane, CH_4, form neutral hydrides. Ammonia is the only common basic hydride:

$$NH_3(aq) + H_2O(l) \rightleftharpoons NH_4^+(aq) + OH^-(aq)$$

while hydrogen chloride is strongly acidic:

$$HCl(g) + H_2O(l) \rightarrow H_3O^+(aq) + Cl^-(aq)$$

hydrogen fluoride less so:

$$HF(g) + H_2O(l) \rightleftharpoons H_3O^+(aq) + F^-(aq)$$

and hydrogen sulfide very much less:

$$H_2S(g) + H_2O(l) \rightleftharpoons H_3O^+(aq) + HS^-(aq)$$

Table 9.15   Acid–base patterns in the covalent hydrides of the Periods 2 and 3 elements

Compound	B_2H_6	CH_4	NH_3	H_2O	HF
Acid–base behavior	Neutral	Neutral	Basic	—	Weakly acidic
Compound		SiH_4	PH_3	H_2S	HCl
Acid–base behavior		Neutral	Neutral	Very weakly acidic	Strongly acidic

9.5 *The (n) Group and (n + 10) Group Similarities*

3(13)	4(14)	5(15)	6(16)	7(17)	12(2)
Al	Si	P	S	Cl	Mg
Sc	Ti	V	Cr	Mn	Zn
Y	Zr	Nb	Mo	Tc	Cd
Lu	Hf	Ta	W	Re	Hg
Lr	Rf	Db	Sg	Bh	Uub

Figure 9.4 This portion of the Periodic Table shows d-block Groups 3–7 and 12, together with the third period members from the related main group (group numbers in parentheses).

It was the similarities in formulas between two sets of groups, such as Groups 4 and 14, that led Mendeleev and others to originally construct a simple eight-column Periodic Table. When chemists became aware of the importance of atomic number in determining periodic order, the resulting 18-column table had group labels 1A, 1B etc., to continue to provide a linkage between the two sets. With the newer 1–18 numbering, this linkage is less apparent and is in danger of being forgotten. This is regrettable, as there are some interesting parallels that actually reinforce the concept of periodicity.

The link is predominantly between compounds and ions of the highest oxidation state of the main group elements and those of the corresponding oxidation state of the matching transition elements. In general, the match in formulas and properties is strongest between the Period 3 main group elements and the first row of the transition metals (Figure 9.4). A general definition for this relationship is:

> There are similarities in chemical formulas and structures of the highest oxidation state of some members of the (n) Group elements and of the members of the corresponding (n + 10) Group elements.

One can argue on simple electron configuration or oxidation state grounds that there should be similarities between these two sets, but the similarities go far beyond simple formula resemblances. We find some commonality in melting points, properties, and, in a few cases, unusual corresponding structures.

Aluminum and Scandium

Aluminum and scandium have so many similarities that the Canadian chemist, Fathi Habashi, has suggested that aluminum's place in the Periodic Table should actually be shifted to Group 3. Certainly this makes sense from the perspective of electron configuration, for the aluminum ion has the noble gas configuration like the 3+ ions of the Group 3 elements, whereas the lower Group 13 elements have filled $(n-1)d^{10}$ orbitals.

Group 3 ion	Electron configuration	Group 13 ion	Electron configuration
		Al^{3+}	[Ne]
Sc^{3+}	[Ar]	Ga^{3+}	[Ar]($3d^{10}$)

The closer resemblance of aluminum to Group 3 in terms of melting point and electrode potential can be seen in Table 9.16.

Table 9.16 **A comparison of some properties of Group 3 and Group 13 elements**

Group 3 elements			Group 13 elements		
Element	m.p. (°C)	E° (V)	Element	m.p. (°C)	E° (V)
—	—	—	Al	660	−1.66
Sc	1540	−1.88	Ga	30	−0.53
Y	1500	−2.37	In	160	−0.34
La	920	−2.52	Tl	300	+0.72
Ac	1050	−2.6	—	—	—

In solution, both Al^{3+} and Sc^{3+} cations hydrolyze significantly to give acid solutions containing polymeric hydroxo species. Upon addition of hydroxide ion to the respective cation, the hydroxides of aluminum and scandium are both produced as gelatinous precipitates. The precipitates redissolve in excess base to give anionic species. In another example, both metals form isomorphous compounds of the type Na_3MF_6 where M = Al or Sc. The term *isomorphous* actually means "same shape" and traditionally referred to crystals having the same shape. Now we tend to use it to mean isostructural; that is, the ions are packed in the same way in the crystal lattice.

Not only does aluminum resemble scandium but also reactive aluminum metal differs significantly in its chemistry from unreactive gallium metal. Among differences in compound types, gallium, like boron, forms a gaseous hydride, Ga_2H_6. Aluminum, however, forms a polymeric white solid hydride $(AlH_3)_x$. Bubbling hydrogen sulfide gas through a solution of the respective cation also has a very different result. Gallium gives a precipitate of gallium sulfide, Ga_2S_3, while aluminum gives a precipitate of aluminum hydroxide. (Likewise, scandium gives a precipitate of scandium hydroxide.) On the other hand, the structures of the aluminum halides do resemble those of gallium halides more than those of scandium halides.

Group 14 and Titanium(IV)

Though there are similarities between titanium(IV) and silicon(IV), there is a much greater similarity of titanium(IV) and tin(IV)—a lower member of Group 14. In fact, this pair have among the closest similarities of elements in different groups. Starting with the oxides, titanium(IV) oxide and tin(IV) oxide are isomorphous and they share the rare attribute of turning yellow on heating (*thermochromism*). There are very close similarities in the melting and boiling points of the chlorides: titanium(IV) chloride (m.p. $-24°C$, b.p. $136°C$) and tin(IV) chloride (m.p. $-33°C$, b.p. $114°C$). Both chlorides behave as Lewis acids and they hydrolyze in water.

$$TiCl_4(l) + 2\ H_2O(l) \rightarrow TiO_2(s) + 4\ HCl(g)$$

$$SnCl_4(l) + 2\ H_2O(l) \rightarrow SnO_2(s) + 4\ HCl(g)$$

Phosphorus (V) and Vanadium(V)

As phosphorus is a nonmetal and vanadium a metal, there is obviously a limit to the comparisons that can be made. Nevertheless, there are some striking similarities in their $+5$ oxidation states. For example, phosphate, PO_4^{3-}, and vanadate, VO_4^{3-}, ions are both strong bases. In addition, the two elements form a large number of polymeric anions, including the unique matching pair of $P_4O_{12}^{4-}$ and $V_4O_{12}^{4-}$.

Sulfur(VI) and Chromium (VI)

Table 9.17 shows some formula similarities between sulfur(VI) and chromium(VI). The resemblance even extends to the physical properties

Table 9.17 Similarities between chromium(VI) and sulfur(VI) species

Group 6		Group 16	
Formula	Systematic name	Formula	Systematic name
CrO_3	Chromium(VI) oxide	SO_3	Sulfur trioxide
CrO_2Cl_2	Chromyl chloride	SO_2Cl_2	Sulfuryl chloride
CrO_4^{2-}	Chromate ion	SO_4^{2-}	Sulfate ion
$Cr_2O_7^{2-}$	Dichromate ion	$S_2O_7^{2-}$	Pyrosulfate ion

of some of the compounds, for example, the two oxychlorides: sulfuryl chloride, SO_2Cl_2 (m.p. $-54°C$, b.p. $69°C$) and chromyl chloride (m.p. $-96°C$, b.p. $117°C$). These compounds resemble each other chemically, decomposing in water. However, there are major chemical differences between sulfur(VI) and chromium(VI); in particular, chromates and dichromates are strongly oxidizing and colored (chromate, yellow, and dichromate, orange), whereas the sulfates and pyrosulfates are nonoxidizing and white.

Chlorine (VII) and Manganese (VII)

The oxo-anions of chlorine(VII) and manganese(VII), perchlorate, ClO_4^-, and permanganate, MnO_4^-, are both strongly oxidizing and their salts are isomorphous. Their oxides, dichlorine heptaoxide, Cl_2O_7, and manganese(VII) oxide, Mn_2O_7, are highly explosive liquids at room temperature.

Chlorine and manganese show another resemblance by forming oxides in an oxidation state that would not be predicted for either element—that of $+4$ (ClO_2 and MnO_2). Though chlorine dioxide is a gas and manganese(IV) oxide is a solid, it is really curious why the most common oxide of both elements should possess such an unexpected oxidation state.

Xenon(VIII) and Osmium (VIII)

The next link is between lower members of Group 8 and Group 18. The chemistry of the metal osmium and nonmetal xenon have some fascinating parallels—particularly in the $+8$ oxidation state. For example, osmium forms a yellow strongly oxidizing oxide, OsO_4, while xenon forms a pale yellow explosive oxide, XeO_4. There are parallels in the formulas of oxyfluorides, too: XeO_2F_4 and OsO_2F_4; and XeO_3F_2 and OsO_3F_2. There are also similarities in the $+6$ oxidation state. The highest fluorides for both elements (formed by direct reaction of the element with fluorine) are in this oxidation state: XeF_6 and OsF_6, and they both form corresponding fluoro-anions: XeF_7^- and OsF_7^-.

The Alkali Metals (Group 1) and the Coinage Metals (Group 11)

Up to now, we have been extolling the usefulness of the links between the Group (n) and ($n + 10$) elements. By contrast, there are no major similarities between the Group 1 and Group 11 elements. In fact, this pair illustrates the extremes of dissimilarity of metallic behavior! The alkali metals are reactive and all the common salts are soluble; the coinage metals are unreactive and most of their $+1$ oxidation state compounds are insoluble. Some examples of the major differences between the elements of the two groups are shown in Table 9.18.

Magnesium and Zinc

Though Groups 1 and 11 were dissimilar, there are major similarities between magnesium (Group 2) and zinc (Group 12). Table 9.19 compares key points of their chemistry.

Aluminum and Iron(III): A Case of Similarities between ($n + 5$) and ($n + 10$) Species

Earlier in this section, we saw that aluminum more closely resembled scandium in Group 3 than the members of Group 13. The chemistry of aluminum also resembles that of the iron(III) ion. Iron(III) and aluminum ions have the same charge and similar sizes (and hence similar charge densities) leading to several similarities. For example, in the vapor phase, both of these ions form covalent chlorides of the form M_2Cl_6. These (anhydrous) chlorides can be used as Friedel–Crafts catalysts in organic chemistry, where they function by the formation of the $[MCl_4]^-$ ion. In addition, the $[M(OH_2)_6]^{3+}$ ions of both metals are very strongly acidic, another result of their high charge densities.

Table 9.18   Contrast of the alkali metals and the coinage metals

Property	Alkali metals	Coinage metals
Common oxidation numbers	Always +1	Silver +1; but copper and gold rarely +1
Chemical reactivity	Very high; increasing down the group	Very low; decreasing down the group
Density	Very low; increasing down the group (0.5 to 1.9 g·cm^{-3})	High; increasing down the group (9 to 19 g·cm^{-3})
Melting points	Very low; decreasing down the group (181°C to 29°C)	High; all about 1000°C
Aqueous redox chemistry	None	Yes (e.g., $Cu^{2+}(aq) \rightarrow Cu^{+}(aq)$)
Solubilities of common salts	All soluble	+1 Oxidation state compounds insoluble

There are, however, some significant differences. For example, the oxides have different properties: aluminum oxide, Al_2O_3, is an amphoteric oxide, whereas iron(III) oxide, Fe_2O_3, is a basic oxide. This difference is utilized in the separation of pure aluminum oxide from the iron-containing bauxite ore in the production of aluminum (see Chapter 13, Section 13.8). The amphoteric aluminum oxide reacts with hydroxide ion to give the soluble tetrahydroxoaluminate ion, $[Al(OH)_4]^-$, whereas the basic iron(III) oxide remains in the solid phase:

$$Al_2O_3(s) + 2\,OH^-(aq) + 3\,H_2O(l) \rightarrow 2\,[Al(OH)_4]^-(aq)$$

9.6   *Isomorphism in Ionic Compounds*

In the previous section, we mentioned how pairs of compounds can be isomorphous, that is, have analogous crystal structures. The best example of isomorphism occurs in a series of compounds called the *alums*. Alums have the general formula $M^+M^{3+}(SO_4^{2-})_2 \cdot 12H_2O$. (The tripositive ion is

Table 9.19   Comparison of the properties of magnesium and zinc

Property	Magnesium	Zinc
Ionic radius	72 pm	74 pm
Oxidation state	+2	+2
Ion color	Colorless	Colorless
Hydrated ion	$Mg(OH_2)_6^{2+}$	$Zn(OH_2)_6^{2+}$
Soluble salts	Chloride, sulfate	Chloride, sulfate
Insoluble salt	Carbonate	Carbonate
Chloride	Covalent, hygroscopic	Covalent, hygroscopic
Hydroxide	Basic	Amphoteric

Table 9.20   A comparison of cation radii in alums

Monopositive ions		Tripositive ions	
K^+	152 pm	Al^{3+}	68 pm
Rb^+	166 pm	Cr^{3+}	75 pm
NH_4^+	151 pm	Fe^{3+}	78 pm

actually a hexahydrate, so the correct formulation is $M^+[M(OH_2)_6]^{3+}$ $(SO_4^{2-})_2 \cdot 6H_2O$.) Large crystals can be grown simply by mixing equimolar mixtures of the monopositive sulfate and the tripositive sulfate. Thus, alums are favorite compounds for crystal-growing competitions. The monopositive cation can be potassium, rubidium, or ammonium, while the tripositive cation is most commonly aluminum, chromium(III), or iron(III). The name alum specifically pertains to the colorless aluminum-containing compound, the others being named chrome alum (deep purple) and ferric alum (pale violet).

The similarities in the sets of ionic radii can be seen in Table 9.20. The lattice energy of alums is very high, accounting for the high stability of the compounds. In fact, alum is the most convenient water-soluble compound of aluminum, while ferric alum is a stable and convenient compound of iron(III). The formation of alums is another similarity between the chemistry of aluminum and iron(III) ions (see Section 9.5).

What determines the ability of an ion to substitute while maintaining the same crystal structure? There are two principles of isomorphous substitution. The first principle states that one ion may substitute for another in a lattice if the two ions have identical charges and differ in radii by no more than 20 percent. This principle is really a restatement of the radius-ratio rules that were discussed in Chapter 5, Section 5.5. A large difference in ionic radii precludes isomorphism, but close values of ionic radii do not necessarily mean isomorphism will occur.

In many cases, mixed structures can be formed. For example, pale purple crystals can be formed by crystallizing a mixture of alum and chrome alum, the crystals having the lattice sites of the 3+ ions randomly filled by aluminum and chromium(III) ions. Isomorphism is of particular importance in mineral chemistry. For example, many precious gems involve isomorphous substitution. Ruby is aluminum oxide containing some chromium(III) ion in place of the aluminum ion. Thus, its formula is represented as $(Al^{3+}, Cr^{3+})_2 (O^{2-})_3$. Likewise, sapphires contain titanium(III) in aluminum oxide: $(Al^{3+}, Ti^{3+})_2(O^{2-})_3$. The composition of the natural gemstones reflects the composition of the molten rock from which the crystals formed. This is true of many minerals. Olivine, a magnesium silicate, Mg_2SiO_4, often contains iron(II) as an isomorphous sustituted ion. The compound of olivine is sometimes represented as $(Mg^{2+}, Fe^{2+})_2(SiO_4^{4-})$. The classic example of isomorphous substitution is that of the lanthanoid phosphates, MPO_4, where M^{3+} is any of the lanthanoids. These ions are so similar in radii that the naturally occurring phosphate ore, monazite, usually contains a mixture of all the lanthanoids.

A second principle of isomorphous substitution is applicable to compounds containing two different cations. This principle states that isomorphous substitution can occur by ions of different charges but the same radii

Table 9.21   Some comparative ion sizes

Ionic radii	+1 Charge	+2 Charge	+3 Charge	+4 Charge	+5 Charge
Small		Be^{2+}	Al^{3+}, Fe^{3+}, Cr^{3+}	Si^{4+}	P^{5+}
Medium	Li^+	Mg^{2+}, Fe^{2+}		Ti^{4+}	W^{5+}
Large	Na^+	Ca^{2+}	La^{3+}		
Very large	K^+, NH_4^+	Ba^{2+}			

as the ions they replace, provided the sum of the cation charges remains the same. Many examples are found in the important series of minerals called the perovskites (Chapter 16, Section 16.6). The parent compound is $(Ca^{2+})(Ti^{4+})(O^{2-})_3$, while one of the many other compounds adopting this structure is $(Na^+)(W^{5+})(O^{2-})_3$, where the monopositive sodium has replaced the similar-size dipositive calcium ion and the pentapositive tungsten ion has replaced the tetrapositive titanium ion. Some of the common substitution possibilities are shown in Table 9.21. (Note that drawing divisions in ionic radii is somewhat arbitrary; for example, an ion at the high end of the "small" category might well substitute for an ion at the low end of the "medium" category.)

9.7   Diagonal Relationships

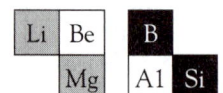

Figure 9.5   Elements commonly considered linked by the diagonal relationship.

Though chemists usually think of periodic trends vertically (down groups) or horizontally (across periods), there are, in fact, other patterns to be found in the periodic table. One of these is the *diagonal relationship*:

> The similarity in chemical properties between an element and that to the lower right of it. This relationship is found for elements in the upper-left corner of the Periodic Table.

We have already seen evidence of a diagonal relationship in the bonding types of the fluorides, oxides, and hydrides, where there is a diagonal of network covalent compounds separating the ionic and covalent bonding types. The diagonal relationship is only chemically significant for three pairs of elements: lithium and magnesium, beryllium and aluminum, and boron and silicon (Figure 9.5).

Similarities of Lithium and Magnesium

The best examples of resemblance between the chemistry of lithium and that of magnesium are:

1. The hardness of lithium metal is greater than that of the alkali metals but similar to that of the alkaline earth metals.

2. Lithium forms a normal oxide, Li_2O, like the alkaline earth metals but unlike the other alkali metals. (Sodium forms Na_2O_2, containing the O_2^{2-} ion, while the heavier alkali metals form compounds containing the O_2^- ion, such as KO_2.)

3. Lithium is the only alkali metal to form a nitride, Li_3N, whereas the alkaline earth metals all form nitrides.

Table 9.22 Charge densities for the alkali metal and alkaline earth metal ions

Group 1 ion	Charge density ($C \cdot mm^{-3}$)	Group 2 ion	Charge density ($C \cdot mm^{-3}$)
Li^+	98	—	—
Na^+	24	Mg^{2+}	120
K^+	11	Ca^{2+}	52
Rb^+	8	Sr^{2+}	33
Cs^+	6	Ba^{2+}	23

4. Three lithium salts—the carbonate, the phosphate, and the fluoride—have very low solubilities. These anions form insoluble salts with the alkaline earth metals.

5. Lithium forms organometallic compounds similar to those of magnesium (the Grignard reagents used in organic chemistry).

6. Many lithium salts exhibit a high degree of covalency in their bonding. This bonding is similar to that of magnesium.

7. The lithium and magnesium carbonates decompose to give the appropriate metal oxide and carbon dioxide. The carbonates of the other alkali metals do not decompose when heated.

How can we explain this? An examination of the charge densities of the elements in Groups 1 and 2 (Table 9.22) reveals that the charge density of lithium is much closer to that of magnesium than to those of the other alkali metals. Hence, similarity in charge density may explain the resemblance in the chemical behaviors of lithium and magnesium.

Similarities of Beryllium and Aluminum

Beryllium and aluminum resemble each other in three ways:

1. In air, both metals form tenacious oxide coatings that protect the interior of the metal sample from attack.

2. Both elements are amphoteric, forming parallel anions—tetrahydroxoberyllates, $[Be(OH)_4]^{2-}$, and tetrahydroxoaluminates, $[Al(OH)_4]^-$—in reactions with concentrated hydroxide ion.

3. Both form carbides (Be_2C and Al_4C_3) containing the C^{4-} ion that react with water to form methane. (The other Group 2 elements form compounds containing the C_2^{2-} ion, such as CaC_2, which react with water to form ethyne.)

However, there are some major differences between the chemical properties of beryllium and aluminum. One of the most apparent differences is in the formula of the hydrated ions each forms. Beryllium forms the $[Be(OH_2)_4]^{2-}$ ion, whereas aluminum forms the $[Al(OH_2)_6]^{3-}$ ion. The lower coordination number of the beryllium may be accounted for in two ways: The beryllium atom has no d orbitals available for bonding, and the ion is physically too small to accommodate six surrounding water molecules at a bonding distance.

Lithium and Mental Health

One important consequence of the diagonal relationship is the treatment of bipolar disorder (commonly called manic depression) by lithium ion. Biochemists have shown that lithium ion functions in part by substituting for magnesium in an enzyme process.

About 1 percent of the population suffer from this debilitating illness in which one's moods oscillate from euphoria and hyperactivity to depression and lethargy. Lithium ion is a mood-stabilizing drug. The discovery of lithium's benefits was made through a combination of accident (serendipity) and observation. In 1938, an Australian psychiatrist, J. Cade, was studying the effects of a large organic anion on animals. To increase the dosage, he needed a more soluble salt. For large anions, the solubilities of the alkali metal ions increase as their radius decreases; hence, he chose the lithium salt. However, when he administered this compound, the animals started to show behavioral changes. He realized that the lithium ion itself must have altered the workings of the brain. Further studies showed that the lithium ion had a profound effect on bipolar disorder patients.

Ironically, the discovery of the health effects of lithium could have been made much earlier, because it had been well known in folk medicine that water from certain lithium-rich British springs helped alleviate the disorder. More recently, a study in Texas showed that locations having lower levels of hospital admissions with manic depression correlated with higher levels of lithium ion in the local drinking water.

It is a particular imbalance of neurotransmitters that gives rise to the bipolar disorder symptoms. This imbalance can be traced back to the enzyme inositolmonophosphatase (abbreviated to IMPase). IMPase converts the monophosphates of inositol, a sugar-like molecule, to free inositol, a process requiring the participation of two magnesium ions. It appears that lithium will readily substitute for one of the magnesium ions in the enzyme pathway, slowing down the process. This alleviates the mood swings, a preferable solution compared to using drug combinations that will subdue the manic phase or counter depressive episodes.

Lithium therapy does have its problems. Side effects include excessive thirst, memory problems, and hand tremor. Additionally, Bertrand's rule curve (Chapter 2) is very narrow; that is, there is only a very small range between theraputic and toxic doses. Despite its problems, lithium therapy, a result of the diagonal relationship, has restored the health of enormous numbers of individuals.

Again we can use charge density arguments to explain this diagonal relationship: that the very high charge density of the beryllium 2+ ion is closer to that of the aluminum 3+ ion than it is to the larger ions of Group 2.

Similarities of Boron and Silicon

A comparison of boron and silicon is our third and final example of the diagonal relationship. This case is very different from the two other examples, for the chemistry of both elements involves covalent bonding. Thus, there can be no justification in terms of ion charge density. In fact, this relationship is not easy to understand except that both elements are on the borderline of the metal–nonmetal divide and have similar electronegativities. Some of the similarities are listed here:

1. Boron forms a solid acidic oxide, B_2O_3, like that of silicon, SiO_2, but unlike that of either aluminum, whose oxide is amphoteric, or carbon, whose oxide, CO_2, is acidic but gaseous.

2. Boric acid, H_3BO_3, is a very weak acid that is similar to silicic acid, H_4SiO_4, in some respects. It bears no resemblance to the amphoteric aluminum hydroxide, $Al(OH)_3$.

3. There are numerous polymeric borates and silicates that are constructed in similar ways, using shared oxygen atoms.

4. Boron forms a range of flammable, gaseous hydrides, just as silicon does. There is only one aluminum hydride—a solid.

9.8 The "Knight's Move" Relationship

Cu	Zn	Ga		
Ag	Cd	In	Sn	Sb
	Tl	Pb	Bi	

Figure 9.6 The elements that seem to exhibit the "knight's move" relationship.

For the later main group elements, the South African chemist, Michael Laing, noticed a relationship between one element and the element one period down and two groups to its right. He called this pattern the *knight's move relationship* from its similarity to the move in the game of chess. This relationship is in the physical properties of compounds, such as melting points and crystal structures. It does not seem to relate to much chemical commonality. The relationship, apparent among the lower members of Groups 11 through 15 (see Figure 9.6), is defined as:

The similarity between an element of Group (n) and Period (m) with the element in Group ($n + 2$) and Period ($m + 1$) in the same oxidation state. This relationship is found among elements in the lower right portion of the Periodic Table.

To illustrate the knight's move relationship, we can use the closeness in the melting point of zinc chloride (275°C) and tin(II) chloride (247°C) and that of cadmium chloride (568°C) and lead(II) chloride (500°C). Although this relationship should not be overemphasized, we find other similarities, such as the identical unusual crystal structures of cadmium iodide and lead(II) iodide.

Similarities of Silver(I), Thallium(I), and Potassium

The most interesting knight's move pair are silver(I) and thallium(I). Both of these very low charge density ions have insoluble halides except for the fluorides. Again we find matches in the melting points, some examples being shown in Table 9.23.

Yet in other ways thallium(I) ion behaves more like potassium ion, particularly in its biochemistry as will be discussed in Section 9.13. Table 9.24 shows the chemical similarities and differences between thallium(I) ion and the ions of potassium and silver.

Table 9.23 Some similarities in melting points between silver(I) and thallium(I) compounds

	Chloride (°C)	Nitrate (°C)
Silver(I)	455	212
Thallium(I)	430	206

Table 9.24   A comparison of the properties of thallium(I) ion to those of silver(I) and potassium ions

Properties of thallium(I)	Properties of silver(I)	Properties of potassium
Forms normal oxide, Tl_2O	Forms normal oxide, Ag_2O	Forms KO_2 not normal oxide
Soluble, very basic hydroxide	Insoluble hydroxide	Soluble, very basic hydroxide
Hydroxide reacts with carbon dioxide to form carbonate	Unreactive hydroxide	Hydroxide reacts with carbon dioxide to form carbonate
Fluoride soluble, other halides insoluble	Fluoride soluble, other halides insoluble	All halides soluble
Chromate brick-red color and insoluble	Chromate brick-red color and insoluble	Chromate yellow and soluble

The Inert Pair Effect

The early chemist, Dumas, called thallium the "duckbill platypus among elements" as a result of thallium more closely resembling elements in groups other than its own.

How can the "knight's move" be explained? Ions of similar charge and size (that is, similar charge densities) are likely to have similarities in their chemistry. For example, silver(I) and thallium(I) both have very low and similar charge densities. The similarities in size of the two ions of different periods is also easy to account for. With the filling of the transition metal series (and the lanthanoids), the d and f electrons are poor shielders of the outer electrons. This contraction results in the ion of the $(n + 2)$ group and $(m + 1)$ period "shrinking" to close to the size of the ion of the same charge of the (n) group and the (m) period.

But this avoids the question of why such elements as thallium, tin, lead, and bismuth form compounds in a lower oxidation state than would be expected from their group number—and why that oxidation state is always less by two (for example, Tl^+, Tl^{3+}; Sn^{2+}, Sn^{4+}; Pb^{2+}, Pb^{4+}; Bi^{3+}, Bi^{5+}). The explanation can be found in the *inert-pair effect*. To illustrate, we will use thallium as an example. All the Group 13 elements have an outer electron configuration of s^2p^1. Thus, formation of a $+1$ ion corresponds to the loss of the single p electron and retention of the two s electrons.

To find a reasonable explanation for the formation of these low-charge ions, we have to consider relativistic effects (mentioned previously in Chapter 2, Section 2.5). The velocities of electrons in outer orbitals, particularly in the $6s$ orbital, become close to that of light. As a result, the mass of these $6s$ electrons increases and, following from this, their mean distance from the nucleus decreases. In other words, the orbital shrinks. This effect is apparent from the successive ionization energies. In Chapter 2, Section 2.6, we saw that ionization energies usually decrease down groups, but a comparison of the first three ionization energies of aluminum and thallium shows that the ionization energy of the outer p electron is greater, and the ionization energies of the pair of s electrons significantly greater, for thallium than for aluminum (Table 9.25).

Recalling the Born–Haber cycles of Chapter 6, Section 6.3, the large energy input needed to form the cation must be balanced by a high lattice

Table 9.25   Comparative ionization energies of aluminum and thallium

Element	Ionization energy (MJ·mol^{-1})		
	First (p)	Second (s)	Third (s)
Aluminum	0.58	1.82	2.74
Thallium	0.59	1.97	2.88

energy (energy output). But the thallium(III) ion is much larger than the aluminum(III) ion; hence, the lattice energy of a thallium(III) ionic compound will be less than that of the aluminum analog. The combination of these two factors, particularly the higher ionization energy, leads to a decreased stability of the thallium(III) ionic state and, hence, the stabilizing of the thallium(I) ionic oxidation state.

9.9   *The Early Actinoid Relationships*

In Chapter 24, we will discuss the chemistry of the two "orphan" series, the lanthanoids and the actinoids. However, the chemistry of the early actinoids actually resembles that of the corresponding member of transition metal series (Figure 9.7).

As an example of this resemblance, we can compare uranium with the Group 6 metals. The most obvious similarity is provided by the oxo-anions: the yellow diuranate ion, $U_2O_7^{2-}$, with the orange dichromate ion, $Cr_2O_7^{2-}$. Uranium forms a uranyl chloride, UO_2Cl_2, matching those of chromyl chloride, CrO_2Cl_2, and molybdenyl chloride, MoO_2Cl_2. In general, as we might expect, uranium bears the closest similarity to tungsten. For example, uranium and tungsten (but not molybdenum and chromium) form stable hexachlorides, UCl_6 and WCl_6, respectively. Just as uranium resembles the Group 6 elements, so protactinium resembles the Group 5 elements and thorium the Group 4 elements.

The similarities in properties result from similarities in outer electron configurations as can be seen from comparisons of molybdenum, tungsten, uranium, and seaborgium, the next "true" member of Group 6.

Ti	V	Cr
Zr	Nb	Mo
Hf	Ta	W
Th	Pa	U

Figure 9.7 The relationship between the early actinoid elements and those of the corresponding transition metal groups.

Atom	Electron configuration	Ion	Electron configuration
Mo	$[Kr]5s^2 4d^4$	Mo^{6+}	$[Kr]$
W	$[Xe]6s^2(4f^{14})5d^4$	W^{6+}	$[Xe](4f^{14})$
U	$[Rn]7s^2 5f^3 6d^1$	U^{6+}	$[Rn]$
Sg	$[Rn]7s^2(5f^{14})6d^4$	Sg^{6+}	$[Rn](5f^{14})$

Whereas the f orbitals in tungsten and seaborgium are "buried," those of uranium are part of the valence shell. In fact, the partial occupancy of $5f$ and $6d$ orbitals in uranium suggests that they are of similar energies. Thus, uranium readily attains a 6+ oxidation state like the members of the Group 6 elements.

9.10   *The Lanthanoid Relationships*

In terms of similarities, the lanthanoids have to be among the most similar of the elements. This might seem surprising, considering that the progression of the elements along the series corresponds to the successive filling of the $4f$ orbitals. There are two reasons for the similarities. The first point is that the common oxidation state results essentially from the balance between the ionization energy of the metal and the lattice energy of the solid salt formed. In the case of each lanthanoid, the optimum reduction in energy corresponds to the formation of the 3+ ion. Second, the

Figure 9.8 The rare earth metals consist of the Group 3 and lanthanoid elements. All of these elements have similar chemical properties and share the common oxidation state of 3+.

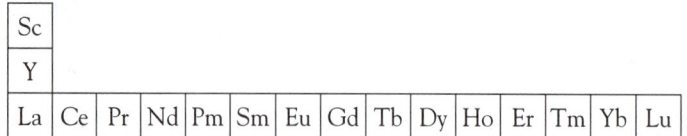

remaining electrons in the $4f$ orbitals are essentially "buried." Thus, the lanthanoids, in general, behave like large main group metals having the one oxidation state of 3+. In this behavior, they resemble the Group 3 elements. As a result, the Group 3 and lanthanoid elements are often classed together as the *rare earth metals* (Figure 9.8)—even though some of them are not rare.

There are two exceptions to the simplicity of the lanthanoids. The first is europium, Eu, which readily forms a 2+ ion. We can see why europium forms a lower oxidation state by looking at the electron configurations of the ions. Formation of the 2+ ion corresponds to the half-filling of the f orbital set and, in fact, the third ionization energy of europium is the highest of the lanthanoids:

Atom	Electron configuration	Ion	Electron configuration
Eu	$[Xe]6s^2 4f^7$	Eu^{2+}	$[Xe]4f^7$
		Eu^{3+}	$[Xe]4f^6$

The europium(II) ion behaves very similarly to an alkaline earth ion; for example, its carbonate, sulfate, and chromate are insoluble as are those of the heavier alkaline earth metals. The ionic radius of europium(II) is actually very similar to that of strontium and, as might be expected, several europium(II) and strontium compounds are isomorphous.

Similarities of Cerium(IV) and Thorium(IV)

Whereas europium has a lower than normal oxidation state, cerium has a higher than normal oxidation state of 4+. Formation of the 4+ ion corresponds to the noble gas configuration and this may account for the fact that cerium has the lowest fourth ionization energy of the lanthanoids.

Atom	Electron configuration	Ion	Electron configuration
Ce	$[Xe]6s^2 4f^1 5d^1$	Ce^{3+}	$[Xe]4f^1$
		Ce^{4+}	$[Xe]$

Cerium(IV) behaves like zirconium(IV) and hafnium(IV) of Group 4 and like thorium(IV) of the corresponding actinoids. For example, all four of these ions form insoluble fluorides and phosphates. There are particularly strong similarities in the chemistry of cerium(IV) and thorium(IV). Cerium(IV) oxide (used in self-cleaning ovens) and thorium(IV) oxide both adopt the fluorite structure. They form isomorphous nitrates, $M(NO_3)_4 \cdot 5H_2O$, where M is Ce or Th, and both form hexanitrato complex ions $[M(NO_3)_6]^{2-}$. The major difference between the two elements in this oxidation state is that thorium(IV) is the thermodynamically stable form of that element while cerium(IV) is strongly oxidizing. It is because of the high redox potential that ammonium hexanitratocerate(IV), $(NH_4)_2[Ce(NO_3)_6]$, is used in redox titrations.

$$Ce^{4+}(aq) + e^- \rightarrow Ce^{3+}(aq) \qquad\qquad E° = +1.44 \text{ V}$$

9.11 "Combo" Elements

The compound carbon monoxide has several similarities to dinitrogen, N_2. For example, they are both triply bonded molecules with similar boiling points: $-196°C$ (N_2) and $-190°C$ (CO). A major reason for the parallel behavior is that the dinitrogen molecule and the carbon monoxide molecule are isoelectronic. This similarity extends to the chemistry of the two molecules. In particular, there are several transition metal compounds where dinitrogen can substitute for a carbon monoxide entity. For example, it is possible to replace one or two carbon monoxides bonded to chromium in $Cr(CO)_6$ to give isoelectronic $Cr(CO)_5(N_2)$ and $Cr(CO)_4(N_2)_2$. The "combo" elements are a subset of isoelectronic behavior in which the sum of the valence electrons of a pair of atoms of one element matches the sum of the valence electrons of two horizontal neighboring elements. A "combo" element can be defined as:

The combination of an $(n - x)$ group element with an $(n + x)$ group element to form compounds that parallel those of the (n) group element.

Boron– Nitrogen Analogs of Carbon Compounds

The best example of a "combo" element is that of the boron and nitrogen combination. Boron has one less valence electron than carbon, and nitrogen has one more. For many years, chemists have tried to make analogs of carbon compounds that contain alternating boron and nitrogen atoms. Included in their successes have been analogs of the pure forms of carbon. The two common allotropes of carbon are graphite, the lubricant, and diamond, the hardest naturally occurring substance known. Unfortunately, both carbon allotropes burn when heated to give carbon dioxide gas, thus precluding the use of either of these substances in high-temperature applications. Boron nitride, BN, however, is the ideal substitute. The simplest method of synthesis involves heating diboron trioxide with ammonia at about $1000°C$:

$$B_2O_3(s) + 2\ NH_3(g) \xrightarrow{\Delta} 2\ BN(s) + 3\ H_2O(g)$$

The product has a graphite-like structure (Figure 9.9) and is an excellent high-temperature, chemically resistant lubricant.

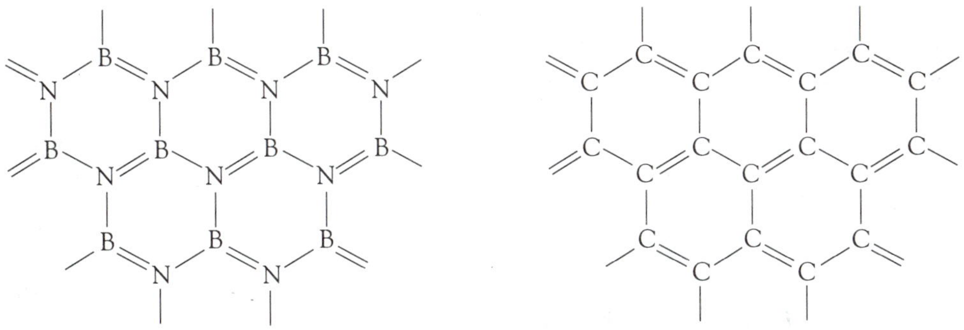

Boron nitride   Graphite

Figure 9.9 Comparative layer structures of boron nitride and graphite.

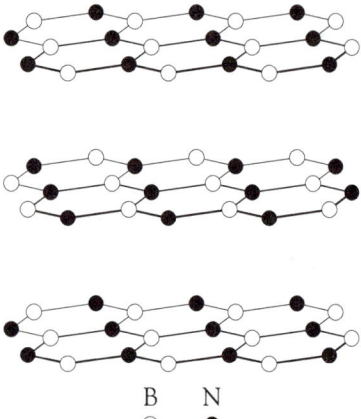

B N

Figure 9.10 Repeating layer structure of boron nitride.

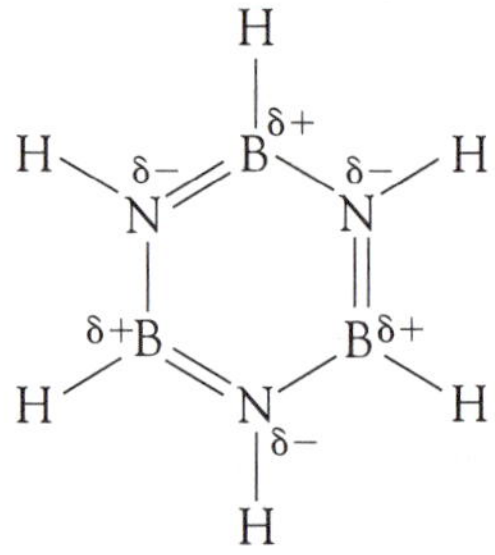

Borazine

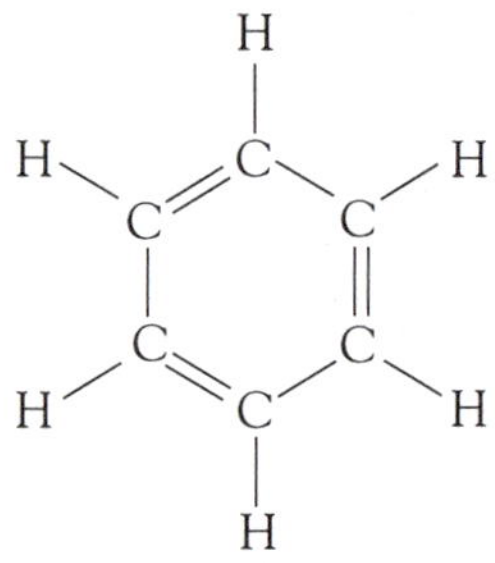

Benzene

Figure 9.11 Comparison of the structures of borazine and benzene.

Unlike graphite, boron nitride is a white solid that does not conduct electricity. This difference is possibly due to differences in the way the layers in the two crystals are stacked. The layers in the graphite-like form of boron nitride are almost exactly the same distance apart as those in graphite, but the boron nitride layers are organized so that the nitrogen atoms in one layer are located directly over boron atoms in the layers above and below, and vice versa. This arrangement is logical, because the partially positive boron atoms and partially negative nitrogen atoms are likely to be electrostatically attracted to each other (Figure 9.10). By contrast, the carbon atoms in one layer of graphite are directly over the center of the carbon rings in the layers above and below. An alternative reason for the lack of electrical conductivity is the weaker aromaticity in the layers due to the differing electronegativities of boron and nitrogen.

In a further analogy to carbon, application of high pressures and high temperatures converts the graphite-like allotrope of boron nitride to a diamond-like form called borazon. This allotrope of boron nitride is similar to diamond in terms of hardness and is far superior in terms of chemical inertness at high temperatures. Hence, borazon is often used in preference to diamond as a grinding agent.

There is another similarity between boron–nitrogen and carbon compounds. The reaction between diborane, B_2H_6, and ammonia gives borazine, $B_3N_3H_6$, a cyclic molecule analogous to benzene, C_6H_6:

$$3\ B_2H_6(g) + 6\ NH_3(g) \rightarrow 2\ B_3N_3H_6(l) + 12\ H_2(g)$$

In fact, borazine is sometimes called "inorganic benzene" (Figure 9.11). This compound is a useful reagent for synthesizing other boron–nitrogen analogs of carbon compounds, but at this time it has no commercial applications. Note that, as one would expect from their comparative electronegativities, the boron atoms bear a slight positive charge; and the nitrogen atoms, a slight negative charge. This assignment is confirmed from the way that electrophilic (electron-loving) reagents bond preferentially to the nitrogen atoms.

Despite similarities in boiling points, densities, and surface tensions, the polarity of the boron–nitrogen bond means that borazine exhibits less aromaticity than benzene. Hence, borazine is much more prone to chemical attack than is the homogeneous ring of carbon atoms in benzene. For example, hydrogen chloride reacts with borazine to give $B_3N_3H_9Cl_3$, in which the chlorine atoms bond to the more electropositive boron atoms:

$$B_3N_3H_6(l) + 3\ HCl(g) \rightarrow B_3N_3H_9Cl_3(s)$$

This compound can be reduced by sodium tetrahydroborate, $NaBH_4$, to give $B_3N_3H_{12}$, an analog of cyclohexane, C_6H_{12}. In fact, like cyclohexane, $B_3N_3H_{12}$ adopts the chair conformation.

"Combo" Elements and Semiconductors

The "combo" element concept is important in the context of semiconductors. Materials scientists realized that by using isoelectronic combinations to match semiconducting elements, it was possible to make semiconductors with desired properties.

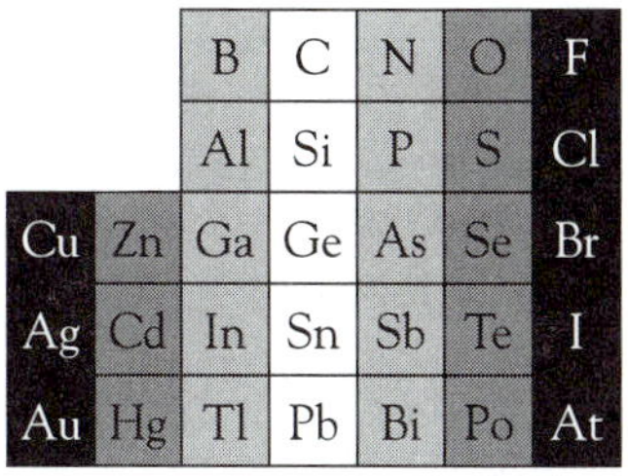

	B	C	N	O	F
	Al	Si	P	S	Cl
Cu	Zn	Ga	Ge	As	Se	Br
Ag	Cd	In	Sn	Sb	Te	I
Au	Hg	Tl	Pb	Bi	Po	At

Figure 9.12 A portion of the periodic table showing the combination of elements that will provide compounds that are isoelectronic with each of the Group 14 elements. Such possible pairs are shown in identical shading.

A particular focus of interest is the Period 4 elements centered on germanium. Germanium adopts the diamond structure, which is the same as the sphalerite structure (Chapter 5, Section 5.5), one of the two packing arrangements of zinc sulfide, but with all of the atoms identical and occupying both anion and cation sites in the lattice. Examples of compounds ("combo" elements) adopting this same sphalerite crystal structure are: gallium arsenide, GaAs; zinc selenide, ZnSe; and copper(I) bromide, CuBr. Though these common "combo" examples are truly isoelectronic, the two elements do not necessarily have to come from the same period; thus, more generally, we can say the sum of the valence electrons must add up to eight. This electron relationship is known as the *Zintl principle*. Figure 9.12 shows the combinations of elements that can result in this series of Zintl solids.

In Chapter 4, Section 4.2, we saw that for metals, there was an overlap between the highest occupied molecular orbitals (HOMOs) and the lowest unoccupied molecular orbitals (LUMOs). It is this overlap that enables electrical conductivity to occur throughout the metal structure. In the case of insulators, the gap between the two energy levels is very large, whereas for semiconductors, the gap is small enough that it is feasible for electrons to be excited to the higher energy state. We can see the trend in Table 9.26 using a family of Zintl (isoelectronic) solids.

It is quite remarkable that the length of side of a unit cell is essentially constant (within experimental error) throughout the series, even though there is a change in bonding from pure covalent (Ge) through polar covalent (GaAs) to substantially ionic (ZnSe, CuBr). As the polarity of the bond increases, so the conductivity of the solid decreases from that of a semiconductor (Ge, GaAs) to that of an insulator (ZnSe, CuBr).

But there is more to the study of this simple series than scientific curiosity. It is these band gaps that allow us to construct light-emitting diodes (LEDs), devices that have been used for many years as indicator lights but which are now finding uses as energy-efficient vehicle rear lights, traffic lights, and night lights. To obtain the chosen color, specific band gaps are needed. These band gaps can be adjusted by substituting proportions of one element for another in a Zintl solid. A particularly useful series is that of $GaP_xAs_{(1-x)}$, where x has a value between zero and one. Gallium phosphide, GaP, itself has a band gap of 222 kJ·mol^{-1}, while as we saw above, that of gallium arsenide is 137 kJ·mol^{-1}. In-between combinations have corresponding intermediate band gaps; for example, $GaP_{0.5}As_{0.5}$ has a band gap of about 200 kJ·mol^{-1}.

Table 9.26 The properties of a series of isoelectronic solids

Solid	Unit–cell dimension (pm)	Electronegativity difference	Energy gap (kJ·mol^{-1})
Ge	566	0.0	64
GaAs	565	0.4	137
ZnSe	567	0.8	261
CuBr	569	0.9	281

9.12   *Pseudo-Elements*

Some polyatomic ions resemble element ions in their behavior, and, in a few cases, there is a corresponding molecule that corresponds to the matching element. We can define this unusual category as:

> A polyatomic ion whose behavior in many ways mimics that of an ion of an element or of a group of elements.

Ammonium as a Pseudo-Alkali Metal Ion

Even though the ammonium ion is a polyatomic cation containing two non-metals, it behaves in many respects like an alkali metal ion. The similarity results from the ammonium ion being a large low-charge cation just like those of the alkali metals. In fact, the radius of the ammonium ion (151 pm) is very close to that of the potassium ion (152 pm). However, the chemistry of ammonium salts more resembles that of rubidium or cesium ions, perhaps because the ammonium ion is not spherical and its realistic radius is larger than its measured value. The similarity to the heavier alkali metals is particularly true of the crystal structures. Ammonium chloride, like rubidium and cesium chloride, has a CsCl structure at high temperatures and a NaCl structure at low temperatures.

The ammonium ion resembles an alkali metal ion in its precipitation reactions. Though all sodium compounds are water-soluble, there are insoluble compounds of the heavier alkali metal ions with very large anions. We find that the ammonium ion gives precipitates with solutions of these same anions. A good example is the hexanitritocobalt(III) ion, $[Co(NO_2)_6]^{3-}$, which is commonly used as a test in qualitative analysis for the heavier alkali metals. A bright yellow precipitate is obtained with the ammonium ion analogously to potassium, rubidium, and cesium ions.

$$3\ NH_4^+(aq) + [Co(NO_2)_6]^{3-}(aq) \rightarrow (NH_4)_3[Co(NO_2)_6](s)$$

However, the similarity does not extend to all chemical reactions that these ions undergo. For example, gentle heating of alkali metal nitrates gives the corresponding nitrite and oxygen gas, but heating ammonium nitrate results in decomposition of the cation and anion to give dinitrogen oxide and water:

$$2\ NaNO_3(s) \xrightarrow{\Delta} 2\ NaNO_2(s) + O_2(g)$$

$$NH_4NO_3(s) \xrightarrow{\Delta} N_2O(g) + 2\ H_2O(g)$$

A major weakness of the parallel between ammonium ion and the heavier alkali metal ions is that the parent pseudo-element of the ammonium ion, "NH$_4$," cannot be isolated.

Cyanide as a Pseudo-Halide Ion

The best example of a pseudo-element ion is cyanide. Not only does it behave very much like a halide ion but also the parent pseudo-halogen, cyanogen, $(CN)_2$, exists. The cyanide ion resembles a halide ion in a remarkable number of ways:

1. Salts of cyanide ion with silver, lead(II) and mercury(I) are insoluble as are those of chloride, bromide and iodide ions. For example:

$$CN^-(aq) + Ag^+(aq) \rightarrow AgCN(s)$$

$$[\text{compared to } Cl^-(aq) + Ag^+(aq) \rightarrow AgCl(s)]$$

2. Like silver chloride, silver cyanide reacts with ammonia to give the diamminesilver(I) cation.

$$AgCN(s) + 2\ NH_3(aq) \rightarrow [Ag(NH_3)_2]^+(aq) + CN^-(aq)$$

$$[\text{compared to } AgCl(s) + 2\ NH_3(aq) \rightarrow [Ag(NH_3)_2]^+(aq) + Cl^-(aq)$$

3. The cyanide ion is the conjugate base of the weak acid, hydrocyanic acid, HCN, parallel to fluoride ion and hydrofluoric acid.

$$HCN(aq) + H_2O(l) \rightleftharpoons H_3O^+(aq) + CN^-(aq)$$

$$[\text{compared to } HF(aq) + H_2O(l) \rightleftharpoons H_3O^+(aq) + F^-(aq)]$$

4. Cyanide ion can be oxidized to the parent pseudo-halogen, cyanogen, similar to the oxidation of halides to halogens. The parallel is particularly close with iodide ion, as they can both be oxidized by very weak oxidizing agents such as the copper(II) ion:

$$2\ Cu^{2+}(aq) + 4\ CN^-(aq) \rightarrow 2\ CuCN(s) + (CN)_2(g)$$

$$[\text{compared to } 2\ Cu^{2+}(aq) + 4\ I^-(aq) \rightarrow 2\ CuI(s) + I_2(s)]$$

5. The parent halogens react with water to form halide and hypohalite ions. In a corresponding manner, cyanogen reacts with base to give the cyanide and cyanate ions:

$$(CN)_2(aq) + 2\ OH^-(aq) \rightarrow CN^-(aq) + OCN^-(aq) + H_2O(l)$$

$$[\text{compared to } Cl_2(aq) + 2\ OH^-(aq) \rightarrow Cl^-(aq) + OCl^-(aq) + H_2O(l)$$

6. Just as the halogens form interhalogen compounds, such as iodine monochloride, ICl, so cyanogen forms pseudointerhalogen compounds such as ICN.

7. The cyanide ion forms numerous complex ions with transition metals, such as $[Cu(CN)_4]^{2-}$, which is similar to its chloride analog, $[CuCl_4]^{2-}$.

9.13 Biological Aspects

In this chapter, we have been looking at patterns in the Periodic Table. Therefore our two examples of biological applications relate to such patterns. First, we explore the importance of the group relationship of strontium to calcium; and second, the link between thallium and potassium which involves the potassium–silver–thallium connection that we discussed previously.

Strontium

Bone and teeth consist of crystals of hydroxyapatite, $Ca_5(PO_4)_3(OH)$, and the fibrous protein, collagen. Two of the reasons why nature chooses hydroxyapatite as a biological structural material is the insolubility of calcium phosphates and the high availability of the calcium ion. Nature might have used strontium instead of calcium except that strontium is about 100 times less abundant in nature than calcium.

When nuclear weapons were first being tested, it was thought that the hazards were localized and of little danger. The 1951 atmospheric tests in Nevada changed that view. Among the fission products was strontium-90, which, as might be expected, substitutes for calcium ion. In fact, strontium is preferentially absorbed over calcium. As a highly radioactive isotope (half-life 29 years),

Thallium Poisoning: Two Case Histories

The most infamous case of thallium poisoning was adapted into a 1995 movie: The Young Poisoner's Handbook. *(Caution: this movie is a black comedy with gruesome incidents.)*

In 1976, a 19-month-old girl from the Middle East country of Qatar was flown to England for treatment of a mysterious illness, initially suspected to be encephalitis. Despite a battery of tests, nothing specific could be found and the girl's condition continued to worsen. One of her intensive care nurses, Marsha Maitland, enjoyed reading Agatha Christie murder-mystery novels, and, at that time, Maitland was reading Christie's *The Pale Horse*. In this novel, the contract killer used a taste-less, water-soluble thallium(I) salt to repeatedly commit "perfect murders." Christie gave accurate descriptions of the symptoms of thallium(I) poisoning, and Maitland noticed how similar they were to the symptoms of the dying girl. She mentioned this fact to the attending physicians who used a forensic pathologist to test urine samples for thallium(I). Maitland was correct, the girl had a very high level of thallium(I) in her body. The soluble Prussian Blue and potassium ion treatments were immediately administered and after 3 weeks, the now-healthy girl was released from the hospital. Where had the thallium come from? Thallium(I) sulfate was used in the Middle East to kill cockroaches and rodents in drains, and it appears that the girl found some of the poison and ingested it.

The most prominent case of recent times occurred in China in 1995. A young chemistry student, Zhu Ling, at Beijing's Tsinghua University had taken seriously ill. Her parents rushed her to one of China's best hospitals, Beijing Union Medical College Hospital, but they were unable to identify the cause of her severe abdominal cramps and burning sensations in her limbs. Also, her hair began to fall out. Her former high school friend, Bei ZhiCheng, was very concerned about her and persuaded his roommate, Cai Quangqing, to help. Cai had access to the internet for research purposes, and he sent an SOS in English, describing Zhu's symptoms to a sci.med. newgroup. This unusual request sparked an enormous worldwide response with over 600 e-mail replies being received over the next 2 weeks. Among suggestions were myesthenia gravis and Guillain–Barré syndrome, but a consensus emerged of deliberate thallium poisoning.

Zhu's doctors initially resented the intrusion of Bei and Cai into the medical diagnosis. However, Zhu's parents took samples of their daughter's blood, urine, hair, and fingernails to the Beijing Institute of Labour, Hygiene, and Occupational Diseases for thallium testing. Analysis showed that Zhu had up to 1000 times the normal levels of thallium in her body. Others following the unfolding drama contacted the Los Angeles County Poison and Drug Information Center about treatment, and they informed the hospital of the importance of immediate soluble Prussian Blue and potassium ion treatment. Within a day, Zhu's thallium levels began to drop, and 10 days later, the thallium(I) concentration was undetectable. Beijing police initiated an investigation of the poisoning attempt.

the strontium-90 irradiates the bone marrow critical to reproduction of cells that mediate immune function. Baby teeth are a convenient way of measuring strontium-90 levels as they are naturally lost and provide measures of strontium-90 levels over the child's lifetime. There are no natural sources of strontium-90. Measurements of strontium-90 levels in baby teeth in St. Louis increased steadily until 1964 when there was a ban on atmospheric weapons tests. Children, with their rapid bone growth, readily absorb strontium-90 and strontium-90 levels have correlated with increases in childhood leukemia rates.

Though atmospheric nuclear testing has long been ended, research has indicated a possible link between radioactive emissions from commercial nuclear power plants and abnormally high levels of childhood leukemia (and other radiation-related diseases) in certain parts of the United States. Thus, concerns about this hazardous isotope still exist.

On the other hand, radioactive strontium-85 is being used to treat extreme bone pain, often resulting from bone cancer. Like the other isotopes of strontium, highly radioactive strontium-85 accumulates in the bones of these patients, the radiation specifically deadening the surrounding nerves that are causing the extreme pain.

Thallium(I)

Just as thallium(I) resembles potassium in its chemistry, so it resembles potassium in its biochemistry. Thallium(I) is a highly toxic ion that accumulates in tissues with high concentrations of potassium ion. Thallium(I) invades cells so readily as it is preferred over potassium by the same cellular transport mechanism. Once in the cell, thallium(I) substitutes for potassium in potassium-activated enzymes and disrupts the functioning of the enzymes. It is also believed that thallium(I), a very soft acid, combines with the soft-base sulfur of thio-aminoacid groups in mitochondria, blocking oxidative phosphorylation. Thallium poisoning causes degenerative changes in all cells, but particularly the nervous system and hair follicles. Unfortunately, many of its symptoms can be diagnosed as other illnesses.

There are two complementary treatments of thallium poisoning. First is the administration of potassium iron(III) hexacyanoferrate(II), $K^+[Fe^{3+}Fe^{2+}(CN^-)_6]$, commonly called soluble Prussian Blue, a nontoxic compound. The use of this compound depends upon the fact that thallium(I) resembles silver(I) in some of its chemistry—which it does in this case. That is, it forms an insoluble compound with the complex ion:

$$[Fe^{3+}Fe^{2+}(CN^-)_6]^-(aq) + Tl^+(aq) \rightarrow Tl[Fe^{3+}Fe^2(CN^-)_6](s)$$

By this means, any thallium(I) in the gastrointestinal tract will be precipitated and excreted. In addition, administration of high potassium ion concentrations can be used to shift the equilibria for tissue-bonded thallium(I).

EXERCISES

9.1 Explain what is meant by: (a) the Zintl principle; (b) the diagonal relationship.

9.2 Explain what is meant by: (a) the knight's move relationship; (b) pseudo-elements.

9.3 What are the common features of an alum?

9.4 What are the similarities and differences of aluminum and iron(III) chemistry?

9.5 Explain the trends in melting point for: (a) the Group 2 elements; (b) the Group 17 elements; (c) the Group 14 elements.

9.6 Write the formulas for the Period 4 main group metal fluorides. Suggest the bonding type in each case.

9.7 Write the formulas for the Period 4 main group metal hydrides. Suggest the bonding type in each case.

9.8 If calcium hydride is melted and electrolyzed, what would you expect will be the products at the (a) anode; (b) cathode?

9.9 Compare and contrast the chemistry of (a) manganese(VII) and chlorine(VII); (b) silver(I) and rubidium.

9.10 There is the following trend in melting points of Period 2 metal oxides: MgO, 2800°C; CaO, 1728°C; SrO, 1635°C; BaO, 1475°C. Suggest a reason for this trend.

9.11 Draw structures for the boron–nitrogen "combo" analog of: (a) naphthalene, $C_{10}H_8$; (b) biphenyl, $C_{12}H_{10}$.

9.12 Suggest which of the following anions is likely to give a precipitate with the ammonium ion: phosphate, PO_4^{3-}, or tetraphenylborate, $[B(C_6H_5)_4]^-$. Give your reason.

9.13 Deduce an electron-dot structure of the cyanogen, $(CN)_2$, molecule and draw the molecular shape. Experimental measurements of the carbon–carbon bond length show it to be shorter than the simple bonding model predicts. Suggest an explanation.

9.14 The thiocyanate ion, SCN^-, also behaves as a pseudo-halide ion.
(a) Write the formula of the parent pseudo-halogen.
(b) Deduce an insoluble compound of the thiocyanate ion.

9.15 Of the alkali metals, the ammonium ion most closely resembles the rubidium or cesium ion. Explain.

9.16 Which metal hydroxide is isostructural with aluminum hydroxide?

9.17 Give one example of how the cyanide ion resembles: (a) the fluoride ion; (b) the chloride ion; (c) the iodide ion.

9.18 Write a chemical equation for the reaction of water with liquid silicon tetrachloride and titanium(IV) chloride.

9.19 Write a chemical equation for the reaction of water with solid sulfur trioxide and with chromium(VI) oxide.

9.20 What are the formulas of the highest oxidation state oxide of chlorine and manganese? What other oxides of the two elements resemble each other in formula?

9.21 (a) Write the formulas of aluminum oxide and scandium oxide.
(b) Write formulas for the oxyanions of phosphorus and vanadium in their highest oxidation states.

9.22 Explain briefly why aluminum might be considered a member of Group 3 instead of Group 13.

9.23 Titanium(IV) nitrate shares many properties, including identical crystal structure, with a metal of a different group. Suggest the identity of the metal.

9.24 Phosphorus forms an oxychloride of formula, $POCl_3$. Which transition metal is likely to form an oxychloride of matching formula?

9.25 In Table 9.6, the melting point of neon is significantly less than that of nitrogen, oxygen, or fluorine. Suggest an explanation.

9.26 One source of scandium is the ore sterrite, $ScPO_4 \cdot 2H_2O$. This ore is isostructural with an ore of a main group metal. Write the formula of that ore.

9.27 Suggest an explanation in terms of orbital occupancy why the oxidation state of +2 is found for the element europium.

9.28 Suggest an explanation in terms of orbital occupancy why the oxidation state of +4 is found for the element cerium.

9.29 Suggest an explanation in energy terms why the oxidation state of +2 is found for the element europium.

9.30 Suggest an explanation in energy terms why the oxidation state of +4 is found for the element cerium.

9.31 "Knight's move" relationships exist when both elements are in the same oxidation states. Which oxidation states would be shared by: (a) copper and indium; (b) cadmium and lead?

9.32 "Knight's move" relationships exist when both elements are in the same oxidation states. Which oxidation states would be shared by: (a) indium and bismuth; (b) zinc and tin?

9.33 Silver bromide has a melting point of 430°C. Which bromide would you expect to have a similar melting point? Check data tables and confirm your answer.

9.34 Write the formulas of two oxo-anions that seaborgium (Sg) might form.

9.35 Carbon and nitrogen form the cyanide ion: $(C\equiv N)^-$. Write the formulas of the corresponding isoelectronic species of: (a) carbon with oxygen; (b) carbon with carbon.

9.36 Sodium is the only alkali metal for which the dioxide(2−), Na_2O_2, is the most stable oxide species. Using Table 9.20, deduce the alkaline earth metal that also forms a stable dioxide(2−) compound.

9.37 Monazite, the lanthanoid phosphate ore, MPO_4, also typically contains about 3 percent of a Group 3 metal ion. Suggest the identity of this ion.

9.38 Phosphorus forms halide and pseudohalide compounds of the form PX_3. Write the formula for the compound with cyanide.

BEYOND THE BASICS

9.39 Write a chemical equation for the reaction of silane, SiH_4, with dioxygen.

9.40 If silane, SiH_4, has a positive free energy of formation, why does it exist at all?

9.41 Magnesium and zinc have similar chemistries as a result of the (n) and $(n + 10)$ relationship. They can be considered related in a different way if the "true" transition metals, Groups 3 to 10, are removed. What would be another way of considering their relationship?

9.42 You have a solution containing either magnesium ion or zinc ion. Suggest a reaction that you could use to identify the cation.

9.43 Moisture/density gauges are used by construction companies to determine the properties of the soil on which they are building. These gauges use two radioactive sources, cesium-137 and americium-241, in their functioning. Many are stolen in the United States, as there is a large black market demand for them. Some are abandoned and/or broken open when the thieves discover they contain radioactive materials. Suggest why the cesium-137 could be a particular hazard.

9.44 There is a compound $Zn_x[P_{12}N_{24}]Cl_2$ that has a similar crystal structure to that of the mineral, sodalite, $Na_8[Al_6Si_6O_{24}]Cl_2$. The total number of valence electrons in the aluminosilicate ion and the phosphonitride ion are the same.
(a) Calculate the charge on the phosphonitride ion.
(b) Calculate x, the number of zinc ions in the zinc phosphonitride.

9.45 Predict the formula for the highest fluoride of iodine. Research whether this compound exists. Suggest why its atom ratio is different from the highest fluoride of chlorine.

9.46 Calculate the oxidation numbers of the other element for each fluoride in Table 9.10 and identify the pattern in these numbers.

9.47 Calculate the oxidation numbers of the other element for each hydride in Table 9.11 and identify the pattern in these numbers.

9.48 Research the formulas of the highest oxides as the Group 8 elements are descended. Calculate the oxidation numbers of the Group 8 element in each case. What do you note about the highest value?

Chapter 10 Hydrogen

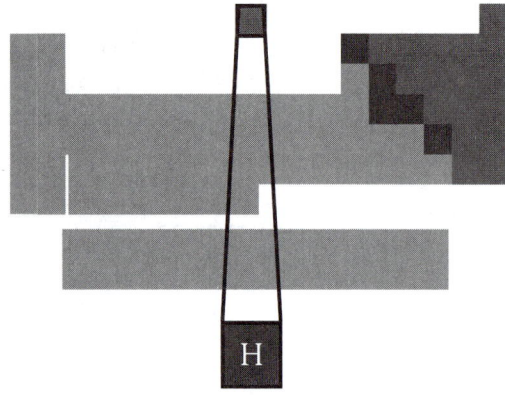

There is only one element in the periodic table that does not belong to any particular group—hydrogen. This element has a unique chemistry. Furthermore, its three isotopes differ so much in their molar masses that the physical and chemical properties of the isotopes are measurably different.

Although hydrogen was described about 200 years ago, the existence of different isotopes of hydrogen is a more recent discovery. In 1931, some very precise measurements of atomic mass indicated that there might be different isotopes of hydrogen. Harold C. Urey at Columbia University decided to try to separate them by applying the concept that the boiling point of a species depends partially on its molar mass. Urey evaporated about 5 L of liquid hydrogen, hoping that the last 2 mL would contain a larger than usual proportion of any higher molar mass isotope. The results proved him correct; the residue had a molar mass double that of normal hydrogen. This form of hydrogen was named deuterium.

Frederick Soddy, who had devised the concept of isotopes, refused to believe that deuterium was an isotope of hydrogen. The reason for his lack of acceptance was his own definition of isotopes: that isotopes are nonseparable. But Urey had separated the two forms of hydrogen, so Soddy argued that they could not be isotopes, preferring to believe that Urey was incorrect rather than question his own definition. Except for Soddy's negative opinion, Urey received considerable recognition for his discovery, culminating in the Nobel Prize for chemistry in 1934. Ironically, the earlier atomic mass measurements were subsequently shown to be in error. In particular, they did not provide any evidence for the existence of hydrogen isotopes. Thus, Urey's search, although successful, was based on erroneous information.

10.1 *Isotopes of Hydrogen*

The isotopes of hydrogen are particularly important in chemistry. Because the relative mass differences between hydrogen's isotopes are so large, there is a significant dissimilarity in physical properties and, to a lesser extent, in chemical behavior among them. Natural hydrogen contains three isotopes: protium, or "common" hydrogen, which contains zero neutrons (abundance 99.985 percent); deuterium, which contains one neutron (abundance 0.015 percent); and radioactive tritium, which contains two neutrons (abundance 10^{-15} percent). In fact, this is the only set of isotopes for which special symbols are used: H for protium, D for deuterium, and T for tritium. As the molar mass of the isotopes increases, there is a significant increase in both the boiling point and the bond energy (Table 10.1).

Table 10.1 Physical properties of the isotopes of hydrogen

Isotope	Molar mass ($g \cdot mol^{-1}$)	Boiling point (K)	Bond energy ($kJ \cdot mol^{-1}$)
H_2	2.02	20.6	436
D_2	4.03	23.9	443
T_2	6.03	25.2	447

The strongest single bond between two atoms of the same element is that between two atoms of tritium in $T_2 - 447\ kJ \cdot mol^{-1}$. That between two hydrogen-1 atoms is $436\ kJ \cdot mol^{-1}$.

The covalent bonds of deuterium and tritium with other elements are also stronger than those of common hydrogen. For example, when water is electrolyzed to give hydrogen gas and oxygen gas, it is the O—H covalent bonds that are broken more readily than O—D bonds. As a result, the remaining liquid contains a higher and higher proportion of "heavy" water, deuterium oxide. When 30 L of water is electrolyzed down to a volume of 1 mL, the remaining liquid is about 99 percent pure deuterium oxide. Normal water and "heavy" water, D_2O, differ in all their physical properties; for example, deuterium oxide melts at 3.8°C and boils at 101.4°C. The density of deuterium oxide is about 10 percent higher than that of protium oxide at all temperatures. As a result, heavy water ice cubes will sink in "light" water at 0°C. Deuterium oxide is used widely as a solvent so that the hydrogen atoms in solute molecules can be studied without their properties being "swamped" by those in the aqueous solvent. Reaction pathways involving hydrogen atoms also can be studied by using deuterium-substituted compounds.

Tritium is a radioactive isotope with a half-life of about 12 years. With such a short half-life, we might expect that none survives naturally; in fact, tritium is constantly being formed by the impact of cosmic rays on atoms in the upper atmosphere. One pathway for its production involves the impact of a neutron on a nitrogen atom:

$$^{14}_{7}N + {}^{1}_{0}n \rightarrow {}^{12}_{6}C + {}^{3}_{1}T$$

The isotope decays to give the rare isotope of helium, helium-3:

$$\underset{1}{\overset{3}{T}} \rightarrow \underset{2}{\overset{3}{He}} + \underset{-1}{\overset{0}{e}}$$

There is a significant demand for tritium. It is sought for medical purposes, where it is useful as a tracer. In its radioactive decay, the isotope emits low-energy electrons (β rays) but no harmful γ rays. The electrons can be tracked by a counter, and cause minimal tissue damage. The most significant consumers of tritium are the military forces of the countries possessing hydrogen (more accurately, tritium) bombs. To extract the traces of tritium that occur in water would require the processing of massive quantities of water. An easier synthetic route entails the bombardment of lithium-6 by neutrons in a nuclear reactor:

$$\underset{3}{\overset{6}{Li}} + \underset{0}{\overset{1}{n}} \rightarrow \underset{2}{\overset{4}{He}} + \underset{1}{\overset{3}{T}}$$

Tritium's short half-life creates a problem for military scientists of nuclear powers because, over time, the tritium content of nuclear warheads diminishes until it is below the critical mass needed for fusion. Hence, warheads have to be periodically "topped up" if they are to remain usable.

10.2   *Nuclear Magnetic Resonance*

One of the most useful tools for studying molecular structure is nuclear magnetic resonance (NMR). This technique involves the study of nuclear spin. As discussed in Chapter 2, Section 2.3, protons and neutrons have spins of ±½. In an atom, there are four possible permutations of nuclear particles: even numbers of both protons and neutrons, odd number of protons and even number of neutrons, even number of protons and odd number of neutrons, and odd numbers of both protons and neutrons. The last three categories, then, will have unpaired nucleons. This condition might be expected to occur in an enormous number of nuclei, but, with spin pairing being a major driving force for the stability of nuclei, only 4 of the 273 stable nuclei have odd numbers of both protons and neutrons.

Unpaired nucleons can have a spin of $+\frac{1}{2}$ or $-\frac{1}{2}$; each spin state has the same energy. However, in a magnetic field, the spin can be either parallel with the field or opposed to it, and the parallel arrangement has lower energy. The splitting of (difference between) the two energy levels is very small and corresponds to the radio frequency range of the electromagnetic spectrum. When we focus a radio wave source on the sample with unpaired nucleons and adjust the frequency of radio waves to the energy level of the splitting, electromagnetic radiation is absorbed by the sample as unpaired nucleons reverse their spins to oppose the field; that is, they move to the higher energy level. In a field of 15 000 gauss, absorption happens at 63.9 MHz (or $6.39 \times 10^7 \ s^{-1}$) for an isolated proton.

The relative intensity of the absorption depends very much on the identity of the nucleus. As it happens, hydrogen-1 gives the most intense absorption among the nuclei (Figure 10.1). This is fortunate, because hydrogen is the most common element in the universe and therefore readily available for study. Even today, years after the discovery of NMR, hydrogen is the element most studied by this technique.

If this were all that NMR could do, it would not be a particularly useful technique. However, the electrons surrounding a nucleus affect the actual

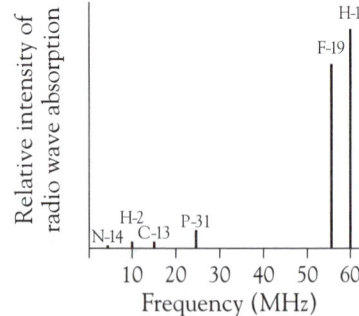

Figure 10.1 Relative intensities of the unique absorption by common isotopes in a magnetic field of 14 000 gauss.

Isotopes in Chemistry

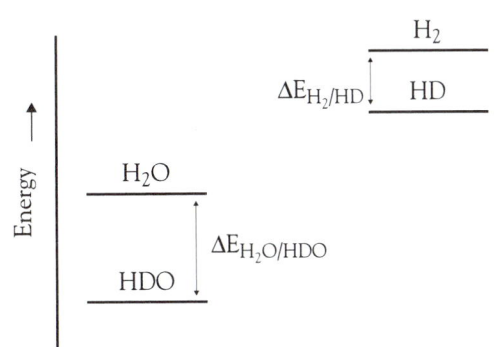

Figure 10.2 A plot of relative energies for the hydrogen gas–water equilibrium for the hydrogen and deuterium isotopes.

In our discussions of the chemistry of elements, we rarely mention the effects of isotopes on chemical reactions, yet such effects are of considerable importance, particularly for hydrogen where isotopic mass differences are so large. The difference in isotope masses can affect reaction rates and in the position of equilibria. We have a better understanding of the role of isotopes as a result of the Bigeleisen–Mayer formulation (co-discovered by the same Maria Goeppert-Mayer, who we mentioned in the Chapter 2 feature on the Origin of the Shell Model of the Nucleus). This relationship showed that bonds to light isotopes are easier to break than to those of heavier isotopes. Thus, the heavy isotope of an element will favor chemical species in which it is bound more strongly. We find, for example, that in the environment, the heavier isotope of sulfur, sulfur-34, is slightly more abundant as sulfate (where sulfur has strong covalent bonds to four oxygen atoms) than as sulfide. It is possible to separate isotopes by means of chemical equilibria. A good example of an isotope effect in a chemical reaction is:

$$HD(g) + H_2O(g) \rightleftharpoons H_2(g) + HDO(g)$$

The plots of energy wells are shown for the four species in Figure 10.2. It can be seen that deuterium forms a proportionally stronger bond with oxygen than with hydrogen. Thus there is an energy preference for the HDO/H_2 combination; or, in other words, the equilibrium lies to the right and it is the water that is enriched in deuterium. It is by means of a series of such equilibria that pure D_2O can be produced.

Carbon is another element for which isotope effects are particularly important—in fact, the proportion of carbon-13 can vary from 0.99 percent to 1.10 percent, depending on the carbon source. When carbon dioxide is absorbed by plants and converted to sugars, different photosynthetic pathways result in different fractionation of the carbon isotopes. For example, from the carbon isotope ratio, we can tell whether a sugar sample is derived from sugar cane or sugar beets. These isotope ratio tests have become invaluable in checking consumer foodstuffs for quality, such as the possible adulteration of honey or wine with low-cost sugar solution. In the chemistry laboratory, there are many applications of isotope effects, including the correlation of infrared absorption spectra with molecular vibrations.

magnetic field experienced by the nucleus. Because the magnetic field for each environment differs from that applied by a magnet, the splitting of the energy levels and the frequency of radiation absorbed are unique for each species. Thus, absorption frequency reflects the atomic environment. The difference in frequency absorbed (called the chemical shift, or simply, shift) is very small—about 10^{-6} of the signal itself. Hence, we report the shifts in terms of parts per million (ppm). In addition, splitting of the transition levels can occur

Table 10.2   Reasons for and against placing hydrogen in Group 1 or 17

	Argument for placement	Argument against placement
Alkali metal group	Forms monopositive ion, $H^+(H_3O^+)$	Is not a metal
	Has a single s electron	Does not react with water
Halogen group	Is a nonmetal	Rarely forms mononegative ion, H^-
	Forms a diatomic molecule	Is comparatively unreactive

through interaction with neighboring odd-spin nuclei. Thus, the relative locations of atoms can often be identified by NMR. This technique is a great aid to chemists, particularly organic chemists, both for the identification of a compound and for the study of electron distributions within molecules. It is also used extensively in the health field under the name of magnetic resonance imaging (MRI).

10.3   Properties of Hydrogen

As stated earlier, hydrogen is a unique element, not belonging to any of the other groups in the periodic table. Some versions of the periodic table place it as a member of the alkali metals, some as a member of the halogens; others place it in both locations; and a few place it on its own. The basic reasons for and against placement of hydrogen in either Group 1 or Group 17 are summarized in Table 10.2. Throughout this book, hydrogen has a place in the periodic table all its own, emphasizing the uniqueness of this element. With an electronegativity higher than those of the alkali metals and lower than those of the halogens, it makes sense to place hydrogen midway between the two groups.

Dihydrogen is a colorless, odorless gas that liquefies at $-253°C$ and solidifies at $-259°C$. Hydrogen gas is not very reactive, partly because of the high H—H covalent bond energy (436 kJ·mol^{-1}). This bond is stronger than the covalent bonds hydrogen forms with most other nonmetals; for example, the H—S bond energy is only 347 kJ·mol^{-1}. Recall that only when the bond energies of the products are similar to or greater than those of the reactants are spontaneous reactions likely. One such reaction is combustion of dihydrogen with dioxygen to produce water. If hydrogen gas and oxygen gas are mixed and sparked, the reaction is explosive:

$$2 H_2(g) + O_2(g) \rightarrow 2 H_2O(g)$$

The reaction has to be enthalpy driven because there is a decrease in entropy (see Chapter 6, Section 6.1). If we add the bond energies, we see that the strong O—H bond (464 kJ·mol^{-1}) makes the reaction thermodynamically feasible (Figure 10.3).

Figure 10.3   Theoretical enthalpy cycle for the formation of water.

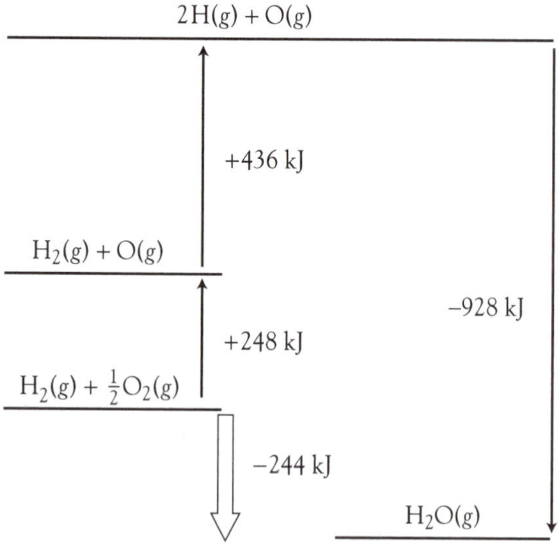

Dihydrogen reacts with the halogens, with the rate of reaction decreasing down the group. It has a violent reaction with difluorine to give hydrogen fluoride:

$$H_2(g) + F_2(g) \rightarrow 2\ HF(g)$$

The reaction of dihydrogen with dinitrogen is very slow in the absence of a catalyst. (This reaction is discussed more fully in Chapter 15.)

$$3\ H_2(g) + N_2(g) \rightleftharpoons 2\ NH_3(g)$$

At high temperatures, dihydrogen reduces many metal oxides to the metallic element. Thus copper(II) oxide is reduced to copper metal:

$$CuO(s) + H_2(g) \xrightarrow{\Delta} Cu(s) + H_2O(g)$$

In the presence of a catalyst (usually powdered palladium or platinum), dihydrogen will reduce carbon–carbon double and triple bonds to single bonds. For example, ethene, C_2H_4, is reduced to ethane, C_2H_6:

$$H_2C{=}CH_2(g) + H_2(g) \rightarrow H_3C{-}CH_3(g)$$

The reduction with dihydrogen is used to convert unsaturated liquid fats (edible oils), which have numerous carbon–carbon double bonds, to higher melting, partially saturated solid fats (margarines), which contain fewer carbon–carbon double bonds.

Preparation of Dihydrogen

In the laboratory, hydrogen gas can be generated by the action of dilute acids on many metals. A particularly convenient reaction is that between zinc and dilute hydrochloric acid:

$$Zn(s) + 2\ HCl(aq) \rightarrow ZnCl_2(aq) + H_2(g)$$

There are several different routes of industrial synthesis, one of these being the *steam reformer process*. In the first step of this process, the endothermic reaction of natural gas (methane) with steam at high temperatures gives carbon monoxide and hydrogen gas. It is difficult to separate the two products because the mixture must be cooled below $-205°C$ before the carbon monoxide will condense. To overcome this problem and to increase the yield of hydrogen gas, the mixture is cooled, additional steam is injected, and the combination is passed over a different catalyst system. Under these conditions, the carbon monoxide is oxidized in an exothermic reaction to carbon dioxide, and the added water is reduced to hydrogen:

$$CH_4(g) + H_2O(g) \xrightarrow[800°C]{Ni} CO(g) + 3\ H_2(g)$$

$$CO(g) + H_2O(g) \xrightarrow[400°C]{Fe_2O_3/Cr_2O_3} CO_2(g) + H_2(g)$$

The carbon dioxide can be separated from hydrogen gas in several ways. One is to cool the products below the condensation temperature of carbon dioxide ($-78°C$), which is much higher than that of dihydrogen ($-253°C$). However, this process still requires large-scale refrigeration systems. Another route involves passage of the gas mixture through a solution of potassium carbonate. As we saw in Chapter 7, Section 7.7 and Chapter 9, Section 9.4, carbon dioxide is an acid oxide, unlike carbon monoxide, which is neutral. Carbon dioxide reacts with the carbonate ion and water to give 2 mol of the hydrogen carbonate ion. When reaction is complete, the potassium hydrogen carbonate solution can be removed and heated to regenerate the

Searching the Depths of Space for the Trihydrogen Ion

We think of chemistry in terms of what occurs at about 100 kPa and 25°C, the conditions on the surface of this planet. But stable chemical species can be very different in other parts of the universe. One of the most interesting is the trihydrogen cation. This ion is formed in the upper levels of planetary atmospheres where solar wind and other sources of high-energy electrons collide with hydrogen molecules.

$$H_2(g) + e^- \rightarrow H_2{}^+(g) + 2\ e^-$$

$$H_2{}^+(g) + H_2(g) \rightarrow H_3{}^+(g) + H(g)$$

The trihydrogen ion is very stable under these conditions of low pressure. The ion has a very characteristic and extremely intense vibrational emission spectrum. Thus, astrochemists have been able to study the trihydrogen ion in the outer planets, Jupiter, Saturn, and Uranus, to give information about the upper levels of the atmospheres of these gas giants.

The ion consists of an equilateral triangle of hydrogen atoms with H–H bond lengths of 87 pm. It is thermodynamically stable in isolation, but in planetary atmospheres, the ion undergoes a variety of ion–ion reactions, the most important decomposition pathway being that of the reaction with electrons:

$$H_3{}^+(g) + e^- \rightarrow H_2(g) + H(g)$$

The most exciting discoveries are yet to come. By measuring gravitational wobbles of nearby stars, astronomers have concluded that many of the stars have planets orbiting them. At the present time, over 40 stars have been shown to have one or more planets by this indirect means. But no-one has yet actually seen one of these planets directly. The reflected sunlight from these planets will be far too weak to detect, but the power of the latest generation of telescopes might be just sensitive enough to detect the trihydrogen emission spectrum. It is probably just a matter of time and the development of even more sensitive telescopes before the telltale trihydrogen spectrum will directly confirm the existence of planets around other stars in our galaxy.

potassium carbonate, while the pure carbon dioxide gas can be collected and pressurized:

$$K_2CO_3(aq) + CO_2(g) + H_2O(l) \rightleftharpoons 2\ KHCO_3(aq)$$

For most purposes, the purity of the dihydrogen (molecular hydrogen) obtained from thermochemical processes is satisfactory. However, very pure hydrogen gas (at least 99.9 percent) is generated by an electrochemical route, the electrolysis of a sodium hydroxide or potassium hydroxide solution. The reaction produces dioxygen at the anode and dihydrogen at the cathode:

Cathode $2\ H_2O(l) + 2\ e^- \rightarrow 2\ OH^-(aq) + H_2(g)$

Anode $2\ OH^-(aq) \rightarrow H_2O(l) + \tfrac{1}{2}\ O_2(g) + 2\ e^-$

10.4 *Hydrides*

Binary compounds of hydrogen are given the generic name of hydrides. Hydrogen, which forms binary compounds with most elements, has an electronegativity only slightly above the median value of all the elements in the Periodic Table. As a result, it behaves as a weakly electronegative nonmetal, forming ionic compounds with very electropositive metals and covalent compounds with nonmetals as we discussed in Chapter 9, Section 9.2. In addition, it forms metallic hydrides with some of the transition metals. The distribution of the three major types of hydrides is shown in Figure 10.4.

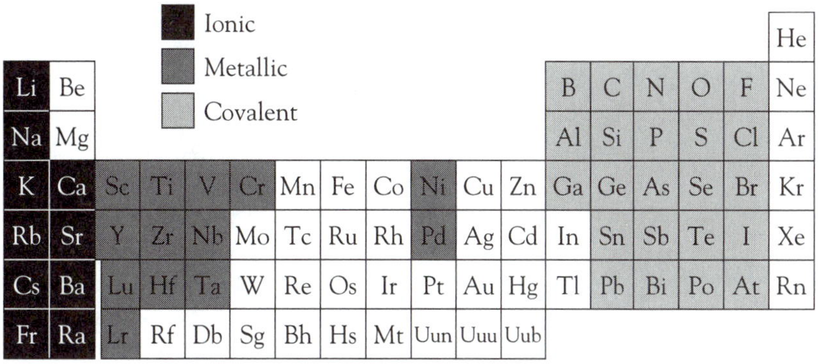

Figure 10.4 Three common types of hydrides: ionic, covalent, and metallic. Only the main and transition groups are shown. A few lanthanoids and actinoids also form metallic hydrides (not shown). For unshaded elements, either hydrides are unknown or poorly characterized.

Ionic Hydrides

All the ionic hydrides are white solids and are formed only by the most electropositive metals. These ionic crystals contain the metal cation and the hydride ion (H^-). Proof of the existence of this anion comes from the electrolysis of lithium hydride in molten lithium chloride. During this process, hydrogen is evolved at the anode, the positive electrode:

$$2\ H^-(\text{LiCl}) \rightarrow H_2(g) + 2\ e^-$$

All these hydrides are very reactive; for example, dihydrogen is produced in the presence of any moisture:

$$\text{LiH}(s) + H_2O(l) \rightarrow \text{LiOH}(aq) + H_2(g)$$

The hydrides can be used as reducing agents. Thus, calcium hydride can be used to reduce water to hydrogen gas. This reaction is sometimes used in organic chemistry as a means to chemically dry organic solvents:

$$\text{CaH}_2(s) + H_2O(l) \rightarrow \text{Ca(OH)}_2(s) + H_2(g)$$

Covalent Hydrides

Hydrogen forms compounds containing covalent bonds with all the nonmetals (except the noble gases) and with very weakly electropositive metals such as gallium and tin. Almost all the simple covalent hydrides are gases at room temperature. There are three subcategories of covalent hydrides:

Those in which the hydrogen atom is nearly neutral

Those in which the hydrogen atom is substantially positive

Those in which the hydrogen atom is slightly negative, including the electron-deficient boron compounds

For the first category, because of their low polarity, the sole intermolecular force between neighboring hydride molecules is dispersion; as a result, these covalent hydrides are gases with low boiling points. Typical examples of these hydrides are stannane (systematic name: tin(IV) hydride), SnH_4 (b.p. $-52°C$), and phosphine, PH_3 (b.p. $-90°C$). There are a few hydrides which have polymeric structures with hydrogen atoms bridging between the metal atoms. It is the "weak" metals: beryllium, magnesium, aluminum, copper, and zinc that form these structures.

The largest group of near-neutral covalent hydrides contains carbon—the hydrocarbons—and comprises the alkanes, the alkenes, the alkynes, and the aromatic hydrocarbons. Many of the hydrocarbons are large molecules in which the intermolecular forces are strong enough to allow them to be liquids or solids at room temperature. All the hydrocarbons are thermodynamically unstable toward oxidation. For example, methane reacts spontaneously with dioxygen to give carbon dioxide and water:

$$CH_4(g) + 2\ O_2(g) \rightarrow CO_2(g) + 2\ H_2O(g) \qquad \Delta G° = -800\ \text{kJ·mol}^{-1}$$

The process is very slow unless the mixture is ignited; that is, the reaction has a high activation energy.

Ammonia, water, and hydrogen fluoride belong to the second category of covalent hydrides—hydrogen compounds containing positively charged hydrogen atoms. These compounds differ from the other covalent hydrides in their abnormally high melting and boiling points. This property is illustrated by the boiling points of the Group 17 hydrides (Figure 10.5).

The positively charged hydrogen in these compounds is attracted by an electron pair on another atom to form a weak bond that is known as a hydrogen bond (discussed in Chapter 3, Section 3.16) but is more accurately called a *protonic bridge*. Even though, as an intermolecular force, protonic bridging is very strong, it is still weak compared to a covalent bond. For example, the $H_2O\cdots HOH$ protonic bridge has a bond energy of 22 kJ·mol^{-1}, compared to 464 kJ·mol^{-1} for the O—H covalent bond. In introductory chemistry texts, protonic bridges are regarded as very strong dipole–dipole interactions occurring as a result of the very polar covalent bonds in the bridged molecules. However, according to this concept of electrostatic attraction, hydrogen chloride should also show this effect, and it does not (to any significant extent).

Hydrogen bonds can also be described in terms of a covalent model using molecular orbital theory. This model utilizes the overlap of a σ orbital on one water molecule with a σ orbital on another molecule. The interaction of these two orbitals results in a bonding/nonbonding pair occupied by one electron pair (in the bonding set). The observation that hydrogen bond lengths are usually much shorter than the sum of the van der Waals radii of the two atoms supports this model. Furthermore, we find that the stronger the protonic bridge, the weaker the O—H covalent bond. Thus, the two bonds are strongly interrelated.

The third category of hydrides includes diborane, B_2H_6, silane, SiH_4, germane, GeH_4, and stannane, SnH_4. The hydrides containing negatively charged hydrogen react violently with oxygen. For example, stannane burns to give tin(IV) oxide and water.

$$SnH_4(g) + 2\ O_2(g) \rightarrow SnO_2(s) + 2\ H_2O(l)$$

A partially negative hydrogen (a *hydridic hydrogen*) is thus much more reactive than a partially positive hydrogen atom.

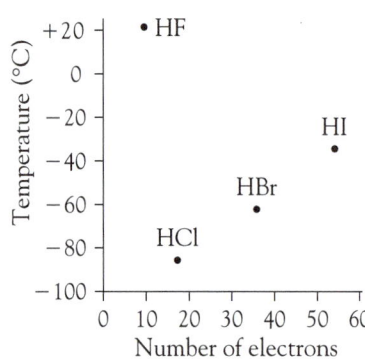

Figure 10.5 Boiling points of the Group 17 hydrides.

The Bonding in Boranes

Three elements form large numbers of hydrides: carbon, boron, and silicon. The hydrides of carbon and silicon contain "normal" covalent bonds, but the bonds in the boron hydrides are unusual because some hydrogen atoms link, or bridge, pairs of boron atoms. Boron has only three outer electrons; hence, the expected formulation of BH_3 would not satisfy the octet rule for boron. The utilization of hydrogen atoms as bridges means, however, that one electron pair can satisfy the bonding requirements of two boron atoms. These bonds are not protonic bridges, but hydridic bridges, because the hydrogen possesses a partial negative charge. The reversed polarity of the bond results in a high chemical reactivity for these compounds. The simplest member of the series is diborane, B_2H_6. Each terminal hydrogen atom forms a normal two-electron bond with a boron atom. Each boron atom then has one electron left, and this is paired with the electron of one of the bridging hydrogen atoms (Figure 10.6).

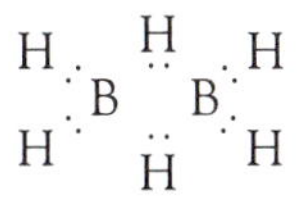

The shape of the molecule can be described as approximately tetrahedral around each boron atom, with the bridging hydrogen atoms in what are sometimes called "banana bonds." The hydridic bonds behave like weak covalent bonds (Figure 10.7).

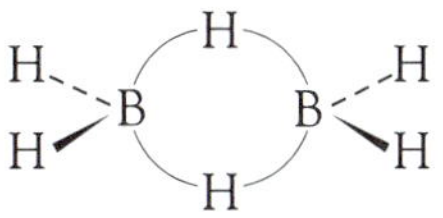

The bonding in a diborane molecule can be described in terms of hybridization concepts. According to these concepts, the four bonds, separated by almost equal angles, would correspond to sp^3 hybridization. Three of the four hybrid orbitals will contain single electrons from the boron atom. Two of these half-filled orbitals would then be involved in bonding with the terminal hydrogen atoms. This arrangement would leave one empty and one half-filled hybrid orbital.

To explain how we make up the eight electrons in the sp^3 orbital set, we consider that the single half-filled hybrid sp^3 orbitals of the two borons overlap with each other and with the $1s$ orbital of a bridging hydrogen atom at the same time. This arrangement will result in a single orbital that encompasses all three atoms (a *three-center bond*). This orbital is capable of containing two electrons (Figure 10.8). An identical arrangement forms the other B—H—B bridge. The bonding electron distribution of diborane is shown in Figure 10.9.

Alternatively, we can consider the molecular orbital explanation (see Chapter 3, Section 3.3). The detailed molecular orbital diagram for this eight-atom molecule is complex. Although molecular orbitals relate to the molecule as a whole, it is sometimes possible to identify molecular orbitals that are involved primarily in one particular bond. In this case, we find that the mixing of the orbital wave functions of the atoms in each bridge bond results in the formation of three molecular orbitals. When we compare the

Figure 10.6 Electron pair arrangement in diborane, B_2H_6.

Figure 10.7 Geometry of the diborane molecule.

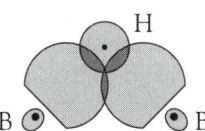

Figure 10.8 Overlap of the sp^3 hybrid orbitals of the two boron atoms with 1s obital of the bridging hydrogen atom.

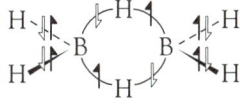

Figure 10.9 The electron pairing that is consistent with sp^3 hybridization for each boron atom and with the two-electron, three-atom B—H—B bridging bonds. The electrons contributed by the hydrogen atoms are the open half-headed arrows.

Molecular
orbitals

Figure 10.10 The molecular orbitals that are involved in the hydridic bridge.

energies of the atomic orbitals with those of the molecular orbitals, we find that one molecular orbital is lower in energy (σ bonding), one is higher in energy (σ antibonding), and the third has an energy level equivalent to the mean energy of the three component atomic orbitals (σ nonbonding). The bridging hydrogen atom contributes one electron, and each boron atom contributes one-half electron. This arrangement fills the bonding orbital between the three atoms (Figure 10.10). Because there is one bonding orbital shared between two pairs of atoms, the bond order for each B—H component must be ½. The same arguments apply to the other bridge. From bond energy measurements, we do indeed find each B—H bridging bond to be about half the strength of a terminal B—H bond, although it is still in the energy range of a true covalent bond, unlike the much weaker protonic bridges in hydrogen-bonded molecules. Of equal importance, the set of molecular orbitals shows that the structure makes maximum use of the few boron electrons. The presence of more electrons would not strengthen the bond because these electrons would enter nonbonding molecular orbitals.

Metallic (*d*-Block) Hydrides

Some transition metals form a third class of hydrides, the metallic hydrides. These compounds are often nonstoichiometric; for example, the highest hydrogen–titanium ratio is found in a compound with the formula $TiH_{1.9}$. The nature of these compounds is complex. Thus, the titanium hydride mentioned previously is now believed to consist of $(Ti^{4+})(H^-)_{1.9}(e^-)_{2.1}$. It is the free electrons that account for the metallic luster and high electrical conductivity of these compounds. The density of the metal hydride is often less than that of the pure metal because of structural changes in the metallic crystal lattice, and the compounds are usually brittle. The electrical conductivity of the metallic hydrides is generally lower than that of the parent metal as well.

Most metallic hydrides can be prepared by warming the metal with hydrogen under high pressure. At high temperatures, the hydrogen is released as dihydrogen gas again. Many alloys (for example, Ni_5La) can absorb and release copious quantities of hydrogen. Their proton densities exceed that of liquid hydrogen, a property that makes them of great interest as a means of hydrogen storage for use in hydrogen-powered vehicles.

A major use of a metal hydride is in the nickel metal hydride batteries that are used in portable computers, cordless vacuum cleaners, cellular phones, and many other cordless electrical devices. The first essential for such a battery is to find a metal alloy that reversibly absorbs and releases hydrogen at ambient temperatures. These hydrogen-absorbing alloys combine a metal, A, whose hydride formation is exothermic with a metal, B, whose hydride formation is endothermic. The alloys exist as four possible ratios: AB (e.g., TiFe), AB_2 (e.g., $ZnMn_2$), AB_5 (e.g., $LaNi_5$), and A_2B (e.g., Mg_2Ni). Combinations are sought that give an essentially energy-neutral hydride formation. It is the combinations $TiNi_2$ and $LaNi_5$ that have proved best suited for the function. In the cells, at the anode, nickel(II) hydroxide is oxidized to nickel(III) oxide hydroxide, while at the cathode, water is reduced to hydrogen atoms which are absorbed into the metal alloy:

$$Ni(OH)_2(s) + OH^-(aq) \rightarrow NiO(OH)(s) + H_2O(l) + e^-$$

$$[\text{Ni-alloy}] + H_2O(l) + e^- \rightarrow [\text{Ni-alloy}]H + OH^-(aq)$$

10.5 *Water and Hydrogen Bonding*

Hydrogen bonding (or more correctly, protonic bridging) between water molecules is of particular importance for life on this planet. Without hydrogen bonding, water would melt at about $-100°C$ and boil at about $-90°C$. Hydrogen bonding results in another very rare property of water—the liquid phase is denser than the solid phase. For most substances, the molecules are packed closer in the solid phase than in the liquid, so the solid has a higher density than the liquid has. Thus, a solid usually settles to the bottom as it starts to crystallize from the liquid phase. Were this to happen with water, those lakes, rivers, and seas in parts of the world where temperatures drop below freezing would freeze from the bottom up. Fish and other marine organisms would be unlikely to survive in such environments.

Fortunately, ice is less dense than liquid water, so in subzero temperatures, a layer of insulating ice forms over the surface of lakes, rivers, and oceans, keeping the water beneath in the liquid phase. The cause of this abnormal behavior is the open structure of ice, which is due to the network of hydrogen bonds (Figure 10.11).

Upon melting, some of these hydrogen bonds are broken and the open structure partially collapses. This change increases the liquid's density. The density reaches a maximum at $4°C$, at which point the increase in density due to the collapsing of the hydrogen-bonded clusters of water molecules is overtaken by the decrease in density due to the increasing molecular motion resulting from the rise in temperature.

To appreciate another unusual property of water, we must look at a phase diagram. A phase diagram displays the thermodynamically stable phase of an element or compound with respect to pressure and temperature. Figure 10.12 shows an idealized phase diagram. The regions between the solid lines represent the phase that is thermodynamically stable under those particular

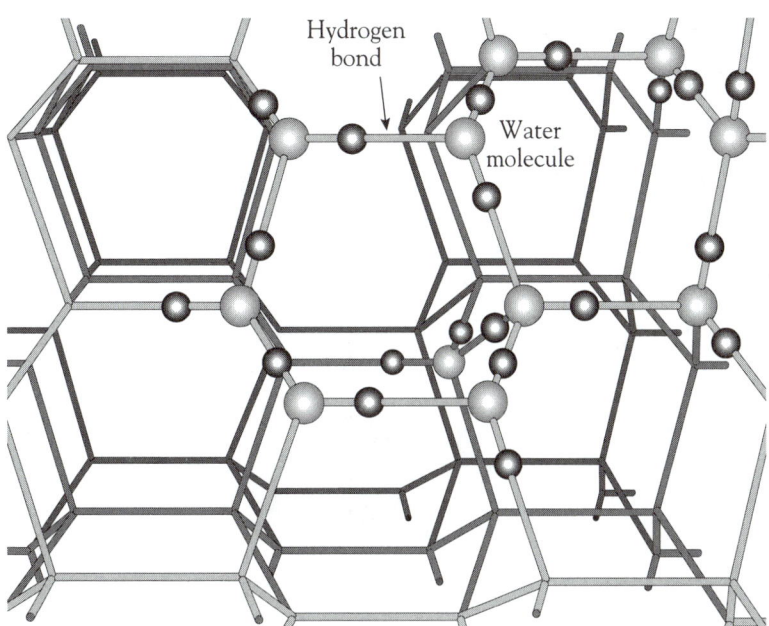

Figure 10.11 A representation of part of the ice structure, showing the open framework. The larger circles represent the oxygen atoms.
(From G. Rayner-Canham et al., Chemistry: A Second Course [Don Mills, ON: Addison–Wesley, 1989] p. 165.)

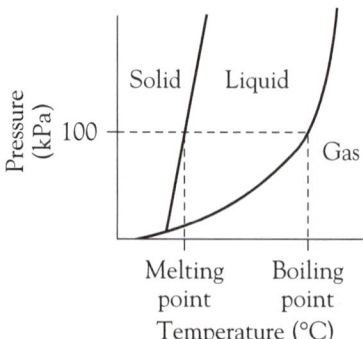

Figure 10.12   An idealized phase diagram.

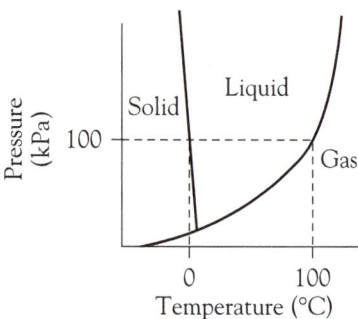

Figure 10.13   Phase diagram for water (not to scale).

combinations of pressure and temperature. The standard melting and boiling points can be determined by considering the phase changes at standard pressure, 100 kPa. When a horizontal dashed line is projected from the point of standard pressure, the temperature at which that line crosses the solid–liquid boundary is the melting point, and the temperature at which the line crosses the liquid–gas line is the boiling point. For almost all substances, the solid–liquid line has a positive slope. This trend means that application of a sufficiently high pressure to the liquid phase will cause the substance to solidify.

Water, however, has an abnormal phase diagram (Figure 10.13) because the density of ice is lower than that of liquid water. The Le Châtelier principle indicates that the denser phase is favored by increasing pressure. So, for water, application of pressure to the less dense solid phase causes it to melt to the denser liquid phase. It is this anomalous behavior that contributes to the feasibility of ice-skating, because the pressure of the blade, together with friction heating, melts the ice surface, providing a low-friction liquid layer down to about $-30°C$.

Furthermore, as a result of hydrogen bonding, solutions of hydronium ion, H_3O^+, or of hydroxide ion, OH^-, have much higher electrical conductivities than do solutions of the same concentration of any other ion. Ionic conductivity is a measure of the rate at which ions travel through the solution. The abnormally high values for these two ions can be explained in terms of hydrogen bonding. If one pictures part of the hydronium ion environment, the ion is hydrogen bonded to neighboring water molecules (Figure 10.14a). When a negative charge is placed in the solution, the hydronium ion does not move; instead, the covalent bonds and hydrogen bonds switch places (Figure 10.14b). Thus, the hydronium ion that conveys the charge to the electrode is not the one that possessed the charge initially (Figure 10.14c).

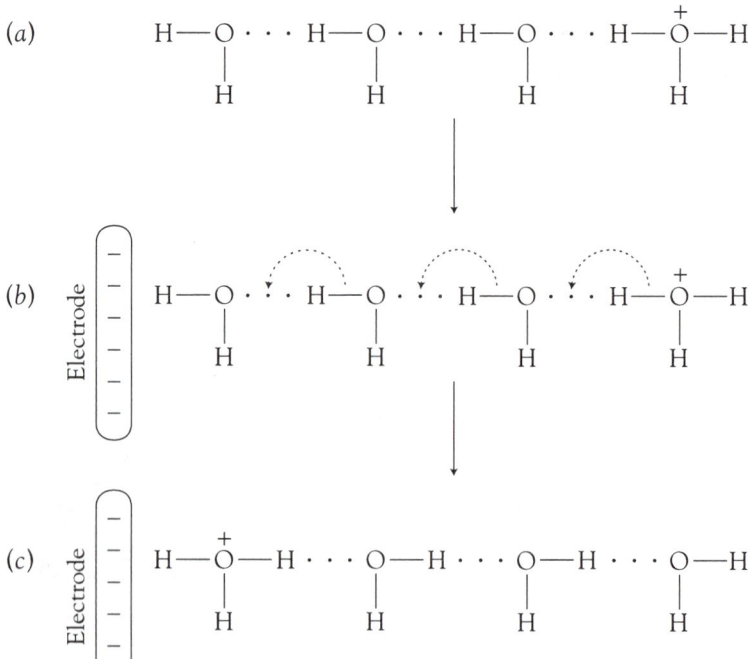

Figure 10.14   The high electrical conductivity of the hydronium ion is explained by the switching of the alternating covalent bonds and hydrogen bonds as the positive charge is attracted toward the negative electrode.

Water: The New Wonder Solvent

As we begin the century of green chemistry, one of the most crucial problems in industry is solvent use. Many solvents, particularly those used in organic synthesis, are toxic to humans and are potential environmental pollutants. One of the best replacements turns out to be our most common solvent, water. Traditionally water was avoided for organic synthesis as it has a low to near-zero solubility for the low- and nonpolar solutes that make up much of organic chemistry. However, at high pressure and temperature, the properties of water change remarkably.

Supercritical water has been promoted for more than 2 decades as a potential agent for waste detoxification. At a temperature of 400–500°C and a pressure of between 20 and 40 MPa, water is miscible with oxygen and with typical environmental toxins. Under such "aggressive" conditions, many toxic organic molecules are decomposed into small molecules such as carbon dioxide, water, and hydrochloric acid.

Under near-critical conditions, water is a more benign solvent and can be used for synthesis rather than decomposition. One of the important differences between normal water and subcritical water is the dielectric constant (a measure of the polarity of the solvent). The increase in pressure and temperature leads to a substantial decrease in the dielectric constant, reducing its polarity to a value close to acetone. For example, heptane is 10^5 times more soluble in near-critical water, while toluene is miscible. Thus, water becomes an excellent solvent for low and nonpolar reactants. In addition, near-critical water is autoionized about 10^3 times more than ordinary water. With the much greater concentrations of hydronium and hydroxide ions, acid- or base-catalyzed reactions can be performed without the need for adding an actual acid or base to the reactants. A wide variety of organic reactions have been shown to proceed in near-critical water cleanly and without need for a catalyst. Table 10.3 shows a comparison of the properties of normal, near-critical and supercritical water.

In industrial processes, between 50 and 80 percent of capital and operating costs are committed to the separation of the products. Reducing the pressure on a near-critical water reaction causes the water dielectric constant to increase. In turn, this renders the organic products insoluble and easily (and cheaply) separable.

Table 10.3 Properties of water

	Ambient	Near-critical	Supercritical
Temperature (°C)	25	275	400
Pressure (kPa)	100	6000	20 000
Density ($g \cdot cm^{-3}$)	1.0	0.7	0.1
Dielectric constant	80	20	2

If near-critical water processes are so wonderful, why are they not yet widely used? To produce the near-critical conditions, thick, expensive, stainless-steel pressure vessels must be used. With some reactions, the more chemically resistant (but extremely expensive) titanium has to be used as a reaction container. Also, there needs to be more research on optimizing conditions before large-scale investment is undertaken. Nevertheless, commercial organic synthesis using near-critical water solvent offers strong advantages such as the avoidance of organic solvent systems, elimination of undesirable catalysts, avoidance of unwanted by-products, and improved reaction selectivity. Surely its time will soon come.

10.6   Clathrates

Until a few years ago, clathrates were a laboratory curiosity. Now the methane and carbon dioxide clathrates in particular are becoming of major environmental interest. A clathrate is defined as a substance in which molecules or atoms are trapped within the crystalline framework of other molecules. The name is derived from the Latin word *clathratus*, which means "enclosed behind bars." In our discussion here we will focus on the gas clathrates of water, sometimes called gas hydrates. Though the latter term is widely used, it is not strictly correct, as the term "hydrate" usually implies some intermolecular attraction between the substance and the surrounding water molecules as, for example, in hydrated metal ions.

The noble-gas clathrates used to be the classic example of a clathrate. In those times, no chemical compounds of the noble gases were known. Thus, clathrates provided the only chemistry of those elements. For example, when xenon dissolves in water under pressure and the solution is cooled below 0°C, crystals with the approximate composition of $Xe \cdot 6H_2O$ are formed. Warming the crystals causes immediate release of the gas. There is no chemical interaction between the noble-gas and the water molecule; the gas atoms are simply locked into cavities in the hydrogen-bonded ice structure. As the ice melts, the cavity structure collapses and the gas atoms are released. However, a significant point is that the presence of a "guest" within the ice structure stabilizes the ice structure and raises the melting point of the ice to several degrees above 0°C.

It was the discovery of methane hydrates (Figure 10.15) on the sea floor that turned clathrates into an issue of major importance. We are now aware that large areas of the ocean floors have thick layers of methane clathrates just beneath the top layer of sediment. It is probable these clathrate layers have formed over eons by the interaction of rising methane from leaking subsurface gas deposits with near-freezing water percolating down through the sediment layers. Provided the water is at a temperature and pressure below those of the melting point of the clathrate, the clathrate will form. Each cubic centimeter of hydrate contains about 175 cm^3 of methane gas measured at SATP (298 K and 100 kPa). The methane content of the clathrate is sufficient that the "ice" will actually burn. It is believed that the total carbon methane clathrate deposits in the world oceans is twice that of the sum of all coal, oil, and natural gas deposits on land.

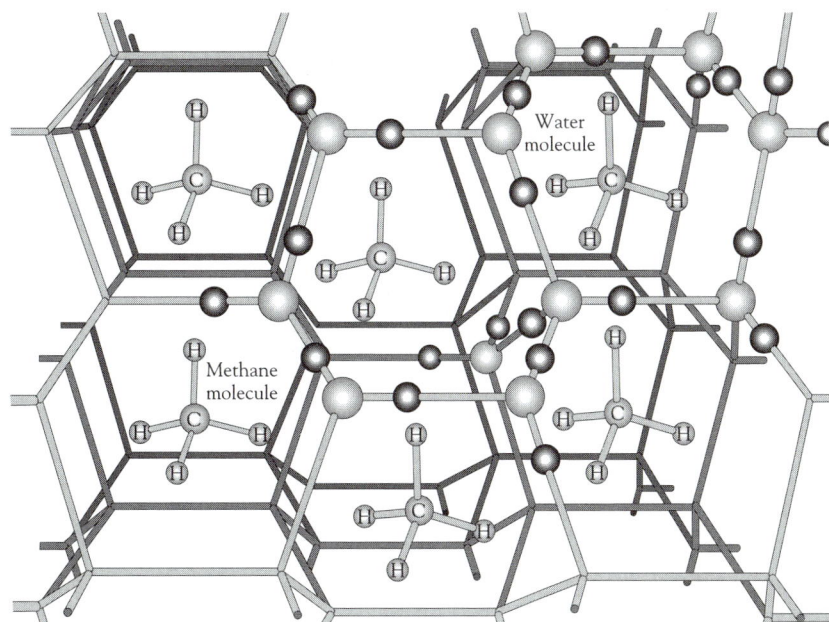

Figure 10.15 A representation of part of the ice structure (see Figure 10.11), showing methane molecules trapped as a clathrate. *(From G. Rayner-Canham et al., Chemistry: A Second Course [Don Mills, ON: Addison–Wesley, 1989], p. 167.)*

As the stability of methane clathrates are so temperature and pressure dependent, there is concern that the warming of oceans may lead to the melting of clathrate deposits, releasing large volumes of methane into the atmosphere. The released methane would then have a significant effect on climate as methane is a potent greenhouse gas (see Chapter 14, Section 14.14). It has been argued that some sudden past changes of climate were triggered by methane release from clathrates. For example, the lowering of water levels during ice ages would have reduced the pressure on seabed deposits, possibly liberating large volumes of gas. The increased methane levels would then have caused global warming, terminating the ice age.

10.7 *Biological Aspects of Hydrogen Bonding*

Hydrogen is a key element in living organisms. In fact, the existence of life depends on two particular properties of hydrogen: the closeness of the electronegativities of carbon and hydrogen and the ability of hydrogen to form hydrogen bonds when covalently bonded to nitrogen or oxygen. The low polarity of the carbon–hydrogen bond contributes to the stability of organic compounds in our chemically reactive world. Biological processes also rely on both polar and nonpolar surfaces, the best example of the latter being the lipids. It is important to realize that nonpolar sections of biological molecules, usually containing just carbon and hydrogen atoms, are just as significant as their polar regions.

Hydrogen bonding is a vital part of all biomolecules. Proteins are held in shape by hydrogen bonds that form cross-links between chains. The strands of DNA and RNA, the genetic material, are held together by hydrogen bonds as well. But more than that, the hydrogen bonds in the double helices are not random; they form between specific pairs of organic bases. These pairs are preferentially hydrogen bonded, because the two components fit together to give particularly close approaches of the hydrogen

Is There Life Elsewhere in Our Solar System?

It is difficult for a chemist or biochemist to visualize any life form that does not depend upon water. Water is an ideal solvent, and it is hydrogen bonds that enable protein and DNA molecules to form their complex structures. This raises the question of where there is liquid water, is there life? Until recently, this seemed to be a theoretical question, as no other planets possessed any surface liquid water. Now our attention is focusing on the larger moons of Jupiter. It was traditionally believed that the other moons in the Solar System would be like our own—rocky, dust-covered, and lifeless. From photos sent back by space probes we now know that the moons of Jupiter and Saturn show some incredible differences in surfaces. The most strikingly different moon was Io with its sulfur volcanoes (see the color feature in the middle of this text) and unique chemistry.

There are, in fact, four large moons of Jupiter: Io, Europa, Ganymede, and Callisto (in order of increasing distance from Jupiter). Europa is the moon of next greatest interest to astrochemists. Our Moon, like other "dead" objects in the Solar System, is covered by craters, mainly formed by impacts of asteroids over the entire geological timescale. Europa, however, is the smoothest surfaced body every discovered. The moon also has a very low average density. The conclusion is that the surface consists of ice. The surface must have been liquid in the geologically recent past to "freeze-over" meteor impact sites. This raises the question as to whether there is liquid water beneath the surface ice. After all, ice is a good insulator (even though the surface is at $-160°C$) and the tidal friction from Jupiter may well provide an energy source. From that question follows another: if there is liquid water, has any life developed in it? A few years ago, this question would have raised laughter, but no more. We are now aware of extremophile bacteria that can thrive in the most unlikely environments. In fact, living bacteria have recently been found in ice layers nearly 4 km beneath the Antarctic ice sheet and only just above a lake, Lake Vostok, that is under the ice sheet itself. To investigate the surface of Europa, a NASA probe, the Europa Ice Clipper, should be returning ice samples to Earth by 2009. This will really be an exciting event. Will there be evidence of recent melting? Will there be ionic solutes dissolved in the ice? And, of course, is there any evidence of organisms?

Attention is now turning to Ganymede and Callisto. Ganymede is the largest of Jupiter's moons; in fact, it is the largest moon in the Solar System, and it is even larger than the planet, Mercury. It, too, has an icy surface, though there is evidence that Ganymede has a core of molten iron. If so, then there may well be a layer of liquid water under the ice. For this moon, the well-cratered ice seems to be about 800 km thick; thus, reaching the liquid layer would be a near impossible task. There is now evidence that Callisto, too, has a salty ocean underneath the old icy surface. However, if there is currently life anywhere else in the Solar System, chemists and biochemists are betting on Europa.

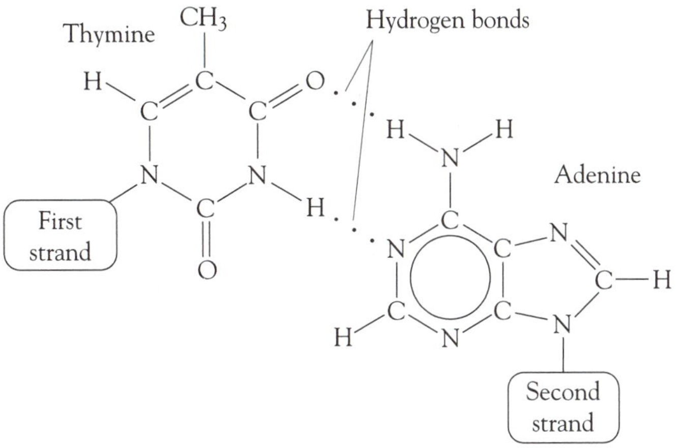

Figure 10.16 Hydrogen bond interaction between thymine and adenine fragments of the two strands in a DNA molecule.

atoms involved in the hydrogen bonding. This bonding is illustrated in Figure 10.16 for the interaction between two particular base units, thymine and adenine. It is the specific matching that results in the precise ordering of the components in the DNA and RNA chains, a system that allows those molecules to reproduce themselves almost completely error-free.

All proteins depend on hydrogen bonding for their function as well. Proteins consist mainly of one or more strands of linked amino acids. But to function, most proteins must form a compact shape. To do this, the protein strand loops and intertwines with itself, being held in place by hydrogen bonds cross-linking one part of the strand to another.

10.8 *Element Reaction Flowchart*

In each chapter discussing chemical elements, a flowchart will be used to display the key reactions of that element. The flowchart for hydrogen is shown in the following.

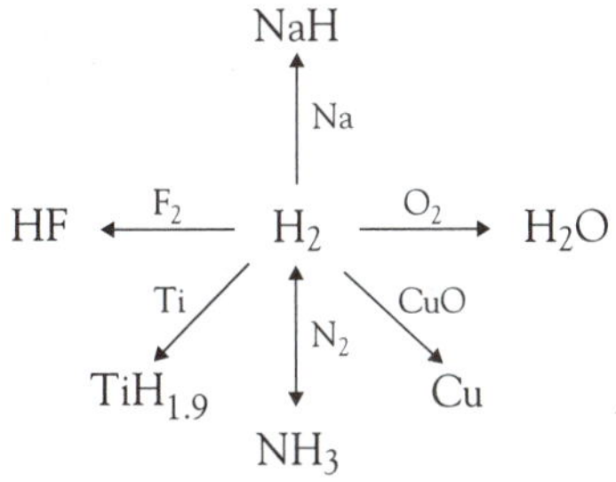

EXERCISES

10.1 Define the following terms: (a) protonic bridge; (b) hydridic bridge.

10.2 Define the following terms: (a) borane; (b) phase diagram.

10.3 An ice cube at 0°C is placed in some liquid water at 0°C. The ice cube sinks. Suggest an explanation.

10.4 Which of the following isotopes can be studied by nuclear magnetic resonance: carbon-12, oxygen-16, oxygen-17?

10.5 When we study the NMR spectrum of a compound, why are the absorption frequencies expressed as ppm?

10.6   Explain why hydrogen is not placed with the alkali metals in the Periodic Table.

10.7   Explain why hydrogen is not placed with the halogens in the Periodic Table.

10.8   Explain why hydrogen gas is comparatively unreactive.

10.9   Is the reaction of dihydrogen with dinitrogen to produce ammonia entropy or enthalpy driven? Do not consult data tables. Explain your reasoning.

10.10 Write chemical equations for the reaction between
(a) tungsten(VI) oxide, WO_3, and dihydrogen with heating
(b) hydrogen gas and chlorine gas
(c) aluminum metal and dilute hydrochloric acid.

10.11 Write chemical equations for the reaction of
(a) potassium hydrogen carbonate on heating
(b) ethyne, $HC{\equiv}CH$, with dihydrogen
(c) lead(IV) oxide with hydrogen gas on heating
(d) calcium hydride and water.

10.12 Show that the combustion of methane:

$$CH_4(g) + 2\,O_2(g) \rightarrow CO_2(g) + 2\,H_2O(l)$$

is indeed spontaneous by calculating the standard molar enthalpy, entropy, free energy of combustion from enthalpy of formation, and absolute entropy values. Use the data tables in the Appendices.

10.13 Construct a theoretical enthalpy cycle (similar to that of Figure 10.3) for the formation of ammonia from its elements. Obtain bond energy information and the standard enthalpy of formation of ammonia from the data tables in the Appendices. Compare your diagram to that in Figure 10.3 and comment on the differences.

10.14 What is the major difference between ionic and covalent hydrides in terms of physical properties?

10.15 Discuss the three types of covalent hydrides.

10.16 Which of the following elements is likely to form an ionic, metallic, or covalent hydride, or no stable hydride: (a) chromium; (b) silver; (c) phosphorus; (d) potassium.

10.17 Write the expected formulas for the hydrides of the Period 4 main group elements from potassium to bromine. What is the trend in the formulas? In what way are the first two members of the series different from the others?

10.18 Construct a plot of the enthalpies of formation of the Group 14 hydrides (Table 10.3) against the element-hydrogen electronegativity difference for each hydride. Suggest an explanation for the general trend.

10.19 Predict which of the following hydrides is a gas or a solid: (a) HCl; (b) NaH. Give your reason in each case.

10.20 Construct a diagram similar to that of Figure 10.14 to explain why the hydroxide ion has a very high electrical conductivity.

10.21 What are the two properties of hydrogen that are crucial to the existence of life?

10.22 Write balanced chemical equations corresponding to each transformation in the Element Reaction Flowchart.

BEYOND THE BASICS

10.23 Predict which of the following hydrides is likely to be strongly hydrogen bonded and hence deduce the likely phases of each of the hydrides at room temperature: (a) H_2O_2; (b) P_2H_4; (c) N_2H_4; (d) B_2H_6.

10.24 Write balanced chemical equations for the air oxidation of: (a) B_2H_6; (b) PbH_4; (c) BiH_3.

10.25 Calculate an approximate value for the bond energy for the bridging B–H bond in diborane using the data tables in the Appendices. In comparison with the normal B–H bond energy, what does this suggest about the bond order? Is this result compatible with the bond order (per bond) deduced from the molecular orbital diagram (Figure 10.10)?

10.26 The hydride ion is sometimes considered as similar to a halide ion; for example, the lattice energies of sodium hydride and sodium chloride are -808 kJ·mol^{-1} and -788 kJ·mol^{-1} respectively. However, the enthalpy of formation of sodium hydride is much less than that of a sodium halide, such as sodium chloride. Use the data tables in the Appendices to calculate enthalpy of formation values for the two compounds and identify the factor(s) that cause the values to be so different.

10.27 Instead of the commonly used reaction of hydrocarbons with air as a high performance aircraft fuel, it was once proposed to use diborane with air. If $\Delta H^\circ_{combustion}(B_2H_6(g)) = -2165$ kJ·mol^{-1} and $\Delta H^\circ_{combustion}(C_2H_6(g)) = -1560$ kJ·mol^{-1}, calculate the comparative energy per gram of diborane compared to per gram of ethane. Use the data tables in the Appendices to calculate the entropy change in each reaction. Suggest why the entropy of combustion values are so different. What would be some practical disadvantages of using diborane?

10.28 Explain why:
(a) Interstitial hydrides have a lower density than that of the parent metal;
(b) Ionic hydrides are more dense than the parent metal.

10.29 Hydrogen gas has been proposed as the best fuel for the 21st century. However, a company in Florida has developed AquaFuel^{TM} as an alternative. This gas mixture is formed by passing a high current through water using carbon electrodes. What are the likely gaseous products of electrolysis? Write a balanced molecular equation for the process. Write a balanced molecular equation for the combustion of the mixture. Calculate the energy released per mole of gas mixture and compare it to the energy released per mole of hydrogen gas combustion.

The Group 1 Elements: The Alkali Metals

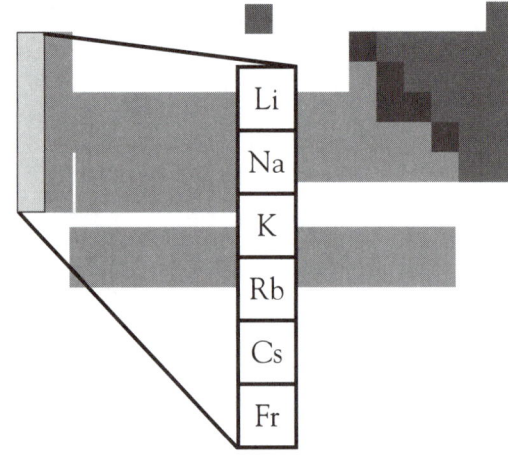

Metals are usually thought of as being dense and nonreactive. The alkali metals, however, are actually the opposite of this characterization, being of both low density and very high chemical reactivity.

Compounds of the alkali metals have been known since ancient times. However, the alkali metal cations are extremely difficult to reduce, and it was not until after electric power was harnessed that the metals themselves could be extracted. A British scientist, Humphry Davy, electrolyzed molten potassium hydroxide in 1807 to extract the first of the alkali metals. Davy obtained such acclaim for his extraction of these metals from their salts that a rhyme was written about him:

Sir Humphry Davy
Abominated gravy
Lived in the odium
Of having discovered sodium.
(E.C. Bentley, 1875–1956)

Nationalism has often become interwoven with chemistry. When Napoleon heard of Davy's discovery, he was extremely angry that the French chemists had not been first. But, by coincidence, it was a French scientist, Marguerite Perey, who in 1939 isolated the one alkali metal that exists only

Table 11.1   Melting points and enthalpies of atomization of the alkali metals

Element	Melting point (°C)	$\Delta H_{atomization}$ (kJ·mol^{-1})
Li	180	162
Na	98	08
K	64	90
Rb	39	82
Cs	29	78

as radioactive isotopes. She named the element francium after her native country—Napoleon would have been delighted!

11.1   Group Trends

Two elements have been named after France: francium and gallium (from the Latin for France, Gaul).

All of the alkali metals are shiny, silver-colored metals. Like the other metals, they have high electrical and thermal conductivities. But in other respects, they are very atypical. For example, the alkali metals are very soft, and they become softer as one progresses down the group. Thus, lithium can be cut with a knife, whereas potassium can be "squashed" like soft butter. Most metals have high melting points, but those of the alkali metals are very low and become lower as the elements in Group 1 become heavier, with cesium melting just above room temperature. In fact, the combination of high thermal conductivity and low melting point makes sodium useful as a heat transfer material in some nuclear reactors. The softness and low melting points of the alkali metals can be attributed to the very weak metallic bonding in these elements. For a "typical" metal, the enthalpy of atomization is in the range of 400 to 600 kJ·mol^{-1}; but as can be seen from Table 11.1, those of the alkali metals are much lower. In fact, there is a correlation between both softness and low melting point and a small enthalpy of atomization.

Even more atypical are the densities of the alkali metals. Most metals have densities between 5 and 15 g·cm^{-3}, but those of the alkali metals are far less (Table 11.2). In fact, lithium has a density one-half that of water!

Lithium is the least dense metal (0.53 g·cm^{-3}) while osmium is the most dense metal (22.6 g·cm^{-3}).

With such a low density, lithium would be ideal for making unsinkable (although soft!) ships, except for one other property of the alkali metals—their high chemical reactivity. The metals are usually stored under oil, because when they are exposed to air, a thick coating of oxidation products covers the lustrous surface of each metal very rapidly. For example, lithium is oxidized to lithium oxide, which in turn reacts with carbon dioxide to give lithium carbonate:

$$4\,Li(s) + O_2(g) \rightarrow 2\,Li_2O(s)$$

$$Li_2O(s) + CO_2(g) \rightarrow Li_2CO_3(s)$$

The alkali metals react with most nonmetals. For example, every molten alkali metal burns in chlorine gas to give off a white smoke of the metal chloride. The reaction of sodium with dichlorine really typifies the wonder of chemistry—that a highly reactive, dangerous metal reacts with a poisonous gas to produce a compound that is essential to life.

$$2\,Na(l) + Cl_2(g) \rightarrow 2\,NaCl(s)$$

Table 11.2   Densities of the alkali metals

Element	Density (g·cm^{-3})
Li	0.53
Na	0.97
K	0.86
Rb	1.53
Cs	1.87

As discussed in Chapter 9, Section 9.1, the reactions of the alkali metals with water are very dramatic, with reactivity increasing down the group. The equation for the reaction of water with potassium is

$$2\,K(s) + 2\,H_2O(l) \rightarrow 2\,KOH(aq) + H_2(g)$$

Because they are so much more reactive than the "average" metal, the alkali metals are sometimes referred to as the supermetals.

11.2   Common Features of Alkali Metal Compounds

All the Group 1 elements are metals. As a result, all the members of the group have common features.

 **Ionic Character**

The alkali metal ions always have an oxidation number of $+1$, and most of their compounds are stable, ionic solids. The compounds are colorless unless they contain a colored anion such as chromate or permanganate. Even for these highly electropositive elements, the bonds in their compounds with nonmetals have a small covalent component.

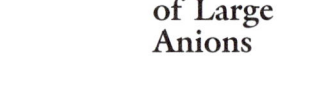 **Stabilization of Large Anions**

Because the cations of the alkali metals (except for that of lithium) have among the lowest charge densities, they are able to stabilize large low-charge anions. For example, the ions of sodium through cesium are the only cations that form solid hydrogen carbonate salts.

 **Ion Hydration**

All ions are hydrated when dissolved in water. However, this is not always true in the solid phase. Hydration in the crystalline solid depends on the balance of lattice energy and ion hydration energies. The lattice energy results from the electrostatic attraction between the cations and anions: the higher the charge density of the ions, the larger the lattice energy. Thus, the lattice energy term favors the loss of an ion's hydration sphere on crystallization to give the small (higher charge density) anhydrous ion. But the hydration energy depends on the attraction between the ion and the surrounding polar water molecules. A major factor contributing to the strength of the ion–dipole attraction is the charge density of the ions. In this ionic tug of war, we find that high charge density usually favors retention of all or part of the hydration sphere in the solid phase, while salts of low-charge ions tend to be anhydrous.

As we mentioned earlier, the alkali metals have very low-charge densities compared to those of other metals. Thus, we would expect—and find—that the majority of solid alkali metal salts are anhydrous. The charge densities of lithium and sodium ions are high enough to favor the formation of a few hydrated salts. An extreme example is lithium hydroxide, which forms an octahydrate, $LiOH \cdot 8H_2O$. With the lowest charge densities of all metals, very few potassium, rubidium, and cesium salts are hydrated.

The low charge densities are reflected in the trend in hydration enthalpy among the alkali metals (Table 11.3). The values are very low (for example, that of the Mg^{2+} ion is 1920 $kJ \cdot mol^{-1}$), and the values decrease as radius increases down the group.

Table 11.3   Hydration enthalpies of the alkali metal ions

Ion	Hydration enthalpy $(kJ \cdot mol^{-1})$
Li^+	519
Na^+	406
K^+	322
Rb^+	301
Cs^+	276

 **Solubility**

Almost all the compounds of the alkali metals are soluble in water, although they are soluble to different extents. For example, a saturated solution of lithium chloride has a concentration of 14 $mol \cdot L^{-1}$, whereas a saturated

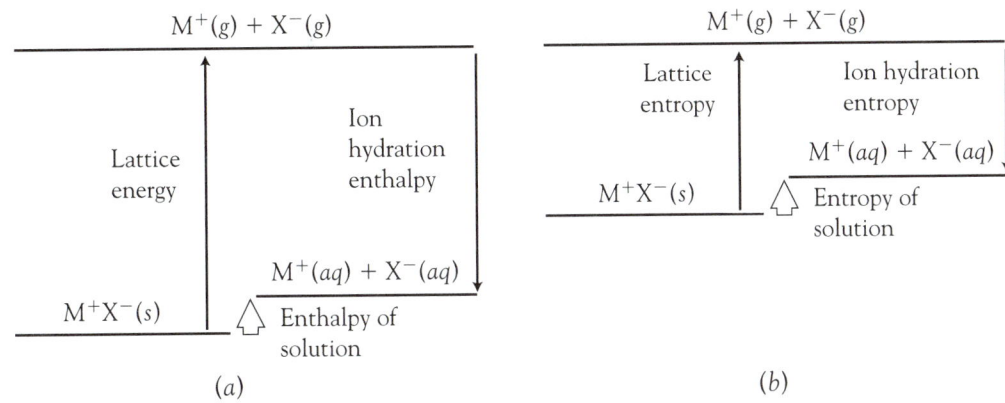

Figure 11.1 Enthalpy cycle (*a*) and entropy cycle (*b*) for the solution of an ionic compound. M^+ is the alkali metal ion and X^- is the anion.

solution of lithium carbonate has a concentration of only $0.18 \ mol \cdot L^{-1}$. We explore the reason for the differences in the next section.

11.3 Solubility of Alkali Metal Salts

It is the solubility of all the common alkali metal salts that makes them so useful as reagents in the chemistry laboratory. Whether it is a nitrate, a phosphate, or a fluoride anion that we need, we can always count on the alkali metal salt to enable us to make a solution of the required anion. Yet the solubilities cover a wide range of values. This variability is illustrated by the solubilities of the sodium halides (Table 11.4).

To explain this solubility trend, we need to look at the energy cycle involved in the formation of a solution from the solid. As we discussed in Chapter 6, Section 6.4, the solubility of a compound is dependent on the enthalpy changes (the lattice energy and the enthalpy of hydration of the cation and anion) together with the corresponding entropy changes. These are shown in Figure 11.1. For the salt to be appreciably soluble, the free energy, $\Delta G°$, should be negative, where

$$\Delta G° = \Delta H° - T\Delta S°$$

If we look at the enthalpy terms (Table 11.5), we see that for each sodium halide, the lattice energy is almost exactly balanced by the sum of the cation and anion hydration enthalpies. In fact, the error in these experimental values is larger than the calculated differences. As a result, we can only say that the lattice energy and hydration enthalpy terms are essentially equal.

Table 11.4 Solubilities of the sodium halides at 25°C

Compound	Solubility $(mol \cdot L^{-1})$
NaF	0.099
NaCl	0.62
NaBr	0.92
NaI	1.23

Table 11.5 Enthalpy factors in the solution process for the sodium halides

Compound	Lattice energy $(kJ \cdot mol^{-1})$	Hydration enthalpy $(kJ \cdot mol^{-1})$	Net enthalpy change $(kJ \cdot mol^{-1})$
NaF	+930	−929	+1
NaCl	+788	−784	+4
NaBr	+752	−753	−1
NaI	+704	−713	−9

Table 11.6   Entropy factors in the solution process for the sodium halides, expressed as $T\Delta S$ values

Compound	Lattice entropy (kJ·mol^{-1})	Hydration entropy (kJ·mol^{-1})	Net entropy change (kJ·mol^{-1})
NaF	+72	−74	−2
NaCl	+68	−55	+13
NaBr	+68	−50	+18
NaI	+68	−45	+23

Table 11.7   Calculated free energy change for the solution process for the sodium halides

Compound	Enthalpy change (kJ·mol^{-1})	Entropy change (kJ·mol^{-1})	Free energy change (kJ·mol^{-1})
NaF	+1	−2	+3
NaCl	+4	+13	−11
NaBr	−1	+18	−19
NaI	−9	+23	−32

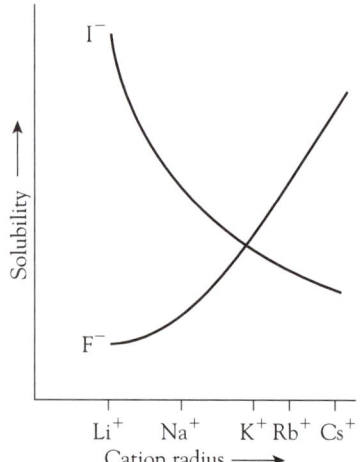

Figure 11.2   Solubility of alkali metal fluorides and iodides as a function of alkali metal ion radius.

When we calculate the entropy changes (Table 11.6), we find that for all the salts except sodium fluoride, the entropy gained by the ions as they are freed from the crystal lattice is numerically larger than the entropy lost when these gaseous ions are hydrated in solution. To obtain the free energy change for the solution process, we combine these two very small net changes in enthalpy and entropy. Amazingly, the calculated free energies provide a trend parallel to that of the measured solubilities (Table 11.7). Furthermore, if we plot the solubilities of the salts one anion forms with different alkali metal cations as a function of the ionic radius of the alkali metal ions, in most cases we get a smooth curve. This curve may have a positive or negative slope (or in some cases, reach a minimum in the middle of the series). To illustrate such trends, the solubilities of alkali metal fluorides and iodides are shown in Figure 11.2.

We can understand the different curves in Figure 11.2 by focusing on the lattice energies. Although there is a strong dependence of lattice energy on ionic charge, there is a secondary relationship to the cation/anion radius ratio; that is, a significant mismatch in ionic sizes will lead to a lower than expected lattice energy. Table 11.8 shows the ionic radii of the cations lithium and cesium and the anions fluoride and iodide. Thus, lithium iodide, the ions of which have very different sizes, is much more soluble than lithium fluoride, the ions of which have similar sizes. Conversely, cesium iodide, the ions of which have similar sizes, is much less soluble than cesium fluoride, in which there is a large mismatch in ionic size.

Table 11.8   Selected ionic radii

Cation	Radius (pm)	Anion	Radius (pm)
Li$^+$	190	F$^-$	119
Cs$^+$	181	I$^-$	206

Mono Lake

This Californian lake, nestled against the Sierra Nevada Mountains, is unique in the world, and this uniqueness derives from its chemistry. Surrounded by mountains and volcanic hills, there is no outlet to the lake, so water loss is by evaporation. The lake is estimated to be among the oldest in North America—at least 760 000 years old. During that time, soluble salts have been leached out of the surrounding rocks by surface waters and underground springs, accumulating in this lake of surface area about 180 km^2 (70 mi^2) with an average depth of about 20 m (60 ft). It has been calculated that the lake contains about 2.8×10^8 tonnes of dissolved salts. Whereas the Great Salt Lake in Utah contains mostly sodium chloride, Mono Lake contains a fascinating mix of soluble sodium and potassium salts, including chloride, sulfate, hydrogen carbonate, carbonate, borate, and a trace of fluoride, iodide, arsenate, and tungstate ions. Among the other metal ions in the lake are calcium, magnesium, and strontium. As a result of the high HCO_3^- and CO_3^{2-} concentrations, the pH of the lake is about 10. Though the bulk composition of the lake is known, there is still much to discover about the ion interactions and the variation in composition with depth and season.

The tufa towers are the most characteristic feature of the lake. They are mainly formed when underwater springs rich in calcium ion mix with the carbonate ion of the lake waters. Hence, the towers identify the location of underwater springs. The water is so saturated in these two ions that calcium carbonate deposits form over the lake bottom, over discarded soft-drink cans and any other debris. The towers only form underwater; thus, those visible today result from a drop in water level since 1941 when feed water was diverted to agricultural and consumable use elsewhere in the state.

Not only is the chemistry of the lake unique, so is its ecology. With the high pH and high soluble salt concentrations, the only water life is algae, brine shrimp, and the black alkali flies. These algae-feeding flies swarm around the lake, spending two of their three life stages entirely underwater. The flies are rich in fat and protein making them an excellent source of food for migrating birds "refueling" for their long journeys. In fact, the lake is one of the most productive ecosystems in the world.

As more and more of the water was diverted, the water level dropped reaching a minimum of about 50 percent of its 1941 volume in 1982, and had the level dropped by much more, the ion concentration would have become too high even for the brine shrimp and alkali flies, killing all life in the lake and the birds that depended upon it. Fortunately, the lake has been designated as an Outstanding National Resource Water. This means that the lake volume must be restored to about 70 percent of its former volume (at present, the figure is about 60 percent), and the volume, allowing for winter–summer variation, must be maintained forever about that value.

Mono Lake is now of interest to NASA's scientists studying Martian chemistry. There are many old Martian lake beds and it is believed that the lakes may have had similar compositions to Mono Lake and perhaps supported related life forms. The astrochemists are eager to obtain ultrahigh resolution photos of the Martian lake beds to see if they, too, have tufa towers which would be an indication of parallel chemical processes.

11.4   Flame Colors

Table 11.9 Alkali metals and their flame colors

Metal	Color
Lithium	Crimson
Sodium	Yellow
Potassium	Lilac
Rubidium	Red-violet
Cesium	Blue

Almost all the alkali metal compounds are water soluble, the exceptions being a very few compounds of potassium, rubidium, and cesium (discussed later). As a result, precipitation tests are not commonly used to detect the presence of an alkali metal. Fortunately, each of the alkali metals produces a characteristic flame color when a sample of an alkali metal salt is placed in a flame (Table 11.9). In the process, energy from the combustion reactions of the fuel is transferred to the metal salt that is placed in the flame. This transfer causes electrons in the alkali metal atoms to be raised to excited states. The energy is released in the form of visible radiation as the electron returns to the ground state. Each alkali metal undergoes its own unique electron transitions. For example, the yellow color of sodium is a result of the energy (photon) emitted when an electron drops from the $3p^1$ orbital to the $3s^1$ orbital of a neutral sodium atom, the ion having acquired its valence electron from the combustion reactions in the flame (Figure 11.3).

11.5   Lithium

With a density of about half that of water, lithium is the least dense of all the elements that are solids at room temperature and pressure. Lithium used to be a laboratory curiosity, but no more. As a result of its very low density, lithium is used in aerospace alloys. For example, alloy LA 141, which consists of 14 percent lithium, 1 percent aluminum, and 85 percent magnesium, has a density of only 1.35 g·cm^{-3}, almost exactly half that of aluminum, the most commonly used low-density metal.

The metal has a bright silvery appearance, but when a surface is exposed to moist air, it very rapidly turns black. Like the other alkali metals, lithium reacts with the dioxygen in air. It is the only alkali metal, and one of a very few elements in the entire Periodic Table, to react with dinitrogen. Breaking the triple bond in the dinitrogen molecule requires an energy input of 945 kJ·mol^{-1}. To balance this energy uptake, the lattice energy of the product must be very high. Of the alkali metals, only the lithium ion, which has the greatest charge density of the group, forms a nitride with a sufficiently high lattice energy:

$$6\,Li(s) + N_2(g) \rightarrow 2\,Li_3N(s)$$

The nitride is reactive, however, forming ammonia when added to water:

$$Li_3N(s) + 3\,H_2O(l) \rightarrow 3\,LiOH(aq) + NH_3(g)$$

Liquid lithium is the most corrosive material known. For example, if a sample of lithium is melted in a glass container, it reacts spontaneously with

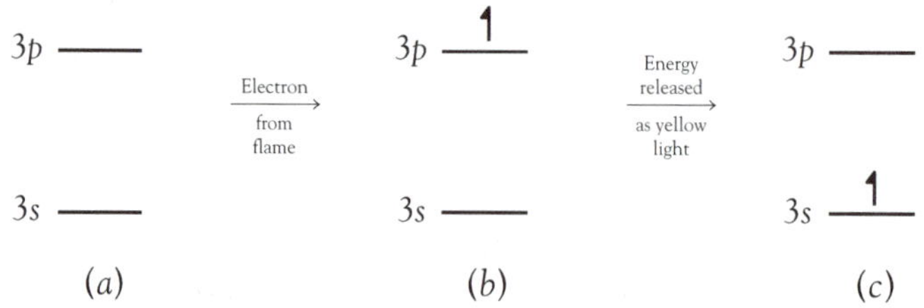

Figure 11.3 In a flame, the sodium ion (a) acquires an electron in the 3p orbital (b). As the electron drops from the excited 3p state to the ground 3s state (c), the energy is released as yellow light.

the glass to produce a hole in the container, the reaction being accompanied by the emission of an intense, greenish white light. In addition, the lithium ion has the most negative standard reduction potential of any element:

$$Li^+(aq) + e^- \rightarrow Li(s) \qquad\qquad E° = -3.05 \text{ V}$$

That is, the metal itself releases more energy than any other element when it is oxidized to its ion (+3.05 V). Yet, of the alkali metals, it has the least spectacular reaction with water. As discussed in Chapter 6, Section 6.6, we must not confuse thermodynamic spontaneity, which depends on the free energy change, with rate of reaction, which is controlled by the height of the activation energy barrier. In this particular case, we must assume that the activation energy for the reaction with water is greater for lithium than for the other alkali metals. Because lithium metal has the greatest lattice energy of the alkali metals and because escape from the lattice must be involved in any oxidation/hydration pathway, it is not really surprising that the activation energy is higher.

The largest industrial use of lithium is in lithium greases—in fact, more than 60 percent of all automotive greases contain lithium. The compound used is lithium stearate, $C_{17}H_{35}COOLi$, which is mixed with oil to give a water-resistant, grease-like material that does not harden at cold temperatures yet is stable at high temperatures.

The comparatively high charge density of the lithium ion is responsible for several other important ways in which lithium's chemistry differs from that of the rest of the alkali metals. In particular, there is an extensive organometallic chemistry of lithium in which the bonding is definitely covalent (see Chapter 22, Section 22.3). Even for common salts, such as lithium chloride, their high solubilities in many solvents of low polarity, particularly ethanol and acetone, indicate a high degree of covalency in the bonding. One specific organometallic compound, butyllithium, LiC_4H_9, is a useful reagent in organic chemistry. It can be prepared by treating lithium metal with chlorobutane, C_4H_9Cl, in a hydrocarbon solvent such as hexane, C_6H_{12}:

$$2 \text{ Li}(s) + C_4H_9Cl(C_6H_{12}) \rightarrow LiC_4H_9(C_6H_{12}) + LiCl(s)$$

After separating the lithium chloride by filtration, the solvent can be removed by distillation; liquid butyllithium remains in the distillation vessel. This compound has to be handled carefully, because it spontaneously burns when exposed to the dioxygen in air.

Lithium Batteries

Lithium is the most common anode material in new battery technology. With its high reduction potential and very low mass per unit of stored energy, it is currently used in compact high-voltage cells. Having a density $\frac{1}{20}$ that of lead, substantial mass savings are possible once the very challenging task of devising an inexpensive reversible (rechargeable) lithium cycle is perfected. Thus, it is strongly favored to replace the lead-acid battery for electric vehicle propulsion.

Lithium batteries are now becoming commonplace, but there are, in fact, many types of lithium battery. The lithium ion rechargeable battery is used in portable computers and cell phones. The anode consists of lithium cobalt(III) oxide, $LiCoO_2$; the cathode is graphite, and an organic liquid is used as the electrolyte. In the charging cycle, at the cathode, lithium ions are released into the solution as electrons are removed from the electrode.

Charge balance is maintained by one cobalt(III) ion being oxidized to cobalt(IV) for each lithium ion released.

$$LiCoO_2(s) \rightarrow Li_{(1-x)}CoO_2 + xLi^+(solvent) + xe^-$$

At the anode, lithium ions enter between the graphite layers and are reduced to lithium metal. This insertion of a "guest" atom into a "host" solid, a process accompanied by only small, reversible changes in structure, is known as intercalation, and the resulting product is called an *intercalation compound*.

$$C(s) + xLi^+(solvent) + xe^- \rightarrow (Li)_xC(s)$$

The discharge of the cell corresponds to the reverse reactions.

There are many other lithium batteries using different electrode materials. Some of the electrode materials and the battery codes are as follows: manganese(IV) oxide (CR), polycarbonmonofluoride (BR), vanadium(V) oxide (VL), and manganese oxide with lithium/aluminum cathode (ML). In most of these cells, the transition metal serves as part of the redox system, oscillating between two oxidation states, $(IV) \leftrightarrow (III)$ for manganese and $(V) \leftrightarrow (IV)$ for vanadium.

An increasingly popular lithium battery system is that using thionyl chloride, $SOCl_2$. This battery provides a high voltage per cell, reliability, long shelf life, low weight, and a constant energy output. However, it is not rechargeable. This class of battery is used in spacecraft, rescue submarines, and submarine torpedoes. A lithium thionyl chloride cell consists of three major components: a metallic lithium or lithium alloy anode, a carbon cathode, and an electrolyte of $Li^+[GaCl_4]^-$ or $Li^+[AlCl_4]^-$ dissolved in thionyl chloride. The anode reaction is that of oxidation of lithium metal to lithium ion:

$$Li(s) \rightarrow Li^+(SOCl_2) + e^-$$

At the cathode, the thionyl chloride is reduced:

$$2\ SOCl_2(l) + 4\ e^- \rightarrow 4\ Cl^-(SOCl_2) + SO_2(SOCl_2) + S(SOCl_2)$$

The lithium ions and chloride ions combine on the surface of the carbon cathode to give lithium chloride which is insoluble in this solvent:

$$Li^+(SOCl_2) + Cl^-(SOCl_2) \rightarrow LiCl(s)$$

It is the sites on the cathode where the lithium chloride is deposited that become inactive. Thus, the cell stops functioning when most sites on the carbon electrode are covered. A related cell design uses sulfuryl chloride, SO_2Cl_2. This solvent has the advantage that upon reduction there are only two products: chloride ion and sulfur dioxide:

$$SO_2Cl_2(l) + 2\ e^- \rightarrow 2\ Cl^-(SO_2Cl_2) + SO_2(SO_2Cl_2)$$

A related battery is the lithium sulfur dioxide cell. This is used in automated external defibrillators (AEDs) that restore normal heart rhythm to victims of sudden cardiac arrest. The cell is able to function down to $-40°C$, so it is also favored for emergency aircraft beacons in cold climates. The sulfur dioxide is dissolved in an organic solvent and is under a pressure of 200–300 kPa. Like all lithium cells, the anode process is the oxidation of lithium, but in this case, the cathode reaction is the reduction of sulfur dioxide to the $S_2O_4^{2-}$ ion.

$$2\ SO_2(solvent) + 2\ e^- \rightarrow S_2O_4^{2-}(solvent)$$

11.6 *Sodium*

Sodium is the alkali metal for which there is the highest industrial demand. Like all the alkali metals, the pure element does not exist naturally because of its very high reactivity. The silvery metal is manufactured by the Downs process, in which sodium chloride (m.p. 801°C) is electrolyzed in the molten state. The electrolysis is done in a cylindrical cell with a central graphite anode and a surrounding steel cathode (Figure 11.4). A mixture of calcium chloride and sodium chloride is used to reduce the melting point and hence lower the temperature at which the cell needs to be operated. Although calcium chloride itself has a melting point of 772°C, a mixture of 33 percent sodium chloride and 67 percent calcium chloride has a melting point of about 580°C. It is the lower melting point of the mixture that makes the process commercially feasible. The two electrodes are separated by a cylindrical steel gauze diaphragm, so that the molten sodium, which floats to the top of the cathode compartment, will be kept away from the gaseous chlorine formed at the anode:

$$Na^+(NaCl) + e^- \rightarrow Na(l)$$

$$2\ Cl^-(NaCl) \rightarrow Cl_2(g) + 2\ e^-$$

The sodium metal produced contains about 0.2 percent calcium metal. Cooling the metal mixture to 110°C allows the calcium impurity (m.p. 842°C) to solidify and sink into the melt. The pure sodium (m.p. 98°C) remains liquid and can be pumped into cooled molds, where it solidifies.

Sodium metal is required for the synthesis of a large number of sodium compounds, but it has two major uses, the first of which is the extraction of other metals. The easiest way to obtain many of the rarer metals such as thorium, zirconium, tantalum, and titanium is by the reduction of their compounds with sodium. For example, titanium can be obtained by reducing titanium(IV) chloride with sodium metal:

$$TiCl_4(l) + 4\ Na(s) \rightarrow Ti(s) + 4\ NaCl(s)$$

The sodium chloride can then be washed away from the pure titanium metal.

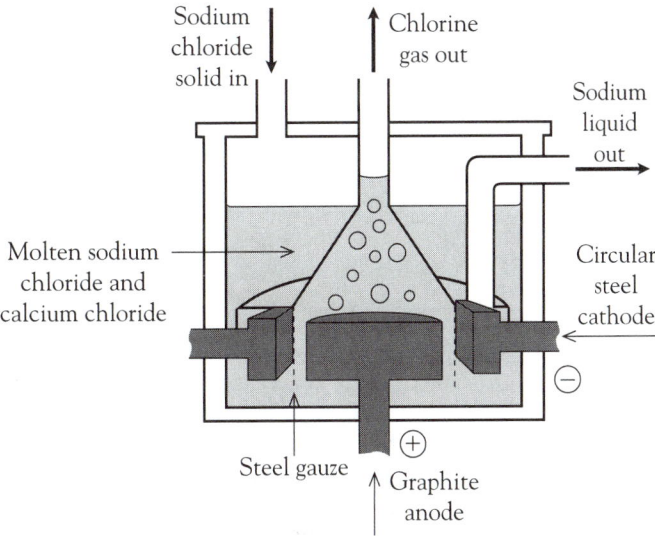

Figure 11.4   Downs cell.

The second major use of sodium metal is in the production of the gasoline additive tetraethyllead (TEL). Although TEL is now banned from gasolines in North America because of its toxicity and the lead pollution resulting from its use, it is still employed throughout much of the world to boost the octane rating of cheap gasolines. The synthesis of TEL uses the reaction between a lead–sodium alloy and ethyl chloride:

$$4 \text{ NaPb}(s) + 4 \text{ C}_2\text{H}_5\text{Cl}(g) \rightarrow (\text{C}_2\text{H}_5)_4\text{Pb}(l) + 3 \text{ Pb}(s) + 4 \text{ NaCl}(s)$$

11.7   Potassium

The potassium found in the natural environment is slightly radioactive because it contains about 0.012 percent of the radioactive isotope potassium-40. In fact, a significant proportion of the radiation generated within our bodies comes from this isotope, which has a half-life of 1.3×10^9 years. Approximately 89 percent of the potassium atoms decay by emitting an electron, while the other 11 percent decay by capturing an electron (evidence that electron density does penetrate the nucleus):

$$^{40}_{19}\text{K} \rightarrow {}^{40}_{20}\text{Ca} + {}^{0}_{-1}\text{e}$$

$$^{40}_{19}\text{K} + {}^{0}_{-1}\text{e} \rightarrow {}^{40}_{18}\text{Ar}$$

The ratio of potassium-40 to argon-40 is one way of dating rocks in that once the magma solidifies, the argon formed will be trapped within the rock structure.

The industrial extraction of potassium metal is accomplished by chemical means. Extraction in an electrolytic cell would be too hazardous because of the extreme reactivity of the metal. The chemical process involves the reaction of sodium metal with molten potassium chloride at 850°C:

$$\text{Na}(l) + \text{KCl}(l) \rightleftharpoons \text{K}(g) + \text{NaCl}(l)$$

Although the equilibrium lies to the left, at this temperature potassium is a gas (b.p. 766°C; b.p. for sodium is 890°C)! Thus, the Le Châtelier principle can be used to drive the reaction to the right by pumping the green potassium gas from the mixture as it is formed.

We have already mentioned that alkali metal salts exhibit a wide range of solubilities. In particular, the least soluble are those with the greatest similarity in ion size. Thus, a very large anion would form the least soluble salts with the larger cations of Group 1. This concept holds for the very large hexanitritocobaltate(III) anion, $[\text{Co(NO}_2)_6]^{3-}$. Its salts with lithium and sodium are soluble, whereas those with potassium, rubidium, and cesium are insoluble. Thus, if a solution is believed to contain either sodium or potassium ion, addition of the hexanitritocobaltate(III) ion can be used as a test. A bright yellow precipitate indicates the presence of potassium ion:

$$3 \text{ K}^+(aq) + [\text{Co(NO}_2)_6]^{3-}(aq) \rightarrow \text{K}_3[\text{Co(NO}_2)_6](s)$$

Another very large anion that can be used in a precipitation test with the larger alkali metals is the tetraphenylborate ion, $[\text{B(C}_6\text{H}_5)_4]^-$:

$$\text{K}^+(aq) + [\text{B(C}_6\text{H}_5)_4]^-(aq) \rightarrow \text{K}[\text{B(C}_6\text{H}_5)_4](s)$$

11.8 Oxides

Most metals in the Periodic Table react with dioxygen gas to form oxides containing the oxide ion, O^{2-}. However, of the alkali metals, only lithium forms a normal oxide when it reacts with oxygen:

$$4\,Li(s) + O_2(g) \rightarrow 2\,Li_2O(s)$$

Sodium reacts with dioxygen to give sodium dioxide(2−), Na_2O_2 (commonly called sodium peroxide),

$$2\,Na(s) + O_2(g) \rightarrow Na_2O_2(s)$$

which contains the dioxide(2−) ion, O_2^{2-} (often called the peroxide ion). The notation "2−" simply indicates the charge on the ion, and it avoids the need for learning the many prefixes that used to be employed for that purpose. We are now using parenthetical Arabic numbers in naming, a method recommended by the American Chemical Society for use whenever there is more than one possible ionic charge (as we shall see shortly).

Sodium dioxide(2−) is diamagnetic, and the oxygen–oxygen bond length is about 149 pm, much longer than the 121 pm in the dioxygen molecule. We can explain the diamagnetism and the weaker bond by constructing the part of the molecular orbital diagram derived from the $2p$ atomic orbitals (Figure 11.5). This diagram shows that three bonding orbitals and two antibonding orbitals are occupied. All the electrons are paired and the net bond order is 1 rather than 2, the bond order in the dioxygen molecule (see Chapter 3, Section 3.4).

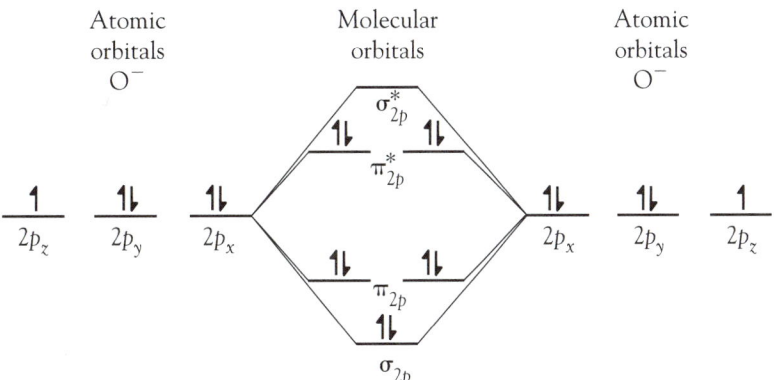

Figure 11.5 Filling of the molecular orbitals derived from the $2p$ orbitals for the dioxide(2−) ion.

The other three alkali metals react with an excess of dioxygen to form dioxides(1−) (traditionally named superoxides) containing the paramagnetic dioxide(1−) ion, O_2^-:

$$K(s) + O_2(g) \rightarrow KO_2(s)$$

The oxygen–oxygen bond length in these ions (133 pm) is less than that in the dioxide(2−) but slightly greater than that in dioxygen itself. We can also explain these different bond lengths in terms of the molecular orbital filling (Figure 11.6). The dioxide(1−) ion possesses three bonding pairs and one and one-half antibonding pairs. The net bond order in a dioxygen(1−) ion is 1½, between the bond order of 1 in the dioxide(2−) ion and the bond order of 2

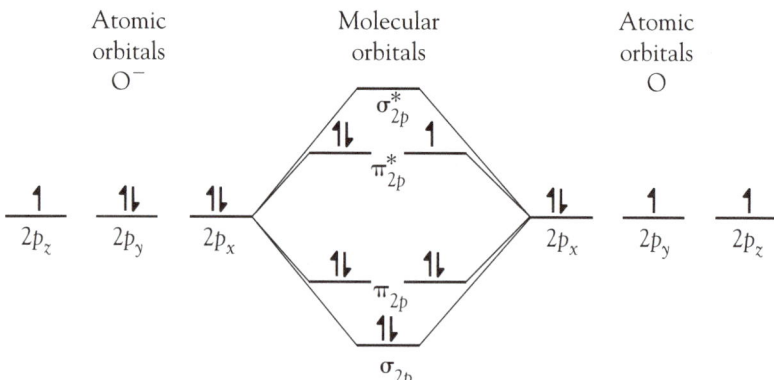

Figure 11.6 Filling of the molecular orbitals derived from the $2p$ orbitals for the dioxide($1-$) ion.

in the dioxygen molecule. We can explain the ready formation of both the dioxide($1-$) and the dioxide($2-$) ions by postulating that the least polarizing cations (those with low charge density) stabilize these large polarizable anions.

All the Group 1 oxides react vigorously with water to give the metal hydroxide solution. In addition, sodium dioxide($2-$) generates hydrogen peroxide, and the dioxides($1-$) produce hydrogen peroxide and oxygen gas:

$$Li_2O(s) + H_2O(l) \rightarrow 2\ LiOH(aq)$$

$$Na_2O_2(s) + 2\ H_2O(l) \rightarrow 2\ NaOH(aq) + H_2O_2(aq)$$

$$2\ KO_2(s) + 2\ H_2O(l) \rightarrow 2\ KOH(aq) + H_2O_2(aq) + O_2(g)$$

Potassium dioxide($1-$) is used in space capsules, submarines, and some types of self-contained breathing equipment because it absorbs exhaled carbon dioxide (and moisture) and releases dioxygen gas:

$$4\ KO_2(s) + 2\ CO_2(g) \rightarrow 2\ K_2CO_3(s) + 3\ O_2(g)$$

$$K_2CO_3(s) + CO_2(g) + H_2O(g) \rightarrow 2\ KHCO_3(s)$$

11.9 Hydroxides

The solid hydroxides are white, translucent solids that absorb moisture from the air until they dissolve in the excess water—a process known as *deliquescence*. The one exception is lithium hydroxide, which forms the stable octahydrate, $LiOH \cdot 8H_2O$. Alkali metal hydroxides are all extremely hazardous because the hydroxide ion reacts with skin protein to destroy the skin surface. Sodium hydroxide and potassium hydroxide are supplied as pellets, and these are produced by filling molds with the molten compound. As solids or in solution, they also absorb carbon dioxide from the atmosphere:

$$2\ NaOH(aq) + CO_2(g) \rightarrow Na_2CO_3(aq) + H_2O(l)$$

The alkali metal hydroxides are convenient sources of the hydroxide ion because they are very water soluble. When hydroxide ion is needed as a reagent, its source is chosen on the basis of either cost or solubility. In inorganic chemistry, sodium hydroxide (caustic soda) is most commonly used as the source of hydroxide ion because it is the least expensive metal hydroxide. Potassium hydroxide (caustic potash) is preferred in organic chemistry because it has a higher solubility in organic solvents than does sodium hydroxide.

Preparation of Sodium Hydroxide

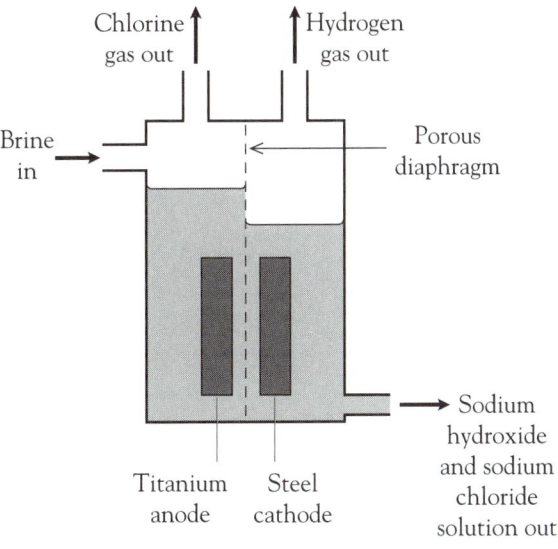

Figure 11.7   Diaphragm cell.

Sodium hydroxide is the sixth most important inorganic chemical in terms of quantity produced. It is prepared by the electrolysis of brine (aqueous sodium chloride). The three common types of electrolytic cells are the diaphragm cell, the membrane cell, and the mercury cathode cell. These cells use prodigious quantities of electricity, the first two types running at between 30 000 and 150 000 A; the mercury cell requires up to 400 000 A.

In the diaphragm cell, water is reduced to hydrogen gas and hydroxide ion at the cathode, and chloride ion is oxidized to chlorine gas at the anode (although some water is oxidized to oxygen gas as well):

$$2\ H_2O(l) + 2\ e^- \rightarrow H_2(g) + 2\ OH^-(aq)$$

$$2\ Cl^-(aq) \rightarrow Cl_2(g) + 2\ e^-$$

The essential design feature (Figure 11.7) is the diaphragm or separator, which prevents the hydroxide ion produced at the cathode from coming into contact with the chlorine gas produced at the anode. This separator, which has pores that are large enough to allow the brine to pass through, used to be made of asbestos, but it is now made of a Teflon® mesh. During the electrolysis, the cathode solution, which consists of a mixture of 11 percent sodium hydroxide and 16 percent sodium chloride, is removed continuously. The harvested solution is evaporated, a process that causes the less soluble sodium chloride to crystallize. The final product is a solution of 50 percent sodium hydroxide and about 1 percent sodium chloride. This composition is quite acceptable for most industrial purposes.

The membrane cell functions like the diaphragm cell except that the anode and cathode solutions are separated by a micropore polymer membrane that is permeable only to cations—in this case, the sodium ion. Thus, the chloride ions from the brine cannot enter the cathode compartment, nor can the hydroxide ions produced in the cathode compartment escape in the opposite direction. As a result, the sodium hydroxide solution produced contains no more than 50 ppm chloride ion contaminant. Unfortunately, the membrane is very expensive, and it can become clogged by trace impurities of calcium ion, which form insoluble calcium hydroxide on the membrane surface.

The mercury cathode cell, as its name implies, uses liquid mercury as the cathode (Figure 11.8). At the anode, chlorine gas is produced, but at the cathode, sodium ion is reduced to sodium metal:

$$Na^+(aq) + e^- \rightarrow Na(Hg)$$

Sodium reduction occurs because the mercury electrode surface inhibits any half-reaction that produces a gas, raising the electrode potential of such a reaction above its standard value. Thus, in this environment the "expected" reduction of hydrogen ion to hydrogen gas actually requires a higher potential than the reduction of sodium ion. The surface phenomenon responsible for this result is called overvoltage.

The sodium—mercury amalgam is pumped to a separate chamber, where water is reacted with the amalgam on a graphite surface:

$$2\ Na(Hg) + 2\ H_2O(l) \rightarrow 2\ NaOH(aq) + H_2(g)$$

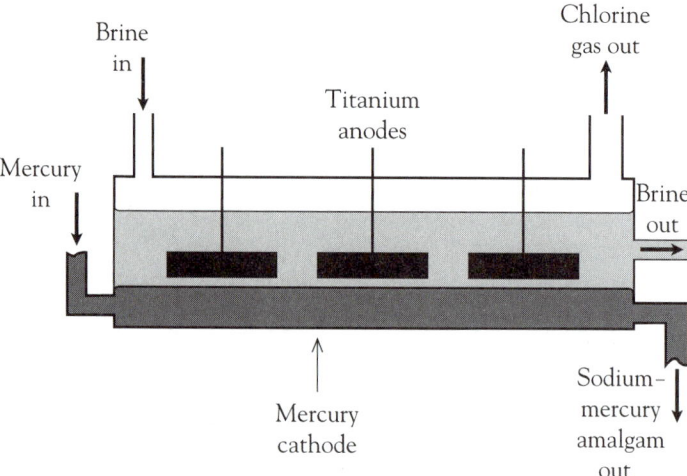

Figure 11.8 Mercury cathode cell.

The sodium hydroxide solution produced by this route is pure and concentrated. Hence, the mercury cathode cell is the preferred source of pure, solid sodium hydroxide. Unfortunately, there is a loss of mercury into the environment during the operation of this process. Traditionally, loss of mercury was regarded as just another business expense, and there was little or no concern about the ultimate fate of the mercury. Much of it ended up in rivers, where it caused severe pollution. Now, of course, we recognize the folly of this approach, and a considerable research effort has been undertaken to completely eliminate mercury losses from the cells. The comparative advantages and disadvantages of the cell types are listed in Table 11.10.

Table 11.10 Advantages and disadvantages of the industrial processes for the production of sodium hydroxide

Process	Advantages	Disadvantages
Diaphragm process	Utilizes less pure brine	Dilute, chloride-contaminated product (11% NaOH)
	Lower electrical consumption, because process is more efficient	Chlorine often oxygen-contaminated
		Asbestos concerns
Membrane process	Produces uncontaminated product	Maximum concentration, 35% NaOH
	Lower electrical consumption	Chlorine often oxygen-contaminated
	No mercury or asbestos problems	Very high-purity brine required
		High cost and short lifetime of membranes
Mercury process	Produces pure, high concentration sodium hydroxide (50%)	Higher electrical consumption
	Produces pure chlorine gas	Needs purer brine than diaphragm process
		Mercury containment problems

Uses of Sodium Hydroxide

About 30 percent of sodium hydroxide production is used as a reagent in organic chemical plants, and about 20 percent is used for the synthesis of other inorganic chemicals. Another 20 percent is consumed in the pulp and paper industry, and the remaining 30 percent is used in hundreds of other ways.

Sodium hydroxide is the most important base in the chemistry laboratory. It also has a number of household uses, where it is commonly referred to as lye. The most direct application takes advantage of its reaction with greases, particularly those in ovens (such as Easy-Off® Oven Cleaner) or those clogging drains (such as Drano®). In some commercial drain-treatment products, aluminum metal is mixed with the sodium hydroxide. When added to water, the following chemical reaction occurs, producing the aluminate ion and hydrogen gas. The hydrogen gas bubbles cause the liquid to churn vigorously, enhancing the contact of grease with fresh sodium hydroxide solution, an action that dissolves the plug more quickly:

$$2 \, Al(s) + 2 \, OH^-(aq) + 6 \, H_2O(l) \rightarrow 2 \, [Al(OH)_4]^-(aq) + 3 \, H_2(g)$$

Sodium hydroxide is also used in the food industry, mainly to provide hydroxide ion for breaking down proteins. For example, potatoes are sprayed with sodium hydroxide solution to soften and remove the skins before processing. (Of course, they are washed thoroughly before the next processing step!) Olives have to be soaked in sodium hydroxide solution to soften the flesh enough to make them edible. Grits, too, are processed with sodium hydroxide solution. The most unusual application is in the manufacture of pretzels. The dough is coated with a thin layer of sodium hydroxide solution before salt crystals are applied. The sodium hydroxide appears to function as a cement, attaching the salt crystals firmly to the dough surface. In the baking process, carbon dioxide is released, thereby converting the sodium hydroxide to harmless sodium carbonate:

$$2 \, NaOH(s) + CO_2(g) \rightarrow Na_2CO_3(s) + H_2O(g)$$

11.10 Sodium Chloride

Salt was one of the earliest commodities to be traded, and 2000 years ago, Roman soldiers were partially paid in salt (sal)—hence our term, salary, for pay.

Seawater is a 3 percent solution of sodium chloride, together with many other minerals. It has been calculated that the sea contains 19 million m^3 of salt—about one and a half times the volume of all North America above sea level. The salt produced by using the Sun's energy to evaporate seawater used to be a major source of income for some Third World countries, such as the Turks and Caicos Islands. Unfortunately, production of salt by this method is no longer economically competitive, and the ensuing loss of income and employment has caused serious economic problems for these countries.

Even today, salt is a vital commodity. More sodium chloride is used for chemical manufacture than any other mineral, with world consumption exceeding 150 million tonnes per year. Today almost all commercially produced sodium chloride is extracted from vast underground deposits, often hundreds of meters thick. These beds were produced when large lakes evaporated to dryness hundreds of millions of years ago. About 40 percent of the rock salt is mined like coal, and the remainder is extracted by pumping water into the deposits and pumping out the saturated brine solution.

Salt Substitutes

We need about 3 g of sodium chloride per day, but in western countries our daily diet usually contains between 8 and 10 g. Provided we have sufficient liquid intake, this level of consumption presents no problem. However, for those with high blood pressure, a decrease in sodium ion intake has been shown to cause a reduction in blood pressure. To minimize sodium ion intake, there are a number of salt substitutes on the market that taste salty but do not contain the sodium ion. Most of these contain potassium chloride and other compounds that mask the bitter, metallic aftertaste of the potassium ion. One enterprising producer of pure household salt claims its product contains "33 percent less sodium." This claim is technically true, and it is accomplished by producing hollow salt crystals. These have a bulk density 33 percent less than the normal cubic crystals. Hence, a spoonful of these salt crystals will contain 33 percent less of both sodium and chloride ions! Provided you sprinkle your food with the same volume of salt, it will obviously have the desired effect; but for the same degree of saltiness, you will need 50 percent more of the product by volume than regular salt.

11.11   *Potassium Chloride*

Like sodium chloride, potassium chloride (commonly called potash) is recovered from ancient dried lake deposits, many of which are now deep underground. About half of the world's reserves of potassium chloride lie under the Canadian provinces of Saskatchewan, Manitoba, and New Brunswick. As the ancient lakes dried, all their soluble salts crystallized. Hence, the deposits are not of pure potassium chloride but also contain crystals of sodium chloride, potassium magnesium chloride hexahydrate, $KMgCl_3·6H_2O$, magnesium sulfate monohydrate, $MgSO_4·H_2O$, and many other salts.

To separate the components, several different routes are used. One employs the differences in solubility: The mixture is dissolved in water and then the salts crystallize out in sequence as the water evaporates. However, this process requires considerable amounts of energy to vaporize the water. A second route involves adding the mixture of crystals to saturated brine. When air is blown through the slurry, the potassium chloride crystals adhere to the bubbles. The potassium chloride froth is then skimmed off the surface. The sodium chloride crystals sink to the bottom and can be dredged out.

The third route is most unusual, because it is an electrostatic process. The solid is ground to a powder, and an electric charge is imparted to the crystals by a friction process. The potassium chloride crystals acquire a charge that is the opposite of that of the other minerals. The powder is then poured down a tower containing two highly charged drums. The potassium chloride adheres to one drum, from which it is continuously removed, and the other salts adhere to the oppositely charged drum. Unfortunately, the reject minerals from potash processing have little use, and their disposal is a significant problem.

There is just one use for all this potassium chloride—as fertilizer. Potassium ion is one of the three essential elements for plant growth (nitrogen and phosphorus being the other two), and about 4.5×10^7 tonnes of potassium chloride is used worldwide for this purpose every year, so it is a major chemical product.

11.12 Sodium Carbonate

Preparation of Sodium Carbonate

The alkali metals (and ammonium ion) form the only soluble carbonates. Sodium carbonate, the most important of the alkali metal carbonates, exists in the anhydrous state (soda ash), as a monohydrate, $Na_2CO_3 \cdot H_2O$, and most commonly as the decahydrate, $Na_2CO_3 \cdot 10H_2O$ (washing soda). The large transparent crystals of the decahydrate *effloresce* (lose water of crystallization) in dry air to form a powdery deposit of the monohydrate:

$$Na_2CO_3 \cdot 10H_2O(s) \rightarrow Na_2CO_3 \cdot H_2O(s) + 9\ H_2O(g)$$

Sodium carbonate is the ninth most important inorganic compound in terms of quantity used. For over half a century, sodium bicarbonate (and from it, the carbonate) was made by the *Solvay*, or ammonia-soda, *process*. However, the environmental problems associated with the reactions used in this process have made it preferable to mine underground deposits. By far the largest quantity in the world, 4.5×10^{10} tonnes, is found in Wyoming. The mineral, known as trona, contains about 90 percent of a mixed carbonate–hydrogen carbonate, $Na_2CO_3 \cdot NaHCO_3 \cdot 2H_2O$, which is commonly called sodium sesquicarbonate. Sesqui means "one and one-half," and it is the number of sodium ions per carbonate unit in the mineral. Sodium sesquicarbonate is not a mixture of the two compounds but a single compound in which the crystal lattice contains alternating carbonate and hydrogen carbonate ions interspersed with sodium ions and water molecules in a $1:1:3:2$ ratio, that is, $Na_3(HCO_3)(CO_3) \cdot 2H_2O$.

In the monohydrate process of extraction, trona is mined like coal about 400 m underground, crushed, and then heated (calcined) in rotary kilns. This treatment converts the sesquicarbonate to carbonate:

$$2\ Na_3(HCO_3)(CO_3)2H_2O(s) \xrightarrow{\Delta} 3\ Na_2CO_3(s) + 5\ H_2O(g) + CO_2(g)$$

The resulting sodium carbonate is dissolved in water and the insoluble impurities are filtered off. The sodium carbonate solution is then evaporated to dryness, thereby producing sodium carbonate monohydrate. Heating this product in a rotary kiln gives the anhydrous sodium carbonate:

$$Na_2CO_3 \cdot H_2O(s) \xrightarrow{\Delta} Na_2CO_3(s) + H_2O(g)$$

The Californian salt lake deposits contain about 6×10^8 tonnes of sodium carbonate, and this also is mined. However, even though the last Solvay plant in the United States closed in 1986, many other countries have to rely on this chemical process for their sodium carbonate supplies. In fact, about 70 percent of the world's supply of this reagent still comes from the Solvay process. For this reason, it is worth discussing. Besides, there is some interesting chemistry involved in the procedure. The overall reaction uses two low-cost reagents, and it appears to be quite simple:

$$2\ NaCl(aq) + CaCO_3(s) \rightleftharpoons Na_2CO_3(aq) + CaCl_2(aq)$$

However, the equilibrium position for this reaction lies far to the left. That is, the converse reaction will occur: Calcium chloride will react with sodium carbonate to give a precipitate of calcium carbonate and a solution of sodium chloride. It is fortunate that the equilibrium does favor the left side. Otherwise the White Cliffs of Dover would have long ago dissolved in the salt water of the English Channel!

To force the reverse reaction and obtain the desired product of sodium carbonate, an indirect, multistep procedure has to be used. In the first step, carbon dioxide is forced into a solution saturated with sodium chloride and ammonia. The carbon dioxide reacts with the ammonia to give ammonium ions and hydrogen carbonate ions:

$$CO_2(g) + NH_3(aq) + H_2O(l) \rightarrow NH_4^+(aq) + HCO_3^-(aq)$$

The solution now contains sodium and ammonium cations and chloride and hydrogen carbonate anions. Upon cooling, the relatively low solubility of sodium hydrogen carbonate in cold water causes it to crystallize:

$$HCO_3^-(aq) + Na^+(aq) \rightarrow NaHCO_3(s)$$

The solid sodium hydrogen carbonate is filtered off and then gently heated to give the carbonate:

$$2\,NaHCO_3(s) \rightarrow Na_2CO_3(s) + H_2O(g) + CO_2(g)$$

The commercial success of the process depends on the recovery of ammonia by the reaction:

$$2\,NH_4^+(aq) + 2\,Cl^-(aq) + Ca(OH)_2(s) \rightarrow$$
$$2\,NH_3(g) + CaCl_2(aq) + 2\,H_2O(l)$$

The calcium hydroxide and carbon dioxide used in the process are both obtained by heating limestone strongly:

$$CaCO_3(s) \xrightarrow{\Delta} CaO(s) + CO_2(g)$$

$$CaO(s) + H_2O(l) \rightarrow Ca(OH)_2(s)$$

Summing these six equations gives the equation for the overall process:

$$2\,NaCl(aq) + CaCO_3(s) \rightarrow Na_2CO_3(aq) + CaCl_2(aq)$$

The problem with the Solvay process is the amount of the by-product calcium chloride that is produced. The demand for calcium chloride is much less than the supply from this reaction. Furthermore, the process is quite energy intensive, making it more expensive than the simple method of extraction from trona.

Uses of Sodium Carbonate

About 50 percent of the U.S. production of sodium carbonate is used in glass manufacture. In the process, the sodium carbonate is reacted with silicon dioxide (sand) and other components at about 1500°C. The actual formula of the product depends on the stoichiometric ratio of reactants (the process is discussed in more detail in Chapter 14, Section 14.18). The key reaction is the formation of a sodium silicate and carbon dioxide:

$$Na_2CO_3(l) + x\,SiO_2(s) \xrightarrow{\Delta} Na_2O \cdot x SiO_2(l) + CO_2(g)$$

Sodium carbonate is also used to remove alkaline earth metal ions from water supplies by converting them to their insoluble carbonates, a process called water "softening." The most common ion that needs to be removed is calcium. Very high concentrations of this ion are found in water supplies that have come from limestone or chalk geological formations:

$$CO_3{}^{2-}(aq) + Ca^{2+}(aq) \rightarrow CaCO_3(s)$$

11.13 Sodium Hydrogen Carbonate

The alkali metals, except for lithium, form the only solid hydrogen carbonates (commonly called bicarbonates). Once again, the notion that low charge density cations stabilize large low charge anions can be used to explain the existence of these hydrogen carbonates.

Sodium hydrogen carbonate is less water-soluble than sodium carbonate. Thus, it can be prepared by bubbling carbon dioxide through a saturated solution of the carbonate:

$$Na_2CO_3(aq) + CO_2(g) + H_2O(l) \rightarrow 2\,NaHCO_3(s)$$

Heating sodium hydrogen carbonate causes it to decompose back to sodium carbonate:

$$2\,NaHCO_3(s) \xrightarrow{\Delta} Na_2CO_3(s) + CO_2(g) + H_2O(g)$$

This reaction provides one application of sodium hydrogen carbonate, the major component in dry powder fire extinguishers. The sodium hydrogen carbonate powder itself smothers the fire, but, in addition, the solid decomposes to give carbon dioxide and water vapor, themselves fire-extinguishing gases.

The main use of sodium hydrogen carbonate is in the food industry, to cause bakery products to rise. It is commonly used as a mixture (baking powder) of sodium hydrogen carbonate and calcium dihydrogen phosphate, $Ca(H_2PO_4)_2$, with some starch added as a filler. The calcium dihydrogen phosphate is acidic and, when moistened, reacts with the sodium hydrogen carbonate to generate carbon dioxide:

$$2\,NaHCO_3(s) + Ca(H_2PO_4)_2(s) \rightarrow$$
$$Na_2HPO_4(s) + CaHPO_4(s) + 2\,CO_2(g) + 2\,H_2O(g)$$

11.14 Ammonia Reaction

The alkali metals themselves have the interesting property of dissolving in liquid ammonia to yield solutions that are deep blue when dilute. These solutions conduct current electrolytically, and the main current carrier in the solution is thought to be the solvated electron that is a product of the ionization of the sodium atoms:

$$Na(s) \rightleftharpoons Na^+(NH_3) + e^-(NH_3)$$

When concentrated by evaporation, the solutions have a bronze color and behave like a liquid metal. On long standing, or more rapidly in the presence of a transition metal catalyst, the solutions decompose to yield the amide salt, $NaNH_2$, and hydrogen gas:

$$2\,Na^+(NH_3) + 2\,NH_3(l) + 2\,e^- \rightarrow 2\,NaNH_2(NH_3) + H_2(g)$$

Table 11.11   Concentrations of ions (mmol·L^{-1})

Ion	[Na$^+$]	[K$^+$]
Red blood cells	11	92
Blood plasma	160	10

11.15   Biological Aspects

Lithium is possibly an essential trace element. Goats raised on a lithium-deficient diet had lower body-weight gain, lower conception rates, and significantly higher mortality rates.

We tend to forget that both sodium and potassium ions are essential to life. For example, we need at least 1 g of sodium ion per day in our diet. However, because of our addiction to salt on foods, the intake of many people is as much as five times that value. Excessive intake of potassium ion is rarely a problem. In fact, potassium deficiency is much more common; thus, it is important to ensure that we include in our diets potassium-rich foods such as bananas and coffee.

The alkali metal ions balance the negative charge associated with many of the protein units in the body. They also help to maintain the osmotic pressure within cells, preventing them from collapsing. In inorganic chemistry, we think of the similarities between sodium and potassium, but in the biological world, it is the difference that is crucial. Cells pump sodium ions out of the cytoplasm and pump potassium ions in (Table 11.11). It is this difference in total alkali metal ion concentrations inside and outside cells that produces an electrical potential across the cell membrane. The potential difference underlies many basic processes, such as the heart's generation of rhythmic electrical signals, the kidney's unceasing separation of vital and toxic solutes in the blood, and the eye's precise control of the lens' refractive index. Most of the 10 W of power produced by the human brain—awake or asleep—results from the Na$^+$/K$^+$-adenosine triphosphatase enzyme pumping potassium ion into and sodium ion out of brain cells. When we "go into shock" as a result of an accident, it is a massive leakage of the alkali metal ions through the cell walls that causes the phenomenon.

The ion-selective enzymes function by having cavities that precisely fit one or another ion size. As well as the difference in ion size, there is also a significant difference in dehydration energy. For the ions to fit in the bonding site, they must lose their hydration sphere. The sodium ion, with its higher charge density, requires 80 kJ·mol^{-1} more energy to release its accompanying water molecules, giving the potassium ion an immediate bonding advantage.

A number of antibiotics seem to be effective because they have the ability to transfer specific ions across cell membranes. These organic molecules, too, have holes in the middle that are just the right size to accommodate an ion with a particular ionic radius. For example, valinomycin has an aperture that is just right for holding a potassium ion but too large for a sodium ion. Thus, the drug functions, at least in part, by selectively transporting potassium ions across biological membranes.

11.16   Element Reaction Flowchart

The three most important elements of this group are lithium, sodium, and potassium; flowcharts are shown for these three elements. Remember, only interrelated reactions are shown on these flowcharts, not all of the important reactions.

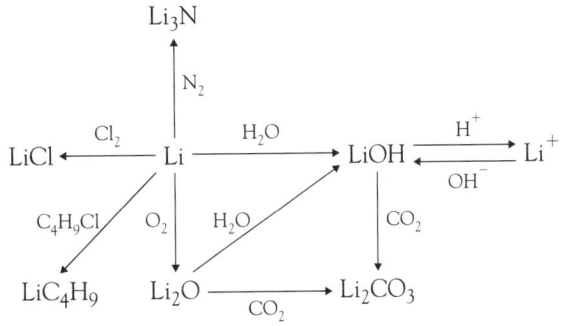

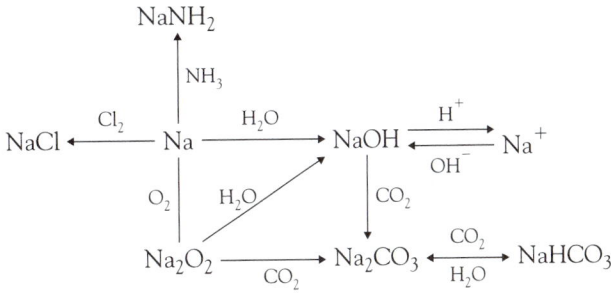

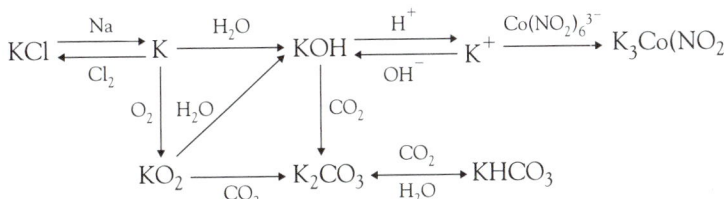

EXERCISES

11.1 Write balanced chemical equations for each of the following reactions:
(a) sodium metal with water
(b) rubidium metal with dioxygen
(c) solid potassium hydroxide with carbon dioxide
(d) heating solid sodium nitrate

11.2 Write balanced chemical equations for each of the following reactions
(a) lithium metal with dinitrogen
(b) solid cesium dioxide(1−) with water
(c) heating solid sodium hydrogen carbonate
(d) heating solid ammonium nitrate

11.3 In what ways do the alkali metals resemble "typical" metals? In what ways are they very different?

11.4 Which is the least reactive alkali metal? Why is this unexpected on the basis of standard oxidation potentials? What explanation can be provided?

11.5 Describe three of the common features of the chemistry of the alkali metals.

11.6 An alkali metal, designated as M, forms a hydrated sulfate, $M_2SO_4 \cdot 10H_2O$. Is the metal more likely to be sodium or potassium? Explain your reasoning.

11.7 Suggest a possible reason why sodium hydroxide is much more water soluble than sodium chloride.

11.8 In the Downs cell for the preparation of sodium metal,
(a) why can't the electrolysis be performed in aqueous solution?
(b) why is calcium chloride added?

11.9 Why is it important to use a temperature of about 850°C in the extraction of potassium metal?

11.10 Describe the advantages and disadvantages of the diaphragm cell for the production of sodium hydroxide.

11.11 Several of the alkali metal compounds have common names. Give the systematic name corresponding to (a) caustic soda; (b) soda ash; (c) washing soda.

11.12 Several of the alkali metal compounds have common names. Give the systematic name corresponding to (a) caustic potash; (b) trona; (c) lye.

11.13 Explain what is meant by (a) efflorescence; (b) intercalation.

11.14 Explain what is meant by (a) supermetals; (b) deliquescent.

11.15 Write the chemical equations for the reactions involved in the Solvay synthesis of sodium carbonate. What are the two major problems with this process?

11.16 Explain briefly why only the alkali metals form solid, stable hydrogen carbonate salts.

11.17 Explain briefly why the ammonium ion is often referred to as a pseudo-alkali metal.

11.18 Construct an approximate molecular orbital diagram to depict the bonding in the gaseous lithium hydride molecule.

11.19 Suggest two reasons why potassium dioxide(1−), not cesium dioxide(1−), is used in spacecraft air recirculation systems.

11.20 Where are the sodium ions and potassium ions located with respect to living cells?

11.21 Write balanced chemical equations corresponding to each transformation in the element reaction flowcharts.

BEYOND THE BASICS

11.22 In this chapter, we have ignored the radioactive member of Group 1, francium. On the basis of group trends, suggest the key properties of francium and its compounds.

11.23 What minimum current at 7.0 V, assuming 100 percent efficiency, would be needed in a Downs cell to produce 1.00 tonne of sodium metal per day? (Passage of 1 mol of electrons requires 9.65×10^4 $A \cdot s^{-1}$ Faraday of electricity.)

11.24 Platinum hexafluoride, PtF_6, has an extremely high electron affinity (772 $kJ \cdot mol^{-1}$). Yet when lithium metal is reacted with platinum hexafluoride, it is lithium fluoride, Li^+F^-, not $Li^+PtF_6^-$ that is formed. Suggest a reason.

11.25 Suggest an explanation why ΔH_f° becomes less negative along the series LiF, NaF, KF, RbF, CsF, while it becomes more negative along the series LiI, NaI, KI, RbI, CsI.

11.26 The atomic mass of lithium is listed as 6.941 $g \cdot mol^{-1}$. However, lithium compounds are not used as primary analytical standards as the atomic mass of the lithium is often about 6.97 $g \cdot mol^{-1}$. Suggest an explanation for this.

11.27 Which compound, sodium fluoride or sodium tetrafluoroborate, $Na[BF_4]$, is likely to be the more soluble in water? Give your reasoning.

11.28 Determine whether the theoretical cesium(II) fluoride, CsF_2, will spontaneously decompose into cesium fluoride:

$$CsF_2(s) \rightarrow CsF(s) + \tfrac{1}{2}\ F_2(g)$$

given that the lattice energy of CsF_2 is 2250 $kJ \cdot mol^{-1}$. The second ionization energy of cesium

is 2.430 $MJ \cdot mol^{-1}$. Obtain all additional data from the Appendices. This calculation will only provide the enthalpy change. For spontaneity we need to find the free energy change from the entropy and enthalpy data. Will the entropy change also favor decomposition? Give your explanation.

11.29 From lattice dimension, the hydride ion appears to have a radius of 130 pm in lithium hydride but 154 pm in cesium hydride. Suggest a reason for the difference in the two values.

11.30 Solid cesium chloride will react with hydrogen chloride gas to give a compound containing a polyatomic anion. Write the formula of the anion. Lithium chloride does not react with hydrogen chloride. Suggest a reason why this reaction does not occur.

11.31 A solution containing equimolar concentrations of the following ions: Li^+, K^+, F^-, and I^-, is evaporated to dryness. Which salts will crystallize out, LiF and KI, or LiI and KF? Check your answer by working out the energetically preferred lattice energies (use the Kapustinskii equation—Chapter 6, Section 6.1).

11.32 In the high temperature/pressure organic substitution reaction of a carbon–chlorine bond with a carbon–fluorine bond using an alkali metal fluoride, MF:

$$R_3C{-}Cl + MF \rightarrow R_3C{-}F + MCl$$

why is the use of potassium fluoride preferred to that of sodium fluoride?

11.33 Suggest why the decay of potassium-40 should most likely lead to the formation of calcium-40.

The Group 2 Elements: The Alkaline Earth Metals

Chapter **12**

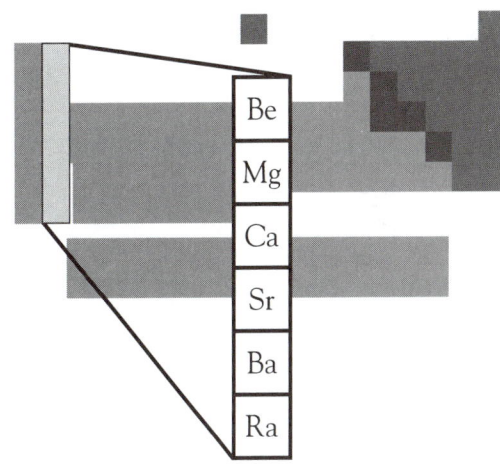

Although harder, denser, and less reactive than the alkali metals, the alkaline earth metals are more reactive and of lower density than a "typical" metal.

The last of the alkaline earth metals to be extracted from its compounds was radium. Marie Curie and André Debierne accomplished this task in 1910, delighting in the bright glow from this element, not realizing that it was the result of the element's intense and dangerous radiation. During the 1930s, cabaret shows sometimes featured dancers painted with radium salts so that they would literally glow in the dark. Some of the dancers may have died of radiation-related diseases, never being aware of the cause. Even quite recently, it was possible to purchase watches with hands and digits painted with radium-containing paint, its glow enabling the owner to read the time in the dark. (Safer substitutes are now available.)

12.1 Group Trends

In this section, we consider the properties of magnesium, calcium, strontium, and barium. Beryllium is discussed separately, because it behaves chemically more like a semimetal. The properties of radium, the radioactive member of the group, are not known in as much detail.

Table 12.1   Densities of the common alkaline earth metals

Element	Density (g·cm^{-3})
Mg	1.74
Ca	1.55
Sr	2.63
Ba	3.62

Table 12.2   Melting points of the common alkaline earth metals

Element	Melting point (°C)	$\Delta H_{atomization}$ (kJ·mol^{-1})
Mg	649	149
Ca	839	177
Sr	768	164
Ba	727	175

The alkaline earth metals are silvery and of fairly low density. As with the alkali metals, density generally increases with increasing atomic number (Table 12.1). The alkaline earth metals have stronger metallic bonding than do the alkali metals, a characteristic that is evident from the significantly greater enthalpies of atomization (Table 12.2). The metallic bonding of the alkaline earth metals is also reflected in both their higher melting points and their greater hardness. Although the density increases down the group (in parallel with that of the alkali metals), the melting points and enthalpies of atomization change very little. The ionic radii increase down the group and are smaller than those of the alkali metals (Figure 12.1).

The alkaline earth metals are less chemically reactive than are the alkali metals, but they are still more reactive than the majority of the other metallic elements. For example, calcium, strontium, and barium all react with cold water, barium reacting the most vigorously of all:

$$Ba(s) + 2\ H_2O(l) \rightarrow Ba(OH)_2(aq) + H_2(g)$$

As with the alkali metals, reactivity increases as mass increases within the group. Thus, magnesium does not react with cold water, but it will react slowly with hot water to produce magnesium hydroxide and hydrogen gas.

They also react with many nonmetals. For example, heated calcium burns in chlorine gas to give calcium chloride.

$$Ca(s) + Cl_2(g) \rightarrow CaCl_2(s)$$

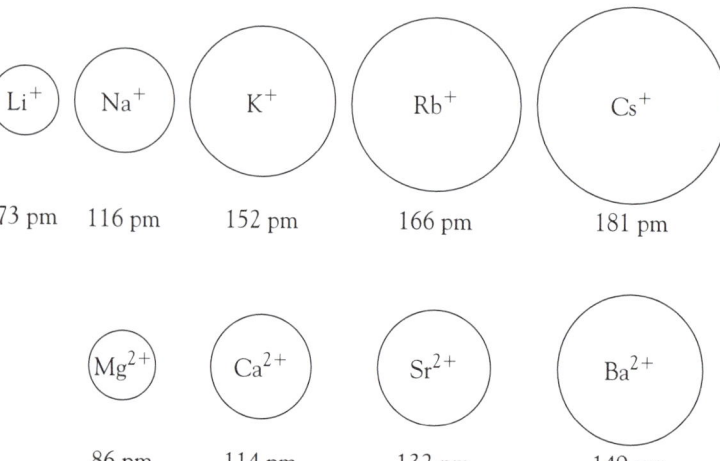

Figure 12.1   Comparison of the ionic radii of the alkali and alkaline earth metals.

Table 12.3 Usual hydration number of common alkaline earth metal salts

Element	MCl_2	$M(NO_3)_2$	MSO_4
Mg	6	6	7
Ca	6	4	2
Sr	6	4	0
Ba	2	0	0

The alkaline earth metals are unusual in a readiness to react with nitrogen gas on heating. Thus, magnesium reacts with dinitrogen to form magnesium nitride.

$$3\ Mg(s) + N_2(g) \rightarrow Mg_3N_2(s)$$

12.2 Common Features of Alkaline Earth Metal Compounds

We again exclude beryllium from the discussion because its properties are very different from those of the other members of Group 2.

Ionic Character

The alkaline earth metal ions always have an oxidation number of $+2$, and their compounds are mainly stable, colorless, ionic solids—unless a colored anion is present. The bonds in alkaline earth metal compounds are mostly ionic in character, but covalent behavior is particularly evident in compounds of magnesium. (Covalency also dominates the chemistry of beryllium.)

Ion Hydration

The salts of the alkaline earth metals are almost always hydrated. For example, calcium chloride can be prepared as the hexahydrate, tetrahydrate, dihydrate, and monohydrate, in addition to its anhydrous form. Table 12.3 shows the usual hydration number (number of molecules of water of crystallization) of some common alkaline earth metal compounds. As the charge density of the metal becomes smaller, so does the hydration number. Paradoxically, the hydroxides of strontium and barium are octahydrates, whereas those of magnesium and calcium are anhydrous.

12.3 Solubility of Alkaline Earth Metal Salts

Whereas all the common Group 1 salts are water soluble, many of those of Group 2 are insoluble. Generally it is the compounds with mononegative anions, such as chloride and nitrate, that are soluble, and those with more than one negative charge, such as carbonate and phosphate, are insoluble. There are also a few salts of anions that show striking trends in solubility. In particular, the sulfates change from soluble to insoluble down the group, whereas the hydroxides change from insoluble to soluble.

In Chapter 11, Section 11.3, we discussed the solubility of alkali metal halides in terms of thermodynamic functions. For the alkaline earth metals, the values of each function differ dramatically from those of the alkali metals, yet the net changes in entropy and enthalpy for the solution process are little different.

First let us consider the enthalpy factors involved. The initial step of our enthalpy cycle is vaporization of the crystal lattice. For a salt of a dipositive cation, about three times the energy will be needed to vaporize the lattice as is needed for a monopositive cation, because there are much greater electrostatic attractions in the dipositive cation salts ($2+$ charge with $1-$,

Table 12.4   Enthalpy factors in the solution process for magnesium chloride and sodium chloride

Compound	Lattice energy ($kJ \cdot mol^{-1}$)	Hydration enthalpy ($kJ \cdot mol^{-1}$)	Net enthalpy change ($kJ \cdot mol^{-1}$)
$MgCl_2$	+2526	−2659	−133
NaCl	+788	−784	+4

versus 1+ with 1−). Furthermore, per mole, three ions must be separated rather than two. However, the enthalpy of hydration of the dipositive ions will also be much greater than those of the monopositive alkali metal ions. As a result of the higher charge densities of the Group 2 ions, the water molecules are more strongly attracted to the "naked" cation, so there is a much greater release of energy when they form a solvation sphere around it. For example, the enthalpy of hydration of the magnesium ion is −1921 $kJ \cdot mol^{-1}$, whereas that of the sodium ion is −435 $kJ \cdot mol^{-1}$ (the ratio of these two values is close to the ratio of their charge densities). Enthalpy data for magnesium chloride and sodium chloride are compared in Table 12.4. As these figures indicate, when (anhydrous) magnesium chloride is dissolved in water, the solution process is noticeably exothermic.

Now let us consider the entropy factors (Table 12.5). The lattice entropy of magnesium chloride is almost exactly one and a half times that of sodium chloride, reflecting the fact that three gaseous ions rather than two are being produced. However, because the magnesium ion has a much higher charge density, the entropy of hydration for the magnesium ion is significantly more negative than that for the sodium ion. There is a much more ordered environment around the magnesium ion, which is surrounded by the strongly held layers of water molecules. Thus, overall, the entropy factors do not favor the solution process for magnesium chloride. Recall that for sodium chloride, it was the entropy factor that favored solution.

When we combine the enthalpy and entropy terms—keeping in mind that all of the data values have associated errors—we see that the solubility process results primarily from very small differences in very large energy terms (Table 12.6). Furthermore, for magnesium chloride, enthalpy factors favor solution and entropy factors oppose them, a situation that is the converse of that for sodium chloride.

It is the much higher lattice energy that partially accounts for the insolubility of the salts containing di- and trinegative ions. As the charge increases, so does the electrostatic attraction that must be overcome in the lattice vaporization step. At the same time, there are fewer ions (two for the metal sulfates compared to three for the metal halides); hence, the total ion hydration enthalpy will be less than that for the salts with mononegative ions. The combination of these two factors, then, is responsible for the low solubility.

Table 12.5   Entropy factors in the solution process for magnesium chloride and sodium chloride, expressed as $T\Delta S$ values

Compound	Lattice entropy ($kJ \cdot mol^{-1}$)	Hydration entropy ($kJ \cdot mol^{-1}$)	Net entropy change ($kJ \cdot mol^{-1}$)
$MgCl_2$	+109	−143	−34
NaCl	+68	−55	+13

Table 12.6 Calculated free energy changes for the solution process for magnesium chloride and sodium chloride

Compound	Enthalpy change (kJ·mol^{-1})	Entropy change (kJ·mol^{-1})	Free energy change (kJ·mol^{-1})
$MgCl_2$	−133	−34	−99
NaCl	+4	+13	−11

12.4 Beryllium

The element beryllium is steel gray and hard; it has a high melting temperature and a low density. It also has a high electrical conductivity, so it is definitely a metal. Because of beryllium's resistance to corrosion, its low density, high strength, and nonmagnetic behavior, beryllium alloys are often used in precision instruments such as gyroscopes. A minor but crucial use is in the windows of X-ray tubes. Absorption of X-rays increases with the square of the atomic number, and beryllium has the lowest atomic number of all the air-stable metals. Hence, it is one of the most transparent materials for the X-ray spectrum.

The sources of beryllium are bertrandite, $Be_4Si_2O_7(OH)_2$, and the gemstone beryl, $Be_3Al_2Si_6O_{18}$, which occurs in various colors because of trace amounts of impurities. When it is a light blue-green, beryl is called aquamarine; when it is deep green, it is called emerald. The green color is due to the presence of about 2 percent chromium(III) ion in the crystal structure. Of course, emeralds are not used for the production of metallic beryllium; the very imperfect crystals of colorless or brown beryl are used instead.

Beryllium compounds have a sweet taste and are extremely poisonous. When new compounds were prepared in the 19th century, it was quite common to report taste as well as melting point and solubility! Inhalation of the dust of beryllium compounds results in a chronic condition known as berylliosis.

The chemistry of beryllium is significantly different from that of the other Group 2 elements because covalent bonding predominates in its compounds. The very small beryllium cation has such a high charge density (1100 C·mm^{-3}) that it polarizes any approaching anion, and overlaps of electron density occur. Hence, the simple ionic compounds of beryllium tend to be found as tetrahydrates, such as $BeCl_2·4H_2O$, which actually consist of $[Be(OH_2)_4]^{2+}·2Cl^-$ ions in the crystal lattice. This tetraaquaberyllium ion, $Be(OH_2)_4^{2+}$, in which the four oxygen atoms of the water molecules are covalently bonded to the beryllium ion, is also the predominant species in aqueous solution. Four coordination is the norm for beryllium, because of the small size of the beryllium ion (Figure 12.2).

Although beryllium is definitely metallic, it has one property that is more characteristic of nonmetals—an ability to form oxyanion species. "Normal" metal oxides generally react with acids to give cations but not with bases to form oxyanions. Thus, beryllium oxide is amphoteric (see Chapter 7, Section 7.7, and Chapter 9, Section 9.4), reacting not only with hydronium ion to form the tetraaquaberyllium ion, $[Be(OH_2)_4]^{2+}$, but also with hydroxide ion to form the tetrahydroxoberyllate ion, $[Be(OH)_4]^{2-}$:

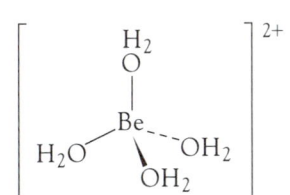

Figure 12.2 Tetrahedral shape of the $[Be(OH_2)_4]^{2+}$ ion.

$$H_2O(l) + BeO(s) + 2\ H_3O^+(aq) \rightarrow [Be(OH_2)_4]^{2+}(aq)$$

$$H_2O(l) + BeO(s) + 2\ OH^-(aq) \rightarrow [Be(OH)_4]^{2-}(aq)$$

Beryllium and the other metals that exhibit amphoteric behavior (including aluminum and zinc) are sometimes called "weak" metals as they tend to be located close to the semimetals and to the semimetal/nonmetal boundary. In the case of beryllium, it is actually located next to boron, on the semimetal/nonmetal boundary. It is only because we fit in the transition metal groups further down into the short form of the Periodic Table that there appears to be a gap between beryllium and boron.

12.5 Magnesium

Magnesium is found in nature as one component in a number of mixed-metal salts such as carnallite, $MgCl_2 \cdot KCl \cdot 6H_2O$, and dolomite, $MgCO_3 \cdot CaCO_3$. These compounds are not simply mixtures of salts but are pure ionic crystals in which the alternating sizes of the cations confer on the crystal lattice a greater stability than that conferred by either cation alone. Thus, carnallite contains arrays of chloride anions with interspersed potassium and magnesium cations and water molecules in a ratio of 3:1:1:6, that is, $KMgCl_3 \cdot 6H_2O$.

Magnesium is the third most common ion in seawater (after sodium and chloride), and seawater is a major industrial source of this metal. In fact, 1 km^3 of seawater contains about 1 million tonnes of magnesium ion. With 10^8 km^3 of seawater on this planet, there is more than enough magnesium for our needs. The Dow Chemical extraction process is based on the fact that magnesium hydroxide has a lower solubility than calcium hydroxide does. Thus, a suspension of finely powdered calcium hydroxide is added to the seawater, causing magnesium hydroxide to form:

$$Ca(OH)_2(s) + Mg^{2+}(aq) \rightarrow Ca^{2+}(aq) + Mg(OH)_2(s)$$

The hydroxide is then filtered off and mixed with hydrochloric acid. The resulting neutralization reaction gives a solution of magnesium chloride:

$$Mg(OH)_2(s) + 2\ HCl(aq) \rightarrow MgCl_2(aq) + 2\ H_2O(l)$$

The solution is evaporated to dryness, and the residue is placed in an electrolytic cell similar to the Downs cell used for the production of sodium. The magnesium collects on the surface of the cathode compartment and is siphoned off. The chlorine gas produced at the anode is reduced back to hydrogen chloride, which is then used to react with more magnesium hydroxide:

$$Mg^{2+}(MgCl_2) + 2\ e^- \rightarrow Mg(l)$$

$$2\ Cl^-(MgCl_2) \rightarrow Cl_2(g) + 2\ e^-$$

Providing an unreactive atmosphere over the molten magnesium is a real problem. For most reactive metal syntheses, such as sodium, the space above the molten metal can be filled with unreactive (and cheap) nitrogen gas. However, as we discussed in Section 12.1, magnesium reacts with dinitrogen. Most plants in western countries currently use expensive sulfur hexafluoride, SF_6, to blanket the molten magnesium to prevent contact with oxygen (or nitrogen). Unfortunately, sulfur hexafluoride is a potent greenhouse gas (see Chapter 16, Section 16.24) and losses to the atmosphere are quite significant. An alternative, used in eastern countries, is sulfur dioxide.

Even though magnesium is a chemically reactive metal, its reactivity is less than would be expected on the basis of its standard reduction potential of -2.37 V, because a thin coating of magnesium oxide rapidly forms over any metal surface exposed to the atmosphere. This coating protects the rest of the metal from attack. Burning magnesium gives an intense white light. The intensity of the light is so great that damage to the retina can occur. The combustion of magnesium powder was used in early photography as a source of illumination:

$$2\ Mg(s) + O_2(g) \rightarrow 2\ MgO(s)$$

The combustion reaction is so vigorous that it cannot be extinguished by using a conventional fire extinguisher material such as carbon dioxide. Burning magnesium even reacts with carbon dioxide to give magnesium oxide and carbon:

$$2\ Mg(s) + CO_2(g) \rightarrow 2\ MgO(s) + C(s)$$

To extinguish reactive metal fires, such as those of magnesium, a class D fire extinguisher must be used. (Classes A, B, and C are used to fight conventional fires.) Class D fire extinguishers contain either graphite or sodium chloride. Graphite produces a solid coating of metal carbide over the combusting surface and effectively smothers the reaction. Sodium chloride melts at the temperature of the burning magnesium and forms an inert liquid layer over the metal surface; it too prevents oxygen from reaching the metal.

Over half of the approximately 4×10^5 tonnes of magnesium metal produced worldwide is used in aluminum–magnesium alloys. The usefulness of these alloys is due primarily to their low density. With a density less than twice that of water (1.74 g·cm^{-3}), magnesium is the lowest density construction metal. Such alloys are particularly important wherever the low density provides significant energy savings: in aircraft, railroad passenger cars, rapid transit vehicles, and bus bodies. For a period of time in the 1970s, these alloys were used in the superstructure of warships because the lower mass of the ship allowed higher speeds. However, during the Falkland Islands War of 1982, the Royal Navy discovered a major disadvantage of this alloy—its flammability when subjected to missile attack. The U.S. Navy had already experienced accidents with the same alloy. An appreciation of the high reactivity of the alkaline earth metals might have prevented these mishaps.

In its chemistry, magnesium differs from the remaining Group 2 metals. For example, heating calcium, strontium, or barium chlorides causes release of the bound water molecules as steam, leaving the anhydrous metal chloride behind. For example,

$$CaCl_2 \cdot 2\ H_2O(s) \xrightarrow{\Delta} CaCl_2(s) + 2\ H_2O(g)$$

However, magnesium chloride monohydrate decomposes when heated to give magnesium chloride hydroxide and hydrogen chloride gas:

$$MgCl_2 \cdot H_2O(s) \xrightarrow{\Delta} Mg(OH)Cl(s) + HCl(g)$$

Magnesium readily forms compounds containing covalent bonds. This behavior can be explained in terms of its comparatively high charge density (120 C·mm^{-3}; calcium's charge density is 52 C·mm^{-3}). For example, magnesium metal reacts with organic compounds called halocarbons (or alkyl halides) such as bromoethane, C_2H_5Br, in a solvent such as ethoxyethane, $(C_2H_5)_2O$, commonly called ether. The magnesium atom inserts itself between the carbon and halogen atoms, forming covalent bonds to its neighbors:

$$C_2H_5Br(ether) + Mg(s) \rightarrow C_2H_5MgBr(ether)$$

These organomagnesium compounds are referred to as *Grignard reagents*, and they are used extensively as intermediates in synthetic organic chemistry. We discuss Grignard reagents in more detail in Chapter 22, Section 22.4.

12.6 Calcium and Barium

Both of these elements are grayish metals that react slowly with the oxygen in air at room temperature but burn vigorously when heated. Calcium burns to give only the oxide:

$$2\ Ca(s) + O_2(g) \rightarrow 2\ CaO(s)$$

whereas barium forms some dioxide(2−) in excess oxygen:

$$2\ Ba(s) + O_2(g) \rightarrow 2\ BaO(s)$$

$$Ba(s) + O_2(g) \rightarrow BaO_2(g)$$

The formation of barium peroxide can be explained in terms of the charge density of barium ion ($23\ C \cdot mm^{-3}$), which is as low as that of sodium ($24\ C \cdot mm^{-3}$). Cations with such a low charge density are able to stabilize polarizable ions like the dioxide(2−) ion.

Whereas beryllium is transparent to X-rays, barium and calcium, both with high atomic numbers, are strong absorbers of this part of the electromagnetic spectrum. It is the calcium ion in bones that causes the skeleton to show up dark on X-ray film. The elements in the soft tissues do not absorb X-rays, a property that presents a problem when one wants to visualize the stomach and intestine. Because barium ion is such a good X-ray absorber, swallowing a solution containing barium should be an obvious way of imaging these organs. There is one disadvantage—barium ion is very poisonous. Fortunately, barium forms an extremely insoluble salt, barium sulfate. This compound is so insoluble ($2.4 \times 10^{-3}\ g \cdot L^{-1}$) that a slurry in water can be safely swallowed, the organs X-rayed, and the compound later excreted.

12.7 Oxides

As mentioned earlier, the Group 2 metals burn in air to yield the normal oxides, except for the member of the group with the lowest charge density—barium—which also forms some barium peroxide. Magnesium oxide is insoluble in water, whereas the other alkaline earth metal oxides react with water to form the respective hydroxides. For example, strontium oxide forms strontium hydroxide:

$$SrO(s) + H_2O(l) \rightarrow Sr(OH)_2(s)$$

Magnesium oxide has a very high melting point, 2825°C, so bricks of this compound are useful as industrial furnace linings. Such high-melting materials are known as *refractory compounds*. Crystalline magnesium oxide is an unusual compound, because it is a good conductor of heat but a very poor conductor of electricity, even at high temperatures. It is this combination of properties that results in its crucial role in electric kitchen range elements. It conducts the heat rapidly from a very hot coil of resistance wire to the metal exterior of the element without allowing any of the electric current to traverse the same route.

Commonly called quicklime, calcium oxide is produced in enormous quantities, particularly for use in steel production (see Chapter 20, Section 20.6).

It is formed by heating calcium carbonate very strongly (over 1170°C):

$$CaCO_3(s) \xrightarrow{\Delta} CaO(s) + CO_2(g)$$

This high-melting oxide is unusual in a different way: When a flame is directed against blocks of calcium oxide, the blocks glow with a bright white light. This phenomenon is called *thermoluminescence*. Before the introduction of electric light, theaters were lighted by these glowing chunks of calcium oxide, hence the origin of the phrase "being in the limelight" for someone who attains a prominent position. Thorium(IV) oxide, ThO_2, exhibits a similar property, hence its use in the mantles of gas-fueled camping lights.

Calcium oxide reacts with water to form calcium hydroxide, a product that is referred to as hydrated lime or slaked lime:

$$CaO(s) + H_2O(l) \rightarrow Ca(OH)_2(s)$$

Hydrated lime is sometimes used in gardening to neutralize acid soils; however, it is not a wise way of accomplishing this because an excess of calcium hydroxide will make the soil too basic:

$$Ca(OH)_2(s) + 2\ H^+(aq) \rightarrow Ca^{2+}(aq) + 2\ H_2O(l)$$

Powdered limestone can be used more safely as a soil neutralizing agent:

$$CaCO_3(s) + 2\ H^+(aq) \rightarrow Ca^{2+}(aq) + CO_2(g) + H_2O(l)$$

12.8 Hydroxides

Whereas magnesium hydroxide is almost completely insoluble in water, calcium and strontium hydroxides are slightly soluble, and barium hydroxide is very soluble (Table 12.7).

A saturated solution of calcium hydroxide is referred to as limewater. The solution is one of the simplest confirmatory tests for carbon dioxide. Bubbling the gas through a calcium hydroxide solution first gives a white precipitate of calcium carbonate. Continued passage of carbon dioxide through the solution causes the precipitate to disappear as a solution of calcium hydrogen carbonate, $Ca(HCO_3)_2$, forms:

$$Ca(OH)_2(aq) + CO_2(g) \rightarrow CaCO_3(s) + H_2O(l)$$

$$CaCO_3(s) + H_2O(l) + CO_2(g) \rightarrow Ca^{2+}(aq) + 2\ HCO_3{}^-(aq)$$

It is this second step that has led to the slow deterioration of marble sculptures in the streets and parks of Europe, a process accelerated by the acid oxides in the industrially polluted air.

Table 12.7 Solubilities of the alkaline earth metal hydroxides

Hydroxide	Solubility (g·L^{-1})
Mg	0.0001
Ca	1.2
Sr	10
Ba	47

12.9 Calcium Carbonate

Calcium is the fifth most abundant element on Earth. It is found largely as calcium carbonate in the massive deposits of chalk, limestone, and marble that occur worldwide. Chalk was formed in the seas, mainly during the Cretaceous period, about 135 million years ago, from the calcium carbonate skeletons of countless marine organisms. Limestone was formed in the same seas, but as a simple precipitate, because the solubility of calcium carbonate was exceeded in those waters:

$$Ca^{2+}(aq) + CO_3{}^{2-}(aq) \rightleftharpoons CaCO_3(s)$$

How Was Dolomite Formed?

One of the great mysteries of geochemistry is how the mineral dolomite was formed. Dolomite is found in vast deposits, including the whole of the Dolomite mountain range in Europe. The chemical structure is $CaMg(CO_3)_2$; that is, it consists of carbonate ions interspersed with alternating calcium and magnesium ions. Of particular interest, many of the world's hydrocarbon (oil) deposits are found in dolomite rock. Yet this composition does not form readily. If you mix solutions of calcium ions, magnesium ions, and carbonate ions in the laboratory, you merely obtain a mixture of calcium carbonate crystals and magnesium carbonate crystals. For 200 years, geochemists have struggled with the problem of how such enormous deposits were formed. To synthesize dolomite, temperatures of over 150°C are required—not typical conditions on the surface of the Earth! Furthermore, magnesium ion concentrations in seawater are far lower than those of calcium ion. The most popular idea is that beds of limestone were formed first and then buried deep in the Earth. Water rich in magnesium ion is then postulated to have circulated through pores in the rock, selectively replacing some of the calcium ions with magnesium ions. For this to happen uniformly throughout thousands of cubic kilometers of rock seems unlikely, but at this time, it is the best explanation that we have.

Some deposits of limestone became buried deep in the Earth's crust, where the combination of heat and pressure caused the limestone to melt. The molten calcium carbonate cooled again as it was pushed back up to the surface, eventually solidifying into the dense solid form that we call marble.

There are two naturally occurring crystalline forms: calcite and aragonite. Calcite is thermodynamically favored at room temperature but by less than 5 kJ·mol^{-1}. Thus, though calcite is by far the most common, aragonite is found in some places. A form of calcite, known as Iceland spar, is unusual in that it transmits two images of any object placed under it. The two images appear because the crystal has two different indices of refraction. Polarizing microscopes rely on Iceland spar (Nicol prisms) for their functioning.

Caves like Carlsbad Caverns and Mammoth Cave occur in beds of limestone. These structures are formed when rainwater seeps into cracks in the limestone. During the descent of rain through the atmosphere, carbon dioxide dissolves in it. The reaction of this dissolved acid oxide with the calcium carbonate produces a solution of calcium hydrogen carbonate:

$$CaCO_3(s) + CO_2(aq) + H_2O(l) \rightleftharpoons Ca^{2+}(aq) + 2\,HCO_3^{-}(aq)$$

The solution is later washed away, leaving a hole in the rock. This is a reversible reaction, and within caves, the evaporation of water from drips of calcium hydrogen carbonate solution results in the formation of calcium carbonate stalagmites and stalactites.

As was mentioned in Chapter 11, Section 11.13, only the alkali metals have a charge density low enough to stabilize the large polarizable hydrogen carbonate ion. Hence, when the water evaporates from the solution of calcium hydrogen carbonate, the compound immediately decomposes back to solid calcium carbonate:

$$Ca(HCO_3)_2(aq) \rightarrow CaCO_3(s) + CO_2(g) + H_2O(l)$$

It is deposited calcium carbonate that forms the stalagmites growing up from a cave floor and the stalactites descending from the roof of the cave.

Calcium carbonate is a common dietary supplement prescribed to help maintain bone density. A major health concern today is the low calcium intake among teenagers. Low levels of calcium lead to larger pores in the bone structure, and these weaker structures mean easier bone fracturing and a higher chance of osteoporosis in later life.

As we mentioned in the Feature "Antacids" in Chapter 7, calcium carbonate is also a popular antacid though it also has a constipative effect. When traveling, it is advisable to drink low-mineral-content bottled water rather than local tap water, because the tap water might well be significantly higher or lower in either calcium (constipative) or magnesium (laxative) ions than your system has become used to at home, thereby causing undesirable effects. (Of course, in certain parts of the world, there is also the danger of more serious health problems from tap water supplies, such as bacterial and viral infections.)

In the form of powdered limestone (commonly called agricultural lime), calcium carbonate is added to farmland to increase the pH by reacting with acids in the soil. Attempts are being made to reduce the effects of acid rain on lake waters by adding large quantities of powdered limestone.

12.10 Cement

About 1500 B.C., it was first realized that a paste of calcium hydroxide and sand (mortar) could be used to bind bricks or stones together in the construction of buildings. The material slowly picked up carbon dioxide from the atmosphere, thereby converting the calcium hydroxide back to the hard calcium carbonate from which it had been made:

$$Ca(OH)_2(s) + CO_2(g) \rightarrow CaCO_3(s) + H_2O(g)$$

Between 100 B.C. and A.D. 400, the Romans perfected the use of lime mortar to construct buildings and aqueducts, many of which are still standing. They also made the next important discovery: that mixing volcanic ash with the lime mortar gave a far superior product. This material was the precursor of our modern cements.

The production of cement is one of the largest modern chemical industries. Worldwide production is about 700 million tonnes, with the United States producing about 10 percent of that figure. Cement is made by grinding together limestone and shales (a mixture of aluminosilicates) and heating the mixture to about 1500°C. The chemical reaction releases carbon dioxide and partially melts the components to form solid lumps called clinker. The clinker is ground to a powder, and a small quantity of calcium sulfate is mixed in. This mixture is known as Portland cement. Chemically, its main components are 26 percent dicalcium silicate, Ca_2SiO_4; 51 percent tricalcium silicate, Ca_3SiO_5; and 11 percent tricalcium aluminate, $Ca_3Al_2O_6$. When water is added, a number of complex hydration reactions take place. A typical idealized reaction can be represented as

$$2\ Ca_2SiO_4(s) + 4\ H_2O(l) \rightarrow Ca_3Si_2O_7 \cdot 3H_2O(s) + Ca(OH)_2(s)$$

The hydrated silicate, called tobermorite gel, forms strong crystals that adhere by means of strong silicon–oxygen bonds to the sand and aggregate (small rocks) that are mixed with the cement. Because the other product in this reaction is calcium hydroxide, the mixture should be treated as a corrosive material while it is hardening.

12.11   Calcium Chloride

Anhydrous calcium chloride is a white solid that absorbs moisture very readily (an example of deliquescence). As a result, it is sometimes used as a drying agent in the chemistry laboratory. The reaction to form the hexahydrate, $CaCl_2 \cdot 6H_2O$, is very exothermic, and this property is exploited commercially. One type of instant hot packs consists of two inner pouches, one containing water and the other, anhydrous calcium chloride. Squeezing the pack breaks the inner partition between the pouches and allows the exothermic hydration reaction to occur. When the dividing partition is broken, calcium chloride solution forms. This process is highly exothermic:

$$CaCl_2(s) \rightarrow Ca^{2+}(aq) + 2\ Cl^-(aq) \qquad \Delta H^\circ = -82\ kJ \cdot mol^{-1}$$

With a 2+ charge cation, the lattice energy is high (about 2200 $kJ \cdot mol^{-1}$); but at the same time, the enthalpy of hydration of the calcium ion is extremely high ($-1560\ kJ \cdot mol^{-1}$) and that of the chloride ion is not insignificant ($-384\ kJ \cdot mol^{-1}$). The sum of these energies yields an exothermic process. By contrast, there is a slight decrease in entropy (256 $J \cdot mol^{-1} \cdot K^{-1}$) as the small, highly charged cation is surrounded in solution by a very ordered sphere of water molecules, thereby diminishing the entropy of the water in the process. This reaction, then, is enthalpy driven.

Anhydrous calcium chloride, instead of sodium chloride, is also used for melting ice. Calcium chloride works in two ways. First, its reaction with water is highly exothermic (as we described previously); and second, calcium chloride forms a freezing mixture that substantially reduces the melting point. Calcium chloride is very water soluble: A mixture of 30 percent calcium chloride and 70 percent water by mass (the eutectic, or minimum, freezing mixture) will remain liquid down to $-55°C$, a temperature much lower than the $-18°C$ produced by the best sodium chloride and water mixture. Another advantage of using the calcium salt is that the calcium ion causes less damage to plants than does the sodium ion.

The concentrated calcium chloride solution has a very "sticky" feel, and this property leads to another of its applications: It is sprayed on unpaved road surfaces to minimize dust problems. It is much less environmentally hazardous than oil, the other substance commonly used. The concentrated solution is also very dense, and for this reason, it is sometimes used to fill tires of earth-moving equipment to give them a higher mass and hence better traction.

12.12   Calcium Sulfate

Calcium sulfate is found as the dihydrate, $CaSO_4 \cdot 2H_2O$, known as gypsum. Mineral deposits of pure, high-density gypsum, called alabaster, have been used for delicate sculptures. Gypsum is also used in some brands of blackboard chalk. When heated to about 100°C, the hemihydrate, plaster of Paris, is formed:

$$CaSO_4 \cdot 2H_2O(s) \rightarrow CaSO_4 \cdot \tfrac{1}{2}H_2O(s) + \tfrac{3}{2}\ H_2O(g)$$

This white powdery solid slowly reacts with water to form long interlocking needles of calcium sulfate dihydrate. It is the strong, meshing crystals of gypsum that give plaster casts their strength. A more correct common name would be "gypsum casts."

One of the major uses of gypsum is in the fire-resistant wallboard used for interior walls in houses and offices. Its nonflammability and low cost are two reasons for choosing this material. But why gypsum and not chalk? The answer lies with the gypsum dehydration reaction that yields the hemihydrate. In a fire, this reaction occurs. Because it is an endothermic process—to the extent of 117 kJ·mol^{-1}—it absorbs energy from the fire. Furthermore, each mole of liquid water produced absorbs the enthalpy of vaporization of

Biomineralization: A New Interdisciplinary "Frontier"

One of the new fields of research is biomineralization: the formation of minerals by biological processes. This interdisciplinary field encompasses inorganic chemistry, biology, geology, biochemistry, and materials science. Though we think of life as being based on organic chemistry, the structures of many living organisms are defined by an inorganic–organic composite material. For example, the bones of vertebrates consist of an organic matrix of elastic fibrous proteins, predominantly collagen, glycoproteins, and mucopolysaccharides. This matrix is filled with inorganic components, about 55 percent being calcium hydroxophosphate, $Ca_5(PO_4)_3(OH)$, the remainder including calcium carbonate, magnesium carbonate, and silicon dioxide. The inorganic filler is hard and pressure-resistant enabling large land-living creatures to exist. The organic components provide elasticity together with tensile, bending, and breaking strength.

The crucial difference between ordinarily minerals and biominerals is that biominerals are grown specifically in the shape for which they are needed. In addition to their use as body frameworks, biominerals have three other roles: as instruments, as parts of sensors, and for mechanical protection. Teeth are the most common inorganic instruments. In vertebrates, the enamel layer of teeth is predominantly hydroxyapatite, but this is not true for all organisms. Marine mollusks of the chiton family synthesize crystals of iron oxides to act in this role.

We have gravity or inertia-sensitive sensors in our inner ear. These are spindle-shaped deposits of the aragonite form of calcium carbonate. Calcium carbonate is relatively dense (2.9 g·cm^{-3}); thus, the movement of this mineral mass with respect to the surrounding sensory cells gives us information on the direction and intensity of the acceleration. One class of bacteria accumulates magnetic iron oxides which they use to orient themselves with respect to the Earth's magnetic field.

Sea urchins provide an example of mechanical protection. They synthesize long, strong needles of calcium carbonate for defense. Such needle crystals are very different from the normal chunky calcite or aragonite crystals. Silicon dioxide crystals are used for defense by several plant species. One of these is the stinging nettle, where the brittle tips of the stinging hairs consist of silicon dioxide (silica).

Table 12.8 shows some of the most important biominerals and their function.

Table 12.8   Some important biominerals and their function

Chemical composition	Mineral name	Occurrence and function
$CaCO_3$	Calcite, aragonite	Exoskeletons (e.g., egg shells, corals, mollusk shells)
$Ca_5(PO_4)_3(OH)$	Hydroxyapatite	Endoskeletons (vertebrate bones and teeth)
$Ca(C_2O_4)$	Whewellite, weddelite	Calcium storage, passive defense of plants
$CaSO_4 \cdot 2H_2O$	Gypsum	Gravity sensor
$SrSO_4$	Celestite	Exoskeleton (some marine unicellular organisms)
$BaSO_4$	Baryte	Gravity sensor
$SiO_2 \cdot nH_2O$	Silica	Exoskeletons, plant defenses
Fe_3O_4	Magnetite	Magnetic sensors, teeth of certain marine organisms
$Fe(O)OH$	Goethite, lepidocrocite	Teeth of certain marine organisms

water—another 144 kJ·mol^{-1}—as it becomes gaseous water. Finally, the gaseous water acts as an inert gas, decreasing the supply of dioxygen to the fire. Chalk, being anhydrous, offers no equivalent reactions.

12.13   Calcium Carbide

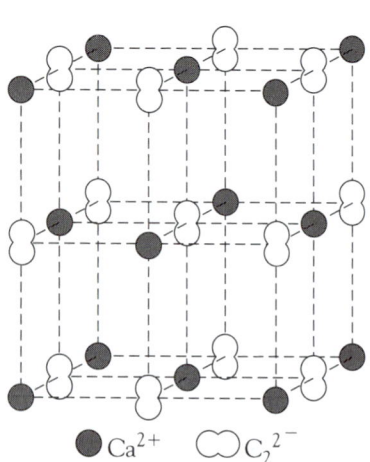

● Ca^{2+} ◯◯ C_2^{2-}

Figure 12.3 Crystal structure of calcium dicarbide(2−), which closely resembles the sodium chloride crystal structure.

This compound was first prepared by accident. Thomas "Carbide" Willson was endeavoring to prepare calcium metal by heating calcium oxide with carbon in an electric furnace. A product was formed and it reacted with water to produce a gas as expected. However, the gas was not the expected hydrogen but acetylene. This synthesis had a major effect on late 19th century life. The solid calcium carbide could be easily stored and transported, and addition of readily available water released a flammable gas. Acetylene lamps enabled automobiles to travel at night and for miners to work more safely underground. Even today, some cave explorers use carbide lamps as they are so reliable and give such an intense light. The acetylene (ethene) proved to be an ideal reactant for the synthesis of numerous organic compounds. In fact, the process was central to the founding of the company Union Carbide.

Even though CaC_2 is commonly called calcium carbide, the compound does not contain the carbide ion, C^{4-}. Instead, it contains the dicarbide(2−) ion, C_2^{2-}, which is commonly called the acetylide ion. The compound adopts the sodium chloride crystal structure, with each anion site being occupied by a dicarbide(2−) unit (Figure 12.3).

Calcium dicarbide(2−) is prepared by heating carbon (coke) and calcium oxide at about 2000°C in an electric furnace:

$$CaO(s) + 3\ C(s) \xrightarrow{\Delta} CaC_2(s) + CO(g)$$

Worldwide production has dropped from about 10 million tonnes in the 1960s to about 5 million tonnes in the 1990s—China is now the main producer—as the chemical industry has shifted to the use of oil and natural gas as the starting point for synthesizing organic compounds. The major use of the carbide process is to produce ethyne (acetylene) for oxyacetylene welding:

$$CaC_2(s) + 2\,H_2O(l) \rightarrow Ca(OH)_2(s) + C_2H_2(g)$$

The very exothermic reaction with dioxygen gives carbon dioxide and water vapor:

$$2\,C_2H_2(g) + 5\,O_2(g) \rightarrow 4\,CO_2(g) + 2\,H_2O(g)$$

Another important reaction of calcium dicarbide($2-$) is that with atmospheric nitrogen, one of the few simple chemical methods of breaking the strong nitrogen–nitrogen triple bond. In the process, calcium dicarbide($2-$) is heated in an electric furnace with nitrogen gas at about 1100°C:

$$CaC_2(s) + N_2(g) \xrightarrow{\Delta} CaCN_2(s) + C(s)$$

The cyanamide ion $[N{=}C{=}N]^{2-}$ is isoelectronic with carbon dioxide, and it also has the same linear structure. Calcium cyanamide is a starting material for the manufacture of several organic compounds, including melamine plastics. It is also used as a slow-release nitrogen-containing fertilizer:

$$CaCN_2(s) + 3\,H_2O(l) \rightarrow CaCO_3(s) + 2\,NH_3(aq)$$

12.14 Biological Aspects

The most important aspect of the biochemistry of magnesium is its role in photosynthesis. Magnesium-containing chlorophyll, using energy from the Sun, converts carbon dioxide and water into sugars and oxygen:

$$6\,CO_2(g) + 6\,H_2O(l) \rightarrow C_6H_{12}O_6(aq) + 6\,O_2(g)$$

Without the oxygen from the chlorophyll reaction, this planet would still be blanketed in a dense layer of carbon dioxide, and without the sugar energy source, it would have been difficult for life to progress from plants to herbivorous animals. Interestingly, the magnesium ion seems to be used for its particular ion size and for its low reactivity. It sits in the middle of the chlorophyll molecule, holding the molecule in a specific configuration. Magnesium has only one possible oxidation number, $+2$. Thus, the electron transfer reactions involved in photosynthesis can proceed without interference from the metal ion.

Both magnesium and calcium ions are present in body fluids. Mirroring the alkali metals, magnesium ions are concentrated within cells, whereas calcium ions are concentrated in the intracellular fluids. Calcium ions are important in blood clotting, and they are required to trigger the contraction of muscles, such as those that control the beating of the heart. In fact, certain types of muscle cramps can be prevented by increasing the intake of calcium ion.

12.15   *Element Reaction Flowchart*

The three most important elements of this group are magnesium, calcium, and barium, and their respective flowcharts are shown in the following.

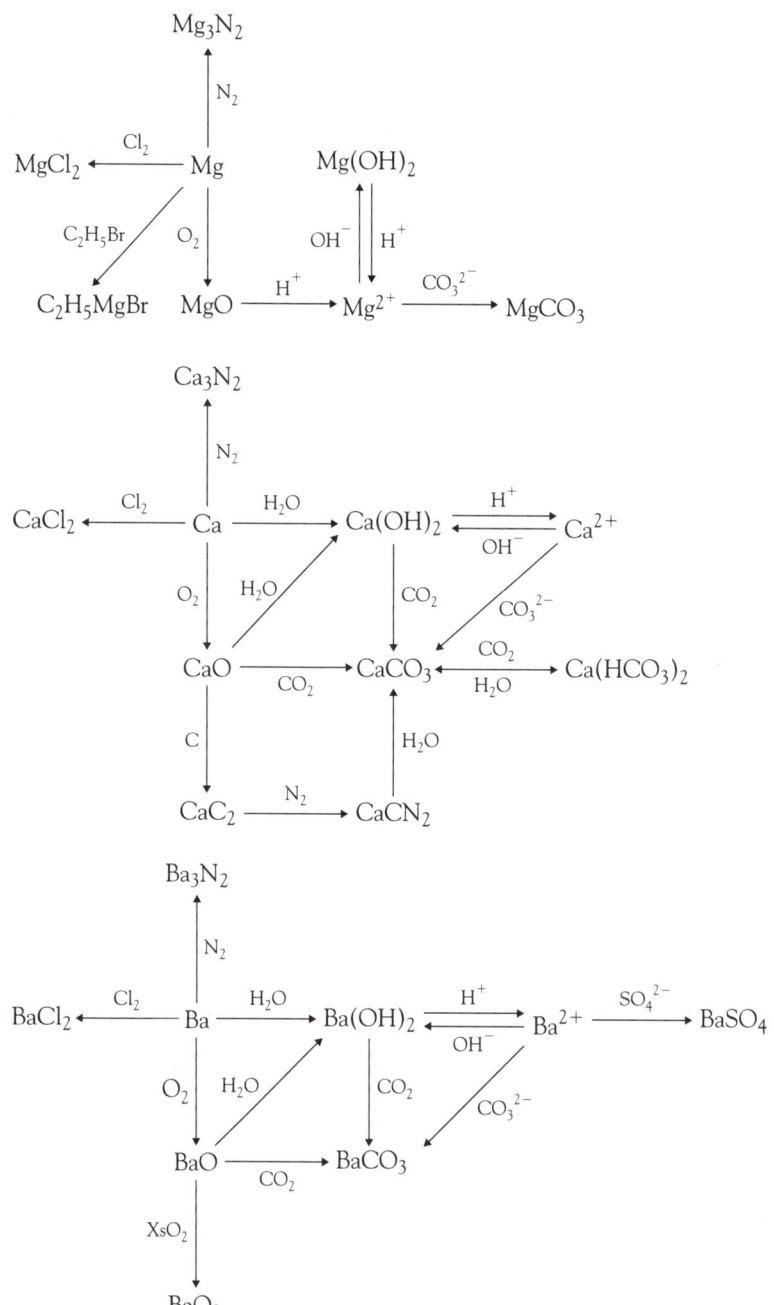

12.1   Write balanced chemical equations for the following processes:
(a) heating calcium in dioxygen
(b) heating calcium carbonate

(c) evaporating a solution of calcium hydrogen carbonate
(d) heating calcium oxide with carbon

12.2 Write balanced chemical equations for the following processes:
(a) adding strontium to water
(b) passing sulfur dioxide over barium oxide
(c) heating calcium sulfate dihydrate
(d) adding strontium dicarbide(2−) to water

12.3 For the alkaline earth elements (except beryllium), which will (a) have the most insoluble sulfate? (b) be the softest metal?

12.4 For the alkaline earth metals (except beryllium), which will (a) have the most insoluble hydroxide? (b) have the greatest density?

12.5 Explain why entropy factors favor the solution of sodium chloride but not that of magnesium chloride.

12.6 Explain why the salts of alkaline earth metals with mononegative ions tend to be soluble, while those with dinegative ions tend to be insoluble.

12.7 What are the two most important common features of the Group 2 elements?

12.8 Explain why the solid salts of magnesium tend to be highly hydrated.

12.9 Suggest why the hydrated beryllium ion has the formula $[Be(OH_2)_4]^{2+}$, while that of magnesium is $[Mg(OH_2)_6]^{2+}$?

12.10 How does the chemistry of magnesium differ from that of the lower members of the Group 2 metals? Suggest an explanation.

12.11 Explain briefly how caves are formed in limestone deposits.

12.12 What are the main raw materials for the manufacture of cement?

12.13 Summarize the industrial process for the extraction of magnesium from seawater.

12.14 How is calcium cyanamide obtained from calcium oxide?

12.15 Several of the alkaline earth metal compounds have common names. Give the systematic name for (a) lime; (b) milk of magnesia; (c) Epsom salts.

12.16 Several of the alkaline earth metal compounds have common names. Give the systematic name for (a) dolomite; (b) marble; (c) gypsum.

12.17 Why is lead commonly used as a shielding material for X-rays?

12.18 The dissolving of anhydrous calcium chloride in water is a very exothermic process. However, dissolving calcium chloride hexahydrate causes a very much smaller heat change. Explain this observation.

12.19 Discuss briefly the similarities between beryllium and aluminum.

12.20 In this chapter, we have ignored the radioactive member of the Group, radium. On the basis of Group trends, suggest the key features of the properties of radium and its compounds.

12.21 Describe briefly the importance of magnesium ion to life on Earth.

12.22 What is the calcium-containing structural material in vertebrates?

12.23 Describe and write corresponding chemical equations showing how you would prepare from magnesium metal each of the following:
(a) magnesium chloride monohydrate
(b) anhydrous magnesium chloride

12.24 Write balanced chemical equations corresponding to each transformation in the Element Reaction Flowcharts.

BEYOND THE BASICS

12.25 From the appropriate data in Appendix 2, calculate the enthalpy and entropy change when plaster of Paris is formed from gypsum. Calculate the temperature at which the process of dehydration becomes significant, that is, when $\Delta G^\circ = 0$.

12.26 Calculate the radius of a dicarbide, C_2^{2-}, ion given that calcium dicarbide(2−), which adopts a sodium chloride lattice, has a lattice energy of $-2911 \text{ kJ·mol}^{-1}$.

12.27 The common hydrate of magnesium sulfate is the heptahydrate, $MgSO_4 \cdot 7H_2O$. In the crystal structure, how many water molecules are likely to

be associated with the cation? With the anion? Give your reasoning.

12.28 Adding powdered limestone (calcium carbonate) to a lake affected by acid rain can decrease the availability of phosphate ion, an important nutrient, but not that of nitrate ion, another nutrient. Write a balanced equation and calculate the standard free energy change for the process to confirm its spontaneity.

12.29 Which of the following gaseous species should be the most stable: BeH, BeH^+, or BeH^-? Show your reasoning.

12.30 Which would you expect to have a higher melting point, magnesium oxide or magnesium fluoride? Explain your reasoning.

12.31 We focus on reactions involving oxygen and water as they are the predominant reactive species on this planet. What would be the parallel reaction to that of calcium oxide with water on a planet whose environment was dominated by gaseous dinitrogen and liquid ammonia?

12.32 Lanthanum, one of the lanthanoid elements, is often regarded by biochemists as a useful analog for calcium, the major difference being the charge of tripositive La^{3+} contrasted to dipositive Ca^{2+}. Assuming this analogy, predict:
  (a) The reaction between lanthanum metal and water.
  (b) Which of the following lanthanum salts are soluble or insoluble in water: sulfate, nitrate, chloride, phosphate, fluoride?

12.33 Beryllium metal can be obtained by the reaction of beryllium fluoride with magnesium metal at 1300°C. Show that the reaction is thermodynamically spontaneous even at 25°C. Is the reaction likely to be more or less favorable at 1300°C? Give your reasoning without a calculation. Why, then, is beryllium commercially synthesized at such a high temperature?

12.34 Suggest why the BeI_4^{2-} ion is not known, even though the $BeCl_4^{2-}$ ion exists.

12.35 Molten beryllium chloride is a poor electrical conductor. However, dissolving sodium chloride in beryllium chloride results in a conducting solution with an optimum conductivity at an $NaCl:BeCl_2$ ratio of $2:1$. Suggest an explanation.

Chapter 13

The Group 13 Elements

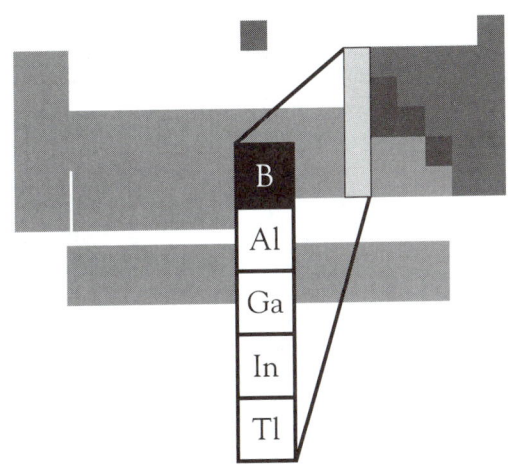

Boron and aluminum are the only members of this group that are of major importance. Boron has some unusual chemistry—particularly its hydrides. Aluminum is one of the most widely used metals, and the properties of its compounds will be the main focus of this chapter.

The German chemist Friedrich Wöhler (better known for his synthesis of urea) was among the first to prepare pure aluminum metal. He did so by heating potassium metal with aluminum chloride; aluminum was produced in a single replacement reaction:

$$3 \text{ K}(l) + \text{AlCl}_3(s) \rightarrow \text{Al}(s) + 3 \text{ KCl}(s)$$

Before he could do this, he had to obtain stocks of the very reactive potassium metal. Because he did not have a battery that was powerful enough to generate the potassium metal electrochemically, he devised a chemical route that used intense heat and a mixture of potassium hydroxide and charcoal. He and his sister, Emilie Wöhler, shared the exhausting work of pumping the bellows to keep the mixture hot enough to produce the potassium. So expensive was aluminum in the mid-19th century that Emperor Napoleon III used aluminum tableware for special state occasions.

13.1   Group Trends

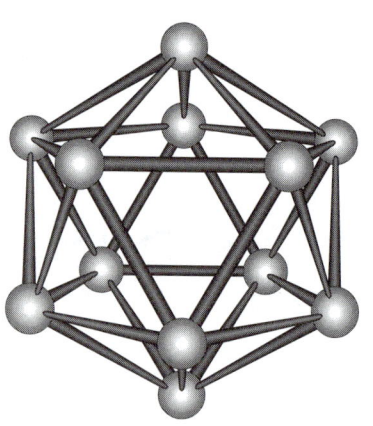

Figure 13.1   Icosahedral arrangement of boron.

Boron exhibits mostly nonmetallic behavior and is classified as a semimetal, whereas the other members of Group 13 are metals. But even the metals have no simple pattern in melting points, although their boiling points do show a decreasing trend as the mass of the elements increases (Table 13.1). The reason for this lack of order is that each element in the group is organized a different way in the solid phase. For example, in one of its four allotropes, boron forms clusters of 12 atoms. Each cluster has a geometric arrangement called an icosahedron (Figure 13.1). Aluminum adopts a face-centered cubic structure, but gallium forms a unique structure containing pairs of atoms. Indium and thallium each form other, different structures. It is only when the elements are melted and the crystal arrangements destroyed that we see, from the decreasing boiling points as the group is descended, that the metallic bond becomes weaker.

Table 13.1   Melting and boiling points of the Group 13 elements

Element	Melting point (°C)	Boiling point (°C)
B	2180	3650
Al	660	2467
Ga	30	2403
In	157	2080
Tl	303	1457

As we would expect, boron, classified as a semimetal, favors covalent bond formation. However, covalency is common among the metallic members of the group as well. The reason for the covalent behavior can be attributed to the high charge and small radius of each metal ion. The resulting high-charge density of Group 13 ions is sufficient to polarize almost any approaching anion enough to produce a covalent bond (Table 13.2). The only way to stabilize the ionic state of Group 13 elements is to hydrate the metal ion. For aluminum, the enormous hydration enthalpy of the tripositive ion, -4665 kJ·mol^{-1}, is almost enough on its own to balance the sum of the three ionization energies, $+5137$ kJ·mol^{-1}. Thus, the hydrated aluminum compounds that we regard as ionic do not contain the aluminum ion, Al^{3+}, as such, but the hexaaquaaluminum ion, $[Al(OH_2)_6]^{3+}$.

It is in Group 13 that we first encounter elements possessing more than one oxidation state. Aluminum has the $+3$ oxidation state, whether the bonding is ionic or covalent. However, gallium, indium, and thallium have

Table 13.2   Charge densities of Period 3 metal ions

Group	Ion	Charge density (C·mm^{-3})
1	Na^+	24
2	Mg^{2+}	120
13	Al^{3+}	364

a second oxidation state of +1. For gallium and indium, the +3 state predominates, whereas the +1 state is most common for thallium (see Chapter 9, Section 9.8). At this point, it is appropriate to note that formulas can sometimes be deceiving. Gallium forms a chloride, $GaCl_2$, a compound implying that a +2 oxidation state exists. However, the actual structure of this compound is now established as $[Ga]^+[GaCl_4]^-$; thus, the compound actually contains gallium in both +1 and +3 oxidation states.

13.2 Boron

Boron is the only element in Group 13 that is not classified as a metal. In Chapter 2, Section 2.4, we classified it as a semimetal. However, on the basis of its extensive oxanion and hydride chemistry, it is equally valid to consider it as a nonmetal. The element can be obtained from its oxide by heating with a reactive metal such as magnesium:

$$B_2O_3(s) + 3\ Mg(l) \xrightarrow{\Delta} 2\ B(s) + 3\ MgO(s)$$

The magnesium oxide can be removed by reaction with acid.

Boron is a rare element in the Earth's crust, but fortunately there are several large deposits of its salts. These deposits, which are found in locations that once had intense volcanic activity, consist of the salts borax and kernite, which are conventionally written as $Na_2B_4O_7 \cdot 10H_2O$ and $Na_2B_4O_7 \cdot 4H_2O$, respectively. Total annual worldwide production of boron compounds amounts to over 3 million tonnes. The world's largest deposit is found at Boron, California; it covers about 10 km^2, with beds of kernite up to 50 m thick. The actual structure of borate ions is much more complex than the simple formulas would indicate. For example, borax actually contains the $[B_4O_5(OH)_4]^{2-}$ ion, shown in Figure 13.2.

About 35 percent of boron production is used in the manufacture of borosilicate glass. Conventional soda glass suffers from thermal shock; that is, when a piece of glass is heated strongly, the outside becomes hot and tries to expand, while the inside is still cold because glass is such a poor conductor of heat. As a result of stress between the outside and the inside, the glass cracks. When the sodium ions in the glass structure are replaced by boron atoms, the degree of glass expansion (more precisely called thermal expansivity) is less than half that of conventional glass. As a result, containers made of borosilicate glass (sold under trademarks such as Pyrex®) are capable of being heated without great danger of cracking. Glass compositions are discussed in Chapter 14, Section 14.18.

In the early part of the 20th century, the major use for boron compounds was as a cleaning agent called borax. This use has now dropped behind that for glassmaking, consuming only 20 percent of production. In detergent formulations, it is no longer borax but sodium peroxoborate, $NaBO_3$, that is used. Once again, the simple formula does not show the true structure of the ion, which is $[B_2(O_2)_2(OH)_4]^{2-}$ (Figure 13.3). The peroxoborate ion is prepared by the reaction of hydrogen peroxide with borax in base:

$$[B_4O_5(OH)_4]^{2-}(aq) + 4\ H_2O_2(aq) + 2\ OH^-(aq) \rightarrow$$
$$2[B_2(O_2)_2(OH)_4]^{2-}(aq) + 3\ H_2O(l)$$

This ion acts as an oxidizing agent as a result of the two peroxo groups (—O—O—) linking the boron atoms. About 5×10^5 tonnes of sodium peroxoborate is produced every year for European detergent manufacturing

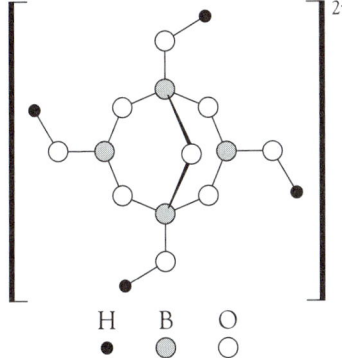

H ●   B ◉   O ○

Figure 13.2 Actual structure of the borate ion in borax.

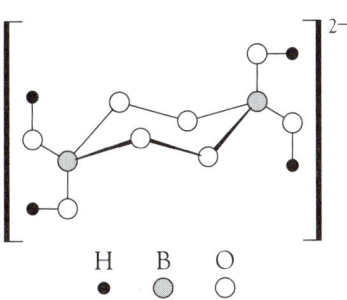

H ●   B ◉   O ○

Figure 13.3 Structure of the peroxoborate ion.

Inorganic Fibers

In our everyday lives, the fibers we encounter are usually organic, for example, nylon and polyester. These materials are fine for clothing and similar purposes, but most organic fibers have the disadvantages of low melting points, flammability, and low strengths. For materials that are strong and unaffected by high temperatures, inorganic materials fit the specifications best. Some inorganic fibers are well known, for example, asbestos and fiberglass. However, it is the elements boron, carbon, and silicon that currently provide some of the toughest materials for our high-technology world. Carbon fiber is the most widely used—not just for tennis rackets and fishing rods but for aircraft parts as well. The Boeing 767 was the first commercial plane to make significant use of carbon fiber; in fact, about 1 tonne is incorporated into the structure of each aircraft. Aircraft constructed with newer technology, such as the Airbus 340, contain a much higher proportion of carbon fibers.

Fibers of boron and silicon carbide, SiC, are becoming increasingly important in the search for tougher, less fatigue-prone materials. The boron fibers are prepared by reducing boron trichloride with hydrogen gas at about 1200°C:

$$2\ BCl_3(g)\ +\ 3\ H_2(g) \rightarrow 2\ B(g)\ +\ 6\ HCl(g)$$

The gaseous boron can then be condensed onto carbon or tungsten microfibers. For example, boron is deposited onto tungsten fibers of 15 μm until the diameters of the coated fibers are about 100 μm. The typical inorganic fiber prices are several hundred dollars per kilogram; so even though production of each type is mostly in the range of hundreds of tonnes, inorganic fiber production is already a billion-dollar business.

companies. It is a particularly effective oxidizing (bleaching) agent at the water temperatures used in European washing machines (90°C), but it is ineffective at the water temperatures usually used in North American washing machines (70°C). In North America, hypochlorites (see Chapter 17, Section 17.9) are used instead.

Boron is a vital component of nuclear power plants because it is a strong absorber of neutrons. Boron-containing control rods are lowered into reactors to maintain the nuclear reaction at a steady rate. Borates are used as wood preservatives and as a fire retardant in fabrics. Borates are also used as a flux in soldering. In this latter application, the borates melt on the hot pipe surface and react with metal oxide coatings, such as copper(II) oxide on copper pipes. The metal borates (such as copper(II) borate) can be easily removed to give a clean metal surface for the soldering.

13.3   Borides

Boron forms a large number of binary compounds. These compounds are all very hard, high melting, and chemically resistant, and they have become of increasing importance as materials that can be used for such purposes as rocket nosecones. However, the stoichiometry of these compounds is far from simple. The most important of the compounds is boron carbide, which has the empirical formula, B_4C. Even though by name it is a carbide, the structure is boron based. The structure is better represented as $B_{12}C_3$ as it consists of B_{12} icosohedra as in the element itself, with carbon atoms linking all the neighboring icosohedra. One preparative method is the reduction of diboron trioxide with carbon.

$$2\ B_2O_3(s) + 7\ C(s) \xrightarrow{\Delta} B_4C(s) + 6\ CO(g)$$

Boron carbide is one of the hardest substances known. Its fibers have enormous tensile strength and are used in bullet-proof clothing. High-density boron carbide armor tiles are placed under the seats of Apache attack helicopters, protecting the occupants from ground fire. A more common use is in some lightweight high-performance bicycle frames where boron carbide is embedded in an aluminum matrix. Boron carbide is also used as a starting material for preparing other tough materials such as titanium boride.

$$2\ TiO_2(s) + B_4C(s) + 3\ C(s) \xrightarrow{\Delta} 2\ TiB_2(s) + 4\ CO(g)$$

Titanium boride belongs to a different class of borides. These borides consist of hexagonal layers of boron ions isoelectronic and isostructural with the graphite allotrope of carbon. The metal ions are located between the boride layers. Each boron atom has a -1 charge, and the stoichiometry of this class of borides corresponds to metals in their $2+$ oxidation state.

Another of the hexagonal layer borides is magnesium boride, MgB_2. This compound is very inexpensive and readily available, yet it was only in 2001 that it was accidentally discovered to be superconducting at low temperatures. Magnesium boride retains its superconductivity up to 39 K, the highest value for a simple (and inexpensive) compound. Research is ongoing to see if there are any close relatives of magnesium boride that exhibit superconductivity to much higher temperatures.

13.4 Boranes

In Chapter 10, Section 10.5, we saw that compounds of boron and hydrogen (boranes) were unusual in terms of their electron-deficient bonding. Here we will explore some borane chemistry. The simplest borane is diborane, B_2H_6. About 200 tonnes per year of this gas is synthesized by the reaction of boron trifluoride with sodium hydride:

$$2\ BF_3(g) + 6\ NaH(s) \rightarrow B_2H_6(g) + 6\ NaF(s)$$

Like most of the boranes, diborane is a highly reactive, toxic, colorless gas. It catches fire in air and explodes when mixed with pure dioxygen. The extremely exothermic reaction produces diboron trioxide and steam:

$$B_2H_6(g) + 3\ O_2(g) \rightarrow B_2O_3(s) + 3\ H_2O(g)$$

The hydride also reacts with any trace of moisture to give boric acid and hydrogen gas:

$$B_2H_6(g) + 6\ H_2O(l) \rightarrow 2\ H_3BO_3(aq) + 3\ H_2(g)$$

Diborane is an important reagent in organic chemistry. The gas reacts with unsaturated hydrocarbons (those containing double or triple carbon–carbon bonds) to form alkylboranes. For example, diborane reacts with propene:

$$B_2H_6(g) + 6\ CH_2{=}CHCH_3(g) \rightarrow 2\ B(CH_2CH_2CH_3)_3(l)$$

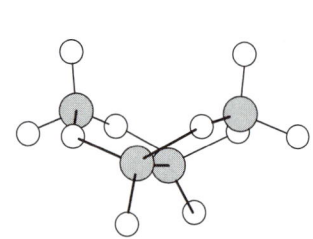

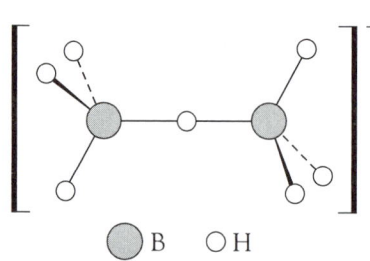

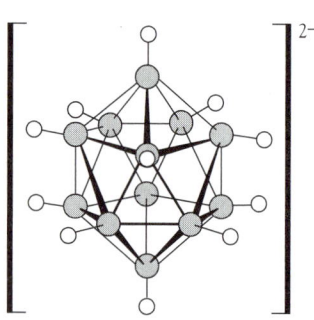

B ◯ H

Figure 13.4 Structure of tetraborane, B_4H_{10}. The boron atoms are shaded.

Figure 13.5   Structure of the $B_2H_7^-$ anion.

Figure 13.6   Structure of the $B_{12}H_{12}^{2-}$ anion.

The product of this *hydroboration* reaction can be reacted with a carboxylic acid to give a saturated hydrocarbon; with hydrogen peroxide to give an alcohol; or with chromic acid to give a ketone or a carboxylic acid. Hydroboration is a favored route of organic synthesis for two reasons: The initial hydride addition is accomplished under very mild conditions, and a wide variety of final products are possible (depending upon the other reagent used).

There are two main series of boranes: one with the generic formula B_nH_{n+4}, the *nido*-boranes, such as $B_{10}H_{14}$; the other with the generic formula B_nH_{n+6}, the *arachno*-boranes, such as B_4H_{10} (Figure 13.4). In each of the boranes, there are bridging hydrogen atoms and, except for diborane, direct boron–boron bonds as well. All of the compounds have positive ΔG_f° values; that is, they are thermodynamically unstable with respect to decomposition into their constituent elements. To name a borane, the numbers of boron atoms are indicated by the normal prefixes, while the numbers of hydrogen atoms are denoted by Arabic numbers in parentheses. Hence, B_4H_{10} is called tetraborane(10) and $B_{10}H_{14}$ is decaborane(14).

Boranes were once considered as possible rocket fuels because they burn very exothermically. In fact, for equal mass, only dihydrogen produces more heat upon combustion. However, the cost of synthesis on a large scale was prohibitive and the solid boron oxides formed clogged the rocket engines. Today the main interest in these compounds is the study of the unique structures that boranes form. In addition to the many boron–hydrogen molecules, there is an equally large array of boron–hydrogen anions. Figure 13.5 shows the structure of the $B_2H_7^-$ ion, which, unlike diborane, has a single boron–hydrogen bridge. One of the most interesting structures is that of the $B_{12}H_{12}^{2-}$ ion, which has the same framework as boron itself, with an icosahedron of boron atoms having a hydrogen atom attached to each boron atom (Figure 13.6).

Boron Neutron Capture Therapy

One of the many avenues under investigation for fighting cancer is boron neutron capture therapy (BNCT). The fundamental principle of this therapy is to have a radioactive source selectively within malignant cells. The radiation would then destroy only those cells and leave the healthy cells untouched. There was a particular interest in this approach in the context of inoperable brain tumors or as a means of killing any tiny clusters of

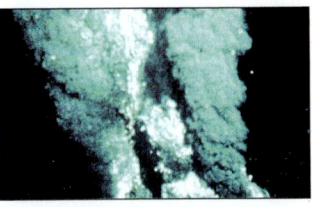

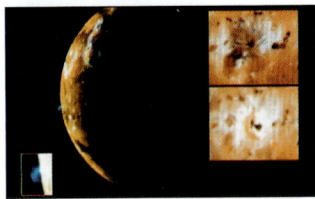
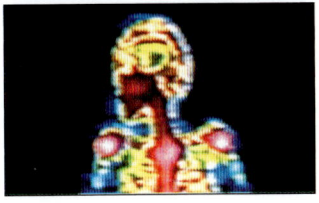

Applications of Inorganic Chemistry

Inorganic chemistry is a wonderous subject in itself, with every element and compound having some unique property. But more than its academic interest, inorganic chemistry is an integral part of the world that we live in: from the air that we breathe, to the waters around us, to the rocks beneath our feet. Each of these topics is primarily a branch of inorganic chemistry with its own name: *atmospheric chemistry, aquatic chemistry,* and *geochemistry.* The living organisms on this planet depend upon many chemical elements for their functioning and this subject, too, has its own identity, that of *bioinorganic chemistry.*

Nature is not the only context in which inorganic chemistry is applied. Our modern civilization depends upon the synthesis of inorganic compounds, from the enormous quantities of such chemicals as ammonia and sulfuric acid to the much smaller quantities of cadmium sulfide, used as a paint pigment. Thus *industrial inorganic chemistry* is one of the cornerstones of our modern economy.

But more than this, inorganic chemistry is part of our future, a part of *materials science,* the synthesis of new and exotic materials for the 21st century that will enable us to live better, more comfortable, environmentally-friendly lives.

In the following features, we have tried to select a few examples of the incredible range of applications of inorganic chemistry. We will see that inorganic chemistry is in space, beneath the seas, in the earth, in medical facilities, and in the materials that surround us.

Memory Metal: The Shape of Things to Come

Though we depict science as progressing through well-thought advances, it is amazing how many of our discoveries take place by accident. One of the most interesting examples of chance is provided by the discovery of memory metal. The story begins with attempts to develop a fatigue-resistant alloy for Navy missile nose cones. The metallurgist, William J. Buehler, discovered that a equi-mole alloy of titanium and nickel had exactly the desired properties. He named the alloy Nitinol (**Ni**ckel **Ti**tanium **N**aval **O**rdnance Laboratory). As demonstrations, he took long straight bands of Nitinol and folded them into an accordion shape. He would show

The Nitinol sculpture by Olivier Deschamps, showing a kneeling woman with a baby on the ground (martensite phase). The woman lifts the baby toward the sky as the metal is transformed by the sun's heat into the austinite phase.
(Photo courtesy of Pascal Goetgheluck.)

how the metal could be stretched repeatedly without breaking. This flexibility, in itself, was a very useful property. At one such demonstration, one of the attendees produced a lighter and idly heated the metal. Much to everyone's amazement, the strip straightened out!—The metal had "remembered" its original preaccordian shape.

With an ordinary metal, bending it causes neighboring crystals to slide over one another. Nitinol has a very unusual crystal structure in that it consists of cubes of nickel atoms with a titanium atom at each cube center, while the titanium atoms themselves are in cubic arrays with a nickel atom at the center of each titanium cube (see Figure). It is this interlocking structure that prevents neighboring crystals from moving relative to each other and imparts the super-elastic properties of the material. At high temperatures, the symmetrical (austenite) phase is stable, but on cooling, the alloy undergoes a phase change to a distorted cubic (martensite) phase. In this phase, the martensite crystals are flexible enough that the bulk metal can be repeatedly bent without fracturing. Gentle warming can then be used to revert the crystals to their original shape. To impart a new shape, the metal must be heated above its phase transition temperature which depends upon the precise mole ratio of the two constituents.

Nitinol has many uses, including that of more comfortable and efficient orthodontic braces. Automatic tweezers provide another use. An ear specialist can bend open the tips of the tweezers, slide them into the patient's ear until the tips surround the foreign object, apply a low current to the wire, warming it-the resulting phase change closing the tips on the object, enabling it to be removed safely.

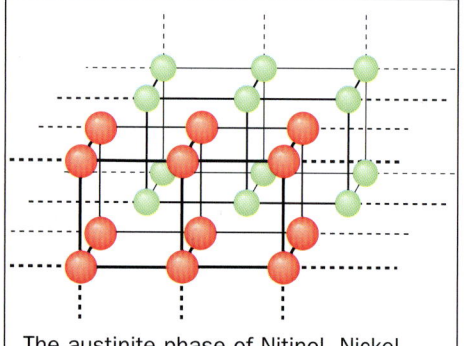

The austinite phase of Nitinol. Nickel atoms (green) adopt a cubic arrangement with a titanium atom at each cube center. The titanium atoms (red) adopt an interlocking cubic arrangement with a nickel ion at each cube center.

Chemosynthesis: Redox Chemistry on the Sea Floor

The bottom of the oceans is a cold and black place. Long thought to be lifeless apart from a few scavenger species living off the dead organisms drifting down from surface waters, we now realize that novel life forms relying on exotic biochemical processes live at these depths. On the surface of the Earth, life relies on photosynthesis to drive the redox cycles that make our life possible. What, then, drives the biological cycles in pitch-black cold environments? Life is not distributed evenly about the ocean floor but it is concentrated about vents, openings in the sea floor from which issue highly toxic plumes of boiling hot water saturated with toxic hydrogen sulfide and equally noxious heavy metal sulfides. On the edge of this unlikely environment, a tremendous range of organisms survive and flourish. The most interesting are the mouthless, gutless tubeworms. These enormous creatures rely on bacteria living inside them to obtain energy by the oxidation of the hydrogen sulfide ion to sulfate ion, the process of *chemosynthesis:*

$$H_2S(aq) + 4\ H_2O(l)$$
$$\rightarrow SO_4{}^{2-}(aq) + 9\ H^+(aq) + 8\ e^-$$

This energy is then used by the worms for the conversion of seawater-dissolved carbon dioxide to the complex carbon-based molecules in their structures. For every mole of hydrogen sulfide ion consumed, nine moles of hydrogen (hydronium) ion are produced. Hence the worms have to possess an efficient biochemical mechanisms to "pump out" the excess acid otherwise they would die from the low pH. The worms survive in the poisonous hydrogen sulfide environment by selectively absorbing the hydrogen sulfide ion rather than hydrogen sulfide itself. The H_2S ion is present at much lower concentrations than the hydrogen sulfide molecule as a result of the acid-base equilibrium below:

$$H_2S(aq) + H_2O(l) \leftrightarrow H_3O^+(aq) + HS^-(aq) \qquad K_{a1} = 9.5 \times 10^{-8}$$

Until recently, our studies of the worms have had to take place on the ocean floor as bringing them to the surface caused instant death. Now researchers at the University of California at Santa Barbara have managed to construct an aquarium habitat of high pressure, low temperature, and approprite levels of carbon dioxide, hydrogen sulfide, and oxygen in which they flourish. Now we can discover much more about the chemistry of these strange organisms. Not only are these of interest in themselves, but some scientists believe that the first living organisms on this planet originated at such sites.

A "black smoker", a vent in the sea floor spewing forth hydrogen sulfide and heavy metal sulfides.
(J. Edmond Whol/Visuals Unlimited)

Concrete: An Old Material with a New Future

Though chemists are devising new and novel materials, there are some timeless materials that will always be the backbone of civilization and concrete is one of these. Most of our larger buildings are of concrete construction as are highways, bridges, dams, tunnels, and so on. In fact, of all the building materials, we probably depend upon concrete the most. In fact, about one tonne is used per person on Earth per year.

The crucial component of concrete is cement. To make cement, limestone (calcium carbonate) is mixed with clay or shale (a mixture of aluminosilicates) and heated to about 2000°C. This process produces lumps of material called "clinker," a mxture of about 50% tricalcium silicate, Ca_3SiO_5, 30% dicalcium silicate, Ca_2SiO_4, with the remainder being calcium aluminate, $Ca_3Al_2O_6$, and calcium ferroaluminate, $Ca_4Al_2Fe_2O_{10}$. The clinker is ground with gypsum (calcium sulfate dihydrate) to give cement powder. Cement reacts with water to form an inorganic "glue" holding together a matrix of sand and aggregate (small rocks). A typical hydration reaction is:

$$2\ Ca_2SiO_4(s) + 4\ H_2O(l) \rightarrow$$
$$Ca_3Si_2O_7{\cdot}3H_2O(s) + Ca(OH)_2(s)$$

The Petronas Towers in Kuala Lumpur, currently the world's tallest building at 452 meters, is constructed of reinforced concrete. (John Mead/Science Photo Library/ Photo Researchers)

Reflected light micrograph of a Portland cement clinker sample. Major compounds are tricalcium silicate (gold crystals), dicalcium silicate (rounded gray, blue, and purple crystals), tricalcium aluminate, and tetracalcium aluminoferrite (bright interstitial material). The width of field is 350 um. (L. Powers, Construction Technology Laboratories, Inc.)

The silicate product, known as tobermorite, forms strong crystals that adhere by means of strong silicon-oxygen bonds to the sand and aggregate.

However, even such a traditional substance as concrete has been reborn as one of our new materials. Autoclaved aerated concrete (AAC) promises to be a major building material of the 21st century. AAC is synthesized by mixing cement with lime (calcium hydroxide), silicate sand (silicon dioxide), water and aluminum power. In addition to the reaction above, the aluminum metal reacts with the hydroxide ion from the hydration reaction and from the added calcium hydroxide to produce hydrogen gas.

$$2\ Al(s) + 2\ OH^-(aq) + 6\ H_2O(l) \rightarrow$$
$$2\ [Al(OH)_4]^-(aq) + 3\ H_2(g)$$

The millions of tiny gas bubbles cause the mixture to swell to five times its original volume. When the concrete has set, it is cut into blocks or slabs of required size then steam-cured in an oven (autoclave). The hydrogen gas diffuses out of the structure, being replaced by air. This low density building material has high thermal insulation properties and it can be made using fly ash, an unwanted product of coal-fired power plants, instead of silica sand. At the end of a building's life, the panels can be disassembled and reused or crushed and remade into new building materials, hence it is probably the most environmentally-friendly construction material.

New Minerals: Going Beyond the Limitations of Geochemistry

Many minerals are formed in the Earth under conditions of high temperature and pressure-and often over hundreds of thousands or millions of years. Now chemists have the ability to synthesize minerals using innovative reaction methods. One particular synthesis is that of the precious stone, lapis lazuli, also called ultramarine. Finely powdered ultramarine is important as an intensely-blue paint and plastics pigment (and eyeshadow) that is non-toxic and stable to light, unlike organic dyes which soon fade in light. This compound has the formula $(Na,Ca)_8[SiAlO_4]_6(S,SO_4)$ where the ratios of sodium to calcium and sulfide to sulfate are variable. Its structure consists of AlO_4 and SiO_4 tetrahedra with the other ions filling holes in the framework. Natural ultramarine suffers from several disadvantages, one of the most important being that it often contains impurities, such as calcium carbonate and iron(II) disulfide. Another problem is its rarity and hence, expense. Two chemists, Sandra Dann and Mark Weller, have produced a synthetic ultramarine that is free of impurities and has a consistent bright blue color. The compound has an exact chemical formula—$Na_8[SiAlO_4]_6(S_2,S_3)$. An advantage of mineral synthesis is that components can be altered to generate minerals that do not exist naturally, a field known as geomimetics. For example, the color of ultramarine is due to the anionic sulfur species. Replacement by chromate gives an equally stable and non-toxic yellow pigment.

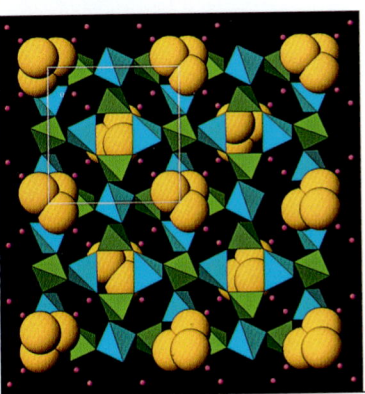

At top, synthetic lapis lazuli (ultramarine). At left, a computer-graphic representation of the structure of lapis lazuli. (Chemistry in Britain, June 1998. Courtesy of Sandra Dann, Loughborough University.)

The minerals produced on Earth largely reflect the abundances of the constituent elements in the Earth's crust. Thus one avenue of research is to synthesize minerals that match the formula of known minerals but with one or two elements substituted by rare elements of the same group in the periodic table. A good example is provided by sodalite $Na_8[SiAlO_4]Cl_2$. Dann and Weller have synthesized the analog $Na_8[GeGaO_4]Br_2$ in which silicon, aluminum, and chlorine are replaced by the element below them in the periodic table. Such a mineral would never be found in nature as the abundances of the replacement elements are of the order of 10_8 less than those in sodalite itself. Such novel compounds have potential as pigments, fluorescers, ferroelectrics, ion exchange materials, catalysts, and magnetic storage devices.

Cosmochemistry: Io, the Sulfur-Rich Moon

Before the exploration of the solar system by spacecraft, many astrochemists believed that all moons were as dull and barren as our own Moon. Since the arrival of Voyagers 1 and 2 at Jupiter in 1979, we now know that the moons of the outer planets have some amazing cosmochemistry. For example, Europa, one of the moons of Jupiter, appears to be covered by ice sheets beneath which there may be oceans of liquid water, perhaps containing some exotic life forms. The strangest body in the solar system is, without doubt, Io, another moon of Jupiter. Io is a brightly-colored moon, much of the color being derived from allotropes of sulfur and from sulfur compounds. Sulfur volcanoes dot the surface of the moon and the fountainlike plumes of the erupting sulfur volcanoes are among the most impressive and beautiful sights in the solar system. In fact, the eruptions resemble giant sulfur geysers, like some fantastic "Old Faithful" jets, spewing sulfur and sulfur compounds far into space before the weak gravity causes it to slowly settle back to the surface. The chemistry of the tenuous atmosphere, too, is unusual, mostly consisting of sulfur dioxide, together with more exotic species such as sulfur monoxide. Why is Io so unique in its chemistry? Sulfur is a common element throughout the solar system but on planets such as Earth, sulfur is found as metal sulfides, particularly iron(II) sulfide. And Io's volcanic activity is a result of the heating generated from enormous gravitational attractions by Jupiter and, to a lesser extent, Europa. We have very little idea of the chemical processes that must occur on this unique body and many new sulfur compounds are probably there waiting to be discovered. A visit to Io would certainly be a top priority for any cosmochemist!

The blue plume at the edge of Io is the eruption of the volcano, Ra Patera, in 1996. The 100 km high geiser-like plume is believed to contain sulfur dioxide and sulfur. (NASA)

Table C6-1 A comparison of Io and our own Moon

	Io	Moon
Mass	9×10^{22} kg	7.3×10^{22} kg
Radius	1.8×10^3 km	1.7×10^3 km
ratio of moon to planet mass	0.00005 (Jupiter)	0.012 (Earth)
Orbital distance from planet	4.2×10^5 km	3.8×10^5 km
Average density	3.6 g·cm^{-3}	3.5 g·cm^{-3}
orbital period	1.8 days	28 days

Technetium: A Rare Element with an Important Medical Use

There is an increasing demand by society that research must be "relevant" or directly applicable to society's needs. Yet relevant or applied science can only develop to the extent that our basic knowledge of science allows. There have been many examples where the original research appeared, at the time, to be of only theoretical interest and yet, subsequently, that discovery proved to have vital applications to our lives and to society. The greatest example was probably the laser, which, during development, seemed to have little use outside the physics laboratory, but has now become a major surgical tool as well as part of our everyday lives. Similarly, when the final gaps were filled in the periodic table, no-one would have expected that technetium, the "missing" element of group 7, would one day have medical applications.

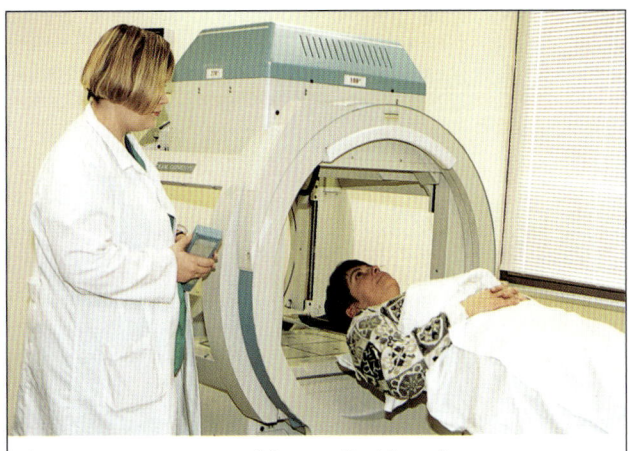

A gamma camera used for medical imaging.
(M.J. Welch, Mallinckrodt Institute of Radiology, Washington University Medical School.)

Technetium is used as a diagnostic radiopharmaceutical, that is, its compounds are given to patients to provide an image of a particular organ or tissue. Why technetium? To provide the image, an isotope is needed that is a γ-ray emitter [Greek gamma]. The radioisotope must have a short half-life, so that after injection there is a high enough level of γ-radiation that the detectors can provide a high-resolution image, yet there is a rapid decay so that there is a minimal radiation damage to the patient. But a short half-life means that there is little time to deliver the radioactive isotope to the hospital for use. Fortunately, the commonly-used isotope of technetium, known as ^{99m}Tc, which has a half-life of 6 hours is produced itself from a molybdenum isotope, ^{99}Mo, of half-life 67 hours. The molybdenum is supplied as molybdate ion, MoO_4^{2-}, absorbed on the surface of an alumina (Al_2O_3) column. When a dilute sodium chloride solution is passed down the column, the insoluble molybdate ion remains bound to the alumina while the pertechnate ion, TcO_4^-, formed by the decay (and analog of the permanganate ion, MnO_4^-) washes off. The ion is then converted to an appropriate complex that binds to a particular site or organ. For example, a technetium phosphate complex is bone-absorbed, finding use in imaging bone tumors, while a different technetium compound has been used for cardiac imaging.

Glass: Ancient Bottles and Modern Lenses

Glass is an unusual substance in that it is not crystalline but an amorphous (shapeless) solid. The oldest glass objects, glass beads found in the near East, date back to about 3000 B.C. but the first glass industry was founded about 1500 B.C. in ancient Egypt. The glass workers quickly developed sophistication to make useful and beautiful glass objects. From Egypt, the skills passed to the Roman Empire and then it was the Islamic world that kept the craft alive during the western dark ages (the 8th to 14th centuries). During

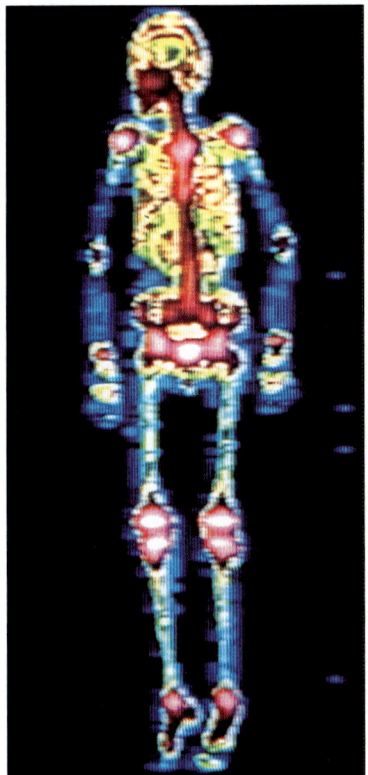

Typical front image from a gamma camera using technetium phosphate to delineate bone.
(Alexander Tsiaras/Science Source/Photo Researchers)

the Renaissance, glass composition and working techniques became so important to the economy of the city state of Venice that glass workers were forbidden by law to leave Venice or divulge the secrets that they knew. Even today, Venetian glass is prized for its outstanding beauty and its complexity of decoration.

Glass science and technology is always at the "cutting-edge," providing us with new and novel transparent materials. The best example is that of photogrey glasses, glasses that darken on exposure to high light levels. This glass contains crystals of colorless silver chloride trapped between the silicon dioxide tetrahedrons. In the presence of ultraviolet light, a charge transfer reaction occurs:

$$Ag^+Cl^- + h\nu \rightarrow Ag^0Cl^0$$

Ancient Egyptian glass vase.
(The Corning Museum of Glass)

The reaction would immediately reverse except for the presence of copper(I) chloride as a deliberate impurity within the silver chloride crystal. The copper(I) reduces the chlorine atom back to chloride ion:

$$Cu^+Cl^0 \rightarrow Cu^{2+} + Cl^-$$

The silver atoms migrate to the surface of the silver chloride crystal and aggregate into small, colloidal particles of silver metal. These particles absorb and reflect the ultraviolet and visible light, lowering the glass transmittance from about 85% to 22%. When the user returns indoors, the copper(II) ions slowly migrate to the crystal surface where they re-oxidize the silver metal to silver ions.

$$Cu^{2+} + Ag^0 \rightarrow Cu^+ + Ag^+$$

The silver ions, in turn, migrate back into the structure of the silver chloride crystal. And so the cycle is ready to begin again.

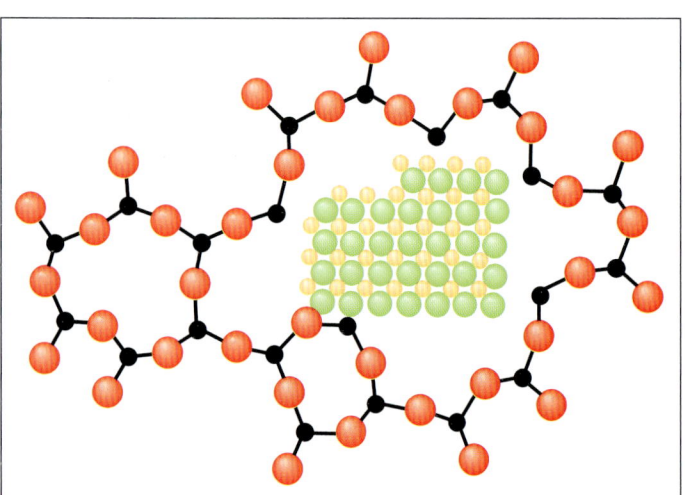

The molecular structure of photogrey glass. The tetrahedral silicate, SiO_4, framwork encloses tiny silver chloride crystals.

Table 13.3 Some effective neutron cross-sectional areas

Isotope	Hydrogen-1	Boron-10	Carbon-12	Nitrogen-14	Oxygen-16
Neutron cross-section (barn)	0.33	3.8×10^3	3.4×10^{-3}	1.8	1.8×10^{-4}

tumor cells following surgical removal of the main tumor. BNCT is a simple and promising concept but a difficult one to turn to reality.

It was in the 1950s that boron was proposed as the key. It was suggested that a stable boron compound be infiltrated into the tumor and irradiated with neutrons converting it to a radioactive isotope, and then letting the radiation destroy the malignant cell. Why boron? is an obvious question. This relates to nuclear chemistry. The ability of an atomic nucleus to capture a particle depends not upon the size of the nucleus but on the nuclear structure. Thus, each nucleus is assigned an effective cross-sectional area; that is, the larger the effective cross-sectional area, the easier it is for a neutron to impact the nucleus. The area is expressed in units of barns, where 1 barn = 10^{-24} cm^2. Boron-10 (with odd numbers of protons and neutrons) has an exceptionally large effective cross-sectional area, while those of hydrogen, carbon, and nitrogen, the major cell components, are quite small (Table 13.3). Thus, boron would be an ideal target.

When a neutron impacts a boron-10 nucleus, boron-11 is initially formed:

$$^{10}_{5}\text{B} + ^{1}_{0}n \rightarrow ^{11}_{5}\text{B}$$

This species is radioactive with a very short lifetime, fissioning to helium-4 and lithium-7 with the release of energy:

$$^{11}_{5}\text{B} \rightarrow ^{7}_{3}\text{Li} + ^{4}_{2}\text{He}$$

This fission energy is enough to propel the two particles about one cell width, damaging whatever molecules they encounter. About 1 billion boron atoms would be enough to completely destroy the cell.

And this is the challenge, to deliver high concentrations of a boron compound specifically to the malignant cells. Early medical research used borate ion, but this proved ineffective. It was the progress in boron hydride chemistry that proved to be the breakthrough. These boron hydrides have a high boron content, are kinetically stable, and can be linked to organic units. One of the simplest "second-generation" ions was $(\text{B}_{12}\text{H}_{11}\text{SH})^{2-}$ (Figure 13.7), commonly called BSH. This was the first family of boron species that showed a significantly greater boron concentration in the tumor compared to that in the blood. The third generation of compounds, currently under development, having four cage-type boranes linked to one large organic molecule, shows even more promise. Nevertheless, it will take many more years of research before BNCT becomes a simple safe means of fighting these small nests of malignant cells that cannot be destroyed by any other means.

Figure 13.7 Structure of the $(\text{B}_{12}\text{H}_{11}\text{SH})^{2-}$ ion.

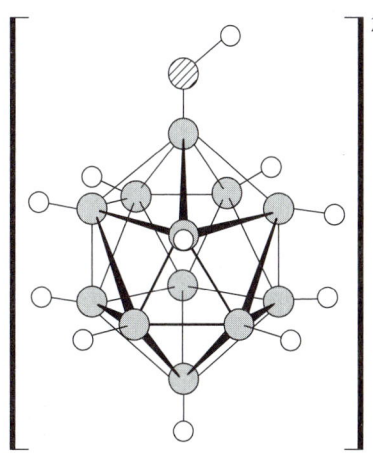

13.5   *Sodium Tetrahydridoborate*

The only other species of boron used on a large scale is the tetrahydridoborate ion, BH_4^-. Most hydrides, except for those of carbon, are spontaneously flammable, unstable compounds. This anion, however, can even be recrystallized from cold water as the sodium salt. The crystal structure of sodium tetrahydridoborate is interesting because it adopts the sodium chloride structure, with the whole BH_4^- ion occupying the same sites as the chloride ion does. Sodium tetrahydridoborate is of major importance as a mild reducing agent, particularly in organic chemistry, where it is used to reduce aldehydes to primary alcohols and ketones to secondary alcohols without reducing other functional groups such as carboxylic groups. The reaction of diborane with sodium hydride is used to produce sodium tetrahydridoborate:

$$2\,NaH(s) + B_2H_6(g) \rightarrow 2\,NaBH_4(s)$$

13.6   *Boron Trifluoride*

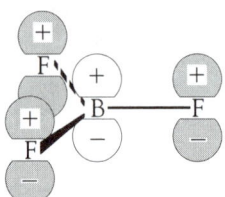

Figure 13.8   Proposed π bonding in boron trifluoride, involving the full p orbitals (shaded) on the fluorine atoms and the empty p_z orbital on the boron atom.

Boron has only three valence electrons, so any boron compound that has simple covalent bonding will be electron deficient with respect to the octet rule. Thus, we saw that the simplest boron hydride dimerizes to give B_2H_6, in which there are two hydridic bridge bonds. Boron trifluoride, however, does not dimerize; it remains as the simple trigonal planar compound, BF_3. A study of the molecule shows that the boron–fluorine bond energy is extremely high ($613\ kJ\cdot mol^{-1}$). This bond energy is far higher than that for any conventional single bond; for example, the carbon–fluorine bond energy is $485\ kJ\cdot mol^{-1}$. To explain the surprising stability of the electron-deficient molecule and the strong covalent bond, it is postulated that there is π bonding as well as σ bonding in the compound. The boron atom has an empty $2p_z$ orbital at right angles to the three σ bonds with the fluorine atoms. Each fluorine atom has a full $2p$ orbital parallel to the boron $2p_z$ orbital. A delocalized π system involving the empty p orbital on the boron and one full p orbital on each of the fluorine atoms can be formed (Figure 13.8).

There is experimental evidence to support this explanation: When boron trifluoride reacts with a fluoride ion to form the tetrahedral tetrafluoroborate ion, BF_4^-, the B—F bond length increases from 130 pm in boron trifluoride to 145 pm in the tetrafluoroborate ion. This lengthening would be expected because the $2s$ and three $2p$ orbitals of the boron in the tetrafluoroborate ion are used to form four σ bonds. Hence, there are no orbitals available for π bonding in the tetrafluoroborate ion, and so the B—F bond in this ion would be a "pure" single bond.

By using the vacant $2p_z$ orbital, boron trifluoride can behave as a powerful Lewis acid. The classic illustration of this behavior is the reaction between boron trifluoride and ammonia, where the nitrogen lone pair acts as the electron pair donor:

$$BF_3(g) + {:}NH_3(g) \rightarrow F_3B{:}NH_3(s)$$

About 4000 tonnes of boron trifluoride is used industrially in the United States every year as both a Lewis acid and a catalyst in organic reactions.

13.7 Boron Trichloride

As we cross the Periodic Table, the chloride of boron is the first chloride that we encounter to exist as small covalently bonded molecules. As such, it is quite typical. Ionic chlorides are solids that dissolve in water to form hydrated cations and anions. However, the typical small-molecule covalent chloride is a gas or liquid at room temperature and reacts violently with water. For example, bubbling boron trichloride (a gas above 12°C) into water produces boric acid and hydrochloric acid:

$$BCl_3(g) + 3\ H_2O(l) \rightarrow H_3BO_3(aq) + 3\ HCl(aq)$$

We can predict the products of these reactions in terms of the relative electronegativities of the two atoms. In this case, the electronegativity of chlorine is much greater than that of boron. Hence, as a water molecule approaches the boron trichloride molecule, we can picture the partially positive hydrogen being attracted to the partially negative chlorine atom, while the partially negative oxygen atom is attracted to the partially positive boron atom (Figure 13.9). A bond shift occurs, and one chlorine atom is replaced by a hydroxyl group. When this process happens two more times, the result is boric acid.

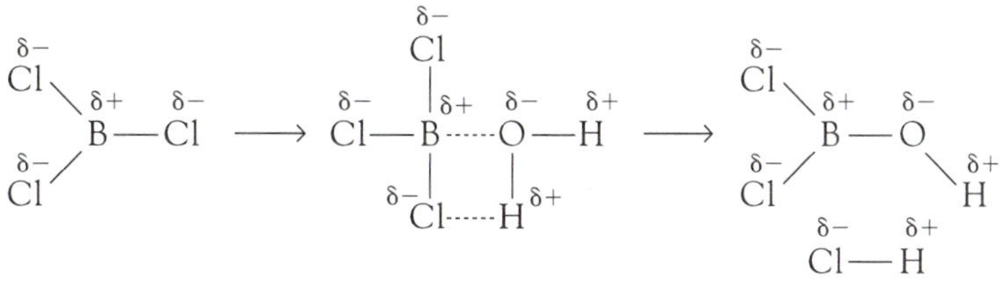

Figure 13.9 First step of the postulated mechanism for hydrolysis of boron trichloride.

13.8 Aluminum

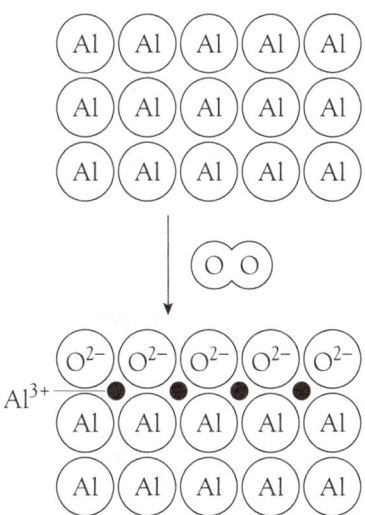

Figure 13.10 Formation of a single oxide layer on the surface of aluminum metal. The small aluminum 3+ ions are indicated by the solid circles.

Because aluminum is a metal with a high negative standard reduction potential, it might be expected to be very reactive. This is indeed the case. Why, then, can aluminum be used as an everyday metal rather than be consigned to the chemistry laboratory like sodium? The answer is found in its reaction with oxygen gas. Any exposed surface of aluminum metal rapidly reacts with dioxygen to form aluminum oxide, Al_2O_3. An impermeable oxide layer, between 10^{-4} and 10^{-6} mm thick, then protects the layers of aluminum atoms underneath. This can happen because the oxide ion has an ionic radius (124 pm) similar to the metallic radius of the aluminum atom (143 pm). As a result, the surface packing is almost unchanged because the small aluminum ions (68 pm) fit into interstices in the oxide surface structure. The process is shown in Figure 13.10.

To increase their corrosion resistance, aluminum products are "anodized." In other words, the aluminum product is used as the anode in an electrochemical cell, and additional aluminum oxide is deposited as an electrolytic product over the naturally formed layers. This anodized aluminum possesses an oxide layer about 0.01 mm thick, and this very thick oxide coating has the useful property of absorbing dyes and pigments so that a colored surface can be produced.

The particular attraction of aluminum as a construction metal is its low density ($2.7\ \text{g·cm}^{-3}$), second only to that of magnesium ($1.7\ \text{g·cm}^{-3}$)—disregarding the very reactive alkali metals. For instance, compare the density of aluminum with that of either iron (7.9 g·cm^{-3}) or gold (19.3 g·cm^{-3}). Aluminum is a good conductor of heat, a property accounting for its role in cookware. It is not as good as copper, however. To spread heat more evenly from the electrical element (or gas flame), higher priced pans have a copper-coated bottom. Aluminum also is exceptional as a conductor of electricity, hence, its major role in electric power lines and home wiring. The major problem with using aluminum wiring occurs at the connections. If aluminum is joined to an electrochemically dissimilar metal, such as copper, an electrochemical cell will be established under damp conditions. This development causes oxidation (corrosion) of the aluminum. For this reason, use of aluminum in home wiring is now discouraged.

Chemical Properties of Aluminum

Like other powdered metals, aluminum powder will burn in a flame to give a dust cloud of aluminum oxide:

$$4 \text{ Al}(s) + 3 \text{ O}_2(g) \rightarrow 2 \text{ Al}_2\text{O}_3(s)$$

and aluminum will burn very exothermically with halogens, such as dichlorine:

$$2 \text{ Al}(s) + 3 \text{ Cl}_2(g) \rightarrow 2 \text{ AlCl}_3(s)$$

Aluminum, like beryllium, is an amphoteric metal, reacting with both acid and base:

$$2 \text{ Al}(s) + 6 \text{ H}^+(aq) \rightarrow 2 \text{ Al}^{3+}(aq) + 3 \text{ H}_2(g)$$

$$2 \text{ Al}(s) + 2 \text{ OH}^-(aq) + 6 \text{ H}_2\text{O}(l) \rightarrow 2 \text{ [Al(OH)}_4]^-(aq) + 3 \text{ H}_2(g)$$

In aqueous solution, the aluminum ion is present as the hexaaquaaluminum ion, $[\text{Al(OH}_2)_6]^{3+}$, but it undergoes a hydrolysis reaction to give a solution of the pentaaquahydroxoaluminum ion, $[\text{Al(OH}_2)_5(\text{OH})]^{2+}$, and the hydronium ion, and then to the tetraaquadihydroxoaluminum ion:

$$[\text{Al(OH}_2)_6]^{3+}(aq) + \text{H}_2\text{O}(l) \rightleftharpoons [\text{Al(OH}_2)_5(\text{OH})]^{2+}(aq) + \text{H}_3\text{O}^+(aq)$$

$$[\text{Al(OH}_2)_5(\text{OH})]^{2+}(aq) + \text{H}_2\text{O}(l) \rightleftharpoons [\text{Al(OH}_2)_4(\text{OH})_2]^+(aq) + \text{H}_3\text{O}^+(aq)$$

Thus, solutions of aluminum salts are acidic, with almost the same acid ionization constant as ethanoic (acetic) acid. The mixture in antiperspirants commonly called aluminum chlorhydrate is, in fact, a mixture of the chloride salts of these two hydroxo ions. It is the aluminum ion in these compounds that acts to constrict pores on the surface of the skin.

Addition of hydroxide ion to aluminum ion first gives a gelatinous precipitate of aluminum hydroxide, but this product redissolves in excess hydroxide ion to give the aluminate ion (more precisely called the tetrahydroxoaluminate ion):

$$[\text{Al(OH}_2)_6]^{3+}(aq) \xrightarrow{\text{OH}^-} \text{Al(OH)}_3(s) \xrightarrow{\text{OH}^-} [\text{Al(OH)}_4]^-(aq)$$

As a result, aluminum 3+ is soluble at low and high pH's but insoluble under neutral conditions (Figure 13.11). Aluminum hydroxide is used in a number of antacid formulations. Like other antacids, the compound is an insoluble base that will neutralize excess stomach acid:

$$\text{Al(OH)}_3(s) + 3 \text{ H}^+(aq) \rightarrow \text{Al}^{3+}(aq) + 3 \text{ H}_2\text{O}(l)$$

Figure 13.11 Aluminum ion solubility as a function of pH.

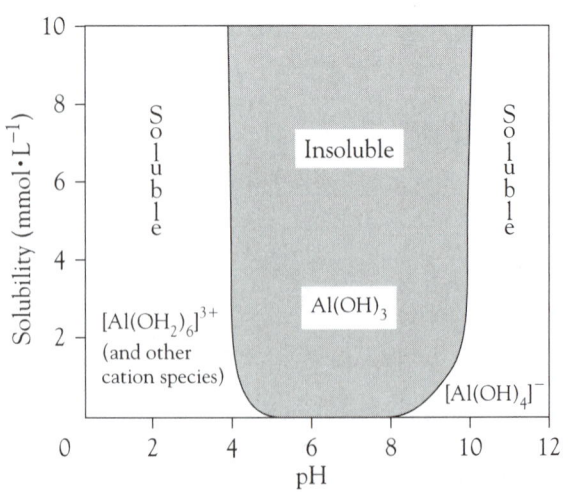

As we discussed in Chapter 9, Section 9.5, much of aluminum chemistry resembles that of scandium more than it does the lower members of Group 13.

Production of Aluminum

The discovery of an electrolytic reduction method by the French chemist Henri Sainte-Claire Deville and the decreasing cost of electricity caused the price of metallic aluminum to drop dramatically in the late 19th century. However, the production of the metal on a large scale required a method that would use an inexpensive, readily available ore. This route was found independently in 1886 by two young chemists; one in France, Paul Héroult, and one in the United States, Charles Hall. Hence, the process is known as the Hall–Héroult process. Charles was assisted by his sister, Julia, who maintained detailed notes of the experiments, though it is now known that she played a quite minor role in the discovery.

Aluminum is the most abundant metal in the Earth's crust, mostly in the form of clays. To this day, there is no economical route for the extraction of aluminum from clay. However, in hot, humid environments, the more soluble ions are leached from the clay structure to leave the ore bauxite (impure hydrated aluminum oxide). Thus, the countries producing bauxite are mainly those near the equator, Australia being the largest source, followed by Guinea, Brazil, Jamaica, and Suriname.

The first step in the extraction process is the purification of bauxite. This step is accomplished by digesting (heating and dissolving) the crushed ore with hot sodium hydroxide solution to give the soluble aluminate ion:

$$Al_2O_3(s) + 2\ OH^-(aq) + 3\ H_2O(l) \rightarrow 2\ [Al(OH)_4]^-(aq)$$

The insoluble materials, particularly iron(III) oxide, are filtered off as "red mud." As we discussed in Chapter 9, Section 9.5, iron(III) ion and aluminum ion have many similarities, but they differ in that aluminum is amphoteric, reacting with hydroxide ion, whereas iron(III) oxide does not react with hydroxide ion. Upon cooling, the equilibrium in the solution shifts to the left, and white aluminum oxide trihydrate precipitates, leaving soluble impurities in solution:

$$2\ [Al(OH)_4]^-(aq) \rightarrow Al_2O_3{\cdot}3H_2O(s) + 2\ OH^-(aq)$$

The hydrate is heated strongly in a rotary kiln (similar to that used in cement production) to give anhydrous aluminum oxide:

$$Al_2O_3{\cdot}3H_2O(s) \xrightarrow{\Delta} Al_2O_3(s) + 3\ H_2O(g)$$

With its high ion charges, aluminum oxide has a very large lattice energy and hence a high melting point (2040°C). However, to electrolyze the aluminum oxide, it was necessary to find an aluminum compound with a much lower melting point. Hall and Héroult simultaneously announced the discovery of this lower melting aluminum compound, the mineral cryolite, whose chemical name is sodium hexafluoroaluminate, Na_3AlF_6. There are few naturally occurring deposits of this mineral; Greenland has the largest deposit. As a result of its rarity, almost all cryolite is manufactured. This in itself is an interesting process, because the starting point is usually a waste material, silicon tetrafluoride, SiF_4, which is produced in the synthesis of hydrogen fluoride. Silicon tetrafluoride gas reacts with water to give insoluble silicon dioxide and a solution of hexafluorosilicic acid, H_2SiF_6, a relatively safe fluorine-containing compound:

$$3\ SiF_4(g) + 2\ H_2O(l) \rightarrow 2\ H_2SiF_6(aq) + SiO_2(s)$$

The acid is then treated with ammonia to give ammonium fluoride:

$$H_2SiF_6(aq) + 6\ NH_3(aq) + 2\ H_2O(l) \rightarrow 6\ NH_4F(aq) + SiO_2(s)$$

Finally, the ammonium fluoride solution is mixed with a solution of sodium aluminate to give the cryolite and ammonia, which can be recycled:

$$6\ NH_4F(aq) + Na[Al(OH)_4](aq) + 2\ NaOH(aq) \rightarrow$$
$$Na_3AlF_6(s) + 6\ NH_3(aq) + 6\ H_2O(l)$$

The detailed chemistry that occurs in the electrolytic cell is still poorly understood, but the cryolite acts as the electrolyte (Figure 13.12). The aluminum oxide is dissolved in molten cryolite at about 950°C. Molten aluminum is produced at the cathode, and the oxygen that is produced at the anode oxidizes the carbon to carbon monoxide (and some carbon dioxide):

$$Al^{3+}(Na_3AlF_6) + 3\ e^- \rightarrow Al(l)$$
$$O^{2-}(Na_3AlF_6) + C(s) \rightarrow CO(g) + 2\ e^-$$

The process is very energy intensive, requiring currents of about 3.5×10^4 A at 6 V. In fact, about 25 percent of the cost of aluminum metal is derived from its energy consumption. The production of 1 kg of aluminum consumes about 2 kg of aluminum oxide, 0.6 kg of anodic carbon, 0.1 kg of cryolite, and 16 kWh of electricity.

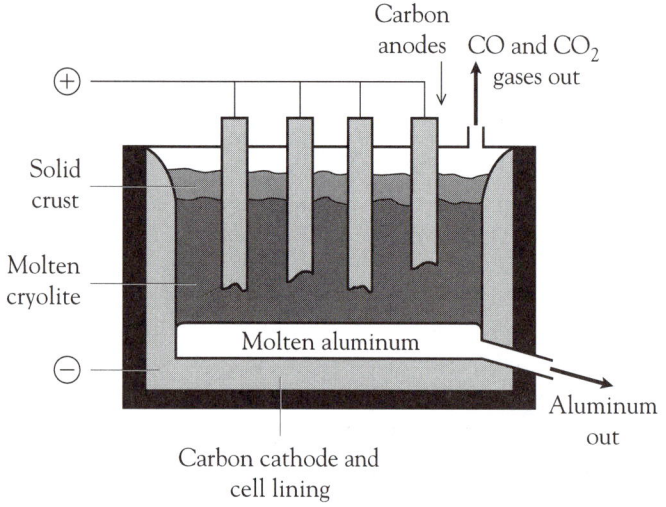

Figure 13.12 Electrolytic cell for aluminum production.

Aluminum production yields four by-products that create major pollution problems:

1. Red mud, which is produced from the bauxite purification and is highly basic

2. Hydrogen fluoride gas, which is produced when cryolite reacts with traces of moisture in the aluminum oxide

3. Oxides of carbon, which are produced at the anode

4. Fluorocarbons, which are produced by reaction of fluorine with the carbon anode

To reduce the red mud disposal problem, the slurry is poured into settling tanks, from which the liquid component, mainly sodium hydroxide solution, is removed and recycled or neutralized. The solid, mostly iron(III) oxide, can then be used as landfill or shipped to iron smelters for extraction of the iron.

The problem of what to do with the emissions of hydrogen fluoride gas has been solved to a large extent by absorbing the hydrogen fluoride in a filter bed of aluminum oxide. The product of this process is aluminum fluoride:

$$Al_2O_3(s) + 6\ HF(g) \rightarrow 2\ AlF_3(s) + 3\ H_2O(g)$$

This fluoride can be added periodically to the melt, thereby recycling the hydrogen fluoride.

A partial solution to the problem of disposing of the large volumes of the oxides of carbon that are produced is to burn the poisonous carbon monoxide, a process giving carbon dioxide and providing some of the heat required to operate the aluminum plant. However, the electrolytic method inevitably produces these two gases, and until an alternative, economical process is devised, aluminum production will continue to contribute carbon dioxide to the atmosphere.

For every tonne of aluminum, about 1 kg of tetrafluoromethane, CF_4, and about 0.1 kg of hexafluoroethane, C_2F_6, are produced. These compounds are significant "greenhouse" gases. The fluorocarbon problem has not yet been solved, and it is the focus of a major research effort by aluminum companies. One advance has been the addition of lithium carbonate to the molten mixture in the electrolytic cell. The presence of lithium carbonate lowers the melting point of the mixture, resulting in a higher current and hence a more efficient cell. At the same time, the presence of the compound reduces fluorine emissions by 25 to 50 percent, thus reducing the production of fluorocarbons.

Fluorosilicic acid is another by-product of the electrolysis process. Until recently, there was little use for this very weak acid. Now, however, it has become a favored source of fluoride ion in the fluoridation of domestic water supplies (see Chapter 17, Section 17.2). At the 1 ppm concentrations in the water supply, the hexafluorosilicate ion will be predominantly hydrolyzed to silicic acid, hydronium ion, and fluoride ion:

$$SiF_6{}^{2-}(aq) + 8\ H_2O(l) \rightleftharpoons H_4SiO_4(aq) + 4\ H_3O^+(aq) + 6\ F^-(aq)$$

The major producers of aluminum metal and the suppliers of bauxite are different. The largest producer of aluminum metal is the United States. The large energy requirement of the production process favors those countries with inexpensive energy sources. Thus, Canada and Norway, neither of which is a bauxite producer or large aluminum consumer, make the top five of aluminum metal producers. Both countries have low-cost hydroelectric power and deep-water ports favoring easy import of ore and export of aluminum metal. The bulk of the value added to the material comes through the processing steps. Even though the developed world relies heavily on third-world countries for the raw material, the third world receives comparatively little in the way of income from the mining phase.

About 25 percent of the output of aluminum metal is used in the construction industry, and lesser proportions are used to manufacture aircraft, buses, and railroad passenger cars (18 percent); containers and packaging (17 percent); and electric power lines (14 percent). Aluminum is becoming increasingly favored for automobile construction. With its lower density, fuel consumption for the same size vehicle is reduced significantly. For example, every tonne of steel replaced by a tonne of aluminum results in a decrease of the emission of 20 tonnes of carbon dioxide over the life of the vehicle. In 1960, the average North American vehicle contained about 2.5 kg of aluminum, while in 2000, the figure is about 110 kg (though some vehicles have over twice that figure).

Aluminum recycling is key to the increased use of aluminum. The recycling process uses only a small fraction of the energy needed to extract aluminum from its ore. Recycling also avoids the environmental problems of the smelting process. Thus, of all the metals, aluminum reclamation is probably the most important for the environment.

13.9 Aluminum Halides

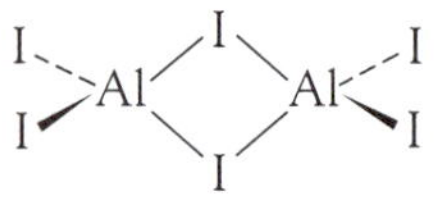

Figure 13.13 Structure of aluminum iodide.

The aluminum halides constitute an interesting series of compounds: Aluminum fluoride melts at 1290°C; aluminum chloride sublimes at 180°C; and aluminum bromide and iodide melt at 97.5 and 190°C, respectively. Thus the fluoride has the characteristic high melting point of an ionic compound, whereas the melting points of the bromide and iodide are typical of covalent compounds. The aluminum ion has a charge density of 364 C·mm^{-3}, so we expect all anions, except the small fluoride ion, to be polarized to the point of covalent bond formation with aluminum. In fact, aluminum fluoride does have a typically ionic crystal structure with arrays of alternating cations and anions. But the bromide and iodide both exist as dimers, Al_2Br_6 and Al_2I_6, analogous to diborane, with two bridging halogen atoms (Figure 13.13). The chloride forms an ionic-type lattice structure in the solid, which collapses in the liquid phase to give molecular Al_2Cl_6 dimers. Thus, the ionic and covalent forms must be almost equal in energy. These dimers are also formed when solid aluminum chloride is dissolved in low-polarity solvents.

Even though anhydrous aluminum chloride appears to adopt an ionic structure in the solid phase, its reactions are more typical of a covalent chloride. This covalent behavior is particularly apparent in the solution processes of the anhydrous aluminum chloride. As mentioned previously, the hexahydrate actually contains the hexaaquaaluminum ion, $[Al(OH_2)_6]^{3+}$. It dissolves quietly in water, although the solution is acidic as a result of hydrolysis. Anhydrous aluminum chloride, however, reacts very exothermically with water in the typical manner of a covalent chloride, producing a hydrochloric acid mist:

$$AlCl_3(s) + 3\ H_2O(l) \rightarrow Al(OH)_3(s) + 3\ HCl(g)$$

Anhydrous aluminum chloride is an important reagent in organic chemistry. In particular, it is used as a catalyst for the substitution of aromatic rings in the *Friedel–Crafts reaction*. The overall reaction can be written as the reaction between an aromatic compound, Ar—H, and an organochloro compound, R—Cl. The aluminum chloride reacts as a strong Lewis acid with the organochloro compound to give the tetrachloroaluminate ion, $AlCl_4^-$, and the carbocation. The carbocation then reacts with the aromatic compound to give the substituted aromatic compound, Ar—R, and a hydrogen ion. The latter decomposes the tetrachloroaluminate ion, regenerating aluminum chloride:

$$R\text{—}Cl + AlCl_3 \rightarrow R^+ + [AlCl_4]^-$$
$$Ar\text{—}H + R^+ \rightarrow Ar\text{—}R + H^+$$
$$H^+ + [AlCl_4]^- \rightarrow HCl + AlCl_3$$

13.10 Aluminum Potassium Sulfate

In Chapter 9, Section 9.6, we discussed the family of compounds called the alums, $M^+M^{3+}(SO_4^{2-})_2 \cdot 12H_2O$. The compound that gave its name to the series is alum, $KAl(SO_4)_2 \cdot 12H_2O$, the only common water-soluble mineral of aluminum. As such, it has played an important role in the dyeing industry. To adsorb a dye permanently onto cloth, the cloth is first soaked in a solution of alum. A layer of aluminum hydroxide is deposited on the cloth's surface, to which dye molecules readily bond. Because of its usefulness, alum has been a valuable import item from Asia since the time of the Romans. Alum crystallizes from an "equimolar" mixture of potassium sulfate and aluminum sulfate to give the formulation of $KAl(SO_4)_2 \cdot 12H_2O$. Alum crystals have a very high lattice stability because the sulfate anions are packed between alternating potassium and hexaaquaaluminum ions. The compound is sometimes used to stop minor bleeding (such as accidental cuts during shaving) because it causes coagulation of proteins on the surface of cells without killing the cells themselves.

13.11 Spinels

Spinel itself is magnesium aluminum oxide, $MgAl_2O_4$; but of more importance are the enormous number of compounds that adopt the same crystal structure and are also called *spinels*. Many of these compounds have unique properties that will make them important in the chemistry of the 21st century. The general formula of a spinel is AB_2X_4, where A is usually a dipositive metal ion; B, usually a tripositive metal ion; and X, a dinegative anion, usually oxygen.

The framework of the unit cell of a spinel consists of 32 oxide ions in an almost perfect cubic close-packed arrangement. Thus, the unit cell composition is actually $A_8B_{16}O_{32}$. Figure 13.14 shows part of the unit cell. The oxide ions form a face-centered cubic array, and there are octahedral sites at the center of the cube and in the middle of each cube edge, and tetrahedral sites in the middle of each "cubelet." In the normal spinel structure, the 8 A cations occupy one-eighth of the tetrahedral holes and the 16 B cations occupy one-half of the octahedral holes. Thus, the unit cell can be considered to consist of "cubelets" of zinc sulfide-type tetrahedral units interspersed among "cubelets" of sodium chloride-type octahedral units.

To indicate site occupancy, we can use subscripts t and o to represent tetrahedral and octahedral cation sites; thus, spinel itself can be written as $(Mg^{2+})_t(2\,Al^{3+})_o(O^{2-})_4$. There are some spinels in which the dipositive ions are located in the octahedral sites. Because there are twice as many available octahedral (B) sites as available tetrahedral (A) sites in the spinel structure, only half of the tripositive ions can be placed in tetrahedral sites; the remainder must occupy octahedral sites. Such compounds are called *inverse spinels*. The most common example is magnetite, Fe_3O_4, or more accurately, $Fe^{2+}(Fe^{3+})_2(O^{2-})_4$. The arrangement here is $(Fe^{3+})_t(Fe^{2+},Fe^{3+})_o(O^{2-})_4$.

We might expect that all spinels would adopt the inverse structure, for the tetrahedral holes are smaller than the octahedral holes and the tripositive cations are smaller than the dipositive cations. However, in addition to size factors, we have to consider energy factors. Because lattice energy depends on the size of the ionic charge, it is the location of the 3+ ion that is responsible for the majority of the energy. Lattice energy will be higher when the 3+ ion is an octahedral site surrounded by six anions than when it occupies a tetrahedral site and is surrounded by only four anions. Nevertheless, the inverse spinel structure is preferred by many transition metal ions because the d-orbital occupancy affects the energy preferences, as we will see in Chapter 19, Section 19.8.

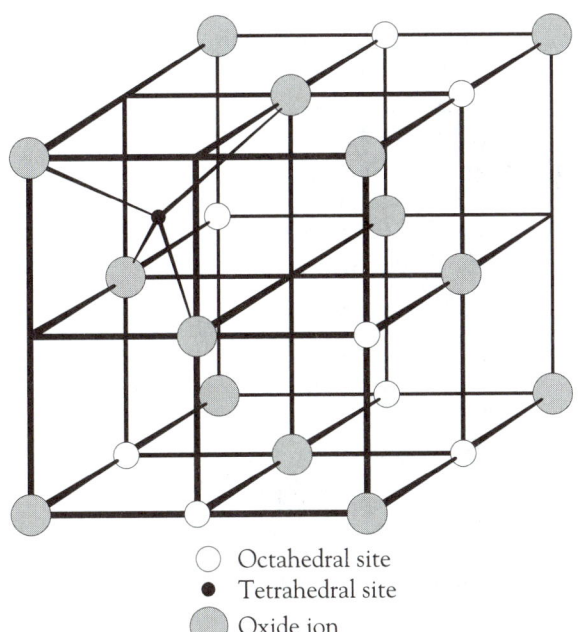

○ Octahedral site
● Tetrahedral site
◯ Oxide ion

Figure 13.14 Part of the unit cell of the spinel structure showing the occupied lattice sites. Of the eight "cubelets" shown, the upper left front cubelet shows an occupied tetrahedral cation site (zinc sulfide-type), whereas the other seven cubelets have some occupied octahedral cation sites (sodium chloride-type).

The interest in spinels derives from their unusual electrical and magnetic properties, particularly those in which the tripositive ion is Fe^{3+}. These compounds are known as *ferrites*. For example, it is possible to synthesize a series of compounds MFe_2O_4, where M is any combination of zinc ions and manganese ions, provided the formula $Zn_xMn_{1-x}Fe_2O_4$ is obeyed. By choosing the appropriate ratio, very specific magnetic properties can be obtained for these zinc ferrites. We discuss ferrites in more detail in Chapter 20, Section 20.6.

Even more peculiar is sodium-β-alumina, $NaAl_{11}O_{17}$. Although its formula does not look like that of a spinel, most of the ions fit the spinel lattice sites. The sodium ions, however, are free to roam throughout the structure. It is this property that makes the compound so interesting, because its electrical conductivity is very high, and it can act as a solid-phase electrolyte. This type of structure offers great potential for low-mass storage batteries.

13.12   Biological Aspects

The Essentiality of Boron

Boron is an essential micronutrient in plants. The element is believed to play a major role in the synthesis of one of the bases for RNA formation and in cellular activities, such as carbohydrate synthesis. After zinc, boron is the most common soil deficiency, worldwide. The class of plants known as dicots have much higher boron requirements than monocots. Crops most susceptible to boron deficiency and which often require boron supplements are alfalfa, carrot, coffee, cotton, peanut, sugar beet, sunflower, swede, and turnip. There is growing evidence that boron is an essential element for mammals, possibly in bone formation.

The Toxicity of Aluminum

Aluminum is the third most abundant element in the lithosphere. Despite its ubiquitousness in the environment, it is a highly toxic metal. Fortunately, under near-neutral conditions aluminum ion forms insoluble compounds, minimizing its bioavailability. Fishes are particularly at risk from aluminum toxicity. Research has shown that the damage to fish stocks in acidified lakes is not due to the lower pH but to the higher concentrations of aluminum ion in the water that result from the lower pH (see Figure 13.13). In fact, an aluminum ion concentration of 5×10^{-6} mol·L^{-1} is sufficient to kill fish.

Human tolerance is greater, but we should still be particularly cautious of aluminum intake. Part of our dietary intake comes from aluminum-containing antacids. Tea is high in aluminum ion, but the aluminum ions form inert compounds when milk or lemon is added. It is advisable not to inhale the spray from aluminum-containing antiperspirants because the metal ion is believed to be absorbed easily from the nasal passages directly into the bloodstream. In Chapter 14, Section 14.27 we will discuss the preventative role of silicon in preventing absorption of aluminum.

Aluminum is the most common metal ion in soils; hence, it is also a concern on the 30 to 40 percent of the world's arable soils where acid soil releases aluminum ions. For some crops, such as corn, it is second only to drought as a factor decreasing crop yields—sometimes by as much as 80 percent. The aluminum ion enters the plant root cells, inhibiting cell metabolism. Farmers in poorer countries cannot afford the regular application

Indium compounds have been found to be effective against sleeping sickness.

of powdered limestone to increase soil pH and immobilize the aluminum as an insoluble hydroxo compound. Some plants are naturally resistant to the aluminum as their roots excrete citric or malic acids into the surrounding soil. These acids form complexes with the aluminum ion, preventing it from being absorbed into the roots. Genetic engineers are now working on the introduction of citric acid-generating genes into important food crop species, hopefully leading to better crop yields.

The Hazard of Thallium

As we mentioned in Chapter 9, Section 9.13, thallium is a highly toxic element as thallium(I), its most common form, and mimics potassium in its biochemical behavior. Thallium is widely distributed in the lithosphere, and it enters the environment primarily from coal burning and cement manufacture. In the smelting of lead from its ores, thallium is a dangerous by-product. For example, in the summer of 2001 at the giant lead and zinc smelter in Trail, British Columbia, Canada, dozens of maintenance workers became ill following exposure to thallium dust during the cleaning of the inside of the smelter ducts. The company owning the plant also admitted to allowing thallium waste to run into the Columbia River.

13.13 *Element Reaction Flowchart*

Boron and aluminum are the only two Group 13 elements that we have discussed in depth.

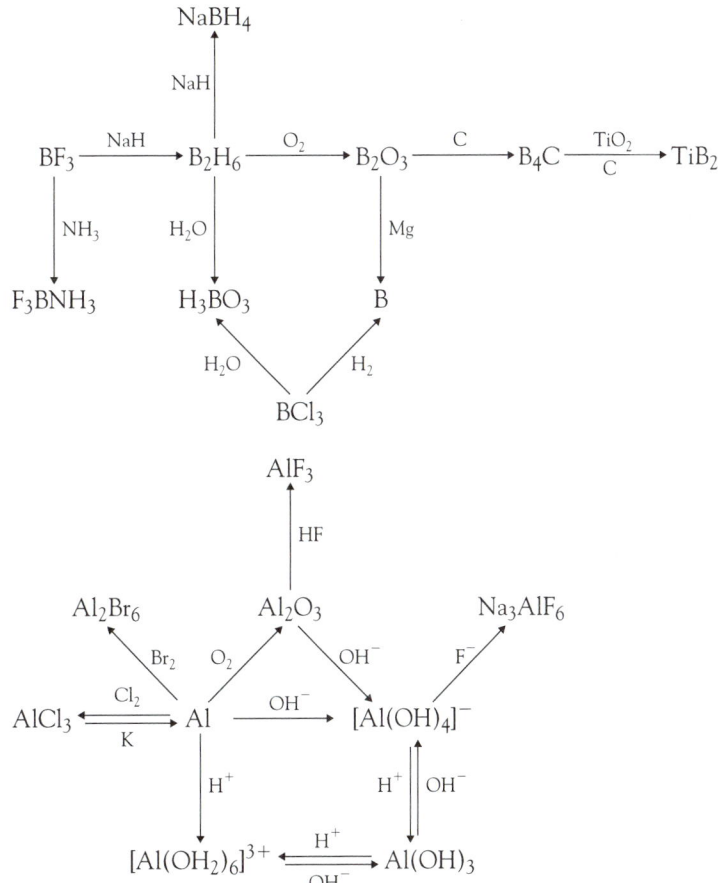

EXERCISES

13.1   Write balanced chemical equations for the following chemical reactions:
   (a) liquid potassium metal with solid aluminum chloride
   (b) solid diboron trioxide with ammonia gas at high temperature
   (c) aluminum metal with hydroxide ion
   (d) tetraborane, B_4H_{10}, and dioxygen

13.2   Write balanced chemical equations for the following chemical reactions:
   (a) liquid boron tribromide with water
   (b) aluminum metal with hydrogen ion
   (c) thallium(I) hydroxide solution with carbon dioxide gas

13.3   With a very high charge density, aluminum would not be expected to exist widely as a free 3+ ion, yet it does exist in the form of a hydrated 3+ ion. Explain why.

13.4   Construct an electron-dot structure for the peroxoborate ion. Thus, deduce the oxidation number of the bridging oxygen atoms.

13.5   From bond energy data, calculate the enthalpy of formation of boron trifluoride. What two factors result in its particularly high value?

13.6   From bond energy data calculate the enthalpy of formation of boron trichloride (gaseous). Why is the value so different from that of boron trifluoride?

13.7   Explain briefly why sheets of aluminum do not oxidize completely to aluminum oxide even though aluminum is a highly reactive metal.

13.8   Explain briefly why solutions of aluminum chloride are strongly acidic.

13.9   Magnesium metal only reacts with acids, whereas aluminum reacts with both acids and bases. What does this behavior tell you about aluminum?

13.10   Describe briefly the steps in the industrial extraction of aluminum from bauxite.

13.11   Explain the potential environmental hazards from aluminum smelting.

13.12   Why are aluminum smelters sometimes located in countries other than those that produce the ore or consume much of the metal?

13.13   Contrast the bonding in the different aluminum halides.

13.14   Why is alum a commonly used salt of aluminum?

13.15   Explain the difference between a spinel and an inverse spinel.

13.16   Explain why thallium(I) compounds are usually ionic species while thallium(III) compounds are more covalent in their behavior.

13.17   Gallium(III) fluoride, GaF_3, sublimes at 950°C, while gallium(III) chloride, $GaCl_3$, melts at 78°C. Suggest an explanation for the significant difference.

13.18   Compare and contrast the chemistry of boron and silicon.

13.19   Why is aluminum a particular environmental problem in the context of acid rain?

13.20   Write balanced chemical equations corresponding to each transformation in the Element Reaction Flowchart.

BEYOND THE BASICS

13.21   The metallic, covalent, and ionic (six-coordinate) radii for aluminum are 143 pm, 130 pm, and 54 pm. Explain why these values are different.

13.22   Aluminum fluoride, AlF_3, is insoluble in pure liquid hydrogen fluoride but dissolves readily in liquid hydrogen fluoride-containing sodium fluoride. When boron trifluoride is bubbled into the solution, aluminum fluoride precipitates. Write two equations to represent these observations and suggest what is happening in each case using an appropriate acid–base concept.

13.23   When aluminum chloride is dissolved in benzene, C_6H_6, a dimer, Al_2Cl_6 is obtained. However,

when the compound is dissolved in diethylether, $(C_2H_5)_2O$, a chemical reaction occurs to give a species containing one aluminum atom. Suggest the identity of the compound.

13.24   When beryllium chloride is vaporized, a dimer of formula Be_2Cl_4 is formed. Suggest a structure for the dimer. Explain your reasoning.

13.25   A solution of beryllium ion, $Be(OH_2)_4^{2+}(aq)$, is strongly acidic. Write a balanced chemical equation for the first step in the process. Explain your reasoning why you would expect this ion to be acidic.

13.26 Zeolite-A, $Na_{12}[(AlO_2)_{12}(SiO_2)_{12}]\cdot27\ H_2O$, is a good ion exchanger, removing such ions as calcium and magnesium from water supplies. What mass of zeolite should a home water softener unit contain if it is to completely remove calcium and magnesium ions at a total concentration of 2.0×10^{-3} mol·L^{-1} from a flowthrough of 1.0×10^6 L of water before it needs recharging?

13.27 The mineral, phlogopite has the formula $KMg_x[AlSi_3O_{10}](OH)_2$. Determine the value of x.

13.28 Construct a Pourbaix diagram for aluminum, showing the species $Al(s)$, $Al^{3+}(aq)$, $Al(OH)_3(s)$, and $Al(OH)_4^{-}(aq)$. $K_{sp}(Al(OH)_3(s)) = 1 \times 10^{33}$. $Al(OH)_3(s) + OH^{-}(aq) \rightleftharpoons Al(OH)_4^{-}(aq)$; $K = 40$. In the range of possible pH and $E°$ values for natural waters, what are the only species likely? Why is this diagram relevant to the acid rain problem?

13.29 The enthalpy of formation of gallium(I) chloride is $+38$ kJ·mol^{-1}, while that of gallium(III) chloride is -525 kJ·mol^{-1}. Show why gallium(I) chloride should be thermodynamically unstable.

13.30 When damp, aluminum sulfide, Al_2S_3, produces the "rotten egg" smell of hydrogen sulfide. Write a balanced equation for the reaction and suggest an explanation for it.

13.31 Aluminum chloride dissolves in the basic solvent, CH_3CN, to give a $1:1$ (cation:anion) conducting solution. The cation has the formula $[Al(NCCH_3)_6]^{3+}$. Suggest the formula of the anion and write a balanced chemical equation for the reaction.

13.32 Boron forms a compound of formula $B_2H_2(CH_3)_4$. Draw a probable structure for this compound.

13.33 When gallium(III) salts are dissolved in water, the $[Ga(OH_2)_6]^{3+}(aq)$ ion is initially formed, but a white precipitate of $GaO(OH)$ slowly forms. Write a balanced chemical equation for the process and suggest how the gallium(III) ion can be kept in solution.

13.34 Thallium forms a selenide of formula $TlSe$. What does the oxidation state of thallium appear to be? What is a more likely structure of the compound?

13.35 Gallium dichloride, $GaCl_2$, is a diamagnetic compound that is a $1:1$ electrolyte in solution containing a simple cation and a tetrachloroanion. Suggest a possible structure for the compound.

13.36 At very low temperatures, the compound B_3F_5 can be synthesized. Spectroscopic evidence shows that the molecule contains two types of fluorine environments in a ratio of $4:1$ and two types of boron environments in a ratio of $2:1$. Suggest a structure for this molecule.

13.37 Boric acid, H_3BO_3, also written as $B(OH)_3$, acts as a weak acid in water. However, it does not do so by loss of a hydrogen ion. Instead, it acts as a Lewis acid toward the hydroxide ion. Write a balanced equation for the reaction of boric acid with water.

13.38 Calculate the standard enthalpy of formation of diboron trioxide, given that $\Delta H_{combustion}(B_2H_6(g)) = -2165$ kJ·mol^{-1}. Use the data tables in the Appendices for the other values required.

13.39 Boron forms two isoelectronic anions: BO_2^{-} and BC_2^{5-}. Construct an electron-dot structure for each ion. There is a third member of this series: BN_2^{n-}. Predict the charge on this ion.

13.40 There is a significant difference in the B—F bond length between that in boron trifluoride (130 pm) and that in the tetrafluoroborate ion (145 pm). One possible explanation was given in this chapter, and an alternative explanation in the feature "An Alternative View of Bonding in Boron Trifluoride" in Chapter 5. Contrast these two explanations and give your perspective.

13.41 Zirconium forms a boride of formula ZrB_{12} which adopts a sodium chloride lattice structure. Is it more likely that the compound is ionic— $[Zr^{4+}][B_{12}^{4-}]$—or simply based on zirconium atoms and a neutral B_{12} cluster? Metallic radius, $Zr = 159$ pm; ionic radius, $Zr^{4+} = 72$ pm; covalent radius, $B = 88$ pm. Explain your reasoning.

The Group 14 Elements

Chapter **14**

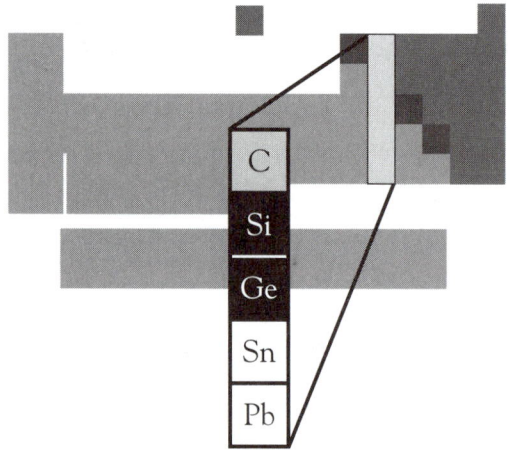

This group contains a nonmetal (carbon), two semimetals (silicon and germanium), and two weakly electropositive metals (tin and lead). Carbon has the most important chemistry of the group. It is the variety of oxyanions, many of which are found in minerals, that makes silicon chemistry interesting. The weakly metallic properties of tin and lead contrast sharply with the properties of the alkali supermetals.

It is likely that no inorganic compound has played a greater role in changing history than lead(II) ethanoate, $Pb(CH_3CO_2)_2$, also called lead(II) acetate. Lead was one of the most important elements during the Roman Empire, 2000 years ago, when about 60 000 tonnes of lead was being smelted annually to provide the elaborate piped water and plumbing system to the Romans. This level of sophistication in lifestyle was not regained until the late 19th century.

We know from the significant levels of lead in human bones from the Roman Empire that the inhabitants were exposed to high concentrations of this element. But it was not the plumbing that presented the major hazard to the Romans. Because they used natural yeasts, the wine that they prepared was quite acidic. To remedy this, winemakers added a sweetener, *sapa*, produced by boiling grape juice in lead pots. The sweet flavor was the result of the formation of "sugar of lead," what we now call lead(II) ethanoate. This sweetener was also used in food preparation, and about 20 percent of the recipes from the period required the addition of sapa. The mental instability of the Roman emperors (many of whom were excessive wine drinkers) was a major contributor to the decline and fall of the Roman Empire. In fact, the idiosyncrasies of many of the emperors match the known symptoms of lead poisoning. Unfortunately, the governing class of Romans never correlated their use of sapa with the sterility and mental disorders that plagued the rulers. Thus, the course of history was probably changed by this sweet-tasting but deadly compound.

14.1 *Group Trends*

The first three elements of Group 14 have very high melting points, a characteristic of network covalent bonding for nonmetals and semimetals, whereas the two metals in the group have low melting points and, common to all metals, long liquid ranges (Table 14.1). All the Group 14 elements form compounds in which they *catenate* (form chains of atoms with themselves). The ability to catenate decreases down the group.

Table 14.1 Melting and boiling points of the Group 14 elements

Element	Melting point (°C)	Boiling point (°C)
C	Sublimes at 4100	
Si	1420	3280
Ge	945	2850
Sn	232	2623
Pb	327	1751

Figure 14.1 Frost diagram in acid solution for the Group 14 elements.

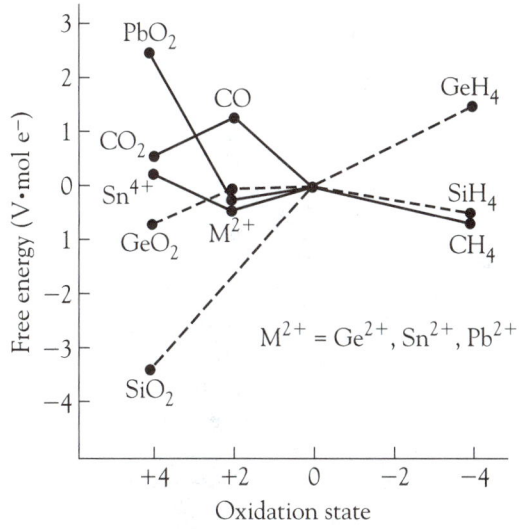

Now that we have reached the middle of the main groups, the nonmetallic properties are starting to predominate. In particular, multiple oxidation states become common. All members of Group 14 form compounds in which they have an oxidation number of +4. This oxidation state involves covalent bonding, even for the two metals of the group. In addition, an oxidation state of −4 exists for the three nonmetals/semimetals when they are bonded to more electropositive elements. Tin and lead also have an oxidation state of +2, which is the only oxidation state in which they form ionic compounds.

The Frost diagram shown in Figure 14.1 provides a summary of the relative stabilities of Group 14 oxidation states. For carbon and germanium, the +4 oxidation state is more thermodynamically stable than is the +2 state, whereas for tin and lead, the +2 state is more stable than the +4 state. For silicon, there is no common compound in which silicon exists in a +2 oxidation state; in contrast, the +2 oxidation state of lead is the most stable and in the +4 state lead is strongly oxidizing. One of the few common examples of carbon in the +2 oxidation state is the reducing compound, carbon monoxide. Among the hydrides, the −4 oxidation state becomes less stable and more strongly reducing as one goes down the group.

14.2 Carbon

Two common allotropes of carbon have been known throughout much of recorded history, but recently, a whole new family of allotropes has been identified.

Diamond

In the diamond form of carbon, there is a network of single, tetrahedrally arranged covalent bonds (Figure 14.2). Diamond is an electrical insulator but an excellent thermal conductor, being about five times better than copper. We can understand the thermal conductivity in terms of the diamond structure. Because the giant molecule is held together by a continuous network of covalent bonds, little movement of individual carbon atoms can occur. Hence, any added heat energy will be transferred as molecular motion directly across the whole diamond. Diamond is a solid to over 4000°C, because an enormous amount of energy is needed to break these strong covalent bonds.

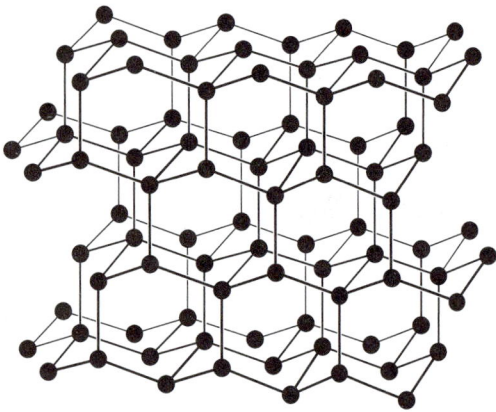

In "normal" diamond, the arrangement of the tetrahedra is the same as that in the cubic ZnS-sphalerite ionic structure (Chapter 5, Section 5.5). There is also an extremely rare form, lonsdaleite (named after the famous crystallographer Kathleen Lonsdale), in which the tetrahedra are arranged in the hexagonal ZnS-wurtzite structure (also shown in Chapter 5, Section 5.5). A crystal of lonsdaleite was first found in the Canyon Diablo meteorite in Arizona, and since then it has been synthesized by a route in which graphite is subjected to high pressure and temperature.

Figure 14.2   Structure of diamond.

Natural (sphalerite-type) diamonds are found predominantly in Africa. Zaire is the largest producer (29 percent), but South Africa (17 percent of the production) still produces the most gem-quality stones. Russia is in second place with 22 percent of world production. In North America, diamonds are found in Crater of Diamonds State Park, Arkansas, but no large-scale mining operations occur there.

The density of diamond ($3.5 \ \text{g·cm}^{-3}$) is much greater than that of graphite ($2.2 \ \text{g·cm}^{-3}$), so a simple application of the Le Châtelier principle indicates that diamond formation from graphite is favored under conditions of high pressure. Furthermore, to overcome the considerable activation energy barrier accompanying the rearrangement of covalent bonds, high temperatures also are required. The lure of enormous profits resulted in many attempts to perform this transformation. The first bulk production of diamonds was accomplished by the General Electric Company in the 1940s, using high temperatures (about 1600°C) and extremely high pressures (about 5 GPa, that is, about 50 000 times atmospheric pressure). The diamonds produced by this method are small and not of gem quality, although they are ideal for drill bits and as grinding material.

The free energy of diamond is $2.9 \ \text{kJ·mol}^{-1}$ higher than that of graphite. Thus, it is only the very slow kinetics of the process that prevents diamonds from crumbling into graphite. For this reason, Western scientists were skeptical when Soviet scientists claimed to have found a method of making layers of diamonds at low temperatures and pressures from a chemical reaction in the gas phase. It was about 10 years before the claims were investigated and shown to be true. We are now aware of the tremendous potential of diamond films as a means of providing very hard coatings—on surgical knives, for example. Diamond films are also promising coatings for computer microprocessor chips. A continuing problem associated with computer chips is their exposure to high temperatures generated by excess heat resulting from electric resistance in the computer's electric circuits. Diamond has a very high thermal conductivity; hence, chips with a diamond coating will be undamaged by the heat produced by high-density circuitry. Diamond film technology is predicted to be a major growth industry over the next decade.

Until the 19th century, it was thought that graphite and diamond were two different substances. It was Humphry Davy—by borrowing one of his spouse's diamonds and setting fire to it—who showed that carbon dioxide is the only product when diamond burns:

$$C(s) + O_2(g) \rightarrow CO_2(g)$$

Fortunately, Davy's wife was rich enough not to be too upset about the loss of one of her gems to the cause of science. This is one of the more expensive chemical methods of testing whether you really have a diamond.

Graphite

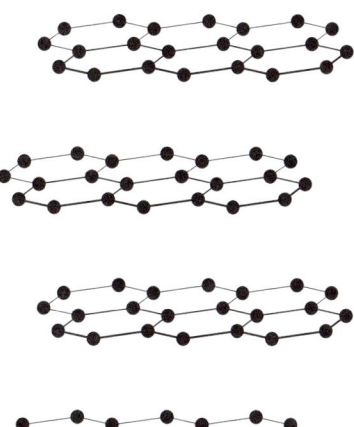

Figure 14.3   Structure of graphite.

The structure of graphite is quite different from that of diamond. Graphite consists of layers of carbon atoms (Figure 14.3). Within the layers, covalent bonds hold the carbon atoms in six-membered rings. The carbon–carbon bond length in graphite is 141 pm. These bonds are much shorter than those in diamond (154 pm) but very similar to the 140 pm bonds in benzene, C_6H_6, a compound that was mentioned in Chapter 9, Section 9.11. This similarity in bond lengths suggests a possible explanation for the short interatomic distance in graphite—there is multiple bonding between the carbon atoms within layers. Like benzene, graphite is assumed to have a delocalized π electron system throughout the plane of the carbon rings resulting from overlap of the $2p_z$ orbitals at right angles to the plane of the rings. This arrangement would result in a net $1\frac{1}{3}$ bonds between each pair of carbon atoms. The measured bond length is consistent with this assumption.

The distance between the carbon layers is very large (335 pm) and is more than twice the value of the van der Waals radius of a carbon atom. Hence, the attraction between layers is very weak. In the common hexagonal form of graphite (see Figure 14.3), alternating layers are aligned to give an *abab* arrangement. Looking at the sequential layers, one-half of the carbon atoms are located in line with carbon atoms in the planes above and below, and the other half are located above and below the centers of the rings.

The layered structure of graphite accounts for one of its most interesting properties—its ability to conduct electricity. More specifically, the conductivity in the plane of the sheets is about 5000 times greater than that at right angles to the sheets. Graphite is also an excellent lubricant by virtue of the ability of sheets of carbon atoms to slide over one another. However, this is not quite the whole story. Graphite also adsorbs gas molecules between its layers. Thus, many chemists argue that the graphite sheets are gliding on molecular "ball bearings," namely, the adsorbed gas molecules.

Even though graphite is thermodynamically more stable than diamond, it is kinetically more reactive as a result of the separation of the carbon sheets. A wide range of substances from alkali metals through the halogens to metal halide compounds are known to react with graphite. In the resulting products, the graphite structure is essentially preserved, with the intruding atoms or ions fitting between the layers in a fairly stoichiometric ratio. We encountered these graphite intercalation compounds earlier in Chapter 11, Section 11.4.

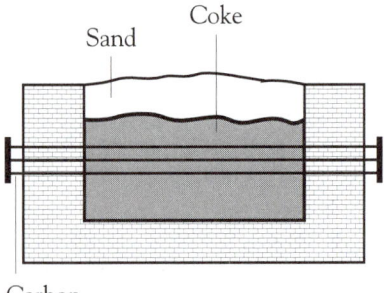

Figure 14.4   Acheson furnace.

Most of the mined graphite comes from the Far East, with China, Siberia, and the two Koreas being the major producers. In North America, Ontario, Canada, has significant deposits. Graphite is also manufactured from amorphous carbon, the most reliable method being the *Acheson process.* In this procedure, powdered coke (amorphous carbon) is heated at 2500°C for about 30 hours. This temperature is produced in an electric furnace that has carbon rods as heating elements (Figure 14.4). The method is rather similar to sublimation in that pure crystalline material is obtained from an impure powder. The amorphous carbon is covered by a layer of sand to prevent it from oxidizing to carbon dioxide. The process is not very energy efficient, but this furnace has fewer operational problems than other types. Thanks to advances in chemical technology, the newer units produce fewer pollutants and are more energy efficient than their predecessors.

Graphite is used in lubricants, as electrodes, and as graphite–clay mixtures in lead pencils. The higher the proportion of clay, the "harder" the pencil. The common mixture is designated "HB." The higher clay (harder) mixtures are designated by various "H" numbers, for example, "2H"; and the higher graphite (softer) mixtures are designated by various "B" numbers. Hence, there is no lead in a lead pencil. The term originated from the similarity between the streak left on a surface from a soft lead object and that from graphite.

Fullerenes

Chemistry is full of surprises, and the discovery of a new series of allotropes of carbon must rank as one of the most unexpected findings of all. The problem with all science is that we are limited by our own imaginations. It has been pointed out that if diamonds did not exist naturally on Earth, it would be very unlikely that any chemist would "waste time" trying to change the structure of graphite by using extremely high pressures. It would be even more unlikely for any agency to advance funding for such a "bizarre" project.

Fullerenes constitute a family of structures in which the carbon atoms are arranged in a spherical or ellipsoidal structure. To make such a structure, the carbon atoms form five- and six-membered rings, similar to the pattern of lines on a soccer ball (the early name for C_{60} was soccerane). The 60-member allotrope, C_{60}, buckminsterfullerene, is the easiest to prepare, and it is aesthetically the nicest, being spherical (Figure 14.5). The 70-member allotrope, C_{70}, is the next most commonly available fullerene. The ellipsoidal structure of this allotrope resembles an American football or a rugby ball.

Figure 14.5   Structure of buckminsterfullerene, C_{60}.

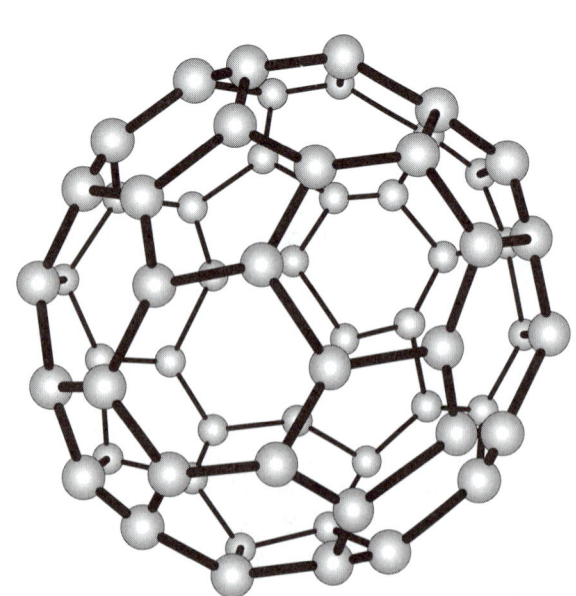

This allotrope family was named after R. Buckminster Fuller, a genius of the 20th century. His name is particularly associated with the geodesic dome, an architectural design of tremendous strength, that has the same structural arrangement as the C_{60} molecule. Contrary to general belief, however, he did not invent the dome. This was done by Walter Bauersfield in Germany, but Buckminster Fuller did make major improvements to the design and popularized it.

One method of manufacturing the fullerenes is to use an intense laser beam to heat graphite to temperatures of over 10 000°C. At these temperatures, sections of the hexagonal planes of carbon atoms peel off the surface and wrap themselves into balls. Now that we know of these molecules, they are turning up everywhere. Common soot contains fullerenes, and they have been found in naturally occurring graphite deposits. Some astrochemists argue that these molecules exist extensively in interstellar space.

Diamond and graphite are insoluble in all solvents because they have network covalent structures. The fullerenes have covalent bonds within the units, but only dispersion forces hold the units together in the solid phase. As a result, they are very soluble in nonpolar solvents such as hexane and toluene. Although black in the solid phase, fullerenes display a wide range of colors in solution: C_{60} gives an intense magenta-purple color; C_{70} is wine red; and C_{76} is bright yellow-green. All the fullerenes sublime when heated, a property providing further evidence of the weak intermolecular forces.

The C_{60} molecules pack together in the same way metal atoms do, forming a face-centered cubic arrangement. The fullerenes have low densities (about 1.5 g·cm^{-3}), and they are nonconductors of electricity. Molecules of C_{60} (and those of other fullerenes) absorb visible light to produce an unstable excited form, which is represented by the symbol $*C_{60}$. The excited form absorbs light many times more efficiently than normal C_{60}, converting the electromagnetic energy to heat. This is a very important property, because it means that as the intensity of light passed into a solution of C_{60} increases, more $*C_{60}$ will be produced and hence more of the light will be absorbed. The intensity of light leaving the solution will be correspondingly reduced. The solution, therefore, acts as an optical limiter. Coating glasses with this material could prevent eye damage in people working with high-intensity lasers; and in the more common world, such coatings could be used to create instant-response sunglasses.

The chemistry of these novel molecules is still a field of intense research. The fullerenes are easily reduced to anions by reaction with Group 1 and Group 2 metals. For example, rubidium fits within the interstices in the C_{60} lattice to give Rb_3C_{60}. This compound is a superconductor at temperatures below 28 K because its structure is actually $[Rb^+]_3[C_{60}^{3-}]$. The extra electrons associated with the fullerenes are free to move throughout the crystal, just like those in a metal. As the cavities in the fullerenes are quite large, it is possible to fit a metal ion within the structure. An example of this is $La@C_{82}$, where the @ symbol is used to indicate that the $3+$ metal ion is within the fullerene. Chemical reaction with the surface of the fullerenes is also possible; thus, reaction with fluorine results in the formation of colorless $C_{60}F_{60}$.

Of the fullerene series, C_{60} is the easiest to produce, followed by C_{70}. Even-numbered fullerenes from C_{70} to well over C_{100} are known. Only recently has any stable fullerene with fewer than 60 carbons been established. This is C_{36}, a black solid that gives a golden-yellow solution. Now that the "less than 60" barrier has been overcome, it is expected that other small fullerenes may soon be synthesized—particularly C_{50}. C_{36} is expected to be the smallest fullerene feasible on the basis of the strained bonds needed to close such a small sphere. The degree of bond strain is apparent from the chemistry of C_{36}, for it is the most reactive of the fullerenes, decomposing even in air.

Carbon Nanotubes

Though fullerenes have gained the greatest recognition, it is nanotubes that chemists believe hold the greater promise as new materials. Nanotubes were first discovered in 1991 by the Japanese scientist, Sumio Iijima. Nanotubes are essentially tiny strips of graphite sheet, rolled into tubes and capped with half a fullerene at each end. They can be made by heating graphite in an inert atmosphere under patented conditions to about 1200°C.

The Discovery of Buckminsterfullerene

Discoveries in science are almost always convoluted affairs, not the sudden "Eureka!" of fiction. W.E. Addison had predicted in 1964 that other allotropes of carbon might exist; and David Jones in 1966 actually proposed the existence of "hollow graphite spheroids."

It was not chemists, however, but two astrophysicists, Donald Huffman of the University of Arizona at Tucson and Wolfgang Krätschmer of the Max Planck Institute for Nuclear Physics at Heidelberg, Germany, who are credited with the first synthesis of fullerenes in 1982. They were interested in the forms of carbon that could exist in interstellar space. They heated graphite rods in a low-pressure atmosphere and obtained a soot. It appeared to have an unusual spectrum, but they attributed that to contamination by oil vapor from the equipment. As a result, they lost interest in the experiment. Two years later, an Australian medical researcher, Bill Burch, at the Australian National University, produced a sublimable form of carbon that he patented as "Technogas." This, too, was probably buckminsterfullerene.

Harold Kroto of the University of Sussex, England, and Richard Smalley of Rice University, Texas, performed the crucial experiment. They were also interested in the nature of carbon in space. When Kroto visited Smalley, the former proposed that they use Smalley's high-powered laser to blast fragments off a graphite surface and then identify the products. Between 4 September and 6 September 1985, they found one batch of products that had a very high proportion of a molecule containing 60 carbon atoms. Over the weekend, two research students, Jim Heath and Sean O'Brien, altered the conditions of the experiment until, time after time, they could consistently obtain this unexpected product.

How could the formula C_{60} be explained? Kroto recalled the geodesic dome that housed the U.S. pavilion at Expo 67 in Montreal. However, he thought the structure consisted of hexagonal shapes, like those making up graphite. The chemists were unaware of the work by the 18th-century mathematician Leonhard Euler, who had shown that it was impossible to construct closed figures out of hexagons alone. Smalley and Kroto now disagree vehemently over which one of them first realized that a spherical structure could be constructed using 20 hexagons and 12 pentagons. Nevertheless, on 10 September, this was the structure that the group postulated for the mysterious molecule. As a result of the disagreement, relations between the two research groups are acrimonious.

The Kroto–Smalley method produced quantities of buckminsterfullerene that were too small for chemical studies. The discovery cycle was completed in 1988 when Huffman realized that the method that he and Krätschmer had used several years earlier must have been forming large quantities of these molecules. These two physicists resumed their production of the soot and developed methods for producing consistently high yields of the allotropes. From subsequent studies, the chemical evidence proving the structures of C_{60} and C_{70} was independently and almost simultaneously produced by the Kroto and the Smalley groups.

Because the carbon atoms in nanotubes are held together by covalent bonds, the tubes are immensely strong—about 100 times that of an equivalent strand of steel. Thus, there are projected uses as a super-strong material. Providing the carbon hexagons are aligned precisely with the long axis of the nanotube, the material is an excellent electrical conductor. This behavior opens the possibility of bundles of nanotubes being the electrical equivalent of optical fibers. However, if there is a "twist" in the hexagons, giving a spiral arrangement, the material behaves as a semiconductor. Open-ended nanotubes also have potential for the reversible storage of hydrogen gas, suggesting they may play a role in the future hydrogen-based economy.

There are two classes of nanotubes: the single-walled nanotubes (SWNTs) and the multiwalled nanotubes (MWNTs). The SWNTs consist of simple carbon nanotubes, whereas the MWNTs consist of concentric layers of nanotubes like a coaxial cable. It is the SWNTs that have the greater promise, but at the present time they are very expensive to synthesize.

Impure Carbon

The major uses of carbon are as an energy source and as a reducing agent. For reducing purposes, an impure form of carbon (coke) is used. This material is produced by heating coal in the absence of air. In this process, the complex coal structure breaks down, boiling off hydrocarbons and leaving behind a porous, low-density, silvery, almost metallic-looking solid. Essentially, coke is composed of microcrystals of graphite that have small proportions of some other elements, particularly hydrogen, bonded in their structure. Much of the distillate produced by the coking process can be used as raw materials in the chemical industry, but the oily and watery wastes are a nightmare of carcinogens. Coke is utilized in the production of iron from iron ore and in other pyrometallurgical processes. Coke production is considerable, and about 5×10^8 tonnes is used worldwide every year.

Carbon black is a very finely powdered form of carbon. This impure micrographite is produced by incomplete combustion of organic materials. It is used in extremely large quantities—about 1×10^7 tonnes per day. It is mixed with rubber to strengthen tires and reduce wear. About 3 kg is used for the average automobile tire, and it is the carbon content that gives a tire its black color.

Another form of carbon known as activated carbon has a very high surface area—typically 10^3 $m^2 \cdot g^{-1}$. This material is used for the industrial decolorizing of sugar and in gas filters, as well as for removing impurities from organic reactions in the university laboratory. The physical chemistry of the absorption process is complex, but in part it works by the attraction of polar molecules to the carbon surface.

Blocks of carbon are industrially important as electrodes in electrochemical and thermochemical processes. For example, about 7.5 million tonnes of carbon is used each year just in aluminum smelters. And, of course, the summer season always increases the consumption of carbon in home barbecues.

14.3 Isotopes of Carbon

Natural carbon contains three isotopes: carbon-12 (98.89 percent), the most prevalent isotope; a small proportion of carbon-13 (1.11 percent); and a trace of carbon-14. Carbon-14 is a radioactive isotope with a half-life of 5.7×10^3 years. With such a short half-life, we would expect little sign of this isotope on Earth. Yet it is prevalent in all living tissue, because the isotope is constantly being produced by reactions between cosmic ray neutrons and nitrogen atoms in the upper atmosphere:

$$^{14}_{7}N + ^{1}_{0}n \rightarrow ^{14}_{6}C + ^{1}_{1}H$$

The carbon atoms react with oxygen gas to form radioactive molecules of carbon dioxide. These are absorbed by plants in photosynthesis. Creatures that eat plants and creatures that eat the creatures that eat plants will all contain the same proportion of radioactive carbon. After the death of the organism, there is no further intake of carbon, and the carbon-14 already present in the body decays. Thus, the age of an object can be determined by measuring the amount of carbon-14 present in a sample. This method provides an absolute scale of dating objects that are between 1000 and 20 000 years old. W.F. Libby was awarded the Nobel Prize for chemistry in 1960 for developing the radiocarbon dating technique.

14.4 The Extensive Chemistry of Carbon

Carbon has two properties that enable it to form such an extensive range of compounds: *catenation* (the ability to form chains of atoms) and multiple bonding (that is, the ability to form double and triple bonds). Extensive use of multiple bonding is found in compounds of carbon, nitrogen, and oxygen. Carbon shows the greatest propensity for catenation of all elements. For catenation, three conditions are necessary:

1. A bonding capacity (valence) greater than or equal to 2.

2. An ability of the element to bond with itself; the self-bond must be about as strong as its bonds with other elements.

3. A kinetic inertness of the catenated compound toward other molecules and ions.

We can see why catenation is frequently found in carbon compounds but only rarely in silicon compounds by comparing bond energy data for these two elements (Table 14.2). Notice that the energies of the carbon–carbon and carbon–oxygen bonds are very similar. However, the silicon–oxygen bond is much stronger than that between two silicon atoms. Thus, in the presence of oxygen, silicon will form —Si—O—Si—O— chains rather than —Si—Si— linkages. We will see later that the silicon–oxygen chains dominate the chemistry of silicon. There is much less of an energy "incentive" to break carbon–carbon bonds in favor of the formation of carbon–oxygen bonds.

It is sobering to realize that two "quirks" of the chemical world make life possible: the hydrogen bond and the catenation of carbon. Without these two phenomena, life of any form (that we can imagine) could not exist.

Table 14.2 Bond energies of various carbon and silicon bonds

Carbon bonds	Bond energy (kJ·mol^{-1})	Silicon bonds	Bond energy (kJ·mol^{-1})
C—C	346	Si—Si	222
C—O	358	Si—O	452

14.5 Carbides

Binary compounds of carbon with less electronegative elements (except hydrogen) are called carbides. Carbides are hard solids with high melting points. Despite this commonality of properties, there are, in fact, three types of bonding in carbides: ionic, covalent, and metallic.

Ionic Carbides

Ionic carbides are formed by the most electropositive elements, the alkali and alkaline earth metals and aluminum. Most of these ionic compounds contain the dicarbide(2−) ion that we discussed in the context of calcium dicarbide (Chapter 12, Section 12.13). The ionic carbides are the only carbides to show much chemical reactivity. In particular, they react with water to produce ethyne, C_2H_2, formerly called acetylene:

$$Na_2C_2(s) + 2\ H_2O(l) \rightarrow 2\ NaOH(aq) + C_2H_2(g)$$

From their formulas, the red beryllium carbide, Be_2C, and the yellow aluminum carbide, Al_4C_3, appear to contain the C^{4-} ion. The high charge density cations Be^{2+} and Al^{3+} are the only ions that can form stable lattices with such a highly charged anion. However, the cations are so small and highly charged and the anion so large that we assume there must be a large degree of covalency in the bonding. Nevertheless, these two carbides do produce methane, CH_4, when they react with water, as would be expected if the C^{4-} ion were present:

$$Al_4C_3(s) + 12\ H_2O(l) \rightarrow 4\ Al(OH)_3(s) + 3\ CH_4(g)$$

Covalent Carbides

Because most nonmetals are more electronegative than carbon, there are few covalent carbides. Silicon carbide, SiC, and boron carbide, B_4C (discussed in Chapter 13, Section 13.3), are the common examples; both are very hard and have high melting points. Silicon carbide is used as a grinding and polishing agent in metallurgical applications, and it is the only nonoxide ceramic product of large-scale industrial importance. Worldwide production of this compound is about 7×10^5 tonnes. Silicon carbide, which is bright green when pure, is produced in an Acheson furnace, like that used to convert coke to graphite. The furnace takes about 18 hours to heat electrically to the reaction temperature of about 2300°C, and it takes another 18 hours for the optimum yield to be formed. The production of this compound is extremely energy intensive, and between 6 and 12 kWh of electricity are needed to produce 1 kg of silicon carbide:

$$SiO_2(s) + 3\ C(s) \xrightarrow{\Delta} SiC(s) + 2\ CO(g)$$

There is tremendous interest in silicon carbide as a material for high-temperature turbine blades, which operate at temperatures that cause metals to lose their strength. Silicon carbide is also being used for the backing of high-precision mirrors because it has a very low coefficient of expansion. This property minimizes distortion problems because silicon carbide mirrors undergo only a negligible change in shape as temperatures fluctuate. A key to the wider use of silicon carbide is the ability to form it into shapes (for such uses as long-lasting engine blocks or replacement human joints). With its very high melting point, that has been an impossibility until now. A considerable amount of current research is focusing on the synthesis of liquid organosilicon compounds. These can be poured into molds and the compound heated to a high enough temperature that decomposition occurs, leaving silicon carbide in the required shape. Silicon carbide is certainly a promising material for the 21st century.

Moissanite: The Diamond Substitute

Until 1998, the only lower cost substitute for diamond was cubic zirconia, ZrO_2 (see Chapter 20, Section 20.2). Now, the development of a commercial synthesis of a hexagonal form of silicon carbide has lead to the introduction of a new gemstone, moissanite. Moissanite had been discovered in the 1890s in the Diablo Canyon, Arizona, meteorite impact crater and it was named after the chemist Henri Moissan. But until recently, it could not be synthesized in large crystals. The synthesis resulted from research into new semiconductor materials for LED and computer use. Though most of the output of pure silicon carbide is still for the high-tech industry, an increasing proportion is for the gemstone market.

Moissanite is an analog of lonsdaleite (discussed in Section 14.2) in which alternate carbon atoms are replaced by silicon atoms. With its close similarity to diamond in composition and structure, it is not surprising that its properties, too, resemble those of diamond (Table 14.3).

Moissanite is almost as hard as diamond, is about the same density and, with a higher refractive index than diamond, actually "sparkles" more than a diamond. Thus, in a comparison between a diamond and a moissonite, most people believe the moissanite to be the real diamond. The standard technique to identify a diamond is to measure its thermal conductivity. Though a nonmetal, it has an extremely high thermal conductivity, similar to that of many metals. The conductivity results from the strong covalent bonds throughout the lattice structure; hence, any molecular vibrations (heat) will be transmitted rapidly through the structure. As moissanite has a similar structure, it, too, has high thermal conductivity. So the traditional method of diamond identification will not eliminate a moissanite. However, examination of a moissonite crystal under a microscope reveals characteristic double refraction for moissonite together with differences in surface polish and inclusions from diamonds.

Table 14.3   A comparison of the properties of diamond, moissanite, and cubic zirconia

	Hardness (Moh's scale)	Refractive index	Density $(g \cdot cm^{-3})$
C, diamond	10	2.24	3.5
SiC, moissonite	9.25–9.5	2.65–2.69	3.2
ZrO_2, cubic zirconia	8.5	2.15	5.8

Metallic Carbides

Metallic carbides are compounds in which the carbon atoms fit within the crystal structure of the metal itself, and they are usually formed by the transition metals. To form a metallic carbide, the metal must assume a close-packed structure, and the atoms usually have a metallic radius greater than 130 pm. The carbon atoms can then fit into the octahedral holes (interstices) in the structure; hence, metallic carbides are also called *interstitial carbides*. If all the octahedral holes are filled, the stoichiometry of these compounds is 1:1.

Because metallic carbides retain the metallic crystal structure, they look metallic and conduct electricity. They are important because they have very high melting points, show considerable resistance to chemical attack, and are extremely hard. The most important of these compounds is tungsten carbide (WC) of which about 20 000 tonnes is produced annually worldwide. Most of the material is used in cutting tools.

Some metals with a radius below 130 pm form metallic carbides, but their metal lattices are distorted. As a result, such compounds are more reactive than true interstitial carbides. The most important of these almost-interstitial carbides is Fe_3C, commonly called cementite. It is microcrystals of cementite that cause carbon steel to be harder than pure iron.

14.6   *Carbon Monoxide*

Carbon monoxide is a colorless, odorless gas. It is very poisonous because it has a 300-fold greater affinity for blood hemoglobin than does oxygen; thus, quite low concentrations of carbon monoxide in air are sufficient to prevent oxygen absorption in the lungs. Without a continuous supply of oxygen, the brain loses consciousness, and death follows unless the supply of oxygenated hemoglobin is restored. Curiously, there is now evidence that carbon monoxide is a messenger molecule in some neurons in the brain. Thus, what is toxic in large quantities is necessary in tiny amounts for the correct functioning of the brain.

The triple bond between carbon and oxygen in carbon monoxide is the strongest bond known with a bond energy of 1070 kJ·mol^{-1}.

The carbon–oxygen bond in carbon monoxide is very short, about the length that would be expected for a triple bond. Figure 14.6 shows a simplified energy level diagram for carbon monoxide for the molecular orbitals derived from $2p$ atomic orbitals. This model gives a bond order of 3 from the filling of one σ-bonding orbital and two π-bonding orbitals.

Carbon monoxide is produced when any carbon-containing compound, including carbon itself, is burned with an amount of oxygen insufficient for complete combustion:

$$2\,C(s) + O_2(g) \rightarrow 2\,CO(g)$$

As automobile engines have become more efficient, carbon monoxide production has diminished substantially. Thus, in that respect the air inhaled on city streets is less harmful than it used to be because its carbon monoxide content is lower.

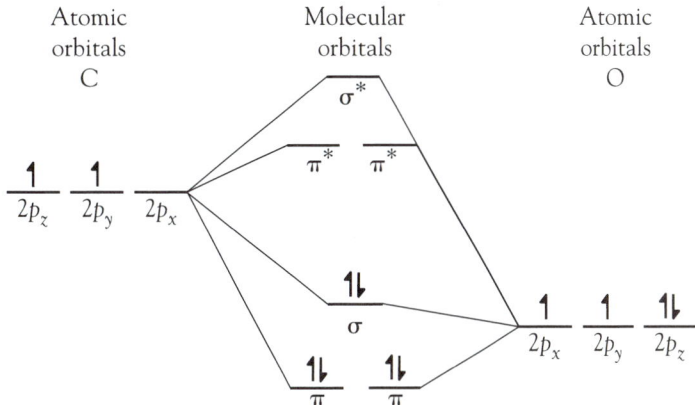

Figure 14.6   Partial simplified molecular orbital energy level diagram for carbon monoxide.

The pure gas is prepared in the laboratory by warming methanoic (formic) acid with concentrated sulfuric acid. In this decomposition, the sulfuric acid acts as a dehydrating agent:

$$HCOOH(l) + H_2SO_4(l) \rightarrow H_2O(l) + CO(g) + H_2SO_4(aq)$$

Carbon monoxide is quite reactive; for example, it burns with a blue flame to carbon dioxide:

$$2\ CO(g) + O_2(g) \rightarrow 2\ CO_2(g)$$

It reacts with chlorine gas in the presence of light or hot charcoal, which serve as catalysts, to give carbonyl chloride, $COCl_2$, a compound better known as the poison gas phosgene:

$$CO(g) + Cl_2(g) \rightarrow COCl_2(g)$$

Although carbonyl chloride is usually recalled as one of the first gases used in warfare, it is actually an industrial chemical produced on the scale of millions of tonnes per year. Carbonyl chloride is especially useful as a starting material for the synthesis of many important compounds such as the polycarbonates that are used widely as tough low-density transparent materials.

When carbon monoxide is passed over heated sulfur, the compound carbonyl sulfide, COS, a promising low-hazard fungicide, is formed:

$$CO(g) + S(s) \rightarrow COS(g)$$

As the Frost diagram in Figure 14.1 shows, carbon monoxide is a strong reducing agent. It is used industrially in this role, for example, in the smelting of iron(III) oxide to iron metal (see Chapter 20, Section 20.6):

$$Fe_2O_{3}(s) + 3\ CO(g) \xrightarrow{\Delta} 2\ Fe(l) + 3\ CO_2(g)$$

It is also an important starting material in industrial organic chemistry. Under high temperatures and pressures, carbon monoxide will combine with hydrogen gas (a mixture known as synthesis gas) to give methanol, CH_3OH. Mixing carbon monoxide with ethene, C_2H_4, and hydrogen gas produces propanal, CH_3CH_2CHO, a reaction known as the *OXO process*:

$$CO(g) + 2\ H_2(g) \xrightarrow{\Delta} CH_3OH(g)$$

$$CO(g) + C_2H_4(g) + H_2(g) \xrightarrow{\Delta} C_2H_5CHO(g)$$

The active catalytic species in this process is a cobalt compound containing covalent bonds to both hydrogen and carbon monoxide, $HCo(CO)_4$; and it is similar compounds of metals with carbon monoxide that we consider next.

Carbon monoxide forms numerous compounds with transition metals. In these highly toxic, volatile compounds, the metal is considered to have an oxidation number of zero. Among the simple carbonyls are tetracarbonylnickel(0), $Ni(CO)_4$, pentacarbonyliron(0), $Fe(CO)_5$, and hexacarbonylchromium(0), $Cr(CO)_6$. Many of the metal carbonyls can be prepared simply by heating the metal with carbon monoxide under pressure. For example, when heated, nickel reacts with carbon monoxide to give the colorless gas tetracarbonylnickel(0):

$$Ni(s) + 4\ CO(g) \xrightarrow{\Delta} Ni(CO)_4(g)$$

These compounds are often used as reagents for the preparation of other low oxidation number compounds of transition metals. The chemistry of carbonyl compounds is discussed more fully in Chapter 22, Section 22.6.

14.7   *Carbon Dioxide*

At pressures above 40 MPa and 1500°C, carbon dioxide can be converted into a polymeric solid with a structure similar to silicon dioxide. This form is believed to be very hard with a high thermal conductivity. The quartz-like form remains stable at room temperature and a pressure of 1 GPa.

Carbon dioxide is a dense, colorless, odorless gas, which does not burn or, normally, support combustion. The combination of high density and inertness has led to its use for extinguishing fires. Because it is about one and a half times denser than air under the same conditions of temperature and pressure, it flows, almost like a liquid, until air currents mix it with the gases of the atmosphere. Thus, it is effective at fighting floor-level fires but almost useless for fighting fires in ceilings. However, carbon dioxide will react with burning metals, such as calcium:

$$2\ Ca_{(s)} + CO_{2(g)} \xrightarrow{\Delta} 2\ CaO_{(s)} + C_{(s)}$$

The inertness of carbon dioxide is leading to an important use in farming. Millions of tonnes of grain are stored until sold. Under such conditions, it is easy for the grain to become infested with insects such as the rusty grain beetle. To prevent infestation, fumigants such as ethylene dibromide (dibromoethane), $C_2H_4Br_2$, methyl bromide (bromomethane), CH_3Br, and phosphine, PH_3, have been used. Ethylene dibromide has been banned as it is carcinogenic to mammals and thus presents a health risk. Methyl bromide is being phased out as it is an ozone depleter. Insects are developing a resistance to phosphine (which is a highly hazardous compound). Carbon dioxide seems to be a perfect replacement. Insects cannot live in a carbon dioxide atmosphere, yet the compound is innocuous once diluted in air.

Carbon dioxide is unusual because it has no liquid phase at normal atmospheric pressures. Instead, the solid sublimes directly to the gas phase. To obtain the liquid phase at room temperature, a pressure of 6.7 MPa (67 times standard atmospheric pressure) must be applied, as shown in the phase diagram in Figure 14.7. Carbon dioxide is usually conveyed in tank cars and cylinders in the liquid form. When the pressure is released, some of the liquid carbon dioxide vaporizes, but the heat absorbed in the expansion process (overcoming the dispersion forces between molecules) is enough to cool the remaining liquid below its sublimation point, −78°C at atmospheric pressure. By inverting the cylinder and opening the valve, solid carbon dioxide, "dry ice," can be collected in a gauze bag or in a CO_2 "patty maker" at room temperature.

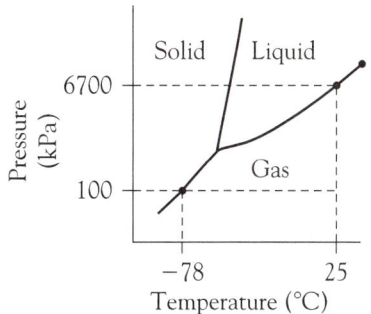

Figure 14.7 Phase diagram for carbon dioxide.

Carbon dioxide is an important industrial chemical. Each year, over 40 million tonnes is used in the United States alone. Half of this quantity is needed as a refrigerant, and another 25 percent is used to carbonate soft drinks. It is also used as a propellant in some aerosol cans, as a pressurized gas to inflate life rafts and life vests, and as a fire-extinguishing material.

There are a number of sources of industrial carbon dioxide: as a byproduct in the manufacture of ammonia, molten metals, cement, and sugar fermentation products. And, of course, we exhaust carbon dioxide into the atmosphere during the complete combustion of any carbon-containing substance: wood, natural gas, gasoline, and oil.

In the laboratory, carbon dioxide is most conveniently prepared by adding dilute hydrochloric acid to marble chips (chunks of impure calcium carbonate), although any dilute acid with a carbonate or hydrogen carbonate can be used (for example, Alka-Selzer® tablets or baking powder):

$$2\ HCl_{(aq)} + CaCO_{3(s)} \rightarrow CaCl_{2(aq)} + H_2O_{(l)} + CO_{2(g)}$$

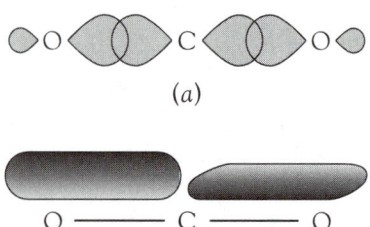

(a)

(b)

Figure 14.8 (a) The σ bonds between the atoms of a carbon dioxide molecule. (b) The two π bonds between the same atoms.

To identify carbon dioxide, the *limewater test* is used. In this test, a gas is bubbled into a saturated solution of calcium hydroxide. When the gas is carbon dioxide, a white precipitate of calcium carbonate forms. Addition of more carbon dioxide results in the disappearance of the precipitate as soluble calcium hydrogen carbonate forms:

$$CO_2(g) + Ca(OH)_2(aq) \rightarrow CaCO_3(s) + H_2O(l)$$

$$CO_2(g) + CaCO_3(s) + H_2O(l) \rightarrow Ca^{2+}(aq) + 2\,HCO_3^{-}(aq)$$

Bond lengths and bond strengths indicate that there are double bonds between the carbon and oxygen atoms in the carbon dioxide molecule. We would predict this bonding pattern both from an electron-dot representation and from simple hybridization theory. On the basis of hybridization theory, we assume that σ bonds are formed from *sp* hybrid orbitals. The remaining *p* orbitals, which are at right angles to the bond direction, then overlap to form two π molecular orbitals (see Figure 14.8).

Carbon dioxide's multiple bonding can also be discussed in terms of molecular orbital theory. Figure 14.9 shows a simplified molecular orbital diagram for carbon dioxide. Because of the differences in the atomic orbital energies of the two elements, it is the $2p$ orbitals of the oxygen atoms that interact with the $2s$ and $2p$ orbitals of the carbon atom. When this happens, two σ, two π, two $π_{NB}$, two π*, and two σ* orbitals are formed. The 12 available electrons fill all the bonding and nonbonding orbitals. Hence, for the molecule as a whole, the net bonding is two σ and two π, which amounts to one σ and one π bond per pair of atoms—the same result we obtained with the hybridization approach.

In aqueous solution, almost all the carbon dioxide is present as $CO_2(aq)$; only 0.37 percent is present as carbonic acid, $H_2CO_3(aq)$:

$$CO_2(aq) + H_2O(l) \rightleftharpoons H_2CO_3(aq)$$

It is fortunate for us that the equilibrium lies to the left, not to the right, and that carbonic acid is a weak acid, because it means that carbonation

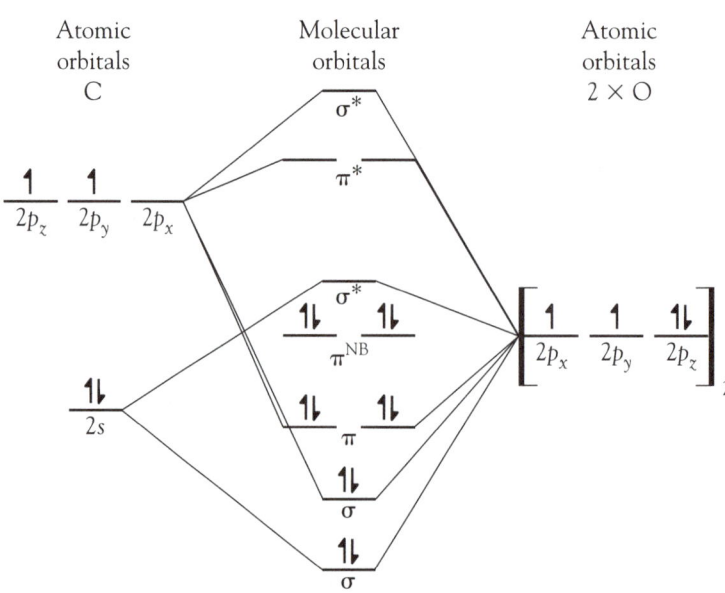

Figure 14.9 A simplified molecular orbital energy level diagram for carbon dioxide.

Carbon Dioxide, Supercritical Fluid

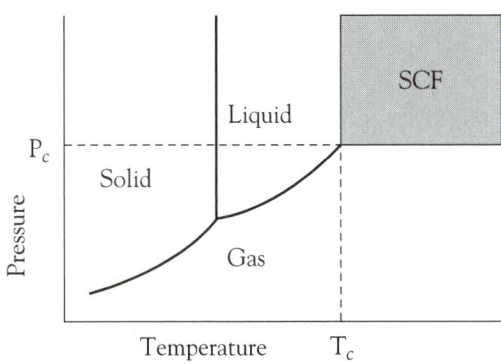

Figure 14.10 A generic phase diagram illustrating the location of the supercritical region (SCF).

In a phase diagram, all the lines are continuous except one, the liquid–gas curve. This line ends abruptly at a particular point known as the critical point. Above the critical pressure and temperature, the properties of the substance are no longer those of a gas or a liquid, but of a unique phase known as a supercritical fluid (Figure 14.10). For carbon dioxide, the critical pressure is about 7.4 MPa (almost 73 times atmospheric pressure) and 30°C.

When a substance reaches the supercritical state, the physical properties of the fluid become intermediate between those of a gas and a liquid. The fluid's solvating powers resemble those of a liquid, while its diffusivity and viscosity resemble those of a gas. Further, the solvation properties of the fluid, particularly its ability to mimic polar or nonpolar solvents, can be altered by changing the temperature and pressure.

A boost to research into supercritical fluids came in 1976. At the time, decaffeinated coffee was produced using dichloromethane, CH_2Cl_2, to extract the caffeine. However, traces of toxic dichloromethane were found to remain in the coffee. Researchers at the Max Planck Institute in Germany discovered that supercritical carbon dioxide was an excellent solvent for caffeine. In addition, the high diffusivity and low viscosity of the supercritic fluid enabled the solvent to rapidly penetrate deep into the coffee beans, extracting close to 100 percent of the caffeine. The majority of decaffeinated coffee is now produced this way, with one plant in Texas alone processing about 25 000 tonnes of coffee beans per year.

The use of supercritical carbon dioxide as a solvent has now become widespread. It is used for the extraction of specific components from tobacco (nicotine), hops, red peppers, and spices, among many others. The technique is also being used for the treatment of waste waters, solid wastes, and refinery wastes. Another application is in natural products chemistry: the extraction of pharmaceuticals from botanicals. Supercritical fluid technology has thus become an integral part of industrial chemistry.

The array of new solvent possibilities, both supercritical fluids and ionic liquids (Chapter 6, Section 6.1) are together called neoteric solvents—*solvents that "break new ground."*

of beverages will not cause them to become unpleasantly acidic. The high solubility of carbon dioxide in water has been explained in terms of carbon dioxide molecules being trapped inside clusters of hydrogen-bonded water molecules—rather like the clathrates (Chapter 10, Section 10.7) in the solid phase.

Carbonic acid has recently been isolated at low temperature and in the absence of water. In the presence of water, dissociation to carbon dioxide and water occurs rapidly. Carbonic acid is an extremely weak diprotic acid, as can be seen from the pK_a values corresponding to each of the ionization steps:

$$H_2CO_3(aq) + H_2O(l) \rightleftharpoons H_3O^+(aq) + HCO_3^-(aq) \qquad pK_{a1} = 6.37$$

$$HCO_3^-(aq) + H_2O(l) \rightleftharpoons H_3O^+(aq) + CO_3^{2-}(aq) \qquad pK_{a2} = 10.33$$

Because it is an acid oxide, carbon dioxide reacts with bases to give carbonates. The presence of excess carbon dioxide results in the formation of the hydrogen carbonates of the alkali and alkaline earth elements:

$$2\, KOH(aq) + CO_2(g) \rightarrow K_2CO_3(aq) + H_2O(l)$$

$$K_2CO_3(aq) + CO_2(g) + H_2O(l) \rightarrow 2\, KHCO_3(aq)$$

14.8   The Greenhouse Effect

Most of us are aware of the so-called greenhouse effect; yet only a small proportion of the population really understand the issues. A tremendous quantity of research is currently being done by chemists, biologists, physicists, geologists, climatologists, ecologists, and geographers to gain information from which we can make more informed judgments.

The greenhouse effect—more correctly called *radiation trapping*—can be described as follows: The energy from the Sun reaches the Earth's surface as electromagnetic radiation, particularly in the visible, ultraviolet, and infrared regions. This energy is absorbed by the Earth's surface and atmosphere. It is re-emitted mainly as infrared radiation ("heat" rays). If all the incoming energy were lost back into space as infrared radiation, the temperature of the Earth's surface would be between $-20°C$ and $-40°C$. Fortunately, there is a reflecting layer in the atmosphere. Many small molecules in the atmosphere, particularly water and carbon dioxide, absorb certain wavelengths of the escaping radiation, the absorbed energy corresponding to the energy of their molecular vibrations. These small molecules act like the glass in a greenhouse. It is the re-radiation of their absorbed energy back to Earth that warms the oceans, land, and air. As a result, the average temperature of the Earth's surface is about $+14°C$. In other words, it is the greenhouse effect that makes this planet habitable.

The wavelengths of the energy absorbed by the molecules in the "greenhouse" layer correspond to certain vibrations of atoms within molecules. All polyatomic molecules, except homonuclear diatomic molecules, absorb energy in the infrared region. Diatomic molecules only absorb infrared radiation if they contain two different atoms (such as carbon monoxide), and no monatomic gases absorb infrared radiation. Thus, dinitrogen, dioxygen, and argon, the three major constituents of the atmosphere, are essentially transparent to infrared radiation. The vibrations of the water molecule and the carbon dioxide molecule that are responsible for the absorption of infrared radiation are shown in Figure 14.11, together with the corresponding wavelength of the energy absorbed. These molecules will also absorb energy with wavelengths that are multiples of the frequencies of these vibrations, but to a lesser extent.

Figure 14.12 shows the infrared spectrum of our atmosphere. The absorptions at wavelengths corresponding to the frequencies of the vibrations of the water and carbon dioxide molecules (see Figure 14.16) are the dominant features of the longer wavelengths, whereas at shorter wavelengths, most of the absorption occurs at wavelengths corresponding to the multiples (harmonics) of the frequencies of these vibrations.

Water vapor is the predominant greenhouse gas, and its atmospheric concentration of about 1 percent has remained fairly constant over geological time

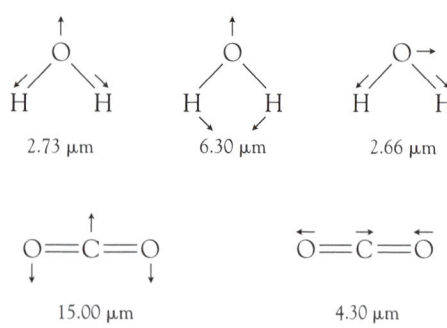

Figure 14.11 The vibrations of water and carbon dioxide and the wavelengths corresponding to the frequencies of these vibrations.

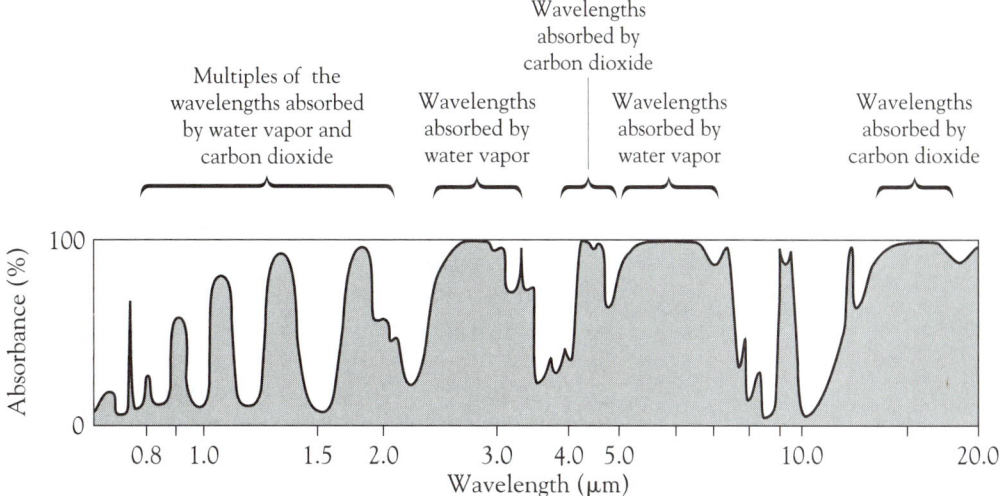

Figure 14.12   Infrared spectrum showing absorption of infrared radiation by various components of the atmosphere.

as a result of the large bodies of water on this planet. However, as far as we can tell from the geological record, the levels of carbon dioxide have fluctuated dramatically. In the warm Carboniferous era, from 350 to 270 million years ago, the carbon dioxide level seems to have been over six times higher than it is now. It is fortunate that the carbon dioxide levels have dropped over geological time, because the Sun has been getting hotter over the same period (known as the faint young Sun phenomenon). Hence, the decreasing greenhouse effect has approximately balanced the increasing solar radiation.

There are three major sources of carbon dioxide: natural outpourings of volcanoes, burning of fossil fuels, and burning of vegetation such as the rain forests. Over geological time, there has been an approximate balance between the emissions of carbon dioxide from volcanoes and the removal of carbon dioxide from the atmosphere by photosynthesis and silicate rock weathering. Now, however, by burning fossil fuels, we are injecting additional carbon dioxide into the atmosphere much faster than it can be removed by natural processes. It is this rapid increase in atmospheric carbon dioxide levels that is causing concern among climatologists (Figure 14.13).

As it is a fact that carbon dioxide is an infrared absorber, it follows logically that more carbon dioxide in the atmosphere will cause an atmospheric warming. It is the degree of warming that is contentious. The amount of carbon dioxide already in the atmosphere is sufficient to absorb most of the outgoing energy of the appropriate wavelengths. Addition of more carbon dioxide will broaden the absorption region. Thus, though doubling the carbon dioxide level will increase the heat retained, it is not a linear relationship. We also have to take into account contributions to global climate change from increases in atmospheric concentration of the other greenhouse gases.

Figure 14.13   The variation of tropospheric carbon dioxide concentrations over the past 50 years. Continuous data recorded at Mauna Loa, Hawaii (D. Keeling and T. Worf, Scripps Institute of Oceanography).

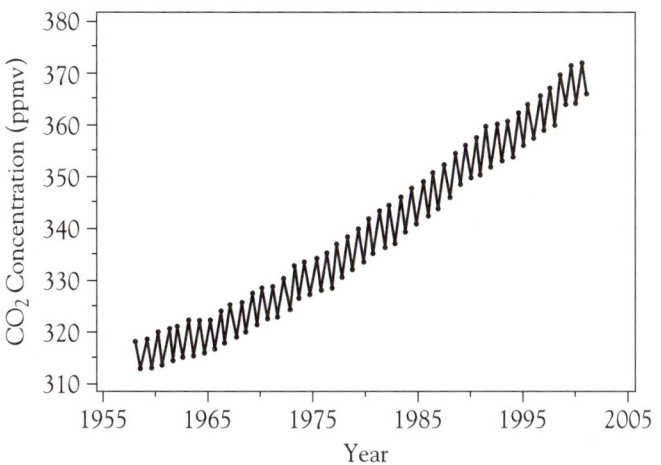

It was from concerns about the rising level of greenhouse gases and their effect on the planet that a meeting of representatives of 161 countries took place in Kyoto, Japan, in 1998. There were two particular concerns resulting from increased levels of greenhouse gases. First, the potential rapid increase in surface temperature will cause significant climate changes and accompanying ecological problems. Second, arising from the first, the melting of the Antarctic and Greenland icecaps will result in a sea level increase that will swamp coastal areas and actually wipe out many island nations completely. The island state of Tuvalu is already suffering from increasing flooding such that the entire population will soon have to find refuge in other countries. Among many vulnerable locations, coastal Florida and all of the Mississippi delta is at long-term risk from rising water levels.

The aim of the resulting Kyoto Protocol was to commit countries to reduce emissions of the following greenhouse gases: carbon dioxide, methane, dinitrogen oxide, hydrofluorocarbons, perfluorocarbons, and sulfur hexafluoride. The goal was to reduce emissions of these gases worldwide by the years 2008–2012 by 5 percent of those measured in 1990. This goal by no means solves the problem, but it did represent a step toward the long-term reduction of greenhouse gas production. Most European countries agreed to aim for an 8 percent reduction while the United States, at the time, committed to aim for a 7 percent reduction. As of 2002, the United States has rejected the Kyoto Protocol, the only country of the 161 signatory nations to do so. There are two obvious ways of complying with the Protocol:

1. Placing a greater dependence upon the generation of power from non-carbon-based fuels, such as wind, wave, and nuclear power.

2. Using carbon resources in a more efficient manner, such as the use of hybrid-fuel passenger vehicles and biodiesel fuels for trucks and other heavy vehicles.

In addition to actually reducing emissions of gases, there are three other routes countries can take to reduce their registered production levels:

1. If an industrialized country helps a developing country reduce its emissions, then that industrialized country can count part of the benefits toward its own reduction goal.

2. Emission–reduction credits are tradable like stocks. Therefore, any country (or company) that reduces its emissions below its own target can "sell" the extra reduction to a country (or company) that did not meet its objectives.

3. Removing greenhouse gases from the atmosphere by increasing the planting of forests can be counted as credit.

Concerned about the implications of climate change, many countries are moving ahead to fulfil their obligations under the Protocol even though it is not yet legally binding on those countries. With the withdrawal of the United States, it becomes more difficult for the Protocol to be approved as a majority of the major-emitting countries must ratify. As the United States alone produces 25 percent of global emissions, most of the other major carbon dioxide producing countries, such as Japan, Canada, Australia, the U.K, and Germany, must sign off on the treaty for it to legally come into effect.

14.9   Hydrogen Carbonates and Carbonates

Hydrogen Carbonates

As we discussed in Chapter 11, Section 11.2, only the alkali metals (except lithium) form solid compounds with the hydrogen carbonate ion, HCO_3^-, and even these decompose to the carbonate when heated:

$$2\ NaHCO_{3(s)} \xrightarrow{\Delta} Na_2CO_{3(s)} + H_2O_{(l)} + CO_{2(g)}$$

Solutions of lithium and Group 2 metal hydrogen carbonates can be prepared, but even in solution the hydrogen carbonates decompose to carbonates when heated. Calcium hydrogen carbonate is formed during the dissolution of carbonate rocks; the dissolution process creates caves, and the subsequent decomposition of the hydrogen carbonate produces stalagmites and stalactites within caves (Chapter 12, Section 12.9):

$$CaCO_{3(s)} + CO_{2(aq)} + H_2O_{(l)} \rightleftharpoons Ca(HCO_3)_{2(aq)}$$

Household water supplies that come from chalk or limestone regions contain calcium hydrogen carbonate. Heating such water in a hot water tank or a kettle shifts the equilibrium in the direction of increasing entropy, precipitating out the calcium carbonate as a solid that is often called "scale":

$$Ca(HCO_3)_{2(aq)} \rightleftharpoons CaCO_{3(s)} + CO_{2(aq)} + H_2O_{(l)}$$

The hydrogen carbonate ion reacts with acids to give carbon dioxide and water and with bases to give the carbonate ion:

$$HCO_3^-{}_{(aq)} + H^+{}_{(aq)} \rightarrow CO_{2(g)} + H_2O_{(l)}$$

$$HCO_3^-{}_{(aq)} + OH^-{}_{(aq)} \rightarrow CO_3^{2-}{}_{(aq)} + H_2O_{(l)}$$

Carbonates

The carbonate ion is very basic in aqueous solution as a result of a hydrolysis reaction that gives hydrogen carbonate and hydroxide ion:

$$CO_3^{2-}{}_{(aq)} + H_2O_{(l)} \rightleftharpoons HCO_3^-{}_{(aq)} + OH^-{}_{(aq)}$$

Thus, concentrated solutions of even that "harmless" household substance, sodium carbonate, commonly called washing soda, should be treated with respect (though not fear).

The carbon–oxygen bonds in the carbonate ion are all the same length and are significantly shorter than a single bond. We can consider the bonding in terms of a σ framework that is centered on the carbon atom and uses sp^2 hybrid orbitals. The electron pair in the remaining *p* orbital of the carbon atom can then form a π bond that is delocalized (shared) over the whole ion (Figure 14.14). With the π bond shared three ways, each carbon–oxygen bond would have a bond order of 1⅓. The notation used to represent this bond order is shown in Figure 14.15.

We can also consider the π system in simplified molecular orbital terms. The four *p* orbitals at right angles to the plane of the molecule interact to form four π molecular orbitals. Appropriate calculations indicate that one will be bonding, two nonbonding, and one antibonding. The three electron pairs that are available to fill these orbitals fill the bonding and nonbonding orbitals (Figure 14.16). Hence, like the hybridization model, molecular orbital theory predicts one π bond shared throughout the four-atom system—or one-third of a π bond for each bonded pair.

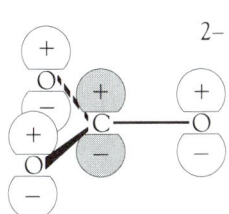

Figure 14.14   Delocalized π bonding in the carbonate ion.

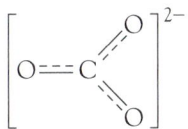

Figure 14.15   Representation of the partial bonds of the carbonate ion.

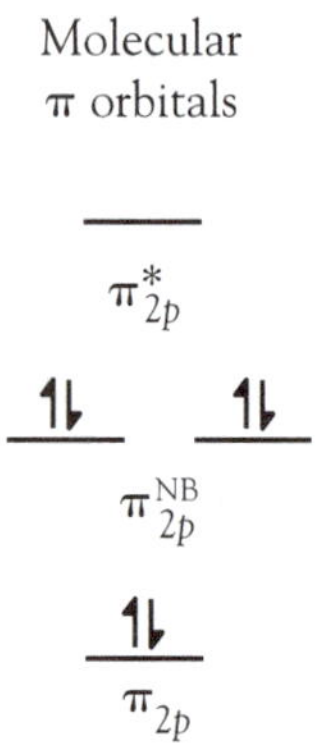

Molecular
π orbitals

π^*_{2p}

⇅   ⇅

π^{NB}_{2p}

⇅

π_{2p}

Figure 14.16 Energy levels of the molecular orbitals involved in π bond formation in the carbonate ion.

Most carbonates are insoluble, the exceptions being ammonium carbonate and the alkali metal carbonates. The alkali metal carbonates (except that of lithium) do not decompose when heated. Lithium carbonate and carbonates of other moderately electropositive metals, such as calcium, give the metal oxide and carbon dioxide when heated:

$$CaCO_3(s) \xrightarrow{\Delta} CaO(s) + CO_2(g)$$

For the carbonates of weakly electropositive metals, such as silver, the metal oxide is itself decomposed by heat. Thus, the final products are the metal, carbon dioxide, and oxygen:

$$Ag_2CO_3(s) \xrightarrow{\Delta} Ag_2O(s) + CO_2(g)$$

$$Ag_2O(s) \xrightarrow{\Delta} 2\,Ag(s) + \tfrac{1}{2}\,O_2(g)$$

As is typical of the behavior of the ammonium ion in oxysalts, both the anion and the cation of ammonium carbonate decompose when heated; the products are ammonia, water, and carbon dioxide:

$$(NH_4)_2CO_3(s) \xrightarrow{\Delta} 2\,NH_3(g) + H_2O(g) + CO_2(g)$$

Sequestration of Carbon Dioxide

Though past efforts at controlling the increase in carbon dioxide levels have focused on reducing emissions, the United States, in particular, is increasing spending into sequestration research. Sequestration involves the storage of emitted carbon dioxide. Sequestering can be accomplished in four different ways:

1. Increasing photosynthetic absorption. The most obvious route involves the planting of more trees. There are limitations to the amount of trees that can be planted. The idea is also being tested of using iron enrichment of seawater to enhance algae production and thus increase marine sequestering. The overall ecological effect of adding megatonnes of iron to ocean basins has yet to be determined.

2. Developing novel chemical technology to convert emitted carbon dioxide into useful products. This avenue seems to have limited potential from the perspective of the quantities of carbon dioxide emitted and also the energy needs of any processes.

3. Storing the gas in underground geological formations. This is already being undertaken. In the Sleipner gas field in the North Sea, carbon dioxide is separated from methane in the natural gas, and the carbon dioxide is pumped into a saline aquifer deep under the ocean floor. Again, this option is limited, this time by the number of suitable burial sites.

4. Pumping the gas into the oceans. According to the U.S. Department of Energy, this is the one solution that offers a virtually limitless depository of excess carbon dioxide.

It is option 4 that relates to the chemistry of carbon dioxide and carbonates. Shallow injection of carbon dioxide into seas will affect the pH of the seawater according to:

$$CO_2(aq) + H_2O(l) \rightleftharpoons H_3O^+(aq) + HCO_3{}^-(aq)$$

Such an increase will have an effect on marine organisms. Of particular concern are the coral reefs. The coral reefs are sometimes called the "rain forests of the seas," as they represent such a concentration of biological diversity. The increase in dissolved carbon dioxide will also cause a shift in the carbonate/hydrogen carbonate equilibrium, such that the carbonate concentration will be reduced below that necessary for sustaining coral growth:

$$CO_2(aq) + CO_3^{2-}(aq) \rightarrow 2\ HCO_3^-(aq)$$

Deep ocean sequestration is becoming of greater interest. It is argued that carbon dioxide can be safely injected into deep water layers without harm to the environment. When carbon dioxide is released into the deep ocean, under the conditions of temperature and pressure, it initially forms a clathrate (see Chapter 10, Section 10.7), also called a hydrate. The hydrate has a high stability, for example, at a depth of 250 m where the pressure is about 2.7 MPa, the clathrate is stable at +5°C. Whereas "normal" ice is less dense than liquid water, the carbon dioxide clathrate has a density of about 1.1 g·cm^{-3} and sinks to the ocean floor. It is envisaged that megatonnes of excess carbon dioxide would be disposed of in this way. There are three major concerns with this concept. First, the layer of carbon dioxide clathrate will smother the bottom life of the deep oceans where the clathrates are deposited. Second, experiments have already shown that fishes show respiratory distress when they approach the carbon dioxide-saturated water around experimental clathrate deposits. Third, over an extended period of time—perhaps hundreds of thousands of years—the clathrates will probably release their captive carbon dioxide to the surrounding waters causing a pH decrease of the oceans. The pH change would obviously have an effect on the ecological balance of marine life.

14.10 Carbon Disulfide

Carbon disulfide is the sulfur analog of carbon dioxide, and it has the same linear geometry. The compound is a colorless, highly flammable, low-boiling liquid with a pleasant smell when pure, but the commercial grade of the compound usually contains very foul-smelling impurities. It is highly toxic, causing damage to the brain and nervous system and, eventually, death. Carbon disulfide is prepared industrially by passing methane gas over molten sulfur at about 700°C, then cooling the products, from which carbon disulfide condenses:

$$CH_4(g) + 4\ S(l) \xrightarrow{\Delta} CS_2(g) + 2\ H_2S(g)$$

Over 1 million tonnes of this reagent is consumed each year, mainly in the production of cellophane and viscose rayon polymers. It is also the starting material for the manufacture of carbon tetrachloride. We tend to forget that industrial chemistry is rarely the conversion of one naturally occurring substance directly to some required product. More often, the product is, itself, only a reagent in the production of many other compounds.

14.11 Carbonyl Sulfide

We are becoming increasingly aware of the complexity of the Earth's atmosphere and of the role of some inorganic compounds which we have ignored in the past as laboratory curiosities without any real-world use. One of these compounds is carbonyl sulfide, written as COS, although its double-bonded structure of S=C=O, resembles that of carbon dioxide. Carbonyl sulfide is the most abundant sulfur-containing gas in the global background atmosphere as a result of its low chemical reactivity, the total amount being estimated as about 5×10^6 tonnes. It is the only sulfur-containing gas to penetrate the stratosphere (except when very powerful volcanic eruptions directly inject sulfur dioxide into the upper atmosphere). The gas is one of the several sulfur-containing compounds produced by soil and marine organisms, another important one being dimethyl sulfide, $(CH_3)_2S$.

14.12   Carbon Tetrahalides

The major divisions of chemistry—inorganic, organic, physical, and analytical—are inventions of chemists attempting to organize this vast and continually growing science. Yet chemistry does not fit into neat little compartments, and the carbon tetrahalides are compounds that belong to the realms of both organic and inorganic chemistry. As a result, they have two sets of names: carbon tetrahalides, according to inorganic nomenclature, and tetrahalomethanes, according to organic nomenclature.

All of the tetrahalides contain the carbon atom tetrahedrally coordinated to the four halogen atoms. The phases of the tetrahalides at room temperature reflect the increasing strength of the intermolecular dispersion forces. Thus, carbon tetrafluoride is a colorless gas; carbon tetrachloride is a dense, almost oily liquid; carbon tetrabromide, a pale yellow solid; and carbon tetraiodide, a bright red solid.

Carbon tetrachloride is an excellent nonpolar solvent. In recent years, the discovery of its cancer-causing ability has made it a solvent of last resort. It was formerly used as a fire-extinguishing material, particularly where water could not be used, for example, around electrical wiring and restaurant deep fat fryers. The liquid vaporizes to form a gas that is five times denser than air, effectively blanketing the fire with an inert gas. However, in addition to its carcinogenic nature, it does oxidize in the flames of a fire to give the poison gas carbonyl chloride, $COCl_2$. Carbon tetrachloride is also a "greenhouse" gas and, in the upper atmosphere, a potent ozone destroyer. It is therefore important to minimize emissions of this compound from industrial plants.

The major industrial route for the synthesis of carbon tetrachloride involves the reaction of carbon disulfide with chlorine. In this reaction, iron(III) chloride is the catalyst. In the first step, the products are carbon tetrachloride and disulfur dichloride. Then at a higher temperature, addition of more carbon disulfide produces additional carbon tetrachloride and sulfur. The sulfur can be reused in the production of a new batch of carbon disulfide:

$$CS_2(g) + 3\ Cl_2(g) \xrightarrow{FeCl_3} CCl_4(g) + S_2Cl_2(l)$$

$$CS_2(g) + 2\ S_2Cl_2(l) \xrightarrow{\Delta} CCl_4(g) + 6\ S(s)$$

The reaction between methane and chlorine is also used to produce carbon tetrachloride:

$$CH_4(g) + 4\ Cl_2(g) \rightarrow CCl_4(l) + 4\ HCl(g)$$

For chemists, one of the interesting properties of carbon tetrachloride is its chemical inertness. For example, it does not react with water, yet silicon tetrachloride does—violently. If we look at the comparative free energies of

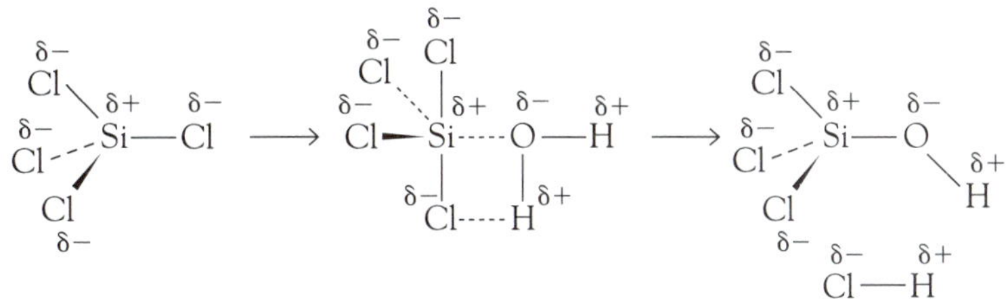

Figure 14.17 First step of the postulated mechanism of hydrolysis of silicon tetrachloride.

reaction with water, we see that, if anything, carbon tetrachloride should be more reactive than silicon tetrachloride:

$$CCl_4(l) + 3\,H_2O(l) \rightarrow H_2CO_3(aq) + 4\,HCl(g) \quad \Delta G° = -380\ \text{kJ·mol}^{-1}$$

$$SiCl_4(l) + 3\,H_2O(l) \rightarrow H_2SiO_3(s) + 4\,HCl(g) \quad \Delta G° = -289\ \text{kJ·mol}^{-1}$$

The answer must lie with kinetic factors, that is, the lack of an available pathway for the reaction. We postulate that the reaction between silicon tetrachloride and water occurs in a manner similar to that described for boron trichloride (see Chapter 13, Figure 13.9): an attack by the polar water molecule in which the partially positive hydrogen bonds to the partially negative chlorine while the partially negative oxygen is bonding to the partially positive silicon (Figure 14.17). This process is repeated three more times, thereby replacing all of the chlorines with hydroxyl groups. The transition states for this process involve a five-coordinate silicon atom. For five bonds, hybridization theory requires the use of at least one *d* orbital of the silicon, as well as the *s* and *p* orbitals. However, carbon, which is a second period element, has no available *d* orbitals. (One can alternatively argue that the cluster of four large chlorine atoms around the small carbon atom block any potential reaction.) Thus, this reaction pathway cannot exist for carbon.

14.13 *Chlorofluorocarbons*

It was Thomas Midgley, Jr., a General Motors chemist, who in 1928 first synthesized dichlorodifluoromethane, CCl_2F_2. This discovery was made as part of a search to find a good, safe refrigerant material. A refrigerant is a compound that, at room temperature, is a gas at low pressures but a liquid at high pressures. Reducing the pressure on the liquid causes it to boil and absorb heat from the surroundings (such as the inside of a refrigerator). The gas is then conveyed outside the cooled container, where it is compressed. Under these conditions, it reliquefies, releasing the enthalpy of vaporization to the surroundings as it does so.

At the time they were discovered, the chlorofluorocarbon family (CFCs) appeared to be a chemist's dream. They were almost completely unreactive and they were nontoxic. As a result, they were soon used in air conditioning systems, as blowing agents for plastic foams, as aerosol propellants, as fire-extinguishing materials, as degreasing agents of electronic circuits, and as anesthetics—to name but a few uses. Annual production amounted to nearly 700 000 tonnes in the peak years. Their lack of reactivity is partly due to the lack of a hydrolysis pathway, but, in addition, the high strength of the carbon–fluorine bond confers extra protection against oxidation.

Chlorofluorocarbons have a specialized and arcane naming system:

1. The first digit represents the number of carbon atoms minus one. For the one-carbon CFCs, the zero is eliminated.

2. The second digit represents the number of hydrogen atoms plus one.

3. The third digit represents the number of fluorine atoms.

4. Structural isomers are distinguished by "a," "b," and so on.

Of the simple CFCs, $CFCl_3$ (CFC-11) and CF_2Cl_2 (CFC-12) were the most widely used.

It was not until the 1970s that the great stability of these compounds—their "best" property—was recognized as a threat to the environment. These compounds are so stable that they will remain in the atmosphere for hundreds of years. Some of these molecules were diffusing into the upper atmosphere (stratosphere), where ultraviolet light cleaved a chlorine atom from each of them. The chlorine atom then reacted with ozone molecules in a series of steps that can be represented in a simplified and not wholly accurate form as

$$Cl + O_3 \rightarrow O_2 + ClO$$

$$ClO \rightarrow Cl + O$$

$$Cl + O_3 \rightarrow O_2 + ClO$$

$$ClO + O \rightarrow Cl + O_2$$

The chlorine atom is then free to repeat the cycle time and time again, destroying enormous numbers of ozone molecules. Incidentally, these chemical species are not those that you can find in a chemistry laboratory, but at the low pressures existing in the upper atmosphere, even free chlorine and oxygen atoms can exist for measurable periods of time.

It was in 1987 that an international meeting in Montreal, Canada, developed the Montreal Protocol. According to the agreement, CFCs would be phased out, in the short term to be replaced by the less harmful HCFCs and ultimately by nonchlorinated compounds. The search was rapidly undertaken to find substitutes for CFCs. This has not been easy, because most potential replacements are flammable, toxic, or suffer from some other major problem. For example, the most promising alternative to the CFC-12 refrigerant is a hydrofluorocarbon (HFC), CF_3—CH_2F (HFC-134a). The absence of a chlorine atom in its structure means that it cannot wreak havoc in the ozone layer. Nevertheless, there are three major problems with its widespread adoption. First, CFC-12 is manufactured in a simple, one-step process from carbon tetrachloride and hydrogen fluoride. The synthesis of HFC-134a, however, requires a costly, multistep procedure. Second, existing refrigeration units have to be altered to operate with the new compound. The major reason is that the pressures used to condense HFC-134a are higher than those used with CFC-12. The costs of building the chemical factories and of modifying refrigerator pumps are a tolerable burden for Western countries but are beyond the means of less developed countries. The final problem is that all the fluorocarbon compounds are also excellent "greenhouse" gases; that is, like carbon dioxide, they absorb infrared radiation and can contribute to global warming. Thus, we must make sure that the CFCs and HFCs are used in closed systems, with zero leakage, and with a legal requirement to return the unit to the manufacturer at the end of its operating life. The manufacturer will then recycle the fluorocarbon refrigerant.

The lesson, of course, is clear. Just because a compound is chemically inert in the laboratory does not mean that it is harmless. No product of the chemical industry can be released into the environment without first considering its impact. At the same time, hindsight is a wonderful thing. It is only in recent years that we have become aware of the importance of the ozone layer and of the many chemical reactions that occur in it (discussed in Chapter 16, Section 16.3). The more crucial moral is that research into the chemical cycles in nature must continue to be funded. If we do not know how the world works at the chemical level, then it will be impossible to predict the effect of any human perturbation.

14.14 Methane

The simplest compound of carbon and hydrogen is methane, CH_4, a colorless, odorless gas. There are enormous quantities of this gas, commonly called natural gas, in underground deposits and in deposits under the seabed (see Chapter 10, Section 10.6). Methane is one of the major sources of thermal energy in use today because it undergoes an exothermic combustion reaction:

$$CH_4(g) + 2\ O_2(g) \rightarrow CO_2(g) + 2\ H_2O(g)$$

Because methane is undetectable by our basic senses (sight and smell), a strong-smelling, organic sulfur-containing compound is added to the gas before it is supplied to customers. We can therefore detect any methane leakage by the odor of the additive.

Many scientists are particularly concerned by the rising concentration of methane. Though it is only present at the ppb level, compared to ppm for carbon dioxide, methane is currently the gas whose concentration in the atmosphere is rising most rapidly. The methane molecule absorbs infrared wavelengths different from those currently being absorbed by carbon dioxide and water vapor, particularly those between 3.4 and 3.5 μm. There have always been traces of methane in the atmosphere as a result of vegetation decay in marshes. However, the proportion in the atmosphere has risen drastically over the last century. The rise is due in part to the rapid growth in the number of cattle and sheep (members of the grazing animals, the ruminants). All ruminants produce large quantities of methane in their digestive tracts, and this gas is expelled into the atmosphere. In addition, methane is generated in the wet soils of paddy fields in which rice is grown.

14.15 Cyanides

Most people are aware of the toxicity of hydrogen cyanide, HCN, and the cyanide ion, CN^-, yet few realize their industrial importance. In fact, over 1 million tonnes of hydrogen cyanide is manufactured each year. There are two modern methods of synthesis of hydrogen cyanide. The *Degussa process* involves the reaction of methane with ammonia at high temperature with a platinum catalyst:

$$CH_4(g) + NH_3(g) \xrightarrow[1200°C]{Pt} CN(g) + 3\ H_2(g)$$

while the *Andrussow process* is similar but requires the presence of dioxygen:

$$2\ CH_4(g) + 2\ NH_3(g) + 3\ O_2(g) \xrightarrow[1100°C]{Pt/Rh} 2\ HCN(g) + 6\ H_2O(g)$$

Hydrogen cyanide is a liquid at room temperature as a result of strong hydrogen bonding between each hydrogen atom and the nitrogen atom of a neighboring molecule. It is extremely toxic, very low concentrations having a faint almond-like odor. The liquid is miscible with water to form an extremely weak acid:

$$HCN(aq) + H_2O(l) \rightleftharpoons H_3O^+(aq)1 + CN^-(aq)$$

About 70 percent of hydrogen cyanide is used to produce many important polymers, including nylon, melamine, and the family of acrylic plastics. Of the remainder, close to 15 percent is converted to sodium cyanide by simple neutralization:

$$NaOH(aq) + HCN(aq) \rightarrow NaCN(aq) + H_2O(l)$$

the salt being obtained by crystallization from the solution. The cyanide ion is used in the extraction of gold and silver from their ores (see Chapter 20, Section 20.9).

The cyanide ion is isoelectronic with carbon monoxide, and both these species react readily with hemoglobin, blocking uptake of dioxygen. The cyanide ion interferes with enzyme processes as well. The most tragic case of cyanide poisoning occurred on November 18, 1978. However, the background story commenced in the late 1960s and early 1970s. A charismatic preacher by the name of Jim Jones organized his own church—the People's Temple. Initially founded as an egalitarian and populist creed appealing to a wide spectrum of altruistic and idealistic people, the People's Temple became very active in the San Francisco area. Convinced that agents of the Federal Government were attempting to undermine the church, Jones moved the Temple with about 1000 followers to the remote jungles of Guyana, near the border with Venezuela. Here, Jones' paranoia grew. Life for its devotees became terrifying, with physical and sexual abuse becoming part of daily rituals. Added to the prison camp-like environment, tropical diseases were rampant. Rehearsals of mass suicides became regular occurrences. The apocalyptic end came when they were visited by a group of relatives of cult members together with reporters and led by Congressman Ryan of California. When the group were about to depart, they were gunned down, and the members of the cult were ordered to drink a solution of sodium cyanide in a soft drink. Those who refused were forced to consume the poisonous beverage (or were shot). In all, over 900 people died. Not only was the cyanide solution poisonous, the cyanide ion is the conjugate base of a very weak acid; hence, its solutions are very basic.

$$CN^-(aq) + H_2O(l) \rightleftharpoons HCN(aq) + OH^-(aq)$$

We cannot imagine the excruciating pain of drinking a solution whose pH was little different from that of the same concentration of sodium hydroxide. Death must have been a welcome relief from the agony.

Hydrogen cyanide, itself, is poisonous. It was in the late 1920s that many U.S. states introduced poisoning with hydrogen cyanide as a method of execution. The gas chamber was a sealed room containing a chair. Beneath the chair hung glass containers filled with sodium or potassium cyanide. At the warden's signal, the containers were released by remote control into a bath of sulfuric acid. The hydrogen cyanide produced was of high enough concentration to cause unconsciousness within seconds and death within 5 minutes.

$$2\ NaCN(aq) + H_2SO_4(aq) \rightarrow Na_2SO_4(aq) + 2\ HCN(g)$$

14.16 *Silicon*

About 27 percent by mass of the Earth's crust is silicon. However, silicon itself is never found in nature as the free element but only in compounds containing oxygen–silicon bonds. The element is a gray, metallic-looking, crystalline solid. Although it looks metallic, it is not classified as a metal because it has a low electrical conductivity.

About half a million tonnes per year of silicon is used in the preparation of metal alloys. Although alloy manufacture is the major use, silicon plays a crucial role in our lives as the semiconductor that enables computers to function. The purity level of the silicon used in the electronics industry has to be exceedingly high. For example, the presence of only 1 ppb of phosphorus is enough to drop the specific resistance of silicon from 150 to 0.1 kΩ·cm. As a result of the expensive purification process, ultrapure electronic grade silicon sells for over 1000 times the price of metallurgical grade (98 percent pure) silicon.

The element is prepared by heating silicon dioxide (quartz) with coke at over 2000°C in an electrical furnace similar to that used for the Acheson process of calcium carbide synthesis. Liquid silicon (melting point 1400°C) is drained from the furnace:

$$SiO_2(s) + 2\ C(s) \xrightarrow{\Delta} Si(l) + 2\ CO(g)$$

To obtain ultrapure silicon, the crude silicon is heated at 300°C in a current of hydrogen chloride gas. The trichlorosilane product, $SiHCl_3$, can be distilled and redistilled until the impurity levels are below the parts per billion level:

$$Si(s) + 3\ HCl(g) \rightarrow SiHCl_3(g) + H_2(g)$$

The reverse reaction is spontaneous at 1000°C, depositing ultrapure silicon. The hydrogen chloride can be reused in the first part of the process:

$$SiHCl_3(g) + H_2(g) \xrightarrow{\Delta} Si(s) + 3\ HCl(g)$$

The ultrapure single crystals needed for solar cells are produced by zone refining (Figure 14.18). This process depends on the fact that the impurities are more soluble in the liquid phase than in its solid phase. A rod of silicon is moved through a high-temperature electric coil that partially melts the silicon. As part of the rod moves beyond the coil, the silicon resolidifies; during the solidification process, the impurities diffuse into the portion of the rod that is still molten. After the entire rod has passed through the coil, the impurity-rich top portion can be removed. The procedure can be repeated until the desired level of purity (less than 0.1 ppb impurity) is obtained.

Computing devices can function because the silicon chips have been selectively "doped." That is, controlled levels of other elements are introduced. If traces of a Group 15 element, such as phosphorus, are mixed into the silicon, then the extra valence electron of the added element is free to roam throughout the material. Conversely, doping with a Group 3 element will result in electron "holes." These holes act as sinks for electrons. A combination of electron-rich and electron-deficient silicon layers makes up the basic electronic circuit. This is a somewhat simplified description of how such devices function, and any interested reader should consult a text on the physics and technology of semiconductors for a more detailed discussion.

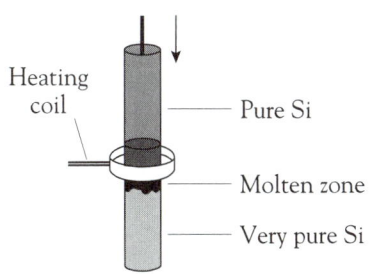

Heating coil — Pure Si

— Molten zone

— Very pure Si

Figure 14.18 Zone refining method for the purification of silicon.

14.17   Silicon Dioxide

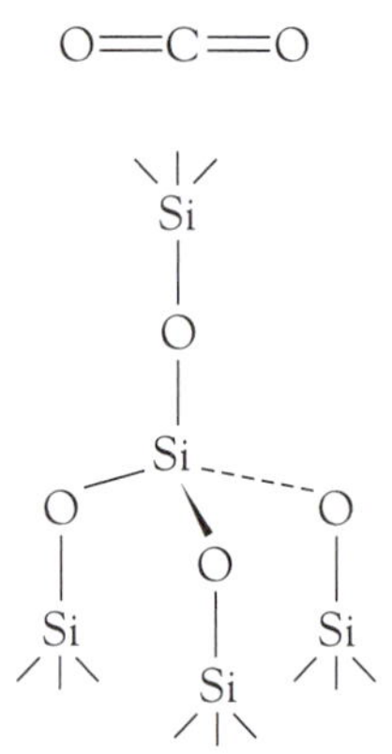

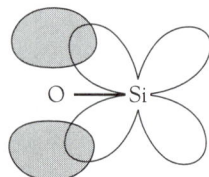

Figure 14.19   Structures of carbon dioxide and silicon dioxide.

Figure 14.20   Overlap of a full oxygen p orbital with an empty silicon d orbital.

The most common crystalline form of silicon dioxide, SiO_2, commonly called silica, is the mineral quartz. Most sands consist of particles of silica that usually contain impurities such as iron oxides. It is interesting to note that carbon dioxide and silicon dioxide share the same type of formula, yet their properties are very different. Carbon dioxide is a colorless gas at room temperature, whereas solid silicon dioxide melts at 1600°C and boils at 2230°C. The difference in boiling points is due to bonding factors. Carbon dioxide consists of small, triatomic, nonpolar molecular units whose attraction to one another is due to dispersion forces. By contrast, silicon dioxide contains a network of silicon–oxygen covalent bonds in a giant molecular lattice. Each silicon atom is bonded to four oxygen atoms, and each oxygen atom is bonded to two silicon atoms, an arrangement consistent with the SiO_2 stoichiometry of the compound (Figure 14.19).

How can we explain this difference? First, the carbon–oxygen single bond (C—O bond energy = 358 kJ·mol^{-1}) is much weaker than the carbon–oxygen double bond (C=O bond energy = 799 kJ·mol^{-1}). So formation of a p_π–p_π bond between the carbon and oxygen atoms more than doubles the strength of the bond. Hence, it is energetically more favorable to form two C=O double bonds than four C—O single bonds, which would be analogous to the silicon dioxide structure. In the case of silicon, silicon–oxygen single bonds can be considered to have a partial double bond character as a result of the overlap of the empty d orbitals on the silicon with the full p orbitals of the oxygen (Figure 14.20). Because of this delocalized bonding, the Si—O bond energy is 452 kJ·mol^{-1}. And it is known that multiple bonds in compounds of elements from Period 3 and higher periods have energies that are not much greater than those of the corresponding single bonds (due to poor p_π–p_π overlap). Thus, for silicon, four single bonds (with partial multiple bond character) are much more preferable than two conventional double bonds.

Silicon dioxide is very unreactive; it reacts only with hydrofluoric acid (or wet fluorine) and molten sodium hydroxide. The reaction with hydrofluoric acid is used to etch designs on glass:

$$SiO_2(s) + 6\ HF(aq) \rightarrow SiF_6{}^{2-}(aq) + 2\ H^+(aq) + 2\ H_2O(l)$$

$$SiO_2(s) + 2\ NaOH(l) \xrightarrow{\Delta} Na_2SiO_3(s) + H_2O(g)$$

Silicon dioxide is mainly used as an optical material. It is hard, strong, and transparent to visible and ultraviolet light, and it has a very low coefficient of expansion. Thus, lenses constructed from it do not warp as the temperature changes.

Silica Gel

Silica gel is a hydrated form of silicon dioxide, $SiO_2 \cdot xH_2O$. It is used as a desiccant (drying agent) in the laboratory and also for keeping electronics and even prescription drugs dry. You may have noticed the packets of a grainy material enclosed with electronic equipment or the little cylinders placed in some drug vials by pharmacists. These enclosures keep the product dry even in humid climates. Commercial silica gel contains about 4 percent water by mass, but it will absorb very high numbers of water molecules over the crystal surface. And it has the particular advantage that it can be reused after heating for several hours; the high temperature drives off the water molecules, enabling the gel to function effectively once more.

Aerogels

In the 1930s, an American chemist, Samuel Kistler, devised a way of drying wet silica gel without causing it to shrink and crack, like mud on a dried riverbank. At the time, there was little interest in the product. Furthermore, the procedure required extremely high pressures, and one laboratory was destroyed by an explosion during the preparation of this material. Now, about 70 years later, chemists have discovered new and safer synthetic routes to this rediscovered family of materials, called *aerogels*. The basic aerogel is silicon dioxide in which a large number of pores exist—so many, in fact, that 99 percent of an aerogel block consists of air. As a result, the material has an extremely low density, yet is quite strong. The translucent solid is also an excellent thermal insulator, and it promises to be a useful fireproof insulating material. Aerogels also have some unique properties. For example, sound travels through aerogels more slowly than through any other medium. Chemists have now prepared aerogels that incorporate other elements, a technique that enables the chemists to vary the characteristics of the aerogels. One type of "kitty litter" is an aerogel.

14.18 Glasses

Glasses are noncrystalline materials. The cooling of molten glass results in an increasingly viscous liquid until it finally becomes infinitely viscous at its solidification point without change into an ordered crystalline structure. Glass has been used as a material for at least 5000 years. It is difficult to obtain a precise figure of current annual production, but it must be about 100 million tonnes.

Almost all glass is silicate glass; it is based on the three-dimensional network of silicon dioxide. Quartz glass is made simply by heating pure silicon dioxide above 2000°C and then pouring the viscous liquid into molds. The product has great strength and low thermal expansion, and it is highly transparent in the ultraviolet region. However, the high melting point precludes the use of quartz glass for most everyday glassware.

The properties of the glass can be altered by mixing in other oxides. The compositions of three common glasses are shown in Table 14.4. About 90 percent of glass used today is soda-lime glass. It has a low melting point, so it is very easy to form soda-lime glass into containers, such as soft-drink bottles. In the chemistry laboratory, we need a glass that will not crack from thermal stress when heated; borosilicate glass (discussed in Chapter 13,

Table 14.4 Approximate compositions of common glasses

Component	Composition (%)		
	Soda-lime glass	Borosilicate glass	Lead glass
SiO_2	73	81	60
CaO	11	—	—
PbO	—	—	24
Na_2O	13	5	1
K_2O	1	—	15
B_2O_3	—	11	—
Other	2	3	<1

Section 13.2) is used for this purpose. Lead glasses have a high refractive index; as a result, cut glass surfaces sparkle like gemstones, and these glasses are used for fine glassware. The element lead is a strong absorber of radiation; hence, a very different use for lead glass is in radiation shields, such as those over cathode ray tubes.

14.19   Silicates

About 95 percent of the rocks of the Earth's crust are silicates, and there is a tremendous variety of silicate minerals. The simplest silicate ion has the formula SiO_4^{4-}; zirconium silicate, $ZrSiO_4$, the gemstone zircon, being one of the few minerals to contain this ion. Silicates are generally very insoluble, as one might expect of rocks that have resisted rain for millions of years. The one common exception is sodium silicate, which can be prepared by reacting solid silicon dioxide with molten sodium carbonate:

$$SiO_2(s) + 2\ Na_2CO_3(l) \xrightarrow{\Delta} Na_4SiO_4(s) + 2\ CO_2(g)$$

A concentrated solution of sodium (ortho)silicate is called water glass, and it is extremely basic as a result of hydrolysis reactions of the silicate anion. Before modern refrigeration became available, the water glass solution was used to preserve eggs, the soft porous shell of calcium carbonate being replaced by a tough, impervious layer of calcium silicate that seals in the egg contents:

$$2\ CaCO_3(s) + SiO_4^{4-}(aq) \rightarrow Ca_2SiO_4(s) + 2\ CO_3^{2-}(aq)$$

Now the sodium silicate solution is used in the "crystal garden" toy. Addition of crystals of colored transition metal salts results in the formation of the appropriate insoluble silicate. For example, adding a crystal of nickel(II) chloride gives a large green "plume" of nickel(II) silicate:

$$2\ NiCl_2(s) + SiO_4^{4-}(aq) \rightarrow Ni_2SiO_4(s) + 2\ Cl^-(aq)$$

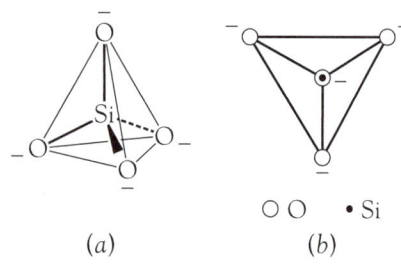

(a)   (b)

○ O   • Si

Figure 14.21   Depiction of the tetrahedral shape of the silicate ion in the conventional (a) and the silicate chemistry (b) forms.

This is not the total extent of silicate chemistry. Oxygen atoms can be shared by different silicon atoms. To show these different structures, silicate chemists depict the units in a manner different from that used in conventional molecular geometry. We can illustrate the approach with the silicate ion itself. Most chemists look at an ion from the side, a perspective giving the arrangement depicted in Figure 14.21a. Silicate chemists look down on a silicate ion, sighting along the axis of a Si—O bond (Figure 14.21b). The corner spheres represent three of the oxygen atoms; the central black dot represents the silicon atom; and the circle around the black dot represents the oxygen atom vertically above it. Instead of drawing the covalent bonds, the edges of the tetrahedron are marked with solid lines.

When a small amount of acid is added to the (ortho)silicate ion, the pyrosilicate ion, $Si_2O_7^{6-}$, is formed:

$$2\ SiO_4^{4-}(aq) + 2\ H^+(aq) \rightarrow Si_2O_7^{6-}(aq) + H_2O(l)$$

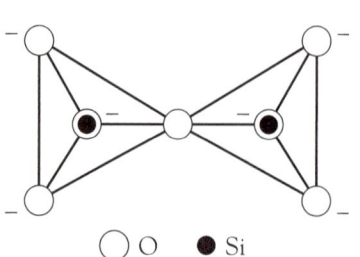

○ O   ● Si

Figure 14.22   Depiction of the $Si_2O_7^{6-}$ ion.

In the pyrosilicate ion, the two silicate ions are linked together by one shared oxygen atom (Figure 14.22). This ion is, itself, not of great importance. However, these silicate units can join to form long chains, and they can cross-link to form a double chain. A polymeric structure with an empirical formula

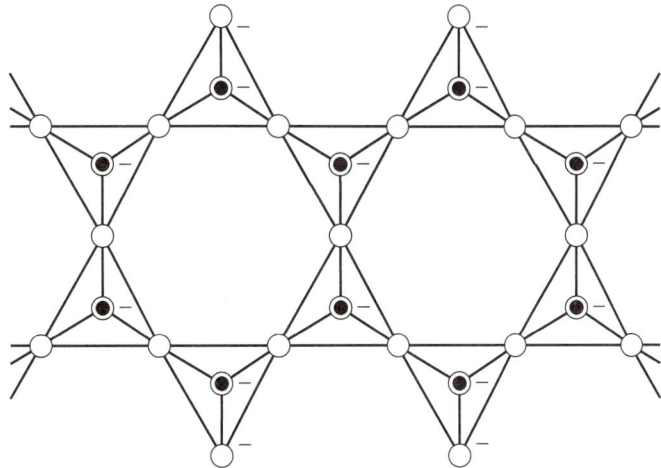

Figure 14.23 Depiction of a section of the $Si_4O_{11}^{6-}$ repeating double chain.

of $Si_4O_{11}^{6-}$ is formed in this way (Figure 14.23). The double chain is an important structure, and it gives rise to a whole family of minerals called the *amphiboles*. The cations that are packed in among these chains determine the identity of the mineral formed. For example, $Na_2Fe_5(Si_4O_{11})_2(OH)_2$ is the mineral crocidolite, more commonly known as blue asbestos.

The double chains of silicate units can link side by side to give sheets of empirical formula $Si_2O_5^{2-}$ (Figure 14.24). One of the sheet silicates is $Mg_3(Si_2O_5)(OH)_4$, chrysotile. This compound, also known as white asbestos, has alternating layers of silicate ions and hydroxide ions, with magnesium ions filling in available holes. Asbestos has been used for thousands of years. For example, the ancient Greeks used it as wicks for their lamps; and the European king Charlemagne astounded his guests in about A.D. 800 by throwing his dirty asbestos tablecloth into a fire and retrieving it unburned

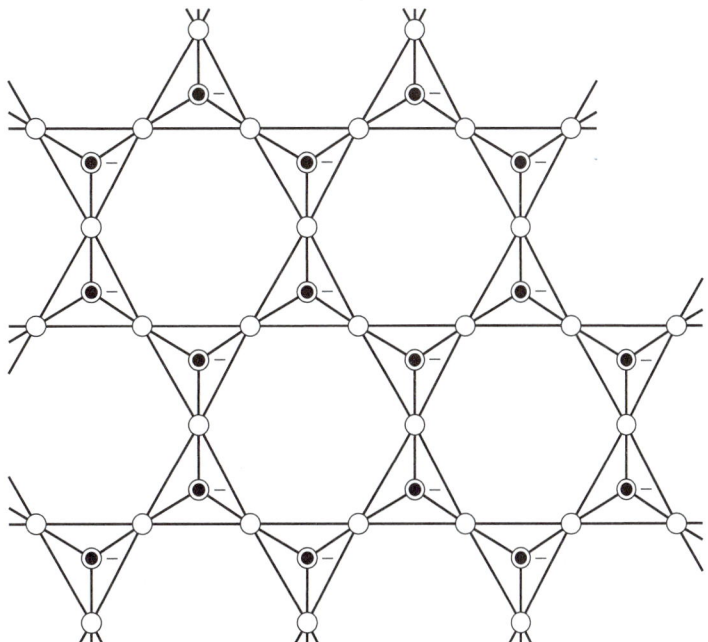

Figure 14.24 Depiction of a section of the sheet silicate, $Si_2O_5^{2-}$.

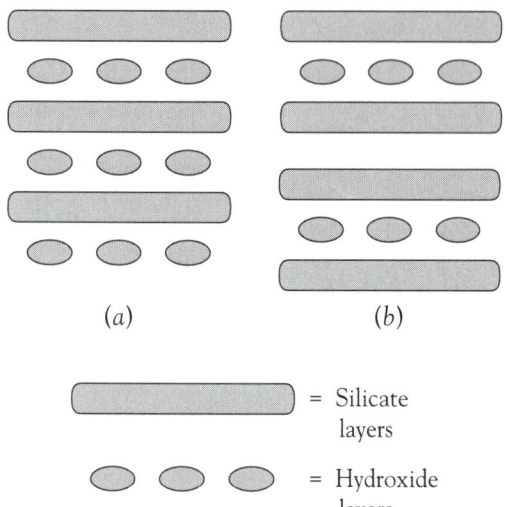

(a)   (b)

⬭ = Silicate
layers

⬬ ⬬ ⬬ = Hydroxide
layers

Figure 14.25 The layer structures of (a) white asbestos and (b) talc.

and clean. The use of asbestos is declining rapidly, now that we are aware of the health risks from embedded asbestos fibers on the lung surface.

Very few nonchemists realize that there are two common forms of this fibrous mineral and that they have very different chemical structures and different degrees of hazard. In fact, about 95 percent of asbestos currently used is the less harmful white asbestos, and only about 5 percent is the more dangerous blue asbestos. Asbestos is a very convenient and inexpensive fireproof material, and it has been extremely difficult for chemists to find hazard-free replacements for all 3000 uses of asbestos. In fact, there is still a significant consumption of asbestos for products such as brake linings, engine gaskets, and even as a filter material for wine.

It is fascinating how minor changes in structure can cause major changes in properties. If, instead of alternating layers of magnesium silicate and magnesium hydroxide, we have a layer of hydroxide ions sandwiched between pairs of layers of silicate ions, we get a different formula, $Mg_3(Si_2O_5)_2(OH)_2$, and a different name, talc (Figure 14.25). Because each sandwich is electrically neutral, it is almost as slippery as graphite (but white instead of black). Talc is used on a very large scale—about 8 million tonnes worldwide—for ceramics, fine paper, paint, and the cosmetic product talcum powder. Clays are mixtures of these sheet silicates. One particular clay mineral, kaolinite, $Al_2Si_2O_5(OH)_4$, is important for the manufacture of ceramics.

The Replacement of Silicon by Aluminum

At first thought, one might consider that aluminum, an ionic metal, and silicon, a covalent semimetal/nonmetal, have little in common. However, there are a large number of mineral structures in which aluminum partially replaces silicon. This should not be too surprising, for as we showed in Chapter 9, Table 9.25, aluminum and silicon fit in similar-sized cation lattice sites. Of course, this is presuming the bonding is ionic. In fact, it is equally valid—and often more useful—to look at these compounds as charged polymeric covalent clusters with cations fitting in the lattice interstices.

The large range of aluminosilicates is actually derived from the basic silicon dioxide structure of a three-dimensional array of SiO_4 units linked by the corner oxygen atoms. In silicon dioxide, the structure will be neutral. By substituting Al^{3+} for Si^{4+}, the lattice acquires one net negative charge for every replacement. For example, replacement of one-fourth of the silicon atoms by aluminum results in an anion of empirical formula $[AlSi_3O_8]^-$; replacement of one-half the silicon atoms gives the formula $[Al_2Si_2O_8]^{2-}$. The charge is counterbalanced by Group 1 or 2 cations. This particular family of minerals comprises the *feldspars*, components of granite. Typical examples are orthoclase, $K[AlSi_3O_8]$, and anorthite, $Ca[Al_2Si_2O_8]$.

14.20 *Zeolites*

One three-dimensional aluminosilicate structure has open channels throughout the network. Compounds with this structure are known as *zeolites*, and their industrial importance is skyrocketing. A number of zeolites exist in nature, but chemists have mounted a massive search for zeolites with novel cavities throughout their structures. There are four major uses for zeolites: as ion exchangers, as adsorption agents, for gas separation, and as catalysts.

Zeolites as Ion Exchangers

If "hard" water (water high in calcium and magnesium ion concentration) is passed through a column containing pellets of a sodium ion zeolite, the Group 2 ions push out the sodium ions. The "soft" water emerging from the column requires less detergent for washing purposes and produces less solid deposit (scum) when soap is used. When the cation sites have been fully exchanged, passage of a saturated salt solution through the column pushes out the alkaline earth metal ions by a process based on the Le Châtelier principle.

Zeolites as Adsorption Agents

The pores in a zeolite are just about the right size for holding small covalent molecules, so one major application is the use of zeolite to dry organic liquids. The water molecule is small enough to fit into a cavity of the zeolite, so it remains in the zeolite, which has effectively "dried" the organic liquid. Strong heating of the "wet" zeolite causes expulsion of the water, so the zeolite can be used again. By choosing a zeolite with a particular pore size, the process can be made quite specific for the removal of particular molecules. These zeolites are called *molecular sieves*. For example, the zeolite of formula $Na_{12}[(AlO_2)_{12}(SiO_2)_{12}] \cdot xH_2O$ has pores that are 400 pm in diameter; pores of this size can accommodate small molecules. The zeolite of formula $Na_{86}[(AlO_2)_{86}(SiO_2)_{106}] \cdot xH_2O$ has holes that are 800 pm in diameter and can accommodate larger molecules.

Zeolites for Gas Separation

Zeolites are very selective in their absorption of gases. In particular, they have a total preference for dinitrogen over dioxygen; 1 L of a typical zeolite contains about 5 L of nitrogen gas. This gas is released when the zeolite is heated. The selective absorption of dinitrogen makes zeolites of great use in the inexpensive separation of the two major components of the atmosphere. For example, a major cost in sewage treatment and steelworks has been the provision of oxygen-enriched air. Traditionally, the only route to oxygen enrichment was liquefying the air and distilling the components. Now, by cycling air through beds of zeolites, the components can be separated inexpensively.

Why is dinitrogen selectively absorbed? After all, both dioxygen and dinitrogen are nonpolar molecules of about the same size. To answer this question, we have to look at the atomic nuclei rather than at the electrons. Nuclei can be spherical or ellipsoidal. If they are ellipsoidal (football-shaped), like odd–odd nitrogen-14, then the nuclei possess an unevenly distributed nuclear charge, known as an electric quadrupole moment. Even–even oxygen-16, however, contains spherical nuclei and thus does not have an electric quadrupole moment. The interior of a zeolite cavity contains an extremely high electric charge, which attracts nuclei with electric quadrupole moments, such as those in the dinitrogen molecule. The effect is much smaller than the electron dipole moment and, apart from this instance, is of little importance in terms of chemical properties.

Zeolites as Catalysts

The modern oil industry depends on zeolites. Crude oil from the ground does not meet many of our modern requirements. It has a high proportion of long-chain molecules, whereas the fuels we need are short-chain molecules with low boiling points. Furthermore, the long-chain molecules in crude oil have straight chains, which is fine for diesel engines, but the gasoline engine

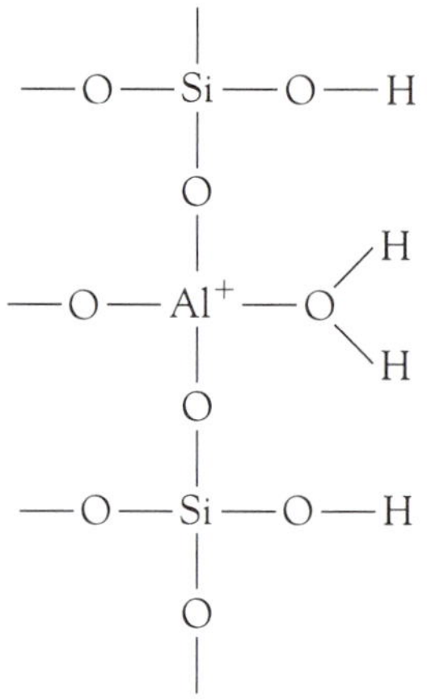

Figure 14.26 Acidic hydrogen on the surface of zeolite ZSM-5.

needs branched-chain molecules for optimum performance. Zeolite catalysts can, under specific conditions, convert straight-chain molecules to branched-chain isomers. The cavities in the zeolite structure act as molecular templates, rearranging the molecular structure to match the cavity shape. In addition to the oil industry, several industrial organic syntheses employ zeolite catalysts to convert a starting material to a very specific product. Such "clean" reactions are rare in conventional organic chemistry; in fact, side reactions giving unwanted products are very common.

One of the most important catalysts is $Na_3[(AlO_2)_3(SiO_2)] \cdot xH_2O$, commonly called ZSM-5. This compound does not occur in nature; it was first synthesized by research chemists at Mobil Oil. It is higher in aluminum than most naturally occurring zeolites, and its ability to function depends on the high acidity of water molecules bound to the high charge density aluminum ions (Figure 14.26). In fact, the hydrogen in ZSM-5 is as strong a Brønsted–Lowry acid as that in sulfuric acid.

The zeolite ZSM-5 catalyzes reactions by admitting molecules of the appropriate size and shape into its pores and then acting as a strong Brønsted–Lowry acid. This process can be illustrated by the synthesis of ethylbenzene, an important organic reagent, from ethene, C_2H_4, and benzene, C_6H_6. It is believed that the ethene is protonated within the zeolite:

$$H_2C\!\!=\!\!CH_{2}(g) + H_2O\!\!-\!\!Al^+(s) \rightarrow H_3C\!\!-\!\!CH_2^{\,+}(g) + HO\!\!-\!\!Al(s)$$

The carbocation can then attack a benzene molecule to give ethylbenzene:

$$H_3C\!\!-\!\!CH_2^{\,+}(g) + C_6H_6(g) + HO\!\!-\!\!Al(s) \rightarrow$$
$$C_6H_5\!\!-\!\!CH_2\!\!-\!\!CH_3(g) + H_2O\!\!-\!\!Al^+(s)$$

14.21 Ceramics

The term *ceramics* describes nonmetallic, inorganic compounds that are prepared by high-temperature treatment. The properties of ceramic materials are a function not only of their chemical composition but also of the conditions of their synthesis. Typically, the components are finely ground and mixed to a paste with water. The paste is then formed into the desired shape and heated to about 900°C. At these temperatures, all the water molecules are lost, and numerous high-temperature chemical reactions occur. In particular, long needle crystals of mullite, $Al_6Si_2O_{13}$, are formed. These make a major contribution to the strength of the ceramic material.

Conventional ceramics are made from a combination of quartz with two-dimensional silicates (clays) and three-dimensional silicates (feldspars). Thus, a stoneware used for household plates will have a composition of about 45 percent clay, 20 percent feldspar, and 35 percent quartz. By contrast, a dental ceramic for tooth caps is made from about 80 percent feldspar, 15 percent clay, and 5 percent quartz.

The major interest today, however, is in nontraditional ceramics, particularly metal oxides. To form a solid ceramic, the microcrystalline powder is heated to just below its melting point, sometimes under pressure as well. Under these conditions, bonding between the crystal surfaces occurs, a process known as *sintering*. Aluminum oxide is a typical example. Aluminum oxide ceramic is used as an insulator in automobile spark plugs and as a replacement for bone tissue, such as in artificial hips. The most widely used non-oxide ceramic, silicon carbide was discussed earlier in this chapter.

As the search for new materials intensifies, boundaries between compound classifications are disappearing. *Cermets* are materials containing cemented grains of metals and ceramic compounds; *glassy ceramics* are glasses in which a carefully controlled proportion of crystals has been grown. Two examples of compounds that can be formed into glassy ceramics are lithium aluminum silicate, $Li_2Al_2Si_4O_{12}$, and magnesium aluminum silicate, $Mg_2Al_4Si_5O_{18}$. These materials are nonporous and are known for their extreme resistance to thermal shock. That is, they can be heated to red heat and then be plunged into cold water without shattering. The major use of this material is in cooking utensils and heat-resistant cooking surfaces. Many of these glassy ceramics are produced by Corning.

Another use for glassy ceramics is in saving lives. Flying food and supplies into troubled parts of the world has been very hazardous for the flight crews. Their low-flying aircraft are easy targets for small-arms ground fire. In the past, all the crew had for protection was relatively ineffective titanium sheets under their seats. The U.S. and British air forces have now equipped some of their Hercules aircraft with glass ceramic tiles underneath and around the flight deck. A high-velocity round hitting the ceramic layer breaks into fragments as it plows into the glassy ceramic, losing most of its kinetic energy in the process. This low-mass, low-cost material has the potential to enable emergency agencies to bring food into war-ravaged areas much more safely.

14.22 Silicones

Silicones, more correctly called polysiloxanes, constitute an enormous family of polymers, and they all contain a chain of alternating silicon and oxygen atoms. Attached to the silicon atoms are pairs of organic groups, such as the methyl group, CH_3. The structure of this simplest silicone is shown in Figure 14.27, where the number of repeating units, n, is very large. To synthesize this compound, chloromethane, CH_3Cl, is passed over a copper–silicon alloy at 300°C. A mixture of compounds is produced, including $(CH_3)_2SiCl_2$:

$$2\ CH_3Cl(g) + Si(s) \xrightarrow{\Delta} (CH_3)_2SiCl_2(l)$$

Water is added, causing hydrolysis:

$$(CH_3)_2SiCl_2(l) + 2\ H_2O(l) \rightarrow (CH_3)_2Si(OH)_2(l) + 2\ HCl(g)$$

The hydroxo compound then polymerizes, with loss of water:

$$n(CH_3)_2Si(OH)_2(l) \rightarrow [\text{—O—Si}(CH_3)_2\text{—}]_n(l) + H_2O(l)$$

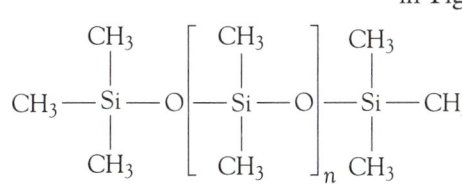

Figure 14.27 Structure of the simplest silicone, catena-poly-[(dimethylsilicon)-μ-oxo]. The number of repeating units, n, is very large.

Silicones are used for a wide variety of purposes. The liquid silicones are more stable than hydrocarbon oils. In addition, their viscosity changes little with temperature, whereas the viscosity of hydrocarbon oils changes dramatically with temperature. Thus, silicones are used as lubricants and wherever inert fluids are needed, for example, in hydraulic braking systems. Silicones are very hydrophobic (nonwetting); hence, they are used in water-repellent sprays for shoes and other items.

By cross-linking chains, silicone rubbers can be produced. Like the silicone oils, the rubbers show great stability to high temperature and to chemical attack. Their multitudinous uses include the face-fitting edges for snorkel and scuba masks. The rubbers also are very useful in medical applications, such as transfusion tubes. However, silicone gels have attained notoriety in

Inorganic Polymers

Seeing how many types of organic polymers are known, why are there not at least as many inorganic polymers? This is a very good question. First of all, the element(s) that comprise the backbone of the polymer must show a tendency to catenate; that is, they must readily form chains. And to be useful, the chains must be stable in the presence of atmospheric oxygen. Second, at least one of the backbone elements must form more than two covalent bonds; otherwise side chain substituents would not be possible. It is the variations in side chain substituents that enables synthetic chemists to "fine-tune" the properties of a polymer to match the needs of a particular application.

Organic polymer chemistry is a well-established branch of chemistry, but the study of inorganic polymers is still in its infancy. A major reason is the lack of equivalent synthetic pathways to that of organic chemistry. Many organic polymers are synthesized by taking multiple bonded monomers and linking them. Multiple bonded inorganic compounds are much harder to synthesize than alkenes and alkynes. As a result, different synthetic routes have had to be devised and more routes are still needed.

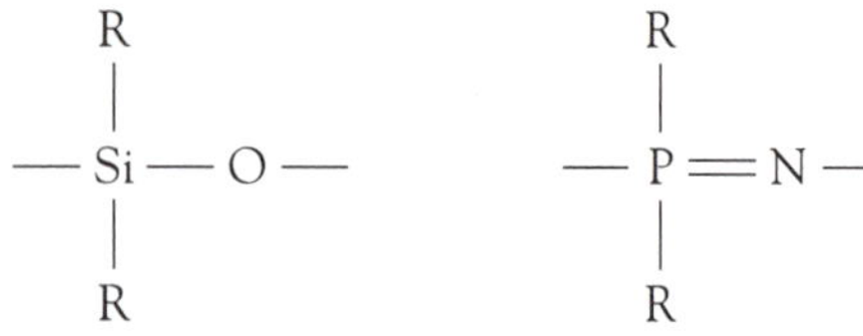

As a result of the synthetic difficulties, until recently, there were only three well-developed families of inorganic polymers, the polysiloxanes (silicones), the polyphosphazenes, and the polysilanes. Figure 14.28 shows the repeating units of the polysiloxanes and the polyphosphazenes. Notice that, by using the phosphorus–nitrogen combination instead of silicon–oxygen, the two series are isoelectronic. However, the bonding is different. Silicon forms four single bonds and oxygen two (with two lone pairs), whereas phosphorus forms five single bonds and nitrogen three (with one lone pair). Thus, the phosphazines have alternating double bonds along the chain length. Like the polysiloxanes, the polyphosphazene polymers are flexible polymers (elastomers) that are superior to organic polymers in their resistance to degradation. As a result, polyphosphazenes are used in aerospace and automobile applications.

The polysilanes, consisting simply of repeating $(-SiR_2-)_n$ belong to a different family, that of the Group 14 polymers. In fact, all of the Group 14 elements form simple polymer chains; thus, there are also polygermanes $(-GeR_2-)_n$ and polystannanes $(-SnR_2-)_n$. Polystannanes are unique in having a backbone solely of metal atoms. It is the polysilanes that have proved the most interesting. Unlike the carbon-based polymers, the electrons in the silicon backbone are delocalized along the chain. As a result, polysilanes are photosensitive and have potential as electrically conducting polymers.

Among the new polymers being studied are those involving three different elements in their backbone, particularly sulfur, nitrogen, and phosphorus. Another bi-element polymer field is that of the boron–nitrogen polymers. As we mentioned in Chapter 9, Section 9.11, "combo" elements provide interesting parallels with the element they mimic—in this case, carbon. For example, the polyiminoborane shown

$$
\begin{array}{c}
\text{R} \\
| \\
-\,\text{Si}-\text{O}- \\
| \\
\text{R}
\end{array}
\qquad
\begin{array}{c}
\text{R} \\
| \\
-\,\text{P}=\text{N}- \\
| \\
\text{R}
\end{array}
$$

    (a)   (b)

Figure 14.28 The repeating units of (a) the polysiloxanes and (b) the polyphosphazenes.

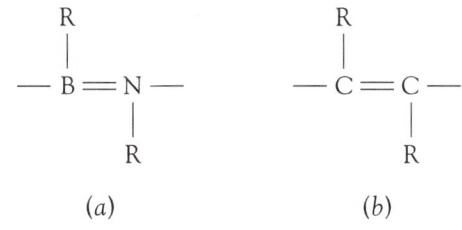

Figure 14.29 The repeating units of (a) the polyiminoboranes and (b) polyacetylene.

in Figure 14.29*a* has been pre-pared. This is analogous in structure to polyacetylene (Figure 14.29*b*).

Though most interest is in the polymers themselves, they are also seen as routes to ceramic materials through pyrolysis at very high temperatures. Thus, a shape can be molded in the polymer, the polymer heated, and the side chains vaporized, with the backbone ceramic retaining the shape of the object. For example, heating polyborazine ($-B_3N_3H_4-)_n$ results in high-purity boron nitride (BN).

their role as a breast implant material. While sealed in a polymer sack, they are believed to be harmless. The major problem arises when the container walls leak or break. The silicone gel can then diffuse into surrounding tissues. The chemical inertness of silicones turns from a benefit to a problem, because the body has no mechanism for breaking down the polymer molecules. Many medical personnel believe that these alien gel fragments trigger the immune system, thereby causing a number of medical problems.

The advantages of the silicone polymers over carbon-based polymers results from several factors. First, the silicon–oxygen bond in the backbone of the molecule is stronger than the carbon–carbon bond in the organic polymers (452 kJ·mol^{-1} compared to about 346 kJ·mol^{-1}) making the silicon-based polymers more resistant to oxidation at high temperatures. It is for this reason that high temperature oil baths always utilize silicone oils, not hydrocarbon oils. The absence of substituents on the oxygen atoms in the chain and the wider bond angle (Si—O—Si being $143°$ compared with $109°$ for C—C—C) results in the greater flexibility of a silicone polymer.

14.23 Tin and Lead

Tin forms two common allotropes: the shiny metallic allotrope, which is thermodynamically stable above $13°C$, and the gray, nonmetallic diamond-structure allotrope, which is stable below that temperature. The change at low temperatures to microcrystals of the gray allotrope is slow at first but accelerates rapidly. This transition is a particular problem in poorly heated museums, where priceless historical artifacts can crumble into a pile of tin powder. The effect can spread from one object to another in contact with it, and this lifelike behavior has been referred to as "tin plague" or "museum disease." The soldiers of Napoleon's army had tin buttons fastening their clothes, and they used tin cooking utensils. It is believed by some that, during the bitterly cold winter invasion of Russia, the crumbling of buttons, plates, and pans contributed to the low morale and hence to the ultimate defeat of the Imperial French troops.

The existence of both a metallic and a nonmetallic allotrope identifies tin as a real "borderline" or weak metal. Tin is also amphoteric, another of its weak metallic properties. Thus, tin(II) oxide reacts with acid to give (covalent) tin(II) salts and with bases to form the stannite ion, $[Sn(OH)_3]^-$

$$SnO(s) + 2\ HCl(aq) \rightarrow SnCl_2(aq) + H_2O(l)$$

$$SnO(s) + NaOH(aq) + H_2O(l) \rightarrow Na^+(aq) + [Sn(OH)_3]^-(aq)$$

Lead, the more economically important of the two metals, is a soft, gray-black, dense solid found almost exclusively as lead(II) sulfide, the mineral

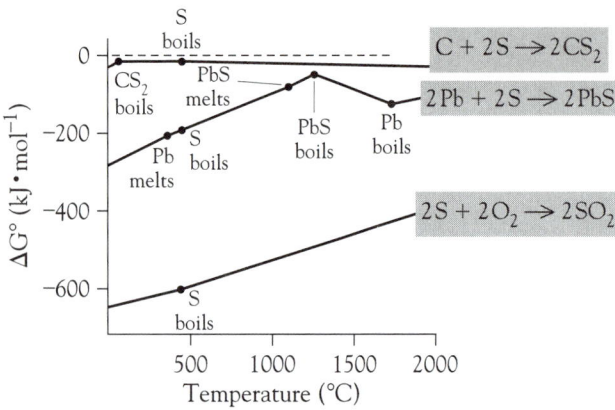

Figure 14.30 Ellingham diagram for lead(II) sulfide, carbon disulfide, and sulfur dioxide.

galena. Lead(II) sulfide cannot be directly reduced with carbon because that reaction has an unfavorable free energy change. This relation is shown in Figure 14.30; note that the lead(II) sulfide line does not cross that of carbon disulfide (the product of oxidation of carbon with sulfur). Instead, the compound is heated with air to oxidize the sulfide ions to sulfur dioxide (which the Ellingham diagram shows to be feasible). The lead(II) oxide can then be reduced with coke to lead metal:

$$2 \; PbS(s) + 3 \; O_2(g) \xrightarrow{\Delta} 2 \; PbO(s) + 2 \; SO_2(g)$$

$$PbO(s) + C(s) \xrightarrow{\Delta} Pb(l) + CO(g)$$

There are two major environmental concerns that arise in connection with this lead extraction process. First, the sulfur dioxide produced contributes to atmospheric pollution unless it is utilized in another process; second, lead dust must not be permitted to escape during the smelting. Lead is highly toxic, so the best solution is to recycle the metal. At the present time, close to half of the 6 million tonnes of lead used annually comes from recycling. The aim must be to substantially increase this proportion. In particular, it would help if all defunct lead–acid batteries were returned for disassembly and reuse of the lead contained in them. Of course, such a move would have a negative economic effect as the result of a decline in employment in the lead mining industry. There would, however, be an increase in employment in the labor-intensive recycling and reprocessing sector.

Tin and lead exist in two oxidation states, +4 and +2. It is possible to explain the existence of the +2 oxidation state in terms of the inert-pair effect, as we did for the +1 oxidation state of thallium. The formation of ions of these metals is rare. Tin and lead compounds in which the metals are in the +4 oxidation state are covalent, except for a few solid-phase compounds. Even when in the +2 oxidation state, tin generally forms covalent bonds, with ionic bonds only being present in compounds in the solid phase. Conversely, lead forms the 2+ ion in solid and in solution. Table 14.5 shows that the charge density for Pb^{2+} is relatively low, whereas that for 4+ ions is extremely high—high enough to cause the formation of covalent bonds with all but the least polarizable anion, fluoride.

Table 14.5 Charge densities of lead ions

Ion	Charge density ($C \cdot mm^{-3}$)
Pb^{2+}	32
Pb^{4+}	196

14.24 Tin and Lead Oxides

The oxides of the heavier members of Group 14 can be regarded as ionic solids. Tin(IV) oxide, SnO_2, is the stable oxide of tin, whereas lead(II) oxide, PbO, is the stable oxide of lead. Lead(II) oxide exists in two crystalline forms, one yellow and the other red. There is also a mixed oxide of lead, Pb_3O_4 (red lead), which behaves chemically as $PbO_2 \cdot 2PbO$; hence, its systematic name is lead(II) lead(IV) oxide. The chocolate-brown lead(IV) oxide, PbO_2, is quite stable, and it is a good oxidizing agent. We can see the differences between tin and lead in this respect from the Frost diagram shown in Figure 14.31.

Tin(IV) oxide is incorporated in glazes used in the ceramics industry. About 3500 tonnes is used annually for this purpose. The consumption of lead(II) oxide is much higher, of the order of 250 000 tonnes annually, because it is used to make lead glass and for the production of the electrode

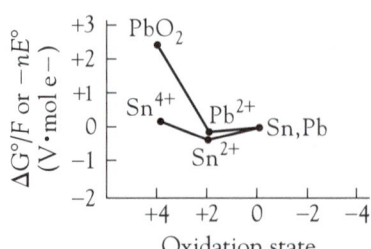

Figure 14.31 Frost diagram for tin and lead.

surfaces in lead–acid batteries. In these batteries, both electrodes are formed by pressing lead(II) oxide into a frame of lead metal. The cathode is formed by oxidizing lead(II) oxide to lead(IV) oxide, and the anode is produced by reducing lead(II) oxide to lead metal. The electric current arises when lead(IV) oxide is reduced to insoluble lead(II) sulfate in the sulfuric acid electrolyte while the lead metal is oxidized to lead(II) sulfate on the other electrode:

$$PbO_2(s) + 4\ H^+(aq) + SO_4{}^{2-}(aq) + 2\ e^- \rightarrow PbSO_4(s) + 2\ H_2O(l)$$

$$Pb(s) + SO_4{}^{2-}(aq) \rightarrow PbSO_4(s) + 2\ e^-$$

These two half-reactions are reversible. Hence, the battery can be recharged by applying an electric current in the reverse direction. In spite of a tremendous quantity of research, it has been very difficult to develop a low-cost, lead-free, heavy-duty battery that can perform as well as the lead–acid battery.

Red lead, Pb_3O_4, has been used on a large scale as a rust-resistant surface coating for iron and steel. Mixed metal oxides, such as calcium lead(IV) oxide, $CaPbO_3$, are now being used as an even more effective protection against salt water for steel structures. The structure of $CaPbO_3$ is discussed in Chapter 16, Section 16.6.

As mentioned earlier, the lead(IV) ion is too polarizing to exist in aqueous solution. Oxygen can often be used to stabilize the highest oxidation number of an element, and this phenomenon is true for lead. Lead(IV) oxide is an insoluble solid in which the Pb^{4+} ions are stabilized in the lattice by the high lattice energy. Even then, one can argue that there is considerable covalent character in the structure. Addition of an acid, such as nitric acid, gives immediate reduction to the lead(II) ion and the production of oxygen gas:

$$2\ PbO_2(s) + 4\ HNO_3(aq) \rightarrow 2\ Pb(NO_3)_2(aq) + 2\ H_2O(l) + O_2(g)$$

In the cold, lead(IV) oxide undergoes a double-replacement reaction with concentrated hydrochloric acid to give covalently bonded lead(IV) chloride. When warmed, the unstable lead(IV) chloride decomposes to give lead(II) chloride and chlorine gas:

$$PbO_2(s) + 4\ HCl(aq) \rightarrow PbCl_4(aq) + 2\ H_2O(l)$$

$$PbCl_4(aq) \rightarrow PbCl_2(s) + Cl_2(g)$$

14.25 *Tin and Lead Chlorides*

Tin(IV) chloride is a typical covalent metal chloride. It is an oily liquid that fumes in moist air to give a gelatinous tin(IV) hydroxide, which we represent as $Sn(OH)_4$ (although it is actually more of a hydrated oxide), and hydrogen chloride gas:

$$SnCl_4(l) + 4\ H_2O(l) \rightarrow Sn(OH)_4(s) + 4\ HCl(g)$$

Like so many compounds, tin(IV) chloride has a small but important role in our lives. The vapor of this compound is applied to freshly formed glass, where it reacts with water molecules on the glass surface to form a layer of tin(IV) oxide. This very thin layer substantially improves the strength of the glass, a property particularly important in eyeglasses. A thicker coating of tin(IV) oxide acts as an electrically conducting layer. Aircraft cockpit windows use such a coating. An electric current is applied across the conducting glass surface, and the resistive heat that is generated prevents frost formation when the aircraft descends from the cold upper atmosphere.

Lead(IV) chloride is a yellow oil that, like its tin analog, decomposes in the presence of moisture and explodes when heated. Lead(IV) bromide and iodide do not exist, because the oxidation potential of these two halogens is sufficient to reduce lead(IV) to lead(II).

Lead(II) chloride is a white insoluble solid that forms an ionic-type crystal lattice. Tin(II) chloride, however, is a white water-soluble solid that forms a solid-phase structure in which there are covalently bridging chloride ions. Its covalent nature is also apparent from its relatively high solubility in low-polarity organic solvents. In the gas phase, tin(II) chloride is a V-shaped molecule. According to VSEPR theory, this molecular shape is due to the presence of a lone pair (Figure 14.32).

With a lone pair, tin(II) chloride would be expected to be a Lewis base. But, as has been said before, chemistry is full of surprises: The compound generally behaves like a Lewis acid. Thus, the lone pair appears to be unreactive—a truer use of the term inert-pair effect. For example, tin(II) chloride reacts with chloride ion to form the trichlorostannate(II) ion, $SnCl_3^-$ (Figure 14.33). The bond angle of about 90° suggests an alternative explanation for the bonding; that is, tin uses pure p orbitals. This model would explain the 90° bond angles and the lack of Lewis base behavior. In other words, the electron pair on the tin atom in tin(II) chloride is in a spherical s orbital rather than being part of a directional sp^2 or sp^3 hybrid. The tin then uses an empty p orbital to bond with the lone pair of the chloride ion.

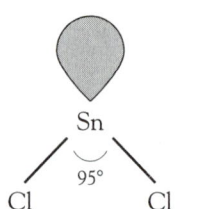

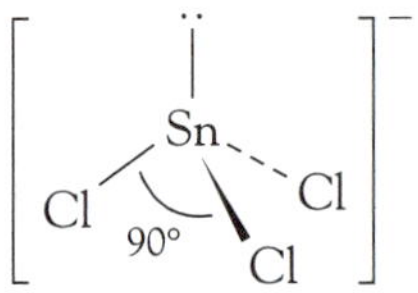

Figure 14.32   The tin(II) chloride molecule in the gas phase.

Figure 14.33   The trichlorostannate(II) ion, $SnCl_3^-$.

14.26   Tetraethyllead

The less electropositive (more weakly metallic) metals form an extensive range of compounds containing metal–carbon bonds. The metal–carbon compound that has been produced on the largest scale is tetraethyllead, $Pb(C_2H_5)_4$, known as TEL. Tetraethyllead is a stable compound that has a low boiling point and at one time was produced on a vast scale as a gasoline additive. One method of synthesis involves the reaction of a sodium–lead alloy with chloroethane (ethyl chloride):

$$4\,NaPb(s) + 4\,C_2H_5Cl(l) \xrightarrow[\Delta]{\text{high pressure}} Pb(C_2H_5)_4(l) + 3\,Pb(s) + 4\,NaCl(s)$$

In a gasoline engine, a spark is used to ignite the mixture of fuel and air. However, straight-chain hydrocarbons will burn simply when compressed with air—the mode of operation of a diesel engine. This reactivity is responsible for the phenomenon of premature ignition (commonly called knocking or pinging), and in addition to making the engine sound as if it is about to fall apart, it can cause severe damage. Branched-chain molecules, however, because of their kinetic inertness, require a spark to initiate combustion (Figure 14.34).

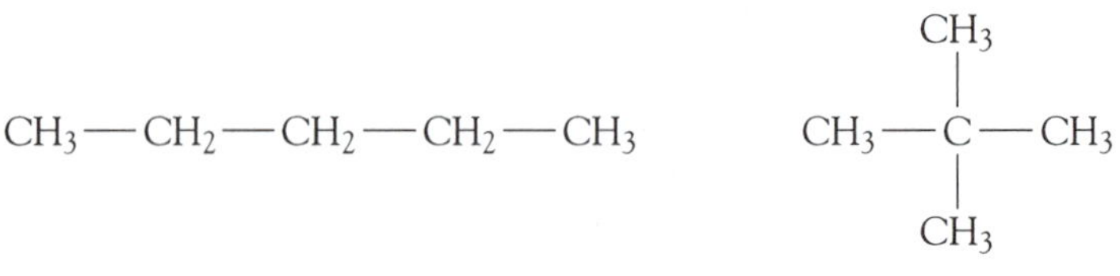

$$CH_3-CH_2-CH_2-CH_2-CH_3$$

$$CH_3-\overset{\overset{\displaystyle CH_3}{|}}{\underset{\underset{\displaystyle CH_3}{|}}{C}}-CH_3$$

(a)   (b)

Figure 14.34   Two hydrocarbons of the same formula, C_5H_{12}: (a) a straight-chain isomer and (b) a branched-chain isomer.

TEL: A Case History

The story of the use of TEL is a prime example of the dominance of economic benefit over health issues and of the control of information and research. The health hazards of lead and, in particular, TEL were known in the early part of the 20th century, yet the chemical corporations, the gasoline companies, and the auto manufacturers colluded to promote TEL, to support research that promoted TEL, and to discredit those warning of health problems. Possible alternative additives, particularly the low-cost ethanol that was popular at the time, were suppressed. In fact, Midgely himself had patented ethanol as a means of enhancing the octane rating of gasoline before he became enamored with TEL. Leaded gasoline first went on sale in 1923, though it was called Ethyl gasoline to hide the fact that it contained lead. It was in the same year that the first (of several) deaths occurred at TEL manufacturing plants. Even in those days there were concerns about the lead released into the environment by the combustion of TEL. For example, the New York Board of Health banned the sale of TEL-enhanced gasoline in 1924, a ban that was lifted in 1926.

One of the pioneer fighters against the use of TEL was Alice Hamilton. Hamilton, the first female faculty member at Harvard Medical School, was the foremost American industrial toxicologist of her time. She expressed her concerns in 1925, the year that the U.S. Surgeon General convened a conference to assess the hazards of TEL. The position of the automobile industry and that of the gasoline manufacturers (who closely colluded on the issue) was that (1) leaded gasoline was essential to the progress of America, (2) any innovation entailed certain risks, and (3) deaths in TEL manufacturing plants was due to carelessness. Dr. Yandell Henderson, a physiologist at Yale University, severely criticized the use of leaded gasoline. However, a committee set up following the conference concluded that there was no good grounds for "prohibiting the use of ethyl gasoline" but suggested further investigations were necessary. However, no funding for these investigations was approved by Congress.

Though evidence of the toxicity of lead accumulated through the 1930s and 1940s, TEL was safe from criticism. Responding to a complaint from Ethyl Gasoline Corporation, manufacturer of TEL (and owned by General Motors and Standard Oil of New Jersey), the Federal Trade Commission (FTC) issued a restraining order preventing competitors from criticizing leaded gasoline in the commercial marketplace. Ethyl gasoline, the FTC order read, "is entirely safe to the health of motorists and the public."

It was the passage of Clean Air legislation in 1970 that largely forced the demise of TEL. The platinum used in catalytic converters is "poisoned" by lead. Even then, Ethyl Corporation sued the Environmental Protection Agency for denying a market for their product. Ethyl claimed that the case against lead was not proven, despite the many studies on its toxicity. Though a lower court upheld Ethyl's claim, this decision was reversed by the U.S. Court of Appeals. In 1982, the then-Administration's Task Force on Regulatory Relief planned to relax or eliminate the lead phaseout, but under political and public pressure, the Government reversed its opposition to lead phaseout. By 1986, the primary phaseout of leaded gasoline in the United States was completed.

The measure of the proportion of branched-chain molecules in gasoline is the *octane rating*; the higher the proportion of branched-chain molecules, the higher the octane rating of the fuel. With the demand for higher performance, higher compression engines, the need for higher octane rated gasoline became acute. The addition of TEL to low octane-rated gasoline increases the octane rating; that is, it prevents premature ignition. In the early 1970s, about 500 000 tonnes of TEL is produced annually for addition to gasoline. In fact, the U.S. Environmental Protection Agency allowed up to 3 g of TEL per gallon of gasoline until 1976.

Thomas Midgley discovered both the chlorofluorocarbons and the role of TEL in improving gasolines. The irony is that both discoveries were designed to make life better through progress in chemistry, and yet both have had quite the opposite long-term effect. Tetraethyllead poses both direct and indirect hazards. The direct hazard has been to people working with gasoline, such as gas station attendants. Because it has a low boiling point, the TEL added to gasoline vaporizes readily; hence, persons exposed to TEL vapor absorb this neurotoxic lead compound through the lining of their lungs; they develop headaches, tremors, and increasingly severe neurologic disorders. The more widespread problem is the lead particulates in the automobile exhausts. In urban areas this is lung absorbed by the inhabitants, while in rural areas near major highways, crops absorb lead, and those consuming the crops will, in turn, experience increased lead intake. A significant proportion of lead in the environment has come from the use of leaded gasoline. To illustrate how the use of TEL has become a global issue, increased lead levels have even been found in the ice cap of Greenland.

Germany, Japan, and the former USSR were quick to outlaw TEL; other countries (such as the United States) followed more slowly. One of the problems of eliminating TEL from gasolines was simply that modern vehicles need high octane rated gasoline. Two solutions have been found: the development of the zeolite catalysts that enable oil companies to convert straight-chain molecules to the required branched-chain molecules and the addition of oxygenated compounds, such as ethanol, to fuels. Thus, the need for octane boosters has been eliminated. More and more countries around the world are phasing out TEL, but it will be many years before the planet will be TEL-free.

14.27 *Biological Aspects*

The Carbon Cycle

There are many biogeochemical cycles on this planet. The largest scale process is the carbon cycle. Of the 2×10^{16} tonnes of carbon, most of it is "locked away" in the Earth's crust as carbonates, coal, and oil. Only about 2.5×10^{12} tonnes is available as carbon dioxide. Every year, about 15 percent of this total is absorbed by plants and algae in the process of photosynthesis, which uses energy from the Sun to synthesize complex molecules such as sucrose.

Some plants are eaten by animals (such as humans), and a part of the stored chemical energy is released during their decomposition to carbon dioxide and water. These two products are returned to the atmosphere by the process of respiration. However, the majority of the carbon dioxide incorporated into plants is returned to the atmosphere only after the death and subsequent decomposition of the plant organisms. Another portion of the plant material is buried, thereby contributing to the soil humus or the formation of peat bogs. The carbon cycle is partially balanced by the copious output of carbon dioxide by volcanoes. The demand for energy has led to the burning of coal and oil, which were formed mainly in the Carboniferous era. This combustion adds about 2.5×10^{10} tonnes of carbon dioxide to the atmosphere each year, in addition to that from natural cycles. Although we are just returning to the atmosphere carbon dioxide that came from there, we are doing so at a very rapid rate, and many scientists are concerned that the rate of return will overwhelm the Earth's absorption mechanisms. This topic is currently being studied in many laboratories.

The Essentiality of Silicon

Silicon is the second most abundant element in the Earth's crust, yet its biological role is limited by the low-water solubility of its common forms, silicon dioxide and silicic acid, H_4SiO_4. At about neutral pH, silicic acid is uncharged and has a solubility of about 2×10^{-3} mol·L^{-1}. As the pH increases, polysilicic acids predominate, then colloidal particles of hydrated silicon dioxide. Though the solubility of silicic acid is low, on the global scale it is enormous, with about 2×10^{11} tonnes of silicic acid entering the sea per year. It is the continuous supply of silicic acid into the sea that enables marine organisms such as diatoms and radiolaria to construct their exoskeletons of hydrated silica.

On a smaller scale, plants require the absorption of about 600 L of water to form about 1 kg of dry mass; thus, plants consist of about 0.15% silicon. The silica is used by the plants to stiffen leaves and stalks. In some plants, it is also used for defense (see the Feature on Biomineralization in Chapter 12). Further up the food chain, herbivores ingest considerable amounts of silica. A sheep consumes about 30 g of silicon per day, though almost all is excreted. Humans are estimated to consume about 30 mg per day, about 60 percent from breakfast cereal and 20 percent from water and drinks. It is the water-dissolved silicic acid that is bioavailable to our bodies.

The most convincing way to illustrate the essentiality of an element is to grow an organism in the total absence of that element. This is a very difficult, but not impossible task. Studies with both rats and chicks showed that silicon-free diets resulted in stunted growths for both animals. Addition of silicic acid to the diet rapidly restored natural growth. The question obviously arose as to the function of the silicon. Chemical studies showed that silicic acid did not react or bind with organic molecules. Thus, incorporation into some essential biosynthetic pathway seemed highly unlikely. The answer seems to lie with its inorganic chemistry. As we saw in Chapter 13, Section 13.12, aluminum is ubiquitous in the environment and this element is highly toxic to organisms. Addition of silicic acid to a saturated neutral solution of aluminum ion causes almost complete precipitation of the aluminum in the form of insoluble hydrated aluminosilicates.

Evidence that silicon did act in a preventative role was provided by a study of young salmon. Those in water containing aluminum ion died within 48 hours. Those in water containing the same concentration of aluminum plus silicic acid thrived. It is now generally accepted that indeed silicon is essential to our diet to inhibit the toxicity of the naturally present aluminum in our foodstuffs.

Though silicon is an essential element, lung-absorbed silica is highly toxic. We have already mentioned the hazards of asbestos. It can cause two serious lung diseases: asbestosis and mesothelioma. The dust of any silicate rock will also cause lung damage, in this case, silicosis. The fundamental cause of the lung problems is due to the total insolubility of the silicates. Once the particles stick in the lungs, they are there for life. The irritation they cause produces scarring and immune responses that lead to the disease state.

The Toxicity of Tin

Although the element and its simple inorganic compounds have a fairly low toxicity, its organometallic compounds are very toxic. Compounds such as hydroxotributyltin, $(C_4H_9)_3SnOH$, are effective against fungal infections in potatoes, grapevines, and rice plants. For many years, organotin compounds were incorporated into the paints used on ships' hulls. The compound would kill the larvae of mollusks, such as barnacles, that tend to attach themselves to a ship's hull, slowing the vessel considerably. However, the organotin compound slowly leaches into the surrounding waters, where, particularly within the confines of a harbor, it destroys other marine organisms. For this reason, its marine use has been curtailed.

The Severe Hazard of Lead

Why is there such a concern about lead poisoning? For many elements the levels to which we are naturally exposed is many times smaller than toxic levels. For lead, however, there is a relatively small safety margin between unavoidable ingestion from our food, water, and air, and the level at which toxic symptoms become apparent. Lead is ubiquitous in our environment. Plants absorb lead from the soil, and water dissolves up traces of lead compounds.

In addition to what our environment contains, humans have used lead products throughout history. Although we no longer use "sugar of lead" as a sweetener, in more recent times lead from a number of other sources has become a hazard. Basic lead carbonate, $Pb_3(CO_3)_2(OH)_2$, was one of the few easily obtainable white substances. Thus, it was used until recent times as a paint pigment, and many old houses have unacceptably high levels of lead resulting from the use of lead-based paints on the walls and ceilings. The same compound was used by women as a cosmetic. The factories producing the lead compound were known as "white cemeteries." Working in such a plant was a last resort. Yet circumstances of illness or death in the family or idleness or drunkenness by a husband gave some women little option even though it was a virtual death sentence.

The shift away from leaded gasolines has led to a drastic reduction in airborne lead particles, but lead is entering the environment from other sources. The most commonly recognized source is the lead battery industry which today comprises about 85 percent of lead consumption. The lead–acid battery is still the most efficient and cost-effective method of storing energy. Lead is the most recycled element—particularly from defunct batteries. The recycling process in the United States is generally carried out using extremely safe conditions with due safeguards for the workers and the environment. However, such facilities are expensive to build and operate. Much of the world's lead is recycled in low-income countries, particularly in Asia, where safety and environmental concerns are lesser priorities. With the cheapness of overseas recycling, much of the lead in the United States, Japan, and other developed countries is shipped to these Far Eastern recycling plants. Though it is a very laudable aim to reduce poverty in such countries by bringing increased employment, in this specific example, the developed nations are exporting pollution.

About 95 percent of absorbed lead substitutes for calcium in the hydroxyapatite of bone. This can be explained as the lead(II) ion is only slightly larger than and has a similar charge density to that of calcium. Thus, the

Cathode-ray-based televisions and computer CRT monitors contain significant quantities of lead together with other hazardous elements. It is important to recycle these just as much as batteries.

body "stores" lead. In the bone, the half-life is about 25 years, so lead poisoning is a very long-term problem. Lead interferes with the synthesis of hemoglobin; thus, it can indirectly cause anemia. At high concentrations, kidney failure, convulsions, brain damage, and then death ensue. There is also strong evidence of neurological effects, including reduced IQ in children exposed to more than minimal lead levels. As part of a program to minimize the hazard, playgrounds built on old industrial sites are checked for lead levels and, if necessary, closed or resurfaced.

14.28   *Element Reaction Flowchart*

Flowcharts are shown for both carbon and silicon.

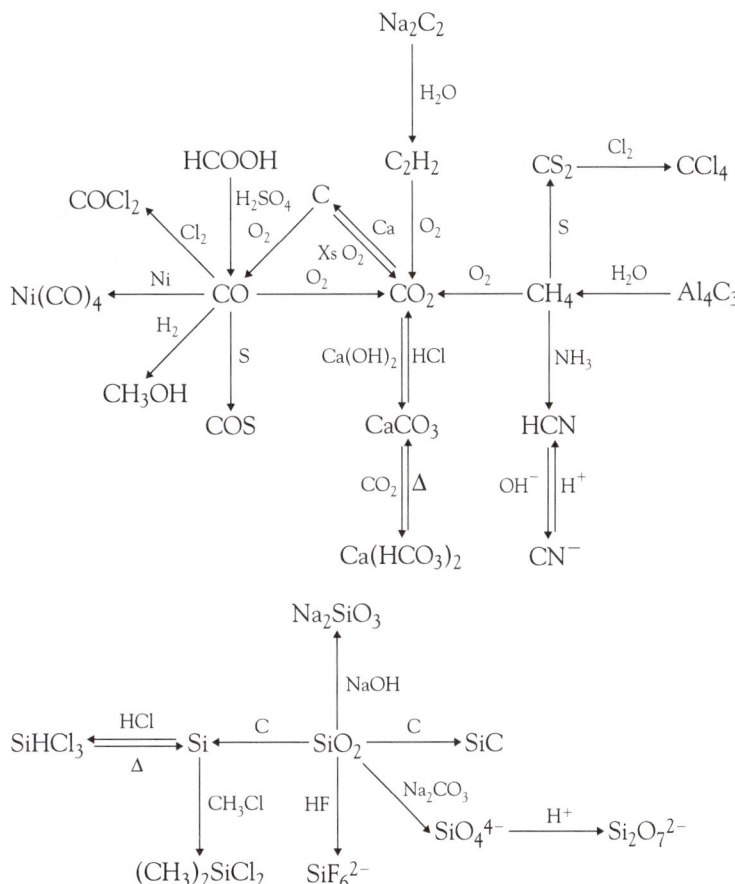

EXERCISES

14.1   Write balanced chemical equations corresponding to the following chemical reactions:
(a) solid lithium dicarbide($2-$) with water
(b) silicon dioxide with carbon
(c) copper(II) oxide heated with carbon monoxide
(d) calcium hydroxide solution with carbon dioxide (two equations)
(e) methane with molten sulfur
(f) silicon dioxide with molten sodium carbonate

(g) lead(IV) oxide with concentrated hydrochloric acid (two equations)

14.2   Write balanced chemical equations corresponding to the following chemical reactions:
(a) solid beryllium carbide with water
(b) carbon monoxide with dichlorine
(c) hot magnesium metal with carbon dioxide
(d) solid sodium carbonate with hydrochloric acid

(e) heating barium carbonate

(f) carbon disulfide gas and chlorine gas

(g) tin(II) oxide with hydrochloric acid

14.3 Define the following terms: (a) catenation; (b) aerogel; (c) ceramic; (d) silicone.

14.4 Define the following terms: (a) glass; (b) molecular sieves; (c) cermet; (d) galena.

14.5 Contrast the properties of the three main allotropes of carbon—diamond, graphite, and C_{60}.

14.6 Explain why (a) diamond has a very high thermal conductivity; (b) high pressure and temperature are required for the traditional method of diamond synthesis.

14.7 Why are fullerenes soluble in many solvents even though both graphite and diamond are insoluble in all solvents?

14.8 Explain why catenation is common for carbon but not for silicon.

14.9 Compare and contrast the three classes of carbides.

14.10 Calcium carbide forms an NaCl structure with a density of 2.22 g·cm^{-3}. Assuming the carbide ion is spherical and taking the ionic radius of calcium as 114 pm, what is the radius of the carbide ion?

14.11 Write the chemical equation for the reaction used in the commercial production of silicon carbide. Is it enthalpy or entropy driven? Explain your reasoning. Calculate the values of ΔH° and ΔS° for the process to confirm your deduction, then calculate ΔG° at 2000°C.

14.12 In compounds of carbon monoxide with metals, it is the carbon atom that acts as a Lewis base. Depict why this is expected using a formal charge representation of the carbon monoxide molecule.

14.13 Carbon dioxide has a negative enthalpy of formation, whereas that of carbon disulfide is positive. Using bond energy data, construct a pair of enthalpy of formation diagrams and identify the reason(s) for such different values.

14.14 Contrast the properties of carbon monoxide and carbon dioxide.

14.15 Discuss the bonding in carbon disulfide in terms of hybridization theory.

14.16 From data tables in the Appendix of ΔH_f° and S° values, show that the combustion of methane is a spontaneous process.

14.17 Explain why silane burns in contact with air, whereas methane requires a spark before it will combust.

14.18 Describe why the CFCs were once thought to be ideal refrigerants.

14.19 Why is HFC-134a, a less than ideal replacement for CFC-12?

14.20 What would be the chemical formula of HFC-134b?

14.21 Why does methane represent a particular concern as a potential greenhouse gas?

14.22 Contrast the properties of carbon dioxide and silicon dioxide and explain these differences in terms of the bond types. Suggest an explanation as to why the two oxides adopt such dissimilar bonding.

14.23 The ion CO_2^- can be prepared using ultraviolet irradiation. Whereas the carbon dioxide molecule is linear, this ion is V-shaped with a bond angle of about 127°. Use an electron-dot diagram to aid your explanation. Also, estimate an average carbon–oxygen bond order for the ion and contrast to that in the carbon dioxide molecule.

14.24 Draw the electron-dot diagram for the symmetrical cyanamide ion, CN_2^{2-}. Then, deduce the bond angle in the ion.

14.25 What geometry would you expect for the ion $:C(CN)_3^-$? In fact, it is trigonal planar. Construct one of the three resonance forms to depict the probable electron arrangement and deduce an average carbon–carbon bond order.

14.26 Ultramarine, the beautiful blue pigment used in oil-based paints, has the formula $Na_x[Al_6Si_6O_{24}]S_2$ where the sulfur is present as the disulfide ion, S_2^{2-}. Determine the value of x.

14.27 In crocidolite, $Na_2Fe_5(Si_4O_{11})_2(OH)_2$, how many of the iron ions must have a 2+ charge and how many a 3+ charge?

14.28 Describe the difference in structure between white asbestos and talc.

14.29 Describe the major uses of zeolites.

14.30 If the water in a zeolite is expelled by strong heating, must the absorption of water by the zeolite be an endo- or exothermic process?

14.31 What advantage of silicone polymers becomes a problem when they are used as breast implants?

14.32 Contrast the properties of the oxides of tin and lead.

14.33 Construct the electron-dot structures of tin(IV) chloride and gaseous tin(II) chloride. Draw the corresponding molecular shapes.

14.34 Lead(IV) fluoride melts at 600°C, whereas lead(IV) chloride melts at 215°C. Interpret the values in relation to the probable bonding in the compounds.

14.35 To form the electrodes in the lead–acid battery, the cathode is produced by oxidizing lead(II) oxide to lead(IV) oxide, and the anode is produced by reducing the lead(II) oxide to lead metal. Write half-equations to represent the two processes.

14.36 Suggest the probable products formed when $CaCS_3$ is heated.

14.37 Write the formulas of two carbon-containing species that are isoelectronic with the C_2^{2-} ion.

14.38 There are two carbides that appear to contain the C^{4-} ion. What are they and how are they related?

14.39 Discuss why inorganic polymer chemistry is much less developed than organic polymer chemistry.

14.40 Discuss the introduction of tetraethyllead and why its use in gasoline continues today.

14.41 Write balanced chemical equations corresponding to each transformation in the Element Reaction Flowcharts.

BEYOND THE BASICS

14.42 Show from the standard reduction potentials in the Appendix that lead(IV) iodide would not be thermodynamically stable in aqueous solution.

14.43 Our evidence that the Romans ingested high levels of lead(II) comes from examination of skeletons. Suggest why the lead ions would be present in bone tissues.

14.44 What are the main sources of lead in the environment today?

14.45 Conventional soda glass, when washed frequently in hot water, tends to become opaque and rough, while pure silica (SiO_2) glass does not lose its brilliance. Suggest an explanation.

14.46 One route for the formation of the trace atmospheric gas, carbonyl sulfide, is hydrolysis of carbon disulfide. Write a chemical equation for this reaction. How would a water molecule attack a molecule of carbon disulfide? Draw a transition state for the attack, showing the bond polarities. Thus, deduce a possible intermediate for the reaction and suggest why it is feasible.

14.47 There is a trimeric silicate ion, $Si_3O_9^{6-}$.
(a) Draw a probable structure for the ion;
(b) phosphorus forms an isoelectronic and isostructural ion. What would be its formula?
(c) another element forms an isoelectronic and isostructural neutral compound. What would be its formula?

14.48 Methyl isocyanate, H_3CNCO has a bent C—N—C bond while silyl isocyanate, H_3SiNCO has a linear Si—N—C bond. Suggest an explanation for the difference.

14.49 In the following reaction, identify which is the Lewis acid and which the Lewis base. Give your reasoning.

$$Cl^-(aq) + SnCl_2(aq) \rightarrow SnCl_3^-(aq)$$

14.50 Tin reacts with both acids and bases. With dilute nitric acid, the metal gives a solution of tin(II) nitrate and ammonium nitrate; with concentrated sulfuric acid, solid tin(II) sulfate, and gaseous sulfur dioxide; with potassium hydroxide solution, a solution of potassium hexahydroxostannate(IV), $K_2Sn(OH)_6$, and hydrogen gas. Write balanced net ionic equations for these reactions.

14.51 Silicon dioxide is a weaker acid than carbon dioxide. Write a balanced chemical equation to show how silicate rocks, such as Mg_2SiO_4, might, in the presence of "carbonic acid," be a partial sink for atmospheric carbon dioxide.

14.52 When aqueous solutions of aluminum ion and carbonate ion are mixed, a precipitate of aluminum hydroxide is formed. Suggest an explanation for this using net ionic equations.

14.53 A flammable gas (A) is reacted at high temperature with a molten yellow element (B) to give compounds (C) and (D). Compound (D) has the odor of rotten eggs. Compound (C) reacts with a pale green gas (E) to give as final product, compound (F) and element (B). Compound (F) can also be produced by the direct reaction of (A) with (E). Identify each species and write balanced chemical equations for each step.

14.54 Magnesium silicide, Mg_2Si, reacts with hydronium ion to give magnesium ion and a reactive gas (X). A mass of 0.620 g of gas (X) occupied a volume of 244 mL at a temperature of 25°C and a pressure of 100 kPa. The sample of gas decomposed in aqueous hydroxide ion solution to give 0.730 L of hydrogen gas and 1.200 g of silicon dioxide. What is the molecular formula of (X)? Write a balanced chemical equation for the reaction of (X) with water.

14.55 Tin(IV) chloride reacts with an excess of ethyl magnesium bromide, $(C_2H_5)MgBr$, to give two products, one of which is a liquid (Y). Compound (Y) contains only carbon, hydrogen, and tin. 0.1935 g of (Y) was oxidized to give 0.1240 g of tin(IV) oxide. Heating 1.41 g of (Y) with 0.52 g of tin(IV) chloride gives 1.93 g of liquid (Z). When 0.2240 g of (Z) was reacted with silver nitrate solution, 0.1332 g of silver chloride was formed. Oxidation of 0.1865 g of (Z) gave 0.1164 g of tin(IV) oxide. Deduce the empirical formulas of (Y) and (Z). Write a balanced chemical equation for the reaction of (Y) with tin(IV) chloride to give (Z).

14.56 The solid compound aluminum phosphate, $AlPO_4$, adopts a quartz-like structure. Suggest why this occurs.

14.57 Use thermodynamic calculations to show that decomposition of calcium hydrogen carbonate is favored at 80°C.

$$CaCO_3(s) + CO_2(aq) + H_2O(l) \rightleftharpoons Ca(HCO_3)_2(aq)$$

14.58 Suggest why the density of moissanite is slightly less than that of diamond, even though the atomic mass of silicon is much greater than that of carbon.

Chapter 15

The Group 15 Elements

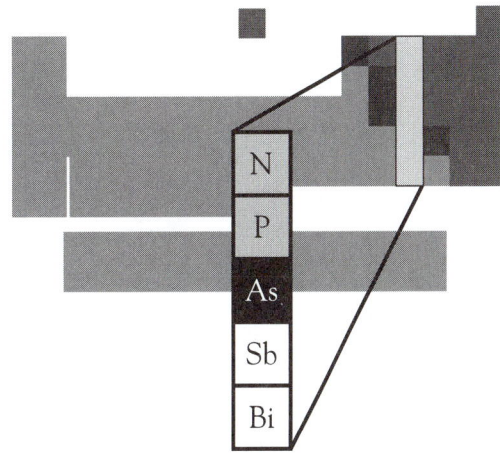

Two of the most dissimilar nonmetallic elements are in the same group: reactive phosphorus and unreactive nitrogen. Of the other members of the group, arsenic is really a semimetal, and the two lower members of the group, antimony and bismuth, exhibit weakly metallic behavior.

The discovery of phosphorus by the German alchemist Hennig Brand in 1669 provides the most interesting saga of the members of this group. The discovery occurred by accident during his investigation of urine. Urine was a favorite topic of research in the 17th century, for it was believed anything gold colored, such as urine, had to contain gold! However, when Brand fermented urine and distilled the product, he obtained a white, waxy, flammable solid with a low melting point—white phosphorus. One hundred years later, a route to extract phosphorus from phosphate rock was devised, and chemists no longer needed buckets of urine to synthesize the element.

In these days of pocket butane lighters, we forget how difficult it used to be to generate a flame. So in 1833, people were delighted to find how easily fire could be produced by using white phosphorus matches. This convenience came at a horrendous human cost, because white phosphorus is extremely toxic. The young women who worked in the match factories died in staggering numbers from phosphorus poisoning. This occupational hazard manifested itself as "phossy jaw," a disintegration of the lower jaw, followed by an agonizing death.

In 1845, the air-stable red phosphorus was shown to be chemically identical to white phosphorus. The British industrial chemist Arthur Albright, who had been troubled by the enormous number of deaths in his match factory, learned of this safer allotrope and determined to produce matches bearing red phosphorus. But mixing the inert red phosphorus with an oxidizing agent gave an instant explosion. Prizes were offered for the development of a safe match, and finally in 1848 some now unknown genius proposed to put half the ingredients on the match tip and the remainder on a strip attached to the matchbox. Only when the two surfaces were brought into contact did ignition of the match head occur—so science and technology moved forward together. (For more details, see the section Matches later in the chapter.)

15.1   Group Trends

The first two members of Group 15, nitrogen and phosphorus, are nonmetals; the remaining three members, arsenic, antimony, and bismuth, have some metallic character. Scientists like to categorize things, but in this group their efforts are frustrated because there is no clear division of properties between nonmetals and metals. Two characteristic properties that we can study are the electrical resistivity of the elements and the acid–base behavior of the oxides (Table 15.1).

Nitrogen and phosphorus are both nonconductors of electricity, and both form acidic oxides, so they are unambiguously classified as nonmetals. The problems start with arsenic. Even though the common allotrope of arsenic looks metallic, subliming and recondensing the solid produce a second allotrope that is a yellow powder. Because it has both metallic-looking and nonmetallic allotropes and forms amphoteric oxides, arsenic can be classified as a semimetal. However, much of its chemistry parallels that of phosphorus, so there is a good case for considering it as a nonmetal.

Antimony and bismuth are almost as borderline as arsenic. Their electrical resistivities are much higher than those of a "true" metal, such as aluminum (2.8 $\mu\Omega\cdot$cm), and even higher than a typical "weak" metal, such as lead (22 $\mu\Omega\cdot$cm). Generally, however, these two elements are categorized as metals.

Table 15.1   Properties of the Group 15 elements

Element	Appearance at SATP	Electrical resistivity ($\mu\Omega\cdot$cm)	Acid–base properties of oxides
Nitrogen	Colorless gas	—	Acidic and neutral
Phosphorus	White, waxy solid	10^{17}	Acidic
Arsenic	Brittle, metallic solid	33	Amphoteric
Antimony	Brittle, metallic solid	42	Amphoretic
Bismuth	Brittle, metallic solid	120	Basic

Table 15.2 Melting and boiling points of the Group 15 elements

Element	Melting point (°C)	Boiling point (°C)
N_2	−210	−196
P_4	44	281
As	Sublimes at 615	
Sb	631	1387
Bi	271	1564

All three of these borderline elements form covalent compounds almost exclusively.

If we want to decide where to draw the vague border between metals and semimetals, the melting and boiling points are as good an indicator as any. In Group 15, these parameters increase as we descend the group, except for a decrease in melting point from antimony to bismuth (Table 15.2). As noted for the alkali metals, the melting points of main group metals tend to decrease down a group, whereas those of nonmetals tend to increase down a group. (We will encounter the latter behavior most clearly with the halogens.) Thus, the increase–decrease pattern shown in Table 15.2 indicates that the lighter members of Group 15 follow the typical nonmetal trend, and the shift to the metallic decreasing trend starts at bismuth. In fact, it is only antimony and bismuth that have the fairly characteristic long liquid range of metals. Thus, we will refer to arsenic as a semimetal and consider antimony and bismuth to be metals, although very "weak" ones.

15.2 *Anomalous Nature of Nitrogen*

In general, the differences between the chemistry of nitrogen and those of the other members of Group 15 relate to some unique features of nitrogen. The major differences are discussed in the following sections.

High Stability of Multiple Bonds

The dinitrogen molecule is a very stable species. Figure 15.1 shows the molecular orbital energy levels involved in the triple bond of that species. The $N{\equiv}N$ bond energy is 942 kJ·mol^{-1}, far greater than that for the triple phosphorus–phosphorus bond (481 kJ·mol^{-1}) and greater even than that for the triple carbon–carbon bond (835 kJ·mol^{-1}). The accepted explanation is

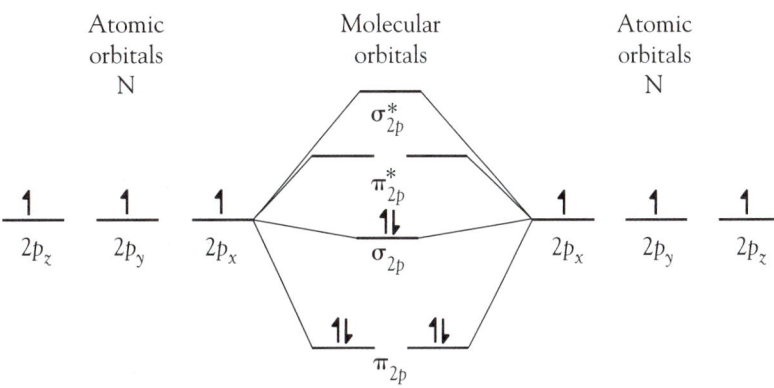

Figure 15.1 Molecular orbital diagram for the 2*p* atomic orbitals of the dinitrogen molecule.

that the p orbitals involved in forming the two π bonds overlap better in the molecule containing the small nitrogen atoms than they do in the other species.

Conversely, nitrogen forms very weak single bonds. The typical nitrogen–nitrogen single bond energy is only 200 $kJ\cdot mol^{-1}$, considerably lower than the bond energy of 346 $kJ\cdot mol^{-1}$ for a carbon–carbon single bond. The argument is made that as Period 2 is traversed from left to right, the atoms become smaller and smaller. At nitrogen, the atoms become so small that electronic repulsions between the nonbonding electrons "force" the atoms farther apart. Thus the nitrogen–nitrogen triple bond is particularly strong, whereas the single bond is comparatively weak. It is this large difference between $N\equiv N$ and $N-N$ bond strengths (742 $kJ\cdot mol^{-1}$) that contributes to the preference in nitrogen chemistry for the formation of the dinitrogen molecule in a reaction rather than chains of nitrogen–nitrogen single bonds, as occurs in carbon chemistry. Furthermore, the fact that dinitrogen is a gas means that an entropy factor also favors the formation of the dinitrogen molecule in chemical reactions.

We can see the difference in behavior between nitrogen and carbon by comparing the combustion of hydrazine, N_2H_4, to that of ethene, C_2H_4. The nitrogen compound burns to produce dinitrogen, whereas the carbon compound gives carbon dioxide:

$$N_2H_4(g) + O_2(g) \rightarrow N_2(g) + 2\ H_2O(g)$$

$$C_2H_4(g) + 3\ O_2(g) \rightarrow 2\ CO_2(g) + 2\ H_2O(g)$$

Curiously, in Groups 15 and 16, it is the second members—phosphorus and sulfur—that are prone to catenation.

Bonding Limitations

Nitrogen forms only a trifluoride, NF_3, whereas phosphorus forms two common fluorides, the pentafluoride, PF_5, and the trifluoride, PF_3. It is argued that the nitrogen atom is simply too small to accommodate more than the three fluorine atoms around it, while the (larger) lower members of the group can manage five (or even six) nearest neighbors. These molecules, such as phosphorus pentafluoride, in which the octet is exceeded for the central atom, are sometimes called *hypervalent compounds*. Traditionally, the bonding model for these compounds assumed that the $3d$ orbitals of the phosphorus played a major role in the bonding. Theoretical studies now suggest that participation of d orbitals is much less than that formerly assumed. However, the only alternative bonding approach is the use of complex molecular orbital diagrams, and these diagrams are more appropriate to upper-level theoretically based inorganic chemistry courses. Like so many aspects of science, we sometimes find it convenient to use a predictive model (such as VSEPR) even when we know it is simplistic and untenable in some respects. Thus, in a course such as this, many chemists continue to explain the bonding in hypervalent compounds in terms of d orbital involvement.

Another example that illustrates the difference in bonding behavior between nitrogen and phosphorus is the pair of compounds, NF_3O and PF_3O. The former contains a weak nitrogen–oxygen bond, whereas the latter contains a fairly strong phosphorus–oxygen bond. For the nitrogen compound, we assume the oxygen is bonded through a coordinate covalent bond, with the nitrogen donating its lone pair in an sp^3 hybrid orbital to a p orbital of the oxygen atom. The phosphorus bonding is analogous to that of silicon

Figure 15.2 Electron-dot representations of the bonding in NF_3O and PF_3O.

Figure 15.3 Representations of the σ and π bonds between the phosphorus and oxygen atoms in PF_3O.

(Chapter 14, Section 14.17), so we can propose that there is some double bond character to the phosphorus–oxygen bond through the overlap of a full p orbital on the oxygen with an empty d orbital on the phosphorus. Figure 15.2 shows possible electron-dot representations for the two compounds, and Figure 15.3 shows an orbital representation of the bonding in the phosphorus compound if $d_\pi–p_\pi$ bonding is invoked.

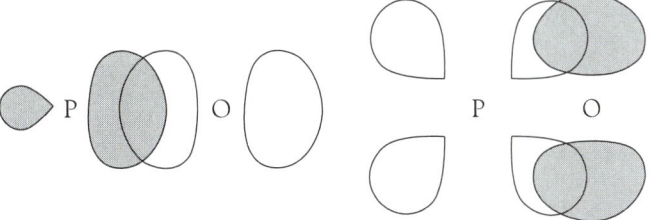

Higher Electronegativity

Nitrogen has a much higher electronegativity than the other members of Group 15. As a result, the polarity of the bonds in nitrogen compounds is often the reverse of that in phosphorus and the other heavier members of the group. For example, the different polarities of the N—Cl and P—Cl bonds result in different hydrolysis products of the respective trichlorides:

$$NCl_3(l) + 3\ H_2O(l) \rightarrow NH_3(g) + 3\ HClO(aq)$$

$$PCl_3(l) + 3\ H_2O(l) \rightarrow H_3PO_3(aq) + 3\ HCl(g)$$

Because the nitrogen–hydrogen covalent bond is strongly polar, ammonia is basic, whereas the hydrides of the other Group 15 elements—phosphine, PH_3, arsine, AsH_3, and stibine, SbH_3—are essentially neutral.

15.3 Nitrogen

The element nitrogen has only one allotrope: the colorless, odorless gas, dinitrogen. Dinitrogen makes up 78 percent by moles of the dry atmosphere at the Earth's surface. Apart from its role in the nitrogen cycle, which we will discuss later, it is very important as an inert diluent for the highly reactive gas in our atmosphere, dioxygen. Without the dinitrogen, every spark in our atmosphere would cause a massive fire. The tragic deaths in 1967 of the astronauts Grissom, White, and Chaffee in an Apollo space capsule were a result of the use of a pure oxygen cabin atmosphere (since discontinued). An accidental electrical spark caused a raging inferno within seconds, killing all of the occupants.

Dinitrogen is not very soluble in water, although like most gases, its solubility increases rapidly with increasing pressure. This is a major problem for deep-sea divers. As they dive, additional dinitrogen dissolves in their bloodstream; as they return to the surface, the decreasing pressure brings the dinitrogen out of solution, and it forms tiny bubbles, particularly around the joints. Prevention of this painful and sometimes fatal problem—called the bends—required divers to return to the surface very slowly. In emergency situations, they were placed in decompression chambers, where the pressure was reapplied and then reduced carefully over hours or days. To avoid this hazard, oxygen–helium gas mixtures are now used for deep diving, because helium has a much lower blood solubility than does dinitrogen.

Industrially, dinitrogen is prepared by liquefying air and then slowly warming the liquid mixture. The dinitrogen boils at −196°C, leaving behind the dioxygen, b.p. $-183°C$. On a smaller scale, dinitrogen can be separated from the other atmospheric gases by using a zeolite, as discussed in Chapter 14, Section 14.20. In the laboratory, dinitrogen can be prepared by gently warming a solution of ammonium nitrite:

$$NH_4NO_2(aq) \rightarrow N_2(g) + 2 H_2O(l)$$

Dinitrogen does not burn or support combustion. It is extremely unreactive toward most elements and compounds. Hence, it is commonly used to

Propellants and Explosives

Propellants and explosives share many common properties. They function by means of a rapid, exothermic reaction that produces a large volume of gas. It is the expulsion of this gas that causes a rocket to be propelled forward (according to Newton's third law of motion), but for the explosive, it is mostly the shock wave from the gas production that causes the damage. There are three factors that make a compound (or a pair of compounds) a potential propellant or explosive:

1. The reaction must be thermodynamically spontaneous and very exothermic, so that a great deal of energy is released in the process.

2. The reaction must be very rapid; in other words, it must be kinetically favorable.

3. The reaction must produce small gaseous molecules, because (according to kinetic theory) small molecules will have high average velocities and hence high momenta.

Although the chemistry of propellants and explosives is a whole science in itself, most of the candidates contain (singly bonded) nitrogen because of the exothermic formation of the dinitrogen molecule. This feature has been of great help in trying to discover terrorist-set explosives in luggage and carry-ons, in that any bags containing abnormally high proportions of nitrogen compounds are suspect.

To illustrate the workings of a propellant, we consider the propellant used in the first rocket-powered aircraft—a mixture of hydrogen peroxide, H_2O_2, and hydrazine, N_2H_4. These combine to give dinitrogen gas and water (as steam):

$$2 H_2O_2(l) + N_2H_4(l) \rightarrow N_2(g) + 4 H_2O(g)$$

The bond energies of the reactants are $O-H = 460$ kJ·mol^{-1}, $O-O = 142$ kJ·mol^{-1}, $N-H = 386$ kJ·mol^{-1}, and $N-N = 247$ kJ·mol^{-1}. Those of the products are $N\equiv N = 942$ kJ·mol^{-1} and $O-H = 460$ kJ·mol^{-1}. Adding the bond energies on each side and finding their difference give the result that 707 kJ·mol^{-1} of heat is released for every 32 g (1 mol) of hydrazine consumed—a very exothermic reaction. And 695 of that 707 kJ·mol^{-1} can be attributed to the conversion of the nitrogen–nitrogen single bond to the nitrogen–nitrogen triple bond.

The largest ever peacetime explosion was the use in 1958 of over 1200 tonnes of explosive to destroy Ripple Rock, a shipping hazard off the coast of British Columbia, Canada. The fragmentation of about 330 000 tonnes of rock eliminated this undersea pinnacle which had ripped open the hulls and sunk at least 119 ships.

This mixture clearly satisfies our first criterion for a propellant. Experimentation showed that the reaction is, indeed, very rapid, and it is obvious from the equation and the application of the ideal gas law that very large volumes of gas will be produced from a very small volume of the two liquid reagents. Because these particular reagents are very corrosive and extremely hazardous, safer mixtures have since been devised by using the same criteria of propellant feasibility.

There is still much research being done on new explosives and propellants. One of the most promising is ammonium dinitramide, $(NH_4)^+[N(NO_2)_2]^-$, known as ADN. From an environmental perspective, unlike the chlorine-containing propellant mixtures, ADN decomposition does not produce pollutants such as chlorine and hydrogen chloride or even carbon dioxide. As ADN is oxygen rich, it can be mixed with reducing agents, such as aluminum powder, to produce even more energy.

provide an inert atmosphere when highly reactive compounds are being handled or stored. About 60 million tonnes of dinitrogen is used every year, worldwide. A high proportion is used in steel production as an inert atmosphere and in oil refineries to purge the flammable hydrocarbons from the pipes and reactor vessels when they need maintenance. Liquid nitrogen is used as a safe refrigerant where very rapid cooling is required. Finally, a significant proportion is employed in the manufacture of ammonia and other nitrogen-containing compounds.

There are a few chemical reactions in which dinitrogen is a reactant. For example, dinitrogen combines on heating with the Group 2 metals and lithium to form ionic nitrides, such as lithium nitride, containing the N^{3-} ion:

$$6\ Li(s)\ +\ N_2(g) \rightarrow 2\ Li_3N(s)$$

If a mixture of dinitrogen and dioxygen is sparked, nitrogen dioxide is formed:

$$N_2(g)\ +\ 2\ O_2(g) \rightleftharpoons 2\ NO_2(g)$$

On a large scale, this reaction takes place in lightning flashes, where it contributes to the biologically available nitrogen in the biosphere. However, it also occurs under the conditions of high pressure and sparking found in modern high-compression gasoline engines. Local concentrations of nitrogen dioxide may be so high that they become a significant component of urban pollution. The equilibrium position for this reaction actually lies far to the left; or, to express this idea another way, nitrogen dioxide has a positive free energy of formation. Its continued existence depends on its extremely slow decomposition rate. Thus, it is kinetically stable. It is one of the roles of the automobile catalytic converter to accelerate the rate of decomposition back to dinitrogen and dioxygen.

Finally, dinitrogen participates in an equilibrium reaction with hydrogen, one that under normal conditions does not occur to any significant extent because of the high activation energy of the reaction (in particular, a single-step reaction cannot occur because it would require a simultaneous four-molecule collision):

$$N_2(g)\ +\ 3\ H_2(g) \rightleftharpoons 2\ NH_3(g)$$

We will discuss this reaction in much more detail shortly.

15.4 Overview of Nitrogen Chemistry

Nitrogen chemistry is complex. For an overview, consider the oxidation state diagram in Figure 15.4. The first thing we notice is that nitrogen can assume formal oxidation states that range from +5 to −3. Second, because it behaves so differently under acidic and basic conditions, we can conclude that the relative stability of an oxidation state is very dependent on pH. Let us look at some specific features of the chemistry of nitrogen.

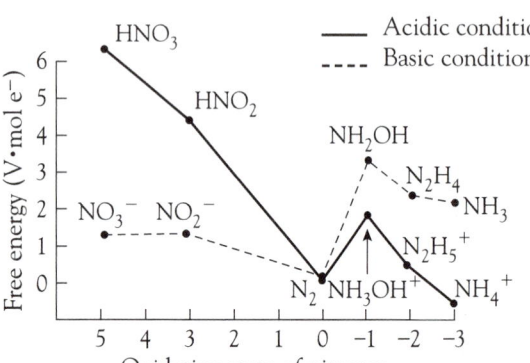

1. Molecular dinitrogen is found at a deep minimum on the Frost diagram. Hence, it is a thermodynamically very stable species. In acidic solution, ammonium ion, NH_4^+, is slightly lower; thus, we might expect that a strong reducing agent would cause dinitrogen to be reduced to the ammonium ion. However, the diagram does not reveal anything about the kinetics of the process, and it is, in fact, kinetically very slow.

Figure 15.4 Frost diagram for the common nitrogen species under acidic and basic conditions.

2. Species that have a high free energy to the left of N_2 are strongly oxidizing. Thus, nitric acid, HNO_3, is a very strong oxidant, although the nitrate ion, NO_3^-, the conjugate base of nitric acid, is not significantly oxidizing.

3. Species that have a high free energy to the right of N_2 tend to be strong reducing agents. Thus, in basic solution, hydroxylamine, NH_2OH, hydrazine, N_2H_4, and ammonia, NH_3, tend to be reducing in their chemical behavior.

4. Both hydroxylamine and its conjugate acid, the hydroxylammonium ion, NH_3OH^+, should readily disproportionate, because they are at convex locations on the diagram. Experimentally, we find that they do disproportionate, but the products are not always those resulting in the greatest decrease in free energy; instead, kinetic factors select the products. Hydroxylamine disproportionates to give dinitrogen and ammonia, whereas the hydroxylammonium ion produces dinitrogen oxide and the ammonium ion:

$$3\ NH_2OH(aq) \rightarrow N_2(g) + NH_3(aq) + 3\ H_2O(l)$$

$$4\ NH_3OH^+(aq) \rightarrow N_2O(g) + 2\ NH_4^+(aq) + 2\ H^+(aq) + 3\ H_2O(l)$$

The First Dinitrogen Compound

Time and time again, chemists fall into the trap of simplistic thinking. As we have said, dinitrogen is very unreactive, but this does not mean that it is totally unreactive. In Chapter 14, Section 14.6, we noted that carbon monoxide could bond to metals (a topic we discuss in more detail in Chapter 22). Dinitrogen is isoelectronic with carbon monoxide, although there is the important difference that dinitrogen is nonpolar, whereas carbon monoxide is polar. Nevertheless, the isoelectronic concept is useful for predicting the possible formation of a compound.

In early 1964 Caesar Senoff, a Canadian chemistry student at the University of Toronto, was working with compounds of ruthenium. He synthesized a brown compound whose composition he was unable to explain. Time passed, and in May 1965, during a discussion with another chemist, it dawned on him that the only feasible explanation was that the molecule contained the N_2 unit bound to the metal in a manner analogous to the carbon monoxide–metal bond. Excitedly, he told his very skeptical supervisor, Bert Allen. After several months, Allen finally agreed to submit the findings to a journal for publication. The manuscript was rejected—a common occurrence when a discovery contradicts accepted thought. After Allen and Senoff rebutted the criticisms, the journal sent the revised manuscript to 16 other chemists for comment and approval before publishing it. Finally, the article appeared in print, and the world of inorganic chemistry was changed yet again.

Since then, transition metal compounds containing the N_2 unit have become quite well known, and some can be made by simply bubbling dinitrogen gas through the solution of a metal compound. (As a consequence, research chemists no longer use dinitrogen as an inert atmosphere for all their reactions.) Some of the compounds are of interest because they are analogs of compounds soil bacteria produce when they convert dinitrogen to ammonia. None of the compounds, however, has become of great practical significance, although they serve as a reminder to inorganic chemists to never say, "Impossible!"

15.5 *Ammonia*

Ammonia is a colorless, poisonous gas with a very strong characteristic smell. It is the only common gas that is basic. Ammonia dissolves readily in water: At room temperature, over 50 g of ammonia will dissolve in 100 g of water, giving a solution of density $0.880\ \text{g·mL}^{-1}$ (known as 880 ammonia). The solution is most accurately called "aqueous ammonia" but is often misleadingly called "ammonium hydroxide." A small proportion does, in fact, react with the water to give ammonium and hydroxide ions:

$$NH_3(aq) + H_2O(l) \rightleftharpoons NH_4^+(aq) + OH^-(aq)$$

This reaction is analogous to the reaction of carbon dioxide with water, and the equilibrium lies to the left. And, like the carbon dioxide and water reaction, evaporating the solution shifts the equilibrium farther to the left. Thus, there is no such thing as pure "ammonium hydroxide."

Ammonia is prepared in the laboratory by mixing an ammonium salt and a hydroxide, for example, ammonium chloride and calcium hydroxide:

$$2\ NH_4Cl(s) + Ca(OH)_2(s) \xrightarrow{\Delta} CaCl_2(s) + 2\ H_2O(l) + 2\ NH_3(g)$$

It is a reactive gas, burning in air when ignited to give water and nitrogen gas:

$$4\ NH_3(g) + 3\ O_2(g) \rightarrow 2\ N_2(g) + 6\ H_2O(l) \quad \Delta G° = -1305\ \text{kJ·mol}^{-1}$$

There is an alternative decomposition route that is thermodynamically less favored, but in the presence of a platinum catalyst, is kinetically preferred; that is, the (catalyzed) activation energy for this alternative route becomes lower than that for the combustion to nitrogen gas:

$$4\ NH_3(g) + 5\ O_2(g) \xrightarrow{Pt} 4\ NO(g) + 6\ H_2O(l) \quad \Delta G° = -1132\ \text{kJ·mol}^{-1}$$

Ammonia acts as a reducing agent in its reactions with chlorine. There are two pathways. With excess ammonia, nitrogen gas is formed, and the excess ammonia reacts with the hydrogen chloride gas produced to give clouds of white, solid ammonium chloride:

$$2\ NH_3(g) + 3\ Cl_2(g) \rightarrow N_2(g) + 6\ HCl(g)$$

$$HCl(g) + NH_3(g) \rightarrow NH_4Cl(s)$$

With excess chlorine, a very different reaction occurs. In this case, the product is nitrogen trichloride, a colorless, explosive, oily liquid:

$$NH_3(g) + 3\ Cl_2(g) \rightarrow 3\ HCl(g) + NCl_3(l)$$

As a base, ammonia reacts with acids in solution to give its conjugate acid, the ammonium ion. For example, when ammonia is mixed with sulfuric acid, ammonium sulfate is formed:

$$2\ NH_3(aq) + H_2SO_4(aq) \rightarrow (NH_4)_2SO_4(aq)$$

Ammonia reacts in the gas phase with hydrogen chloride to give a white smoke of solid ammonium chloride:

$$NH_3(g) + HCl(g) \rightarrow NH_4Cl(g)$$

The formation of a white film over glass objects in a chemistry laboratory is usually caused by the reaction of ammonia escaping from reagent bottles with acid vapors, particularly hydrogen chloride.

Ammonia condenses to a liquid at $-35°C$. This boiling point is much higher than that of phosphine, PH_3 ($-134°C$), because ammonia molecules form strong hydrogen bonds with their neighbors. Liquid ammonia is a good polar solvent because it autoionizes just as water does to produce the ammonium cation and the amide anion, NH_2^-:

$$2\ NH_3(l) \rightleftharpoons NH_4^+(NH_3) + NH_2^-(NH_3)$$

Recall that the autoionization of water gives

$$2\ H_2O(l) \rightleftharpoons H_3O^+(aq) + OH^-(aq)$$

Thus, a whole range of ammonia acid–base chemistry exists in which the ammonium ion is the conjugate acid and the amide ion the conjugate base of ammonia. For example, there is a neutralization reaction,

$$NH_4^+(NH_3) + NH_2^-(NH_3) \rightarrow 2\ NH_3(l)$$

that is analogous to that between the hydronium ion and the hydroxide ion in aqueous solution:

$$H_3O^+(aq) + OH^-(aq) \rightarrow 2\ H_2O(l)$$

With its lone electron pair, ammonia is also a strong Lewis base. One of the "classic" Lewis acid–base reactions involves that between the gaseous electron-deficient boron trifluoride molecule and ammonia to give the white solid compound in which the lone pair on the ammonia is shared with the boron:

$$:NH_3(g) + BF_3(g) \rightarrow F_3B:NH_3(s)$$

Ammonia also acts like a Lewis base when it coordinates to metal ions. For example, it will displace the six water molecules that surround a nickel(II) ion, because it is a stronger Lewis base than water:

$$[Ni(OH_2)_6]^{2+}(aq) + 6\ NH_3(aq) \rightarrow [Ni(NH_3)_6]^{2+}(aq) + 6\ H_2O(l)$$

15.6 Nitrogen Fertilizers and the Industrial Synthesis of Ammonia

It has been known for hundreds of years that nitrogen compounds are essential for plant growth. Manure was once the main source of this ingredient for soil enrichment. But the rapidly growing population in Europe during the 19th century necessitated a corresponding increase in food production. The solution, at the time, was found in the sodium nitrate (Chile saltpeter) deposits in Chile. This compound was mined in vast quantities and shipped around Cape Horn to Europe. The use of sodium nitrate fertilizer prevented famine in Europe and provided Chile with its main income, turning it into an extremely prosperous nation. However, it was clear that the sodium nitrate deposits would one day be exhausted. Thus, chemists rushed to find some method of forming nitrogen compounds from the unlimited resource of unreactive nitrogen gas.

Discovery of the Haber Process

It was Fritz Haber, a German chemist, who showed in 1908 that at about 1000°C, traces of ammonia are formed when nitrogen gas and hydrogen gas are mixed:

$$N_2(g) + 3\,H_2(g) \rightleftharpoons 2\,NH_3(g)$$

A physical chemist, Walther Nernst (of Nernst equation fame), pointed out that a study of the thermodynamics of the system would enable the conditions of maximum yield to be found. In fact, the conversion of dinitrogen and dihydrogen into ammonia is exothermic and results in a decrease in gas volume and a resulting decrease in entropy. To "force" the reaction to the right, the Le Châtelier principle suggests that the maximum yield of ammonia would be at low temperature and high pressure. However, the lower the temperature, the slower the rate at which equilibrium is reached. A catalyst might help, but even then there are limits to the most practical minimum temperature. Furthermore, there are limits to how high the pressure can go, simply in terms of the cost of thick-walled containers and pumping systems.

Haber found that adequate yields could be obtained in reasonable time by using a pressure of 20 MPa (200 atm) and a temperature of 500°C. However, it took 5 years for a chemical engineer, Carl Bosch, to actually design an industrial-size plant for the chemical company BASF that could work with gases at this pressure and temperature. Unfortunately, completion of the plant coincided with the start of World War I. With Germany blockaded by the Allies, supplies of Chile saltpeter were no longer available; nevertheless, the ammonia produced was used for the synthesis of explosives rather than for crop production. Without the *Haber–Bosch process*, the German and Austro-Hungarian armies might well have been forced to capitulate earlier than 1918, simply because of a lack of explosives.

The Modern Haber–Bosch Process

To prepare ammonia in the laboratory, we can simply take cylinders of nitrogen gas and hydrogen gas and pass them into a reaction vessel at appropriate conditions of temperature, pressure, and catalyst. But neither dinitrogen nor dihydrogen is a naturally occurring pure reagent. Thus, for the industrial chemist, obtaining the reagents inexpensively, on a large scale, with no useless by-products, is a challenge.

The first step is to obtain the dihydrogen gas. This is accomplished by the *steam reforming process*, where a hydrocarbon, such as methane, is mixed with steam at high temperatures (about 750°C) and at high pressures (about 4 MPa). This process is endothermic, so high temperatures would favor product formation on thermodynamic grounds, but high pressure is used for kinetic reasons to increase the collision frequency (reaction rate). A catalyst, usually nickel, is present for the same reason:

$$CH_4(g) + H_2O(g) \rightarrow CO(g) + 3\ H_2(g)$$

Catalysts are easily inactivated (*poisoned*) by impurities. Thus, it is crucial to remove impurities from the reactants (*feedstock*). Sulfur compounds are particularly effective at reacting with the catalyst surface and deactivating it by forming a layer of metal sulfide. Thus, before the methane is used, it is pretreated to convert contaminating sulfur compounds to hydrogen sulfide. The hydrogen sulfide is then removed by passing the impure methane over zinc oxide:

$$ZnO(s) + H_2S(g) \rightarrow ZnS(s) + H_2O(g)$$

Next, air is added to the mixture of carbon monoxide and dihydrogen, which still contains some methane—deliberately. The methane burns to give carbon monoxide, but, by controlling how much methane is present, the amount of dinitrogen left in the deoxygenated area should be that required to achieve the 1:3 stoichiometry of the Haber–Bosch reaction:

$$CH_4(g) + \tfrac{1}{2}\ O_2(g) + 2\ N_2(g) \rightarrow CO(g) + 2\ H_2(g) + 2\ N_2(g)$$

There is no simple way of removing carbon monoxide from the mixture of gases. For this reason, and to produce an additional quantity of hydrogen, the third step involves the oxidation of the carbon monoxide to carbon dioxide by using steam. This *water gas shift process* is performed at fairly low temperatures (350°C) because it is exothermic. Even though a catalyst of iron and chromium oxides is used, the temperature cannot be any lower without reducing the rate of reaction to an unacceptable level:

$$CO(g) + H_2O(g) \rightleftharpoons CO_2(g) + H_2(g)$$

The carbon dioxide can be removed by a number of different methods. Carbon dioxide has a high solubility in water and in many other solvents. Alternatively, it can be removed by a chemical process such as the reversible reaction with potassium carbonate:

$$CO_2(g) + K_2CO_3(aq) + H_2O(l) \rightleftharpoons 2\ KHCO_3(aq)$$

The potassium hydrogen carbonate solution is pumped into tanks where it is heated to generate pure carbon dioxide gas and potassium carbonate solution:

$$2\ KHCO_3(aq) \rightleftharpoons CO_2(g) + K_2CO_3(aq) + H_2O(l)$$

The carbon dioxide is liquefied under pressure and sold, and the potassium carbonate is returned to the ammonia processing plant for reuse.

Now that a mixture of the pure reagents of dinitrogen and dihydrogen gas has been obtained, the conditions are appropriate for the simple reaction that gives ammonia:

$$N_2(g) + 3\ H_2(g) \rightarrow 2\ NH_3(g)$$

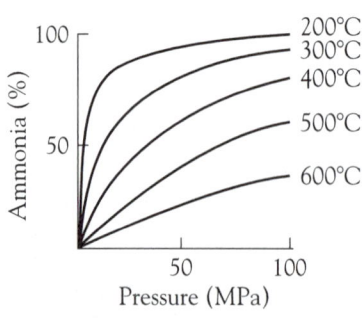

Figure 15.5 Percentage yields of ammonia as a function of pressure, at various temperatures.

The practical thermodynamic range of conditions is shown in Figure 15.5. As mentioned earlier, to "force" the reaction to the right, high pressures are

Haber and Scientific Morality

It has been said that many scientists are amoral because they fail to consider the applications to which their work can be put. The life of Fritz Haber presents a real dilemma: Should we regard him as a hero or as a villain? As discussed earlier, Haber devised the process of ammonia synthesis, which he intended to be used to help feed the world; yet the process was turned into a source of materials to kill millions. He cannot easily be faulted for this, but his other interest is more controversial. Haber argued that it was better to incapacitate the enemy during warfare than to kill them. Thus, he worked enthusiastically on poison gas research during the First World War. His first wife, Clara Immerwahr Haber, a talented chemist, pleaded with him to desist and when he did not, she committed suicide.

In 1918, Haber was awarded the Nobel Prize for his work on ammonia synthesis, but many chemists opposed the award on the basis of his poison gas research. After that war, Haber was a key figure in the rebuilding of Germany's chemical research community. Then in 1933, the National Socialist government took power, and Haber, of Jewish origin himself, was told to fire all of the Jewish workers at his institute. He refused and resigned instead, bravely writing: "For more than 40 years I have selected my collaborators on the basis of their intelligence and their character and not on the basis of their grandmothers, and I am not willing to change this method which I have found so good."

This action infuriated the Nazi leaders, but in view of Haber's international reputation, they did not act against him at that time. In 1934, the year after his death, the German Chemical Society held a memorial service for him. The German government was so angered by this tribute to someone who had stood up against their regime that they threatened arrest of all those chemists who attended. But their threat was hollow. The turnout of so many famous chemists for the event caused the Gestapo to back down.

used. But the higher the pressure, the thicker the reaction vessels and piping required to prevent an explosion—and the thicker the containers, the higher the cost of construction.

Today's ammonia plants utilize pressures between 10 and 100 MPa (100 and 1000 atm). The lower the temperature, the higher the yield but the slower the rate. With current high-performance catalysts, the optimum conditions are 400°C to 500°C. The catalyst is the heart of every ammonia plant. The most common catalyst is specially prepared high-surface-area iron containing traces of potassium, aluminum, calcium, magnesium, silicon, and oxygen. About 100 tonnes of catalyst is used in a typical reactor vessel, and, provided all potential "poisons" are removed from the incoming gases, the catalyst will have a working life of about 10 years. The mechanism of the reaction is known to involve the dissociation of dinitrogen to atomic nitrogen on the crystal face of the iron catalyst, followed by reaction with atomic hydrogen, similarly bonded to the iron surface.

After leaving the reactor vessel, the ammonia is condensed. The remaining dinitrogen and dihydrogen are then recycled back through the plant to be mixed with the fresh incoming gas. A typical ammonia plant produces about 1000 tonne·day^{-1}. The most crucial concern is to minimize energy consumption. A traditional Haber–Bosch plant consumed about 85 GJ·tonne^{-1} of ammonia produced, whereas a modern plant, built to facilitate energy recycling, uses only about 30 GJ·tonne^{-1}.

Even today, the most important use of ammonia itself is in the fertilizer industry. The ammonia is often applied to fields as ammonia gas. Ammonium sulfate and ammonium phosphate also are common solid fertilizers. These are simply prepared by passing the ammonia into sulfuric acid and phosphoric acid, respectively:

$$2 \, NH_3(g) + H_2SO_4(aq) \rightarrow (NH_4)_2SO_4(aq)$$

$$3 \, NH_3(g) + H_3PO_4(aq) \rightarrow (NH_4)_3PO_4(aq)$$

Ammonia is also used in a number of industrial syntheses, particularly that of nitric acid, as we will discuss shortly.

15.7   *The Ammonium Ion*

The colorless ammonium ion is the most common nonmetallic cation used in the chemistry laboratory. As we discussed in Chapter 9, Section 9.12, this tetrahedral polyatomic ion can be thought of as a pseudo-alkali metal ion, close in size to the potassium ion. Having covered the similarities with alkali metals in that section, here we will focus on the unique features of the ion. In particular, unlike the alkali metal ions, the ammonium ion does not always remain intact: It can be hydrolyzed, dissociated, or oxidized.

The ammonium ion is hydrolyzed in water to give its conjugate base, ammonia:

$$NH_4^{+}(aq) + H_2O(l) \rightleftharpoons H_3O^{+}(aq) + NH_3(aq)$$

As a result, solutions of ammonium salts of strong acids, such as ammonium chloride, are slightly acidic.

Ammonium salts can volatilize (vaporize) by dissociation. The classic example of this is ammonium chloride:

$$NH_4Cl(s) \rightleftharpoons NH_3(g) + HCl(g)$$

If a sample of ammonium chloride is left open to the atmosphere, it will "disappear." It is this same decomposition reaction that is used in "smelling salts." The pungent ammonia odor, which masks the sharper smell of the hydrogen chloride, has a considerable effect on a semicomatose individual (although it should be noted that the use of smelling salts except by medical personnel is now deemed to be unwise and potentially dangerous).

Finally, the ammonium ion can be oxidized by the anion in the ammonium salt. These are reactions that occur when an ammonium salt is heated, and each one is unique. The three most common examples are the thermal decomposition of ammonium nitrite, ammonium nitrate, and ammonium dichromate. The reaction of ammonium dichromate is often referred to as the "volcano" reaction. A source of heat, such as a lighted match, will cause the orange crystals to decompose, producing sparks and a large volume of dark green chromium(III) oxide. Although this is a very spectacular decomposition reaction, it needs to be performed in a fume hood, because a little ammonium dichromate dust usually is dispersed by the reaction, and this highly carcinogenic material can be absorbed through the lungs:

$$NH_4NO_2(aq) \xrightarrow{\Delta} N_2(g) + 2\ H_2O(l)$$

$$NH_4NO_3(s) \xrightarrow{\Delta} N_2O(g) + 2\ H_2O(l)$$

$$(NH_4)_2Cr_2O_7(s) \xrightarrow{\Delta} N_2(g) + Cr_2O_3(s) + 4\ H_2O(g)$$

15.8 Other Hydrides of Nitrogen

Besides ammonia, nitrogen forms two other compounds with hydrogen: hydrazine, N_2H_4, and hydrogen azide, HN_3.

Hydrazine

Hydrazine is a fuming, colorless liquid. It is a weak base, forming two series of salts, in which it is either monoprotonated or diprotonated:

$$N_2H_4(aq) + H_3O^+(aq) \rightleftharpoons N_2H_5^+(aq) + H_2O(l)$$

$$N_2H_5^+(aq) + H_3O^+(aq) \rightleftharpoons N_2H_6^{2+}(aq) + H_2O(l)$$

However, hydrazine is a strong reducing agent, reducing iodine to hydrogen iodide and copper(II) ion to copper metal:

$$N_2H_4(aq) + 2\ I_2(aq) \rightarrow 4\ HI(aq) + N_2(g)$$

$$N_2H_4(aq) + 2\ Cu^{2+}(aq) \rightarrow 2\ Cu(s) + N_2(g) + 4\ H^+(aq)$$

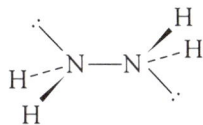

Figure 15.6 The hydrazine molecule.

Most of the 20 000 tonnes produced worldwide annually is used as the reducing component of a rocket fuel, usually in the form of assymetrical dimethyl hydrazine, $(CH_3)_2NNH_2$. Another derivative, dinitrophenylhydrazine, $H_2NNHC_6H_3(NO_2)_2$, is used in organic chemistry to identify carbon compounds containing the $C{=}O$ grouping. The structure of hydrazine is like that of ethane, except that two ethane hydrogens are replaced by lone pairs of electrons, one pair on each nitrogen atom (Figure 15.6)

Hydrogen Azide

Hydrogen azide, a colorless liquid, is quite different from the other nitrogen hydrides. It is acidic, with a pK_a similar to that of acetic acid:

$$HN_3(aq) + H_2O(l) \rightleftharpoons H_3O^+(aq) + N_3^-(aq)$$

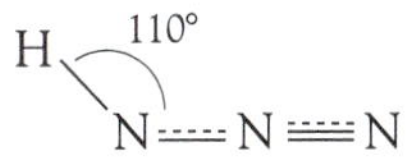

The compound has a repulsive, irritating odor and is extremely poisonous. It is highly explosive, producing hydrogen gas and nitrogen gas:

$$2\ HN_3(l) \rightarrow H_2(g) + 3\ N_2(g)$$

Figure 15.7 The hydrogen azide molecule. The bond orders of the two nitrogen–nitrogen bonds are about 1½ and 2½.

The three nitrogen atoms in a hydrogen azide molecule are colinear, with the hydrogen at a 110° angle (Figure 15.7). The nitrogen–nitrogen bond lengths in hydrogen azide are 124 pm and 113 pm (the end N—N bond is shorter). A typical N=N bond is 120 pm, and the N≡N bond in the dinitrogen molecule is 110 pm. Thus, the bond orders in hydrogen azide must be approximately 1½ and 2½, respectively. The bonding can be pictured simply as an equal resonance mixture of the two electron-dot structures shown in Figure 15.8, one of which contains two N—N double bonds and the other, a N—N single bond and a triple N—N bond.

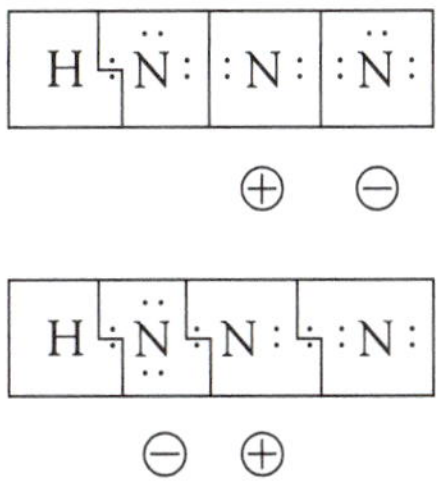

The azide ion, N_3^-, is isoelectronic with carbon dioxide, and it is presumed to have an identical electronic structure. The nitrogen–nitrogen bonds are of equal length (116 pm), an observation that reinforces the concept that the presence of the hydrogen atom in hydrogen azide causes the neighboring N—N bond to weaken (and lengthen to 124 pm) and the more

Figure 15.8 The bonding in a hydrogen azide can be pictured as a resonance mixture of these two structures.

distant one to strengthen (and shorten to 113 pm). In its chemistry, the azide ion behaves as a pseudo-halide ion (Chapter 9, Section 9.12). For example, mixing a solution of azide ion with silver ion gives a precipitate of silver azide, AgN_3, analogous to silver chloride, $AgCl$. Azide ion also forms parallel complex ions to those of chloride ion, such as $[Sn(N_3)_6]^{2-}$, the analog of $[SnCl_6]^{2-}$.

It is interesting how so much of chemistry can be used either destructively or constructively. The azide ion is now used to save lives—by the automobile air bag. It is crucial that an air bag inflate extremely rapidly, before the victim is thrown forward after impact. The only way to produce such a fast response is through a controlled chemical explosion that produces a large volume of gas. For this purpose, sodium azide is preferred: It is about 65 percent nitrogen by mass, can be routinely manufactured to a high purity (at least 99.5 percent), and decomposes cleanly to sodium metal and dinitrogen at 350°C:

$$2\ NaN_3(s) \rightarrow 2\ Na(l)\ +\ 3\ N_2(g)$$

In an air bag, this reaction takes place in about 40 ms. Obviously, we would not want the occupants to be saved from a crash and then have to face molten sodium metal. There are a variety of reactions that can be used to immobilize the liquid product. One of these involves the addition of potassium nitrate and silicon dioxide to the mixture. The sodium metal is oxidized by the potassium nitrate to sodium oxide, producing more nitrogen gas. The alkali metal oxides then react with the silicon dioxide to give inert glassy metal silicates.

$$10\ Na(l)\ +\ 2\ KNO_3(s) \rightarrow K_2O(s)\ +\ 5\ Na_2O(s)\ +\ N_2(g)$$

$$2\ K_2O(s)\ +\ SiO_2(s) \rightarrow K_4SiO_4(s)$$

$$2\ Na_2O(s)\ +\ SiO_2(s) \rightarrow Na_4SiO_4(s)$$

Lead(II) azide is important as a detonator: It is a fairly safe compound unless it is impacted, in which case it explosively decomposes. The shock wave produced is usually sufficient to detonate a more stable explosive such as dynamite:

$$Pb(N_3)_2(s) \rightarrow Pb(s)\ +\ 3\ N_2(g)$$

15.9 The Pentanitrogen Cation

Though most simple inorganic compounds have been known for over 100 years, there are still new compounds being discovered. One of the most interesting is the pentanitrogen cation, N_5^+, the first known stable cation of the element, and only the third all-nitrogen species known (the others being N_2 and N_3^-). A salt of this cation was first synthesized in 1999 as part of a research program into high-energy materials at the U.S. Edwards Air Force Base in California. To stabilize the large cation, the large hexafluoroarsenate(V) anion was used. The actual synthesis reaction was:

$$(N_2F)^+(AsF_6)^-(HF)\ +\ HN_3(HF) \xrightarrow{-78°C} (N_5)^+(AsF_6)^-(s)\ +\ HF(l)$$

The pentanitrogen ion is an extremely strong oxidizing agent and will explosively oxidize water to oxygen gas. One potential use of this compound is to prepare yet other species that nobody thinks can be made.

15.10 *Nitrogen Oxides*

Nitrogen forms a plethora of common oxides: dinitrogen oxide, N_2O; nitrogen monoxide, NO; dinitrogen trioxide, N_2O_3; nitrogen dioxide, NO_2; dinitrogen tetroxide, N_2O_4; and dinitrogen pentoxide, N_2O_5. Each of the oxides is actually thermodynamically unstable with respect to decomposition to its elements, but all are kinetically stabilized.

Dinitrogen Oxide

The sweet-smelling, gaseous dinitrogen oxide is also known as nitrous oxide or, more commonly, laughing gas. This name results from the intoxicating effect of low concentrations. It is sometimes used as an anesthetic, although the high concentrations needed to cause unconsciousness make it unsuitable for more than brief operations such as tooth extraction. Anesthetists have been known to become addicted to the narcotic gas. Because the gas is very soluble in fats, tasteless, and nontoxic, its major use is as a propellant in pressurized cans of whipped cream.

Dinitrogen oxide is a fairly unreactive, neutral gas, although it is the only common gas other than oxygen to support combustion. For example, magnesium burns in dinitrogen oxide to give magnesium oxide and nitrogen gas:

Figure 15.9 The dinitrogen oxide molecule. The N–N bond order is about 2½, and the N–O bond order is about 1½.

$$N_2O(g) + Mg(s) \rightarrow MgO(s) + N_2(g)$$

The standard method of preparation of dinitrogen oxide involves the thermal decomposition of ammonium nitrate. This reaction can be accomplished by heating the molten solid to about 280°C. An explosion can ensue from strong heating, however, so a safer route is to heat an ammonium nitrate solution that has been acidified with hydrochloric acid:

$$NH_4NO_3(aq) \xrightarrow{H^+} N_2O(g) + 2\ H_2O(l)$$

Dinitrogen oxide is isoelectronic with carbon dioxide and the azide ion. However, contrary to what one might expect, the atoms are arranged asymmetrically, with a N—N bond length of 113 pm and a N—O bond length of 119 pm. These values indicate a nitrogen–nitrogen bond order of close to 2½ and a nitrogen–oxygen bond order close to 1½ (Figure 15.9).

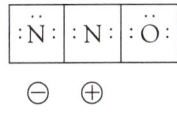

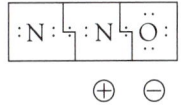

Figure 15.10 The bonding in dinitrogen oxide can be pictured as a resonance mixture of these two structures.

Like hydrogen azide, dinitrogen oxide can be pictured simply as a molecule that resonates between two electron-dot structures, one of which contains a N—O double bond and a N—N double bond and the other, a N—O single bond and a N—N triple bond (Figure 15.10).

Nitrogen Monoxide

One of the most curious simple molecules is nitrogen monoxide, also called nitric oxide. It is a colorless, neutral, paramagnetic gas. Its molecular orbital diagram resembles that of carbon monoxide, but with one additional electron that occupies an antibonding orbital (Figure 15.11). Hence, the predicted net bond order is 2½.

Chemists expect molecules containing unpaired electrons to be very reactive. Yet nitrogen monoxide in a sealed container is quite stable. Only when it is cooled to form the colorless liquid or solid does it show a tendency to form a dimer, N_2O_2, in which the two nitrogen atoms are joined by a single bond.

Consistent with the molecular orbital representation, nitrogen monoxide readily loses its electron from the antibonding orbital to form the nitrosyl ion, NO^+, which is diamagnetic and has a shorter N–O bond length (106 pm) than that of the parent molecule (115 pm). This triple-bonded

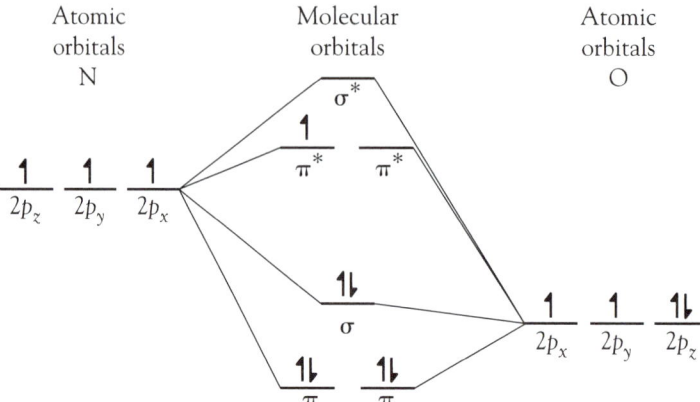

Figure 15.11 Molecular orbital energy level diagram for the $2p$ atomic orbitals of the nitrogen monoxide molecule.

ion is isoelectronic with carbon monoxide, and it forms many analogous metal complexes.

Nitrogen monoxide does show a high reactivity toward dioxygen, and once a sample of colorless nitrogen monoxide is opened to the air, brown clouds of nitrogen dioxide form:

$$2\,NO(g) + O_2(g) \rightleftharpoons 2\,NO_2(g)$$

The molecule is an atmospheric pollutant, commonly formed as a side reaction in high-compression internal combustion engines when dinitrogen and dioxygen are compressed and sparked:

$$N_2(g) + O_2(g) \rightleftharpoons 2\,NO(g)$$

The easiest method for preparing the gas in the laboratory involves the reaction between copper and 50 percent nitric acid:

$$3\,Cu(s) + 8\,HNO_3(aq) \rightarrow 3\,Cu(NO_3)_2(aq) + 4\,H_2O(l) + 2\,NO(g)$$

However, the product is always contaminated by nitrogen dioxide. This contaminant can be removed by bubbling the gas through water, because the nitrogen dioxide reacts rapidly with water.

Until recently, a discussion of simple nitrogen monoxide chemistry would have ended here. Now we realize that this little molecule plays a vital role in our bodies and those of all mammals. In fact, the prestigious journal *Science* called it the 1992 Molecule of the Year. It has been known since 1867 that organic nitro compounds, such as nitroglycerine, can relieve angina, lower blood pressure, and relax smooth muscle tissue. Yet it was not until 1987 that Salvador Moncada and his team of scientists at the Wellcome Research Laboratories in Britain identified the crucial factor in blood vessel dilation as nitrogen monoxide gas. That is, organic nitro compounds were broken down to produce this gas in the organs.

Since this initial work, we have come to realize that nitrogen monoxide is crucial in controlling blood pressure. There is even an enzyme (nitric oxide synthase) whose sole task is the production of nitrogen monoxide. At this point, a tremendous quantity of biochemical research is concerned with the role of this molecule in the body. A lack of nitrogen monoxide is implicated as a cause of high blood pressure, whereas septic shock, a leading cause of death in intensive care wards, is ascribed to an excess of nitrogen monoxide. The gas appears to have a function in memory and in the stomach. Male erections have been proved to depend on production of nitrogen monoxide, and there are claims of important roles for nitrogen monoxide in female uterine contractions. One question still to be answered concerns the life span of these molecules, considering the ease with which they react with oxygen gas.

Dinitrogen Trioxide

Dinitrogen trioxide, the least stable of the common oxides of nitrogen, is a dark blue liquid that decomposes above 230°C. It is prepared by cooling a stoichiometric mixture of nitrogen monoxide and nitrogen dioxide:

$$NO(g) + NO_2(g) \rightleftharpoons N_2O_3(l)$$

Dinitrogen trioxide is the first of the acidic oxides of nitrogen. In fact, it is the acid anhydride of nitrous acid. Thus, when dinitrogen trioxide is mixed with water, nitrous acid is formed; and when it is mixed with hydroxide ion, the nitrite ion is produced:

$$N_2O_3(l) + H_2O(l) \rightarrow 2\ HNO_2(aq)$$

$$N_2O_3(l) + 2\ OH^-(aq) \rightarrow 2\ NO_2^-(aq) + H_2O(l)$$

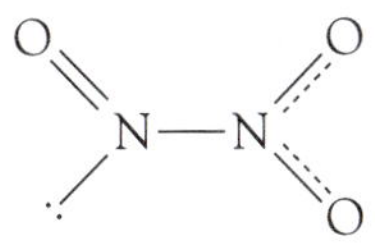

Figure 15.12 The dinitrogen trioxide molecule.

Although, simplistically, dinitrogen trioxide can be considered to contain two nitrogen atoms in the +3 oxidation state, the structure is asymmetric (Figure 15.12), an arrangement that shows it to be a simple combination of the two molecules with unpaired electrons from which it is prepared (nitrogen monoxide and nitrogen dioxide). In fact, the nitrogen–nitrogen bond length in dinitrogen trioxide is abnormally long (186 pm) relative to the length of the single bond in hydrazine (145 pm).

Bond length data indicate that the single oxygen is bonded to the nitrogen with a double bond, whereas the other two oxygen–nitrogen bonds each have a bond order of about 1½. This value is the average of the single and double bond forms that can be constructed with electron-dot formulas.

Nitrogen Dioxide and Dinitrogen Tetroxide

These two toxic oxides coexist in a state of dynamic equilibrium. Low temperatures favor the formation of the colorless dinitrogen tetroxide, whereas high temperatures favor the formation of the dark red-brown nitrogen dioxide:

$$\underset{\text{colorless}}{N_2O_4(g)} \rightleftharpoons \underset{\text{red-brown}}{2\ NO_2(g)}$$

At the normal boiling point of -1°C, the mixture contains 16 percent nitrogen dioxide, but the proportion of nitrogen dioxide rises to 99 percent at 135°C.

Nitrogen dioxide is prepared by reacting copper metal with concentrated nitric acid:

$$Cu(s) + 4\ HNO_3(l) \rightarrow Cu(NO_3)_2(aq) + 2\ H_2O(l) + 2\ NO_2(g)$$

It is also formed by heating heavy metal nitrates, a reaction that produces a mixture of nitrogen dioxide and oxygen gases:

$$Cu(NO_3)_2(s) \xrightarrow{\Delta} CuO(s) + 2\ NO_2(g) + ½\ O_2(g)$$

And, of course, it is formed when nitrogen monoxide reacts with dioxygen:

$$2\ NO(g) + O_2(g) \rightarrow 2\ NO_2(g)$$

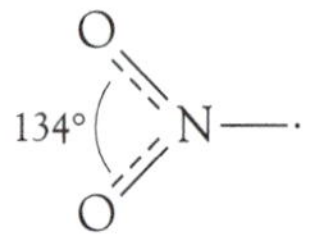

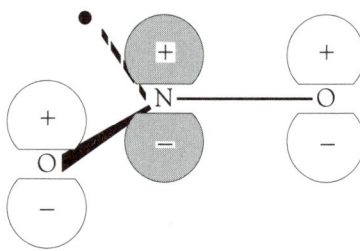

Figure 15.13   The nitrogen dioxide molecule.

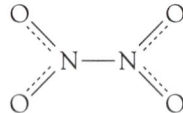

Figure 15.14   Overlap of the p orbitals at right angles to the molecular plane of nitrogen dioxide.

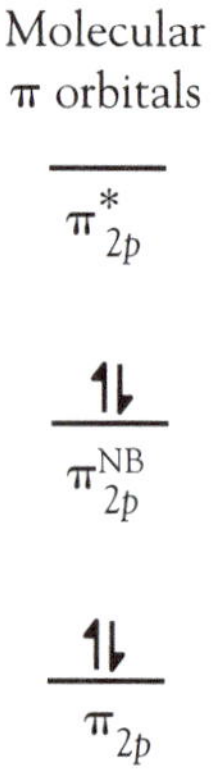

Molecular
π orbitals

$\overline{}$
π^*_{2p}

$\underline{\uparrow\downarrow}$
π^{NB}_{2p}

$\underline{\uparrow\downarrow}$
π_{2p}

Figure 15.15   Molecular orbital energy level diagram of the orbitals involved in π bond formation in nitrogen dioxide.

Figure 15.16   The dinitrogen tetroxide molecule.

■■■■■ **Dinitrogen Pentoxide**

Nitrogen dioxide is an acid oxide, dissolving in water to give nitric acid and nitrous acid:

$$2\,NO_2(g) + H_2O(l) \rightleftharpoons HNO_3(aq) + HNO_2(aq)$$

This potent mixture of corrosive, oxidizing acids is produced when nitrogen dioxide, formed from automobile pollution, reacts with rain. It is a major damaging component of urban precipitation.

Nitrogen dioxide is a V-shaped molecule with an O—N—O angle of 134°, an angle slightly larger than the true trigonal planar angle of 120°. Because the third bonding site is occupied by a single electron rather than by a lone pair, it is not unreasonable for the bonding angle to be opened up (Figure 15.13). The oxygen–nitrogen bond length indicates a 1½ bond order, like that in the NO_2 half of dinitrogen trioxide.

It is useful to compare the π bonding in nitrogen dioxide to that in carbon dioxide. The linear structure of carbon dioxide allows both sets of p orbitals that are at right angles to the bonding direction to overlap and participate in π bonding. In the bent nitrogen dioxide molecule, the p orbitals are still at right angles to the bonding direction, but, in the plane of the molecule, they are skewed with respect to one another and cannot overlap to form a π system. As a result, the only π bond that can form is at right angles to the plane of the molecule (Figure 15.14). Combining this set of three p orbitals results in the formation of three π molecular orbitals: one bonding, one nonbonding, and one antibonding (Figure 15.15). Of the four electrons available, two electrons enter the bonding molecular orbital; and two, the nonbonding orbital. This arrangement gives a net single π bond. However, this single π bond is shared between two bonded pairs; hence, each pair has one-half a π bond.

The "odd" electron is believed to occupy a weakly antibonding σ molecular orbital derived from the nitrogen p orbital in the molecular plane. (In hybridization terms, the single electron occupies an sp^2-type orbital on the nitrogen atom, and the other two hybrid orbitals form σ bonds with the oxygen atoms.)

The O—N—O bond angle in the dinitrogen tetroxide molecule is almost identical to that in nitrogen dioxide (Figure 15.16). But dinitrogen tetroxide has an abnormally long (and hence weak) nitrogen–nitrogen bond, although at 175 pm, it is not as weak as the N—N bond in dinitrogen trioxide. The N—N bond is formed by the combination of the weakly antibonding σ orbitals of the two NO_2 units (overlap of the sp^2 hybrid orbitals containing the "odd" electrons, in hybridization terminology). The resulting N—N bonding molecular orbital will have correspondingly weak bonding character. In fact, the N—N bond energy is only about 60 kJ·mol⁻¹.

This colorless, solid, deliquescent oxide is the most strongly oxidizing of the nitrogen oxides. It is also strongly acidic, reacting with water to form nitric acid:

$$N_2O_5(s) + H_2O(l) \rightarrow 2\,HNO_3(aq)$$

In the liquid and gas phases, the molecule has a structure related to those of the other dinitrogen oxides, N_2O_3, and N_2O_4, except that an oxygen atom links the two NO_2 units (Figure 15.17). Once again, the two pairs of p

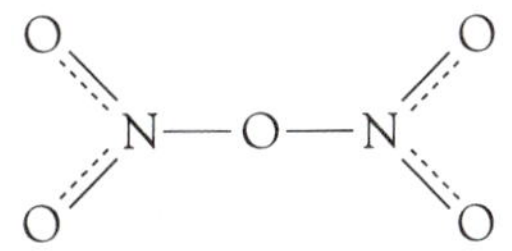

Figure 15.17 The dinitrogen pentoxide molecule.

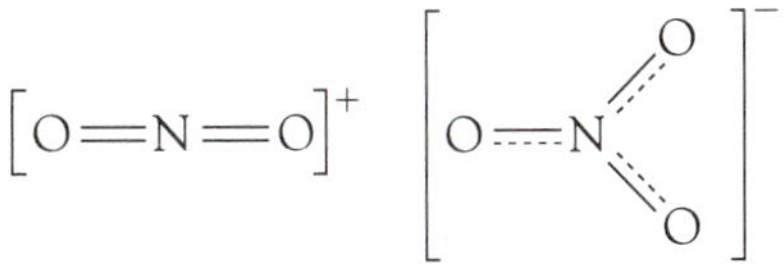

Figure 15.18 The nitryl cation and nitrate anion present in solid-phase dinitrogen pentoxide.

electrons provide a half π bond to each oxygen–nitrogen pair. Of more interest, however, is the bonding in the solid phase. We have already seen that compounds of metals and nonmetals can be covalently bonded. Here we have a case of a compound of two nonmetals that contains ions! In fact, the crystal structure consists of alternating nitryl cations, NO_2^+, and nitrate anions, NO_3^- (Figure 15.18).

15.11 *The Nitrate Radical*

Most people are aware that the Earth's atmosphere is predominantly dinitrogen and dioxygen and that trioxygen and carbon dioxide are also important atmospheric gases. What very few realize is the crucial role of certain trace gases, one of which is the nitrate radical, NO_3. This highly reactive free radical was first identified in the troposphere in 1980 where it is now known to play a major role in night-time atmospheric chemistry.

The nitrate radical is formed by the reaction of nitrogen dioxide with ozone.

$$NO_2(g) + O_3(g) \rightarrow NO_3(g) + O_2(g)$$

During the day, it is decomposed by light (photolysed), the product depending upon the wavelength of light:

$$NO_3(g) \xrightarrow{h\nu} NO(g) + O_2(g)$$
$$NO_3(g) \xrightarrow{h\nu} NO_2(g) + O(g)$$

However at night, the nitrate radical is the predominant oxidizing species on the Earth's surface, even though its concentration is usually in the 0.1 to 1 ppb range. This role is crucial in urban environments where there are high levels of hydrocarbons. Thus, it will remove a hydrogen atom from an alkane (represented as RH in the following equation) to give a reactive alkyl radical and hydrogen nitrate, the latter reacting with water to give nitric acid.

$$NO_3(g) + RH(g) \rightarrow R(g) + HNO_3(g)$$

With alkenes, addition occurs to the double bond to form highly oxidizing and reactive organo nitrogen and peroxy compounds, including the infamous peroxyacetyl nitrate, $CH_3COO_2NO_2$, known as PAN, a major eye irritant in the photochemical smog found in many city atmospheres.

15.12 *Nitrogen Halides*

Nitrogen trichloride is a typical covalent chloride. It is a yellow, oily liquid that reacts with water to form ammonia and hypochlorous acid:

$$NCl_3(aq) + 3\ H_2O(l) \rightarrow NH_3(g) + 3\ HClO(aq)$$

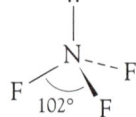

Figure 15.19 The nitrogen trifluoride molecule.

The compound is highly explosive when pure, because it has a positive free energy of formation. However, nitrogen trichloride vapor is used quite extensively (and safely) to bleach flour.

By contrast, nitrogen trifluoride is a thermodynamically stable, colorless, odorless gas of low chemical reactivity. For example, it does not react with water at all. Such stability and low reactivity are quite common among covalent fluorides. Despite having a lone pair like ammonia (Figure 15.19), it is a weak Lewis base. The F—N—F bond angle in nitrogen trifluoride (102°) is significantly less than the tetrahedral angle. One explanation for the weak Lewis base behavior and the decrease in bond angle from 109½° is that the nitrogen–fluorine bond has predominantly p orbital character (for which 90° would be the optimum angle), and the lone pair is in the nitrogen s orbital rather than in a more directional sp^3 hybrid.

There is one unusual reaction in which nitrogen trifluoride does act as a Lewis base: It forms the stable compound nitrogen oxide trifluoride, NF_3O, when an electric discharge provides the energy for its reaction with oxygen gas at very low temperature:

$$2\,NF_3(g) + O_2(g) \rightarrow 2\,NF_3O(g)$$

Nitrogen oxide trifluoride is often used as the classic example of a compound with a coordinate covalent bond between the nitrogen and oxygen atoms.

15.13 Nitrous Acid

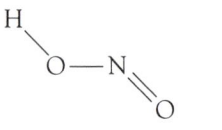

Figure 15.20 The nitrous acid molecule.

Nitrous acid is a weak acid that is unstable, except in solution. It can be prepared by mixing a metal nitrite and a solution of a dilute acid at 0°C in a double replacement reaction. Barium nitrite and sulfuric acid give a pure solution of nitrous acid, because the barium sulfate that is formed has a very low solubility:

$$Ba(NO_2)_2(aq) + H_2SO_4(aq) \rightarrow 2\,HNO_2(aq) + BaSO_4(s)$$

The shape of the nitrous acid molecule is shown in Figure 15.20.

Even at room temperature, disproportionation of aqueous nitrous acid occurs to give nitric acid and bubbles of nitrogen monoxide. The latter reacts rapidly with the oxygen gas in the air to produce brown fumes of nitrogen dioxide:

$$3\,HNO_2(aq) \rightarrow HNO_3(aq) + 2\,NO(g) + H_2O(l)$$

$$2\,NO(g) + O_2(g) \rightarrow 2\,NO_2(g)$$

Nitrous acid is used as a reagent in organic chemistry; for example, diazonium salts are produced when nitrous acid is mixed with an organic amine (in this case, aniline, $C_6H_5NH_2$):

$$C_6H_5NH_2(aq) + HNO_2(aq) + HCl(aq) \rightarrow C_6H_5N_2{}^+Cl^-(s) + 2\,H_2O(l)$$

The diazonium salts are used, in turn, to synthesize a wide range of organic compounds.

15.14 Nitric Acid

A colorless, oily liquid when pure, nitric acid is extremely hazardous. It is obviously dangerous as an acid, but, as can be seen from the Frost diagram (see Figure 15.4), it is a very strong oxidizing agent, making it a potential danger in the presence of any oxidizable material. The acid, which melts at $-42°C$ and boils at $183°C$, is usually slightly yellow as a result of a light-induced decomposition reaction:

$$4\ HNO_3(aq) \rightarrow 4\ NO_2(g) + O_2(g) + 2\ H_2O(l)$$

When pure, liquid nitric acid is almost completely nonconducting. A small proportion ionizes as follows (all species exist in nitric acid solvent):

$$2\ HNO_3(l) \rightleftharpoons H_2NO_3^+ + NO_3^-$$

$$H_2NO_3^+ \rightleftharpoons H_2O + NO_2^+$$

$$H_2O + HNO_3 \rightleftharpoons H_3O^+ + NO_3^-$$

giving an overall reaction of

$$3\ HNO_3 \rightleftharpoons NO_2^+ + H_3O^+ + 2\ NO_3^-$$

The nitryl cation is important in the nitration of organic molecules; for example, the conversion of benzene, C_6H_6, to nitrobenzene, $C_6H_5NO_2$, an important step in numerous organic industrial processes.

Concentrated nitric acid is actually a 70 percent solution in water (corresponding to a concentration of about $16\ mol \cdot L^{-1}$), whereas "fuming nitric acid," an extremely powerful oxidant, is a red solution of nitrogen dioxide in pure nitric acid. Even when dilute, it is such a strong oxidizing agent that the acid rarely evolves hydrogen when mixed with metals; instead, a mixture of nitrogen oxides is produced and the metal is oxidized to its cation.

The terminal O—N bonds are much shorter (121 pm) than the O—N bond attached to the hydrogen atom (141 pm). This bond length indicates multiple bonding between the nitrogen and the two terminal oxygen atoms. In addition to the electrons in the σ system, there are four electrons involved in the O—N—O π system, two in a bonding orbital and two in a nonbonding orbital, a system giving a bond order of $1\frac{1}{2}$ for each of those nitrogen–oxygen bonds (Figure 15.21).

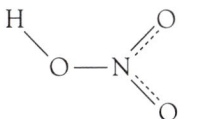

Figure 15.21 The nitric acid molecule.

We sometimes forget how our existence depends on thermodynamics and kinetics. For example, the first industrial synthesis of nitric acid was performed by simple combination of dinitrogen, dioxygen, and water, using an electric arc furnace to provide the energy to overcome the very high kinetic barrier. Although industrially the kinetic factor was a problem, in reality it is a blessing, for without the energy barrier, all the dioxygen in the atmosphere would long ago have been converted to nitric acid and our seas would have become dilute nitric acid:

$$\tfrac{1}{2}\ N_2(g) + \tfrac{5}{4}\ O_2(g) + \tfrac{1}{2}\ H_2O(g) \xrightarrow{\Delta} HNO_3(g)$$

Now we use the *Ostwald process* for nitric acid synthesis. This process utilizes much of the ammonia produced by the Haber process. The process is performed in three steps. First, a mixture of ammonia and dioxygen (or air) is passed through a platinum metal gauze. This is a very efficient, highly exothermic process that causes the gauze to glow red-hot. Contact time with the catalyst is limited to about 1 ms to minimize unwanted side reactions. The step is performed at low pressures to take advantage of the entropy effect; that is, the formation of 10 gas moles from 9 gas moles (an application of the Le Châtelier principle) to shift the equilibrium to the right:

$$4\ NH_3(g) + 5\ O_2(g) \rightarrow 4\ NO(g) + 6\ H_2O(g)$$

Additional oxygen is added to oxidize the nitrogen monoxide to nitrogen dioxide. To improve the yield of this exothermic reaction, heat is removed from the gases, and the mixture is placed under pressure:

$$2 \, NO(g) + O_2(g) \rightarrow 2 \, NO_2(g)$$

Finally, the nitrogen dioxide is mixed with water to give a solution of nitric acid:

$$3 \, NO_2(g) + H_2O(l) \rightarrow 2 \, HNO_3(l) + NO(g)$$

This reaction also is exothermic. Again, cooling and high pressures are used to maximize yield. The nitrogen monoxide is returned to the second stage for reoxidation.

Pollution used to be a major problem for nitric acid plants. The older plants were quite identifiable by the plume of yellow-brown gas—escaping nitrogen dioxide. State-of-the-art plants have little trouble in meeting the current emission standards of less than 200 parts per million (ppm) nitrogen oxides in their flue gases. Older plants now mix stoichiometric quantities of ammonia into the nitrogen oxides, a mixture producing harmless dinitrogen and water vapor:

$$NO(g) + NO_2(g) + 2 \, NH_3(g) \rightarrow 2 \, N_2(g) + 3 \, H_2O(g)$$

Worldwide, about 80 percent of the nitric acid is used in fertilizer production. This proportion is only about 65 percent in the United States, because about 20 percent is required for explosives production.

15.15 *Nitrites*

The colorless nitrite ion reacts with acids to form nitrous acid. At room temperature, the nitrous acid rapidly decomposes, as noted earlier in this chapter:

$$NO_2{}^-(aq) + H^+(aq) \rightarrow HNO_2(aq)$$

$$3 \, HNO_2(aq) \rightarrow HNO_3(aq) + 2 \, NO(g) + H_2O(l)$$

$$2 \, NO(g) + O_2(g) \rightarrow 2 \, NO_2(g)$$

The nitrite ion is a weak oxidizing agent; hence, nitrites of metals in their lower oxidation states cannot be prepared. For example, nitrite will oxidize iron(II) ion to iron(III) ion and is simultaneously reduced to lower oxides of nitrogen.

Sodium nitrite is a commonly used meat preservative, particularly in cured meats such as ham, hot dogs, sausages, and bacon. The nitrite ion inhibits the growth of bacteria, particularly *Clostridium botulinum*, an organism that produces the deadly botulism toxin. Sodium nitrite is also used to treat packages of red meat, such as beef. Blood exposed to the air rapidly produces a brown color, but shoppers much prefer their meat purchases to look bright red. Thus, the meat is treated with sodium nitrite; the nitrite ion is reduced to nitrogen monoxide, which then reacts with the hemoglobin to form a very stable bright red compound. It is true that the nitrite will prevent bacterial growth in this circumstance as well, but these days, the meat is kept at temperatures low enough to inhibit bacteria. To persuade shoppers to prefer brownish rather than red meat will require a lot of re-education. Now that all meats are treated with sodium nitrite, there is concern

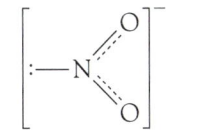

Figure 15.22 The nitrite ion.

that the cooking process will cause the nitrite ion to react with amines in the meat to produce nitrosamines, compounds containing the –NNO functional group. These compounds are known to be carcinogenic. However, as long as preserved meats are consumed in moderation, it is generally believed that the cancer risk is minimal.

The nitrite ion is V-shaped, as a result of the lone pair on the central nitrogen (Figure 15.22). The N—O bond length is 124 pm, longer than that in nitrogen dioxide (120 pm), but still much shorter than the N—O single bond (143 pm). As mentioned earlier, the lone electron in nitrogen dioxide is believed to occupy a weakly antibonding orbital. With analogous orbitals, the additional electron in the nitrite ion should also enter the antibonding orbital, thus leading to a bond weaker than that in nitrogen dioxide.

15.16 Nitrates

Nitrates of almost every metal ion in its common oxidation states are known, and, of particular note, all are water soluble. For this reason, nitrates tend to be used whenever a solution of a cation is required. Although nitric acid is strongly oxidizing, the colorless nitrate ion is not, under normal conditions (see Figure 15.4). Hence, one can obtain nitrates of metals in their lower oxidation states, such as iron(II).

The most important nitrate is ammonium nitrate; in fact, this one chemical accounts for the major use of nitric acid. About 1.5×10^7 tonnes is produced annually worldwide. It is prepared simply by the reaction of ammonia with nitric acid:

$$\text{NH}_3(g) + \text{HNO}_3(aq) \rightarrow \text{NH}_4\text{NO}_3(aq)$$

One of the common cold packs utilizes solid ammonium nitrate and water. When the dividing partition is broken, ammonium nitrate solution forms. This process is highly endothermic:

$$\text{NH}_4\text{NO}_3(s) \rightarrow \text{NH}_4^+(aq) + \text{NO}_3^-(aq) \qquad \Delta H^\circ = +26 \text{ kJ·mol}^{-1}$$

The endothermicity must be a result of comparatively strong cation–anion attractions in the crystal lattice and comparatively weak ion–dipole attractions to the water molecules in solution. If the enthalpy factor is positive but the compound is still very soluble, the driving force must be a large increase in entropy. In fact, there is such an increase: $+110 \text{ J·mol}^{-1}\text{·K}^{-1}$. In the solid, the ions have a low entropy, whereas in solution the ions are mobile. At the same time, the large ion size and low charge result in little ordering of the surrounding water molecules. Thus it is an increase in entropy that drives the endothermic solution process of ammonium nitrate.

Ammonium nitrate is a convenient and concentrated source of nitrogen fertilizer, although it has to be handled with care. At low temperatures, it decomposes to dinitrogen oxide; but at higher temperatures, explosive decomposition to dinitrogen, dioxygen, and water vapor occurs:

$$\text{NH}_4\text{NO}_3(s) \xrightarrow{\Delta} 2 \text{ H}_2\text{O}(g) + \text{N}_2\text{O}(g)$$

$$2 \text{ NH}_4\text{NO}_3(s) \xrightarrow{\Delta} 2 \text{ N}_2(g) + \text{O}_2(g) + 4 \text{ H}_2\text{O}(g)$$

About 1955, the North American blasting explosives industry recognized the potential of an ammonium nitrate–hydrocarbon mixture. As a result, a

The accidental fire on a ship carrying wax-coated pellets of ammonium nitrate killed at least 500 people in Texas City, Texas, in 1947.

mixture of ammonium nitrate with fuel oil has become very popular with the industry. It is actually quite safe, because the ammonium nitrate and fuel oil can be stored separately until use, and a detonator is then employed to initiate the explosion. This mixture was put to more tragic use in the Oklahoma City, Oklahoma, bombing in 1995.

Other nitrates decompose by different routes when heated. Sodium nitrate melts, and then, when strongly heated, bubbles of oxygen gas are produced, leaving behind sodium nitrite:

$$2\ NaNO_3(l) \xrightarrow{\Delta} 2\ NaNO_2(s) + O_2(g)$$

Most other metal nitrates give the metal oxide, nitrogen dioxide, and oxygen. For example, heating blue crystals of copper(II) nitrate heptahydrate first yields a green liquid, as the water of hydration is released and dissolves the copper(II) nitrate itself. Continued heating boils off the water, and the green solid then starts to release dioxygen and the brown fumes of nitrogen dioxide, leaving the black residue of copper(II) oxide:

$$2\ Cu(NO_3)_2(s) \xrightarrow{\Delta} 2\ CuO(s) + 4\ NO_2(g) + O_2(g)$$

Both nitrates and nitrites can be reduced to ammonia in basic solution with zinc or Devarda's alloy (a combination of aluminum, zinc, and copper). This reaction is a common test for nitrates and nitrites, because there is no characteristic precipitation test for this ion. The ammonia is usually detected by odor or with damp red litmus paper (which will turn blue):

$$NO_3{}^-(aq) + 6\ H_2O(l) + 8\ e^- \rightarrow NH_3(g) + 9\ OH^-(aq)$$

$$Al(s) + 4\ OH^-(aq) \rightarrow Al(OH)_4{}^-(aq) + 3\ e^-$$

The "brown ring test" for nitrate involves the reduction of nitrate with iron(II) in very acidic solution, followed by replacement of one of the coordinated water molecules of the remaining iron(II) to give a brown complex ion:

$$[Fe(OH_2)_6]^{2+}(aq) \rightarrow [Fe(OH_2)_6]^{3+}(aq) + e^-$$

$$NO_3{}^-(aq) + 4\ H^+(aq) + 3\ e^- \rightarrow NO(g) + 2\ H_2O(l)$$

$$[Fe(OH_2)_6]^{2+}(aq) + NO(g) \rightarrow [Fe(OH_2)_5NO]^{2+}(aq) + H_2O(l)$$

The nitrate ion is trigonal planar and has short nitrogen–oxygen bonds (122 pm)—bonds slightly shorter than those in the nitrite ion. Figure 15.23 shows an electron-dot diagram of one of the three possible resonance structures. If each contributes equally, the bond order will be 1⅓. The partial bonds of this ion are shown in Figure 15.24.

In molecular orbital terms, we envisage a π electron system involving one p orbital of the nitrogen atom and one p orbital of each of the three oxygen atoms. Molecular orbital calculations show that combining the atomic orbitals of four planar atoms gives four π molecular orbitals: one bonding, two nonbonding, and one antibonding. Six electrons are available; thus, two electrons enter the bonding molecular orbital; and four, the nonbonding orbital. This arrangement gives a net single π bond (Figure 15.25). However, this π system is shared over three bonds; hence, each bond possesses one-third of a π bond—the same result as that derived from the simple resonance structures.

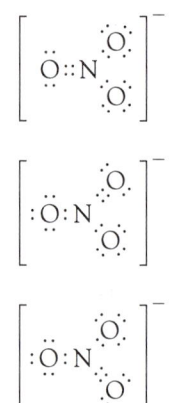

Figure 15.23 The three resonance forms of the nitrate ion.

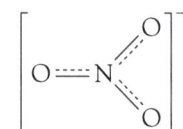

Figure 15.24 The nitrate ion.

Molecular
π orbitals

$$\overline{}$$
π^*_{2p}

$\underline{\text{1⬇}}\qquad\underline{\text{1⬇}}$
π^{NB}_{2p}

$\underline{\text{1⬇}}$
π_{2p}

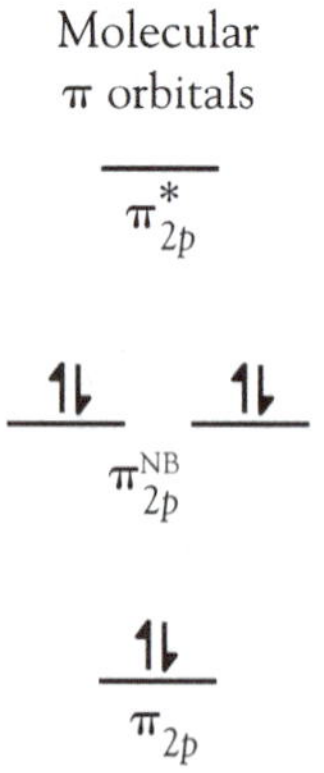

Figure 15.25 Molecular orbital energy level diagram of the orbitals involved in π bond formation in the nitrate ion.

15.17   *Overview of Phosphorus Chemistry*

Although they are neighbors in the Periodic Table, the redox behavior of nitrogen and phosphorus could not be more different (Figure 15.26). Whereas the higher oxidation states of nitrogen are strongly oxidizing in acidic solution, those of phosphorus are quite stable. In fact, the highest oxidation state of phosphorus is the most thermodynamically stable; and the lowest oxidation state, the least stable—the converse of nitrogen chemistry.

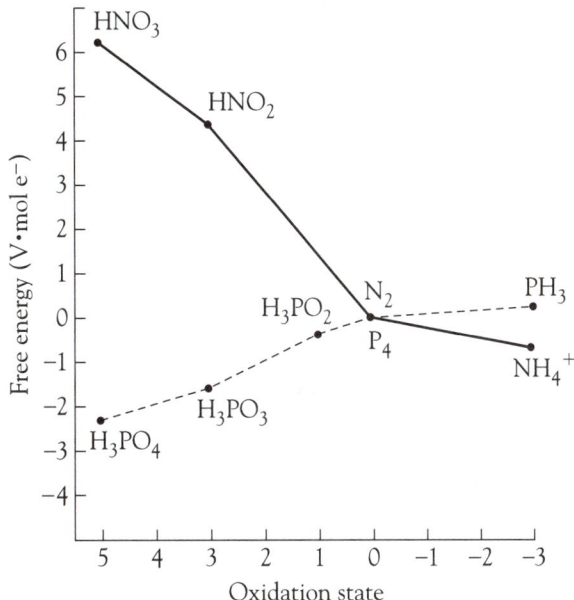

Figure 15.26 Frost diagram comparing the stability of the oxidation states of phosphorus and nitrogen in acidic solution.

15.18   *Allotropes of Phosphorus*

Phosphorus has several allotropes. The most common is white phosphorus (sometimes called yellow phosphorus); the other common one is red phosphorus. White phosphorus is a very poisonous, white, waxy-looking substance. It is a tetratomic molecule with the phosphorus atoms at the corners of a tetrahedron (Figure 15.27). Tetraphosphorus is an extremely reactive substance, possibly because of its highly strained bond structure. It burns vigorously in air to give tetraphosphorus decaoxide:

$$P_4(s) + 5\ O_2(g) \rightarrow P_4O_{10}(s)$$

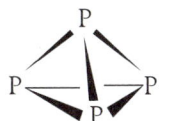

Figure 15.27 The white phosphorus molecule.

The oxide is formed in an electronically excited state, and as the electrons fall to the lowest energy state, visible light is released. In fact, the name phosphorus is derived from the phosphorescent glow when white phosphorus is exposed to air in the dark.

Because it is so reactive toward oxygen, white phosphorus has to be stored under water. This allotrope, having only weak dispersion forces between neighboring molecules, melts at 44°C. Although insoluble in hydrogen-bonding solvents such as water, it is extremely soluble in nonpolar organic solvents such as carbon disulfide.

Even though white phosphorus is formed when liquid phosphorus solidifies, it is the least thermodynamically stable of the allotropes. When

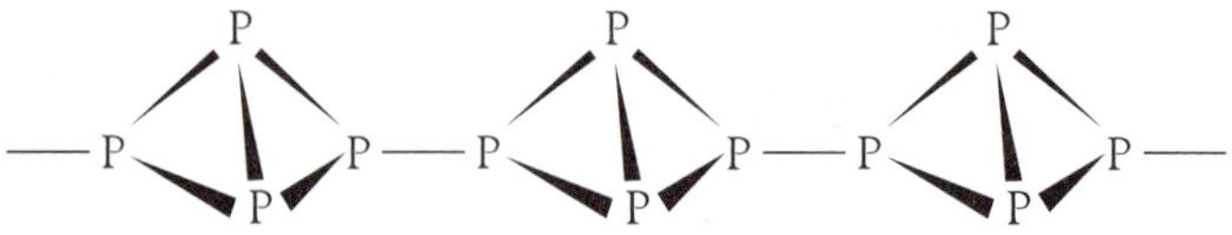

Figure 15.28 The arrangement of atoms in red phosphorus.

exposed to ultraviolet radiation (for example, from fluorescent lights), the white phosphorus slowly turns to red phosphorus. In this allotrope, one of the bonds in the tetrahedral structure of white phosphorus has broken open and joined to a neighboring unit (Figure 15.28). Thus, red phosphorus is a polymer with bonds less strained than those of the white allotrope.

The more thermodynamically stable red phosphorus has properties that are completely different from those of the white allotrope. It is stable in air, reacting with the dioxygen in air only above about 400°C. The melting point of the red allotrope is about 600°C, at which temperature the polymer chains break to give the same P_4 units contained in white phosphorus. As we would expect for a covalently bonded polymer, red phosphorus is insoluble in all solvents.

Curiously, the most thermodynamically stable form of phosphorus, black phosphorus, is the hardest to prepare. To make black phosphorus, white phosphorus is heated under pressures of about 1.2 GPa! This densest allotrope (as might be expected from the preferred reaction conditions) has a complex polymeric structure.

15.19   *Industrial Extraction of Phosphorus*

Phosphorus is such a reactive element that quite extreme methods have to be used to extract it from its compounds. The raw material is calcium phosphate. This compound is found in large deposits in central Florida, the Morocco-Sahara region, and the Pacific island of Nauru. The origins of these deposits are not well understood, although they may have been the result of the interaction of the calcium carbonate of coral reefs with the phosphate-rich droppings of seabirds over a period of hundreds of thousands of years.

Processing of phosphate rock is highly dependent on electric energy. As a result, the ore is usually shipped to countries where electric power is abundant and inexpensive, such as North America and Europe. The conversion of phosphate rock to the element is accomplished in a very large electric furnace containing 60-tonne carbon electrodes. In this electrothermal process, the furnace is filled with a mixture of ore, sand, and coke, and a current of about 180 000 A (at 500 V) is applied across the electrodes. At the 1500°C operating temperature of the furnace, the calcium phosphate reacts with carbon monoxide to give calcium oxide, carbon dioxide, and gaseous tetraphosphorus:

$$2\ Ca_3(PO_4)_2(s) + 10\ CO(g) \xrightarrow{\Delta} 6\ CaO(s) + 10\ CO_2(g) + P_4(g)$$

The carbon dioxide is then reduced back to carbon monoxide by the coke:

$$CO_2(g) + C(s) \xrightarrow{\Delta} 2\ CO(g)$$

Some of the gas is reused, but the remainder escapes from the furnace. The calcium oxide reacts with silicon dioxide (sand) to give calcium silicate (slag):

$$CaO(s) + SiO_2(s) \rightarrow CaSiO_3(l)$$

Table 15.3 Materials consumed and produced in the extraction of 1 tonne of phosphorus

Required	Produced
10 tonne calcium phosphate (phosphate rock)	1 tonne white phosphorus 8 tonne calcium silicate (slag)
3 tonne silicon dioxide (sand)	¼ tonne iron phosphides
1½ tonne carbon (coke)	0.1 tonne filter dust
14 MWh electrical energy	2500 m^3 flue gas

The escaping carbon monoxide is burned and the heat is used to dry the three raw materials:

$$2\ CO(g) + O_2(g) \rightarrow 2\ CO_2(g)$$

To condense the gaseous tetraphosphorus, it is pumped into a tower and sprayed with water. The liquefied phosphorus collects at the bottom of the tower and is drained into holding tanks. The average furnace produces about 5 tonnes of tetraphosphorus per hour.

There are two common impurities in the phosphate ore. First, there are traces of fluorapatite, $Ca_5(PO_4)_3F$, that react at the high temperatures to produce toxic and corrosive silicon tetrafluoride. This contaminant is removed from the effluent gases by treating them with sodium carbonate solution. The process produces sodium hexafluorosilicate, Na_2SiF_6, which is a commercially useful product. The second impurity is iron(III) oxide, which reacts with the tetraphosphorus to form ferrophosphorus (mainly Fe_2P, one of the several interstitial iron phosphides), a dense liquid that can be tapped from the bottom of the furnace below the liquid slag layer. Ferrophosphorus can be used in specialty steel products such as railroad brake shoes. The other by-product from the process, calcium silicate (slag), has little value apart from road fill. The cost of this whole process is staggering, not only for its energy consumption but also for the total mass of materials. These are listed in Table 15.3.

The major pollutants from the process are dusts, flue gases, phosphorus-containing sludge, and process water from the cooling towers. Older plants had very bad environmental records. In fact, the technology has changed to such an extent that it is now more economical to abandon an old plant and build a new one that will produce as little pollution as is possible using modern technology. However, the abandoned plant may become a severe environmental problem for the community in which it is located as a result of leaching from the waste material dumps.

The need for pure phosphorus is in decline because the energy costs of its production are too high to make it an economical source for most phosphorus compounds. Furthermore, demand for phosphate-based detergents has dropped because of ecologic concerns. Nevertheless, elemental phosphorus is still the preferred route for the preparation of high-purity phosphorus compounds, such as phosphorus-based insecticides and match materials.

15.20 Matches

Despite the prevalence of cheap butane lighters, match consumption is still between 10^{12} and 10^{13} per year. As mentioned at the beginning of this chapter, the modern safety match depends on a chemical reaction between the match head and the strip on the matchbox. The head of the match is mostly potassium chlorate, $KClO_3$, an oxidizing agent, whereas the strip contains red phosphorus and antimony sulfide, Sb_2S_3, both of which oxidize very exothermically when brought in contact with the potassium chlorate.

In addition to the safety match, there is also the "strike-anywhere" match. In this case, the two chemical components, the oxidizing agent (potassium chlorate) and the reducing agent (tetraphosphorus trisulfide, P_4S_3), are mixed in the match head. Any source of friction, such as the glass-paper strip on the matchbox or a brick wall, can provide the activation energy necessary to start the reaction.

15.21 Phosphine

The analog of ammonia—phosphine, PH_3—is a colorless, highly poisonous gas. The two hydrides differ substantially because the P—H bond is much less polar than the N—H bond. Thus, phosphine is a very weak base, and it does not form hydrogen bonds. In fact, the phosphonium ion, PH_4^+, the equivalent of the ammonium ion, is difficult to prepare. Phosphine itself can be prepared by mixing a phosphide of a very electropositive metal with water:

$$Ca_3P_2(s) + 6\ H_2O(l) \rightarrow 2\ PH_3(g) + 3\ Ca(OH)_2(aq)$$

In phosphine, the P–H bond angle is only 93° rather than 107°, the angle of the N–H bond in ammonia. The phosphine angle suggests that the phosphorus atom is using p orbitals rather than sp^3 hybrids for bonding.

Although phosphine itself has few uses, substituted phosphines are important reagents in organometallic chemistry as we shall see in Chapter 22, Section 22.10. The most common of the substituted phosphines is triphenylphosphine, $P(C_6H_5)_3$, often abbreviated to PPh_3.

15.22 Phosphorus Oxides

Phosphorus forms two oxides: tetraphosphorus hexaoxide, P_4O_6, and tetraphosphorus decaoxide, P_4O_{10}. They are both white solids at room temperature. Tetraphosphorus hexaoxide is formed by heating white phosphorus in an environment with a shortage of oxygen:

$$P_4(s) + 3\ O_2(g) \xrightarrow{\Delta} P_4O_6(s)$$

Conversely, tetraphosphorus decaoxide, the more common and more important oxide, is formed by heating white phosphorus in the presence of an excess of oxygen:

$$P_4(s) + 5\ O_2(g) \xrightarrow{\Delta} P_4O_{10}(s)$$

Tetraphosphorus decaoxide can be used as a dehydrating agent because it reacts vigorously with water in a number of steps to give, ultimately, phosphoric acid:

$$P_4O_{10}(s) + 6\ H_2O(l) \rightarrow 4\ H_3PO_4(l)$$

Tetraphosphorus decaoxide will dehydrate many compounds, for example, nitric acid to dinitrogen pentoxide and organic amides, $RCONH_2$, to nitriles, RCN.

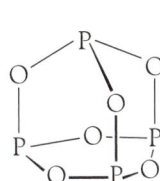

Figure 15.29 The tetraphosphorus hexaoxide molecule.

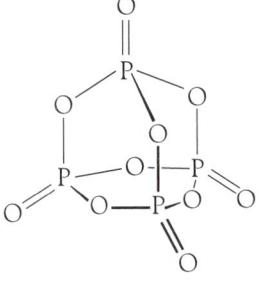

Figure 15.30 The tetraphosphorus decaoxide molecule.

The structure of both these oxides is based on the tetrahedron of white phosphorus (tetraphosphorus) itself. In tetraphosphorus hexaoxide, oxygen atoms have inserted themselves in all the phosphorus–phosphorus bonds (Figure 15.29). In tetraphosphorus decaoxide, four additional oxygen atoms form coordinate covalent bonds to the phosphorus atoms, extending out from the corners of the tetrahedron (Figure 15.30). These bonds have multiple bond character, which we can explain as a partial contribution to the bonding by an empty d orbital on each phosphorus atom forming a π bond with a full p orbital on each corner oxygen atom.

15.23 *Phosphorus Chlorides*

In parallel with the oxides, there are two chlorides: phosphorus trichloride, PCl_3, a colorless liquid, and phosphorus pentachloride, PCl_5, a white solid. Phosphorus trichloride is produced when chlorine gas reacts with an excess of phosphorus:

$$P_4(s) + 6\ Cl_2(g) \rightarrow 4\ PCl_3(l)$$

An excess of chlorine results in phosphorus pentachloride:

$$P_4(s) + 10\ Cl_2(g) \rightarrow 4\ PCl_5(l)$$

Phosphorus trichloride reacts with water to give phosphonic acid, H_3PO_3 (commonly called phosphorous acid), and hydrogen chloride gas (Figure 15.31):

$$PCl_3(l) + 3\ H_2O(l) \rightarrow H_3PO_3(l) + 3\ HCl(g)$$

This behavior contrasts with that of nitrogen trichloride, which, as mentioned earlier, hydrolyzes to give ammonia and hypochlorous acid (Figure 15.32):

$$NCl_3(l) + 3\ H_2O(l) \rightarrow NH_3(g) + 3\ HClO(aq)$$

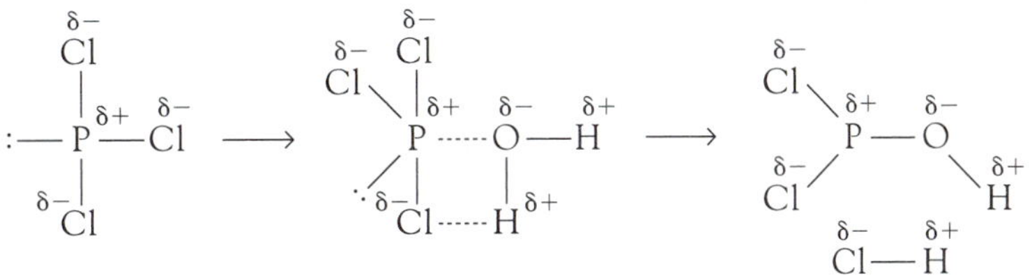

Figure 15.31 The proposed mechanism for the first step in the reaction between phosphorus trichloride and water.

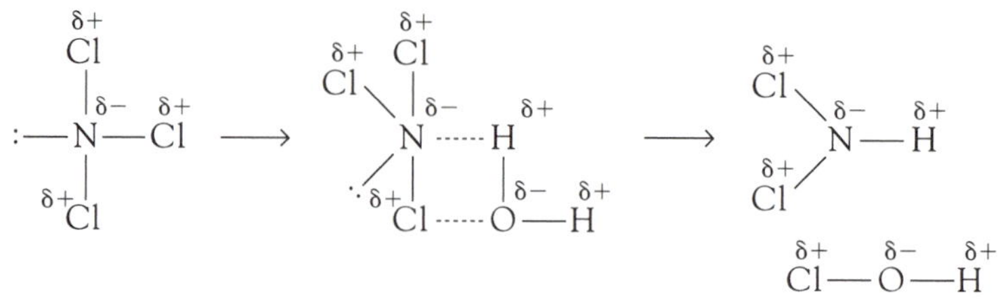

Figure 15.32 The proposed mechanism for the first step in the reaction between nitrogen trichloride and water.

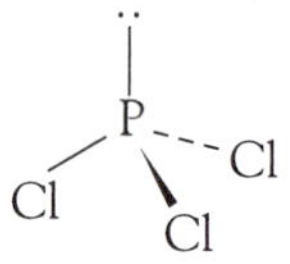

Figure 15.33 The phosphorus trichloride molecule.

Phosphorus trichloride is an important reagent in organic chemistry, and its worldwide production amounts to about 250 000 tonnes. For example, it can be used to convert alcohols to chloro compounds. Thus, 1-propanol is converted to 1-chloropropane by phosphorus trichloride:

$$PCl_3(l) + 3\ C_3H_7OH(l) \rightarrow 3\ C_3H_7Cl(g) + H_3PO_3(l)$$

Phosphorus trichloride has a trigonal shape that is explained by the lone pair on the phosphorus atom (Figure 15.33).

Phosphorus pentachloride is also used as an organic reagent, but it is less important, annual production being only about 20 000 tonnes worldwide. Like phosphorus trichloride, it reacts with water, but in a two-step process, the first step yielding phosphoryl chloride, $POCl_3$:

$$PCl_5(s) + H_2O(l) \rightarrow POCl_3(l) + 2\ HCl(g)$$

$$POCl_3(l) + 3\ H_2O(l) \rightarrow H_3PO_4(l) + 3\ HCl(g)$$

In the gas phase, phosphorus pentachloride is a trigonal bipyramidal covalent molecule (Figure 15.34), but in the solid phase, phosphorus pentachloride adopts the ionic structure $PCl_4^+PCl_6^-$ (Figure 15.35).

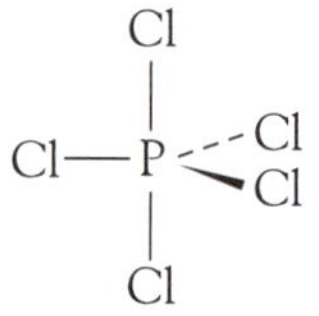

Figure 15.34 The shape of the phosphorus pentachloride molecule in liquid and gas phases.

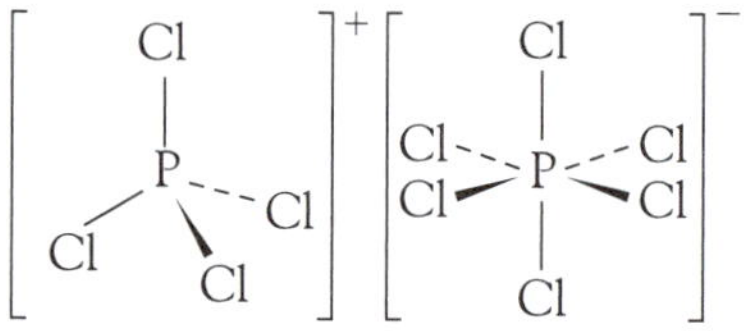

Figure 15.35 The two ions present in solid-phase phosphorus pentachloride.

15.24   *Phosphorus Oxychloride*

One of the most important industrial phosphorus compounds is phosphorus oxychloride, $POCl_3$. This dense toxic liquid, which fumes in moist air, is produced industrially by the catalytic oxidation of phosphorus trichloride:

$$2\ PCl_3(l) + O_2(g) \rightarrow 2\ POCl_3(l)$$

There is an extensive range of chemicals made from phosphorus oxychloride. Tri-*n*-butyl phosphate, $(C_5H_{11}O)_3PO$, commonly abbreviated to TBP, is a useful selective solvent, such as for separating uranium and plutonium compounds. Similar compounds are crucial to our lives as fire retardants that are sprayed on children's clothing, aircraft and train seats, curtain materials, and many more items that we encounter in our everyday lives.

15.25 *Common Oxyacids of Phosphorus*

There are three oxyacids of phosphorus that we will mention here: phosphoric acid, H_3PO_4; phosphonic acid, H_3PO_3 (commonly called phosphorous acid); and phosphinic acid, H_3PO_2 (commonly called hypophosphorous acid). The first, phosphoric acid, is really the only oxyacid of phosphorus that is important. However, the other two acids are useful for making a point about the character of oxyacids.

In an oxyacid, for the hydrogen to be significantly acidic, it must be attached to an oxygen atom—and this is normally the case. In general, as we progress through a series of oxyacids—for example, nitric acid, $(HO)NO_2$, and nitrous acid, $(HO)NO$—it is one of the terminal oxygen atoms that is lost as the oxidation state of the central element is reduced. Phosphorus is almost unique in that the oxygens linking a hydrogen to the phosphorus are the ones that are lost. Thus, phosphoric acid possesses three ionizable hydrogen atoms; phosphonic acid, two; and phosphinic acid, only one (Figure 15.36).

Figure 15.36 The bonding in the three common oxyacids of phosphorus: (a) phosphoric acid, (b) phosphonic acid, (c) phosphinic acid.

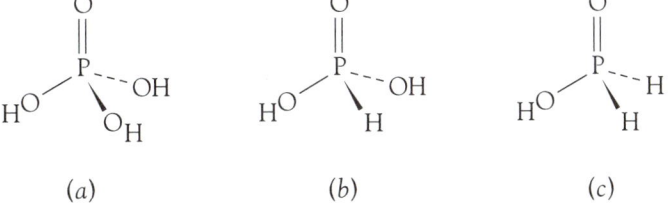

(a) (b) (c)

Phosphoric Acid

Pure (ortho)phosphoric acid is a colorless solid, melting at 42°C. A concentrated aqueous solution of the acid (85 percent by mass and having a concentration of 14.7 mol·L^{-1}) is called "syrupy" phosphoric acid, its viscous nature being caused by extensive hydrogen bonding. As discussed earlier, the acid is essentially nonoxidizing. In solution, phosphoric acid is a weak acid, undergoing three ionization steps:

$$H_3PO_4(aq) + H_2O(l) \rightleftharpoons H_3O^+(aq) + H_2PO_4^-(aq)$$

$$H_2PO_4^-(aq) + H_2O(l) \rightleftharpoons H_3O^+(aq) + HPO_4^{2-}(aq)$$

$$HPO_4^{2-}(aq) + H_2O(l) \rightleftharpoons H_3O^+(aq) + PO_4^{3-}(aq)$$

The pure acid is prepared by burning white phosphorus to give tetraphosphorus decaoxide, then treating the oxide with water:

$$P_4(s) + 5\ O_2(g) \rightarrow P_4O_{10}(s)$$

$$P_4O_{10}(s) + 6\ H_2O(l) \rightarrow 4\ H_3PO_4(l)$$

This is known as the "thermal" process. Such high purity is not required for most purposes, so where trace impurities can be tolerated, it is much more energy efficient to treat calcium phosphate with sulfuric acid to give a solution of phosphoric acid and a precipitate of calcium sulfate:

$$Ca_3(PO_4)_2(s) + 3\ H_2SO_4(aq) \rightarrow 3\ CaSO_4(s) + 2\ H_3PO_4(aq)$$

The only problem associated with this process (known as the "wet" process) is the disposal of the calcium sulfate. Some of this product is used in the building industry, but production of calcium sulfate exceeds the uses; hence,

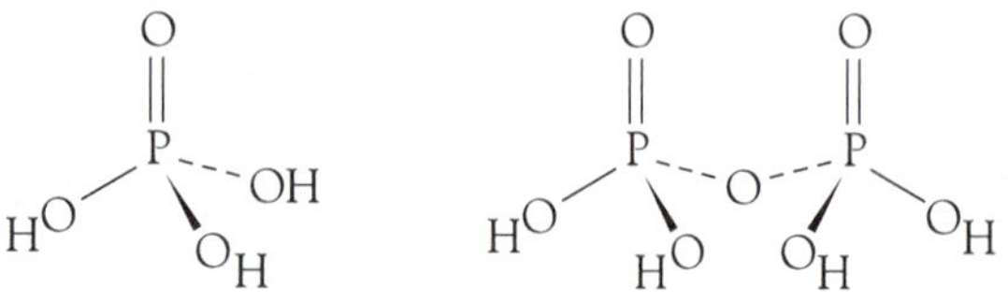

(a)

(b)

Figure 15.37 The shape of (a) (ortho)phosphoric acid and (b) pyrophosphoric acid.

most of it must be dumped. Furthermore, when the phosphoric acid is concentrated at the end of the process, many of the impurities precipitate out. This "slime" must be disposed of in an environmentally safe manner.

Heating phosphoric acid causes a stepwise loss of water; in other words, the phosphoric acid molecules undergo condensation. The first product is pyrophosphoric acid, $H_4P_2O_7$. As in phosphoric acid, each phosphorus atom is tetrahedrally coordinated (Figure 15.37). The next product is tripolyphosphoric acid, $H_5P_3O_{10}$:

$$2\ H_3PO_4(l) \xrightarrow{\Delta} H_4P_2O_7(l) + H_2O(l)$$

$$3\ H_4P_2O_7(l) \xrightarrow{\Delta} 2\ H_5P_3O_{10}(l) + H_2O(l)$$

Subsequent condensations give products with even greater degrees of polymerization.

Most of the phosphoric acid is used for fertilizer production. Phosphoric acid also is a common additive to soft drinks, its weak acidity preventing bacterial growth in the bottled solutions. It often serves a second purpose in metal containers. Metal ions may be leached from the container walls, but the phosphate ions will react with the metal ions to give an inert phosphate compound, thus preventing any potential metal poisoning. Phosphoric acid is also used on steel surfaces as a rust remover, both in industry and for home automobile repairs.

15.26   Phosphates

With high lattice energies resulting from the high anion charge, most phosphates are insoluble. The alkali metals and ammonium phosphates are the only common exceptions to this rule. There are three series of phosphate salts: the phosphates, containing the PO_4^{3-} ion; the hydrogen phosphates, containing the HPO_4^{2-} ion; and the dihydrogen phosphates, containing the $H_2PO_4^-$ ion. In solution, there is an equilibrium between the three species and phosphoric acid itself. For example, a solution of phosphate ion will hydrolyze:

$$PO_4^{3-}(aq) + H_2O(l) \rightleftharpoons HPO_4^{2-}(aq) + OH^-(aq)$$

$$HPO_4^{2-}(aq) + H_2O(l) \rightleftharpoons H_2PO_4^-(aq) + OH^-(aq)$$

$$H_2PO_4^-(aq) + H_2O(l) \rightleftharpoons H_3PO_4(aq) + OH^-(aq)$$

Each successive equilibrium lies more and more to the left. Thus the concentration of actual phosphoric acid is minuscule. A solution of sodium phosphate, then, will be quite basic, almost entirely as a result of the first equilibrium. It is this basicity (good for reaction with greases) and the complexing ability of the phosphate ion that make a solution of sodium phosphate a common kitchen cleaning solution (known as TSP, trisodium phosphate).

Figure 15.38 shows how the proportions of the phosphate species depend on pH. A solution of sodium hydrogen phosphate, $Na_2(HPO_4)$, will be basic as a result of the second step of the preceding sequence:

$$HPO_4^{2-}(aq) + H_2O(l) \rightleftharpoons H_2PO_4^-(aq) + OH^-(aq)$$

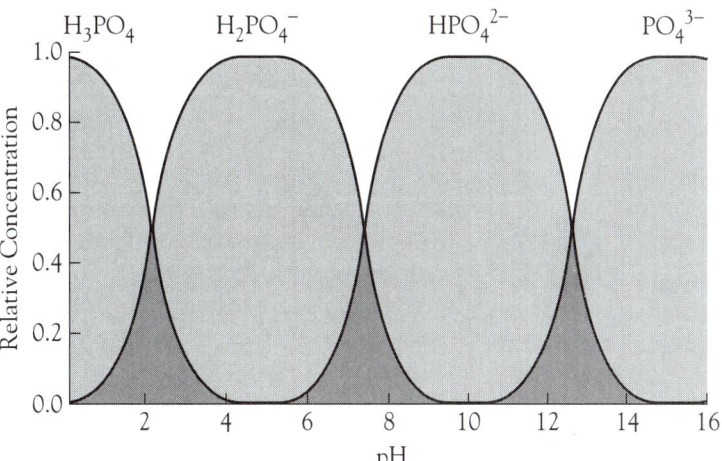

Figure 15.38 The relative concentrations of phosphate species at different values of pH.

A solution of sodium dihydrogen phosphate, $Na(H_2PO_4)$, however, is slightly acidic as a result of the following reaction predominating:

$$H_2PO_4^-(aq) + H_2O(l) \rightleftharpoons H_3O^+(aq) + HPO_4^{2-}(aq)$$

Solid hydrogen phosphates and dihydrogen phosphates are only known for most of the monopositive cations (the alkali metals ions and the ammonium ion) and a few of the dipositive cations, such as calcium ion. As we have seen before, to stabilize a large low-charge anion, a low charge density cation is required. For most dipositive and all tripositive metal ions, the metal ion precipitates the small proportion of phosphate ion from a solution of one of the acid ions. The Le Châtelier principle then drives the equilibria to produce more phosphate ion, causing additional metal phosphate to precipitate.

The phosphates have a tremendous range of uses. As mentioned earlier, trisodium phosphate is used as a household cleaner. Other sodium phosphates, such as sodium pyrophosphate, $Na_4P_2O_7$, and sodium tripolyphosphate, $Na_5P_3O_{10}$, are often added to detergents, because they react with calcium and magnesium ions in the tap water to form soluble compounds, preventing the deposition of scum in the washing. However, when the phosphate-rich wastewater reaches lakes, it can cause a rapid growth of algae and other simple plant life. This formation of green, murky lakes is called *eutrophication*. Phosphates are also added to detergents as fillers. Fillers are required because we are used to pouring cupfuls of solid detergent into a washing machine. However, only small volumes of cleaning agents are actually needed, so most of the detergent is simply inert materials. In more frugal societies, such as Japan, people are used to adding spoonfuls of detergent—hence, less wasteful filler is required.

Disodium hydrogen phosphate is used in the preparation of pasteurized processed cheese, although even today, the reason why this ion aids in the cheese-making process is not well understood. The ammonium salts, diammonium hydrogen phosphate and ammonium dihydrogen phosphate, are useful nitrogen–phosphorus combination fertilizers. Ammonium phosphates also make excellent flame retardants for drapes, theater scenery, and disposable paper clothing and costumes.

The calcium phosphates are used in many circumstances. For example, "combination baking powder" relies on the reaction between calcium dihydrogen phosphate and sodium hydrogen carbonate to produce the carbon dioxide gas so essential for baking. The reaction can be simplistically represented as

$$Ca(H_2PO_4)_2(aq) + NaHCO_3(aq) \rightleftharpoons CaHPO_4(aq)$$
$$+ NaH_2PO_4(aq) + CO_2(g) + H_2O(g)$$

Other calcium phosphates are used as mild abrasives and polishing agents in toothpaste. Finally, calcium dihydrogen phosphate is used as a fertilizer. The calcium phosphate rock is too insoluble to provide phosphate for plant growth, so it is treated with sulfuric acid to produce calcium dihydrogen phosphate:

$$Ca_3(PO_4)_2(s) + 2\ H_2SO_4(aq) \rightleftharpoons Ca(H_2PO_4)_2(s) + 2\ CaSO_4(s)$$

This compound is only slightly soluble in water, but soluble enough to release a steady flow of phosphate ions into the surrounding soil, where they can be absorbed by plant roots.

It was an English farmer, John Lawes, who discovered in 1842 that phosphate fertilizer could be made more soluble by the reaction with sulfuric acid. This was the synthesis of the first artificial fertilizer.

Nauru, the World's Richest Island

In our studies of industrial chemistry, or any branch of science, it is important to consider the human element. The extraction of phosphate rock from the island of Nauru is an illustrative case history.

The Republic of Nauru, located in the Pacific Ocean, has an area of 21 km^2, yet it is one of the world's major suppliers of calcium phosphate. Between 1 and 2 million tonnes of phosphate rock is mined there each year, thus providing that small nation with a gross national product of about $200 million per year. For the approximately 5000 native Nauruans, this income provides an opulent lifestyle, with all the conveniences of modern living from washing machines to VCRs. In addition, they hire servants and maids from Asia and from other island countries to perform the work around the house and garden.

The downside of this "idyllic" life is that there is little incentive to work or study. Obesity, heart disease, and alcohol abuse have suddenly become major problems. The long-term effects on the environment have also been catastrophic. Extraction of the phosphate rock is like large-scale dentistry. The ore is scooped up from between enormous toothlike stalks of coral limestone, some of which are 25 m high. These barren pinnacles of limestone will be all that is left of 80 percent of the island when the deposits are exhausted. Furthermore, silt runoff from the mining operation has damaged the offshore coral reefs that once provided abundant fishing resources.

To provide a future for the island and the islanders, the majority of the mining royalties are now placed in the Phosphate Royalties Trust. The money is intended to give the islanders a long-term benefit. This may involve the importation of millions of tonnes of soil to resurface the island. Alternatively, some planners envisage the purchase of some sparsely populated island for relocation of the entire population and an abandonment of Nauru. Whatever the future, phosphate mining has changed the lives of these islanders forever.

15.27 Biological Aspects

Nitrogen: Inert but Essential

Just as there is a carbon cycle, there is also a nitrogen cycle, for all plant life requires nitrogen for growth and survival. Between 10^8 and 10^9 tonnes of nitrogen is cycled between the atmosphere and the lithosphere in a 1-year period. The dinitrogen in the atmosphere is converted by bacteria to compounds of nitrogen. Some of the bacteria exist free in the soil, but members of the most important group, the *Rhizobium*, form nodules on the roots of pea, bean, alder, and clover plants. This is a symbiotic relationship, with the bacteria providing the nitrogen compounds to the plants, and the plants providing a stream of nutrients to the bacteria. To do this at a high rate at normal soil temperatures, the bacteria use enzymes such as nitrogenase.

Nitrogenase contains two components: a large protein containing two metals, iron and molybdenum, and a smaller one containing iron. The bioinorganic chemistry involved in this process is still not well understood, but it is believed that one of the crucial steps involves the formation of a molybdenum bond with the dinitrogen molecule. It is hoped that an understanding of the route used by bacteria will enable us to produce ammonia for fertilizers by a room-temperature process rather than the energy-intensive Haber–Bosch procedure.

Phosphorus: The Phosphorus Cycle

Phosphorus is another element essential for life. For example, the free hydrogen phosphate and dihydrogen phosphate ions are involved in the blood buffering system. More important, phosphate is the linking unit in the sugar esters of DNA and RNA, and phosphate units make up part of ATP, the essential energy storage unit in living organisms. Finally, bone is a phosphate mineral, calcium hydroxide phosphate, $Ca_5(OH)(PO_4)_3$, commonly called apatite.

Just as there are carbon and nitrogen cycles, there is an active phosphorus cycle. Phosphorus is an essential element for plant growth though its soil concentration is usually quite low—between 1 mg$\cdot$L^{-1} and 0.001 mg$\cdot$L^{-1}. One major difference from the carbon and nitrogen cycles is that phosphorus has no major gas-phase compound; thus, cycling is almost totally between aqueous and solid phases involving the phosphate ion. Organic phosphates are the most soluble in soils. These are phosphates covalently bound as phospholipids, nucleic acids, and inositol phosphates where inositol is a sugar-like molecule $C_6H_6(OH)_6$.

At the normal pH range of soils, inorganic phosphate is present as the hydrogen phosphate and dihydrogen phosphate ions. However, the equilibrium is shifted by the presence of high-charge metal ions that force the equilibria toward insoluble, and hence unavailable, compounds. Of particular importance, calcium ion forms insoluble calcium phosphate. As the pH increases, so the equilibrium favors formation of the insoluble compound.

$$3\ Ca^{2+}(aq) + 2\ HPO_4{}^{2-}(aq) \rightleftharpoons Ca_3(PO_4)_2(s) + 2\ H^+(aq)$$

Solution availability of phosphate is also limited at the acid end of the scale. As we mentioned in Chapter 13, Section 13.8, aluminum ion becomes more soluble as the pH decreases. The solubilized aluminum ion species will then react with the phosphate species to give insoluble aluminophosphate compounds. For example:

$$[Al(OH_2)_4(OH)_2]^+(aq) + H_2PO_4{}^-(aq) \rightleftharpoons Al(OH)_2H_2PO_4(s) + 2\ H_2O(l)$$

As a result of these two reaction types, the level of inorganic phosphate in soils is very low except in a narrow window from pH 6 to pH 7.

Arsenic: The Water Poisoner

Amazing as it may seem, arsenic also is an element essential to life. But we only need trace amounts of this element, whose role is still unknown. Anything more than a tiny amount of an inorganic compound of arsenic causes acute poisoning. In part, poisoning is believed to occur by the reaction of the soft-acid arsenic(III) with the soft-base sulfide of thio-aminoacids blocking the sulfur atoms from crosslinking forming disulfide bridges. In fact, one of the treatments for arsenic poisoning is the administration of thiol compounds to "mop up" the arsenite ions.

Many people believe that well water is, by definition, "pure" water. In fact, the composition of well water reflects the soluble (and possibly toxic) components of the water-laden strata. It is always important to check rural well-water supplies for trace elements such as arsenic. Arsenic poisoning from well water is now a serious health problem in the Asian country of Bangladesh. In rural Bangladesh, villagers had traditionally relied on disease-carrying surface ponds for their water supply. The government, together with many international aid agencies carried out a highly successful campaign of drilling over 18 000 wells to provide safe, deep well water.

However, in their enthusiasm for nonsurface water, they did not check on the underlying geological strata. These happened to be high in leachable arsenic compounds. As a result, in many places, illness from water-born diseases was replaced by illness and death from arsenic poisoning and arsenic-induced tumors. The immediate task was to identify which of the 18 000 wells exceeds the limit of 0.05 ppm. In a developed country such as the United States, the test samples would be rushed to a laboratory with state-of-the-art instrumentation. This is not an option for a poor country such as Bangladesh. Instead, a low-cost kit was developed by a Japanese company using the traditional Marsh test for arsenic. This kit could be carried by

Paul Erhlich and His "Magic Bullet"

The story of the founding of chemotherapy is told in a classic movie, Dr. Erhlich Magic Bullet, released in 1940.

Arsenic has also been used as a lifesaver. In the 19th century, physicians had no means of combating infections, and patients usually died. The whole nature of medicine changed in 1863 when a French scientist, Béchamps, noticed that an arsenic compound was toxic to some microorganisms. A German, Paul Erhlich, decided to synthesize new arsenic compounds, testing each one for its organism-killing ability. In 1909, with his 606th compound of arsenic, he found a substance that selectively killed the syphilis organism. At the time, syphilis was a feared and widespread disease for which there was no cure, only suffering, dementia, and death. Erhlich's arsenic compound, what he dubbed a "magic bullet," provided miraculous cures; and a search for other chemical compounds that could be used in the treatment of disease was launched.

This field, chemotherapy, has produced one of the most effective tools of controlling bacterial infections and those of many other microorganisms. Chemotherapy also provides one of the lines of attack against cancerous tissues. And it all started with an arsenic compound.

community health workers from village to village and used by relatively unskilled personnel. With the identification of the most hazardous wells, the next step is to find a low-cost method of arsenic removal. This solution has to be found very rapidly as each day results in more deaths.

The most famous case of arsenic poisoning is believed to be that of Napoleon. Modern chemical analysis has shown very high levels of arsenic in Napoleon's hair. Many have argued that his British captors (or perhaps French rivals) poisoned him to remove him as a threat to European stability were he to escape again from captivity. However, chemical research has turned up an alternative explanation: that he was poisoned by his wallpaper. At that time, copper(II) hydrogen arsenite, $CuHAsO_3$, was used as a pigment to give a beautiful green color in wallpapers. In a dry climate, this pigment was quite safe, but in the chronically damp house in which Napoleon was held on the island of St. Helena, molds grew on the walls. Many of these molds metabolize arsenic compounds to trimethylarsenic, $(CH_3)_3As$, a gas. Amazingly, a sample of his bedroom wallpaper has survived and it does indeed use the arsenite pigment. Thus, Napoleon quite possibly inhaled this toxic gas while in bed, and the sicker he became, the more time he spent in his bedroom, thus hastening his death from this toxic element.

Bismuth: The Medicinal Element

Though bismuth is not known to serve any essential role in biological systems, bismuth compounds have been used in the treatment of bacteria-related illnesses in Western medicine for over 250 years. Bismuth compounds are to be found in traditional Chinese medicine and in some medicinal plants (herbal remedies) that are bismuth accumulators.

Though bismuth compounds are used in the treatment of syphilis and certain types of tumors, the major use of bismuth-containing compounds is in the treatment of gastrointestinal (stomach) disorders. Commercial preparations include Pepto-Bismol™, which contains bismuth sub-salicylate (BSS) and De-Nol™, which contains colloidal bismuth sub-citrate (CBS). These compounds are antimicrobial but also appear to fortify the gastric mucus and stimulate cytoprotective processes. BSS and CBS have been found to be effective against the bacterium *Helicobacter pylori,* now known to be a causative factor of most types of gastroduodenal ulcers. There is also strong evidence of the effectiveness of BSS against traveller's diarrhea. In fact, Canadian health authorities advise visitors to regions of the world where bacterial food and water levels are high to take one or two BSS tablets prior to each meal as a preventative measure. (Though not a scientifically meaningful sample, one of the co-authors of this text has found this advice to be very effective.)

The efficacy of bismuth as a bacterial killer must relate to some unique aspect of its chemistry. Bismuth's chemistry is defined by its weakly metallic properties. The predominant oxidation state is +3, but the free cation does not exist and covalent behavior predominates. Instead, BiO^+, is the only common ionic species. This is known as the bismuthyl ion and its compounds are called "basic" or "oxy-" or "sub-"salts. In fact, the aqueous chemistry is dominated by the formation of clusters containing six bismuth and eight oxygen atoms, such as $[Bi_6O_4(OH)_4]^{6+}$. As a result of the complexity of its solution chemistry, even the stoichiometry of compounds such as BSS and CBS is variable, depending upon the precise conditions of synthesis. Much research still needs to be done to discover how bismuth compounds function as bactericides.

15.28 *Element Reaction Flowchart*

Flowcharts are shown for both nitrogen and phosphorus, the two key elements in Group 15.

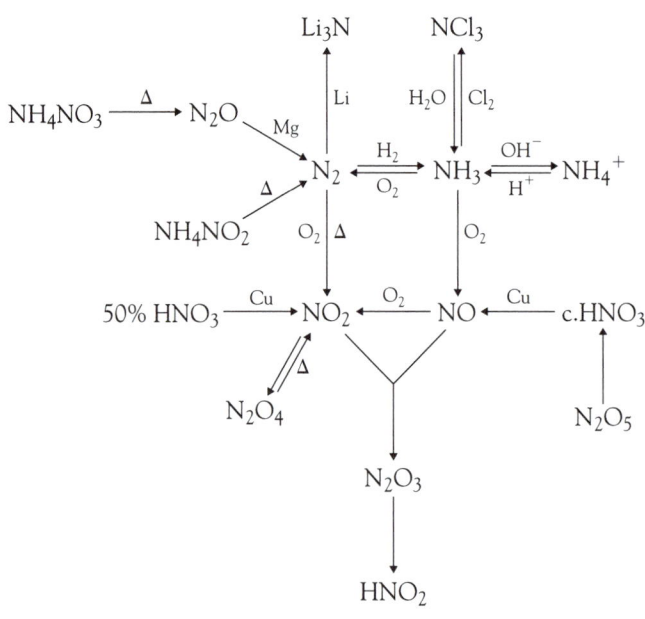

15.1 Write balanced chemical equations for the following chemical reactions:
(a) arsenic trichloride with water
(b) magnesium with dinitrogen
(c) ammonia with excess chlorine
(d) methane with steam
(e) hydrazine and oxygen
(f) heating a solution of ammonium nitrate
(g) sodium hydroxide solution with dinitrogen trioxide
(h) heating sodium nitrate
(i) heating tetraphosphorus decaoxide with carbon.

15.2 Write balanced chemical equations for the following chemical reactions:
(a) heating a solution of ammonium nitrite
(b) solutions of ammonium sulfate with sodium hydroxide
(c) ammonia with phosphoric acid

(d) decomposition of sodium azide
(e) nitrogen monoxide and nitrogen dioxide
(f) heating solid lead nitrate
(g) tetraphosphorus with an excess of dioxygen
(h) calcium phosphide with water
(i) hydrazine solution and dilute hydrochloric acid.

15.3 Why is it hard to categorize arsenic as either a metal or a nonmetal?

15.4 What are the factors that distinguish the chemistry of nitrogen from that of the other members of Group 15?

15.5 Contrast the behavior of nitrogen and carbon by comparing the properties of (a) methane and ammonia; (b) ethene and hydrazine.

15.6 Contrast the bonding to oxygen in the two compounds, NF_3O and PF_3O.

15.7 (a) Why is dinitrogen very stable? (b) Yet why is dinitrogen not always the product during redox reactions involving nitrogen compounds?

15.8 When ammonia is dissolved in water, the solution is often referred to as "ammonium hydroxide." Discuss whether this terminology is appropriate.

15.9 In the Haber process for ammonia synthesis, the recycled gases contain increasing proportions of argon gas. Where does the argon come from? Suggest how it might be removed.

15.10 Why is it surprising that high pressure is used in the steam reforming process during ammonia synthesis?

15.11 Discuss the differences between the ammonium ion and the alkali metal ions.

15.12 Using bond energies, calculate the heat released when gaseous hydrazine burns in air (oxygen) to give water vapor and nitrogen gas.

15.13 Construct a possible electron-dot structure for the azide ion. Identify the location of the formal charges.

15.14 Construct three possible electron-dot structures for the theoretical molecule, N—O—N. By assignment of formal charges, suggest why the actual dinitrogen oxide molecule has its asymmetrical structure.

15.15 Taking into account the nitrogen gas produced in both steps of the airbag reaction, calculate the mass of sodium azide needed to fill a 70-L airbag with dinitrogen at 298 K and 100 kPa pressure.

15.16 Nitrogen monoxide can form a cation, NO^+, and an anion, NO^-. Calculate the bond order in each of these species.

15.17 Nitrogen trifluoride boils at $-129°C$ while ammonia boils at $-33°C$. Account for the difference in these values.

15.18 Draw the shape of each of the following molecules: (a) dinitrogen trioxide; (b) dinitrogen pentoxide (solid and gas phases); (c) phosphorus pentafluoride.

15.19 Draw the shape of each of the following molecules: (a) dinitrogen oxide; (b) dinitrogen tetroxide; (c) phosphorus trifluoride; (d) phosphonic acid.

15.20 Describe the physical properties of: (a) nitric acid; (b) ammonia.

15.21 Explain why, in the synthesis of nitric acid, the reaction of nitrogen monoxide with dioxygen is performed at high pressure and with cooling.

15.22 Write balanced equations for the following reactions:
  (a) the reduction of nitric acid to the ammonium ion by zinc metal
  (b) the reaction of solid silver sulfide with nitric acid to give silver ion solution, elemental sulfur, and nitrogen monoxide

15.23 Contrast the properties of the two common allotropes of phosphorus.

15.24 Contrast the properties of ammonia and phosphine.

15.25 Phosphine, PH_3, dissolves in liquid ammonia to give $NH_4^+PH_2^-$. What does this tell you about the relative acid–base strengths of the two Group 15 hydrides?

15.26 In the "strike-anywhere" match, assume that the potassium chlorate is reduced to potassium chloride and the tetraphosphorus trisulfide is oxidized to tetraphosphorus decaoxide and sulfur dioxide. Write a balanced chemical equation for the process and identify the oxidation number changes that have occurred.

15.27 A compound is known to have the formula NOCl (nitrosyl chloride).
  (a) Construct an electron-dot diagram for the molecule and identify the oxidation number of nitrogen.
  (b) What is the anticipated nitrogen–oxygen bond order?
  (c) From the $\Delta H_f°$ value for this compound of $+52.6$ kJ·mol^{-1} and appropriate bond energy data, calculate the N—O bond energy in this compound. Compare it with values for N—O single and double bonds.

15.28 (a) Construct an electron-dot structure for $POCl_3$ (assume it to be similar to PF_3O) and hence draw its molecular shape.
  (b) According to the hybridization concept, what is the likely hybridization of the central phosphorus atom?
  (c) The phosphorus–oxygen distance is very short; how would you explain this?

15.29 Another compound of phosphorus and chlorine is P_2Cl_4. Construct an electron-dot structure for this compound and then draw its molecular shape.

15.30 When gaseous dinitrogen tetraoxide is bubbled into liquid nitric acid solvent, the N_2O_4 ionizes to form a conducting solution. Suggest the identity of the products on the basis of known positive and negative ions containing only nitrogen and oxygen. Write a balanced equation for the reaction.

15.31 In the solid phase, PCl_5 forms $PCl_4^+PCl_6^-$. However, PBr_5 forms $PBr_4^+Br^-$. Suggest a reason why the bromine compound has a different structure.

15.32 The experimentally determined bond angles for arsine (AsH_3), arsenic trifluoride, and arsenic trichloride are 92°, 96°, and 98½°, respectively. Offer explanations for the trends in the values.

15.33 Figure 15.7 depicts the nitrogen–nitrogen bond orders in hydrogen azide, while Figure 15.8 shows the contributing resonance structures. How does the bonding differ in the azide ion, N_3^-?

15.34 Write the formula of two ions that are isoelectronic with the dinitrogen molecule.

15.35 Hydroxylamine, NH_2OH, can be oxidized to nitrate ion by bromate ion, BrO_3^-, which is itself reduced to bromide ion. Write a balanced chemical equation for the reaction.

15.36 Suggest why acidification promotes the decomposition of ammonium nitrate to dinitrogen oxide and water. Hint: Consult Figure 15.4.

15.37 Gaseous NOF reacts with liquid SbF_5 to give an electrically conducting solution. Write a balanced chemical equation for the reaction.

15.38 A solution of the hydrogen phosphate ion is basic, whereas the dihydrogen phosphate ion is acidic. Write chemical equilibria for the predominant reactions that account for this difference in behavior.

15.39 In Section 15.25, we mentioned pyrophosphoric acid, $H_4P_2O_7$. What are the formulas of the equivalent isoelectronic acids of sulfur and silicon?

15.40 There is an oxyanion of a transition metal that is isostructural with the pyrophosphate ion, $P_2O_7^{4-}$. Write the formula of the corresponding ion and explain your reasoning.

15.41 Explain the terms (a) eutrophication; (b) symbiotic relationship; (c) chemotherapy; (d) apatite.

15.42 Write balanced chemical equations corresponding to each transformation in the Element Reaction Flowchart.

BEYOND THE BASICS

15.43 When gaseous phosphine is bubbled into liquid hydrogen chloride, a conducting solution is formed. The product reacts with boron trichloride to give another ionic compound. Suggest the identity of each product. Write a balanced equation for each reaction and identify each reactant as a Lewis acid or base.

15.44 A possible mechanism for the formation of ammonia is:

$$N_2(g) + H_2(g) \rightarrow N_2H_2(g)$$

$$N_2H_2(g) + H_2(g) \rightarrow N_2H_4(g)$$

$$N_2H_4(g) + H_2(g) \rightarrow 2\ NH_3(g)$$

Use bond energy calculations to determine ΔH for each step. Then suggest the major weakness with this possible route. Suggest another possible mechanism that would account for the slowness of the reaction. Determine the ΔH value for the replacement step in the mechanism and show that it is indeed feasible.

15.45 Deduce the shape of the nitrate radical. Suggest approximate values for the bond angles. What would be the expected average value of the N—O bond order?

15.46 When phosphorus oxychloride fumes in moist air, what is the likely chemical reaction?

15.47 Suggest a structure for the compound $P_4O_6S_4$.

15.48 An alternative route for the synthesis of phosphorus oxychloride involves the reaction of phosphorus pentachloride with tetraphosphorus decaoxide:

$$6\ PCl_5(s) + P_4O_{10}(s) \rightarrow 10\ POCl_3(l)$$

However, the reaction is usually performed by bubbling chlorine gas though a mixture of phosphorus trichloride and tetraphosphorus decaoxide. Suggest the reason for this.

15.49 Suggest why sodium azide is comparatively stable while heavy metal azides, such as copper(II) azide are much more likely to explode.

15.50 The nitrite ion is an ambidentate ion, bonding through oxygen or nitrogen. Which category of base (hard, borderline, soft) would you expect it to be in each of the two bonding types?

15.51 The equilibrium constant, K, for a reaction can be found from the expression $\Delta G° = -RT\ln K$ where R is the Ideal Gas constant ($8.31\ J \cdot mol^{-1} \cdot K^{-1}$).
(a) Determine the equilibrium constant for the formation of ammonia from its elements at 298 K.
(b) Assuming $\Delta H°$ and $S°$ are temperature independent, calculate the equilibrium constant at 775 K.

(c) 775 K is a common operating temperature for the dinitrogen-dihydrogen reaction in an ammonia synthesis plant. In view of your answers to (a) and (b), why would such a high temperature be used for the reaction?

15.52 The potentials of most nitrogen redox reactions cannot be measured directly. Instead, the values are obtained from free energy values. If $\Delta G_f^\circ(NH_3(aq))$ is -26.5 kJ·mol^{-1}, what is the standard potential for the $N_{2(g)}/NH_3(aq)$ half-reaction in basic solution?

15.53 Determine the N—N bond energy in the dinitrogen tetraoxide molecule from appropriate thermodynamic data. Assume that the bonding within the NO_2 units is the same as that in the nitrogen dioxide molecule itself.

15.54 The energy needed to break a particular bond type differs from compound to compound. For example, the breaking of a nitrogen–chlorine bond requires much more energy for:

$$NCl_3(g) \rightarrow NCl_2(g) + Cl(g)$$

than for:

$$NOCl(g) \rightarrow NO(g) + Cl(g)$$

Suggest an explanation for this.

15.55 Combinations of hydrogen phosphate and dihydrogen phosphate ions are often used as buffer mixtures. To prepare 1.0 L of a pH 6.80 buffer mixture, what masses of Na_2HPO_4 and NaH_2PO_4 would be needed to give a total buffer concentration of 0.10 mol·L^{-1}. The $pK_{a2}(H_3PO_4(aq))$ is 7.21.

15.56 Liquid phosphorus oxychloride is a useful nonaqueous solvent. Suggest the formulas of the cation and anion that would form on self-ionization. Identify which one would be the solvent's acid and which one the conjugate base.

15.57 On warming, phosphorus pentachloride dissociates into phosphorus trichloride and dichlorine; however, phosphorus pentafluoride does not dissociate. Use bond energy arguments to explain the different behavior of the two pentahalides.

15.58 One mole of phosphinic acid, H_3PO_2, requires 2 moles of diiodine to be oxidized to phosphoric acid, the diiodine being reduced to iodide. Determine the formal oxidation number of phosphorus in phosphinic acid. By comparison, what would be the oxidation number of phosphorus in phosphonic acid, H_3PO_3?

15.59 In the text, we mention that nitrogen only forms a trifluoride, whereas phosphorus forms both a trifluoride and a pentafluoride. However, nitrogen does form a compound of empirical formula NF_5. Suggest a possible structure for this compound.

15.60 When nitrogen monoxide and air are mixed, dinitrogen tetraoxide and nitrogen dioxide are formed. However, nitrogen monoxide produced in automobile exhausts at parts per million concentration only reacts very slowly with the dioxygen in the atmosphere. Suggest a possible mechanism for the reaction and explain why the reaction at low nitrogen monoxide concentrations is so slow.

15.61 A red substance (A), when heated in the absence of air, vaporized and recondensed to give a yellow waxy substance (B). (A) did not react with air at room temperature, but (B) burned spontaneously to give clouds of a white solid (C). (C) dissolved exothermically in water to give a solution containing a triprotic acid (D). (B) reacted with a limited amount of chlorine to give a colorless fuming liquid (E), which in turn reacted further with chlorine to give a white solid (F). (F) gave a mixture of (D) and hydrochloric acid when treated with water. When water was added to (E), a diprotic acid (G) and hydrochloric acid were produced. Identify substances (A) to (G) and write equations for all reactions.

15.62 When magnesium metal is heated in nitrogen gas, a pale gray compound (A) is formed. Reaction of (A) with water gives a precipitate of (B) and a gas (C). Gas (C) reacts with hypochlorite ion to form a colorless liquid (D) of empirical formula NH_2. Liquid (D) reacts in a 1:1 ratio with sulfuric acid to produce the ionic compound (E) $N_2H_6SO_4$. An aqueous solution of (E) reacts with nitrous acid to give a solution which, after neutralization with ammonia, produces a salt (F), with empirical formula NH. The compound (F) contains one cation and one anion per formula unit. The gas (C) reacts with heated sodium metal to give a solid (G) and hydrogen gas. When the solid (G) is heated with dinitrogen oxide in a 1:1 mole ratio, a solid (H) and water are produced. The anion in (H) is the same as that in (F). Identify substances (A) to (H).

15.63 An alternative to the Haber process for the synthesis of ammonia is the reaction between lithium nitride and water. Write a balanced equation for this reaction and suggest why it is not commercially viable.

15.64 Hydrogen azide reacts with diiodine in a 2:1 mole ratio. Deduce the products using a balanced chemical equation.

15.65 There are numerous analogs between water and ammonia chemistry; for example, the base $K^+NH_2^-$ in the ammonia system parallels K^+OH^- in the water system. Deduce the ammonia-system analog for each of the following compounds: H_2O_2, HNO_3, H_2CO_3. Hint: Not all oxygen atoms need be replaced.

15.66 Nitrogen forms two compounds with hydrogen and oxygen that are superficially similar: hydroxylamine, NH_2OH, and ammonium hydroxide, NH_4OH. However, the former is purely covalently bonded, while the latter consists of two separate ions. Use your knowledge of bonding to draw the shapes of each of these compounds.

15.67 Nitrogen trichloride is an explosive compound. Write a balanced chemical equation for its decomposition to its constituent elements and use bond energy data from the Appendices to explain the exothermicity of the reaction.

15.68 When methylammonium chloride, $CH_3NH_3^+Cl^-$ is dissolved in heavy water, D_2O, only half of the hydrogen atoms in the compound are replaced by deuterium. Explain why this happens.

15.69 When phosphinic acid, H_3PO_2, is dissolved in heavy water, D_2O, HD_2PO_2 is formed. Suggest an explanation.

15.70 If a strip of magnesium is ignited and placed in a container of dinitrogen oxide, it continues to burn brightly. Write a balanced chemical equation for the reaction. If the products are as you predict, how does this confirm the structure of the dinitrogen oxide molecule?

15.71 The azide ion, N_3^-, is isoelectronic with carbon dioxide, CO_2. Deduce the formulas of two other nitrogen-containing ions that are isoelectronic with them.

15.72 Hyponitrous acid and nitroamide (nitramide) have the same formulas of $N_2O_2H_2$. Draw structures of these two compounds and contrast their acid–base behavior.

15.73 The pentanitrogen hexafluoroarsenate(V) salt, $(N_5)^+(AsF_6)^-$, is only marginally stable at room temperature. Suggest how a more stable salt of this cation might be synthesized.

15.74 Phosphorus and nitrogen form a polyatomic ion, $(P_4N_{10})^{n-}$. Deduce the charge, $n-$, on this ion.

15.75 The azide ion can be considered as a pseudo-halide ion. Identify:
(a) an aqueous cation which would give a precipitate on addition to azide ion solution
(b) the reaction of azide ion with water
(c) a way in which azide ion does not resemble a halide ion

15.76 The pentanitrogen, N_5^+, ion is V-shaped. Construct the electron-dot diagram and determine the bond order for each bond in the ion.

15.77 Write a balanced chemical equation for the reaction of the explosive, ADN, $(NH_4)^+[N(NO_2)_2]^-$ with aluminum powder. Explain the three main reasons why you would expect this reaction to be highly exothermic. What other aspect of the reaction would make it a good rocket propellant?

15.78 An isomer of hydrazoic acid, HN_3, called cyclotriazine, contains a triangle of nitrogen atoms. Draw an electron-dot diagram for cyclotriazine. Do you expect all three nitrogen–nitrogen bond lengths to be equivalent? Give your reasoning.

15.79 Research the traditional Marsh test used for arsenic. Write balanced chemical equations for each step in the procedure.

Chapter 16

The Group 16 Elements

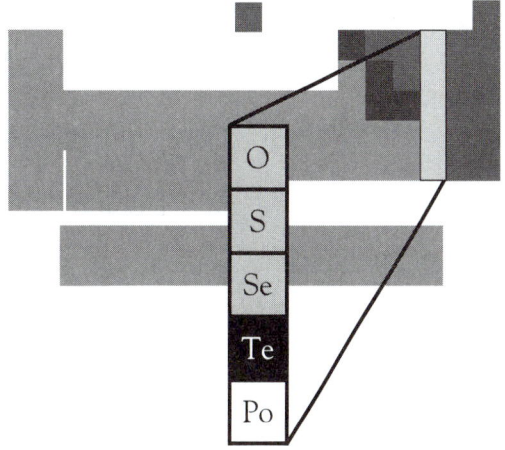

Again, it is the first two members of the group that have the most significant chemistry: oxygen and sulfur. The differences between the first and second members that we saw for the Group 15 elements (nitrogen and phosphorus) are repeated in this group, except it is the oxygen that is the more reactive. Selenium and tellurium both possess some semimetallic behavior, and it is only the radioactive element, polonium, that can be said to exhibit metallic character.

It is a common fallacy to link a particular discovery with a specific name. Our modern perception of the progress of science shows that discoveries usually involve the work of many individuals. For example, the discovery of oxygen is credited to the 18th-century chemist Joseph Priestley, when, in fact, it was a long-forgotten Dutch inventor, Cornelius Drebble, who first reported the preparation of the gas about 150 years earlier. Nevertheless, Priestley does deserve the bulk of the credit, for he made extensive studies of pure oxygen gas and, very bravely, breathed this gas, then known as "de-phlogisticated air." Priestley performed these experiments in Birmingham, England, where he was a Nonconformist minister. He was known for his "leftist" views on politics and religion—for example, he supported the French and American revolutions. A mob burned his church, home, and library. He fled to the United States, where he dedicated one of his books to Vice President John Adams, noting, "It is happy that, in this country, religion has no connection with civil power."

The discovery of oxygen marked the end of the phlogiston theory of combustion. According to that theory, burning involved the loss of phlogiston. However, the French scientist Guyton de Morveau (see Introduction to Chapter 8) showed that burning a metal gave a product that showed a gain in weight. His colleague Antoine Lavoisier realized that something had to be added in the combustion process. It was oxygen. But revolutionary concepts are often slow to be accepted in science, and this was true of the idea that combustion is linked to the addition of oxygen. In fact, many chemists of the time, including Joseph Priestley, never did accept this idea.

16.1 Group Trends

This group is sometimes called the *chalcogens*. Oxygen, sulfur, and selenium are the nonmetals of the group; tellurium is generally regarded as a semimetal; polonium is considered to be the only true metal in Group 16. Certainly the melting and boiling points show the rising trend characteristic of nonmetals, followed by the falling trend at polonium, a trend characteristic of metals (Table 16.1). Our categorization of polonium as a metal is supported by its low electrical resistivity of 43 $\mu\Omega\cdot$cm. On the basis of electrical resistivity, the common allotrope of selenium is a nonmetal (10^{16} $\mu\Omega\cdot$cm), whereas tellurium is usually classified as a semimetal (10^6 $\mu\Omega\cdot$cm).

Except for oxygen, of course, there are patterns in the oxidation states of the Group 16 elements. We find all of the even-numbered oxidation states from +6, through +4 and +2, to −2. The stability of the −2 and +6 oxidation states decreases down the group, whereas that of the +4 state increases. As happens in many groups, the trends are not as regular as we would like. For example, the acids containing atoms in the +6 oxidation state are sulfuric acid, which we can represent as $(HO)_2SO_2$ to indicate the bonding, and selenic acid, $(HO)_2SeO_2$; but telluric acid has the formula $(HO)_6Te$, or H_6TeO_6.

16.2 Anomalous Nature of Oxygen

The anomalies of oxygen chemistry are similar to those of nitrogen; that is, the formation of strong π bonds using the $2p$ atomic orbitals and a small size, precluding the possibility of more than four covalent bonds. To illustrate the latter point, oxygen forms only one normal oxide with fluorine, OF_2, whereas sulfur forms several compounds with fluorine, including SF_6.

Table 16.1 Melting and boiling points of the Group 16 elements

Element	Melting point (°C)	Boiling point (°C)
O_2	−219	−183
S_8	119	445
Se_8	221	685
Te	452	987
Po	254	962

Table 16.2 Bond energies for the Group 16 elements

Bond	σ Bond energy (kJ·mol^{-1})	π Bond energy (kJ·mol^{-1})
Oxygen–oxygen	142	350
Sulfur–sulfur	270	155
Selenium–selenium	210	125

High Stability of Multiple Bonds

Like nitrogen, the oxygen–oxygen double bond (494 kJ·mol^{-1}) is much stronger than the oxygen–oxygen single bond (142 kJ·mol^{-1}). The oxygen–oxygen single bond is particularly weak; the carbon–carbon single bond energy is 335 kJ·mol^{-1}.

If we consider a double bond energy to consist of the single (σ) bond energy plus the energy of the second (π) bond, we can see from Table 16.2 that double bond formation results in a considerable energy gain for oxygen but very little for sulfur and selenium. This difference accounts for the lack of multiple bond formation by the other members of the group.

Lack of Catenated Compounds

In Group 14, the ability to catenate decreases down the group. However, in Group 16, sulfur forms the longest chains. In fact, compounds containing two oxygen atoms bonded together are usually strong oxidizing agents,

Oxygen Isotopes in Geology

Although we usually consider oxygen atoms to have eight neutrons (oxygen-16), there are in fact two other stable isotopes of the element. The isotopes and their abundances are

Isotope	Abundance (%)
Oxygen-16	99.763
Oxygen-17	0.037
Oxygen-18	0.200

Thus, one oxygen atom in every 500 has a mass that is 12 percent greater than the other 499. This "heavy" oxygen will have slightly different physical properties, both as the element and in its compounds. In particular, $H_2{}^{18}O$ has a vapor pressure significantly lower than that of $H_2{}^{16}O$. Hence, in an equilibrium between liquid and gaseous water, the gas phase will be deficient in oxygen-18. Because the most evaporation occurs in tropical waters, it is those waters that will have a higher concentration of oxygen-18. This increased proportion of oxygen-18 will be found in all the marine equilibria that involve oxygen.

We can use the ratio of the two isotopes of oxygen to determine the temperature of the seas in which shells were formed millions of years ago simply by determining the oxygen isotopic ratio in the calcium carbonate of the shells. The higher the proportion of oxygen-18, the warmer the waters of those ancient seas.

and compounds containing three oxygen atoms bonded together are virtually unknown. Such behavior can be explained by postulating that the oxygen–oxygen bond is weaker than its bonds to other elements. For example, the oxygen–sulfur single bond energy of 275 kJ·mol^{-1} is almost twice as strong as the oxygen–oxygen single bond. Thus, oxygen will endeavor to bond to other elements rather than to itself. Conversely, the sulfur–sulfur single bond energy of 270 kJ·mol^{-1} is only slightly lower than that of its bonds to other elements, thereby stabilizing catenation in sulfur compounds.

16.3   *Oxygen*

Oxygen exists in two allotropic forms: the common dioxygen and the less common trioxygen, commonly called ozone.

Dioxygen

Dioxygen is a colorless, odorless gas that condenses to a pale blue liquid. Because it has a low molar mass and forms a nonpolar molecule, it has very low melting and boiling points. The gas does not burn, but it does support combustion. In fact, almost all elements will react with oxygen at room temperature or when heated. The main exceptions are the "noble" metals, such as platinum, and the noble gases. For a reaction to occur, the state of division of the reactant is often important. For example, very finely powdered metals such as iron, zinc, and even lead will catch fire in air at room temperature. These finely divided forms of metals are sometimes called *pyrophoric*, a term reflecting their ability to catch fire. For example, zinc dust will inflame to give white zinc oxide:

$$2\ Zn(s) + O_2(g) \rightarrow 2\ ZnO(s)$$

Dioxygen is the reactive gas that makes up 21 percent of Earth's atmosphere. This oxidizing gas is not naturally occurring in planetary atmospheres. The "normal" atmosphere of a planet is reducing, containing hydrogen, methane, ammonia, and carbon dioxide. It was the process of photosynthesis that started to convert the carbon dioxide component of Earth's early atmosphere to dioxygen about 2.5×10^9 years ago; its present oxygen-rich state was attained about 5×10^8 years ago. Thus, we can look for signs of life similar to our own on planets around other stars just by sending dioxygen detectors.

Dioxygen is not very soluble in water, about 2×10^{-5} mole fraction at 25°C, compared to 6×10^{-4} for carbon dioxide. Nevertheless, the concentration of oxygen in natural waters is high enough to support marine organisms. The solubility of dioxygen decreases with increasing temperature; hence, it is the cold waters, such as the Labrador and Humboldt currents, that are capable of supporting the largest fish stocks—and have been the focus of the most severe overfishing. Even though the solubility of dioxygen is low, it is twice that of dinitrogen. Hence, the gas mixture released by heating air-saturated water will actually be enriched in dioxygen.

The measurement of dissolved oxygen (sometimes called DO) is one of the crucial determinants of the health of a river or lake. Low levels of dissolved oxygen can be caused by eutrophication (excessive algae and plant growth) or an input of high-temperature water from an industrial cooling system. As a temporary expedient, air-bubbling river barges can be used to increase dissolved dioxygen levels. This has been done in London, England, to help bring back game fish to the river Thames. Almost the opposite of DO is BOD—biological oxygen demand; this measure indexes the potential for oxygen consumption by aquatic organisms. Thus, a high BOD can indicate potential problems in a lake or river.

Dioxygen is a major industrial reagent; about 10^9 tonnes is used worldwide every year, most in the steel industry. Dioxygen is also used in the synthesis of nitric acid from ammonia (Chapter 15, Section 15.14). Almost all the oxygen is obtained by fractional distillation of liquid air. Dioxygen is also consumed in large quantities by hospital facilities. In that context, it is mostly used to raise the dioxygen partial pressure in gas mixtures given to people with respiratory problems, making absorption of oxygen gas easier for poorly functioning lungs.

In the laboratory, there are a number of ways of making dioxygen gas. For example, strong heating of potassium chlorate in the presence of manganese(IV) oxide gives potassium chloride and oxygen gas:

$$2 \ KClO_3(l) \xrightarrow[\Delta]{MnO_2} 2 \ KCl(s) + 3 \ O_2(g)$$

The catalytic decomposition of potassium chlorate is the source of emergency oxygen in commercial aircraft. This is a much more compact source of oxygen than high-pressure cylinders of oxygen gas.

However, a much safer route in the laboratory is the catalytic decomposition of aqueous hydrogen peroxide. Again, manganese(IV) oxide can be used as the catalyst:

$$2 \ H_2O_2(aq) \rightarrow 2 \ H_2O(l) + O_2(g)$$

As discussed in Chapter 3, Section 3.4, the molecular orbital model is the only representation of the bonding in dioxygen that fits with experimental evidence. Figure 16.1 indicates that the net bond order is 2 (six bonding electrons and two antibonding electrons), with the two antibonding electrons having parallel spins. The molecule, therefore, is paramagnetic.

However, an energy input of only 95 kJ·mol^{-1} is required to cause one of the antibonding electrons to "flip" and pair with the other antibonding electron (Figure 16.2). This spin-paired (diamagnetic) form of dioxygen reverts to the paramagnetic form in seconds or minutes, depending on both the concentration and the environment of the molecule. The diamagnetic form can be prepared by the reaction of hydrogen peroxide with sodium hypochlorite:

$$H_2O_2(aq) + ClO_2(aq) \rightarrow O_2(g)[\text{diamagnetic}] + H_2O(l) + Cl_2(aq)$$

or, alternatively, by shining ultraviolet radiation on paramagnetic oxygen in the presence of a sensitizing dye.

Diamagnetic dioxygen is an important reagent in organic chemistry and gives products different from those of the paramagnetic form. Furthermore, diamagnetic oxygen, which is very reactive and formed by ultraviolet radiation,

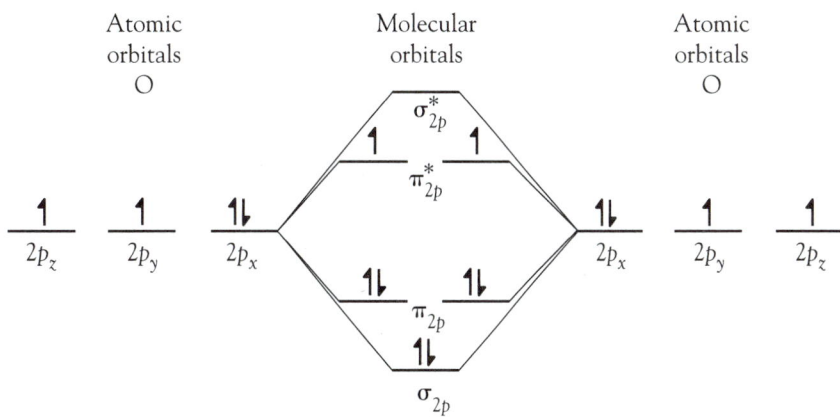

Figure 16.1 Molecular orbital energy level diagram showing the combination of the 2p atomic orbitals in the dioxygen molecule.

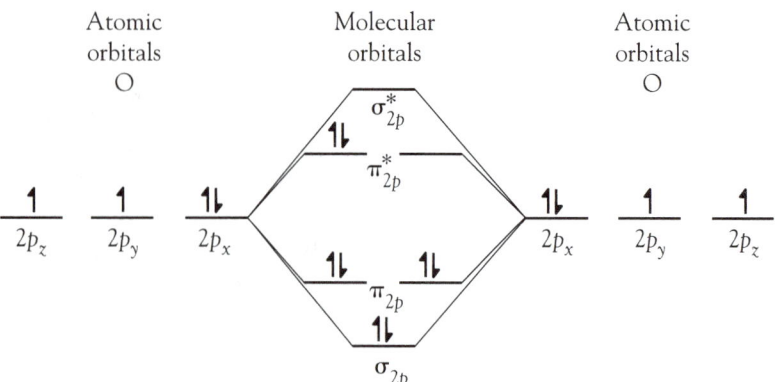

Figure 16.2 Molecular orbital energy level diagram showing the combination of the $2p$ atomic orbitals in the more common of the two diamagnetic forms of the dioxygen molecule.

has been implicated in skin cancer induction. Diamagnetic oxygen is often referred to as singlet oxygen, and the paramagnetic form is called triplet oxygen.

It requires 95 kJ·mol^{-1} to pair up the electrons in the antibonding orbital. There is a second singlet form of dioxygen in which the spin of one electron is simply flipped over, so the resulting unpaired electrons have opposite spins (Figure 16.3). Surprisingly, this arrangement requires much more energy to attain, about 158 kJ·mol^{-1}. As a result, this other singlet form is of little laboratory importance. In Chapter 14, Section 14.8, we showed that most of the absorption features in the infrared spectrum of the atmosphere could be explained in terms of vibrations of the water and carbon dioxide molecules. However, there was an absorption at the precise wavelength of 0.76 μm that we did not explain (Figure 16.4). This wavelength, in fact, represents the electronic absorption of energy corresponding to the production of the higher energy singlet form of dioxygen from the normal triplet oxygen. The absorption corresponding to the formation of the lower energy singlet form is hidden under a massive carbon dioxide absorption at 1.27 μm (shown in Figure 14.12).

Trioxygen (Ozone)

This thermodynamically unstable allotrope of oxygen is a diamagnetic gas with a strong odor. In fact, the "metallic" smell of ozone can be detected in concentrations as low as 0.01 ppm. The gas is extremely toxic; the maximum permitted concentration for extended exposure is 0.1 ppm. The gas is produced in regions of high voltages; thus, photocopying machines and laser printers have been responsible for high levels of ozone in many office environments.

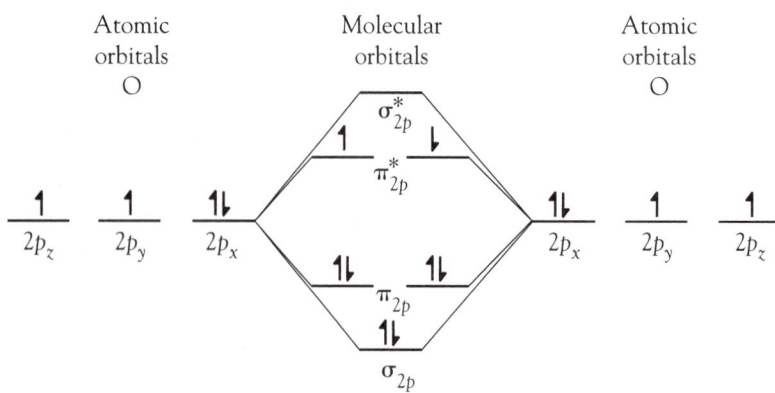

Figure 16.3 Molecular orbital energy level diagram showing the combination of the $2p$ atomic orbitals in the less common of the two diamagnetic forms of the dioxygen molecule.

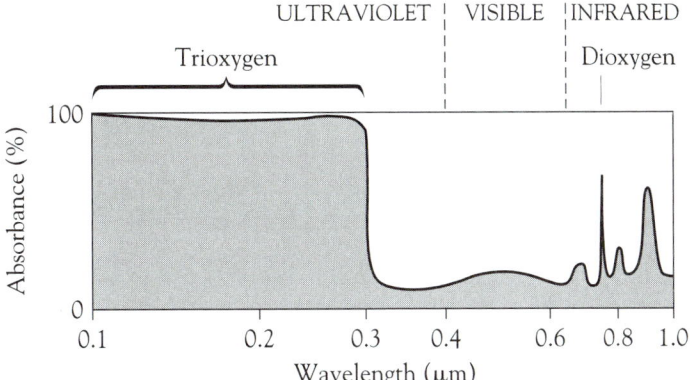

Figure 16.4 The infrared, visible, and ultraviolet portions of the electromagnetic spectrum of the atmosphere.

The ozone generated may well have been one cause of headaches and other complaints by office workers. Some machines have carbon filters on the air exhaust to minimize the trioxygen emissions, and these need to be replaced periodically according to the manufacturer's recommendations. However, technological advances have enabled the development of duplicators and printers that produce very low levels of trioxygen.

A convenient way to generate trioxygen is to pass a stream of dioxygen through a 10- to 20-kV electric field. This field provides the energy necessary for the reaction:

$$3\ O_2(g) \rightarrow 2\ O_3(g) \qquad\qquad \Delta H_f^\circ = +143\ \mathrm{kJ \cdot mol^{-1}}$$

At equilibrium, the concentration of trioxygen is about 10 percent. The trioxygen slowly decomposes to dioxygen, although the rate of conversion depends on the phase (gas or aqueous solution).

Trioxygen is a very powerful oxidizing agent, much more powerful than dioxygen, as can be seen from a comparison of reduction potentials in acid solution:

$$O_3(g) + 2\ H^+(aq) + 2\ e^- \rightarrow O_2(g) + H_2O(l) \qquad E^\circ = +2.07\ \mathrm{V}$$

$$O_2(g) + 4\ H^+(aq) + 4\ e^- \rightarrow 2\ H_2O(l) \qquad E^\circ = +1.23\ \mathrm{V}$$

In fact, in acid solution, fluorine and the perxenate ion, XeO_6^{4-}, are the only common oxidizing agents that are stronger than trioxygen. Its range of oxidizing ability is illustrated by the following reactions: one in the gas phase, one in aqueous solution, and the third with a solid:

$$2\ NO_2(g) + O_3(g) \rightarrow N_2O_5(g) + O_2(g)$$

$$CN^-(aq) + O_3(g) \rightarrow OCN^-(aq) + O_2(g)$$

$$PbS(s) + 4\ O_3(g) \rightarrow PbSO_4(s) + 4\ O_2(g)$$

It is the strongly oxidizing nature of trioxygen that enables it to be used as a bactericide. For example, it is used to kill bacteria in bottled waters; and in France, particularly, it is used to kill organisms in municipal water supplies and public swimming pools. However, water purification experts in North America have preferred the use of chlorine gas for water purification. There are advantages and disadvantages of both bactericides. Ozone changes to dioxygen over a fairly short period of time; thus, its bactericidal action is not long lasting. However, ozone is chemically innocuous in water supplies.

Dichlorine remains in the water supply to ensure bactericidal action, but it reacts with any organic contaminants in the water supply to form hazardous organochlorine compounds (see Chapter 17, Section 17.11).

On the surface of the Earth, ozone is a dangerous compound, a major atmospheric pollutant in urban areas. In addition to its damaging effect on lung tissue and even on exposed skin surfaces, ozone reacts with the rubber of tires, causing them to become brittle and crack. The ozone is produced by the photolysis of nitrogen dioxide, itself formed mainly from internal combustion engines:

$$NO_2(g) \xrightarrow{uv} NO(g) + O(g)$$

$$O(g) + O_2(g) \rightarrow O_3(g)$$

However, as most people now know, it is a different story in the upper atmosphere, where the ozone in the stratosphere provides a vital protective layer for life on Earth. The absorption region that results from atmospheric ozone absorption is shown in Figure 16.4. The process is quite complex, but the main steps are as follows: First, the shorter wavelength ultraviolet radiation reacts with dioxygen to produce atomic oxygen:

$$O_2(g) \xrightarrow{uv} 2\, O(g)$$

The atomic oxygen reacts with dioxygen to give trioxygen:

$$O(g) + O_2(g) \rightarrow O_3(g)$$

The trioxygen absorbs longer wavelength ultraviolet radiation and decomposes back to dioxygen:

$$O_3(g) \xrightarrow{uv} O_2(g) + O(g)$$

$$O_3(g) + O(g) \rightarrow 2\, O_2(g)$$

Chemistry is rarely this simple, and the chemistry of the stratosphere is no exception. There are alternative routes for the destruction of ozone that involve trace components of the stratosphere, including hydrogen atoms, the hydroxyl radical, nitrogen monoxide, and chlorine atoms. These species, represented here by X, undergo a catalytic cycle for the decomposition of ozone without ultraviolet ray absorption:

$$X(g) + O_3(g) \rightarrow XO(g) + O_2(g)$$

$$XO(g) + O(g) \rightarrow X(g) + O_2(g)$$

to give the net reaction:

$$O(g) + O_3(g) \rightarrow 2\, O_2(g)$$

There is a particular reason why these four species act in this role—apart from the fact that they are known trace constituents of the upper atmosphere. Because there are equal numbers of gas molecules, entropy cannot be a significant driving factor for either reaction step; both steps must be exothermic. This places limits on X. Thus, for the first step, the X—O bond energy must be greater than the difference between the O_3 and O_2 enthalpies of formation (107 kJ·mol^{-1}). For the second step, the X—O bond energy must be less than the dioxygen bond energy (498 kJ·mol^{-1}). And these conditions hold when X is H, OH, NO, or Cl.

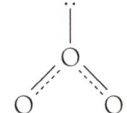

Figure 16.5 The ozone molecule.

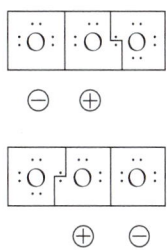

Figure 16.6 The bonding of trioxygen can be interpreted as the mean of these two resonance structures.

Trioxygen is a V-shaped molecule with a bond angle of 117°. Its oxygen–oxygen bonds are of equal length and have a bond order of about 1½ (Figure 16.5). The bond angles and bond lengths of trioxygen are very similar to those of the isoelectronic nitrite ion.

We can use resonance structures to account for the bond order (Figure 16.6). However, a molecular orbital approach provides a better representation (Figure 16.7). Like the nitrogen dioxide arrangement, we envisage the $2p$ orbitals to be those responsible for the bonding. A set of σ bonds is formed by overlap of the four p orbitals along the bonding axes: two of the p orbitals on the central oxygen plus one each on the terminal oxygen atoms. This arrangement results in the formation of four molecular orbitals: two bonding and two antibonding. The three p orbitals at right angles to the molecular plane provide three π molecular orbitals: one bonding, one nonbonding, and one antibonding. Finally, the remaining p orbital (in the molecular plane but at right angles to the bonding direction) on each of the terminal oxygen atoms will be nonbonding. Filling these orbitals with the twelve $2p$ electrons results in two net σ bonds and one π bond, the latter being shared by the three atoms. Like the resonance approach, the 1½ bond order matches the experimental bond information.

Ozone forms compounds with the alkali and alkaline earth metals. These compounds contain the trioxide(1−) ion, O_3^-. As we would expect from lattice stability arguments, it is the larger cations, such as cesium, that form the most stable trioxides. It has been shown that the trioxide(1−) is also V-shaped, so the molecular orbital diagram should be similar to that of trioxygen itself (Figure 16.8). Thus, the additional electron in the anion should enter the π antibonding orbital. This arrangement would reduce the π bonding to one-half, or one-fourth, per oxygen–oxygen bond. Experimental measurements have shown this to be the case. The oxygen–oxygen bond length is 135 pm in the trioxide (1−) ion, slightly longer than the 128-pm bond in trioxygen itself.

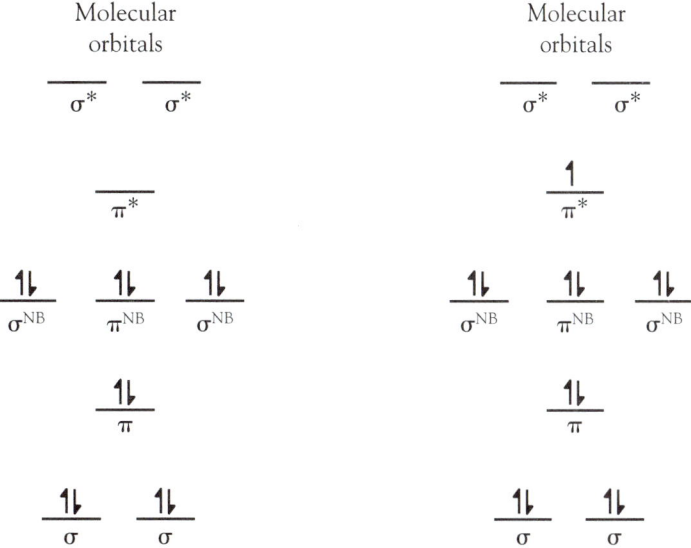

Figure 16.7 The molecular orbital energy level diagram for orbitals derived from the $2p$ atomic orbitals in trioxygen.

Figure 16.8 The molecular orbital energy level diagram for orbitals derived from the $2p$ atomic orbitals in the trioxide(1−) ion.

16.4 *Bonding in Covalent Oxygen Compounds*

An oxygen atom will usually form two single covalent bonds or one multiple bond, ordinarily a double bond. When two single bonds are formed, the angle between the bonds can be significantly different from the 109½° tetrahedral angle. The traditional explanation for the bond angle of 104½° in water asserts that the lone pairs occupy more space than bonding pairs, thus "squashing" the bond angle in the water molecule.

However, when we compare the two halogen–oxygen compounds—oxygen difluoride, OF_2 (with a bond angle of 103°), and dichlorine oxide, Cl_2O (with a bond angle of 111°)—we have to look for a different explanation. The best explanation relates to the degree of orbital mixing. In Chapter 3, Section 3.10, we considered the hybridization model of bonding, where orbital characters mixed. At the time, we considered integral values of mixing: for example, one *s* orbital with three *p* orbitals to form four sp^3 hybrid orbitals. However, there is no reason why the mixing cannot be fractional. Thus, some covalent bonds can have more *s* character; and others, more *p* character. Also, beyond Period 2, we have to keep in mind that *d* orbitals might be mixed in as well. This approach of partial orbital mixing is, in fact, moving toward the more realistic molecular orbital representation of bonding.

It was Henry A. Bent who proposed an empirical rule—the *Bent rule*—to explain, among other things, the variation in bond angles of oxygen compounds. The rule states: More electronegative substituents "prefer" hybrid orbitals with less *s* character, and more electropositive substituents "prefer" hybrid orbitals with more *s* character. Thus, with fluorine (more electronegative), the bond angle tends toward the 90° angle of two "pure" *p* orbitals on the oxygen atom. Conversely, the angle for chlorine (less electronegative) is greater than that for an sp^3 hybrid orbital, somewhere between the 109½° angle of sp^3 hybridization and the 120° angle of sp^2 hybridization. An alternative explanation for the larger angle in dichlorine oxide is simply that there is a steric repulsion between the two large chlorine atoms, thus increasing the angle.

Oxygen can form coordinate covalent bonds, either as a Lewis acid or as a Lewis base. The former is very rare; the compound NF_3O (mentioned in Chapter 15, Section 15.2) is one such case. However, oxygen readily acts as a Lewis base, for example, in the bonding of water molecules to transition metal ions through a lone pair on the oxygen. There are also numerous examples of double bonded oxygen, such as that in PF_3O. In these cases, there is a π bond that involves a lone pair on the oxygen and an empty *d* orbital on the other element.

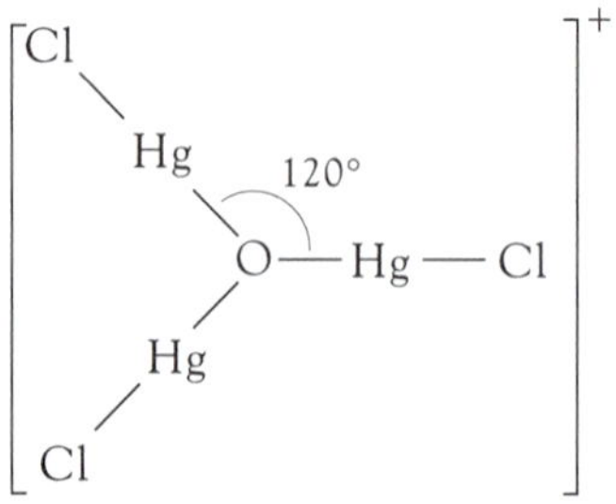

There are also other bonding modes of oxygen. In particular, oxygen can form three equivalent covalent bonds. The classic example is the hydronium ion, in which each bond angle is close to that of the tetrahedral value of 109½° (Figure 16.9). However, such oxygen-containing molecules are not always tetrahedral. In the unusual cation $[O(HgCl)_3]^+$, the atoms are all coplanar and the Hg—O—Hg bond angle is 120° (Figure 16.10). To explain this, we must assume that the lone pair on the oxygen atom is not in its usual sp^3 hybrid orbital but in a *p* orbital, where it can form a π bond with empty $6p$ orbitals of the mercury atom.

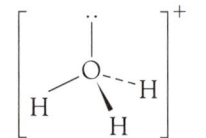

Figure 16.9 The hydronium ion.

Figure 16.10 The $[O(HgCl)_3]^+$ ion.

Table 16.3 Highest stable oxidation states of oxides and fluorides of three elements

Element	Highest oxide	Highest fluoride
Chromium	CrO_3 (+6)	CrF_5 (+5)
Xenon	XeO_4 (+8)	XeF_6 (+6)
Osmium	OsO_4 (+8)	OsF_7 (+7)

Oxygen often "brings out" a higher oxidation state than does fluorine. This may be a result of the ability of oxygen to form a π bond, using one of its own full p orbitals and an empty orbital on the other element, or it may simply be steric grounds. For example, osmium can fit four oxygen atoms to form OsO_4 but not eight fluorine atoms to form OsF_8 (Table 16.3).

16.5 Trends in Oxides

The properties of an oxide depend upon the oxidation number of the other element. For metal oxides, we find there is a transition in bonding types, as discussed in Chapter 9, Section 9.2. For example, chromium(III) oxide, Cr_2O_3, has a melting point of 2266°C, a value typical of ionic compounds, while chromium(VI) oxide, CrO_3, has a melting point of 196°C, a value typical of covalent compounds. This shift from ionic to covalent behavior can be linked to the increase in charge density of the metal ion.

This change in bonding type can be used to explain the difference in acid–base behavior among metal oxides. If the metal is in a low oxidation state, typically +2, the oxide is basic (and sometimes reducing); for example, ionic manganese(II) oxide reacts with acid to give the aqueous manganese(II) ion:

$$MnO(s) + 2\ H^+(aq) \rightarrow Mn^{2+}(aq) + H_2O(l)$$

If the metal is in the 3+ oxidation state, the metal is often amphoteric. Chromium(III) oxide, for example, reacts with acids to give the chromium(III) ion and with a strong base to give the chromite anion, CrO_2^-. The oxide of a high oxidation state metal is often acidic and oxidizing. Thus, covalently bonded chromium(VI) oxide reacts with water to give chromic acid:

$$CrO_3(s) + H_2O(l) \rightarrow H_2CrO_4(aq)$$

The oxides of nonmetals are always covalently bonded. Those with the element in a low oxidation state tend to be neutral, whereas those with the element in the higher oxidation states tend to be acidic. For example, dinitrogen oxide, N_2O, is neutral, whereas dinitrogen pentoxide dissolves in water to give nitric acid:

$$N_2O_5(g) + H_2O(l) \rightarrow 2\ HNO_3(l)$$

The higher the oxidation state of the other element, the more acidic the properties. For example, sulfur dioxide is weakly acidic, while sulfur trioxide is a strongly acidic oxide.

It is always necessary to be cautious when assigning oxidation states in oxygen compounds, for, as we have seen, there are oxygen ions in which oxygen itself has an abnormal oxidation state. These include the dioxide(2−) ion, O_2^{2-}, the dioxide(1−) ion, O_2^-, and the trioxide(1−) ion, O_3^-. These exist only in solid-phase compounds, specifically those in which the metal cation has a charge density low enough to stabilize these large, low-charge anions.

Compounds in which the other element appears to have an abnormally high oxidation number usually contain the peroxide —O—O— linkage in which each oxygen atom has the oxidation number of -1. Recalculating the oxidation numbers in such cases gives the other element a normal oxidation number. For example, in K_2O_2, it is the oxygen that has the peroxide oxidation number of -1 rather than the oxide oxidation number of -2, not potassium having an abnormal value of $+2$.

16.6 Mixed Metal Oxides

In Chapter 13, Section 13.11, we discussed a family of mixed metal oxides, the spinels. These compounds have empirical formulas of AB_2X_4, where A is usually a dipositive metal ion; B, usually a tripositive metal ion; and X, a dinegative anion, usually oxygen. The crystal lattice consists of a framework

New Pigments through Perovskites

Inorganic compounds are the mainstay of the pigment industry. Whenever permanent colors are required, inorganic compounds are preferred over organic ones for their long-term stability. Traditionally, lead(II) carbonate was used for white; yellow to red is provided by cadmium sulfide; green, by chromium(III) oxide; brown, by iron(III) oxide; and blue, by complex copper(II) compounds. A replacement for lead(II) carbonate has been titanium(IV) oxide. Cadmium sulfide has been particularly useful as the compound is very insoluble and extremely light-stable. By altering the particle size, it is possible to produce any color between yellow and red. However, a replacement for cadmium sulfide would remove one more toxic compound from the artist's palette and from the commercial paint industry.

To find a brilliant and pure color, it is necessary to find a compound in which there is a very sharp electronic transition. For gases, we can get very pure transitions, but the atoms and ions in solids usually give very "fuzzy" absorptions. That is, because of the atomic vibrations in the solid, there are a range of ground and excited states. As a result, very broad absorption spectra are obtained. The answer has been found in a series of perovskites. This series has calcium (Ca^{2+}) and lanthanum(III) (La^{3+}) as the large low-charge cations, tantalum(V) (Ta^{5+}) as the small cation, and oxide (O^{2-}) and nitride (N^{3-}) as the anions. The generic structure is:

$$(Ca^{2+})_{(1-x)}(La^{3+})_x(Ta^{5+})(O^{2-})_{(2-x)}(N^{3-})_{(1+x)}$$

As the ratio of the 2+ calcium and the 3+ lanthanum alter, so the ratio of the 2− oxide and the 3− nitride must change in step, so that electrical neutrality is maintained. By varying the proportion of the cations and anions, it is possible to synthesize pure colors ranging from a bright yellow, where $x = 0.15$, to an intense red, where $x = 0.90$. This is a great advance in color chemistry, for it is now possible to produce a very precise color just by adjusting the element ratio.

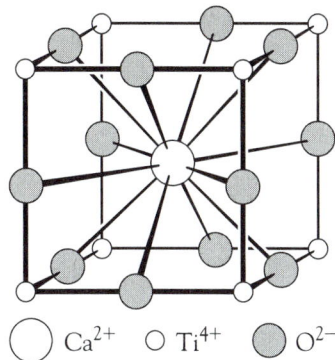

○ Ca²⁺ ○ Ti⁴⁺ ○ O²⁻

Figure 16.11 The perovskite unit cell.

of oxide ions with metal ions in octahedral and tetrahedral sites. This is not the only possible structure of mixed metal oxides: There are many more, one of which is the *perovskite* structure; and, like the spinels, perovskites are of great interest to materials scientists.

Perovskites have the general formula ABO_3, where A is usually a large dipositive metal ion and B is generally a small tetrapositive metal ion. It is important to distinguish these mixed metal oxides from the oxyanion salts that we generally study in inorganic chemistry. The metal salts of oxyanions can have a formula parallel to that of a perovskite: AXO_3, involving a metal (A), a nonmetal (X), and oxygen. In these compounds, the XO_3 consists of a covalently bonded polyatomic ion. For example, sodium nitrate consists of the Na^+ and NO_3^- ions arrayed in a sodium chloride structure, in which each nitrate ion occupies the site equivalent to the chloride ion site. However, in perovskites, such as the parent compound, calcium titanate, $CaTiO_3$, there is no such thing as a "titanate ion." Instead, the crystal lattice consists of independent Ca^{2+}, Ti^{4+}, and O^{2-} ions.

The packing in the perovskite unit cell is shown in Figure 16.11. The large calcium ion occupies the center of the cube; it is surrounded by 12 oxide ions. Eight titanium(IV) ions are located at the cube corners; each has six oxide neighbors (three being in adjacent unit cells). Many perovskites are ferroelectric materials (although they contain no iron). Such compounds can convert a mechanical pulse into an electrical signal (and vice versa), a property that is important for many electronic devices. In Chapter 14, Section 14.24, another perovskite, $CaPbO_3$, was mentioned as a rust-resistant coating for metal surfaces.

On this planet, the most important perovskite is the mixed magnesium iron silicate $(Mg,Fe)SiO_3$. It is the major component of the lower mantle, the layer of the Earth between a depth of 670 and 2900 km. This compound consists of Mg^{2+} and Fe^{2+} ions alternating in the M^{2+} sites at the center of the unit cubes, Si^{4+} "ions" at the corners, and O^{2-} ions occupying the anion sites. (In the silicate perovskites, there is probably some degree of covalency to the silicon–oxygen bonds.)

16.7 Water

Water is the only common liquid on this planet. Without water as a solvent for our chemical and biochemical reactions, life would be impossible. Yet a comparison of water to the other hydrides of Group 16 would lead us to expect it to be a gas at the common range of temperatures found on Earth. In fact, on the basis of a comparison with the other Group 16 hydrides, we would expect water to melt at about $-100°C$ and boil at about $-90°C$ (Figure 16.12). As discussed in Chapter 10, Section 10.6, the cause of its abnormally high melting and boiling points is the strength of the hydrogen bonding between water molecules. To break these bonds, more energy (and hence a higher temperature) is necessary for the phase change.

Liquid water has formed and re-formed Earth's surface over geological time. It has been able to do this because it has the ability to dissolve ionic substances, particularly the alkali and alkaline earth metals and the common anions such as chloride and sulfate. Thus, the composition of seawater reflects the leaching of ions from minerals since the time that Earth cooled enough to form liquid water. The current composition of this solution—the oceans, which make up 97 percent of the water on Earth—is shown in Table 16.4.

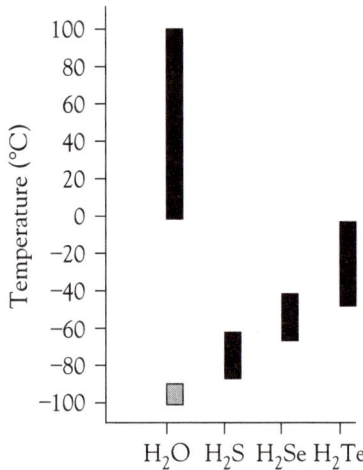

Figure 16.12 The liquid ranges of the Group 16 hydrides. If water molecules were not bound to one another by hydrogen bonds, then its liquid range would be between −90 and −100°C (shaded rectangle).

Table 16.4 Ion proportions of the principal ionic constituents in seawater

Cation	Concentration (%)	Anion	Concentration (%)
Sodium	86.3	Chloride	94.5
Magnesium	10.0	Sulfate	4.9
Calcium	1.9	Hydrogen carbonate	0.4
Potassium	1.8	Bromide	0.1

Many of our mineral deposits were formed by aqueous processes. We assume that the massive deposits of the alkali and alkaline earth minerals were formed by deposition from ancient seas and lakes. Less obvious is the mechanism of formation of heavy metal sulfide deposits, such as lead(II) sulfide. In fact, they also are the result of aqueous solution processes. Even though these minerals are extremely insoluble at common temperatures and pressures, this is not the case under the extremely high pressures and very high temperatures that exist deep under the Earth's surface. Under those conditions, many ions dissolve and are transported near to the surface, where reductions in temperature and pressure cause precipitation to occur.

Water acts as a solvent for these ionic compounds by the interaction of ions in the crystal lattice with the dipoles of the water molecules. Thus, the partially negative oxygen atoms of water molecules are attracted to the cations, and the partially positive hydrogen atoms of water molecules are attracted toward the anions. As discussed in Chapter 6, Section 6.4, it is the balance of the enthalpy and entropy terms for the breakup of the crystal lattice and those terms for the formation of the hydrated ions that determines solubility. It is generally true that the interaction between the anion and water is simply that of ion–dipole electrostatic forces. For the cation, the picture is murkier. Certainly for the low charge density alkali metals, the interaction can still be attributed to ion–dipole electrostatic forces. When alkali metal salts crystallize, either the water molecules are lost in the process (as in sodium chloride) or the water molecules simply fill holes in the crystal lattice (as in sodium sulfate decahydrate).

When we consider higher charge density cations, particularly those of the transition metals, the interaction is much stronger and more closely resembles that of a covalent bond. For example, the iron(III) ion crystallizes as the hexaaquairon(III) ion, $[Fe(H_2O)_6]^{3+}$, with the oxygen atoms of the water molecules organized in a precise octahedral arrangement around the metal ion. In these hydrates we consider the water molecules to be essentially bonded to the cations by using an oxygen lone pair to form a coordinate covalent bond. In this way, the metal ion is acting as a Lewis acid, and the water molecule is the Lewis base. To indicate that it is the oxygen end of the water molecule that is bonded to the metal ion, we conventionally write such a hydrated ion as $[Fe(OH_2)_6]^{3+}$, reversing the order of the hydrogen and oxygen. The interaction between the water molecules and transition metal ions is discussed in detail in Chapter 19, Section 19.8.

The hydration process is usually exothermic. Anhydrous calcium chloride, for example, reacts with water to form the hexahydrate:

$$CaCl_2(s) + 6\ H_2O(l) \rightarrow [Ca(OH_2)_6]^{2+}(aq) + 2\ Cl^-(aq)$$

Anhydrous calcium chloride is sometimes used to melt ice, where part of its benefits result from the exothermic hydration as well as its effect of lowering the freezing point of water.

Water is also the basis of our acid–base system, autoionizing to give the hydronium and hydroxide ions, the strongest acid and base, respectively, in our aqueous world:

$$2\ H_2O(l) \rightleftharpoons H_3O^+(aq) + OH^-(aq)$$

Although in chemical equations we often write the hydronium ion as $H^+(aq)$, such a tiny high charge density ion could not exist in solution. There is good evidence of the existence of the hydronium ion: When hydrochloric acid freezes, it gives crystals of hydronium chloride, $H_3O^+Cl^-$. The hydronium ion resembles the ammonium ion in some respects. For example, perchloric acid forms a solid, $H_3O^+ClO_4^-$, whose crystals are the same shape as ammonium perchlorate, $NH_4^+ClO_4^-$. Actually, even the hydronium ion is only an approximation of the actual hydrogen ion environment, because there is evidence that four water molecules hydrogen-bond to the hydrogen ion to give an ion of formula $H_9O_4^+$.

16.8 Hydrogen Peroxide

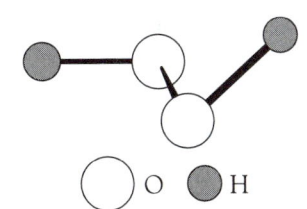

Figure 16.13 The hydrogen peroxide molecule.

$$\bigcirc\ O\quad \bullet\ H$$

Pure hydrogen peroxide is an almost colorless (slightly bluish), viscous liquid; its high viscosity is a result of the high degree of hydrogen bonding. It is an extremely corrosive substance that should always be handled with great care. The shape of the molecule is unexpected; the H—O—O bond angle in the gas phase is only 94½° (about 10° less than the H—O—H bond angle in water), and the two H—O units form a dihedral angle of 111° with each other (Figure 16.13).

Hydrogen peroxide is thermodynamically very unstable with respect to disproportionation:

$$H_2O_2(l) \rightarrow H_2O(l) + \tfrac{1}{2}\ O_2(g) \qquad \Delta G^\circ = -119.2\ kJ{\cdot}mol^{-1}$$

However, when pure, it is slow to decompose because of kinetic factors. (The reaction pathway must have a high activation energy.) Almost anything—transition metal ions, metals, blood, dust—will catalyze the decomposition.

A solution of hydrogen peroxide can be prepared in the laboratory by the reaction of sodium peroxide with water:

$$Na_2O_2(s) + 2\ H_2O(l) \rightarrow 2\ NaOH(aq) + H_2O_2(aq)$$

It is advisable to handle even dilute solutions of hydrogen peroxide with gloves and eye protection, because it attacks the skin. Hydrogen peroxide can act as an oxidizing or reducing agent in both acidic and basic solutions. Oxidations are usually performed in acidic solution; reductions, in basic solution:

$$H_2O_2(aq) + 2\ H^+(aq) + 2\ e^- \rightarrow 2\ H_2O(l) \qquad E^\circ = +1.77\ V$$

$$HO_2^-(aq) + OH^-(aq) \rightarrow O_2(g) + H_2O(l) + 2\ e^- \qquad E^\circ = +0.08\ V$$

Hydrogen peroxide will oxidize iodide ion to iodine and reduce permanganate ion in acid solution to manganese(II) ion. Hydrogen peroxide has an important application to the restoration of antique paintings. One of the favored white pigments was white lead, a mixed carbonate-hydroxide,

$Pb_3(OH)_2(CO_3)_2$. Traces of hydrogen sulfide cause the conversion of this white compound to black lead(II) sulfide, which discolors the painting. Application of hydrogen peroxide oxidizes the lead(II) sulfide to white lead(II) sulfate, thereby restoring the correct color of the painting:

$$PbS(s) + 4\,H_2O_2(aq) \rightarrow PbSO_4(s) + 4\,H_2O(l)$$

Apart from being a convenient redox reagent, hydrogen peroxide is used to test for the presence of chromium ions. Addition of hydrogen peroxide to dichromate ion solutions results in the formation of blue chromium(VI) oxide peroxide, $CrO(O_2)_2$. This covalent compound can be extracted into a low-polarity organic solvent, such as ethoxyethane (diethyl ether).

Hydrogen peroxide is a major industrial chemical; about 10^6 tonnes is produced worldwide every year. Its uses are highly varied, from paper bleaching to household products, particularly hair bleaches. Hydrogen peroxide is also used as an industrial reagent, for example, in the synthesis of sodium peroxoborate (Chapter 13, Section 13.2).

16.9   Hydroxides

Almost every metallic element forms a hydroxide. The colorless hydroxide ion is the strongest base in aqueous solution. It is very hazardous, because it reacts with the proteins of the skin, producing a white opaque layer. For this reason, it is particularly hazardous to the eyes. Despite its dangerous nature, many household products, particularly oven and drain cleaners, utilize solid or concentrated solutions of sodium hydroxide. It is also important to realize that, through hydrolysis, very high levels of hydroxide ion are present in products that do not appear to contain them. For example, the phosphate ion, used in sodium phosphate-containing cleansers, reacts with water to give hydroxide ion and the hydrogen phosphate ion:

$$PO_4^{3-}(aq) + H_2O(l) \rightarrow HPO_4^{2-}(aq) + OH^-(aq)$$

Sodium hydroxide is prepared by electrolysis of aqueous brine (Chapter 11, Section 11.9). Calcium hydroxide is obtained by heating calcium carbonate to give calcium oxide, which is then mixed with water to give calcium hydroxide:

$$CaCO_3(s) \xrightarrow{\Delta} CaO(s) + CO_2(g)$$
$$CaO(s) + H_2O(l) \rightarrow Ca(OH)_2(s)$$

Calcium hydroxide is actually somewhat soluble in water—soluble enough to give a significantly basic solution. A mixture of the saturated solution with a suspension of excess solid calcium hydroxide was referred to as "whitewash," and it was used as a low-cost white coating for household painting.

Many metal hydroxides can be prepared by adding a metal ion solution to a hydroxide ion solution. Thus, green-blue copper(II) hydroxide can be prepared by mixing solutions of copper(II) chloride with sodium hydroxide:

$$CuCl_2(aq) + 2\,NaOH(aq) \rightarrow Cu(OH)_2(s) + 2\,NaCl(aq)$$

Many of the insoluble hydroxides precipitate out of solution as *gelatinous* (jellylike) solids, making them difficult to filter. Furthermore, some of the metal hydroxides are very unstable; they lose water to form the oxide, which, with its higher charge, forms a more stable lattice. For example, even gentle warming of the green-blue copper(II) hydroxide gel produces the black solid copper(II) oxide:

$$Cu(OH)_2(s) \xrightarrow{\Delta} CuO(s) + H_2O(l)$$

Solutions of the soluble hydroxides (the alkali metals, barium, and ammonium) react with the acidic oxide, carbon dioxide, present in the air to give the metal carbonate. For example, sodium hydroxide reacts with carbon dioxide to give sodium carbonate solution:

$$2\,NaOH(aq) + CO_2(g) \rightarrow Na_2CO_3(aq) + H_2O(l)$$

For this reason, solutions of hydroxides should be kept sealed except while being used. It is also one of the reasons why sodium hydroxide contained in glass bottles should be sealed with a rubber stopper rather than with a glass stopper. Some of the solution in the neck of the bottle will react to form crystals of sodium carbonate—enough to effectively "glue" the glass stopper into the neck of the bottle.

Alkali and alkaline earth metal hydroxides react with carbon dioxide, even when they are in the solid phase. In fact, "whitewashing" involves the penetration of the partially soluble calcium hydroxide into the wood or plaster surface. (Often the hydroxide ion acts additionally as a degreasing agent.) Over the following hours and days, it reacts with the carbon dioxide in the air to give microcrystalline, very insoluble, intensely white calcium carbonate:

$$Ca(OH)_2(aq) + CO_2(g) \rightarrow CaCO_3(s) + H_2O(l)$$

This process, performed by many of our ancestors, involved some very practical chemistry!

16.10 *The Hydroxyl Radical*

In Chapter 15, Section 15.11, we saw that the chemical reactions of the troposphere were dominated at night by the nitrate radical. In the daytime, it is another radical, the hydroxyl radical, OH, that is the most important reactive species.

The hydroxyl radical is present during the day at concentrations of between 10^5 and 10^6 molecules·cm^{-3}. It is most commonly formed by the photodissociation of ozone by light of wavelengths less than 319 nm to give atomic and molecular oxygen (oxygen and dioxygen). The two species produced are in excited states, the atomic oxygen having two of its p electrons paired instead of the ground state condition of all the p electrons being unpaired. The excited state can be indicated by an asterisk, $*$, following the relevant chemical formula. Thus, the reaction is represented as:

$$O_3(g) \xrightarrow{h\nu} O^*(g) + O_2(g)$$

About 80 percent of the excited oxygen atoms collide with a dinitrogen or dioxygen molecule (represented as M), lose their excitation energy, and become ground-state atoms.

$$O^*(g) + M(g) \rightarrow O(g) + M^*(g)$$

The ground-state atomic oxygen can then recombine with dioxygen to give ozone (trioxygen):

$$O(g) + O_2(g) \rightarrow O_3(g)$$

The ozone can then photodissociate, and the cycle begins again. However, the other 20 percent of excited oxygen atoms do not collide with dinitrogen or dioxygen molecules; instead, they collide with water molecules to form two hydroxyl radicals:

$$O^\star(g) + H_2O(g) \rightarrow OH(g) + OH(g)$$

These hydroxyl radicals are potent oxidizing agents. On the whole, this is good, for they oxidize, fragment, and destroy gas-phase organic molecules. For example, methane is oxidized to methyl hydroperoxide, CH_3OOH, then to methanal, $HCHO$, and finally to carbon dioxide. Hydroxyl radicals oxidize atmospheric nitrogen dioxide to nitric acid and hydrogen sulfide to sulfur dioxide.

There is now concern about hydroxyl radical concentrations in indoor air. These radicals are produced in part by the ozone emissions of such equipment as photocopiers and laser printers. The hydroxyl radicals react with the "cocktail" of organic compounds that we find in poorly ventilated indoor air such as exhaled breath, deodorants, perfumes and gas emissions from room furniture to produce a wide range of toxic oxidized fragments.

16.11   Sulfur

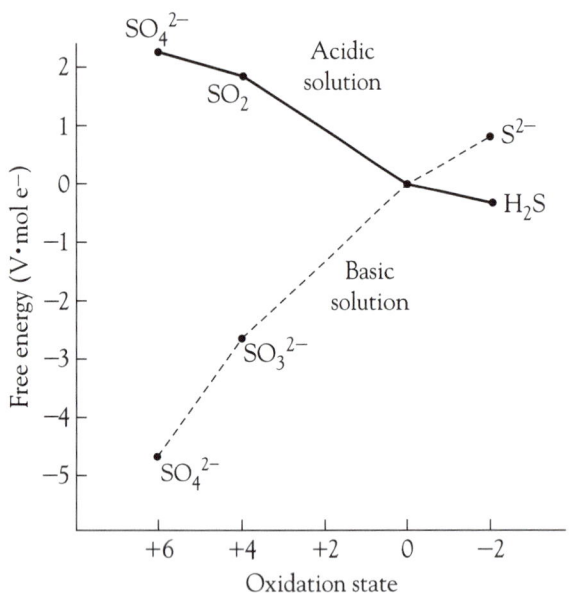

Figure 16.14   Frost diagram for sulfur in acidic and basic solutions.

Sulfur (the other nonmetal in Group 16) has a range of even oxidation states from +6 through +4 and +2 to −2. The oxidation state diagram for sulfur in acidic and basic solutions is shown in Figure 16.14. The comparatively low free energy of the sulfate ion in acidic solution indicates that the ion is only weakly oxidizing. In basic solution, the sulfate ion is completely nonoxidizing, and it is the most thermodynamically stable sulfur species. Although on a convex curve, the +4 oxidation state is actually quite kinetically stable. The Frost diagram does show that in acidic solution the +4 state tends to be reduced, whereas in basic solution it tends to be oxidized. The element itself is usually reduced in acidic environments but oxidized in base. Figure 16.14 also shows that the sulfide ion (basic solution) is a fairly strong reducing agent but that hydrogen sulfide is a thermodynamically stable species.

After carbon, sulfur is the element most prone to catenate. However, there are only two available bonds. Thus, the structures are typically chains of sulfur atoms with some other element or group of elements at each end: Dihydrogen polysulfides have the formula, $HS–S_n–SH$; and the polysulfur dichlorides, $ClS–S_n–SCl$, where n can have any value between 0 and 20.

16.12   Allotropes of Sulfur

Although elemental sulfur has been known from the earliest times, it is only in the last 20 years that the allotropy of sulfur has been clarified. The most common naturally occurring allotrope, S_8, cyclooctasulfur, has a zigzag

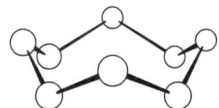

Figure 16.15 The cyclooctasulfur molecule.

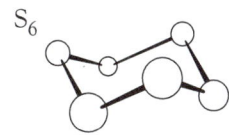

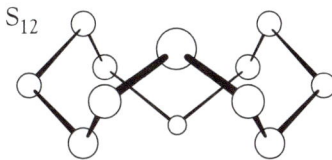

Figure 16.16 The cyclohexasulfur and cyclododecasulfur molecules.

arrangement of the atoms around the ring (Figure 16.15). This allotrope crystallizes to form needle crystals above 95°C; but below that temperature, "chunky" crystals are formed. The crystals, which are referred to as monoclinic and rhombic forms, differ simply in the way in which the molecules pack. These two forms are *polymorphs* of each other, not allotropes. Polymorphs are defined as different crystal forms in which the identical units of the same compound are packed differently.

In 1891, a sulfur allotrope with a ring size other than 8 was first synthesized. This allotrope, S_6, cyclohexasulfur, was the second of many true allotropes of sulfur to be discovered. To distinguish allotropes and polymorphs, we can more correctly define *allotropes* as forms of the same element that contain different molecular units. Sulfur allotropes with ring sizes that range from 6 to 20 have been definitely synthesized, and there is evidence that allotropes with much larger rings exist. The most stable, apart from cyclooctasulfur, is S_{12}, cyclododecasulfur. The structures of cyclohexasulfur and cyclododecasulfur are shown in Figure 16.16.

Cyclohexasulfur can be synthesized by mixing sodium thiosulfate, $Na_2S_2O_3$, and concentrated hydrochloric acid:

$$6\,Na_2S_2O_3(aq) + 12\,HCl(aq) \rightarrow S_6(s) + 6\,SO_2(g) + 12\,NaCl(aq) + 6\,H_2O(l)$$

However, there is now a fairly logical synthesis of the even-numbered rings (which are more stable than the odd-numbered rings). The method involves the reaction of the appropriate hydrogen polysulfide, H_2S_x, with the appropriate polysulfur dichloride, S_yCl_2, such that $(x + y)$ equals the desired ring size. Thus cyclododecasulfur can be prepared by mixing dihydrogen octasulfide, H_2S_8, and tetrasulfur dichloride, S_4Cl_2, in ethoxyethane, $(C_2H_5)_2O$, a solvent:

$$H_2S_8(eth.) + S_4Cl_2(eth.) \rightarrow S_{12}(s) + 2\,HCl(g)$$

However, cyclo octasulfur is the allotrope found almost exclusively in nature and as the product of almost all chemical reactions, so we will concentrate on the properties of this allotrope. At its melting point, cyclooctasulfur forms a low-viscosity, straw-colored liquid. But when the liquid is heated, there is an abrupt change in properties at 159°C. The most dramatic transformation is a 10^4-fold increase in viscosity. The liquid also darkens considerably. We can explain these changes in terms of a rupture of the rings. The octasulfur chains then link one to another to form polymers containing as many as 20 000 sulfur atoms. The rise in viscosity, then, is explained by a replacement of the free-moving S_8 molecules by these intertwined chains, which have strong dispersion force interactions.

As the temperature increases toward the boiling point of sulfur (444°C), the viscosity slowly drops as the polymer units start to fragment as a result of the greater thermal motion. If this liquid is poured into cold water, a brown transparent rubbery solid—plastic sulfur—is formed. This material slowly changes to microcrystals of rhombic sulfur.

Boiling sulfur produces a green gaseous phase, most of which consists of cyclooctasulfur. Raising the temperature even more causes the rings to fragment; and by 700°C, a violet gas is observed. This gas contains disulfur molecules, S_2, analogous to dioxygen.

16.13   *Industrial Extraction of Sulfur*

Elemental sulfur is found in large underground deposits in both the United States and Poland. It is believed they were formed by the action of anaerobic bacteria on lake-bottom deposits of sulfate minerals. A discovery causing great excitement among planetary scientists is the evidence for large deposits of sulfur on Jupiter's moon, Io (see color feature).

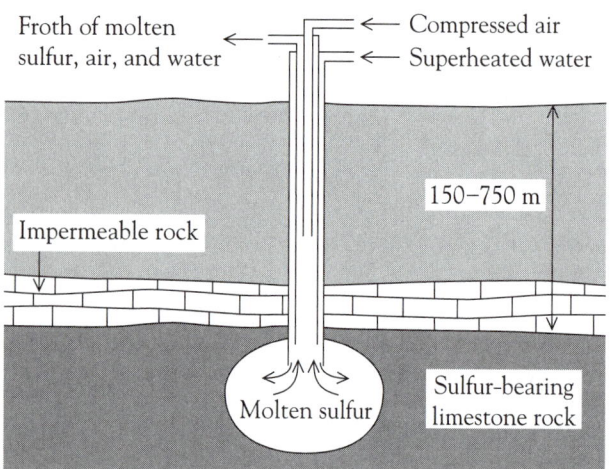

The method of extraction, the *Frasch process*, was devised by Canadian scientist, Herman Frasch (Figure 16.17). The sulfur deposits are between 150 and 750 m underground and are typically about 30 m thick. A pipe, 20 cm in diameter, is sunk almost to the bottom of the deposit. Then a 10-cm pipe is inserted inside the larger one; this pipe is a little shorter than the outer pipe. Finally, a 2.5-cm pipe is inserted into the middle pipe but ends about halfway down the length of the outer pipes.

Figure 16.17   The Frasch method of extraction of sulfur.

Water at 165°C is initially pumped down both outer pipes; this water melts the surrounding sulfur. The flow of superheated water down the 10-cm pipe is discontinued, and liquid pressure starts to force the dense liquid sulfur up that pipe. Compressed air is pumped down the 2.5-cm pipe, producing a low-density froth that flows freely up the 10-cm pipe to the surface. At the surface, the sulfur–water–air mixture is pumped into gigantic tanks, where it cools and the violet sulfur liquid crystallizes to a solid yellow block. The retaining walls of the tank are then removed, and dynamite is used to break up the blocks to a size that can be transported by railcar.

As we mentioned previously, the United States and Poland are the only countries fortunate enough to have large underground deposits of elemental sulfur. For their sulfur needs, other nations must resort to natural gas deposits, many of which contain high levels of hydrogen sulfide. Deposits that are low in hydrogen sulfide are known as "sweet gas," and those containing high levels, between 15 and 20 percent typically, are known as "sour gas." Gas producers are obviously only too pleased to find a market for this contaminant of the hydrocarbon mixtures.

The production of elemental sulfur from the hydrogen sulfide in natural gas is accomplished by using the *Claus process*. The hydrogen sulfide is first extracted from the sour natural gas by bubbling the gas through ethanolamine, $HOCH_2CH_2NH_2$, a basic organic solvent, the hydrogen sulfide acting as a Brønsted–Lowry acid:

$$HOCH_2CH_2NH_2(l) + H_2S(g) \rightarrow HOCH_2CH_2NH_3{}^+(solvent) + HS^-(solvent)$$

The solution is removed and warmed, causing the hydrogen sulfide gas to be released. The hydrogen sulfide is then mixed with dioxygen in a 2:1 mole ratio rather than the 2:3 mole ratio that would be needed to oxidize all the hydrogen sulfide to water and sulfur dioxide. One-third of the hydrogen sulfide burns to give sulfur dioxide gas, and the sulfur dioxide produced then reacts with the remaining two-thirds of the hydrogen sulfide to give elemental sulfur. To meet modern emissions standards, the process has been improved; in modern plants, 99 percent conversion occurs, an extraction much better than the 95 percent conversion achieved in older plants.

$$2\ H_2S(g) + 3\ O_2(g) \rightarrow 2\ SO_2(g) + 2\ H_2O(g)$$
$$\underline{4\ H_2S(g) + 2\ SO_2(g) \rightarrow 6\ S(s) + 4\ H_2O(g)}$$
$$6\ H_2S(g) + 3\ O_2(g) \rightarrow 6\ S(s) + 6\ H_2O(g)$$

About 53 percent of world sulfur production comes from by-product sulfur produced in the Claus (or a related) process; about 23 percent comes from the Frasch process; about 18 percent is obtained by heating the mineral iron pyrite, FeS_2 (iron(II) disulfide). Heating this compound in the absence of air decomposes the S_2^{2-} ion to elemental sulfur and iron(II) sulfide:

$$FeS_2(s) \xrightarrow{\Delta} FeS(s) + S(g)$$

Most of the world's sulfur production is needed for the synthesis of sulfuric acid, a process discussed later in this chapter. The remainder is used to synthesize sulfur chemicals such as carbon disulfide, for the vulcanization (hardening) of rubber, and for the synthesis of sulfur-containing organic dyes. Some of the elemental sulfur is added to asphalt mixes to make more frost-resistant highway surfaces.

16.14 Hydrogen Sulfide

Most people have heard of the gas that smells like "rotten eggs," although not as many could identify which gas it is. In fact, the obnoxious odor of hydrogen sulfide is almost unique. More important, this colorless gas is extremely toxic—more toxic than hydrogen cyanide. Because it is much more common, hydrogen sulfide presents a much greater hazard. As mentioned earlier, it sometimes is a component of the natural gas that issues from the ground; thus, gas leaks from natural gas wellheads can be dangerous.

Hydrogen sulfide is used in enormous quantities in the separation of "heavy water" from regular water. There is an equilibrium between water and hydrogen sulfide that energetically slightly favors the deuterium isotope in the water component:

$$HDS(g) + H_2O(l) \rightleftharpoons H_2S(g) + HDO(l)$$

$$HDS(g) + HDO(l) \rightleftharpoons H_2S(g) + D_2O(l)$$

By this means, the deuterium isotope can be concentrated (enriched) from 0.016 to 15 percent in the water. Fractional distillation to $\frac{1}{40}$th of the volume leaves a residue that is 99 percent deuterium oxide (D_2O having a slightly higher boiling point than H_2O). Communities located near such plants usually have quick-response evacuation procedures to minimize the inherent dangers of this industry. The odor can be detected at levels as low as 0.02 ppm; headaches and nausea occur at about 10 ppm, and death, at 100 ppm. Using smell to detect the gas is not completely effective, because it kills by affecting the central nervous system, including the sense of smell.

Hydrogen sulfide is produced naturally by anaerobic bacteria. In fact, this process, which occurs in rotting vegetation and bogs and elsewhere, is the source of most of the natural-origin sulfur in the atmosphere. The gas can be prepared in the laboratory by reacting a metal sulfide with a dilute acid, such as iron(II) sulfide with dilute hydrochloric acid:

$$FeS(s) + 2\ HCl(aq) \rightarrow FeCl_2(aq) + H_2S(g)$$

Hydrogen sulfide burns in air to give sulfur or sulfur dioxide, depending on the gas to air ratio:

$$2 \, H_2S(g) + O_2(g) \rightarrow 2 \, H_2O(l) + 2 \, S(s)$$

$$2 \, H_2S(g) + 3 \, O_2(g) \rightarrow 2 \, H_2O(l) + 2 \, SO_2(g)$$

In solution, it is oxidized to sulfur by almost any oxidizing agent:

$$H_2S(aq) \rightarrow 2 \, H^+(aq) + S(s) + 2 \, e^- \qquad E° = +0.141 \text{ V}$$

The common test for the presence of significant concentrations of hydrogen sulfide utilizes lead(II) acetate paper (or a filter paper soaked in any soluble lead(II) salt, such as the nitrate). In the presence of hydrogen sulfide, the colorless lead(II) acetate is converted to black lead(II) sulfide:

$$Pb(CH_3CO_2)_2(s) + H_2S(g) \rightarrow PbS(s) + 2 \, CH_3CO_2H(g)$$

In an analogous reaction, the blackening of silver tableware is usually attributed to the formation of black silver(I) sulfide.

The hydrogen sulfide molecule has a V-shaped structure, as would be expected for an analog of the water molecule. However, as we descend the group, the bond angles in their hydrides decrease (Table 16.5). The variation of bond angle can be explained in terms of a decreasing use of hybrid orbitals by elements beyond Period 2. Hence, it can be argued that the bonding in hydrogen selenide involves p orbitals only. This reasoning is the most commonly accepted explanation, because it is consistent with observed bond angles in other sets of compounds.

Table 16.5 Bond angles of three Group 16 hydrides

Hydride	Bond angle
H_2O	104½°
H_2S	92½°
H_2Se	90°

16.15   Sulfides

Only the Groups 1 and 2 metals and aluminum form soluble sulfides. These readily hydrolyze in water, and, as a result, solutions of sulfides are very basic:

$$S^{2-}(aq) + H_2O(l) \rightleftharpoons HS^-(aq) + OH^-(aq)$$

There is enough hydrolysis of the hydrogen sulfide ion, in turn, to give the solution a strong odor of hydrogen sulfide:

$$HS^-(aq) + H_2O(l) \rightleftharpoons H_2S(g) + OH^-(aq)$$

The sodium–sulfur system provides the basis for a high-performance battery. In most batteries, the electrodes are solids and the electrolyte a liquid. In this battery, however, the two electrodes, sodium and sulfur, are liquids and the electrolyte, $NaAl_{11}O_{17}$, is a solid. The electrode processes are

$$Na(l) \rightarrow Na^+(NaAl_{11}O_{17}) + e^-$$

$$n \, S(l) + 2 \, e^- \rightarrow S_n^{2-}(NaAl_{11}O_{17})$$

The battery is extremely powerful and it can be recharged readily. It shows great promise for industrial use, particularly for commercial electricity-driven delivery vehicles. However, adoption of this battery for household purposes is unlikely, because it operates at about 300°C. Of course, it has to remain sealed to prevent reaction of the sodium and sulfur with the oxygen or water vapor in the air.

All other metal sulfides are very insoluble. Many minerals are sulfide ores. The most common of these are listed in Table 16.6. Sulfides tend to be used

Table 16.6 Common sulfide minerals

Common name	Formula	Systematic name
Cinnabar	HgS	Mercury(II) sulfide
Galena	PbS	Lead(II) sulfide
Pyrite	FeS$_2$	Iron(II) disulfide
Sphalerite	ZnS	Zinc sulfide
Orpiment	As$_2$S$_3$	Diarsenic trisulfide
Stibnite	Sb$_2$S$_3$	Diantimony trisulfide
Chalcopyrite	CuFeS$_2$	Copper(II) iron(II) sulfide'

for specialized purposes. The intense black diantimony trisulfide was one of the first cosmetics, being used as eye shadow from earliest recorded times.

Today, sodium sulfide is the sulfide in highest demand. Between 10^5 and 10^6 tonnes is produced every year by the high-temperature reduction of sodium sulfate with coke:

$$Na_2SO_4(s) + 2\ C(s) \rightarrow Na_2S(l) + 2\ CO_2(g)$$

Sodium sulfide is used to remove hair from hides in the tanning of leather. It is also used in ore separation by flotation, for the manufacture of sulfur-containing dyes, and in the chemical industry, such as the precipitation of toxic metal ions, particularly lead.

Other sulfides in commercial use are selenium disulfide, SeS$_2$, a common additive to antidandruff hair shampoos, and molybdenum(IV) sulfide, MoS$_2$, an excellent lubricant for metal surfaces, either on its own or suspended in oil. Metal sulfides tend to be dense, opaque solids, and it is this property that makes the bright yellow cadmium sulfide, CdS, a popular pigment for oil painting.

Disulfide Bonds and Hair

Hair consists of amino acid polymers (proteins) cross-linked by disulfide units. In about 1930, it was shown by researchers at the Rockefeller Institute that these links could be broken by sulfides or molecules containing —SH groups in slightly basic solution. This discovery proved to be the key to the present-day method for "permanently" changing the shape of hair, from curly to straight or vice versa.

In the process, a solution of the thioglycollate ion, HSCH$_2$CO$_2^-$, is poured on the hair, reducing the —S—S— cross-links to —SH groups:

$$2\ HSCH_2CO_2^-(aq) + \text{---S---S---}(hair) \rightarrow [SCH_2CO_2^-]_2(aq) + 2\ \text{---S---H}(hair)$$

By using curlers or straighteners, the protein chains can then be mechanically shifted with respect to their neighbors. Application of a solution of hydrogen peroxide then reoxidizes the —SH groups to re-form new cross-links of —S—S—, thus holding the hair in the new orientation:

$$2\ \text{---S---H}(hair) + H_2O_2(aq) \rightarrow \text{---S---S---}(hair) + 2\ H_2O(l)$$

The formation of insoluble metal sulfides used to be common in inorganic qualitative analysis. Hydrogen sulfide is bubbled through an acid solution containing unknown metal ions. The presence of the high hydrogen ion concentration reduces the sulfide ion concentration to extremely low levels:

$$H_2S(aq) + 2\ H_2O(l) \rightleftharpoons 2\ H_3O^+(aq) + S^{2-}(aq)$$

This very low level of sulfide ion is still enough to precipitate the most insoluble metal sulfides, those with a solubility product, K_{sp}, smaller than 10^{-30}. These metal sulfides are separated by filtration or centrifugation. The pH of the filtrate is increased by adding base. This increase shifts the sulfide equilibrium to the right, thereby raising the concentration of sulfide ions to the point where those metal sulfides with a solubility product between 10^{-20} and 10^{-30} (mainly the transition metals of Period 4) precipitate. Specific tests could then be used to identify which metal ions are present. More recently, thio-acetamide, a reagent that hydrolyzes to hydrogen sulfide, has been used for such tests.

In addition to forming conventional sulfides, some elements form disulfides, S_2^{2-}, ions analogous to dioxide($2-$). Thus, FeS_2 does not contain iron in a high oxidation state, but the disulfide ion and Fe^{2+}. Also, the alkali and alkaline earth metals form polysulfides, which contain the S_n^{2-} ion, where n has values between 2 and 6.

16.16   Sulfur Oxides

Sulfur Dioxide

The common oxide of sulfur, sulfur dioxide, is a colorless, dense, toxic gas with an acid "taste." The maximum tolerable levels for humans is about 5 ppm, but plants begin to suffer in concentrations as low as 1 ppm. The taste is a result of the reaction of sulfur dioxide with water on the tongue to give the weak acid, sulfurous acid:

$$SO_2(g) + H_2O(l) \rightleftharpoons H_2SO_3(aq)$$

Sulfur dioxide is very water soluble, but like ammonia and carbon dioxide, almost all the dissolved gas is present as the sulfur dioxide molecule; only a very small proportion forms sulfurous acid. To prepare the gas in the laboratory, a dilute acid is added to a solution of a sulfite or a hydrogen sulfite:

$$SO_3^{2-}(aq) + 2\ H^+(aq) \rightarrow H_2O(l) + SO_2(g)$$

$$HSO_3^-(aq) + H^+(aq) \rightarrow H_2O(l) + SO_2(g)$$

Sulfur dioxide is one of the few common gases that is a reducing agent, itself being easily oxidized to the sulfate ion:

$$SO_2(aq) + 2\ H_2O(l) \rightarrow SO_4^{2-}(aq) + 4\ H^+(aq) + 2\ e^-$$

To test for reducing gases, such as sulfur dioxide, we can use an oxidizing agent that undergoes a color change, the most convenient one being the dichromate ion. A filter paper soaked in acidified orange dichromate ion will turn green as a result of the formation of the chromium(III) ion:

$$Cr_2O_7^{2-}(aq) + 14\ H^+(aq) + 6\ e^- \rightarrow 2\ Cr^{3+}(aq) + 7\ H_2O(l)$$

Since Earth first solidified, sulfur dioxide has been produced by volcanoes in large quantities. However, we are now adding additional, enormous quantities of this gas to the atmosphere. Combustion of coal is the worst offender, because most coals contain significant levels of sulfur compounds. In London, the yellow smog of the 1950s caused by home coal fires led to thousands of premature deaths. Currently, coal-fired electric power stations are the major sources of unnatural sulfur dioxide in the atmosphere. Oil, too, contributes to the atmospheric burden of sulfur dioxide, for the lowest cost heating oil is sulfur rich. Thus, many schools and hospitals, in their need to conserve finances, become the cause of poorer air quality when they choose the lowest cost oil for their heating purposes. Finally, many metals are extracted from sulfide ores, and the traditional smelting process involves the oxidation of sulfide to sulfur dioxide, thereby providing an additional source of the gas. Copper is one such metal that is extracted from sulfide ores (see Chapter 20, Section 20.9).

In the past, the easiest solution to industrial air pollution problems was to provide ever taller smokestacks so that the sulfur dioxide would travel appreciable distances from the source. But, during its time in the upper atmosphere, the sulfur dioxide is oxidized by the hydroxyl radical and hydrated to give droplets of sulfuric acid, a much stronger acid than sulfurous acid:

$$SO_2(g) + OH(g) \rightarrow HOSO_2(g)$$

$$HOSO_2(g) + O_2(g) \rightarrow HO_2(g) + SO_3(g)$$

$$SO_3(g) + H_2O(g) \rightarrow H_2SO_4(aq)$$

This product precipitates as acid rain, many hundreds or thousands of kilometers away, making the problem "someone else's." Currently, researchers are studying methods to minimize sulfur dioxide emissions. One of these involves the conversion of sulfur dioxide to solid calcium sulfate. In a modern coal-burning power plant, powdered limestone (calcium carbonate) is mixed with the powdered coal. The coal burns, producing a flame at about 1000°C, a temperature high enough to decompose the calcium carbonate:

$$CaCO_3(s) \xrightarrow{\Delta} CaO(s) + CO_2(g)$$

Then the calcium oxide reacts with sulfur dioxide and oxygen gas to give calcium sulfate:

$$2\,CaO(s) + 2\,SO_2(g) + O_2(g) \xrightarrow{\Delta} 2\,CaSO_4(s)$$

Because the second step is about as exothermic as the first step is endothermic, no heat is lost in the overall process. The fine dust of calcium sulfate is captured by electrostatic precipitators.

The solid calcium sulfate can be used for fireproof insulation and roadbed cement. However, as this process becomes more and more widely used, the supply of calcium sulfate will outstrip demand and using it for landfill will become more and more common. Thus, we have replaced a gaseous waste by a less harmful solid waste, but we have not eliminated the waste problem completely.

Sulfur dioxide does have some positive uses. It is used as a bleach and as a preservative, particularly for fruits. In this latter role, it is very effective at killing molds and other fruit-destroying organisms. Unfortunately, some people are sensitive to traces of the substance.

The sulfur dioxide molecule is V-shaped, with an S—O bond length of 143 pm and a O—S—O bond angle of 119°. The bond length is much shorter than that of a sulfur–oxygen single bond (163 pm) and very close to that of a typical sulfur-oxygen double bond (140 pm). The bonding and shape

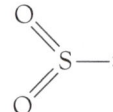

Figure 16.18   A possible representation of the bonding in sulfur dioxide.

is shown in Figure 16.18. The similarity of the sulfur dioxide bond angle to the trigonal angle of 120° (sp^2 hybridization) can be explained in terms of a σ bond between each sulfur–oxygen pair and a lone pair of electrons on the sulfur atom. We might expect the π bond system to resemble that of the nitrite ion; but, in addition, we could invoke some contribution to multiple bonding by interaction between the empty $3d$ orbitals of sulfur and full p orbitals of the oxygen atoms.

Sulfur Trioxide

Most people have heard of sulfur dioxide, but few have heard of the other important oxide, sulfur trioxide, a colorless liquid at room temperature. The liquid and gas phases contain a mixture of sulfur trioxide, SO_3, and a trimer, S_3O_9 (Figure 16.19). The liquid freezes at 16°C to give crystals of trisulfur nonaoxide. In the presence of moisture, long-chain solid polymers are formed and have the structure $HO(SO_3)_nOH$, where n is about 10^5. Sulfur trioxide is a very acidic, deliquescent oxide, reacting with water to form sulfuric acid:

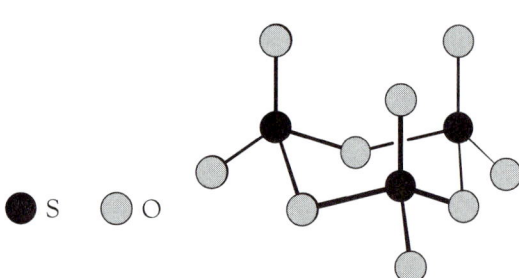

● S   ○ O

Figure 16.19   The trisulfur nonaoxide molecule, a solid allotrope of sulfur trioxide.

$$SO_3(s) + H_2O(l) \rightarrow H_2SO_4(l)$$

This oxide is so little known because oxidation of sulfur almost always gives sulfur dioxide, not sulfur trioxide. Even though the formation of sulfur trioxide is even more thermodynamically favored than that of sulfur dioxide (-370 kJ·mol^{-1} for sulfur trioxide, -300 kJ·mol^{-1} for sulfur dioxide), the oxidation has a high activation energy. Thus, the pathway from sulfur dioxide to sulfur trioxide is kinetically controlled:

$$2\,SO_2(g) + O_2(g) \rightarrow 2\,SO_3(g)$$

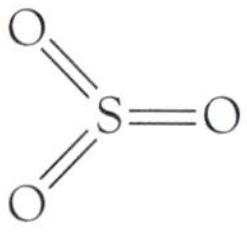

When liquid sulfur trioxide boils, the gaseous molecules formed are planar SO_3 (Figure 16.20). Like sulfur dioxide, all the sulfur–oxygen bond lengths are equally short (142 pm) and very close to the typical double bond value. Again, this bonding is best interpreted by means of a π system involving the sulfur $3d$ orbitals.

Figure 16.20   A possible representation of the bonding in sulfur trioxide.

16.17   Sulfuric Acid

Sulfuric acid is an oily, dense liquid that freezes at 10°C. It mixes with water very exothermically. For this reason, the concentrated acid should be slowly added to water, not the reverse process, and the mixture should be stirred continuously. The molecule contains a tetrahedral arrangement of oxygen atoms around the central sulfur atom (Figure 16.21). The short bond lengths and the high bond energies suggest that there must be double bond character in the sulfur bonds to each terminal oxygen atom.

The pure liquid has a significant electrical conductivity as a result of self-ionization reactions:

$$2\,H_2SO_4(l) \rightleftharpoons H_2O(H_2SO_4) + H_2S_2O_7(H_2SO_4)$$

$$H_2O(H_2SO_4) + H_2SO_4(l) \rightleftharpoons H_3O^+(H_2SO_4) + HSO_4^-(H_2SO_4)$$

$$H_2S_2O_7(H_2SO_4) + H_2SO_4(l) \rightleftharpoons H_3SO_4^+(H_2SO_4) + HS_2O_7^-(H_2SO_4)$$

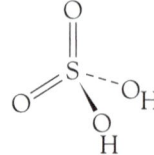

Figure 16.21   A possible representation of the bonding in sulfuric acid.

Concentrated sulfuric acid is a water mixture with an acid concentration of 18 mol·L^{-1}. We usually think of sulfuric acid as just an acid, but in fact it can react in five different ways.

Sulfuric Acid as an Acid

Dilute sulfuric acid is used most often as an acid. It is a strong, diprotic acid, forming two ions, the hydrogen sulfate ion and the sulfate ion:

$$H_2SO_4(aq) + H_2O(l) \rightleftharpoons H_3O^+(aq) + HSO_4^-(aq)$$

$$HSO_4^-(aq) + H_2O(l) \rightleftharpoons H_3O^+(aq) + SO_4^{2-}(aq)$$

The first equilibrium lies far to the right, but the second one, less so. Thus, the predominant species in a solution of sulfuric acid are the hydronium ion and the hydrogen sulfate ion.

Sulfuric Acid as a Dehydrating Agent

The concentrated acid will remove the elements of water from a number of compounds. For example, sugar is converted to carbon and water. This exothermic reaction is spectacular:

$$C_{12}H_{22}O_{11}(s) + H_2SO_4(l) \rightarrow 12\ C(s) + 11\ H_2O(g) + H_2SO_4(aq)$$

The acid serves this function in a number of important organic reactions. For example, addition of concentrated sulfuric acid to ethanol produces ethene, C_2H_4, or ethoxyethane, $(C_2H_5)_2O$, depending on the reaction conditions:

$$C_2H_5OH(l) + H_2SO_4(l) \rightarrow C_2H_5OSO_3H(aq) + H_2O(l)$$

$$C_2H_5OSO_3H(aq) \rightarrow C_2H_4(g) + H_2SO_4(aq) \qquad \text{[excess acid]}$$

$$C_2H_5OSO_3H(aq) + C_2H_5OH(l) \rightarrow (C_2H_5)_2O(l) + H_2SO_4(aq)$$
$$\text{[excess ethanol]}$$

Sulfuric Acid as an Oxidizing Agent

Although sulfuric acid is not as strongly oxidizing as nitric acid, if it is hot and concentrated, it will function as an oxidizing agent. For example, hot concentrated sulfuric acid reacts with copper metal to give the copper(II) ion, and the sulfuric acid itself is reduced to sulfur dioxide and water:

$$Cu(s) \rightarrow Cu^{2+}(aq) + 2\ e^-$$

$$2\ H_2SO_4(l) + 2\ e^- \rightarrow SO_2(g) + 2\ H_2O(l) + SO_4^{2-}(aq)$$

Sulfuric Acid as a Sulfonating Agent

The concentrated acid is used in organic chemistry to replace a hydrogen atom by the sulfonic acid group ($-SO_3H$):

$$H_2SO_4(l) + CH_3C_6H_5(l) \rightarrow CH_3C_6H_4SO_3H(s) + H_2O(l)$$

Sulfuric Acid as a Base

A Brønsted–Lowry acid can only act as a base if it is added to a stronger proton donor. Sulfuric acid is a very strong acid; hence, only extremely strong acids such as fluorosulfonic acid (see feature "Superacids" in Chapter 7) can cause it to behave as a base:

$$H_2SO_4(l) + HSO_3F(l) \rightleftharpoons H_3SO_4^+(H_2SO_4) + SO_3F^-(H_2SO_4)$$

16.18 Industrial Synthesis of Sulfuric Acid

Sulfuric acid is synthesized in larger quantities than any other chemical. In the United States alone, production is about 165 kg per person per year! All synthetic routes use sulfur dioxide, and in some plants this reactant is obtained directly from the flue gases of smelting processes. However, in North America most of the sulfur dioxide is produced by burning molten sulfur in dry air:

$$S(l) + O_2(g) \xrightarrow{\ \Delta\ } SO_2(g)$$

It is more difficult to oxidize sulfur further. As we mentioned earlier, there is a kinetic barrier to the formation of sulfur trioxide. Thus, an effective catalyst must be used to obtain commercially acceptable rates of reaction. We also need to ensure that the position of equilibrium is to the right side of the equation. To accomplish this, we invoke the Le Châtelier principle, which predicts that an increase in pressure will favor the side of the equation with the fewer moles of gas—in this case, the product side. This reaction is also exothermic; thus, the choice of temperature must be high enough to produce a reasonable rate of reaction, even though these conditions will result in a decreased yield:

$$2\ SO_2(g)\ +\ O_2(g)\ \xrightarrow[\Delta]{V_2O_5}\ 2\ SO_3(g)$$

In the *Contact process*, pure, dry sulfur dioxide and dry air are passed through a catalyst of vanadium(V) oxide on an inert support. The gas mixture is heated to between 400 and 500°C, which is the optimum temperature for conversion to sulfur trioxide with a reasonable yield at an acceptable rate.

Sulfur trioxide reacts violently with water—not a process that is acceptable in industry. However, it does react more controllably with concentrated sulfuric acid itself to give pyrosulfuric acid, $H_2S_2O_7$ (Figure 16.22):

$$SO_3(g)\ +\ H_2SO_4(l) \rightarrow H_2S_2O_7(l)$$

The pyrosulfuric acid is then diluted with water to produce an additional mole of sulfuric acid:

$$H_2S_2O_7(l)\ +\ H_2O(l) \rightarrow 2\ H_2SO_4(l)$$

All steps in the process are exothermic. In fact, the entire process of converting elemental sulfur to sulfuric acid produces 535 kJ·mol^{-1} of heat. An essential feature of any sulfuric acid plant is effective utilization of this waste heat, either as direct heating for some other industrial process or in the production of electricity.

This process is associated with two potential pollution problems. First, some of the sulfur dioxide escapes. Legislation in most pollution-conscious countries limits emissions to less than 0.5 percent of the processed gas. Second, despite using the pyrosulfate route, some of the sulfuric acid escapes as a fine mist. Newer plants have mist eliminators to reduce this problem.

Use of the sulfuric acid varies from country to country. In the United States, the vast majority of acid is employed in the manufacture of fertilizers, such as the conversion of the insoluble calcium phosphate to the more soluble calcium dihydrogen phosphate:

$$Ca_3(PO_4)_2(s)\ +\ 2\ H_2SO_4(aq) \rightarrow Ca(H_2PO_4)_2(s)\ +\ 2\ CaSO_4(s)$$

or the production of ammonium sulfate fertilizer:

$$2\ NH_3(g)\ +\ H_2SO_4(aq) \rightarrow (NH_4)_2SO_4(aq)$$

In Europe, however, a higher proportion of the acid is used for manufacturing other products such as paints, pigments, and sulfonate detergents.

There is an increasing interest in trying to reclaim waste sulfuric acid. At the present time, the cost of removing contaminants and concentrating the dilute acid is greater than the cost of preparing the acid from sulfur. However, recovery is now preferred over dumping. If the acid is pure but too dilute (in other words, the only contaminant is water), then pyrosulfuric

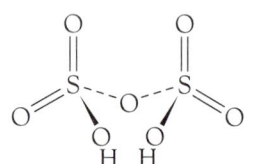

Figure 16.22 A possible representation of the bonding in the pyrosulfuric acid molecule.

acid is added to increase the concentration of acid to usable levels. For contaminated acid, high-temperature decomposition produces gaseous sulfur dioxide, which can be removed and used to synthesize fresh acid:

$$2 \ H_2SO_4(aq) \xrightarrow{\Delta} 2 \ SO_2(g) + 2 \ H_2O(l) + O_2(g)$$

16.19 Sulfites

Although sulfurous acid is mostly an aqueous solution of sulfur dioxide, the sulfite and hydrogen sulfite ions are real entities. In fact, sodium sulfite is a major industrial chemical with annual production being about 10^6 tonnes. It is most commonly prepared by bubbling sulfur dioxide into sodium hydroxide solution:

$$2 \ NaOH(aq) + SO_2(g) \rightarrow Na_2SO_3(aq) + H_2O(l)$$

In the laboratory and in industry, sodium sulfite is used as a reducing agent, itself being oxidized to sodium sulfate:

$$SO_3{}^{2-}(aq) + H_2O(l) \rightarrow SO_4{}^{2-}(aq) + 2 \ H^+(aq) + 2 \ e^-$$

The main use of sodium sulfite is as a bleach in the Kraft process for the production of paper. In this process, the sulfite ion attacks the polymeric material (lignin) that binds the cellulose fibers together (the loose cellulose fibers make up the paper structure). A secondary use, as we will see shortly, is in the manufacture of sodium thiosulfate. Like sulfur dioxide, sodium sulfite can be added to fruit as a preservative.

The sulfur–oxygen bond lengths in the sulfite ion are 151 pm, slightly longer than the 140-pm S—O double bond. Although it is possible to draw electron-dot structures with all single bonds, we can use formal charge representations to see why multiple bonds are preferred (Figure 16.23).

In Figure 16.23(a), the single bond representation has formal charges on each atom, which makes this bonding arrangement unlikely. Figure 16.23(c) shows two double bonds and negative charges on neighboring atoms, again an unlikely scenario. It is the structure in Figure 16.23(b), with one double bond, that has the minimum formal charge arrangement. However, it should be kept in mind that formal charge is a very simplistic method of approaching bonding and that a molecular orbital study provides a much more valid picture. If we take Figure 16.23(b) to represent one of three possible resonance structures, then each sulfur–oxygen bond can be assigned an average bond order of 1⅓.

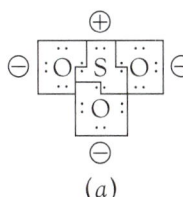

(a)

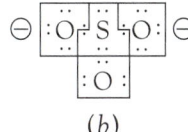

(b)

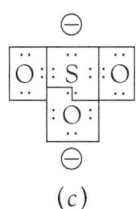

(c)

Figure 16.23 Three formal charge representations for the sulfite ion.

16.20 Sulfates

Sulfates and nitrates are the most commonly encountered metal salts. There are several reasons for the use of sulfates:

1. Most sulfates are water soluble, making them a useful source of the metal cation. Two important exceptions are lead(II) sulfate, which plays an important role in the lead-acid battery, and barium sulfate, used in X-rays of soft tissues such as the stomach.

2. The sulfate ion is not oxidizing or reducing. Hence, the sulfate ion can form salts with metals in both their higher and their lower

common oxidation states, for example, iron(II) sulfate and iron(III) sulfate. Furthermore, when dissolved in water, the sulfate ion will not initiate a redox reaction with any other ion present.

3. The sulfate ion is the conjugate base of a moderately strong acid (the hydrogen sulfate ion), so the anion will not significantly alter the pH of a solution.

4. The sulfates tend to be thermally stable, at least more stable than the equivalent nitrate salts.

Sulfates can be prepared by the reaction between a base, such as sodium hydroxide, and the stoichiometric quantity of dilute sulfuric acid:

$$2\,NaOH(aq) + H_2SO_4(aq) \rightarrow Na_2SO_4(aq) + 2\,H_2O(l)$$

or by the reaction between an electropositive metal, such as zinc, and dilute sulfuric acid:

$$Zn(s) + H_2SO_4(aq) \rightarrow ZnSO_4(aq) + H_2(g)$$

or by the reaction between a metal carbonate, such as copper(II) carbonate, and dilute sulfuric acid:

$$CuCO_3(s) + H_2SO_4(aq) \rightarrow CuSO_4(aq) + CO_2(g) + H_2O(l)$$

The common test for the presence of sulfate ion is the addition of barium ion, which reacts with the anion to give a dense white precipitate, barium sulfate:

$$Ba^{2+}(aq) + SO_4^{\,2-}(aq) \rightarrow BaSO_4(s)$$

Like the sulfite ion, the sulfate ion has a short sulfur–oxygen bond, a characteristic indicating considerable multiple bond character. In fact, at 149 pm, its length is about the same as that in the sulfite ion, within experimental error.

16.21 *Hydrogen Sulfates*

Like the hydrogen carbonates, only the alkali and alkaline earth metals have charge densities low enough to stabilize these large low-charge anions in the solid phase. The value of the second ionization of sulfuric acid is quite large, so the hydrogen sulfates give an acidic solution:

$$HSO_4^{\,-}(aq) + H_2O(l) \rightleftharpoons H_3O^+(aq) + SO_4^{\,2-}(aq)$$

It is the high acidity of the solid sodium hydrogen sulfate that makes it useful as a household cleaning agent, such as Saniflush®.

Hydrogen sulfates can be prepared by mixing the stoichiometric quantities of sodium hydroxide and sulfuric acid and evaporating the solution:

$$NaOH(aq) + H_2SO_4(aq) \rightarrow NaHSO_4(aq) + H_2O(l)$$

They decompose in an unexpected manner when heated, forming the metal pyrosulfate:

$$2\,NaHSO_4(s) \xrightarrow{\Delta} Na_2S_2O_7(s) + H_2O(l)$$

16.22 *Thiosulfates*

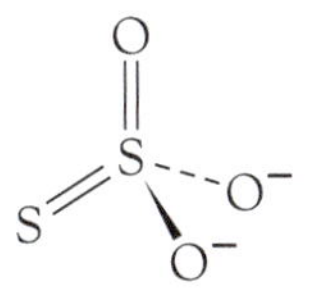

Figure 16.24 A possible representation of the bonding in the thiosulfate ion.

The thiosulfate ion resembles the sulfate ion, except that one oxygen atom has been replaced by a sulfur atom (*thio-* is a prefix meaning "sulfur"). These two sulfur atoms are in completely different environments; the additional sulfur behaves more like a sulfide ion. In fact, a formal assignment of oxidation numbers gives the central sulfur a value of $+5$; and the other one, a value of -1, as discussed in Chapter 8, Section 8.3. The shape of the thiosulfate ion is shown in Figure 16.24. Although the ion is depicted as having two double bonds and two single bonds, the multiple bond character is actually spread more evenly over all the bonds.

Sodium thiosulfate pentahydrate, commonly called "hypo," can easily be prepared by boiling sulfur in a solution of sodium sulfite:

$$SO_3^{2-}(aq) + S(s) \rightarrow S_2O_3^{2-}(aq)$$

The thiosulfate ion is not very thermally stable, first dissolving in the water of crystallization:

$$Na_2S_2O_3 \cdot 5H_2O(s) \xrightarrow{\Delta} Na_2S_2O_3(aq) + 5\ H_2O(l)$$

then, when heated more strongly, disproportionating to give three different oxidation states of sulfur: sodium sulfate, sodium sulfide, and sulfur:

$$4\ Na_2S_2O_3(s) \xrightarrow{\Delta} 3\ Na_2SO_4(s) + Na_2S(s) + 4\ S(s)$$

The gentle warming that causes the sodium thiosulfate pentahydrate to lose the water of crystallization is a reversible endothermic process:

$$Na_2S_2O_3 \cdot 5H_2O(s) \rightleftharpoons Na_2S_2O_3(aq) + 5\ H_2O(l) \qquad \Delta H^\circ = +55\ kJ \cdot mol^{-1}$$

The equilibrium has generated considerable interest as a heat storage system. In this process, heat from the Sun is absorbed by solar panels and transferred to an underground tank of the hydrated compound. This input of heat causes the hydrate to decompose and dissolve in the water produced. Then, in the cool of the night, heat released as the compound crystallizes can be used to heat the dwelling.

When handling solutions of the thiosulfate ion, it is important to avoid the presence of acid. The hydrogen (hydronium) ion first reacts to form thiosulfuric acid, which decomposes rapidly to give a white suspension of sulfur and the characteristic odor-taste of sulfur dioxide. This particular disproportionation is further evidence that the two sulfur atoms are in different oxidation states. Presumably, it is the central sulfur that provides the higher oxidation state sulfur in sulfur dioxide:

$$S_2O_3^{2-}(aq) + 2\ H^+(aq) \rightarrow H_2S_2O_3(aq)$$

$$H_2S_2O_3(aq) \rightarrow H_2O(l) + S(s) + SO_2(g)$$

Sodium thiosulfate is also used in redox titrations. For example, it is used to determine the concentration of iodine in aqueous solutions. During the assay, the iodine is reduced to iodide, and the thiosulfate ion of known concentration is oxidized to the tetrathionate ion:

$$2\ S_2O_3^{2-}(aq) \rightarrow S_4O_6^{2-}(aq) + 2\ e^-$$

$$I_2(aq) + 2\ e^- \rightarrow 2\ I^-(aq)$$

The Chemistry of Photography

The thiosulfate ion is particularly important in traditional film photography. Film is coated with silver bromide. During picture-taking with black-and-white film, the light reduces a few of the silver ions to silver metal. In a silver bromide microcrystal, from 10 to 100 atoms of silver metal are usually produced. This number would be far too few to see; thus, the first processing step is to add a developer, usually hydroquinone, $C_6H_6O_2$. A developer selectively reduces all the silver ions in any crystal that already contains silver atoms. By this means, we enhance the image by a factor of about 10^{10} times.

Next, we must remove the remaining insoluble silver bromide; otherwise the film would all turn black when it is exposed to light. To accomplish this, thiosulfate ion is added. This reagent reacts with the remaining silver ion to form a soluble complex, the tris(thiosulfato)silver(I) ion, $Ag(S_2O_3)_3^{5-}$, which is washed away. The black particles of silver metal are left behind to form the image. The process is essentially the same for color photography, except that organic dyes are involved as well:

$$AgBr(s) + 3\ S_2O_3^{2-}(aq) \rightarrow [Ag(S_2O_3)_3]^{5-}(aq) + Br^-(aq)$$

Although sodium thiosulfate used to be the preferred salt, ammonium thiosulfate is becoming more widely used, because it allows faster processing and silver can be recovered from the waste solution more easily.

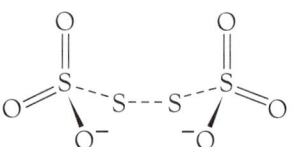

Figure 16.25 A possible representation of the bonding in the tetrathionate ion.

The tetrathionate ion contains bridging sulfur atoms (Figure 16.25).

Mixing cold solutions of thiosulfate ion and iron(III) ion gives a characteristic deep purple complex ion:

$$Fe^{3+}(aq) + 2\ S_2O_3^{2-}(aq) \rightarrow [Fe(S_2O_3)_2]^-(aq)$$

When warmed, this bis(thiosulfato)ferrate(III) ion, $[Fe(S_2O_3)_2]^-$, undergoes a redox reaction to give the iron(II) ion and the tetrathionate ion:

$$[Fe(S_2O_3)_2]^-(aq) + Fe^{3+}(aq) \rightarrow 2\ Fe^{2+}(aq) + S_4O_6^{2-}(aq)$$

16.23   Peroxodisulfates

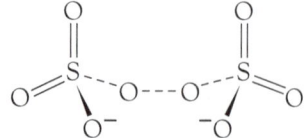

Figure 16.26 A possible representation of the bonding in the peroxodisulfate ion.

Although the sulfate ion contains sulfur in its highest possible oxidation state of $+6$, it can be oxidized electrolytically to the peroxodisulfate ion by using smooth platinum electrodes, acidic solution, and high current densities. These conditions favor oxidations that do not produce gases such as the competing oxidation of water to dioxygen:

$$2\ HSO_4^-(aq) \rightarrow S_2O_8^{2-}(aq) + 2\ H^+(aq) + 2\ e^-$$

This ion contains a dioxo bridge (Figure 16.26) with an analogous structure to that of the tetrathionate ion (see Figure 16.25). Hence, the two sulfur atoms still have formal oxidation states of $+6$, but the bridging oxygen atoms have been oxidized from -2 to -1. The terminal S—O bond lengths are all equivalent at 150 pm; once again, there must be considerable multiple bonding. The parent acid, peroxodisulfuric acid, is a white solid, but it is the two salts, potassium peroxodisulfate and ammonium peroxodisulfate, that are important as powerful, stable oxidizing agents:

$$S_2O_8{}^{2-}(aq) + 2\ e^- \rightarrow 2\ SO_4{}^{2-}(aq) \qquad\qquad E^\circ = +2.01\ V$$

16.24   Sulfur Hexafluoride

The most important compound of sulfur and fluorine is sulfur hexafluoride, SF_6. This compound is a colorless, odorless, unreactive gas. About 6500 tonnes is produced per year by simply burning molten sulfur in fluorine gas:

$$S(l) + 3\ F_2(g) \rightarrow SF_6(g)$$

As would be expected from simple VSEPR theory, the molecule is octahedral (Figure 16.27).

Figure 16.27   The sulfur hexafluoride molecule.

As a result of its stability, low toxicity, and inertness, sulfur hexafluoride is used as an insulating gas in high-voltage electrical systems. At a pressure of about 250 kPa, it will prevent a discharge across a 1 MV potential difference that is separated by only 5 cm. Another major use is to blanket molten magnesium during the refining of the metal. There are many other uses, including filling noise-insulating double- and triple-glazed windows.

The very high molar mass of this gas makes it useful for several scientific applications. For example, air pollution can be tracked for thousands of kilometers by releasing a small amount of sulfur hexafluoride at the pollution source. The extremely high molar mass is so unique that the contaminated air mass can be identified days later by its tiny concentration of sulfur hexafluoride molecules. Similarly, deep ocean currents are being identified by bubbling sulfur hexafluoride into deep-water layers and then tracking the movement of the gas.

However, the very inertness of sulfur hexafluoride makes it a particular problem in the context of climatic impact. The gas absorbs radiation throughout much of the otherwise transparent part of the infrared region in the atmosphere. As a result, it is an extraordinarily effective greenhouse gas; 1 tonne of sulfur hexafluoride being equivalent to 23 900 tonnes of carbon dioxide in terms of infrared absorption. Further, there are no destruction pathways for sulfur hexafluoride, except above 60 km, where it is destroyed by the intense ultraviolet radiation. Thus, the atmospheric lifetime of the gas is estimated as at least 3000 years. Compared to total carbon dioxide emissions, sulfur hexafluoride represents less than 1 percent of contributions to increased energy absorption. However, with the increasing use of this gas, it is imperative that we take all possible steps to ensure that as little sulfur hexafluoride as possible escapes into the atmosphere.

16.25   Other Sulfur Halides

It is interesting that the other common sulfur–fluorine compound, sulfur tetrafluoride, is extremely reactive. It decomposes in the presence of moisture to hydrogen fluoride and sulfur dioxide:

$$SF_4(g) + 2\ H_2O(l) \rightarrow SO_2(g) + 4\ HF(g)$$

Its high reactivity might be due to the "exposed" lone pair site where reaction can take place. The compound is a convenient reagent for the fluorination

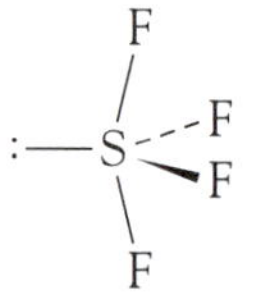

Figure 16.28   The sulfur tetrafluoride molecule.

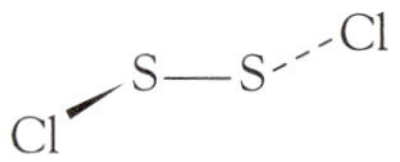

Figure 16.29   The disulfur dichloride molecule.

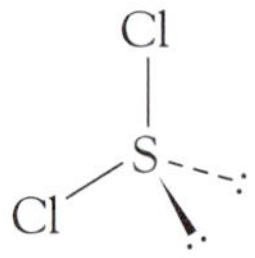

Figure 16.30   The sulfur dichloride molecule.

of organic compounds. For example, it converts ethanol to fluoroethane. As simple VSEPR theory predicts, it has a slightly distorted seesaw shape (Figure 16.28).

Whereas sulfur forms high oxidation state compounds with fluorine, it forms stable low-oxidation-state compounds with chlorine. Bubbling chlorine through molten sulfur produces disulfur dichloride, S_2Cl_2, a toxic, yellow liquid with a revolting odor:

$$2\ S(l) + Cl_2(g) \rightarrow S_2Cl_2(l)$$

The compound is used in the *vulcanization* of rubber; that is, the formation of disulfur cross-links between the carbon chains that make the rubber stronger. The shape of the molecule resembles that of hydrogen peroxide (Figure 16.29).

Surprisingly, no compound of sulfur and chlorine containing a higher sulfur oxidation state than +2 is stable at room temperature. If chlorine is bubbled through disulfur dichloride in the presence of catalytic diiodine, sulfur dichloride, SCl_2, is formed:

$$S_2Cl_2(l) + Cl_2(g) \xrightarrow{\ I_2\ } 2\ SCl_2(l)$$

This foul-smelling red liquid is used in the manufacture of a number of sulfur-containing compounds, including the notorious mustard gas, $S(CH_2CH_2Cl)_2$. Mustard gas was used in the First World War and, more recently, by the Iraqi government against some of its citizens. Liquid droplets containing this gas cause severe blistering of the skin, followed by death. As predicted by VSEPR theory, the sulfur dichloride molecule is V-shaped (Figure 16.30).

16.26   *Sulfur–Nitrogen Compounds*

There are several sulfur–nitrogen compounds. Some of these are of interest because their shapes and bond lengths cannot be explained in terms of simple bonding theory. The classic example is tetrasulfur tetranitride, S_4N_4. Unlike the crown structure of octasulfur, tetrasulfur tetranitride has a closed, basket-like shape, with multiple bonding around the ring and weak bonds cross-linking the pairs of sulfur atoms (Figure 16.31).

Of much more interest, however, is the polymer $(SN)_x$, commonly called polythiazyl. This bronze-colored, metallic-looking compound was first synthesized in 1910; yet it was not until 50 years later that an investigation of its properties showed it to be an excellent electrical conductor. In fact, at very low temperatures (0.26 K), it becomes a superconductor. There is an intense interest in making related nonmetallic compounds that have metallic properties, both because of their potential for use in our everyday lives and because they may help us develop the theory of metals and superconductivity.

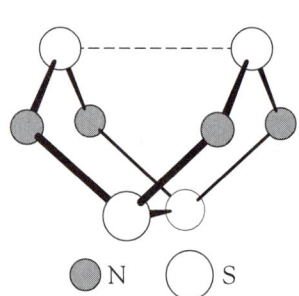

N ◯ S

Figure 16.31   The tetrasulfur tetranitride molecule.

16.27   *Selenium*

Until the 1960s, the only major use of selenium was as a glass additive. Addition of cadmium selenide, CdSe, to a glass mixture results in a pure ruby-red color that is much valued by glass artisans. Cadmium selenide is a semiconductor compound used in photocells because its electrical conductivity varies as a function of the light intensity to which it is exposed.

It was the invention of xerography (from the Greek *xero*, "dry," *graphy*, "writing") as a means for duplicating documents that turned an element of little interest to one that affects everyone's life. Xerography is made feasible by the unique photoconducting properties of selenium. The heart of a photocopier (and a laser printer) is a drum coated with selenium. The surface is charged in an electric field of about 10^5 V·cm^{-1}. The areas exposed to a high light intensity (the white areas of the image) lose their charge as a result of photoconductivity. Toner powder then adheres to the charged areas of the drum (corresponding to the black parts of the image). In the next step, the toner is transferred to the paper, where a heat source melts the particles, bonding them to the paper fibers and producing the photocopy. Tellurium is used in color photocopiers to alter the color sensitivity of the drum.

16.28 *Biological Aspects*

Oxygen: The Most Essential Element

We can live without food for days, without water for hours or days (depending on the temperature); but without dioxygen, life ceases in a very short time. We inhale about 10 000 L of air per day, from which we absorb about 500 L of oxygen gas. The dioxygen molecule bonds at the lung surface to a hemoglobin molecule; in fact, an oxygen molecule covalently bonds to each of the four iron atoms in a hemoglobin molecule. The process of uptake is amazing in that once the first dioxygen molecule is bonded to an iron atom, the ease of bonding of the second dioxygen is increased, as is the third and fourth, in turn. This *cooperative effect* contrasts strikingly with the normal chemical equilibria, in which successive steps are usually less favored.

The hemoglobin transports the dioxygen to the muscles and other energy-utilizing tissues, where it is transferred to myoglobin molecules. The myoglobin molecule (similar to one of the units in hemoglobin) contains one iron ion, and it bonds with the dioxygen molecule even more strongly than the hemoglobin molecule does. Once the first dioxygen molecule is removed from the hemoglobin, the cooperative effect operates again, this time making it easier and easier to remove the remaining dioxygen molecules. The myoglobin molecules store the dioxygen until it is needed in the energy-producing redox reaction with sugars that provide the energy our bodies require to survive and function.

Sulfur: The Importance of Oxidation State

Ethanethiol, CH_3CH_2SH, listed by the Guinness Book of World Records *as the world's most evil-smelling substance, is added to odorless natural gas supplies so that we can discover leaks. Concentrations as low as 50 ppb are detectable by the human nose.*

Sulfur resembles nitrogen in that its biologically important oxidation state is the negative one. Just as aminoacids incorporate —NH$_2$ (Ox. No. $N = -3$), so they incorporate —SH (Ox. No. $S = -2$), the thiol unit in the important amino acid, cysteine, HSCH$_2$(NH$_3^+$)COO$^-$. The presence of sulfur enable this specific amino acid in a protein chain to cross-link to another as we mentioned in the feature "Disulfide Bonds and Hair" earlier in this chapter. Sulfur is also one of the most crucial coordination sites in proteins for metal ions. It bonds to the widest range of metal ions of any amino acid functional group. The metal ions might be expected to be those favoring a soft base, but, in fact, the sulfur is a strong bonding site for some metals one might think prefer hard bases. The metal ions are zinc(II), copper(I), copper(II), iron(II), iron(III), molybdenum(IV)–(VI), and nickel(I)–(III).

Other sulfur-containing biological molecules include vitamin B$_1$ (thiamine) and the coenzyme biotin (which, in spite of its name, does not contain tin). Furthermore, many of our antibiotics, such as penicillin, cephalosporin, and sulfanilamide are sulfur-containing substances. In the -2 oxidation state, the

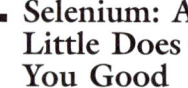

Figure 16.32 The molecule responsible for the eye irritation that accompanies the task of peeling onions.

majority of the simple sulfur-containing compounds have obnoxious odors. For example, the odorous molecules from onion, garlic, and skunk all contain sulfur in this oxidation state. Many of the naturally occurring sulfur-containing molecules involve rather bizarre chemical structures. For example, the lachrymatory (tear-inducing) factor in onions is the molecule depicted in Figure 16.32, containing the unusual C—S—O group.

Just as there are carbon, nitrogen, and phosphorus cycles, so there is a sulfur cycle. As indicated by the Pourbaix diagram (Figure 16.33), under the normal range of potential and pH, sulfate is the thermodynamically favored species.

$$HS^-(aq) + 4\,H_2O(l) \rightarrow SO_4^{2-}(aq) + 9\,H^+(aq) + 8\,e^-$$
$$\Delta G^\circ = -21.4 \text{ kJ·mol}^{-1}$$

If sulfur(VI) is thermodynamically preferred, an obvious question is how sulfur(−II) is such a common oxidation state. Organisms accomplish the reduction by coupling it with a strongly thermodynamically favored oxidation to give a net negative free energy change. A typical example is the oxidation of a carbohydrate to carbon dioxide:

$$C_6H_{12}O_6(aq) + 3\,SO_4^{2-}(aq) + 3\,H^+(aq) \rightarrow 6\,CO_2(g) + 3\,HS^-(aq) + 6\,H_2O(l)$$
$$\Delta G^\circ = -25.6 \text{ kJ·mol}^{-1}$$

Selenium: A Little Does You Good

Selenium is essential to health. It is utilized in enzymes and in amino acids such as selenomethionine. Among other roles, selenium compounds break down peroxides that will damage the cytoplasm in cells. Unfortunately, this element shows one of the narrowest ranges of tolerance. Clinical deficiency sets in below levels of about 0.05 ppm in food intake, but concentrations over 5 ppm cause chronic poisoning. (That is why selenium dietary supplements should be treated with extreme caution.) Sufferers from selenium poisoning produce the garlic-like odor of dimethylselenium, $(CH_3)_2Se$. In parts of the western-central United States selenium levels in the soil are quite high, and animals that graze on selenium-accumulating wild plants can suffer from selenium poisoning, known as "blind staggers" and "alkali disease."

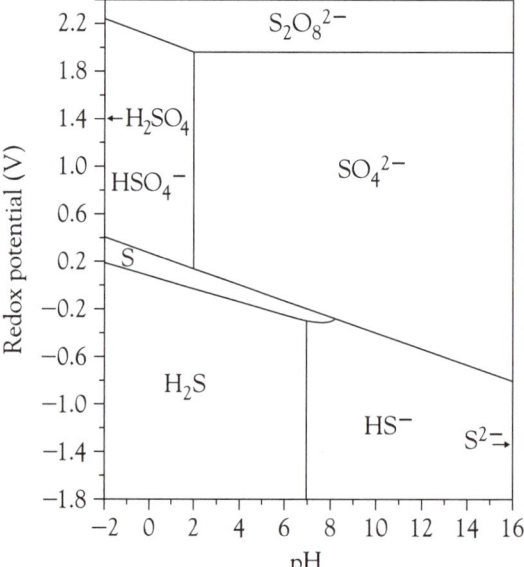

Figure 16.33 The Pourbaix diagram for sulfur. *(From W. Kaim and B. Schwederski, Bioinorganic Chemistry: Inorganic Elements in the Chemistry of Life, John Wiley, Chichester, 1994, p. 324.)*

The active ingredient in one type of antidandruff shampoo is selenium disulfide, SeS_2.

Selenium deficiency is much more prevalent than excess. In the United States, the Pacific northwest, the northeast, and Florida have soils that are very low in selenium. Animals whose diet is low in selenium suffer from muscular degeneration, known as "white muscle disease." For humans, the continentwide movement of foodstuffs normally provides an adequate level of selenium in a balanced diet. To ensure adequate selenium intake (about 100 mg per year), one should ensure having a diet containing selenium-rich foods such as mushrooms, garlic, asparagus, fish, and animal liver or kidney. A correlation has been shown between higher levels of selenium in water supplies and decreased incidence of breast and colon cancer.

It is in parts of China where the soils are almost totally selenium-deficient and where major human health problems exist. In addition to health problems directly attributable to selenium deficiency, the deficiency results in a lowered resistance to viral infection. Kashan disease, fatal and endemic across a broad band of rural China, results from inflammation of heart muscles by the family of Cocksackie viruses. These viruses are normally harmless but mutate into a virulent form in the weakened immune system of a selenium-deficient host. It has been suggested that the Asian origin of most new influenza viruses might have the same origin: that they similarly mutate in the selenium-deficient population of the region. Perhaps a massive international effort to eradicate selenium deficiency in China might have the side benefit to all humanity of reducing the incidence of new flu strains.

Though we talk about a deficiency in selenium, it is important to realize that, in reality, we are referring to compounds of selenium. In the typical biological range of E and pH, the most common species are those of selenium(IV), specifically the selenate ion, SeO_3^{2-} and the hydrogen selenate ion, $HSeO_3^-$.

16.29 Element Reaction Flowchart

Flowcharts are shown for both oxygen and sulfur, the two key elements in Group 16.

EXERCISES

16.1 Write balanced chemical equations for the following chemical reactions:
 (a) finely divided iron with dioxygen
 (b) solid barium sulfide with trioxygen
 (c) solid barium dioxide(2−) and water
 (d) potassium hydroxide solution with carbon dioxide

 (e) sodium sulfide solution with dilute sulfuric acid
 (f) sodium sulfite solution and sulfuric acid
 (g) sodium sulfite solution with cyclo-octasulfur

16.2 Write balanced chemical equations for the following chemical reactions:
 (a) heating potassium chlorate
 (b) solid iron(II) oxide with dilute hydrochloric acid

(c) iron(II) chloride solution with sodium hydroxide solution

(d) dihydrogen octasulfide with octasulfur dichloride in ethoxyethane

(e) heating sodium sulfate with carbon

(f) sulfur trioxide gas and liquid sulfuric acid

(g) peroxodisulfate ion with sulfide ion

16.3   Why is polonium the only element in this group to be classified as a metal?

16.4   Discuss the essential differences between oxygen and the other members of Group 16.

16.5   Define the following terms: (a) pyrophoric; (b) polymorphs; (c) cooperative effect.

16.6   Define the following terms: (a) mixed metal oxide; (b) vulcanization; (c) the Claus process.

16.7   Why is Earth's atmosphere so chemically different from that of Venus?

16.8   River and lake water are commonly used by electrical generating plants for cooling purposes. Why is this a potential problem for wildlife?

16.9   Predict the bond order in the trioxygen cation, O_3^+. Explain your reasoning. Is the ion paramagnetic or diamagnetic?

16.10   As we have seen, dioxygen forms two anions, O_2^- and O_2^{2-} with bond lengths of 133 and 149 pm, respectively; the length of the bond in the dioxygen molecule itself is 121 pm. In addition, dioxygen can form a cation, O_2^+. The bond length in this ion is 112 pm. Use a molecular orbital diagram to deduce the bond order and the number of unpaired electrons in the dioxygen cation. Is the bond order what you would expect for the bond length?

16.11   Dibromine oxide decomposes above 240°C. Would you expect the Br—O—Br bond angle to be larger or smaller than the Cl—O—Cl bond angle in dichlorine oxide? Explain your reasoning.

16.12   Osmium forms osmium(VIII) oxide, OsO_4, but the fluoride with the highest oxidation number of osmium is osmium(VII) fluoride, OsF_7. Suggest an explanation.

16.13   Suggest a structure for the O_2F_2 molecule, explaining your reasoning. Determine the oxidation number of oxygen in this compound and comment on it.

16.14   The mineral thortveitite, $Sc_2Si_2O_7$, contains the $[O_3Si—O—SiO_3]^{6-}$ ion. The Si—O—Si bond angle in this ion has the unusual value of 180°. Use hybridization concepts to account for this.

16.15   The compound $F_3C—O—O—O—CF_3$ is unusual for oxygen chemistry. Explain why.

16.16   Barium forms a sulfide of formula BaS_2. Use an oxidation number approach to account for the structure of this compound. Suggest why this compound exists, but not similar compounds with the other alkaline earth metals.

16.17   Draw structures of the following molecules and ions: (a) sulfuric acid; (b) the SF_5^- ion; (c) sulfur tetrafluoride; (d) the SOF_4 molecule. Hint: The oxygen is in the equatorial plane.

16.18   Draw structures of the following molecules and ions: (a) thiosulfate ion; (b) pyrosulfuric acid; (c) peroxodisulfuric acid; (d) the SO_2Cl_2 molecule.

16.19   Suggest a structure for the $S_4(NH)_4$ molecule. Explain your reasoning.

16.20   Disulfur difluoride, S_2F_2, rapidly converts to thio-thionylfluoride, SSF_2. Construct electron-dot diagrams for these two molecules. Use oxidation numbers to explain why this rearrangement would occur.

16.21   The unstable molecule, SO_4, contains a three-membered ring of the sulfur atom and two oxygen atoms. The other two oxygen atoms are doubly bonded to the sulfur atom. Draw an electron-dot formula for the compound. Then derive the oxidation states of each atom in this molecule and show that no abnormal oxidation states are involved.

16.22   Describe the hazards of (a) trioxygen; (b) hydroxide ion; (c) hydrogen sulfide.

16.23   Describe, using a chemical equation, why "whitewash" was such an effective and inexpensive painting material.

16.24   Although sulfur catenates, it does not have the extensive chemistry that is found for carbon. Explain briefly.

16.25   Describe the changes in cyclooctasulfur as it is heated. Explain the observations in terms of changes in molecular structure.

16.26   Describe the essential features of the Frasch and the Claus processes.

16.27   The bond angle in hydrogen telluride, H_2Te, is 89½°; that in water is 104½°. Suggest an explanation.

16.28   Explain why an aqueous solution of sodium sulfide has an odor of hydrogen sulfide.

16.29   Describe the five ways in which sulfuric acid can behave in chemical reactions.

16.30   Why must the formation of sulfur trioxide from sulfur dioxide be an exothermic reaction?

16.31   Which will have the stronger average sulfur–oxygen bond energy, sulfur trioxide or the sulfite ion? Use formal charges to justify your answer.

16.32   Suggest two alternative explanations why telluric acid has the formula H_6TeO_6 rather than H_2TeO_4, analogous to sulfuric and selenic acids.

16.33   Construct formal charge representations for (a) sulfate ion; (b) sulfurous acid.

16.34   Why is the sulfate anion commonly used in chemistry?

16.35   What are the chemical tests used to identify (a) hydrogen sulfide; (b) sulfate ion?

16.36   What are the major uses for (a) sulfur hexafluoride; (b) sodium thiosulfate?

16.37   Why is sulfur dioxide rather than sulfur trioxide the most common sulfur compound in the oxygen-rich atmosphere?

16.38   What would happen on this planet if hydrogen bonding ceased to exist between water molecules?

16.39   Whereas potassium ozonide, KO_3, is unstable and explosive, tetramethylammonium ozonide, $[(CH_3)N]O_3$, is stable up to 75°C. Suggest an explanation.

16.40   Suggest an explanation why sulfur hexafluoride sublimes at $-64°C$ while sulfur tetrafluoride boils at $-38°C$.

16.41   Draw the structure of the NS_2^+ ion. What neutral molecule is isoelectronic and isostructural with it?

16.42   Given that the S=S bond energy is 427 kJ·mol^{-1}, using data from Appendix 5, calculate the enthalpy of reaction for:

$$2\ X_{(g)} \rightarrow X_{2(g)}$$
$$8\ X_{(g)} \rightarrow X_{8(g)}$$

where X = oxygen and X = sulfur. Hence, show that the formation of the diatomic molecule is energetically preferable for oxygen, while it is the octamer that is favored for sulfur.

16.43   "Selenium is beneficial and toxic to life." Discuss this statement.

16.44   Write balanced chemical equations corresponding to each transformation in the Element Reaction Flowcharts.

BEYOND THE BASICS

16.45   Calculate the enthalpy of formation of (theoretical) gaseous sulfur hexachloride and compare it to the tabulated value for sulfur hexafluoride. Suggest why the values are so different.

16.46 S_2F_{10} is an unusual fluoride of sulfur. It consists of two SF_5 units joined by a sulfur–sulfur bond. Calculate the oxidation number for the sulfur atoms. This molecule disproportionates. Suggest the products and write a balanced equation. Use oxidation numbers to explain why the products would be those that you have selected.

16.47   Ammonium thioglycollate is used for hair straightening or curling, while calcium thioglycollate is used for hair removal. The ammonium glycollate does not act so drastically upon the hair because the solution is less basic than that of the calcium salt. Explain the reason for the difference in solution pH.

16.48   Use an approximate molecular orbital diagram to determine the bond order in the hydroxyl radical.

16.49   If the daytime hydroxyl radical concentration is 5×10^5 molecules·cm^{-3}, what concentration does that represent in ppm or ppb at SATP (25°C and 100 kPa)? The gas constant, R, is 8.31 kPa·L·mol^{-1}·K^{-1}.

16.50   (a) Sulfur tetrafluoride reacts with cesium fluoride to give an electrically conducting solution of a monatomic cation and a polyatomic anion containing the sulfur atom. Write a chemical equation for the reaction.
   (b) Sulfur tetrafluoride reacts with boron trifluoride to give an electrically conducting solution in which the polyatomic cation contains the sulfur. Write a chemical equation for the reaction.
   (c) Suggest an explanation why sulfur tetrafluoride behaved differently in the two reactions.

16.51   Calculate the length of side of the unit cell of the perovskite calcium titanate assuming
   (a) the titanium and oxide ions are in contact, or
   (b) the calcium and oxide ions are in contact.
   The radius of the Ti^{4+} ion is 74 pm.

16.52 The crucial reaction in polluted tropospheric air is:

$$NO_{2(g)} \xrightarrow{h\nu} NO_{(g)} + O_{(g)}$$

It is this energy input and the formation of reactive oxygen atoms that initiate most of the pollution chemistry cycles. Calculate the minimum energy needed for this process and hence the longest wavelength of light that could initiate this process. Also, show that the parallel reaction

$$CO_{2(g)} \xrightarrow{h\nu} CO_{(g)} + O_{(g)}$$

is not feasible in terms of the wavelength of light required.

16.53 Construct two possible electron-dot diagrams for the $SOCl_2$ molecule. Use formal charge principles to deduce which is the more likely. Then draw the structural formula and mark the approximate bond angles.

16.54 The bacteria in hydrothermal vents derive their metabolic energy from the oxidation of the hydrogen sulfide ion to the sulfate ion. Calculate the free energy change for this process at 298 K from standard reduction potentials in Appendix 3.

16.55 Sulfur dioxide is the most important contributor to acid rain. It can be removed from coal-burning power plants using calcium carbonate. Write a balanced equation for the reaction. If a coal contained 3.0 percent sulfur, what mass of calcium carbonate would be needed to react with the sulfur dioxide produced from 1000 tonnes of coal?

16.56 When hydrogen peroxide is added to a basic solution of potassium chromate, a compound of formula K_3CrO_8 is formed. Deduce the oxidation state of chromium in this compound, explaining your reasoning.

16.57 Another oxyacid of sulfur is H_2SO_5. Calculate the apparent oxidation number of sulfur in this compound. Is this oxidation number feasible? Suggest a structure of the compound that would give a common oxidation number for the sulfur atom. Hint: This acid reacts with water to form sulfuric acid and another product.

16.58 Which, dinitrogen trioxide or dinitrogen pentaoxide, is likely to be the more acidic oxide? Give your reasoning.

16.59 Write a balanced equation for the reaction of aqueous sulfite ion with aqueous peroxodisulfate ion.

16.60 The half-reaction for the reduction of dioxygen to water is

$$O_{2(g)} + 4\,H^+(aq) + 4\,e^- \rightarrow 2\,H_2O(l)$$

The standard potential provided in the Appendices is not relevant to normal atmospheric conditions of pO_2 of 20 kPa and a pH of about 7.00. Calculate the reduction potential under these more realistic conditions.

16.61 The standard electrode potential for the reduction of sulfate under acid conditions is given by:

$$SO_4{}^{2-}(aq) + 4\,H^+(aq) + 2\,e^- \rightarrow SO_2(aq) + 2\,H_2O(l)$$
$$E° = +0.17\text{ V}$$

Calculate the potential under basic conditions.

16.62 Identify each of the following reactants, writing balanced chemical equations for each reaction.
(a) A metal (A) reacts with water to give a colorless solution of compound (B) and a colorless gas (C). Common dilute diprotic acid (D) is added to (B), forming a dense white precipitate (E).
(b) A solution of (F) slowly decomposes to give a liquid (G) and a colorless gas (H). Gas (H) reacts with colorless gas (C) to give liquid (G).
(c) Under certain conditions, colorless acidic gas (I) will react with gas (H) to give a white solid (J). Addition of (G) to (J) gives a solution of acid (D).
(d) Metal (A) burns in excess gas (H) to give compound (K). Compound (K) dissolves in water to produce a solution of (B) and (F).

16.63 A gas (A) was bubbled into a solution of a common monopositive hydroxide (B) to give a solution of the salt (C). The cation of (B) gives a precipitate with the tetraphenylborate ion. Heating yellow solid (D) with a solution of (C) and evaporating the water gives crystals containing anion (E). Addition of iodine to a solution of anion (E) gives iodide ion and a solution of anion (F). Addition of hydrogen ion to a solution of anion (E) initially produces acid (G), which decomposes to form solid (D) and gas (A). Identify (A) through (G), writing balanced equations for each step.

16.64 Write a balanced equation for the reaction of pure liquid sulfuric acid with pure liquid perchloric acid (a stronger acid).

16.65 Construct an electron-dot structure for the NSF_3 molecule in which the nitrogen–sulfur bond is (a) double; (b) triple. Decide which structure is most likely the major contributor to the bonding on the basis of formal charge.

16.66 S_4N_4 can be fluorinated to give $S_4N_4F_4$. Are the fluorine atoms bonded to the sulfur or nitrogen atoms of the ring? Suggest an answer and explain your reasoning.

16.67 If you have a solution of sodium hydrogen sulfite and allow it to crystallize, you do not obtain that compound but sodium metabisulfite instead:

$$2\,Na^+(aq) + HSO_3^-(aq) \rightleftharpoons Na_2S_2O_5(s) + H_2O(l)$$

Suggest why this reaction occurs and suggest a cation that would probably enable a solid hydrogen sulfite to be crystallized.

16.68 Sulfur forms two unusual compounds with fluorine that have identical empirical formulas (structural isomers), FSSF and SSF$_2$. Draw electron-dot structures for these molecules and deduce the oxidation state of sulfur in each compound.

16.69 Sulfur forms two other unusual compounds with fluorine that have identical empirical formulas (structural isomers), SF$_2$ and F$_3$SSF. Draw electron-dot structures for these molecules and deduce the oxidation state of sulfur in each compound.

Chapter 17

The Group 17 Elements: The Halogens

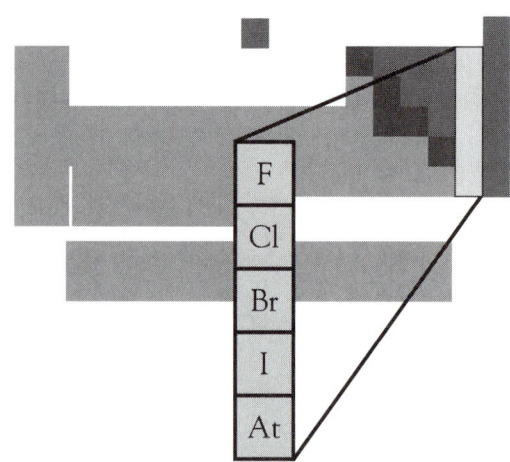

We started with the group containing the most reactive metals, and now we have reached the group containing the most reactive non-metals. Whereas the reactivity of the alkali metals increases down the group, the most reactive halogen is at the top of the group.

Each discovery of a new halogen provided a major advance in our knowledge of chemistry. For example, in the late 1700s, chemists believed that all acids contained oxygen. By this reasoning, hydrochloric acid (then known as muriatic acid) had to contain oxygen. When Scheele, in 1774, prepared a new green gas—dephlogisticated muriatic acid—from hydrochloric acid, it was argued by Lavoisier and most chemists that the substance (chlorine gas) was simply a new compound, containing more oxygen than the muriatic acid itself. This misconception lasted until 1810, when Davy showed that the gas was indeed a new element, chlorine, and in the process overthrew the definition of an acid.

The discovery of iodine involved the field of natural products chemistry, a most important research area today. It had been known for possibly thousands of years that ingestion of burnt sponge was an effective treatment for goiter. Physicians at the time wanted to know what it was in the sponge that actually provided the cure, particularly because ingestion of the entire sponge could also cause the side effect of severe stomach cramps. In 1819, the French chemist J.F. Coindet showed that the beneficial ingredient was iodine and that potassium iodide would produce the same benefits without the side effects. Even today, we consume "iodized salt" as a goiter preventive.

Bromine was the next halogen to be discovered, and the significance of this discovery lies in the fact that three elements—chlorine, bromine, and iodine—were thereby shown to have similar properties. It was one of the first indications that there were patterns in the properties of elements. The identification by the German chemist Döbereiner between 1817 and 1829 of such groups of three elements, or "triads," was one of the first steps toward the discovery of the periodicity of chemical elements.

Fluorine proved to be the most elusive. Many, many attempts were made in the 19th century to obtain this reactive element from its compounds. Hydrogen fluoride, an incredibly poisonous substance, was often used as the starting material. At least two chemists died from inhaling the fumes, and many more suffered lifelong pain from damaged lungs. It was the French chemist Henri Moissan, together with the laboratory assistance of his spouse, Léonie Lugan, who devised an electrolytic apparatus for its synthesis in 1886. Moissan, who received the Nobel Prize for his discovery of fluorine, died prematurely, himself a victim of hydrogen fluoride poisoning.

17.1 Group Trends

Under normal conditions (SATP), fluorine is a pale yellow gas (although many texts erroneously refer to it as pale green or colorless); chlorine is a pale green gas; bromine, an oily, red-brown liquid; and iodine, a black, metallic-looking solid. Bromine is the only nonmetallic element that is liquid at room temperature. The vapor pressures of bromine and iodine are quite high. Thus, toxic, red-brown bromine vapor is apparent when a container of bromine is opened; and toxic violet vapors are produced whenever iodine crystals are gently heated. Although iodine looks metallic, it behaves like a typical nonmetal in most of its chemistry.

As before, we will ignore the chemistry of the radioactive member of the group, in this case, astatine. All of astatine's isotopes have very short half-lives; hence, they emit high-intensity radiation. Nevertheless, the chemistry of this element has been shown to follow the trends seen in the other members of this group. Astatine is formed as a rare decay product from isotopes of uranium. Astatine is probably the rarest element on Earth; the top 1 km of Earth's entire crust is estimated to contain no more than 44 mg of astatine. Despite this incredibly low concentration, one of the long-overlooked pioneers of radiochemistry, the Austrian scientist Berta Karlik and her student Trudy Beinert actually managed to prove the existence of this element in nature.

Table 17.1 shows the smooth increase in the melting and boiling points of these nonmetallic elements. Because the diatomic halogens possess only

Table 17.1 Melting and boiling points of the Group 17 elements

Element	Melting point (°C)	Boiling point (°C)	Number of electrons
F_2	−219	−188	18
Cl_2	−101	−34	34
Br_2	−7	+60	70
I_2	+114	+185	106

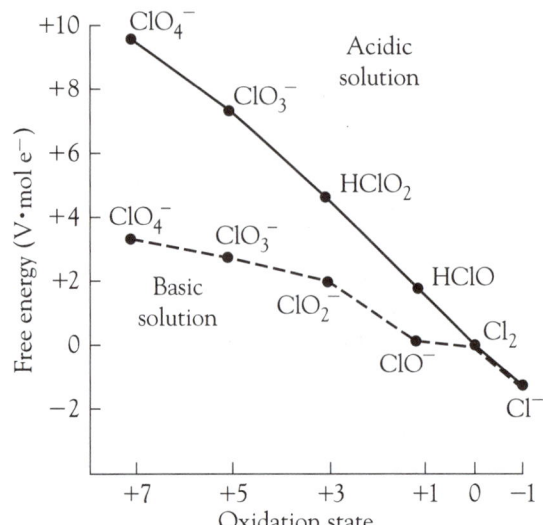

Figure 17.1 Frost diagram for chlorine in acidic and basic solutions.

dispersion forces between molecules, their melting and boiling points depend on the polarizability of the molecules, a property that, in turn, is dependent on the total number of electrons.

The oxidation state of fluorine is always −1, whereas the other halogens have common oxidation numbers of −1, +1, +3, +5, and +7. As the oxidation state diagram in Figure 17.1 shows, the higher the oxidation state, the stronger the oxidizing ability. Whatever its positive oxidation state, a halogen atom is more oxidizing in an acidic solution than in a basic solution. The chloride ion is the most stable chlorine species, for the dichlorine molecule can be reduced to the chloride ion in both acidic and basic solutions. In basic solution, the convex point on which the dichlorine molecule is located indicates that it will undergo disproportionation to the chloride and hypochlorite ions.

All the halogens have odd atomic numbers; hence, as discussed in Chapter 2, Section 2.3, they are expected to have few naturally occurring isotopes. In fact, fluorine and iodine each have only one isotope; chlorine has two (76 percent chlorine-35; 24 percent chlorine-37), and bromine also has two (51 percent bromine-79; 49 percent bromine-81).

The formulas of the oxyacids in which the halogen atom is in its highest oxidation state parallel those of the Group 16 acids, with the Period 5 member (iodine) having a structure different from those of the lighter elements of the group. Thus, the structure of perchloric acid can be represented as $(HO)ClO_3$; and perbromic acid, as $(HO)BrO_3$. Periodic acid, however, has the structure of $(HO)_5IO$ (or H_5IO_6), similar to that of isoelectronic telluric acid, H_6TeO_6.

Table 17.2 Bond energies of the Group 17 elements

Element	Bond energy $(kJ \cdot mol^{-1})$
F_2	155
Cl_2	240
Br_2	190
I_2	149

17.2   *Anomalous Nature of Fluorine*

As previously noted for nitrogen and oxygen, the Period 2 members owe much of their distinctiveness to their bonding limitations. However, fluorine has some additional unique features.

The Weakness of the Fluorine–Fluorine Bond

The bond energies of the halogens from chlorine to iodine show a systematic decrease, but the bond energy of fluorine does not fit the pattern (Table 17.2). To fit the trend, we would expect a fluorine–fluorine bond energy of about 300 $kJ \cdot mol^{-1}$ rather than the actual value of 155 $kJ \cdot mol^{-1}$.

The Fluoridation of Water

A dentist, Frederick McKay, noticed the remarkable lack of cavities in the population in the Colorado Springs, Colorado, area in 1902. He tracked down the apparent cause as being the higher than average levels of fluoride ion in the drinking water. We now know that a concentration of about 1 ppm is required to convert the softer tooth material hydroxyapatite, $Ca_5(PO_4)_3(OH)$, to the tougher fluoroapatite, $Ca_5(PO_4)_3F$. A higher concentration of 2 ppm results in a brown mottling of the teeth; and at 50 ppm, toxic health effects occur. As a means of minimizing dental decay, health authorities around the world, including the American Dental Association, have recommended that fluoride levels in drinking water supplies be optimized to a level of about 1 ppm. The first city in the world to add a controlled level of fluoride ion was Grand Rapids, Michigan, in 1945. Many parts of the world have natural fluoride levels in excess of the recommended value. For example, in parts of Texas, the natural levels are over 2 ppm, while water sources in parts of Africa and Asia have dangerous levels in excess of 20 ppm.

The major concern of most health experts is not fluoridated water but fluoride toothpaste. The toothpaste contains high concentrations of fluoride ion designed to surface-coat the tooth with the fluoro-compound. Unfortunately young children have a predilection for swallowing toothpaste and thus massively exceeding the recommended fluoride intake. It is for this reason that toothpaste tubes carry such warnings as: "To prevent swallowing, children under six years of age should only use a pea-sized amount and be supervised during brushing" and "Warning: Keep [tube] out of reach of children under 6 years of age." It is commonly recommended that fluoride toothpaste not be used with children under 2 years of age as the lower the age, the higher the proportion of toothpaste swallowed. The overconsumption of fluoride toothpaste by children is generally believed to be responsible for a rise in dental fluorosis, a harmless but unsightly mottling of teeth.

Though many countries supplement the fluoride levels in water as a means of reducing dental decay, some countries have added fluoride ion to table salt as a simpler means of increasing fluoride intake. However, this route means that everyone gets supplemental fluoride in addition to whatever is the natural level in the water supply.

The traditional compound used for fluoridation was sodium fluoride; however, this has largely been replaced by fluorosilicic acid, H_2SiF_6, or its sodium salt. Fluorosilicic acid, a by-product of the phosphate mining industry, hydrolyzes in very dilute solution to give silicic acid, H_4SiO_4, hydronium ion and fluoride ion. It is the level of fluoridation that is most questionable. A concentration of 1 ppm was optimum if that was the only source of intake. Now with other sources such as toothpaste and foodstuffs processed in fluoride-supplemented water, it is generally considered that a lower concentration of about 0.7 ppm is preferable.

Although a number of reasons have been suggested, most chemists believe that the weak fluorine–fluorine bond is a result of repulsions between the nonbonded electrons of the two atoms of the molecule. It accounts in part for the high reactivity of fluorine gas.

Bonding Limitations

Like the other Period 2 elements, fluorine is limited to an octet of electrons in its covalent compounds. Thus, fluorine almost always forms just one covalent bond, one of the few exceptions being the H_2F^+ ion.

The High Electronegativity of Fluorine

With its very high electronegativity, the fluorine atom forms the strongest hydrogen bonds of any element. Apart from its major effect on the melting and boiling points of hydrogen fluoride, the hydrogen bonding results in the formation of a very stable polyatomic anion, HF_2^-.

The Ionic Nature of Many Fluorides

Metals in "normal" oxidation states form fluorides that often are ionic, whereas their equivalent compounds with chlorine, bromine, and iodine are covalent. This difference is due to the low polarizability of the small fluoride ion. For example, aluminum fluoride is ionic in its behavior, whereas the other aluminum halides show appreciable covalent behavior.

The High Oxidation States Found in Fluorides

Fluorine, being a very strong oxidizing agent, often "brings out" a higher oxidation state in a metal than the other halogens do. For example, vanadium forms vanadium(V) fluoride, VF_5 (oxidation state of vanadium, +5), but the highest oxidation state that vanadium achieves in compounds with chlorine is that in vanadium(IV) chloride, VCl_4 (oxidation state of vanadium, +4).

Differences in Fluoride Solubilities

Because the fluoride ion is much smaller than the other halide ions, the solubilities of metal fluorides differ from those of the other halides. For example, silver fluoride is soluble, whereas the other silver halides are not. Conversely, calcium fluoride is insoluble, whereas the other calcium halides are soluble. This pattern is a result of the differences in lattice energy between the metal fluorides and the other metal halides. Thus, the large silver ion has a relatively low lattice energy with the small fluoride ion. Conversely, the lattice energy is maximized when the small, high charge density calcium ion is matched with the small fluoride ion.

17.3   Fluorine

Difluorine is the most reactive element in the Periodic Table—in fact, it has been called the "Tyrannosaurus rex" of the elements. Fluorine gas is known to react with every other element in the Periodic Table except helium, neon, and argon. In the formation of fluorides, it is the enthalpy factor that is usually the predominant thermodynamic driving force.

For covalent fluorides, the negative enthalpy of formation is partially due to the weak fluorine–fluorine bond, which is readily broken, and partially to the extremely strong element–fluoride bond that is formed. For example, in Table 17.3 the bond energies of fluorine and chlorine, for bonds with themselves and with carbon, are compared. For ionic compounds, the crucial enthalpy factors are, again, the weak fluorine–fluorine bond that is broken and the high lattice energy due to the small, high charge density fluoride ion. The significant effect of the fluoride ion on lattice energies can be seen in a comparison of lattice energies for the sodium halides (Table 17.4), all of which adopt the NaCl structure.

Table 17.3 Comparison of fluorine and chlorine bond energies

Bond	Bond energy (kJ·mol^{-1})	Bond	Bond energy (kJ·mol^{-1})
F—F	155	C—F	485
Cl—Cl	240	C—Cl	327

Table 17.4 The lattice energies of sodium halides

Halide	Lattice energy (kJ·mol^{-1})
NaF	915
NaCl	781
NaBr	743
NaI	699

The synthesis of fluorides usually produces compounds whose other element has a high oxidation state. Thus, powdered iron burns spectacularly in fluorine gas to form iron(III) fluoride, not iron(II) fluoride:

$$2 \, Fe(s) + 3 \, F_2(g) \rightarrow 2 \, FeF_3(s)$$

Similarly, sulfur burns brightly to give sulfur hexafluoride:

$$S(s) + 3 \, F_2(g) \rightarrow SF_6(g)$$

About 40 percent of the industrially produced fluorine gas is used to prepare sulfur hexafluoride.

Fluorine oxidizes water to oxygen gas while simultaneously being reduced to fluoride ion:

$$F_2(g) + 2 \, e^- \rightarrow 2 \, F^-(aq) \qquad E° = +2.87 \, V$$

$$2 \, H_2O(l) \rightarrow 4 \, H^+(aq) + O_2(g) + 4 \, e^- \qquad E° = -1.23 \, V$$

We can explain the reason for such a high fluorine reduction potential by comparing the free energy cycles for the formation of the hydrated fluoride and chloride ions from their respective gaseous elements (Figure 17.2). The first step is (half of) the bond dissociation free energy, for which chlorine has the slightly higher value. In the next step, the energy required is the electron affinity, and the value for chlorine is again slightly higher, thus almost canceling out the fluorine advantage in the first step. It is the third step, the hydration of the respective ions, for which the free energy change of the fluoride ion is very much greater than that of the chloride ion. This large free energy change results from the strong interactions of the high charge density fluoride ion with surrounding water molecules to give a network of F···H—O hydrogen bonds. Because $\Delta G° = -nFE°$, the large free energy of reduction directly translates to a very positive standard reduction potential, thus accounting for the great strength of difluorine as an oxidizing agent.

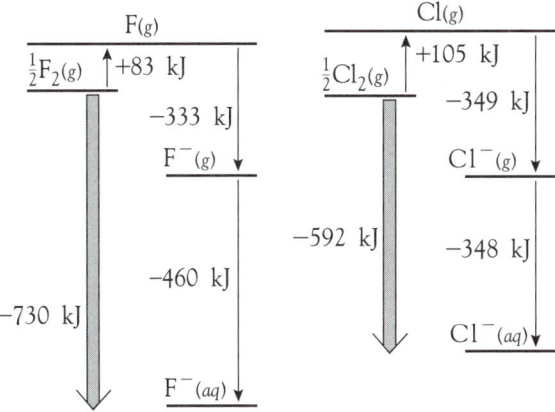

Figure 17.2 The free energy terms in the reduction of difluorine and dichlorine to the aqueous fluoride and chloride ions, respectively.

Difluorine is still produced by the Moissan electrochemical method, a process devised 100 years ago. The cells can be laboratory-size, running at currents between 10 and 50 A, or industrial-size, with currents up to 15 000 A. The cell contains a molten mixture of potassium fluoride and hydrogen fluoride in a 1:2 ratio and operates at about 90°C. The jacket of the apparatus is used to heat up the cell initially and then to cool it as the exothermic electrolysis occurs. At the central carbon anode, fluoride ion is

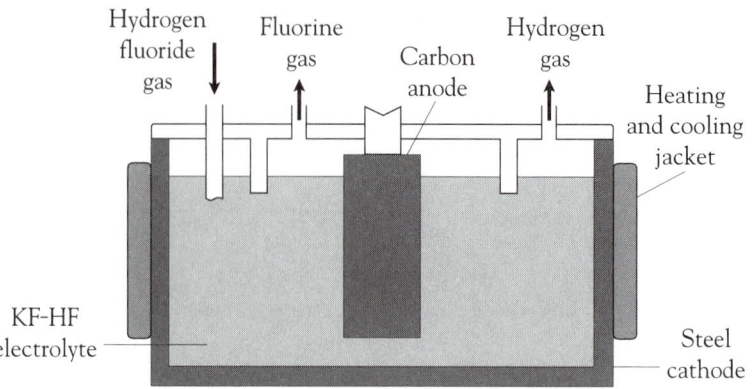

Figure 17.3 The cell used in the production of difluorine.

oxidized to fluorine, and at the steel cathode walls of the container, hydrogen gas is produced (Figure 17.3):

$$2\ F^-(g) \rightarrow F_2(g) + 2\ e^-$$

$$2\ H^+(g) + 2\ e^- \rightarrow H_2(g)$$

Hydrogen fluoride gas must be bubbled into the cell continuously to replace that used in the process. Annual production is at least 10^4 tonnes. We mentioned earlier that a significant proportion of fluorine is used in the synthesis of sulfur hexafluoride. The other major use is the preparation of uranium(VI) fluoride. This low-boiling-point uranium(VI) halide is used to separate the isotopes of uranium; uranium-235 is used in the manufacture of bombs and also in certain types of nuclear reactors. Uranium(VI) fluoride is prepared in two steps. Uranium(IV) oxide, UO_2, reacts with hydrogen fluoride to give uranium(IV) fluoride, UF_4:

$$UO_2(s) + 4\ HF(g) \xrightarrow{\Delta} UF_4(s) + 2\ H_2O(g)$$

As expected, this lower oxidation state fluoride is an ionic compound and thus a solid. The uranium(IV) fluoride is then treated with difluorine gas to oxidize the uranium to the +6 oxidation state:

$$UF_4(s) + F_2(g) \rightarrow UF_6(g)$$

In such a high oxidation state, even this fluorine compound exhibits covalent properties, such as a sublimation temperature of about 60°C. We discuss the extraction of uranium in greater detail in Chapter 23, Section 23.4.

17.4   *Chlorine*

Chlorine gas is very poisonous; a concentration of over 30 ppm is lethal after a 30-minute exposure. It is the dense, toxic nature of dichlorine that led it to be used as the first wartime poison gas. In 1915, as a result of a German chlorine gas attack, 20 000 Allied soldiers were incapacitated and 5000 of them died. By contrast, the toxicity of low concentrations of chlorine toward microorganisms has saved millions of human lives. It is through chlorination that waterborne disease-causing organisms have been virtually eradicated from domestic water supplies in Western countries. Curiously, there used to be a great enthusiasm for the benefits of chlorine gas. President

Calvin Coolidge, among others, used sojourns in a "chlorine chamber" as a means of alleviating his colds. It is probable that many who tried this "cure" finished up with long-term lung damage instead.

This pale green, dense, toxic gas is also very reactive, although not as reactive as fluorine. It reacts with many elements, usually to give the higher common oxidation state of the element. For example, iron burns to give iron(III) chloride, not iron(II) chloride; phosphorus burns in excess chlorine to give phosphorus pentachloride:

$$2\ \text{Fe}(s) + 3\ \text{Cl}_2(g) \rightarrow 2\ \text{FeCl}_3(s)$$

$$2\ \text{P}(s) + 5\ \text{Cl}_2(g) \rightarrow 2\ \text{PCl}_5(s)$$

However, particularly with nonmetals, the highest oxidation state of the nonmetal in chlorides is usually much lower than its oxidation state in the equivalent fluoride. Thus, as discussed in Chapter 16, Section 16.25, sulfur dichloride, SCl_2 (oxidation number of sulfur, $+2$), is the chloride with the highest oxidation state of the sulfur atom at room temperature, whereas sulfur hexafluoride, SF_6 (oxidation number of sulfur, $+6$), is obtainable with the strongly oxidizing fluorine.

Chlorine gas is most easily prepared in the laboratory by adding concentrated hydrochloric acid to solid potassium permanganate. The chloride ion is oxidized to dichlorine and the permanganate ion is reduced to the manganese(II) ion:

$$2\ \text{HCl}(aq) \rightarrow 2\ \text{H}^+(aq) + \text{Cl}_2(g) + 2\ \text{e}^-$$

$$\text{MnO}_4^-(aq) + 8\ \text{H}^+(aq) + 5\ \text{e}^- \rightarrow \text{Mn}^{2+}(aq) + 4\ \text{H}_2\text{O}(l)$$

Dichlorine can act as a chlorinating agent. For example, mixing ethene, C_2H_4, with dichlorine gives 1,2-dichloroethane, $C_2H_4Cl_2$:

$$\text{CH}_2{=}\text{CH}_2(g) + \text{Cl}_2(g) \rightarrow \text{CH}_2\text{Cl}{-}\text{CH}_2\text{Cl}(g)$$

Chlorine is a strong oxidizing agent, having a very positive standard reduction potential (although much less than that of difluorine):

$$\text{Cl}_2(aq) + 2\ \text{e}^- \rightarrow 2\ \text{Cl}^-(aq) \qquad\qquad E° = +1.36\ \text{V}$$

Dichlorine reacts with water to give a mixture of hydrochloric and hypochlorous acids:

$$\text{Cl}_2(aq) + \text{H}_2\text{O}(l) \rightleftharpoons \text{H}^+(aq) + \text{Cl}^-(aq) + \text{HClO}(aq)$$

At room temperature, a saturated solution of chlorine in water contains about two-thirds hydrated dichlorine molecules and one-third of the acid mixture. It is the hypochlorite ion in equilibrium with the hypochlorous acid, rather than chlorine itself, that is used as an active oxidizing (bleaching) agent:

$$\text{HClO}(aq) + \text{H}_2\text{O}(l) \rightleftharpoons \text{H}_3\text{O}^+(aq) + \text{ClO}^-(aq)$$

The industrial preparation of chlorine is accomplished by electrolysis of aqueous sodium chloride solution (brine); the other product is sodium hydroxide. (This process was discussed in Chapter 11.) Chlorine is produced on a vast scale, about 10^8 tonnes annually, worldwide. The majority of the product is used for the synthesis of organochlorine compounds. And appreciable quantities are used in the pulp and paper industry to bleach paper, in water treatment, and in the production of titanium(IV) chloride, $TiCl_4$, an intermediate step in the extraction of titanium from its ores. There is increasing concern about the use of chlorine for both pulp and paper production and for water purification. In the former case, the dichlorine reacts with organic compounds in the wood pulp to produce toxic chloro compounds. These by-products end up in the wastewater and are pumped into rivers and seas. Because of increasingly stringent limits on the organochlorine levels in effluents from pulp mills, companies must now reduce such emissions to near-zero levels.

Chlorine has become a battleground between the chemical industry and the environmental movement. During recent years, there has been a movement to ban chlorine completely as a feedstock in industrial chemistry. Unfortunately, politics generally lead to the polarization of views rather than to a logical case-by-case consideration of whether a specific chlorine compound is preferable for a particular purpose, balancing cost, availability, and environmental impact. We noted previously that dichlorine is not necessary for pulp bleaching or even for water purification, yet it is hard to find a good replacement for polyvinyl chloride and for other chlorine-containing plastics, or for some chlorine-containing pharmaceuticals. In addition, a very large number of non-chlorine-containing chemicals are synthesized industrially from chlorine-containing starting materials. Finally, we have to keep in mind that many organo-chlorine compounds are produced naturally. For example, it is now believed that algae produce about 5 million tonnes of chloromethane per year, while fungi and lichens produce more complex organochlorine compounds.

17.5 *Hydrogen Fluoride and Hydrofluoric Acid*

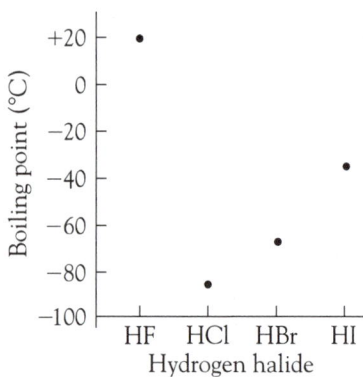

Figure 17.4 Boiling points of the hydrogen fluoride.

Figure 17.5 Hydrogen bonding in hydrogen fluoride.

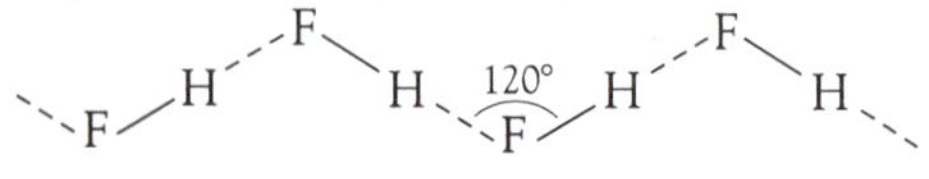

Hydrogen fluoride is a colorless, fuming liquid with a boiling point of 20°C. This is much higher than the boiling points of the other hydrogen halides, as can be seen from Figure 17.4. The high boiling point of hydrogen fluoride is a result of the very strong hydrogen bonding between neighboring hydrogen fluoride molecules. Fluorine has the highest electronegativity of all the elements, so the hydrogen bond formed with fluorine is the strongest of all. The hydrogen bonds are linear with respect to the hydrogen atoms but are oriented at 120° with respect to the fluorine atoms. Thus the molecules adopt a zigzag arrangement (Figure 17.5).

Hydrogen fluoride is miscible with water and forms hydrofluoric acid:

$$HF(aq) + H_2O(l) \rightleftharpoons H_3O^+(aq) + F^-(aq)$$

It is a weak acid with a pK_a of 3.2, unlike the other hydrohalic acids, which are very strong, all having negative pK_a values. As discussed in Chapter 7, Section 7.4, the relative weakness of hydrofluoric acid can be ascribed to the fact that the hydrogen–fluorine bond is much stronger than the other hydrogen–halide bonds. Thus, the dissociation into ions is less energetically favorable.

In more concentrated solutions, hydrofluoric acid ionizes to an even greater degree, the converse of the behavior of other acids. The cause of this behavior is well understood: a second equilibrium step that becomes more important at higher hydrogen fluoride concentrations and gives the linear hydrogen difluoride ion:

$$F^-(aq) + HF(aq) \rightleftharpoons HF_2^-(aq)$$

The hydrogen difluoride ion is so stable that alkali metal salts, such as potassium hydrogen difluoride, KHF_2, can be crystallized from solution.

This acid ion is unique, for it involves a bridging hydrogen atom. It used to be regarded as a hydrogen fluoride molecule with a fluoride ion hydrogen bonded to it. However, recent studies have shown that the hydrogen is centrally located between the two fluorine atoms.

Hydrofluoric acid is very corrosive, even though it is a weak acid. It is one of the few substances to attack glass and for this reason, hydrofluoric acid is always stored in plastic bottles. The reaction with glass produces the hexafluorosilicate ion, SiF_6^{2-}:

$$SiO_2(s) + 6\ HF(aq) \rightarrow SiF_6^{2-}(aq) + 2\ H^+(aq) + 2\ H_2O(l)$$

This property is used in the etching of glass. An object to be etched is dipped in molten wax, and the wax is allowed to harden. The required pattern is then cut through the wax layer. Dipping the glass in hydrofluoric acid enables the hydrofluoric acid to react with only those parts of the glass surface that are exposed. After a suitable depth of glass has been dissolved, the object is removed from the acid bath, rinsed with water, and the wax melted off, leaving the glass with the desired etched pattern.

About 10^6 tonnes of hydrofluoric acid is produced each year, worldwide. Hydrogen fluoride is produced by heating calcium fluoride, the mineral fluorspar, with concentrated sulfuric acid:

$$CaF_2(s) + H_2SO_4(l) \xrightarrow{\Delta} 2\ HF(g) + CaSO_4(s)$$

The product is either liquefied by refrigeration or added to water to give hydrofluoric acid. To lower the cost of this endothermic reaction, plants have been constructed next to sulfuric acid production facilities. The heat from the exothermic reactions in the sulfuric acid plant is then used for the hydrogen fluoride process.

Obviously, in the production of a substance as toxic as hydrofluoric acid, the flue gases have to be carefully "scrubbed" to prevent traces of hydrogen fluoride from escaping into the environment. The other product in the reaction is the ubiquitous calcium sulfate. A simple stoichiometric calculation shows that, for every tonne of hydrogen fluoride, nearly 4 tonnes of calcium sulfate is produced. As with the other industrial processes that produce this by-product, some is utilized, but much of it is used as landfill.

We have discussed the use of hydrofluoric acid for glass etching, but, in fact, almost all the hydrofluoric acid produced commercially is used as the starting material for the synthesis of other fluorine-containing chemicals. For example, sodium fluoride is produced by mixing hydrofluoric acid with sodium hydroxide solution:

$$NaOH(aq) + HF(aq) \rightarrow NaF(aq) + H_2O(l)$$

Evaporation gives sodium fluoride crystals, a compound used for water fluoridation. The reaction between hydrofluoric acid and potassium hydroxide solution, in a $2:1$ mole ratio, gives the acid salt, potassium hydrogen fluoride, which is used in the manufacture of fluorine gas:

$$KOH(aq) + 2\ HF(aq) \rightarrow KHF_2(aq) + H_2O(l)$$

17.6   *Hydrochloric Acid*

Hydrogen chloride is extremely soluble in water; in fact, concentrated hydrochloric acid contains about 38 percent by mass of hydrogen chloride, a concentration of 12 mol·L^{-1}. This acid is a colorless liquid with a pronounced acidic odor, which is due to the equilibrium between gaseous and aqueous hydrogen chloride:

$$HCl(aq) \rightleftharpoons HCl(g)$$

The technical grade reagent often has a yellowish color from an iron(III) ion impurity.

In contrast to hydrofluoric acid, hydrochloric acid is a strong acid ($pK_a = -7$), ionizing almost completely:

$$HCl(aq) + H_2O(l) \rightarrow H_3O^+(aq) + Cl^-(aq)$$

As the oxidation state diagram shows (see Figure 17.1), the chloride ion is a very stable species. Hence, dilute hydrochloric acid is often the acid of choice over the oxidizing nitric acid and, to a lesser extent, sulfuric acid. For example, zinc metal will react with hydrochloric acid to give the zinc ion and hydrogen gas:

$$Zn(s) + 2 \, HCl(aq) \rightarrow ZnCl_2(aq) + H_2(g)$$

When zinc reacts with nitric acid, there is often some reduction of the nitrate ion to give nitrogen dioxide.

The traditional method of producing hydrochloric acid in the laboratory is to generate hydrogen chloride gas from the reaction of sodium chloride with sulfuric acid. In the first step, at 150°C, sodium hydrogen sulfate is formed:

$$NaCl(s) + H_2SO_4(l) \xrightarrow{\Delta} NaHSO_4(s) + HCl(g)$$

The mixture is heated to 550°C, at which temperature the sodium hydrogen sulfate reacts with an excess of sodium chloride to form the sodium sulfate and additional hydrogen chloride gas. The gas is then dissolved in water to form the acid:

$$NaHSO_4(s) + NaCl(s) \xrightarrow{\Delta} Na_2SO_4(s) + HCl(g)$$

This method is used in industry, too, and hydrogen chloride is also produced by the direct combination of dichlorine and dihydrogen gas:

$$H_2(g) + Cl_2(g) \rightarrow 2 \, HCl(g)$$

The largest proportion of hydrogen chloride produced actually is a byproduct from other industrial processes, such as the synthesis of carbon tetrachloride:

$$CH_4(g) + 4 \, Cl_2(g) \rightarrow CCl_4(l) + 4 \, HCl(g)$$

About 10^7 tonnes of hydrochloric acid is used worldwide every year. It has a wide range of uses: as a common acid, for removing rust from steel surfaces (a process called "pickling"); in the purification of glucose and corn syrup; in the acid treatment of oil and gas wells; and in the manufacture of chlorine-containing chemicals. The acid is available in many hardware stores under the archaic name of muriatic acid, its main home uses being for the cleaning of concrete surfaces and rust removal. A very small number of persons are unable to synthesize enough stomach acid, and these individuals must ingest capsules of dilute hydrochloric acid with every meal.

17.7 Halides

Ionic Halides

Most ionic chlorides, bromides, and iodides are soluble in water, dissolving to give the metal ion and the halide ion. However, many metal fluorides are insoluble. For example, as mentioned earlier, calcium chloride is very water soluble, whereas calcium fluoride is insoluble. We explain these observations in terms of the greater lattice energy in crystals containing the small, high charge density anion and the high charge density cation.

Solutions of soluble fluorides are basic because the fluoride ion is the conjugate base of the weak acid hydrofluoric acid:

$$F^-(aq) + H_2O(l) \rightleftharpoons HF(aq) + OH^-(aq)$$

There are two possible ways to form metal halides: combining metal and halogen to give a metal ion with the higher oxidation state and combining metal and hydrogen halide to give a metal ion with the lower oxidation state. The preparations of iron(III) chloride and iron(II) chloride illustrate this point:

$$2\ Fe(s) + 3\ Cl_2(g) \rightarrow 2\ FeCl_3(s)$$

$$Fe(s) + 2\ HCl(g) \rightarrow FeCl_2(s) + H_2(g)$$

In the first case, dichlorine is acting as a strong oxidizing agent, whereas in the second case the hydrogen ion is a weak oxidizing agent.

Hydrated metal halides can be prepared from the metal oxide, carbonate, or hydroxide and the appropriate hydrohalic acid. For example, magnesium chloride hexahydrate can be prepared from magnesium oxide and hydrochloric acid, followed by crystallization of the solution:

$$MgO(s) + 2\ HCl(aq) + 5\ H_2O(l) \rightarrow MgCl_2{\cdot}6H_2O(s)$$

In most cases, the anhydrous salt cannot be prepared by heating the hydrate, because decomposition occurs instead. Thus, magnesium chloride hexahydrate gives magnesium hydroxide chloride, Mg(OH)Cl, when heated:

$$MgCl_2{\cdot}6H_2O(s) \xrightarrow{\Delta} Mg(OH)Cl(s) + HCl(g) + 5\ H_2O(l)$$

To obtain the anhydrous chloride from the hydrate, we have to chemically remove the water. This can be done by using thionyl chloride, $SOCl_2$; the reaction gives sulfur dioxide and hydrogen chloride gases:

$$MgCl_2{\cdot}6H_2O(s) + 6\ SOCl_2(l) \rightarrow MgCl_2(s) + 6\ SO_2(g) + 12\ HCl(g)$$

This is a common way to dehydrate metal chlorides.

Not all metal iodides in which the metal ion takes its higher oxidation state can be prepared, because the iodide ion itself is a reducing agent. For example, iodide ion will reduce copper(II) ion to copper(I):

$$2\ Cu^{2+}(aq) + 4\ I^-(aq) \rightarrow 2\ CuI(s) + I_2(s)$$

As a result, copper(II) iodide does not exist.

The common test for distinguishing chloride, bromide, and iodide ions involves the addition of silver nitrate solution to give a precipitate. Using X^- to represent the halide ion, we can write a general equation:

$$X^-(aq) + Ag^+(aq) \rightarrow AgX(s)$$

Silver chloride is white; silver bromide is cream colored; silver iodide is yellow. Like most silver compounds, they are light sensitive, and, over a period of hours, the solids change to shades of gray as metallic silver forms.

To confirm the identity of the halogen, ammonia solution is added to the silver halide. A precipitate of silver chloride reacts with dilute ammonia to form the soluble diamminesilver(I) ion, $[Ag(NH_3)_2]^+$:

$$AgCl(s) + 2\,NH_3(aq) \rightarrow [Ag(NH_3)_2]^+(aq) + Cl^-(aq)$$

The other two silver halides are not attacked. Silver bromide does react with concentrated ammonia, but silver iodide remains unreactive even under these conditions.

There also are specific tests for each halide ion. The chloride ion test is quite hazardous, because it involves the reaction of the suspected chloride with a mixture of potassium dichromate and concentrated sulfuric acid. When warmed gently, the volatile, red, toxic compound chromyl chloride, CrO_2Cl_2, is produced:

$$K_2Cr_2O_7(s) + 4\,NaCl(s) + 6\,H_2SO_4(l) \rightarrow 2\,CrO_2Cl_2(l) + 2\,KHSO_4(s)$$
$$+ 4\,NaHSO_4(s) + 3\,H_2O(l)$$

The vapor can be bubbled into water, where it forms a yellow solution of chromic acid, H_2CrO_4:

$$CrO_2Cl_2(g) + 2\,H_2O(l) \rightarrow H_2CrO_4(aq) + 2\,HCl(aq)$$

To test for bromide ion and iodide ion, a solution of dichlorine in water (aqueous chlorine) is added to the halide ion solution. The appearance of a yellow to brown color suggests the presence of either of these ions:

$$Cl_2(aq) + 2\,Br^-(aq) \rightarrow Br_2(aq) + 2\,Cl^-(aq)$$

$$Cl_2(aq) + 2\,I^-(aq) \rightarrow I_2(aq) + 2\,Cl^-(aq)$$

To distinguish them, we rely on the fact that the halogens themselves are nonpolar molecules. Thus, they will "prefer" to dissolve in nonpolar or low-polarity solvents, such as carbon tetrachloride. If the brownish aqueous solution is shaken with such a solvent, the halogen should transfer to the low-polarity, nonaqueous layer. If the unknown is dibromine, the color will be brown, whereas that of diiodine will be bright purple.

There is another, very sensitive test for iodine: It reacts with starch to give a blue color (blue-black when concentrated solutions are used). In this unusual interaction, the starch polymer molecules wrap themselves around the iodine molecules. There is no actual chemical bond involved. The equilibrium is employed qualitatively in starch–iodide paper. When the paper is exposed to an oxidizing agent, the iodide is oxidized to iodine. The starch in the paper forms the starch–iodine complex, and the blue-black color is readily observed. Quantitatively, starch is used as the indicator in redox titrations involving the iodide–iodine redox reaction.

Diiodine, as already mentioned, is a nonpolar molecule. Thus, its solubility in water is extremely low. It will, however, "dissolve" in a solution of iodide ion. This is, in fact, a chemical reaction producing the triiodide ion, I_3^- (discussed later):

$$I_2(s) + I^-(aq) \rightleftharpoons I_3^-(aq)$$

The large, low-charge triiodide ion will actually form solid compounds with low charge density cations such as rubidium, with which it forms rubidium triiodide, RbI_3.

Iodide ion will also undergo a redox reaction with iodate ion, IO_3^-, in acid solution to give diiodine:

$$IO_3^-(aq) + 6\ H^+(aq) + 5\ I^-(aq) \rightarrow 3\ I_2(s) + 3\ H_2O(l)$$

This reaction is often used in titrimetric analysis of iodide solutions. The diiodine can then be titrated with thiosulfate ion of known concentration:

$$I_2(s) + 2\ S_2O_3^{2-}(aq) \rightarrow 2\ I^-(aq) + S_4O_6^{2-}(aq)$$

Covalent Halides

As a result of weak intermolecular forces, most covalent halides are gases or liquids with low boiling points. The boiling points of these nonpolar molecules are directly related to the strength of the dispersion forces between the molecules. This intermolecular force, in turn, is dependent on the number of electrons in the molecule. A typical series is that of the boron halides (Table 17.5), which illustrates the relationship between boiling point and number of electrons.

Many covalent halides can be prepared by treating the element with the appropriate halogen. When more than one compound can be formed, the mole ratio can be altered to favor one product over the other. For example, in the presence of excess chlorine, phosphorus forms phosphorus pentachloride, whereas in the presence of excess phosphorus, phosphorus trichloride is formed:

$$2\ P(s) + 5\ Cl_2(g) \rightarrow 2\ PCl_5(s)$$

$$2\ P(s) + 3\ Cl_2(g) \rightarrow 2\ PCl_3(l)$$

If the element itself is inert, such as dinitrogen, an alternative route must be used. For nitrogen, the preferred method is to mix ammonia and chlorine gas:

$$NH_3(g) + 3\ Cl_2(g) \rightarrow NCl_3(l) + 3\ HCl(g)$$

Most covalent halides react vigorously with water. For example, phosphorus trichloride reacts with water to give phosphonic acid and hydrogen chloride:

$$PCl_3(l) + 3\ H_2O(l) \rightarrow H_3PO_3(l) + 3\ HCl(g)$$

Table 17.5 Boiling points of the boron halides

Compound	Boiling point (°C)	Number of electrons
BF_3	-100	32
BCl_3	$+12$	56
BBr_3	$+91$	110
BI_3	$+210$	164

However, some covalent halides are kinetically inert, particularly the fluorides, such as carbon tetrafluoride and sulfur hexafluoride.

It is important to remember that metal halides can contain covalent bonds even when the metal is in a high oxidation state. For example, tin(IV) chloride behaves like a typical covalent halide. It is a liquid at room temperature, and it reacts violently with water:

$$SnCl_4(l) + 2\ H_2O(l) \rightarrow SnO_2(s) + 4\ HCl(g)$$

If a nonmetallic element exists in a number of possible oxidation states, then the highest oxidation state is usually stabilized by fluorine and the lowest by iodine. This pattern reflects the decreasing oxidizing ability of elements in Group 17 as the group is descended. However, we must always be careful with our application of simplistic arguments. For example, the nonexistence of phosphorus pentaiodide, PI_5, is more likely to be due to the fact that the size of the iodine atom limits the number of iodine atoms that will fit around the phosphorus atom rather than to the spontaneous reduction of phosphorus from the +5 to the +3 oxidation state.

17.8 Chlorine Oxides

Chlorine Monoxide

Many species that are too unstable to exist in significant concentrations at ambient temperature and pressure play important roles in atmospheric chemistry. For example, in Chapter 15, Section 15.11, we described how nitrogen trioxide acted as a night-time tropospheric detergent. Another important atmospheric molecule is chlorine monoxide, ClO. Chlorine monoxide is a key stratospheric species responsible for causing the "ozone hole," a decrease in uv-filtering ozone concentration over the south and, to a lesser extent, north polar regions during their respective Spring seasons.

The saga is believed to begin with the build-up of chlorine molecules, predominantly from the breakdown of CFCs, during the dark Winter in the isolated air mass over the Antarctic. With the arrival of Spring, sunlight causes the weakly bonded chlorine molecules (bond energy 242 kJ·mol^{-1}) to dissociate into chlorine atoms:

$$Cl_2 \xrightarrow{h\nu} 2\ Cl$$

The chlorine atoms react with ozone (trioxygen) to give chlorine monoxide and dioxygen:

$$Cl + O_3 \rightarrow ClO + O_2$$

If the reaction terminated here, the damage to the ozone layer would be minimal. However, the chlorine monoxide takes part in a reaction cycle that regenerates the chlorine atoms, causing this process to be catalytic. That is, the chlorine atom acts as a catalyst for the conversion of ozone to dioxygen. The first step in this process is the combination of two chlorine monoxide radicals to form the ClOOCl dimer molecule. However, dissociation would immediately occur unless the two radicals simultaneously collide with a third body, M. It is the role of the species M to remove the excess energy. The identity of M is any molecule that can remove the energy—usually dinitrogen, N_2, or dioxygen, O_2, as these are the most common atmospheric molecules.

$$ClO + ClO + M \rightarrow Cl_2O_2 + M^*$$

Sunlight again becomes involved, this time to fission the Cl_2O_2 molecule assymetrically:

$$Cl_2O_2 \xrightarrow{h\nu} ClOO + Cl$$

The ClOO species is very unstable and rapidly breaks down to give a chlorine atom and a dioxygen molecule:

$$ClOO \rightarrow Cl + O_2$$

Then the chlorine atoms are again available to react with ozone molecules. It is this catalytic cycle that results in the severe polar ozone depletion.

Chlorine Dioxide

Figure 17.6 A possible electron-dot representation of the bonding in chlorine dioxide.

Chlorine forms several isolable oxides, all of which have positive values of free energy of formation. As a result of this and a low activation energy of decomposition, they are very unstable and have a tendency to explode. Chlorine dioxide, ClO_2, is the only oxide of major interest. It is a yellow gas that condenses to a deep red liquid at 11°C. The compound is quite soluble in water, giving a fairly stable, green solution.

Chlorine dioxide is paramagnetic, like nitrogen dioxide. Yet it shows no tendency to dimerize. The chlorine–oxygen bond length is only 140 pm, much shorter than the 170 pm that is typical for a single bond length, and it is very close to that of a typical chlorine–oxygen double bond. A possible electron-dot structure reflecting this bond order is shown in Figure 17.6.

The bond angle in chlorine dioxide is 118°, a value suggesting that the σ bond framework involves sp^2 hybrid orbitals on the chlorine. If this were the case, two of these orbitals would be used in bonding to the oxygen atoms and the third would hold the lone pair (accounting for the approximately 120° bond angle). The odd electron would then be in the unhybridized chlorine p orbital. The technique of electron spin resonance (ESR) can be used to investigate the properties of unpaired electrons. In a magnetic field, unpaired electrons have a preference for one spin direction $(-\frac{1}{2})$ over the other $(+\frac{1}{2})$. By applying energy, in this case, in the microwave region of the spectrum, we can cause the electron to reverse its spin. The energy (or energies) needed to do this gives us information about the electron environment. When we study chlorine dioxide by this means, we find that the odd electron is indeed in a pure p orbital of the chlorine atom at right angles to the molecular plane. In addition, it spends about 15 percent of its time in a p orbital of each oxygen atom.

These findings provide an explanation for why nitrogen dioxide dimerizes and chlorine dioxide does not. The ESR results confirm that the unpaired electron of nitrogen dioxide is largely in an sp-type hybrid orbital of nitrogen (Chapter 15), although it spends more time delocalized on the oxygen atoms (about 25 percent on each) than it does in chlorine dioxide. However, in chlorine dioxide, the unhybridized p orbital has two lobes and is not oriented in a single direction like that of a hybrid orbital. Hence, it can be argued that the lone electron is less available for pairing with another lone electron in a second chlorine dioxide molecule. There are probably some chlorine d orbitals involved in the π bonding to account for the substantial double bond character.

Chlorine dioxide, usually diluted with dinitrogen or carbon dioxide for safety, is a very powerful oxidizing agent. For example, to bleach flour to make white bread, chlorine dioxide is 30 times more effective than dichlorine. Large quantities of chlorine dioxide also are used as dilute aqueous solutions for bleaching wood pulp to make white paper. In this role, it is preferred over dichlorine, because chlorine dioxide bleaches without significant formation of hazardous chlorinated wastes. Similarly, chlorine dioxide is being used increasingly for domestic water treatment, because, in this context too, it does not chlorinate hydrocarbon pollutants that are present in the water to any measurable extent. Hence, use of this reagent avoids the problems discussed earlier. Chlorine dioxide was used to destroy any anthrax spores contaminating Congressional offices during the anthrax-letter scare in 2001.

Thus, even though pure chlorine dioxide is explosive, it is of major industrial importance. About 10^6 tonnes is produced every year worldwide. It is difficult to determine the exact production total, because the gas is so hazardous that it is generally produced in comparatively small quantities at the sites where it is to be used. The synthetic reaction involves the reduction of chlorine in the $+5$ (ClO_3^-) oxidation state by chlorine in the -1 (Cl^-) oxidation state in very acid conditions to give chlorine in the $+4$ (ClO_2) and 0 (Cl_2) oxidation states:

$$2\ ClO_3^-(aq) + 4\ H^+(aq) + 2\ Cl^-(aq) \rightarrow 2\ ClO_2(g) + Cl_2(g) + 2\ H_2O(l)$$

In North America, sulfur dioxide is added to reduce (and remove) the dichlorine gas to chloride ion; the sulfur dioxide is simultaneously oxidized to sulfate:

$$SO_2(g) + 2\ H_2O(l) \rightarrow SO_4^{2-}(aq) + 4\ H^+(aq) + 2\ e^-$$

$$Cl_2(g) + 2\ e^- \rightarrow 2\ Cl^-(aq)$$

However, this process produces sodium sulfate waste. A German process separates the dichlorine from the chlorine dioxide, then reacts the dichlorine with hydrogen gas to produce hydrochloric acid. The acid can then be reused in the synthesis.

17.9   *Chlorine Oxyacids and Oxyanions*

Chlorine forms a series of oxyacids and oxyanions for each of its positive odd oxidation states from $+1$ to $+7$. The shapes of the ions (and related acids) are based on a tetrahedral arrangement around the chlorine atom (Figure 17.7). The short chlorine–oxygen bonds in each of the ions indicate that multiple bonding must be present, possibly involving some contribution to the π bonding by the full p orbitals on the oxygen atoms and empty d orbitals on the chlorine atom.

As discussed in Chapter 7, Section 7.4, acid strength increases as the number of oxygen atoms increases. Thus, hypochlorous acid is very weak;

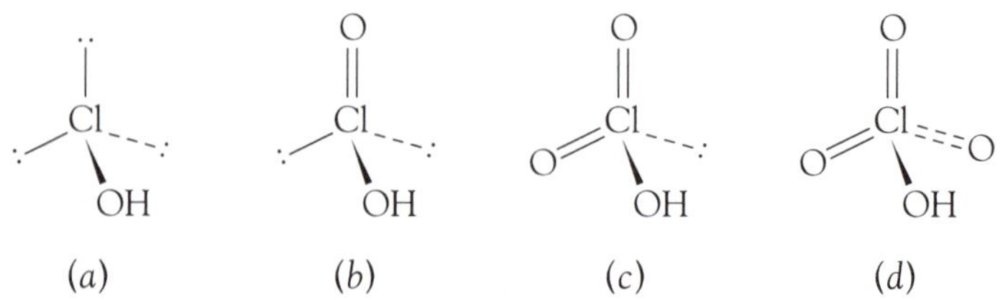

Figure 17.7   A possible representation of the bonding in (*a*) hypochlorous acid; (*b*) chlorous acid; (*c*) chloric acid; and perchloric acid.

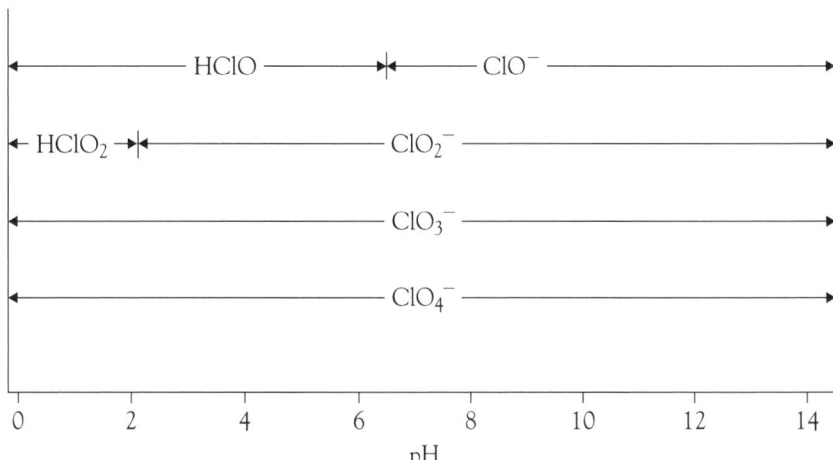

Figure 17.8 The pH predominance diagram for the chlorine oxyacids.

chlorous acid, weak; chloric acid, strong; and perchloric acid, very strong. The relative acid strengths can best be seen from the pH predominance diagram for the chlorine oxyacids (Figure 17.8).

Hypochlorous Acid and the Hypochlorite Ion

Hypochlorous acid is a very weak acid; thus, solutions of hypochlorites are very basic as a result of the hydrolysis reaction:

$$ClO^-(aq) + H_2O(l) \rightleftharpoons HClO(aq) + OH^-(aq)$$

Hypochlorous acid is a strong oxidizing agent and, in the process, is reduced to chlorine gas:

$$2\ HClO(aq) + 2\ H^+(aq) + 2\ e^- \rightarrow Cl_2(g) + 2\ H_2O(l) \quad E° = +1.64\ V$$

The hypochlorite ion, however, is a weaker oxidizing agent that is usually reduced to the chloride ion:

$$ClO^-(aq) + H_2O(l) + 2\ e^- \rightarrow Cl^-(aq) + 2\ OH^-(aq) \quad E° = +0.89\ V$$

It is this oxidizing (bleaching and bactericidal) power that renders the hypochlorite ion useful.

Hypochlorous acid and hydrochloric acid are formed when dichlorine is dissolved in cold water:

$$Cl_2(aq) + H_2O(l) \rightleftharpoons H^+(aq) + Cl^-(aq) + HClO(aq)$$

The two compounds of industrial importance are sodium hypochlorite and calcium hypochlorite. The former is prepared by the electrolysis of brine, in a manner similar to that used for the preparation of sodium hydroxide (see Chapter 11, Section 11.9). However, to produce sodium hypochlorite, the electrolysis is performed in a single chamber with stirrers to mix the hydroxide ion produced at the cathode with the chlorine produced at the anode:

$$2\ Cl^-(aq) \rightarrow Cl_2(aq) + 2\ e^- \qquad \text{(anode)}$$

$$2\ H_2O(l) + 2\ e^- \rightarrow H_2(g) + 2\ OH^-(aq) \qquad \text{(cathode)}$$

$$Cl_2(aq) + 2\ OH^-(aq) \rightarrow Cl^-(aq) + ClO^-(aq) + H_2O(l) \quad \text{(mixing reaction)}$$

The reaction vessel must be cooled, for as we will see in our discussion of the chlorate ion, a different reaction occurs in warm solutions. Commercial hypochlorite solutions such as those in Clorox® or Javex® contain an approximately equimolar mixture of sodium hypochlorite and sodium chloride.

Sodium hypochlorite is not stable in the solid phase; thus, calcium hypochlorite is often used as a solid source of hypochlorite ion. There are several processes for its synthesis, but the most elegant is the reaction between a suspension of calcium hydroxide and chlorine gas:

$$2\ Ca(OH)_2(s) + 2\ Cl_2(g) \rightarrow Ca(ClO)_2 \cdot 2\ H_2O(s) + CaCl_2(aq)$$

The calcium chloride produced is soluble, but calcium hypochlorite dihydrate is insoluble. Hence the latter can be separated by filtration to provide a high-purity product.

Sodium hypochlorite solution is used for bleaching and decolorization of wood pulp and textiles; both sodium and calcium hypochlorites are used in disinfection. The latter process has its hazards. Although the labels on sodium hypochlorite solution containers warn about the hazards of mixing cleansers, a knowledge of chemistry is required to understand the problem. As we mentioned, commercial sodium hypochlorite solution contains chloride ion. In the presence of hydrogen (hydronium) ion such as that in a sodium hydrogen sulfate–based cleanser, the hypochlorous acid reacts with the chloride ion to produce chlorine gas:

$$ClO^-(aq) + H^+(aq) \rightarrow HClO(aq)$$

$$HClO(aq) + Cl^-(aq) + H^+(aq) \rightleftharpoons Cl_2(g) + H_2O(l)$$

Several injuries and deaths have been caused by this simple redox reaction.

The Chlorate Ion

Although chlorites are of little interest, chlorates have several uses. Sodium chlorate can be prepared by bubbling dichlorine into a hot solution of sodium hydroxide. The sodium chloride, which is less soluble than sodium chlorate, precipitates:

$$3\ Cl_2(aq) + 6\ NaOH(aq) \xrightarrow{\Delta} NaClO_3(aq) + 5\ NaCl(s) + 3\ H_2O(l)$$

Industrially, the process is performed in a cell like that used for the production of sodium hypochlorite, except hot solution conditions are employed.

Potassium chlorate is used in large quantities to make matches and fireworks. Like all chlorates, it is a strong oxidizing agent that can explode unpredictably when mixed with reducing agents. Considerable amounts of sodium chlorate are consumed in the production of chlorine dioxide.

Chlorates decompose when heated, although in an unusual manner. The route for producing potassium chlorate has been studied in the most detail. When heated to temperatures below 370°C, disproportionation occurs to give potassium chloride and potassium perchlorate:

$$4\ KClO_3(l) \xrightarrow{\Delta} KCl(s) + 3\ KClO_4(s)$$

This is a synthetic route for the perchlorate. When potassium chlorate is heated above 370°C, the perchlorate that is formed by disproportionation decomposes:

$$KClO_4(s) \xrightarrow{\Delta} KCl(s) + 2\ O_2(g)$$

Swimming Pool Chemistry

In North America, we usually rely on dichlorine or chlorine-based compounds, such as calcium hypochlorite, to destroy the microorganisms in our swimming pools. In fact, the most potent disinfectant is hypochlorous acid. In many public pools, this compound is formed when dichlorine reacts with water:

$$Cl_2(aq) + H_2O(l) \rightleftharpoons H^+(aq) + Cl^-(aq) + HClO(aq)$$

To neutralize the hydronium ion, sodium carbonate (soda ash) is added:

$$CO_3^{2-}(aq) + H^+(aq) \rightarrow HCO_3^-(aq)$$

As a secondary result of this addition, the chlorine equilibrium shifts to the right, thus providing more hypochlorous acid.

In smaller pools, the hydrolysis of the hypochlorite ion provides the hypochlorous acid:

$$ClO^-(aq) + H_2O(l) \rightarrow HClO(aq) + OH^-(aq)$$

Acid must then be added to reduce the pH:

$$H^+(aq) + OH^-(aq) \rightarrow H_2O(l)$$

It is important to keep the hypochlorous acid concentrations at levels high enough to protect against bacteria and other pool organisms. This is a particularly difficult task in outdoor pools because hypochlorous acid decomposes in bright light and at high temperatures:

$$2\ HClO(aq) \rightarrow 2\ HCl(aq) + O_2(g)$$

The production of stinging eyes in a swimming pool is usually blamed on "too much chlorine." In fact, it is the converse problem, for the irritated eyes can be caused by the presence of chloramines in the water, such as NH_2Cl. These nasty compounds are formed through the reaction of hypochlorous acid with ammonia-related compounds, such as urea from urine, provided by the bathers:

$$NH_3(aq) + HClO(aq) \rightarrow NH_2Cl(aq) + H_2O(l)$$

To destroy them, we need to add more dichlorine, a process known as superchlorination. This additional dichlorine will react with the chloramines, decomposing them to give hydrochloric acid and dinitrogen:

$$2\ NH_2Cl(aq) + Cl_2(aq) \rightarrow N_2(g) + 4\ HCl(aq)$$

An increasingly popular approach is to use trioxygen (ozone) as the primary bactericide. This disinfectant causes much less eye irritation. However, because trioxygen slowly decomposes into dioxygen, a low level of chlorine-based compounds has to be added to the water to maintain safe conditions.

The pathway for the slow, uncatalyzed reaction is different from that for the reaction catalyzed by manganese(IV) oxide. When catalyzed, the pathway giving potassium chloride and dioxygen involves potassium permanganate (which produces a purple color) and potassium manganate, K_2MnO_4. The mechanism is a nice illustration of the chemical participation of a catalyst.

$$2\ KClO_3(s) + 2\ MnO_2(s) \rightarrow 2\ KMnO_4(s) + Cl_2(g) + O_2(g)$$
$$2\ KMnO_4(s) \rightarrow K_2MnO_4(s) + MnO_2(s) + O_2(g)$$
$$\underline{K_2MnO_4(s) + Cl_2(g) \rightarrow 2\ KCl(s) + MnO_2(s) + O_2(g)}$$
$$2\ KClO_3(s) \rightarrow 2\ KCl(s) + 3\ O_2(g) \qquad \text{[net equation]}$$

Thus, the oxygen is oxidized from the oxidation state of -2 to 0; the manganese cycles from $+4$ through $+7$ and $+6$ back to $+4$; and the chlorine is reduced from $+5$ to 0 to -1. Additional evidence for this mechanism is a faint odor of dichlorine, which is released during the first step.

Perchloric Acid and the Perchlorate Ion

The strongest simple acid of all is perchloric acid. The pure acid is a colorless liquid that can explode unpredictably. As a result of its oxidizing nature and high oxygen content, contact with organic materials such as wood or paper causes an immediate fire. Concentrated perchloric acid, usually a 60 percent aqueous solution, is rarely used as an acid but is far more often used as a very powerful oxidizing agent, for example, to oxidize metal alloys to the metal ions so that they can be analyzed. Special perchloric acid fume hoods should be used when these oxidations are performed. Cold dilute solutions of perchloric acid are reasonably safe, however.

Sodium perchlorate is prepared industrially by the electrolytic oxidation of sodium chlorate:

$$ClO_3{}^-(aq) + H_2O(l) \rightarrow ClO_4{}^-(aq) + 2\ H^+(aq) + 2\ e^- \qquad \text{(anode)}$$

$$2\ H^+(aq) + 2\ e^- \rightarrow H_2(g) \qquad \text{(cathode)}$$

The solubility of an alkali metal salt decreases with increasing cation size. That is, the increasing size (decreasing charge densities) of the ions will reduce the hydration energies to the point that they are exceeded by the lattice energy. Thus, potassium perchlorate is only slightly soluble ($20\ g{\cdot}L^{-1}$ of water). By contrast, silver perchlorate is amazingly soluble, to the extent of $5\ kg{\cdot}L^{-1}$ of water. The high solubility of silver perchlorate in low-polarity organic solvents as well as water suggests that its bonding in the solid phase is essentially covalent rather than ionic. That is, only dipole attractions need to be overcome to solubilize the compound rather than the much stronger electrostatic attractions in an ionic crystal lattice, which can be overcome only by a very polar solvent.

Potassium perchlorate is used in fireworks and flares, but about half the commercially produced perchlorate is used in the manufacture of ammonium perchlorate. This compound is used as a component, along with the reducing agent, aluminum, in solid booster rockets:

$$6\ NH_4ClO_4(s) + 8\ Al(s) \rightarrow 4\ Al_2O_3(s) + 3\ N_2(g) + 3\ Cl_2(g) + 12\ H_2O(g)$$

Each shuttle launch uses 850 tonnes of the compound, and total U.S. consumption is about 30 000 tonnes. Until recently, the only two U.S. plants manufacturing ammonium perchlorate were located in Henderson, Nevada,

The Discovery of the Perbromate Ion

The perchlorate ion, ClO_4^-, and the periodate ion, IO_6^{5-}, have been known since the 19th century, yet the perbromate ion could never be prepared. Many scientists, including Linus Pauling, devised theories to account for its nonexistence. For example, it was argued that the stability of the perchlorate ion was due to the strong π bonds that involved the chlorine $3d$ orbitals. The claim was made that, for bromine, a very poor overlap of the $4d$ bromine orbitals with the $2p$ orbitals of the oxygen destabilized the theoretical perbromate ion.

The theories had to be rewritten in 1968 when the American chemist E.H. Appelman discovered synthetic routes to this elusive perbromate ion. One of the routes involved another new discovery, a compound of xenon. In this process, xenon difluoride was used as an oxidizing agent.

$$XeF_2(aq) + BrO_3^-(aq) + H_2O(l) \rightarrow Xe(g) + 2\ HF(aq) + BrO_4^-(aq)$$

The second route, which is now used to produce the perbromate ion on a large scale, involves the use of difluorine as the oxidizing agent in basic solution:

$$BrO_3^-(aq) + F_2(g) + 2\ OH^-(aq) \rightarrow BrO_4^-(aq) + 2\ F^-(aq) + H_2O(l)$$

So why is the ion so elusive even though it is thermodynamically stable? The answer lies in the high reduction potential of the ion:

$$BrO_4^-(aq) + 2\ H^+(aq) + 2\ e^- \rightarrow BrO_3^-(aq) + H_2O(l) \quad E° = +1.74\ V$$

By contrast, the reduction potential for perchlorate is $+1.23$ V; and for periodate, $+1.64$ V. Hence, only extremely strong oxidizing agents such as xenon difluoride and difluorine are capable of oxidizing bromate to perbromate. Thus, before dismissing any conjectured compound as impossible to synthesize, we must always be sure to explore all the possible preparative routes and conditions.

a suburb of Las Vegas. The attractions of the site were the cheap electricity from the Hoover Dam and the very dry climate, which make the handling and storage of the hygroscopic ammonium perchlorate much easier.

There is another major problem with ammonium perchlorate—it decomposes when heated above 200°C:

$$2\ NH_4ClO_4(s) \xrightarrow{\Delta} N_2(g) + Cl_2(g) + 2\ O_2(g) + 4\ H_2O(g)$$

On May 4, 1988, this decomposition occurred on a massive scale at one of the manufacturing plants. A series of explosions destroyed half of the nation's ammonium perchlorate production capacity, as well as causing death, injury, and extensive property damage. Several issues were raised by the accident, such as the feasibility of constructing such plants close to residential areas and the dependence of the space and military rocket programs on only two manufacturing facilities for the nation's entire supply of a crucial chemical compound.

17.10 Interhalogen Compounds and Polyhalide Ions

There is an enormous number of combinations of pairs of halogens forming interhalogen compounds and polyhalide ions. The neutral compounds fit the formulas XY, XY_3, XY_5, and XY_7, where X is the halogen of higher atomic mass and Y, that of lower atomic mass. All permutations are known for XY and XY_3, but XY_5 is only known where Y is fluorine. Thus, once again, it is only with fluorine that the highest oxidation states are obtained. The formula XY_7, in which X would have the oxidation state of $+7$, is found only in IF_7. The common argument for the lack of chlorine and bromine analogs is simply that of size: Only the iodine atom is large enough to accommodate seven fluorine atoms.

All of the interhalogens can be prepared by combination reactions of the constituent elements. For example, heating iodine and fluorine in a $1:7$ ratio gives iodine heptafluoride, IF_7:

$$I_2(g) + 7\ F_2(g) \xrightarrow{\Delta} 2\ IF_7(g)$$

While a solution of iodine monochloride can be prepared by bubbling chlorine gas through a solution of iodine in carbon tetrachloride:

$$Cl_2(g) + I_2(CCl_4) \rightarrow 2\ ICl(CCl_4)$$

The simple interhalogens such as iodine monochloride have colors intermediate between those of the constituents. In fact, iodine monochloride can be considered as a "combo" element (Chapter 9, Section 9.11) analog of bromine. However, the melting points and boiling points of the interhalogens are slightly higher than the mean values of the constituents because the interhalogen molecules are polar. More important, the chemical reactivity of an interhalogen compound is usually similar to that of the more reactive parent halogen. To chlorinate an element or compound, it is often more convenient to use solid iodine monochloride than chlorine gas, although sometimes the nonhalogen atom in the two products has different oxidation states. This outcome can be illustrated for the chlorination of vanadium:

$$V(s) + 2\ Cl_2(g) \rightarrow VCl_4(l)$$

$$V(s) + 3\ ICl(s) \rightarrow VCl_3(s) + \tfrac{3}{2}\ I_2(s)$$

In solution, interhalogen molecules are hydrolyzed to the hydrohalic acid of the more electronegative halogen and the hypohalous acid of the less electronegative halogen. For example:

$$BrCl(g) + H_2O(l) \rightarrow HCl(aq) + HBrO(aq)$$

Ruby-red, solid iodine monochloride is used in biochemistry as the *Wij reagent* for the determination of the number of carbon–carbon double bonds in an oil or fat. When we add the brown solution of the interhalogen to the unsaturated fat, decolorization occurs as the halogens add across the double bond:

$$-CH{=}CH{-}\ +\ ICl \rightarrow \underset{\displaystyle \overset{|}{I}\quad\ \overset{|}{Cl}}{-CH{-}CH{-}}$$

When a permanent brown color remains, the reaction has been completed. The results are reported as the iodine number—the volume (milliliters) of a standard iodine monochloride solution needed to react with a fixed mass of fat.

The only interhalogen compound produced on an industrial scale is chlorine trifluoride, a liquid that boils at 11°C. It is a convenient and extremely powerful fluorinating agent, as a result of its high fluorine content and high bond polarity. It is particularly useful in the separation of uranium from most of the fission products in used nuclear fuel. At the reaction temperature of 70°C, uranium forms liquid uranium(VI) fluoride. However, most of the major reactor products, such as plutonium, form solid fluorides:

$$U(s) + 3\ ClF_3(l) \rightarrow UF_6(l) + 3\ ClF(g)$$

$$Pu(s) + 2\ ClF_3(l) \rightarrow PuF_4(s) + 2\ ClF(g)$$

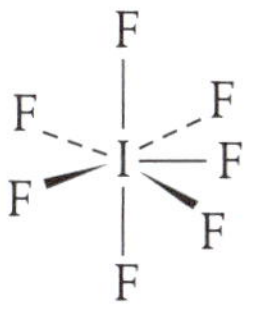

The uranium compound can then be separated from the mixture by distillation.

The interhalogen compounds are of particular interest to inorganic chemists because of their geometries. The shapes of the compounds all follow the VSEPR rules, even iodine heptafluoride, IF_7, which has the rare pentagonal bipyramidal shape of a seven-coordinate species (Figure 17.9).

The halogens also form polyatomic ions. Iodine is the only halogen to readily form polyhalide anions by itself. The triiodide ion, I_3^-, is important because its formation (discussed in Section 17.7) provides a means of "dissolving" molecular iodine in water by using a solution of the iodide ion:

$$I_2(s) + I^-(aq) \rightleftharpoons I_3^-(aq)$$

The ion is linear and has equal iodine–iodine bond lengths of about 293 pm; these bonds are slightly longer than the single bond in the diiodine molecule (272 pm). There are many other polyiodide ions, including I_5^- and I_7^-, but these are less stable than the triiodide ion.

There also are a wide variety of interhalogen cations and anions, for example, the dichloroiodine ion, ICl_2^+, and the tetrachloroiodate ion, ICl_4^-. These are mainly of interest in terms of the match of their actual shape with that predicted by VSEPR theory (Figure 17.10).

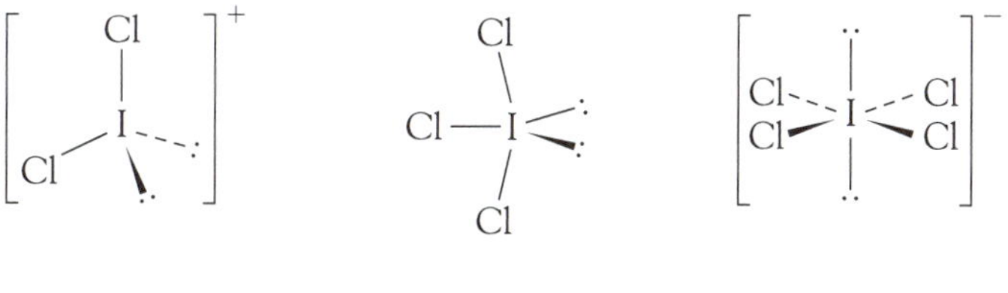

Figure 17.9 The iodine heptafluoride molecule, which has a pentagonal arrangement in the horizontal plane.

Figure 17.10 (a) The dichloroiodine ion, ICl_2^+; (b) the iodine trichloride molecule, ICl_3; and (c) the tetrachloroiodate ion, ICl_4^-.

(a)   (b)   (c)

17.11 Biological Aspects

The halogens are unique in that every stable member of the group has a biological function.

Fluorine and Killer Plants

In several regions of the world, cattle ranchers have a major problem from toxic plants. Australia suffers particularly severely as about a thousand plant species there are known to be toxic to animals and humans. A significant number of these plants produce the fluoroacetate ion, CH_2FCOO^-. Though acetate ion is harmless (unless consumed in enormous quantities), the substitution of one fluorine for a hydrogen changes the properties of the ion substantially. For example, fluoroacetic acid is a strong acid with a pK_a of 2.59 compared to a value of 4.76 for acetic acid. The fluoroacetate ion acts by blocking the Krebs cycle in mammals, causing a build-up of citric acid and resulting in heart failure.

The plants absorb traces of fluoride ion from the soil and then incorporate it into their biochemical pathways. It is thought that these plants produce the fluoroacetate ion as a defense mechanism against predators. Though Australia probably has the most fluoroacetate-producing species, South Africa has a plant that produces the fluoroacetate ion to a concentration of 1%, meaning that the ingestion of one leaf of the plant is suffcient to kill a cow.

Sodium fluoroacetate is used by some in the United States and elsewhere as a poison for unwanted mammals such as coyotes. One of Australia's most famous double-murder cases (the Bogle–Chandler case) is believed to have been caused by deliberate fluoroacetate poisoning.

Chlorine: The Challenge of THMs

The chloride ion has a vital role in the ion balance in our bodies. It does not appear to play an active role but simply acts to balance the positive ions of sodium and potassium. However, covalently bonded chlorine is far less benign. Most of the toxic compounds with which we are currently concerned—for example, DDT and PCBs—are chlorine-containing molecules. The argument has been made to completely ban the production of chlorine-containing covalently bonded compounds. However, this would result in the elimination of many useful materials such as polyvinylchloride (PVC). It is important to recognize that there are organochloro-compounds produced by a variety of organisms, thus banning synthetic chloro-compounds would not totally eliminate chloro-compounds from the planet.

A group of compounds of particular concern are the trihalomethanes (THMs). These are produced when water rich in organic matter is chlorinated to disinfect it for human consumption. The organic matter is found in surface waters that have dissolved humic and fulvic acids from rotting vegetation. The acids themselves are comparatively harmless, though they do cause the water to have a brownish tint. It is the chlorination process that results in the fragmentation of the complex organic molecules to give small chloro-molecules such as trichloromethane, $CHCl_3$. The existence of THMs in some tap-water samples was discovered in the 1970s, and regulations limiting the allowable level were introduced in the early 1980s. At that time, a maximum level of 100 ppb was generally considered safe but now there is evidence of a weak correlation between drinking more than five glasses of water of high THM (over 75 ppb) water per day and miscarriages. For this reason, there are proposals to reduce the permissible THM levels. One way to accomplish this is to use ozone (trioxygen, see Chapter 16, Section 16.3) or chlorine dioxide (see Section 17.8) as the primary disinfectant of the water supply. These compounds decompose the organic acids without significant generation of THMs. However, because ozone decomposes within a short time, a small concentration of chlorine needs to be added to water supplies to maintain potability as it travels many kilometers through the aging, leaky pipes of most of our cities to its destination in your home.

Bromine: The Methyl Bromide Problem

The most contentious bromine-containing compound is bromomethane, more commonly called methyl bromide. Methyl bromide, CH_3Br, is a broad spectrum fumigant used in control of pest insects, nematodes, weeds, pathogens, and rodents. Apart from its unmatched range, it has a high volatility

Potassium bromide has been used in medicine as a sedative and as an anticonvulsant in the treatment of epilepsy.

and hence leaves little residue. About 80 000 tonnes is used every year, about 75%, as soil fumigant; 22%, postharvest; and 3%, structural pest control tool. Methyl bromide is important to the agricultural community because it controls a wide variety of pests at low cost. In fact, there is no one chemical that can currently be found to replace all of its roles.

As sea water contains relatively high concentrations of bromide ion, it is not surprising that many marine organisms synthesize organo-bromine compounds. The function of most of these unique molecules is unknown at present. However, we do know that some of the metabolic pathways results in the production of methyl bromide. Thus, a significant proportion of the atmospheric burden of methyl bromide—probably about half—comes from natural sources.

There are two concerns with methyl bromide. First, methyl bromide is an efficient ozone-depleting substance (ODS), bromine being 50 times more efficient at destroying ozone as chlorine. The first global controls on methyl bromide emissions were established in 1995 though the compound is still in widespread agricultural use. The second concern is that methyl bromide is lethal against nontarget organisms as well. It is a cancer suspect in humans as well as being acutely toxic. It can cause central nervous system and respiratory system failure. In low doses, headaches, nausea, then in higher doses, muscle spasms, coma, convulsions, and death. Thus, farm workers are at risk from handling the compound.

Iodine: The Thyroid Element

About 75 percent of the iodine in the human body is found in one location—the thyroid gland. Iodine is utilized in the synthesis of the hormones thyroxine (Figure 17.11) and triiodothyronine. These hormones are essential for growth, for the regulation of neuromuscular functioning, and for the maintenance of male and female reproductive functions. Yet goiter, the disease resulting from a deficiency of the thyroid hormones, is found throughout the world, including a band across the northern United States, much of South America, and Southeast Asia; there are localized areas of deficiency in most other countries of the world. One common cause of the disease is a lack of iodide ion in the diet. To remedy the iodine deficiency, potassium iodide is added to common household salt (iodized salt).

A symptom of goiter, which can have causes other than simple iodide deficiency, is a swollen lower part of the neck. The enlargement is the attempt of the thyroid gland to maximize absorption of iodide in iodine-deficient circumstances. In previous centuries, women with mild goiter were favored as marriage partners because their swollen necks enabled them to display their expensive and ornate necklaces more effectively.

Why is iodine essential? Bound to carbon, the iodine cannot take part in any redox function, nor is it chemically available for covalent bonding to other molecules. As its electronegativity is close to that of carbon, it is unlikely

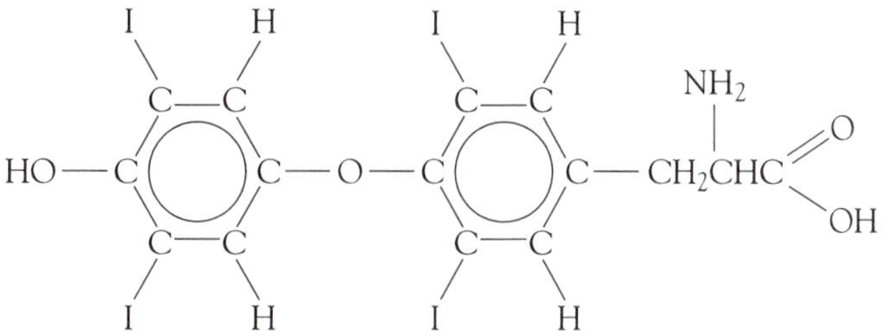

Figure 17.11 The thyroxine molecule.

that the iodine atoms make any significant change to the overall electronic structure of the molecule and hence the molecular properties. The clue seems to be in the very large covalent radius of the iodine atom—133 pm—about twice the radius of carbon, nitrogen, and oxygen. This corresponds to an eightfold greater volume. The iodine atoms of the iodo-organic molecules seem to be designed to fit in certain large cavities in a matching enzyme site, holding the enzyme in a unique conformation. Evidence for this role is provided by the fact that substituting the large isopropyl unit, $(CH_3)_2CH$—, for iodine results in a molecule with similar hormonal activity. Of course, this raises the question as to why a biological system should choose such a rare element for an essential pathway. The only obvious answer relates to presumed marine origin of life. The seas are iodine-rich, and, in such an environment, incorporation of iodine would have provided a simple pathway to addition of a bulky substituent. Land organisms then became stuck with a process that necessitated an element much less common in the nonaquatic environment.

17.12 Element Reaction Flowchart

Flowcharts are shown for fluorine, chlorine, and iodine.

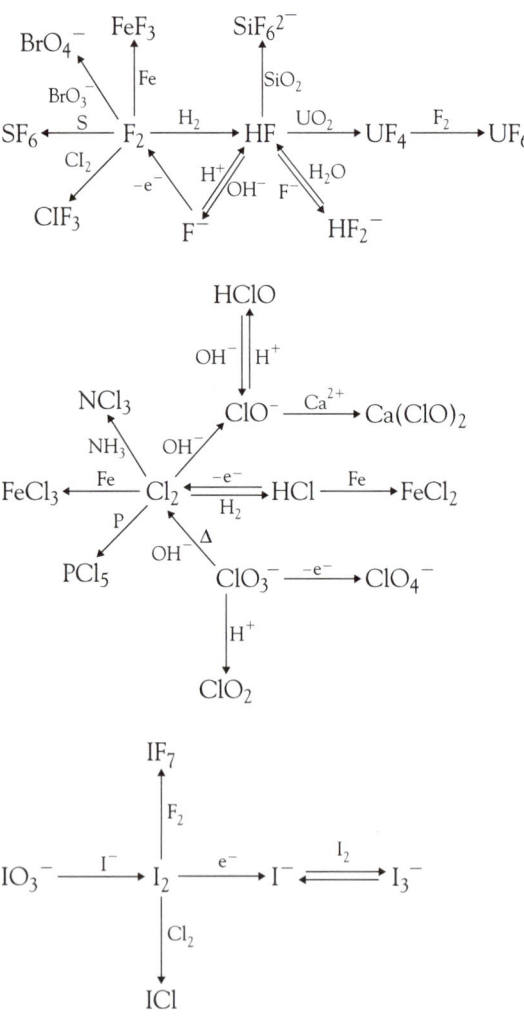

EXERCISES

17.1 Write balanced chemical equations for the following chemical reactions:
(a) uranium(IV) oxide with hydrogen fluoride
(b) calcium fluoride with concentrated sulfuric acid
(c) liquid sulfur tetrachloride with water
(d) aqueous dichlorine and hot sodium hydroxide solution
(e) diiodine with difluorine in a $1:5$ mole ratio
(f) bromine trichloride and water

17.2 Write balanced chemical equations for the following chemical reactions:
(a) lead metal with excess dichlorine
(b) magnesium metal with dilute hydrochloric acid
(c) the hypochlorite ion with sulfur dioxide gas
(d) mild heating of potassium chlorate
(e) solid iodine monobromide with water
(f) phosphorus and iodine monochloride

17.3 Summarize the unique features of fluorine chemistry.

17.4 What are the shapes of the following species: (a) BrF_2^+; (b) BrF_3; (c) BrF_4^-.

17.5 Suggest an explanation for why difluorine is so reactive toward other nonmetals.

17.6 For chlorine, the only two naturally occurring isotopes are chlorine-35 and chlorine-37, while for bromine, they are bromine-79 and bromine-81. Suggest why chlorine-36 and bromine-80 are not stable isotopes.

17.7 Use the formation of solid iodine heptafluoride to indicate why entropy cannot be a driving force in the reactivity of fluorine.

17.8 Why cannot difluorine be produced electrolytically from an aqueous solution of sodium fluoride by a similar process to that used to produce dichlorine from sodium chloride solution?

17.9 In the Frost diagram for chlorine, the Cl_2/Cl^- lines are identical for acidic and basic solution. Explain why.

17.10 Why, in the Frost diagram (Figure 17.1), is the acid species of chloric acid written as ClO_3^-, while that of chlorous acid is written $HClO_2$?

17.11 Suggest a reason why hydrofluoric acid is a weak acid, whereas the binary acids of the other halogens are all strong acids.

17.12 Explain why, as the solution becomes more concentrated, hydrofluoric acid ionizes to a lesser extent at first, then to a greater extent at high concentrations.

17.13 If annual hydrogen fluoride production is 1.2×10^6 tonnes per year, calculate the mass of calcium sulfate produced per annum by this process.

17.14 Why would you expect the hydrogen difluoride ion to form a solid compound with potassium ion?

17.15 Deduce the oxidation number for oxygen in hypofluorous acid, HOF.

17.16 Why is hydrochloric acid used as a common laboratory acid in preference to nitric acid?

17.17 Suggest how you would prepare (a) chromium(III) chloride, $CrCl_3$, from chromium metal; and (b) chromium(II) chloride, $CrCl_2$, from chromium metal.

17.18 Suggest how you would prepare (a) selenium tetrachloride, $SeCl_4$, from selenium; and (b) diselenium dichloride, Se_2Cl_2, from selenium.

17.19 Explain why iron(III) iodide is not a stable compound.

17.20 Describe the tests used to identify each of the halide ions.

17.21 Calculate the enthalpy of reaction of ammonium perchlorate with aluminum metal. Apart from the exothermicity of the reaction, what other factors would make it a good propellant mixture?

17.22 Construct an electron-dot formula for the triiodide ion. Thus, deduce the shape of the ion.

17.23 The concentration of hydrogen sulfide in a gas supply can be measured by passing a measured volume of gas over solid diiodine pentoxide. The hydrogen sulfide reacts with the diiodine pentoxide to give sulfur dioxide, diiodine, and water. The diiodine can then be titrated with thiosulfate ion; and the hydrogen sulfide concentration, calculated. Write chemical equations corresponding to the two reactions.

17.24 Carbon tetrachloride has a melting point of $-23°C$; carbon tetrabromide, $+92°C$; and carbon tetraiodide, $+171°C$. Provide an explanation for this trend. Estimate the melting point of carbon tetrafluoride.

17.25 The highest fluoride of sulfur is sulfur hexafluoride. Suggest why sulfur hexaiodide does not exist.

17.26 Construct electron-dot structures of chlorine dioxide that have zero, one, and two double bonds (one of each) and decide which would be preferred on the basis of formal charge assignments.

17.27 Another compound of chlorine and oxygen, Cl_2O_4, is more accurately represented as chlorine perchlorate, $ClOClO_3$. Draw the electron-dot structure of this compound and determine the oxidation number of each chlorine in the compound.

17.28 Describe the uses of (a) sodium hypochlorite; (b) chlorine dioxide; (c) ammonium perchlorate; (d) iodine monochloride.

17.29 Predict some physical and chemical properties of astatine as an element.

17.30 Explain why the cyanide ion is often considered a pseudohalogen.

17.31 The thiocyanate ion, SCN^-, is linear. Construct reasonable electron-dot representations of this ion by assigning formal charges. The carbon–nitrogen bond length is known to be close to that of a triple bond. What does this tell you about the relative importance of each representation?

17.32 How does fluoride ion affect the composition of teeth?

17.33 Write balanced chemical equations corresponding to each transformation in the Element Reaction Flowcharts.

17.34 Iodine forms an oxide, diiodine pentoxide that resembles dinitrogen pentoxide in structure. Construct the electron-dot structure of the iodine compound and contrast the bonding with that in N_2O_5 (Chapter 15). Explain why the bonding differs. What is the oxidation state of each iodine atom, of the bridging oxygen atom, of each terminal oxygen atom?

17.35 Draw the electron-dot structure of the dichlorine dioxide, ClOOCl, molecule. Deduce the oxidation state of the chlorine atoms and of the oxygen atoms.

17.36 Explain why ammonium fluoride adopts the wurtzite structure whereas ammonium chloride adopts the sodium chloride lattice.

17.37 Explain why, of the tetraphosphonium halides, $[PH_4]I$ is the most stable toward decomposition.

17.38 Tetramethylammonium fluoride, $(CH_3)_4NF$, reacts with iodine heptafluoride, IF_7, to give an electrically conducting solution. Write a chemical equation for the reaction.

17.39 We noted that iodine monochloride can be considered as a "combo" element analog of bromine. Which other halogen pair can be considered as a "combo" element analog of another halogen?

BEYOND THE BASICS

17.40 Iodine pentafluoride undergoes self-ionization. Deduce the formulas of the cation and anion formed in the equilibrium and write a balanced equation for the equilibrium. Construct electron-dot diagrams for the molecule and the two ions. Which ion is the Lewis acid, and which is the Lewis base? Explain your reasoning.

17.41 The melting point of ammonium hydrogen difluoride, $(NH_4)^+(HF_2)^-$ is only +26°C. This is much lower than what one would expect for an ionic lattice. Suggest what might be happening.

17.42 Calculate the enthalpy of formation of the chlorine molecular ion, Cl_2^+, given that the bond energy of molecular chlorine is 240 kJ·mol^{-1}, while the first ionization energy of the chlorine atom is 1250 kJ·mol^{-1}, and that of the chlorine molecule is 1085 kJ·mol^{-1}. Comment on the bond strength in the molecular ion compared to that in the neutral molecule.

17.43 Construct an approximate molecular orbital energy level diagram to depict the bonding in chlorine monofluoride.

17.44 Use the principle of formal charge to determine the average bond order in the phosphate ion and the perchlorate ion. Use these two results to suggest the average bond order in the sulfate ion.

17.45 Which of dichlorine oxide and dichlorine heptaoxide is likely to be the more acidic oxide? Give your reasoning.

17.46 Dichlorine heptoxide, Cl_2O_7, is a colorless oily liquid.
(a) Calculate the oxidation state of chlorine in the compound.
(b) Draw the probable structure of the compound.
(c) Write a balanced chemical equation for the reaction of this compound with water.
(d) Write the formula of an analogous compound of a metallic element. Give your reasoning.

(e) Write the formula of two probable isoelectronic and isostructural ions.

17.47 Draw a probable structure of the dichlorine trioxide molecule. Will it be completely linear or bent? If bent, suggest an approximate bond angle.

17.48 As we discussed, there are strong parallels between the chemistry of the pseudohalogens and the halogens. On this assumption, write balanced chemical equations for the reactions between:
(a) cyanogen, $(CN)_2$, and cold sodium hydroxide solution
(b) thiocyanate ion, NCS^-, and acidified permanganate ion solution

17.49 Why is ammonium perchlorate an explosive hazard while sodium perchlorate is much less hazardous? Use a balanced equation to illustrate your argument and identify which elements undergo a change in oxidation state.

17.50 Diiodine reacts with an excess of dichlorine to form a compound of formula ICl_x. One mole of ICl_x reacts with an excess of iodide ion to produce chlorine gas and two moles of diiodine. What is the empirical formula of ICl_x?

17.51 Fluorine, chlorine, and oxygen form a series of polyatomic ions: $(F_2ClO_2)^-$, $(F_4ClO)^-$, $(F_2ClO)^+$, and $(F_2ClO_2)^+$. Deduce the molecular shape of each of these ions.

17.52 Fluorine forms only one oxide, F_2O. Draw the electron-dot structure of the compound and determine the oxidation state of fluorine and of oxygen in the compound. Explain why the oxidation state of oxygen is unusual. Suggest why one would not expect any other fluorine oxides. Use as comparisons the compounds Cl_2O and Cl_2O_7.

17.53 Thallium forms an iodide, TlI_3. Suggest the actual formulation of the compound given the following information:

$$Tl^{3+}(aq) + 2\,e^- \rightarrow Tl^+(aq) \qquad E^\circ = +1.25\text{ V}$$

$$I_3^-(aq) + 2\,e^- \rightarrow 3\,I_2(aq) \qquad E^\circ = +0.55\text{ V}$$

Why would this formulation be expected?

17.54 Refer back to the section on chlorine monoxide. Sum together the reaction steps for the ozone destruction process and write the overall reaction. Identify catalytic and intermediate species.

17.55 The ion, $I(N_3)_2^-$, is known.
(a) Why should the existence of this ion be expected?
(b) Deduce whether the electronegativity of the azide unit is higher or lower than that of iodine.
(c) Deduce the geometry of the azide units about the iodine atom.
(d) Suggest how the ion was stabilized.

17.56 The standard enthalpy of formation of gaseous iodine monochloride is $+18$ kJ·mol^{-1}.
(a) Write a balanced chemical equation for the process.
(b) Using Cl—Cl and I—I bond energy data and the enthalpy of sublimation of diiodine (62 kJ·mol^{-1}), calculate the I—Cl bond energy.
(c) If the standard free energy of formation is -5 kJ·mol^{-1}, calculate the entropy change for the reaction. Why would you expect the value to have the sign that it does (two reasons).

17.57 (a) Chlorine trifluoride reacts with boron trifluoride to give an ionic product in which the chlorine is in the polyatomic cation. Write a chemical equation for the reaction.
(b) Chlorine trifluoride reacts with potassium fluoride to give an ionic product in which the chlorine is in the polyatomic cation. Write a chemical equation for the reaction.
(c) Comment on the difference in reactions.

The Group 18 Elements: The Noble Gases

Chapter 18

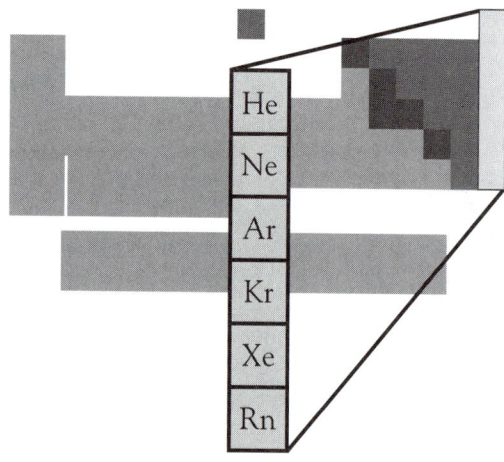

The noble gases make up the least reactive group in the periodic table. In fact, xenon is the only noble gas to form a wide range of compounds—and then only with highly electronegative elements. It is doubtful that stable chemical compounds will ever be made of helium and neon.

Although it had been noted as early as 1785 that there was something else in air besides oxygen and nitrogen, it was not until 100 years later that Sir William Ramsay showed that this other gas produced a previously unknown spectrum when an electric discharge was passed through it. Because every element has a unique spectrum, the gas producing the new spectrum had to be a new element. He named it *argon,* from the Greek word for lazy, because of its unreactive nature; and he suggested that it might be the first member of a new group in the periodic table.

In fact, one other element in this group—helium—had already been discovered in 1868, but not on Earth. Observations of the spectrum of the Sun had shown some lines that did not belong to any element known at that time. The new element was named hel-ium, the first part of the name indicating that it was first discovered in the Sun (Greek, *helios*) and the ending indicating that it was expected to be a metal. The element was first isolated on Earth in 1894 from uranium ores, and a few years later, it was realized that the helium is produced during the radioactive decay of uranium and its

Table 18.1 Melting and boiling points of the noble gases

Noble gas	Melting point (°C)	Boiling point (°C)	Number of electrons
He	—	−269	2
Ne	−249	−245	10
Ar	−189	−186	18
Kr	−157	−152	36
Xe	−112	−109	54
Rn	−71	−62	86

daughter elements. In 1926 it was suggested that the name of the element be changed to helion, to indicate that it was not a metal, but by that time the former name was too well established.

Every one of the noble gases was first identified by its unique emission spectrum. Hence, it was really physical chemists rather than inorganic chemists who founded the study of this group of elements.

18.1 Group Trends

All the elements in Group 18 are colorless, odorless, monatomic gases at room temperature. They neither burn nor support combustion; in fact, they make up the least reactive group in the Periodic Table. The very low melting and boiling points of the noble gases indicate that the dispersion forces holding the atoms together in the solid and liquid phases are very weak. The trend in the melting and boiling points, shown in Table 18.1, corresponds to the increasing number of electrons and, hence, greater polarizability.

Because the elements are all monatomic gases, there is a well-behaved trend in densities at the same pressure and temperature. The trend is a simple reflection of the increase in molar mass (Table 18.2). Air has a density of about 1.3 g·L^{-1}; so, relative to air, helium has an extremely low density. Conversely, radon is among the densest of gases at SATP.

To date, chemical compounds have been isolated at room temperature only for the three heaviest members of the group: krypton, xenon, and radon. Few compounds of krypton are known, whereas xenon has an extensive chemistry. The study of radon chemistry is very difficult because all the isotopes of radon are highly radioactive.

Table 18.2 Densities of the noble gases (at SATP)

Noble gas	Density (g·L^{-1})	Molar mass (g·mol^{-1})
He	0.2	4
Ne	1.0	20
Ar	1.9	39
Kr	4.1	84
Xe	6.4	131
Rn	10.6	222

18.2   Unique Features of Helium

When helium is cooled to as close to absolute zero as we can reach, it is still a liquid. In fact, even at a temperature of 1.0 K, a pressure of about 2.5 MPa is required to cause it to solidify. However, liquid helium is an amazing substance. At a pressure of 100 kPa, the gas condenses at 4.2 K to form an ordinary liquid (referred to as helium I); but when cooled below 2.2 K, the properties of the liquid (now helium II) are dramatically different. For example, helium II is an incredibly good thermal conductor, 10^6 times greater than helium I and much better than even silver, the best metallic conductor at room temperature. Even more amazing, its viscosity drops to close to 0. When helium II is placed in an open container, it literally "climbs the walls" and runs out over the edges. These and many other bizarre phenomena exhibited by helium II are best interpreted in terms of quantum behavior in the lowest possible energy states of the element. A full discussion is in the realm of quantum physics.

18.3   Uses of the Noble Gases

Table 18.3   Abundance of the noble gases in the dry atmosphere

Noble gas	Abundance (% by moles)
He	0.00052
Ne	0.0015
Ar	0.93
Kr	0.00011
Xe	0.0000087
Rn	Trace

All the stable noble gases are found in the atmosphere, although only argon is present in a high proportion (Table 18.3). Helium is found in high concentrations in some underground natural gas deposits, where it has been accumulating from the decay of radioactive elements in the Earth's crust. Gas reservoirs in the southwestern United States are among the largest in the world, and the United States is the world's largest producer of helium. In fact, the discovery of the deposits in the 1920s caused the price of helium gas to drop from $\$88 \cdot L^{-1}$ (1915) to $0.05 \mathcal{c} \cdot L^{-1}$ (1926).

Because it is the gas with the second lowest density (dihydrogen having the lowest), helium is used to fill balloons. Dihydrogen would provide more "lift," but its flammability is a major disadvantage. Almost everyone has heard of the Hindenburg disaster, the burning of a transatlantic airship. Yet few are aware that the airship was designed to use helium. When the National Socialist party came to power in Germany in the 1930s, the U.S. government placed an embargo on helium shipments to Germany, fearing that the gas would be used for military purposes. Thus, when the airship was completed, dihydrogen had to be used. Today, the public thinks of airships solely in their advertising role. However, they have also been used as long-endurance flying radar posts by the U.S. Coast Guard to identify illegal drug-carrying flights. And an airship has been used to study the upper canopy of the rain forest in the Amazon basin, a vital task that would be very difficult to do by any other means. Using modern technology, new designs of airships are being constructed for a variety of tasks, such as ecotourism and heavy lifting.

Helium is used in deep-sea diving gas mixtures as a replacement for the more blood-soluble nitrogen gas in air. The velocity of sound is much greater in low-density helium than in air. As many people are aware, this property gives breathers of helium "Mickey Mouse" voices. It should be added that the combination of dry helium gas and the higher frequency of vibrations in the larynx can cause voice damage to those who frequently indulge in the gas for fun.

Of great scientific importance, liquid helium is the only safe means of cooling scientific apparatus close to 0 K. Many pieces of equipment use superconducting magnets to obtain very high magnetic fields, but at present the coils only become superconducting at extremely low temperatures.

All the other gases are obtained as by-products of the production of dioxygen from air. Some argon also is obtained from industrial ammonia synthesis, where it accumulates during the recycling of the unused atmospheric gases. Argon production is quite large, approaching 10^6 tonnes per year. Its major use is as an inert atmosphere for high-temperature metallurgical processes. Argon and helium are both used as an inert atmosphere in welding; neon, argon, krypton, and xenon are used to provide different colors in "neon" lights. The denser noble gases, particularly argon, have been used to fill the air space between the glass layers of thermal insulating windows. This use is based on the low thermal conductivities of these gases; for example, that of argon at 0°C is 0.017 $J \cdot s^{-1} \cdot m^{-1} \cdot K^{-1}$. Dry air at the same temperature has a thermal conductivity of 0.024 $J \cdot s^{-1} \cdot m^{-1} \cdot K^{-1}$.

The high abundance of argon in the atmosphere is a result of the radioactive decay of potassium-40, the naturally occurring radioactive isotope of potassium. As mentioned in Chapter 11, Section 11.7, this isotope has two decay pathways, one of which involves the capture of a core electron to form argon-40:

$$\ce{^{40}_{19}K} + \ce{^{0}_{-1}e} \rightarrow \ce{^{40}_{18}Ar}$$

18.4   *A Brief History of Noble-Gas Compounds*

The story of the discovery of the noble-gas compounds has become part of the folklore of inorganic chemistry. Unfortunately, like most folklore, the "true" story has been buried by myth. It was in 1924 that the German chemist von Antropoff made the suggestion that is obvious to us today: Because they have eight electrons in their valence level, the noble gases could form compounds with up to eight covalent bonds. Following that, in 1933 the American chemist Linus Pauling predicted the formulas of some possible noble-gas compounds, such as oxides and fluorides. Two chemists at Caltech, Don Yost and Albert Kaye, set out to make compounds of xenon and fluorine. At the time, they thought they had been unsuccessful, but there is evidence that they did, in fact, make the first noble-gas compound.

It was only after Yost and Kaye's admitted failure that the myth of the inertness of the noble gases spread. The "full octet" was claimed to be the reason, even though every inorganic chemist knew that many compounds involving nonmetals beyond the second period violated this "rule." So things remained, with this dogma being accepted by generation after generation of chemistry students, until the upsurge in interest in inorganic chemistry in the 1960s. It was Neil Bartlett, working at the University of British Columbia, who then approached the problem from a different direction in 1962.

Bartlett had been working with platinum(VI) fluoride, which he found was such a strong oxidizing agent that it oxidized dioxygen gas to form the compound $O_2^+PtF_6^-$. While teaching a first-year chemistry class, he noticed that the first ionization energy of xenon was almost identical to that of the dioxygen molecule. Despite the skepticism of his colleagues and students, he managed to synthesize an orange-yellow compound that he claimed was $Xe^+PtF_6^-$. This reaction was the first proven formation of a compound of a noble gas. However, the compound did not have this simplistic formula, and it is now believed to have been a mixture of compounds that contained the XeF^+ ion. Unknown to Bartlett, Rudolf Hoppe, in Germany, had for some years been working with enthalpy cycles, and he had come to the conclusion that, on thermodynamic grounds, xenon fluorides should exist. By passing an electric discharge through a mixture of xenon and difluorine, he was able to prepare xenon difluoride. Unfortunately for Hoppe, this discovery came a few weeks after Bartlett's discovery. Since then, the field of noble-gas chemistry has blossomed. Xenon is the sole noble gas to form a rich diversity of compounds, and then only with electronegative elements, such as fluorine, oxygen, and nitrogen.

Is It Possible To Make Compounds of the Early Noble Gases?

Why are there no chemical compounds of helium, neon, or argon? In fact, this statement is not quite correct. The gas-phase ion HeH^+ was first synthesized in 1925. However, it is true to say that no isolable stable compound of the early noble gases has been synthesized so far.

The known chemistry of krypton might be a guide to the potential chemistry of argon. The only known binary compound of krypton is krypton difluoride, formed by the reaction of krypton and fluorine in ultraviolet light at $-196°C$. The compound decomposes at about $-20°C$. Krypton difluoride is an extremely strong oxidizing agent. For example, it will oxidize/fluorinate metallic gold to give $(KrF)^+(AuF_6)^-$. The fluorokrypton cation is quite stable and will undergo reactions as a Lewis acid. Theoretical calculations have shown that the binding energy of the analogous fluoroargon cation should be similar and that $(ArF)^+(AuF_6)^-$ and similar compounds with large hexafluoroanions should be capable of existence. The challenge is to find a synthetic route.

At the time of writing, there has been one argon compound synthesized and conclusively identified, HArF. It has been formed by irradiating a mixture of argon and hydrogen fluoride at about $-255°C$. Infrared spectroscopy was used to show that there were indeed H—Ar and Ar—F covalent bonds formed. Unfortunately, the compound decomposes above $-245°C$. Nevertheless, this is first step that shows argon chemistry may indeed be possible.

Will stable compounds of helium and neon ever be made? Theoretical calculations show that HHeF might be more likely to form than HNeF as the helium atom is small enough that there may be a three-center bond formed over the three atoms. However, the difficulty will be in finding a reaction pathway. There is a strong possibility that room-temperature-stable helium and neon compounds will never be made. However, in chemistry, one should never say never.

18.5 Xenon Fluorides

Xenon forms three fluorides:

$$Xe(g) + F_2(g) \rightarrow XeF_2(s)$$

$$Xe(g) + 2\,F_2(g) \rightarrow XeF_4(s)$$

$$Xe(g) + 3\,F_2(g) \rightarrow XeF_6(s)$$

The product depends on the mole ratios of the reactants and on the exact reaction conditions of temperature and pressure, although very high partial pressures of difluorine are needed to form the xenon hexafluoride.

All three xenon fluorides are white solids and are stable with respect to dissociation into elements at ordinary temperatures; that is, they have negative free energies of formation at 25°C. As noted earlier, it is not necessary to invoke any novel concepts to explain the bonding; in fact, the three compounds are isoelectronic with well-established iodine polyfluoride anions. Table 18.4 shows the formulas of the compounds and the number of electron pairs around the central atom.

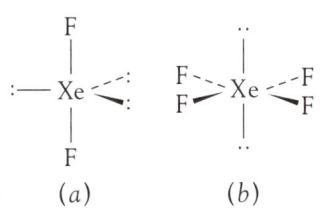

Figure 18.1 (a) Xenon difluoride; (b) xenon tetrafluoride.

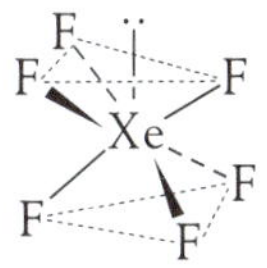

Figure 18.2 Probable capped octahedral structure of xenon hexafluoride in the gas phase.

Table 18.4 Isoelectronic xenon halides and iodine polyfluoride anions

Number of electron pairs	Xenon halides	Iodine polyfluoride anions
5	XeF_2	IF_2^-
6	XeF_4	IF_4^-
7	XeF_6	IF_6^-

The shapes of xenon difluoride and tetrafluoride are exactly those predicted from simple VSEPR theory (Figure 18.1). Xenon hexafluoride, with six bonding pairs and one lone pair around the xenon atom, might be expected to adopt some form of pentagonal bipyramid like iodine heptafluoride. As discussed in Chapter 3, Section 3.9, there are three possible arrangements: pentagonal bipyramid, capped trigonal prism, and capped octahedron. The structural studies of xenon hexafluoride in the gas phase indicate that it adopts the capped octahedral arrangement (Figure 18.2).

What is the driving force in the formation of the xenon fluorides? Let us take xenon tetrafluoride as an example. If we look at the equation for the formation of the compound from its elements, we see that the entropy change must be negative, considering that 1 mole of solid is being formed from 3 moles of gas:

$$Xe(g) + 2\,F_2(g) \rightarrow XeF_4(s)$$

The negative free energy must therefore result from a negative enthalpy change—an exothermic reaction. Figure 18.3 shows the enthalpy cycle for the formation of this compound from its elements. In the cycle, 2 moles of difluorine are dissociated into atoms, then 4 moles of xenon–fluorine bonds are formed, followed by solidification of the product. The stability of this compound clearly depends on the moderately high Xe—F bond energy and the low dissociation energy of the fluorine molecule.

Figure 18.3 Enthalpy cycle for the formation of xenon tetrafluoride.

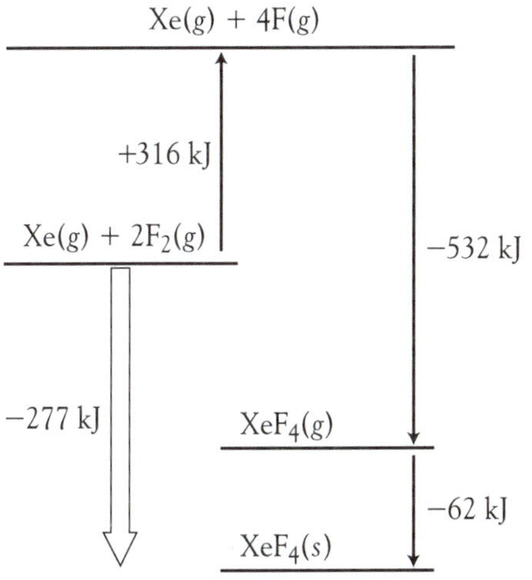

$Xe(g) + 4F(g)$

$+316\ kJ$

$Xe(g) + 2F_2(g)$

$-532\ kJ$

$-277\ kJ$

$XeF_4(g)$

$-62\ kJ$

$XeF_4(s)$

All the fluorides hydrolyze in water; for example, xenon difluoride is reduced to xenon gas:

$$2 \text{ XeF}_2(s) + 2 \text{ H}_2\text{O}(l) \rightarrow 2 \text{ Xe}(g) + \text{O}_2(g) + 4 \text{ HF}(l)$$

Xenon hexafluoride is first hydrolyzed to xenon oxide tetrafluoride, $XeOF_4$, which in turn is hydrolyzed to xenon trioxide:

$$\text{XeF}_6(s) + \text{H}_2\text{O}(l) \rightarrow \text{XeOF}_4(l) + 2 \text{ HF}(l)$$

$$\text{XeOF}_4(l) + 2 \text{ H}_2\text{O}(l) \rightarrow \text{XeO}_3(s) + 4 \text{ HF}(l)$$

The fluorides are strong fluorinating agents. For example, xenon difluoride can be used to fluorinate double bonds in organic compounds. It is a very "clean" fluorinating agent, in that the inert xenon gas can be easily separated from the required product:

$$\text{XeF}_2(s) + \text{CH}_2{=}\text{CH}_2(g) \rightarrow \text{CH}_2\text{FCH}_2\text{F}(g) + \text{Xe}(g)$$

Furthermore, a fluoride in which the other element is in its highest possible oxidation state can be produced by using xenon fluorides as reagents. Thus xenon tetrafluoride will oxidize sulfur tetrafluoride to sulfur hexafluoride:

$$\text{XeF}_4(s) + 2 \text{ SF}_4(g) \rightarrow 2 \text{ SF}_6(g) + \text{Xe}(g)$$

18.6   *Xenon Oxides*

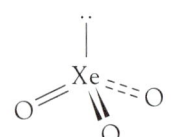

Figure 18.4 Possible representation of the bonding in the xenon trioxide molecule.

Xenon forms two common oxides, xenon trioxide and xenon tetroxide. As mentioned in Chapter 16, Section 16.4, oxygen usually "brings out" a higher oxidation number of an element than does fluorine; and with xenon, this is certainly the case.

Xenon trioxide is a colorless, deliquescent solid that is quite explosive. The oxide is an extremely strong oxidizing agent, although its reactions are often kinetically slow. Because of its lone pair, it is a trigonal pyramidal molecule, as predicted by VSEPR theory (Figure 18.4). The bond length indicates that, as discussed before for silicon, phosphorus, and sulfur, there is some degree of multiple bonding.

Xenon trioxide reacts with dilute base to give the hydrogen xenate ion, $HXeO_4^-$. However, this ion is not stable, and disproportionation to xenon gas and the perxenate ion, XeO_6^{4-}, occurs:

$$\text{XeO}_3(s) + \text{OH}^-(aq) \rightarrow \text{HXeO}_4^-(aq)$$

$$2 \text{ HXeO}_4^-(aq) + 2 \text{ OH}^-(aq) \rightarrow \text{XeO}_6^{4-}(aq) + \text{Xe}(g) + \text{O}_2(g) + 2 \text{ H}_2\text{O}(l)$$

Alkali and alkaline earth metal salts of the perxenate ion can be crystallized; they are all colorless, stable solids. In the perxenate ion, the xenon is surrounded by the six oxygen atoms in an octahedral arrangement. Perxenates are among the most powerful oxidizing agents known, which is not really surprising considering that the xenon is in the formal oxidation state of +8. For example, they rapidly oxidize manganese(II) ion to permanganate, themselves being reduced to the hydrogen xenate ion:

$$\text{XeO}_6^{4-}(aq) + 5 \text{ H}^+(aq) + 2 \text{ e}^- \rightarrow \text{HXeO}_4^-(aq) + 2 \text{ H}_2\text{O}(l)$$

$$4 \text{ H}_2\text{O}(l) + \text{Mn}^{2+}(aq) \rightarrow \text{MnO}_4^-(aq) + 8 \text{ H}^+(aq) + 5 \text{ e}^-$$

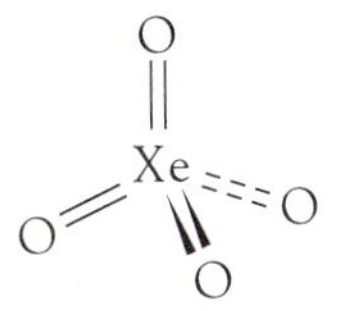

Figure 18.5 A possible representation of the bonding in the xenon tetraoxide molecule.

Xenon tetroxide is prepared by adding concentrated sulfuric acid to solid barium perxenate:

$$Ba_2XeO_6(s) + 2\ H_2SO_4(aq) \rightarrow 2\ BaSO_4(s) + XeO_4(g) + 2\ H_2O(l)$$

This oxide, also with xenon in the oxidation state of $+8$, is an explosive gas. Its structure has been shown to be tetrahedral (Figure 18.5). This geometry is expected from VSEPR arguments. Furthermore, the fact that it is a gas at room temperature suggests that it is probably a nonpolar molecule.

18.7 Biological Aspects

Radon: The Dangerous Gas

None of the noble gases have any positive biological functions. Radon, however, has recently been in the news because it accumulates inside buildings. The radiation it releases as it decays may be a significant health hazard. Radon isotopes are produced during the decay of uranium and thorium. Only one isotope, radon-222, has a half-life long enough (3.8 days) to cause major problems, and this particular isotope is produced in the decay of uranium-238. This process is happening continuously in the rocks and soils, and the radon produced normally escapes into the atmosphere. However, the radon formed beneath dwellings permeates through cracks in concrete floors and basement walls, a process that is enhanced when the pressure inside the house is lower than the external value. This pressure differential occurs when ventilation fans, clothes dryers, and other mechanical devices pump air out of the house. Furthermore, our concern about saving energy has prompted us to build houses that are more airtight, thereby preventing exchange of radon-rich interior air with exterior fresh air.

It is not actually the radon that is the problem but the solid radioactive isotopes produced by its subsequent decay. These solid particles attach themselves to lung tissue, subsequently irradiating it with α particles (helium nuclei) and β particles (electrons), disrupting the cells, and even initiating lung cancer. Notice from Table 18.5 that it is the steps following the formation of radon which have the short half-lives and therefore the high rate of radiation.

Awareness of the problem arose from an incident at the Limerick Nuclear Generating Station, Pennsylvania. When leaving such an installation, workers have to pass through a radiation detector to ensure they have not become contaminated with radioactive materials. By accident, one of the workers, Stanley Watras, entered the plant through the detector, setting it off. Investigators were puzzled until they checked his house, which showed very high levels of radiation to which he and his family were constantly being exposed. The radiation was a result of enormous levels of radon and its decay products leaking into the house from a vein of uranium-bearing ore that lay under the house.

The very unpleasant radioactivity decontamination process is shown vividly in the movie Silkwood.

There is clear evidence that exposure to high levels of radon does increase the probability of lung cancer. The concentration of radon at which significant hazard exists is still under debate. Certainly, cigarette smoking presents a far greater hazard than radon exposure for the average person. However, investigators have discovered houses where the radiation levels are about 100 times greater than normal levels. Usually these houses are built over geological deposits that produce high levels of radon. It is possible to have a home tested for radon by a certified technician or testing process,

Table 18.5 The major radioactive decays and half-lives for the uranium-238 decay chain

Decay step	Half-life
$^{238}_{92}U \rightarrow\, ^{234}_{90}Th + ^{4}_{2}He$	4.5×10^9 years
$^{234}_{90}Th \rightarrow\, ^{234}_{91}Pa + ^{0}_{-1}e$	24 days
$^{234}_{91}Pa \rightarrow\, ^{234}_{92}U + ^{0}_{-1}e$	1.2 min
$^{234}_{92}U \rightarrow\, ^{230}_{90}Th + ^{4}_{2}He$	2.5×10^5 years
$^{230}_{90}Th \rightarrow\, ^{226}_{88}Ra + ^{4}_{2}He$	8.0×10^4 years
$^{226}_{88}Ra \rightarrow\, ^{222}_{86}Rn + ^{4}_{2}He$	1.6×10^3 years
$^{222}_{86}Rn \rightarrow\, ^{218}_{84}Po + ^{4}_{2}He$	3.8 days
$^{218}_{84}Po \rightarrow\, ^{214}_{82}Pb + ^{4}_{2}He$	3.1 min
$^{214}_{82}Pb \rightarrow\, ^{214}_{83}Bi + ^{0}_{-1}e$	27 min
$^{214}_{83}Bi \rightarrow\, ^{214}_{84}Po + ^{0}_{-1}e$	20 min
$^{214}_{84}Po \rightarrow\, ^{210}_{82}Pb + ^{4}_{2}He$	1.6×10^{-4} s
$^{210}_{82}Pb \rightarrow\, ^{210}_{83}Bi + ^{0}_{-1}e$	22 years
$^{210}_{83}Bi \rightarrow\, ^{210}_{84}Po + ^{0}_{-1}e$	5.0 days
$^{210}_{84}Po \rightarrow\, ^{206}_{82}Pb + ^{4}_{2}He$	138 days

but it is always advisable to have a properly ventilated house, to prevent not only possible radon accumulation but also, more generally, to flush out continuously all the air pollutants that are present in most modern well-sealed houses and offices.

18.8 Element Reaction Flowchart

A Flowchart is shown for xenon, the only noble gas with a significant chemistry.

EXERCISES

18.1 Write balanced chemical equations for the following chemical reactions:
(a) xenon with difluorine in a 1:2 mole ratio
(b) xenon tetrafluoride with phosphorus trifluoride

18.2 Write balanced chemical equations for the following chemical reactions:
(a) xenon difluoride with water
(b) solid barium perxenate with sulfuric acid

18.3 Describe the trends in the physical properties of the noble gases.

18.4 Why is argon (thermal conductivity 0.017 $J \cdot s^{-1} \cdot m^{-1} \cdot K^{-1}$ at 0°C) more commonly used as a thermal insulation layer in glass windows than xenon (thermal conductivity 0.005 $J \cdot s^{-1} \cdot m^{-1} \cdot K^{-1}$ at 0°C)?

18.5 What are the unusual features of liquid helium?

18.6 Why would we expect noble-gas compounds to exist?

18.7 A bright green ion, Xe_2^+, has been identified. Suggest the bond order for this ion, showing your reasoning.

18.8 Bartlett's noble-gas compound is now known to contain the XeF^+ ion. Construct the electron-dot formula for this ion. By comparison with interhalogen chemistry, would this ion be predicted to exist?

18.9 What are the key thermodynamic factors in the formation of xenon–fluorine compounds?

18.10 For the formation of xenon tetrafluoride, $\Delta G_f^\circ = -121.3$ kJ·mol^{-1} and $\Delta H_f^\circ = -261.5$ kJ·mol^{-1}. Determine the value for the standard entropy of formation of this compound. Why do you expect the sign of the entropy change to be negative?

18.11 Estimate the enthalpy of formation of xenon tetrachloride from the following data: bond energy (Xe—Cl) [estimated] = 86 kJ·mol^{-1}; enthalpy of sublimation of solid xenon tetrachloride [estimated] = 60 kJ·mol^{-1}. Obtain any other required data from the data tables in the Appendices.

18.12 One of the few krypton compounds known is krypton difluoride, KrF_2. Calculate the enthalpy of formation of this compound using the data in the Appendices (the Kr—F bond energy is 50 kJ·mol^{-1}).

18.13 Construct an electron-dot structure for $XeOF_4$ with a xenon–oxygen (a) single bond; (b) double

bond. Decide which is more significant on the basis of formal charge.

18.14 Determine the shapes of the following ions: (a) XeF_3^+; (b) XeF_5^+; (c) XeO_6^{4-}.

18.15 Determine the oxidation number of xenon in each of the compounds in Exercise 18.14.

18.16 Which of the noble gases would you choose as:
(a) the lowest temperature liquid refrigerant?
(b) the least expensive inert atmosphere?
(c) an electric discharge light source requiring a safe gas with the lowest ionization energy?

18.17 It is possible to prepare a series of compounds of formula $MXeF_7$, where M is an alkali metal ion. Which alkali metal ion should be used in order to prepare the most stable compound?

18.18 Suggest an explanation why xenon forms compounds with oxygen in the +8 oxidation state but with fluorine only up to an oxidation state of +6.

18.19 Write a balanced chemical equation for the reaction of krypton difluoride with gold to give $(KrF)^+(AuF_6)^-$.

18.20 Briefly discuss why radon is a health hazard.

18.21 Write balanced chemical equations corresponding to each transformation in the Element Reaction Flowchart.

BEYOND THE BASICS

18.22 Why is $XeCl_2$ likely to be much less stable than XeF_2?

18.23 Xenon difluoride reacts with antimony pentafluoride, SbF_5, to give an electrically conducting solution. Write a chemical equation for the reaction.

18.24 The reduction potentials for the $H_4XeO_6(aq)/XeO_3(aq)$ reduction half-reaction is $+2.3$ V, whereas that for the $XeO_3(aq)/Xe(g)$ reduction half-reaction is $+1.8$ V. Calculate a value for the

half-cell potential of:

$$8\,H^+(aq) + H_4XeO_6(aq) + 8\,e^- \rightarrow Xe(g) + 6\,H_2O(l)$$

18.25 The fact that argon difluoride has not been prepared despite strenuous efforts suggests that the argon–fluorine bond must be very weak. Use a theoretical enthalpy cycle to determine an approximate maximum value that the Ar—F bond energy could have. Do entropy factors favor or oppose the formation of argon difluoride? Give your reasoning.

Introduction to Transition Metal Complexes

Chapter **19**

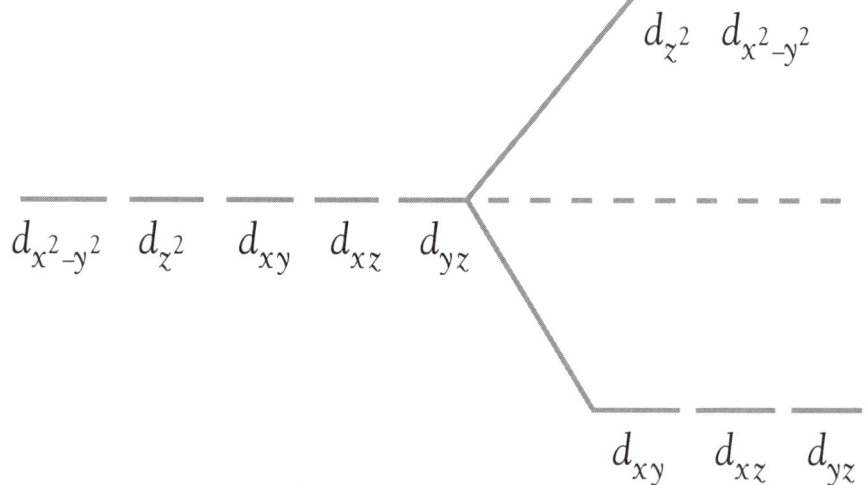

*I*t is the enormity of the number of compounds that is the obvious feature of the transition metals. We will see the ways in which transition metals can form this galaxy of compounds, introduce the naming system used for them, and discuss the modern theories of bonding that are used to explain their diversity. In addition, we will revisit hard–soft acid–base concepts in the context of transition metal compounds.

The compounds of the transition metals have always held a special interest for inorganic chemists. Whereas the compounds of the main group metals are almost always white, the transition metal compounds come in every color of the rainbow. Chemists were fascinated by the fact that it was sometimes possible to make compounds of the same formula but in different colors. For example, chromium(III) chloride hexahydrate, $CrCl_3 \cdot 6H_2O$, can be synthesized in purple, pale green, and dark green forms.

The initial explanation for this multitude of compounds was that, like organic compounds, the components of the transition metal compounds were strung out in chains. It was the Swiss chemist Alfred Werner who, during a restless night in 1893, devised the concept that transition metal compounds consisted of the metal ion surrounded by the other ions and molecules. This novel theory was accepted in Germany, but it received a hostile reception in the English-speaking world. Over the next 8 years, Werner and his students prepared several series of transition metal compounds, searching for proof of his theory. As more and more evidence accumulated, the opposition disintegrated, and he was awarded the Nobel Prize for chemistry in 1913 in recognition of his contribution. Although Werner deserved the credit for devising this theory, we must always keep in mind that the toil at the research bench was done mainly by his research students. In particular, one of the most crucial pieces of evidence was established by a young British student, Edith Humphrey.

19.1 Transition Metals

Though some people use the terms "d-block elements" and "transition metals" interchangeably, this is not strictly true. Inorganic chemists generally restrict the term *transition metal* to that of an element that has at least one simple ion with an incomplete outer set of d electrons. For example, chromium has two common oxidation states (plus several other less common ones). The $+3$ oxidation state has a partially filled d-set even though the $+6$ state has an empty d-set. Thus, chromium is considered to be a transition metal.

Atom	Electron configuration	Ion	Electron configuration
Cr	$[Ar]4s^1 3d^5$	Cr^{3+}	$[Ar]3d^3$
		Cr^{6+}	$[Ar]$

On the other hand, the only common oxidation state of scandium is $+3$. As this state has an empty d-set, scandium (and the other members of Group 3) is excluded from the transition metal designation. In fact, as we mentioned in Chapter 9, Section 9.5, scandium closely resembles the main group metal aluminum in its chemical behavior. The Group 3 elements also resemble the $4f$-block elements in their chemistry and hence are discussed together with them in Chapter 23.

Atom	Electron configuration	Ion	Electron configuration
Sc	$[Ar]4s^2 3d^1$	Sc^{3+}	$[Ar]$

At the other end of the d-block, we have to consider the elements that retain a full d-set in their oxidation states. The Group 12 elements—zinc, cadmium, and mercury—fit this category. Their common oxidation state is $+2$. Thus, these elements are not considered as transition metals. For this reason, the Group 12 elements are discussed in a separate chapter—Chapter 21.

Atom	Electron configuration	Ion	Electron configuration
Zn	$[Ar]4s^2 3d^{10}$	Zn^{2+}	$[Ar]3d^{10}$

The postactinoid metals from Rf to Uuu are transition metals, but because they are all short-lived radioactive elements, it is common to discuss

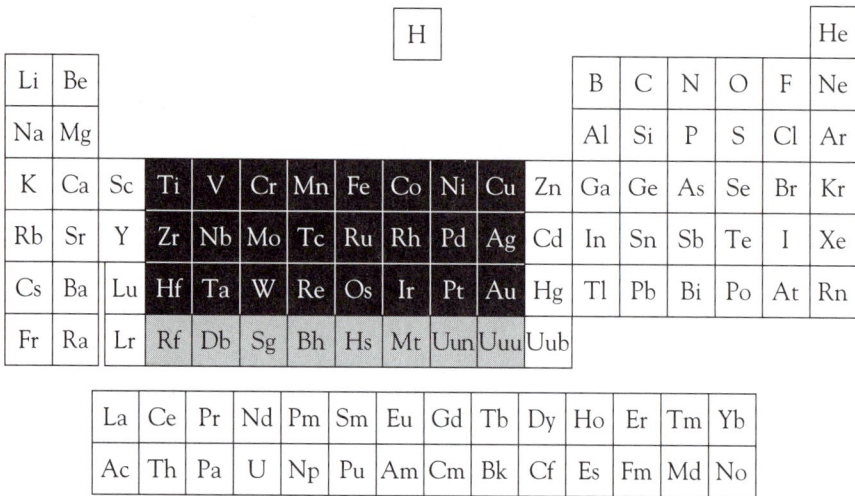

Figure 19.1 A Periodic Table with the elements usually defined as transition metals shown in black and corresponding postactinoid metals shaded.

them together with the actinoid metals (Chapter 23). To summarize, the elements commonly considered as transition metals are shown in Figure 19.1.

19.2   Transition Metal Complexes

We rarely encounter a "naked" transition metal ion, because the ion is usually covalently bonded to other ions or molecules. These groupings are called metal *complexes*, and it is the number and diversity of the metal complexes that provide the wealth of transition metal chemistry.

It was Alfred Werner's proposal that metal ions had not only a particular value of charge but also some characteristic "combining power." That is, there was a specific number of molecules or ions with which a transition metal would combine. We now refer to this number (or numbers) as the *coordination number(s)* of the element, and it is usually 4 or 6. The molecules or ions that are covalently bonded to the central metal ion are called *ligands.*

One of the best illustrations of the concept is shown by the series of compounds that can be prepared from platinum(II) and ammonia, chloride ions, and potassium ions. These compounds are shown in Table 19.1. The key to understanding this multiplicity of compounds was provided by measurements of the electrical conductivity of their solutions and by gravimetric

Table 19.1   Formulas and structures of a series of platinum(II) complexes

Composition	Number of ions	Modern formulation
$PtCl_2 \cdot 4\,NH_3$	3	$[Pt(NH_3)_4]^{2+}\ 2\,Cl^-$
$PtCl_2 \cdot 3\,NH_3$	2	$[PtCl(NH_3)_3]^+\ Cl^-$
$PtCl_2 \cdot 2\,NH_3$	0	$[PtCl_2(NH_3)_2]$ (two forms)
$KPtCl_3 \cdot NH_3$	2	$K^+[PtCl_3(NH_3)]^-$
K_2PtCl_4	3	$2\,K^+\ [PtCl_4]^{2-}$

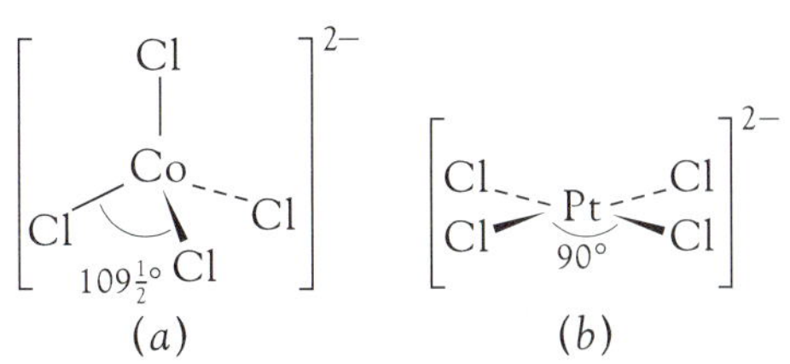

Figure 19.2 (a) The tetrahedral tetra-chlorocobaltate(II) ion and (b) the square planar tetrachloroplatinate(II) ion.

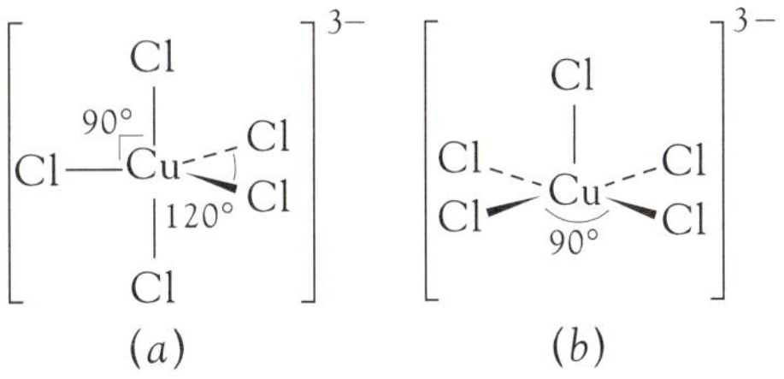

Figure 19.3 The two stereochemical arrange-ments of the pentachlorocuprate(II) ion: (a) trigo-nal bipyramid and (b) square-based pyramid.

analysis using silver nitrate solution. Thus, the presence of three ions in so-lution and 2 moles of precipitating siliver chloride in the first case can only be explained if the two chloride ions are not covalently bonded to the plat-inum. In the second complex, the presence of two ions and only 1 mole of free chloride ion that can be precipitated as silver chloride shows that only one chloride ion is ionic and that the other must be part of the coordina-tion sphere of the platinum. Similar arguments can be made for the other compounds.

The bonding theories will be discussed shortly, but for the moment we can consider complex formation to be the result of coordinate covalent bond formation, the metal ion acting as a Lewis acid and the ligands acting as Lewis bases.

19.3 Stereochemistries

Transition metal complexes have a wide range of shapes. With four ligands, there are two alternatives: tetrahedral and square planar. Tetrahedra are more common in Period 4 transition metals, and square planar complexes are more prevalent among the Periods 5 and 6 transition metal series. Figure 19.2(a) shows the tetrahedral geometry of the tetrachlorocobaltate(II) ion, $[CoCl_4]^{2-}$, and Figure 19.2(b) shows the square planar configuration of the tetrachloroplatinate(II) ion, $[PtCl_4]^{2-}$.

There are few simple complexes with five ligands, but it is interesting to find that, like the four-ligand situation, these complexes have two stereo-chemistries: trigonal bipyramidal, like the main group compounds, and square-based pyramidal (Figure 19.3). The energy difference between these two configurations must be very small, because the pentachlorocuprate(II) ion, $[CuCl_5]^{3-}$, adopts both structures in the solid phase, the preference de-pending on the identity of the cation.

The most common number of simple ligands is 6, and almost all of these complexes adopt the octahedral arrangement. This configuration is shown in Figure 19.4 for the hexafluorocobaltate(IV) ion, $[CoF_6]^{2-}$. Cobalt com-pounds usually have cobalt oxidation states of $+2$ and $+3$; thus, as we discussed in Chapter 17, Section 17.2, it is fluoride that has to be used to attain the unusual higher oxidation state of $+4$.

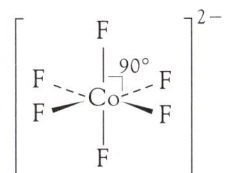

Figure 19.4 The octahedral hexafluorocobaltate(IV) ion.

Ligands

As mentioned earlier, the atoms, molecules, or ions attached to the metal ion are called ligands. For most ligands, such as water or chloride ion,

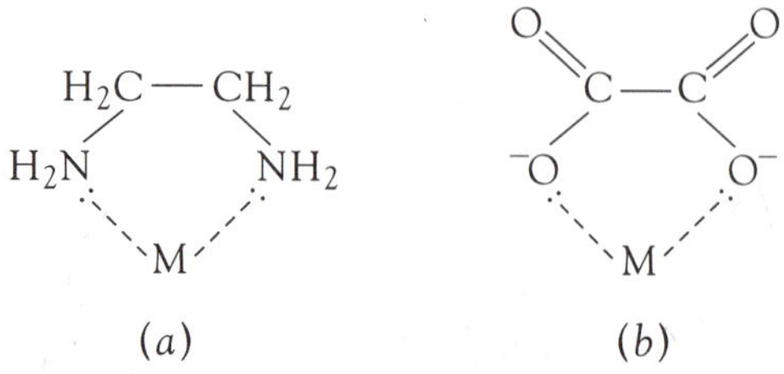

(a)

(b)

Figure 19.5 (a) The 1,2-diaminoethane molecule, $H_2NCH_2CH_2NH_2$, and (b) the oxalate ion, $^-O_2CCO_2^-$. The atoms that coordinate to the metal have dashed lines to the metal ion, M, showing how the bonding will occur.

each occupies one coordination site. These species are known as *monodentate* ligands (from the Greek word meaning "one-toothed"). There are several molecules and ions that take up two bonding sites; common examples are the 1,2-diaminoethane molecule, $H_2NCH_2CH_2NH_2$ (commonly called ethylenediamine and abbreviated "en"), and the oxalate ion, $^-O_2CCO_2^-$. Such groups are called *bidentate* ligands (Figure 19.5). More complex ligands can be synthesized and will bond to three, four, five, and even six coordination sites. These species are called tridentate, tetradentate, pentadentate, and hexadentate ligands, respectively. All ligands that form more than one attachment to a metal ion are called *chelating* ligands (from the Greek *chelos*, meaning "claw-like").

Ligands and Oxidation States of Transition Metals

Another feature common to transition metals is their wide range of oxidation states. The preferred oxidation state is very dependent on the nature of the ligand; that is, various types of ligands stabilize low, normal, or high oxidation states.

Ligands That Tend To Stabilize Low Oxidation States The two common ligands that particularly favor metals in low oxidation states are the carbon monoxide molecule and the isoelectronic cyanide ion. For example, iron has an oxidation number of 0 in $Fe(CO)_5$.

Ligands That Tend To Stabilize "Normal" Oxidation States Most common ligands, such as water, ammonia, and halide ions, fall into this category. For example, iron exhibits its common oxidation states of +2 and +3 with water: $[Fe(OH_2)_6]^{2+}$ and $[Fe(OH_2)_6]^{3+}$. There are many cyanide complexes in normal oxidation states as well. This is not unexpected, for the ion is a pseudo-halide ion (as discussed in Chapter 9) and hence is capable of behaving like a halide ion.

Ligands That Tend To Stabilize High Oxidation States Like nonmetals, transition metals only adopt high oxidation numbers when complexed with fluoride and oxide ions. We have already mentioned the hexafluorocobaltate(IV) ion, $[CoF_6]^{2-}$, as one example. In the tetraoxoferrate(VI) ion, $[FeO_4]^{2-}$, the oxide ions stabilize the abnormal +6 oxidation state of iron.

19.4 *Isomerism in Transition Metal Complexes*

In the early history of coordination chemistry, the existence of pairs of compounds with the same formula yet different properties proved to be very perplexing to inorganic chemists. Werner was among the first to realize that the different properties represented different structural arrangements (isomers). Isomers can be categorized as structural isomers and stereoisomers. For *stereoisomers,* the bonds to the metal ion are identical, whereas the bonds of *structural isomers* are different. These categories can be further subdivided, as shown in Figure 19.6.

Structural Isomerism

Structural isomerism has three common types: linkage isomerism, ionization isomerism, and hydration isomerism. Ionization and hydration isomerism are

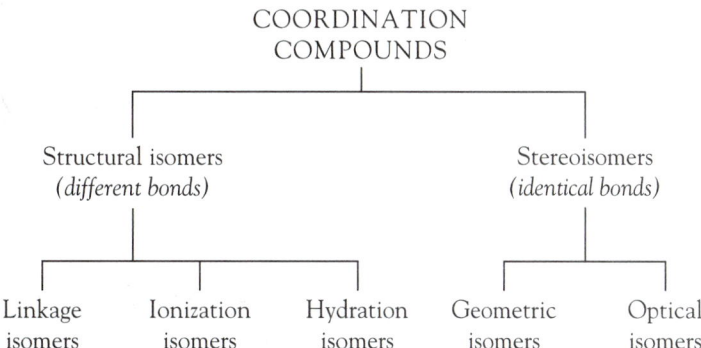

COORDINATION
COMPOUNDS

Structural isomers
(*different bonds*)

Stereoisomers
(*identical bonds*)

Linkage
isomers

Ionization
isomers

Hydration
isomers

Geometric
isomers

Optical
isomers

Figure 19.6   Isomer types.

sometimes categorized together as *coordination-sphere isomerism* because in both cases it is the identity of the ligands that differs.

Linkage Isomerism   Some ligands can form bonds through more than one atom. For example, the thiocyanate ion, NCS^-, can bond through either the nitrogen or the sulfur. This particular ambidentate ligand was discussed in Chapter 7, Section 7.9, under Borderline Bases, as the choice of ligating atom depends in part on the hard–soft acid nature of the metal ion. A classic example of *linkage isomerism* involves the nitrite ion, which can form bonds through the nitrogen atom, —NO_2, or through one of the oxygen atoms, —ONO. A pentamminecobalt(III) complex, $Co(NH_3)_5Cl_2(NO_2)$, conveniently illustrates this isomerism as the two isomers have different colors. One of these, the red form, contains the $[Co(ONO)(NH_3)_5]^{2+}$ ion, in which one of the oxygen atoms of the nitrite ion is bonded to the cobalt(III) ion (Figure 19.7(a)). The other isomer, the yellow form, contains the $[Co(NO_2)(NH_3)_5]^{2+}$ ion, in which the nitrogen atom is bonded to the cobalt(III) ion (Figure 19.7(b)).

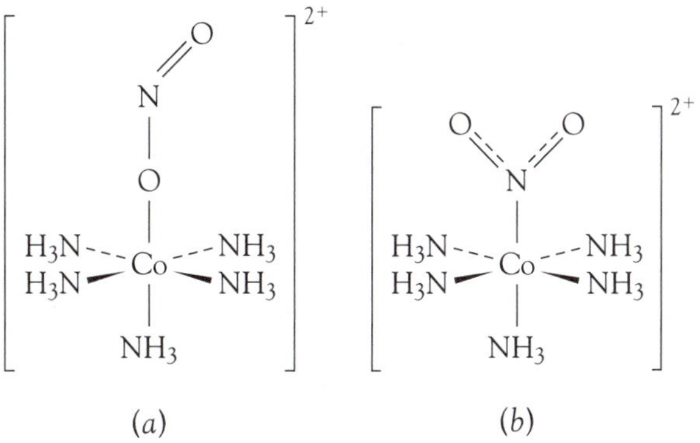

Figure 19.7   The two linkage isomers of the pentam-minecobalt(III) nitrite complex. (a) the nitrito-form, (b) the nitro-form.

Ionization Isomerism   *Ionization isomers* give different ions when dissolved in solution. Again, there is a classic example: $Co(NH_3)_5Br(SO_4)$. If barium ion is added to a solution of the red-violet form, a white precipitate of barium sulfate forms. Addition of silver ion has no effect. Hence, the complex ion must have the formula $[CoBr(NH_3)_5]^{2+}$, with an ionic sulfate ion. A solution of the red form, however, does not give a precipitate with barium ion; instead, a cream-colored precipitate is formed with silver ion. Hence, this complex ion must have the structure of $[CoSO_4(NH_3)_5]^+$, with an ionic bromide ion.

Hydration Isomerism   *Hydration isomerism* is very similar to ionization isomerism in that the identity of the ligand species is different for the two isomers. In this case, rather than different types of ions, it is the proportion of coordinated water molecules that differs between isomers. It is possible to have a series of complexes in which the proportion of coordinated water molecules differs. The three structural isomers of formula $CrCl_3 \cdot 6H_2O$ provide the best example. In the violet form, the six water molecules are coordinated; hence, the formula for this compound is more correctly written as $[Cr(OH_2)_6]^{3+} \, 3Cl^-$. As evidence, all three chloride ions are precipitated from solution by silver ion. In the light green form, one of the chloride ions is not precipitated by silver ion; hence, the complex is assigned the structure $[CrCl(OH_2)_5]^{2+} \, 2Cl^- \cdot H_2O$ (light green). Finally, only one chloride ion can be precipitated by silver ion from a solution of the dark green form; hence, this compound must have the structure $[CrCl_2(OH_2)_4]^+ \, Cl^- \cdot 2H_2O$.

Stereo-isomerism

The two types of inorganic stereoisomers, geometric and optical, are parallel to those found in organic chemistry, except that in inorganic chemistry, optical isomerism is most common for a metal ion in an octahedral environment rather than for the tetrahedral environment of organic carbon compounds.

Geometric Isomerism   Inorganic *geometric isomers* are analogous to organic geometric isomers that contain carbon–carbon double bonds. Geometric isomers must have two different ligands, A and B, attached to the same metal, M. For square planar compounds, geometric isomerism occurs in compounds of the form MA_2B_2. The term *cis* is used for the isomer in which ligands of one kind are neighbors, and *trans* is used to identify the isomer in which ligands of one kind are opposite each other (Figure 19.8). Geometric isomers also exist for square planar complexes of the form MA_2BC, where *cis* refers to ligands A being neighbors; and *trans*, to ligands A being opposite each other.

In the earlier discussion of platinum(II) complexes (see Table 19.1), we noted that there are two chemically different forms of $[PtCl_2(NH_3)_2]$. This observation indicates the probable geometry, even though there are two possibilities: tetrahedral and square planar. If the isomers were tetrahedral, all of the bond directions would be equivalent and only one form of the compound would be possible. However, as Figure 19.8 shows, the square planar arrangement allows two geometric isomers.

There are two formulas of octahedral compounds having only two kinds of ligands for which geometric isomers are possible. Compounds with the formula MA_4B_2 can have the two B ligands on opposite sides or as neighbors. Hence, these, too, are known as *trans* and *cis* isomers (Figure 19.9). Octahedral compounds with the formula MA_3B_3 also can have isomers (Figure 19.10). If one set of ligands (A) occupies three sites of the corners of a triangular face of the octahedron, 90° apart from each other, and the B set, the opposing triangular face, then the prefix *fac-* (for facial) is used. However, if the three A ligands occupy the three sites in the horizontal plane and the B set, three sites in the vertical plane, then the geometry is described by the prefix *mer-* (for meridional).

Optical Isomerism   Again, inorganic *optical isomerism* is analogous to that of organic chemistry. Optical isomers are pairs of compounds in which one isomer is a nonsuperimposable mirror image of the other. One of the characteristics of optical isomers is that they rotate the plane of polarized light, one isomer rotating the light in one direction and the other isomer in the opposite direction. Compounds that exist as optical isomers are called ***chiral compounds***.

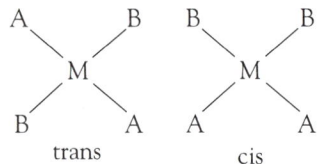

Figure 19.8   The geometric isomers of a square planar MA_2B_2 arrangement.

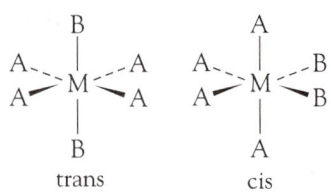

Figure 19.9   The geometric isomers of an octahedral MA_4B_2 arrangement.

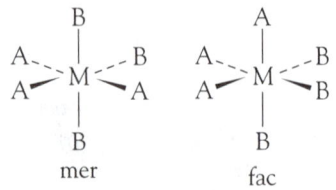

Figure 19.10   The geometric isomers of an octahedral MA_3B_3 arrangement.

This form of isomerism is found most commonly when a metal is surrounded by three bidentate ligands; 1,2-diaminoethane, $H_2NCH_2CH_2NH_2$, en, mentioned earlier, is one such ligand. Hence, the complex ion would have the formula $[M(en)_3]^{n+}$, where $n+$ is the charge of the transition metal ion. The two optical isomers of this complex ion are shown in Figure 19.11, where the 1,2-diaminoethane molecules are depicted schematically as pairs of linked nitrogen atoms.

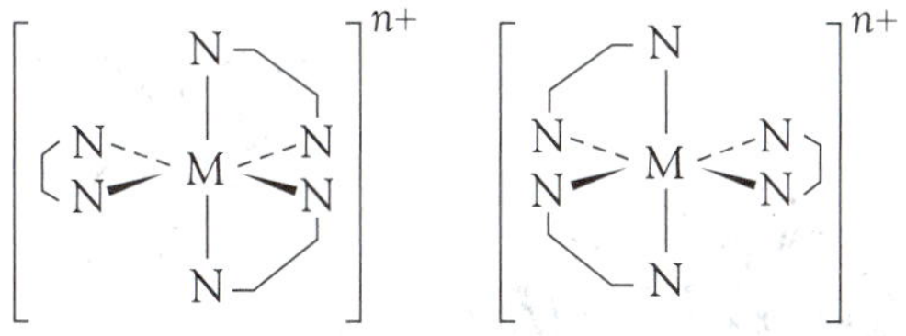

Figure 19.11 The two optical isomers of the $[M(en)_3]^{n+}$ ion. The linked nitrogen atoms represent the 1,2-diaminoethane bidentate ligands.

Platinum Complexes and Cancer Treatment

It is a common misconception that scientific research works the same way as does technology, where goals are set and the appropriate solutions found. In science, however, there is so much that is not known that we still rely to a large extent on observing the unexpected. It was in 1965 that Barnett Rosenberg, of Michigan State University, was studying the rate of bacterial growth in the presence of electric fields. He and his co-researchers were surprised to find that the bacteria in electric fields were growing without dividing. The group spent a considerable amount of time looking for possible causes of this, such as pH changes, temperature changes, and so on. Having excluded every probable cause, they examined the electrodes that they were using to generate the electric charge. These were made of platinum, a metal that was "well known" to be extremely unreactive.

However, their tests showed that some of the platinum metal was being oxidized, and it was the oxidation products, the diamminedichloroplatinum(II), $PtCl_2(NH_3)_2$, and diamminetetrachloroplatinum(IV), $PtCl_4(NH_3)_2$, molecules, that were causing the bacterial abnormalities. Further, only the *cis* geometric isomers were active. This biological activity of platinum compounds was completely unexpected. Because they prevented cell division, the compounds were tested for antitumor activity, and the *cis*-diamminedichloroplatinum(II) compound seemed particularly effective. The compound is now available for cancer treatment under the name cisplatin. It does have side effects, and chemists, now that they are aware of the potential of platinum compounds, are looking for more effective and less toxic analogs. The key to this compound's effectiveness seems to be the ability of the *cis*-$(H_3N)_2Pt$ unit to cross-link DNA units, bending and partially unwinding the double helix, thereby preventing further DNA synthesis. The *trans* isomers show no biological activity. Hence, these compounds demonstrate the influence of isomerism on the chemical behavior of compounds.

19.5   Naming Transition Metal Complexes

Because of the multiplicity of transition metal complexes, the simple system of inorganic nomenclature proved unworkable. As a result, special rules for naming transition metal complexes were devised.

1. Nonionic species are written as one word; ionic species are written as two words with the cation first.

2. The central metal atom is identified by name, which is followed by the formal oxidation number in Roman numerals in parentheses, such as (IV) for a +4 state and (-II) for a 2− state. If the complex is an anion, the ending -*ate* adds to the metal name or replaces any -*ium, -en,* or -*ese* ending. Thus, we have cobaltate and nickelate, but chromate and tungstate (not chromiumate or tungstenate). For a few metals, the anion name is derived from the old Latin name of the element: ferrate (iron), argentate (silver), cuprate (copper), and aurate (gold).

3. The ligands are written as prefixes of the metal name. Neutral ligands are given the same name as the parent molecule, whereas the names of negative ligands are given the ending -*o* instead of -*e*. Thus, sulfate becomes sulfato and nitrite becomes nitrito. Anions with -*ide* endings have them completely replaced by -*o*. Hence, chloride ion becomes chloro; iodide, iodo; cyanide, cyano; and hydroxide, hydroxo. There are three special names: Coordinated water is commonly named *aqua;* ammonia, *ammine;* and carbon monoxide, *carbonyl.*

4. Ligands are always placed in alphabetical order. (In chemical formulas, the symbols of anionic ligands always precede those of neutral ligands.)

5. For multiple ligands, the prefixes *di-, tri-, tetra-, penta-,* and *hexa-* are used for 2, 3, 4, 5, and 6, respectively.

6. For multiple ligands already containing numerical prefixes (such as 1,2-diaminoethane), the prefixes used are *bis-, tris-,* and *tetrakis-* for 2, 3, and 4. This is not a rigid rule. Many chemists use these prefixes for all polysyllabic ligands.

Examples

Let us try naming some of the platinum metal complexes discussed earlier in this chapter. Notice that, in chemical formulas, square brackets, [ ], are used to enclose all units linked together by covalent bonds.

Example 1: $[\text{Pt}(\text{NH}_3)_4]\text{Cl}_2$ Because this compound has separate ions, the name will consist of (at least) two words (rule 1). There are two negative chloride ions outside of the complex, so the complex itself must have the formula $[\text{Pt}(\text{NH}_3)_4]^{2+}$. The ammonia ligands are neutral; hence, the platinum must have an oxidation state of +2. Hence, we start with the stem name platinum(II) (rule 2). The ligand is ammonia; hence, we use the name ammine (rule 3). But there are four ammonia ligands, so we add a prefix and get tetraammine (rule 5). Finally, the chloride anions must be included. They are free, uncoordinated chloride ions, so they are called chloride, not chloro.

We do not identify the number of chloride ions because the oxidation number enables us to deduce it. Hence, the full name is tetraammineplatinum(II) chloride.

Example 2: $[PtCl_2(NH_3)_2]$ This is a nonionic species; hence, it will have a one-word name (rule 1). Again, to balance the two chloride ions, the platinum is in the +2 oxidation state, so we start with platinum(II) (rule 2). The ligands are named ammine for ammonia and chloro for chloride (rule 3). Alphabetically, ammine comes before chloro (rule 4); thus, we have the prefix diamminedichloro (rule 5). The whole name is diamminedichloroplatinum(II). As mentioned earlier, this particular compound is square planar and exists as two geometric isomers. We refer to these isomers as *cis*-diamminedichloroplatinum(II) and *trans*-diamminedichloroplatinum(II).

Example 3: $K_2[PtCl_4]$ Again, two words are needed (rule 1), but in this case, the platinum is in the anion, $[PtCl_4]^{2-}$. The metal is in the +2 oxidation state; thus, the anionic name will be platinate(II) (rule 2). There are four chloride ligands, giving the prefix tetrachloro (rules 3 and 5), and the separate potassium cations. The complete name is potassium tetrachloroplatinate(II).

Example 4: $[Co(en)_3]Cl_3$ The complex ion is $[Co(en)_3]^{3+}$. Because (en), $H_2NCH_2CH_2NH_2$, is a neutral ligand, the cobalt must be in a +3 oxidation state. The metal, then, will be cobalt(III). The full name of the ligand is 1,2-diaminoethane and contains a numerical prefix, so we use the alternate prefix set (rule 6) to give tris(1,2-diaminoethane)—parentheses are used to separate the ligand name from the other parts of the name. Finally, we add the chloride anions. The full name is tris(1,2-diaminoethane)cobalt(III) chloride.

Unfortunately, there are a number of transition metal compounds that have well-known common names in addition to their systematic names. For example, a few complexes are identified by the name of their discoverer, such as Zeise's compound, Wilkinson's catalyst, and Magnus's green salt.

The Ewens–Bassett Nomenclature System

The nomenclature system that we have used to this point was first devised by Alfred Stock in 1919, and it is still the one most widely used. As we have seen, the system uses Roman numerals to indicate the oxidation state of the central atom, and from that, the ion charge can be calculated. An alternative system was devised by R. Ewens and H. Bassett in 1949. According to their rules, the ion charge is bracketed in Arabic numerals. It was, in fact, the Ewens–Bassett system that we used in Chapter 11, Section 11.8, to distinguish the O_2^- ion from the O_2^{2-} ion, the former being called the dioxide(1−) ion; and the latter, the dioxide(2−) ion. (These are much more useful names than the traditional superoxide and peroxide.) The Ewens–Bassett system has advantages for systematic naming of polyatomic ions containing one element: for example, the dioxides mentioned here: C_2^{2-}, dicarbide(2−) (instead of carbide or acetylide); O_3^-, trioxide(1−) (not ozonide); and N_3^-, trinitride(1−) (not azide).

Apart from indicating charge rather than oxidation number, the Stock and Ewens–Bassett systems employ the same nomenclature rules. The use of either Roman or Arabic numerals identifies the method used in a particular name; for example, $K_4[Fe(CN)_6]$ is called potassium hexacyanoferrate(II) by the Stock method, because the iron in the complex ion has a formal oxidation state of $+2$. The Ewens–Bassett name for this compound is potassium hexacyanoferrate$(4-)$, because the complex anion has a $4-$ charge. For neutral molecules, no number is shown; thus, *cis*-$[Pt(NH_3)_2Cl_2]$ would be *cis*-diamminedichloroplatinum rather than the Stock name of *cis*-diamminedichloroplatinum(II). The Ewens–Bassett name, then, can be found just from the charge balance. And this system is also useful when the ion is so complicated that any formal oxidation state is difficult to identify. Conversely, seeing the oxidation state in the name enables us to identify whether the metal is in a typical, high, or low oxidation state. In the following chapters, we will consistently use the Stock system for transition metal complexes.

19.6 An Overview of Bonding Theories of Transition Metal Compounds

For many decades, chemists and physicists struggled with possible explanations to account for the large number of transition metal compounds. Such explanations had to account for the variety of colors found among the compounds, the wide range of stereochemistries, and the paramagnetism of many compounds.

One of the first approaches was to regard the bonding as that between a Lewis acid (the metal ion) and Lewis bases (the ligands). This model produced the 18-electron rule or the effective atomic number (EAN) rule. This model works for many metal compounds in which the metal is in a low oxidation state; but it does not work for most compounds, nor does it explain the color or paramagnetism of many transition metal compounds. It is useful in the context of organometallic compounds; thus, we will defer discussion of this bonding model to Chapter 22, Section 22.4.

Following from this, the American chemist Linus Pauling proposed the valence-bond theory in which he assumed that the bonding of transition metals was similar to that of typical main group elements, assigning different modes of hybridization to the metal ion depending on the known geometry of the compound. This theory did account for the different stereochemistries and formulas, but it, too, failed to account for colors and unpaired electrons. We briefly outline this theory in the following subsection, though it is not widely used.

Two physicists, Hans Bethe and Johannes Van Vleck, approached the problem from a completely different direction. They assumed that the interaction between a metal ion and its ligands was totally electrostatic in nature. Known as *crystal field theory*, it has been remarkably—amazingly—successful in accounting for the properties of transition metal complexes. Through this and the following chapter, most of our explanations of the behavior of transition metal complexes will be based on crystal field theory. Nevertheless, it is clear that there is a component of covalent bonding between ligand and metal. To refine crystal field theory by taking the covalent character into account, an empirical constant (the Racah parameter) can be added to calculations. This modified form of crystal field theory is known as *ligand field theory*.

The most sophisticated approach is molecular orbital theory. This most sophisticated theory is not necessary for the discussion of conventional transition metal complexes.

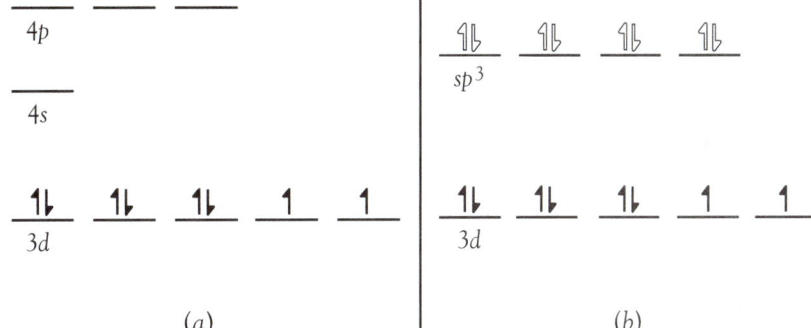

Figure 19.12 (a) The electron distribution of the free nickel(II) ion. (b) The hybridization and occupancy of the higher energy orbitals by electron pairs (open half-headed arrows) of the chloride ligands.

Valence-Bond Theory

In Chapter 3, Section 3.10, we introduced valence-bond theory in the context of the main group elements, together with the concept of orbital hybridization. Valence-bond theory and orbital hybridization can also be used to explain some aspects of bonding in transition metal complexes. Using valence-bond theory, we consider the interaction between the metal ion and its ligands to be that of a Lewis acid with Lewis bases, but in this case, the donated ligand electron pairs are considered to occupy the empty higher orbitals of the metal ion. This arrangement is shown for the tetrahedral tetrachloronickelate(II) ion, $[NiCl_4]^{2-}$, in Figure 19.12. The free nickel(II) ion has an electron configuration of $[Ar]3d^8$ with two unpaired electrons. According to the theory, the 4s and 4p orbitals of the nickel hybridize to form four sp^3 hybrid orbitals, and these are occupied by an electron pair from each chloride ion (the Lewis bases).

This representation accounts for the two unpaired electrons in the complex ion and the tetrahedral shape expected for sp^3 hybridization. However, we can only construct the orbital diagrams once we know from a crystal structure determination and magnetic measurements what the ion's shape is and what the number of unpaired electrons actually are. For chemists, a theory should be predictive, if possible, and valence-bond theory is not. For example, some iron(III) compounds have five unpaired electrons and others have one unpaired electron, but valence-bond theory cannot predict this.

The theory also has some conceptual flaws. In particular, it does not explain why the electron pairs occupy higher orbitals, even though there are vacancies in the 3d orbitals. For some Period 4 transition metal complexes, the ligand electron pairs have to be assigned to 4d orbitals as well as to 4s and 4p orbitals, even though there is room in the 3d orbitals. In addition, the theory fails to account for the color of the transition metal complexes, one of the most obvious features of these compounds. For these reasons, the valence-bond theory has become little more than a historical footnote.

19.7 Crystal Field Theory

As we have just seen, the classic valence-bond theory was unable to explain many of the aspects of transition metal complexes. In particular, valence-bond theory did not satisfactorily explain the different numbers of unpaired electrons that we find among the transition metal ions. For example, the hexaaquairon(II) ion, $[Fe(OH_2)_6]^{2+}$, has four unpaired electrons, whereas the hexacyanoferrate(II) ion, $[Fe(CN)_6]^{4-}$, has no unpaired electrons.

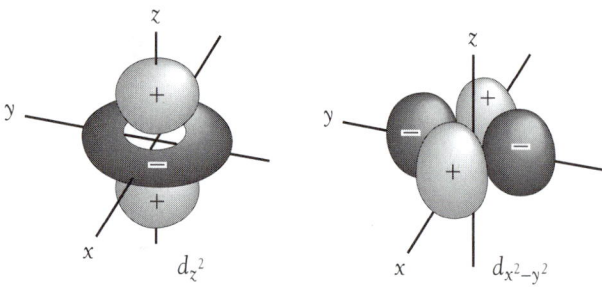

Figure 19.13 Representations of the shapes of the $3d_{z^2}$ and $3d_{x^2-y^2}$ orbitals.

Despite its simplistic nature, crystal field theory has proved remarkably useful for explaining the properties of Period 4 transition metal complexes. The theory assumes that the transition metal ion is free and gaseous, that the ligands behave like point charges, and that there are no interactions between metal d-orbitals and ligand orbitals. The theory also depends upon the probability model of the d-orbitals, that there are two d-orbitals whose lobes are oriented along the Cartesian axes, $d_{x^2-y^2}$ and d_{z^2} (Figure 19.13) and three d-orbitals whose lobes are oriented between the Cartesian axes, d_{xy}, d_{xz}, and d_{yz} (Figure 19.14).

Figure 19.14 Representations of the shapes of the $3d_{xy}$, $3d_{xz}$, and $3d_{yz}$ orbitals. *(Adapted from D.F. Shriver, P. Atkins, and C.A. Langford, Inorganic Chemistry, 2nd ed. [New York: W.H. Freeman, 1994], p. 25.)*

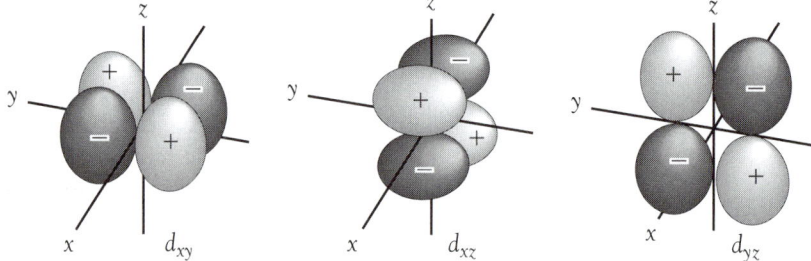

We can consider complex formation as a series of events:

1. The initial approach of the ligand electrons forms a spherical shell around the metal ion. Repulsion between the ligand electrons and the metal ion electrons will cause an increase in energy of the metal ion d orbitals.

2. The ligand electrons rearrange so that they are distributed in pairs along the actual bonding directions (such as octahedral or tetrahedral). The mean d orbital energies will stay the same, but the orbitals oriented along the bonding directions will increase in energy, and those between the bonding directions will decrease in energy. This loss in d orbital degeneracy will be the focus of the crystal field theory discussion.

3. Up to this point, complex formation would not be favored, because there has been a net increase in energy as a result of the ligand electron–metal electron repulsion (step 1). Furthermore, the decrease in the number of free species means that complex formation will generally result in a decrease in entropy. However, there will be an attraction between the ligand electrons and the positively charged metal ion that will result in a net decrease in energy. It is this third step that provides the driving force for complex formation.

These three hypothetical steps are summarized in Figure 19.15.

Octahedral Complexes

Although it is the third step that provides the energy for complex formation, it is the second step—the loss of degeneracy of the d orbitals—that is crucial

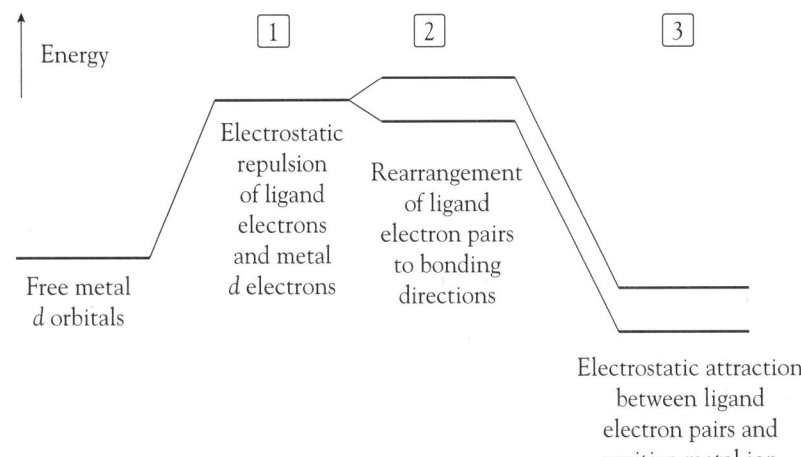

Figure 19.15 The hypothetical steps in complex ion formation according to crystal field theory.

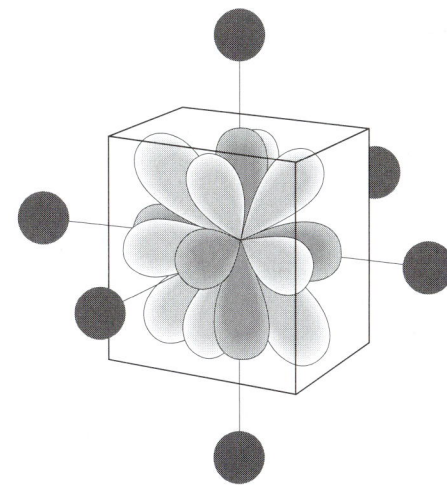

Figure 19.16 The orientation of six ligands with respect to the metal d orbitals. *(Adapted from J.E. Huheey, E.A. Keiter, and R.L. Keiter, Inorganic Chemistry, 4th ed. [New York: HarperCollins, 1993], p. 397.)*

for the explanation of the color and magnetic properties of transition metal complexes. Examining the octahedral situation first, we see that the six ligands are located along the Cartesian axes (Figure 19.16). As a result of these negative charges along the Cartesian axes, the energy of the orbitals aligned along these axes, the $d_{x^2-y^2}$ and d_{z^2} orbitals, will be higher than those of the d_{xy}, d_{xz}, and d_{yz} orbitals. This splitting is represented in Figure 19.17. The energy difference between the two sets of d orbitals in the octahedral field is given the symbol Δ_{oct}. The sum of the orbital energies equals the degenerate energy (sometimes called the baricenter). Thus, the energy of the two higher energy orbitals ($d_{x^2-y^2}$ and d_{z^2}) is $+\frac{3}{5}\Delta_{oct}$, and the energy of the three lower energy orbitals (d_{xy}, d_{xz}, and d_{yz}) is $-\frac{2}{5}\Delta_{oct}$ below the mean.

If we construct energy diagrams for the different numbers of d electrons, we see that for the d^1, d^2, and d^3 configurations, the electrons will all fit into the lower energy set (Figure 19.18). This net energy decrease is known as the *crystal field stabilization energy*, or CFSE.

For the d^4 configuration, there are two possibilities: The fourth d electron can either pair up with an electron in the lower energy level or it can occupy the upper energy level, depending on which situation is more energetically favorable. If the octahedral crystal field splitting, Δ_{oct}, is smaller than the pairing energy, then the fourth electron will occupy the higher orbital. If the pairing energy is less than the crystal field splitting, then it is energetically preferred for the fourth electron to occupy the lower orbital.

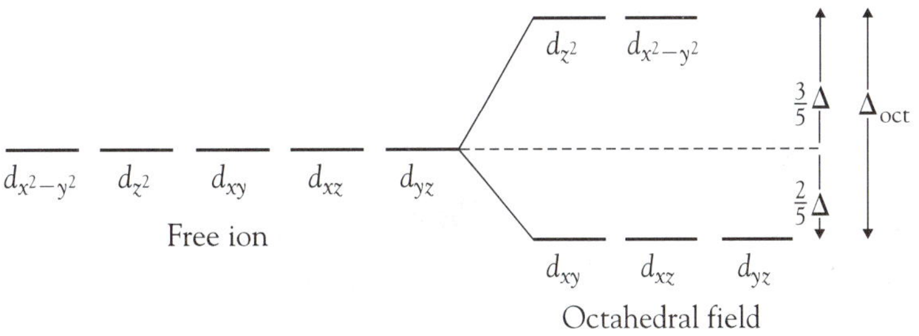

$$d_{z^2} \quad d_{x^2-y^2}$$

$$\frac{3}{5}\Delta \quad \Delta_{oct}$$

$$d_{x^2-y^2} \quad d_{z^2} \quad d_{xy} \quad d_{xz} \quad d_{yz}$$

Free ion

$$\frac{2}{5}\Delta$$

$$d_{xy} \quad d_{xz} \quad d_{yz}$$

Octahedral field

Figure 19.17 The splitting of the d orbital energies that occurs when the metal ion is surrounded by an octahedral array of ligands.

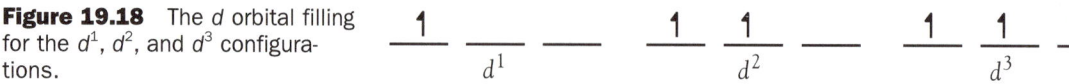

Figure 19.18 The d orbital filling for the d^1, d^2, and d^3 configurations.

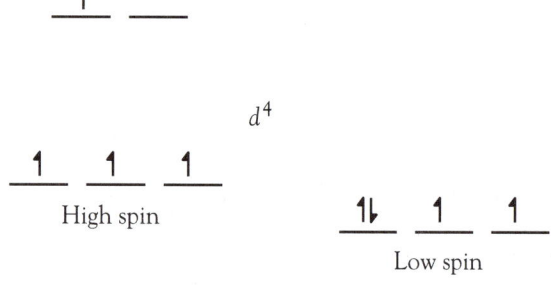

d^4

High spin

Low spin

Figure 19.19 The two possible spin situations for the d^4 configuration.

The two situations are shown in Figure 19.19. The result having the greater number of unpaired electrons is called the high-spin (or weak field) situation, and that having the lesser number of unpaired electrons is called the low-spin (or strong field) situation.

Two possible spin conditions exist for each of the d^4, d^5, d^6, and d^7 electron configurations in an octahedral environment. The number of possible unpaired electrons corresponding to each d electron configuration is shown in Table 19.2, where h.s. and l.s. indicate high-spin and low-spin, respectively.

The energy level splitting depends on four factors:

1. *The identity of the metal.* The crystal field splitting, Δ, is about 50 percent greater for the second transition series compared to the first, whereas the third series is about 25 percent greater than the second. There is a small increase in the crystal field splitting along each series.

2. *The oxidation state of the metal.* Generally, the higher the oxidation state of the metal, the greater the crystal field splitting. Thus, most cobalt(II) complexes are high spin as a result of the small crystal field splitting, whereas almost all cobalt(III) complexes are low spin as a result of the much larger splitting by the 3+ ion.

3. *The number of the ligands.* The crystal field splitting is greater for a larger number of ligands. For example, Δ_{oct}, the splitting for six

Table 19.2 The d electron configurations and corresponding number of unpaired electrons for an octahedral stereochemistry

Configuration	Number of unpaired electrons	Common examples
d^1	1	Ti^{3+}
d^2	2	V^{3+}
d^3	3	Cr^{3+}
d^4	4 (h.s.), 2 (l.s.)	Mn^{3+}
d^5	5 (h.s.), 1 (l.s.)	Mn^{2+}, Fe^{3+}
d^6	4 (h.s.), 0 (l.s.)	Fe^{2+}, Co^{3+}
d^7	3 (h.s.), 1 (l.s.)	Co^{2+}
d^8	2	Ni^{2+}
d^9	1	Cu^{2+}

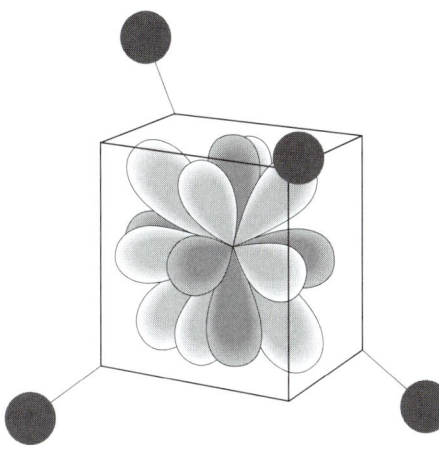

Figure 19.20 The orientation of four ligands with respect to the metal *d* orbitals. *(Adapted from J.E. Huheey, E.A. Keiter, and R.L. Keiter, Inorganic Chemistry, 4th ed. [New York: HarperCollins, 1993], p. 402.)*

ligands in an octahedral environment is much greater than Δ_{tet}, the splitting for four ligands in a tetrahedral environment.

4. *The nature of the ligands.* The common ligands can be ordered on the basis of the effect that they have on the crystal field splitting. This ordered listing is called the *spectrochemical series.* Among the common ligands, the splitting is largest with carbonyl and cyanide and smallest with iodide. The ordering for most metals is

$$I^- < Br^- < Cl^- < F^- < OH^- < OH_2 < NH_3^- < en < CN^- < CO$$

Thus, for a particular metal ion, it is the ligand that determines the value of the crystal field splitting. Consider the d^6 iron(II) ion. According to crystal field theory, there are the two spin possibilities: high spin (weak field) with four unpaired electrons and low spin (strong field) with all electrons paired. We find that the hexaaquairon(II) ion, $[Fe(OH_2)_6]^{2+}$, possesses four unpaired electrons. The water ligands, being low in the spectrochemical series, produce a small Δ_{oct}; hence, the electrons adopt a high-spin configuration. Conversely, the hexacyanoferrate(II) ion, $[Fe(CN)_6]^{4-}$, is found to be diamagnetic (zero unpaired electrons). Cyanide is high in the spectrochemical series and produces a large Δ_{oct}; hence, the electrons adopt a low-spin configuration.

Tetrahedral Complexes

The second most common stereochemistry is tetrahedral. Figure 19.20 shows the tetrahedral arrangement of four ligands around the metal ion. In this case, it is the d_{xy}, d_{xz}, and d_{yz} orbitals that are more in line with the approaching ligands than the $d_{x^2-y^2}$ and d_{z^2} orbitals. As a result, it is the $d_{x^2-y^2}$ and d_{z^2} orbitals that are lower in energy, and the tetrahedral energy diagram is inverted relative to the octahedral diagram (Figure 19.21).

With only four ligands instead of six, and the ligands not quite pointing directly at the three d orbitals, the crystal field splitting is much less than that in the octahedral case; in fact, as we mentioned previously, it is about four-ninths of Δ_{oct}. As a result of the small orbital splitting, the tetrahedral complexes are almost all high spin. Tetrahedral geometries are most commonly found for halogen complexes, such as the tetrachlorocobaltate(II) ion, $[CoCl_4]^{2-}$, and for the oxyanions, such as the tetraoxomolybdate(VI) ion, MoO_4^{2-} (commonly called molybdate).

Square Planar Complexes

For the Period 4 transition metals, it is only nickel that tends to form square planar complexes, such as the tetracyanonickelate(II) ion, $[Ni(CN)_4]^{2-}$. These complexes are diamagnetic. We can develop a crystal field diagram to see why this is so, even though both octahedral and tetrahedral geometries result in two unpaired electrons for the d^8 configuration.

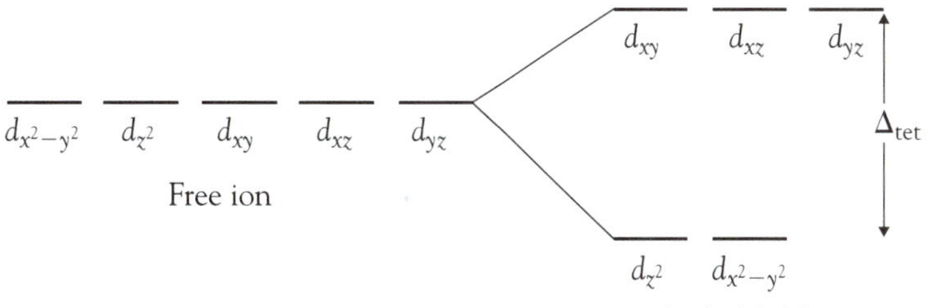

Figure 19.21 The splitting of the *d* orbital energies that occurs when the metal ion is surrounded by a tetrahedral array of ligands.

$d_{x^2-y^2}$ d_{z^2} d_{xy} d_{xz} d_{yz}

Free ion

d_{xy} d_{xz} d_{yz}

Δ_{tet}

d_{z^2} $d_{x^2-y^2}$

Tetrahedral field

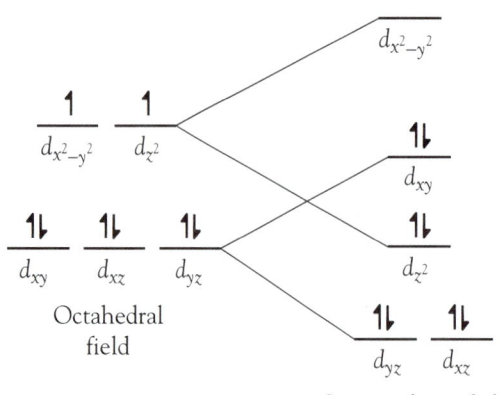

Figure 19.22 The d^8 orbital energy diagram for the square planar environment, as derived from the octahedral diagram.

If we start from the octahedral field and withdraw the ligands from the z-axis, the d_{z^2} orbital will no longer feel the electrostatic repulsion from the axial ligands; hence, it will drop substantially in energy. The other two orbitals with z-axis components—the d_{xz} and d_{yz}—will also undergo a decrease in energy. Conversely, with the withdrawal of the axial ligands, there will be a greater electrostatic attraction on the ligands in the plane, and these will become closer to the metal ion. As a result, the $d_{x^2-y^2}$ and d_{xy} orbitals will increase substantially in energy (Figure 19.22). As the nickel(II) complex with cyanide is diamagnetic, the splitting of the $d_{x^2-y^2}$ and the d_{xy} orbitals must be greater than the pairing energy for this combination.

19.8 Successes of Crystal Field Theory

A good chemical theory is one that can account for many aspects of physical and chemical behavior. By this standard, crystal field theory is remarkably successful, because it can be used to explain most of the properties that are unique to transition metal ions. Here, we will look at a selection of them.

Magnetic Properties

Any theory of transition metal ions has to account for the paramagnetism of many of the compounds. The degree of paramagnetism is dependent on the identity of the metal, its oxidation state, its stereochemistry, and the nature of the ligand. Crystal field theory explains the paramagnetism very well in terms of the splitting of the d orbital energies, at least for the Period 4 transition metals. For example, we have just seen how crystal field theory can explain the diamagnetism of square planar nickel(II) ion, which contrasts with the paramagnetism of the tetrahedral and octahedral geometries.

Colors of Transition Metal Complexes

The most striking feature of transition metal complexes is the range of colors that they exhibit. These colors are the result of absorptions in the visible region of the electromagnetic spectrum. For example, Figure 19.23 shows the visible absorption spectrum of the purple hexaaquatitanium(III) ion, $[Ti(OH_2)_6]^{3+}$. This ion absorbs light in the green part of the spectrum, transmitting blue and red light to give the blended purple color.

The titanium(III) ion has a d^1 electron configuration, and with six water molecules as ligands, we can consider the ion to be in an octahedral field. The resulting d orbital splitting is shown on the left-hand side of Figure 19.24. An absorption of electromagnetic energy causes the electron to shift

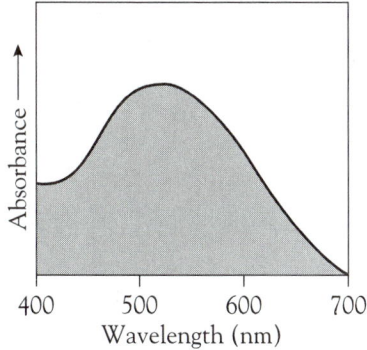

Figure 19.23 The visible absorption spectrum of the hexaaquatitanium(III) ion.

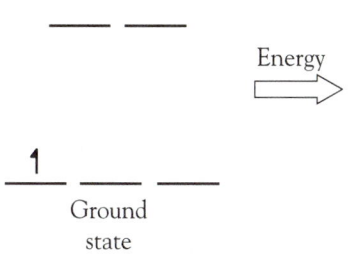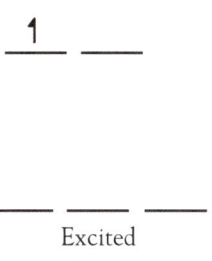

Figure 19.24 The electron transition corresponding to the visible absorption of the titanium(III) ion.

to the upper d orbital set as shown on the right-hand side of Figure 19.24. The electron subsequently returns to the ground state, and the energy is released as thermal motion rather than as electromagnetic radiation. The absorption maximum is at about 520 nm, which represents an energy difference between the upper and lower d orbital sets of about 230 kJ·mol^{-1}. This energy difference represents the value of Δ, the crystal field splitting.

As is apparent from Figure 19.23, the electronic absorption bands are very broad (especially so for d^1 and d^9 configurations). These bands are broad because the electron transition time is much shorter than the vibrations occurring within the molecule. When the ligands are farther away from the metal than the mean bond length, the field is weaker and the splitting is less; hence, the transition energy is smaller than the "normal" value. Conversely, when the ligands are closer to the metal, the field is stronger, the splitting is greater, and the transition energy is larger than the normal value. We can confirm this explanation by cooling the complex to close to absolute zero, thereby reducing the molecular vibrations. When we do, as predicted, the bands in the visible absorption spectrum become much narrower.

The hexachlorotitanate(III) ion, $[TiCl_6]^{3-}$, has an orange color as a result of an absorption centered at 770 nm. This value corresponds to a crystal field splitting of about 160 kJ·mol^{-1}. The lower value reflects the weakness of the chloride ion as a ligand relative to the water molecule; that is, chloride is lower than water in the spectrochemical series.

For most branches of chemistry, we measure energy differences in kilojoules per mole, but transition metal chemists usually report crystal field splittings in a frequency unit called wavenumbers. This is simply the reciprocal of the wavelength expressed as centimeters. Thus, the units of wavenumbers will be cm^{-1}, called reciprocal centimeters. For example, the crystal field splitting for the hexaaquatitanium(III) ion is usually cited as 19 200 cm^{-1} rather than as 230 kJ·mol^{-1}.

We will discuss the visible absorptions of other electron configurations later.

Hydration Enthalpies

Another of the parameters that can be explained by crystal field theory is the enthalpy of hydration of transition metal ions. This is the energy released when gaseous ions are hydrated, a topic discussed in Chapter 6, Section 6.4.

$$M^{n+}(g) + 6\ H_2O(l) \rightarrow [M(OH_2)_6]^{n+}(aq)$$

As the effective nuclear charge of metal ions increases across a period, we expect the electrostatic interaction between the water molecules and the metal ions to increase regularly along the transition metal series. In fact, we find deviations from a linear relationship (Figure 19.25). To explain this observation, we assume that the greater hydration enthalpy is the result of the crystal field stabilization energy, which can be calculated in terms of Δ_{oct}, the crystal field splitting. Recall that for an octahedral field, the d_{xy}, d_{xz}, and d_{yz} orbitals are lowered in energy by $\frac{2}{5}\Delta_{oct}$ and the $d_{x^2-y^2}$ and d_{z^2} orbitals are raised in energy by $\frac{3}{5}\Delta_{oct}$. Thus for a particular electron configuration, it is possible

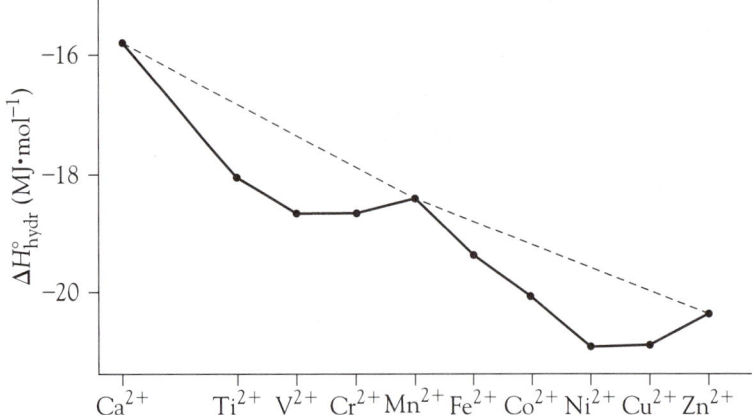

Figure 19.25 Experimental hydration enthalpies of the dipositive ions of the Period 4 transition metals.

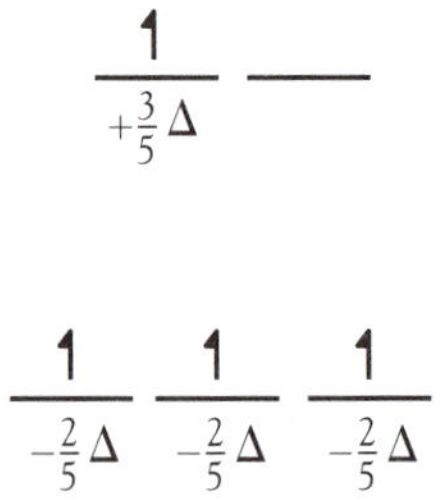

$$\frac{1}{+\frac{3}{5}\Delta} \quad \underline{}$$

$$\frac{1}{-\frac{2}{5}\Delta} \quad \frac{1}{-\frac{2}{5}\Delta} \quad \frac{1}{-\frac{2}{5}\Delta}$$

Figure 19.26 The crystal field stabilization energy for the d^4 high-spin electron configuration.

━━━ **Spinel Structures**

to calculate the net contribution of the crystal field to the hydration enthalpy. Figure 19.26 illustrates the situation for the d^4 high-spin ion. This ion would have a net stabilization energy of

$$-\left[3\left(\tfrac{2}{5}\Delta_{oct}\right)\right] + \left[1\left(\tfrac{3}{5}\Delta_{oct}\right)\right] = -0.6\Delta_{oct}$$

The complete set of crystal field stabilization energies is listed in Table 19.3.

These values correspond remarkably well with the deviations of the hydration enthalpies. Of particular note, it is only the d^0, d^5 (high spin), and d^{10} ions that fit the expected near-linear relationship, and these all have zero crystal field stabilization energy.

Yet another triumph of crystal field theory is the explanation for the transition metal ion arrangements in the spinel structures that we first met in Chapter 13, Section 13.11. The spinel is a mixed oxide, usually of general formula $(M^{2+})(M^{3+})_2(O^{2-})_4$, with the metal ions occupying both octahedral and tetrahedral sites. In a normal spinel, all of the 2+ ions are in the tetrahedral sites and the 3+ ions are in the octahedral sites; whereas in an inverse

Table 19.3 Crystal field stabilization energies (CFSE) for the dipositive, high spin ions of various Period 4 metals

Ion	Configuration	CFSE
Ca^{2+}	d^0	$-0.0\,\Delta_{oct}$
—	d^1	$-0.4\,\Delta_{oct}$
Ti^{2+}	d^2	$-0.8\,\Delta_{oct}$
V^{2+}	d^3	$-1.2\,\Delta_{oct}$
Cr^{2+}	d^4	$-0.6\,\Delta_{oct}$
Mn^{2+}	d^5	$-0.0\,\Delta_{oct}$
Fe^{2+}	d^6	$-0.4\,\Delta_{oct}$
Co^{2+}	d^7	$-0.8\,\Delta_{oct}$
Ni^{2+}	d^8	$-1.2\,\Delta_{oct}$
Cu^{2+}	d^9	$-0.6\,\Delta_{oct}$
Zn^{2+}	d^{10}	$-0.0\,\Delta_{oct}$

spinel, the 2+ ions are in the octahedral sites and the 3+ ions fill the tetrahedral sites and the remaining octahedral sites.

The choice of normal spinel or inverse spinel for mixed transition metal oxides is determined usually (but not always) by the option that will give the greater crystal field stabilization energy. This can be illustrated by a pair of oxides that each contains ions of one metal in two different oxidation states: Fe_3O_4, containing Fe^{2+} and Fe^{3+} and Mn_3O_4 containing Mn^{2+} and Mn^{3+}. The former adopts the inverse spinel structure: $(Fe^{3+})_t(Fe^{2+}, Fe^{3+})_o(O^{2-})_4$. All these ions are high spin, so the Fe^{3+} ion (d^5) has a zero CFSE; but the Fe^{2+} ion (d^6) has a nonzero CFSE. Because a crystal field splitting for the tetrahedral geometry is four-ninths that of the equivalent octahedral environment, the CFSE of an octahedrally coordinated ion will be greater than that of a tetrahedrally coordinated ion. This energy difference accounts for the octahedral site preference of the Fe^{2+} ion. Unlike the mixed iron oxide, the mixed manganese oxide has the normal spinel structure: $(Mn^{2+})_t(Mn^{3+})_o(O^{2-})_4$. In this case, it is the Mn^{2+} ion (d^5) that has a zero CFSE and the Mn^{3+} ion (d^4) that has a nonzero CFSE. Hence it is the Mn^{3+} ion that preferentially occupies the octahedral sites.

The Earth and Crystal Structures

There is still much to be discovered about the structure of the planet that we inhabit. However, we do know that it consists of three regions: the crust, mantle, and core. The crust occupies the surface 30–70 km for continents and 5–15 km for oceanic crust. Then the mantle extends down to a depth of about 2900 km where it meets the core. The core is predominantly iron, molten to a depth of about 5100 km and then solid at the center.

The mantle provides some interesting crystal-phase chemistry. Between about 100 and 400 km depth, the mantle consists of the mineral olivine, $(Mg,Fe)_2SiO_4$, where the composition varies as the two metal ions can both occupy the same lattice site in this simple silicate (Chapter 9). Below the olivine layer, it is a spinel structure, probably mainly $MgFe_2O_4$, that stretches down to the lower mantle. The lower mantle begins at a depth of about 670 km. This layer consists of the denser perovskite structure (Chapter 16, Section 16.6), $(Mg,Fe)SiO_3$.

There is an increasing interest in the chemistry of the deep Earth and phenomena such as earthquakes. There are two types of earthquakes: shallow and deep. Deep earthquakes occur when dense spinel forms around the margins of descending streams of olivine. Shallow earthquakes seem more complex, but part of their origin may be from dehydration reactions. The surface rock, serpentine, is hydrated olivine. As a layer of serpentine is dragged down into the Earth at a trench boundary, the heat will cause the serpentine to dehydrate to olivine. The high-pressure water can act to force cracks open and also as a lubricant. These are only hypotheses at present. Much more research needs to be done on the properties of spinels and perovskites under high pressure and the dehydration of minerals if we are to obtain a more detailed knowledge of our planet.

19.9 More on Electronic Spectra

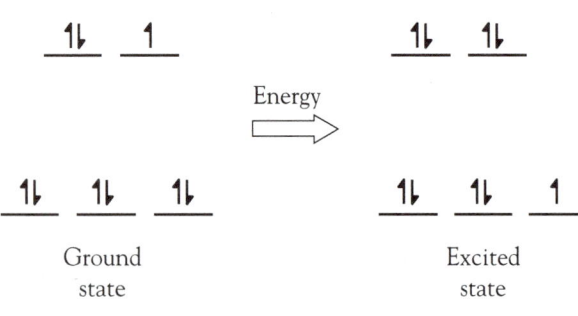

Figure 19.27 The electron transition for the d^9 electron configuration.

Figure 19.28 Three possible electron transitions for the d^2 electron configuration.

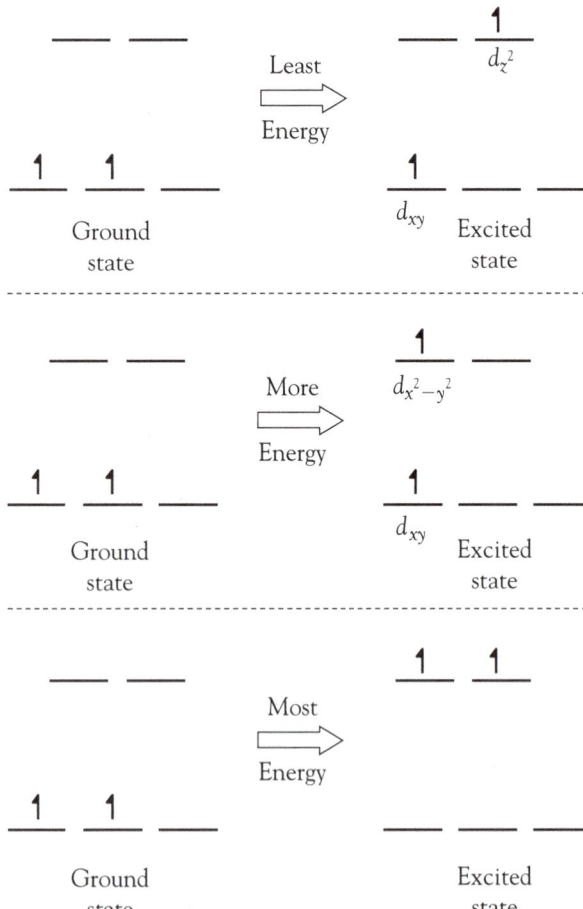

In the previous section, we saw that the single visible absorption of the titanium(III) ion can be explained in terms of a d electron transition from the lower level to the upper level in the crystal field. For a copper(II) ion (d^9) in an octahedral environment, a single broad visible absorption is also observed. As in the d^1 situation, the absorption of this ion can be interpreted to mean that one electron is excited to the upper level in the octahedral crystal field (Figure 19.27).

For d^2 ions, we might expect two absorption peaks, corresponding to the excitation of one or both of the electrons. However, a total of three fairly strong absorptions are observed. To explain this, we have to consider interelectronic repulsions. In the ground state, a d^2 ion, such as the hexaaquavanadium(III) ion, has two electrons with parallel spins in any two of the three lower energy orbitals: d_{xy}, d_{xz}, and d_{yz}. When one electron is excited, the resulting combination can have different energies, depending on whether the two electrons are occupying overlapping orbitals and therefore repelling each other. For example, an excited configuration of $(d_{xy})^1(d_{z^2})^1$ will be lower in energy because the two electrons occupy very different volumes of space, whereas the $(d_{xy})^1(d_{x^2-y^2})^1$ configuration will be higher in energy because both electrons occupy space in the x and y planes.

By calculation, it can be shown that the combinations $(d_{xy})^1(d_{z^2})^1$, $(d_{xz})^1(d_{x^2-y^2})^1$, $(d_{yz})^1(d_{x^2-y^2})^1$ all have the same lower energy, and $(d_{xy})^1(d_{x^2-y^2})^1$, $(d_{xz})^1(d_{z^2})^1$, and $(d_{yz})^1(d_{z^2})^1$ all have the same higher energy. This accounts for two of the transitions, and the third transition corresponds to the excitation of both electrons into the upper levels to give the configuration $(d_{x^2-y^2})^1$ $(d_{z^2})^1$. Three possibilities are shown in Figure 19.28.

Sometimes very weak absorptions also appear in the visible spectrum. These correspond to transitions in which an electron has reversed its spin, such as that shown in Figure 19.29. Such transitions, which involve a change in spin state (*spin-forbidden transitions*), are of low probability and hence have very low intensity in the spectrum.

We find that the visible spectra of d^1, d^4 (high spin), d^6 (high spin), and d^9 can be interpreted in terms of a single transition, whereas d^2, d^3, d^7 (high spin), and d^8 spectra can be interpreted in terms of three transitions. The remaining high spin case, d^5, is unique in that the only possible transitions are spin-forbidden. As a result, complexes such as hexaaquamanganese(II) ion and hexaaquairon(III) ion have very pale colors. A more detailed study of transition metal ion spectra is the domain of theoretical inorganic chemistry.

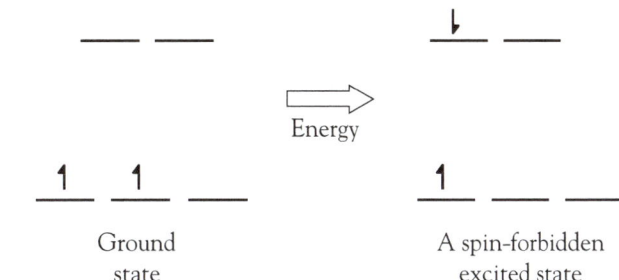

Figure 19.29 A possible spin-forbidden transition for the d^2 electron configuration.

Ground state

A spin-forbidden excited state

19.10 *Thermodynamic versus Kinetic Factors*

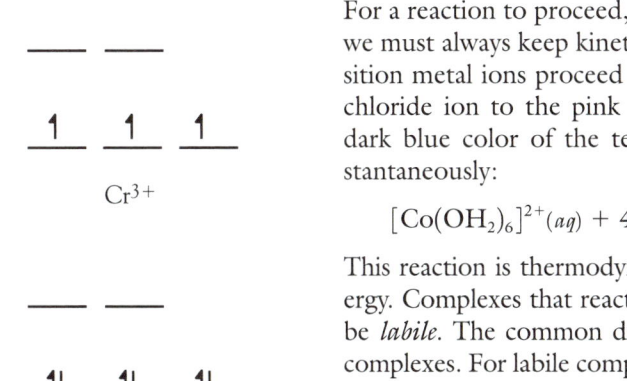

Cr^{3+}

Co^{3+}

Figure 19.30 Inert complexes are formed by metal ions with the d^3 and low spin d^6 electron configurations.

For a reaction to proceed, there must be a decrease in free energy. However, we must always keep kinetic factors in mind. Most solution reactions of transition metal ions proceed rapidly. For example, addition of a large excess of chloride ion to the pink hexaaquacobalt(II) ion, $[Co(OH_2)_6]^{2+}$, gives the dark blue color of the tetrachlorocobaltate(II) ion, $[CoCl_4]^{2-}$, almost instantaneously:

$$[Co(OH_2)_6]^{2+}(aq) + 4\ Cl^-(aq) \rightarrow [CoCl_4]^{2-}(aq) + 6\ H_2O(l)$$

This reaction is thermodynamically favored and also has a low activation energy. Complexes that react quickly (for example, within 1 minute) are said to be *labile*. The common dipositive Period 4 transition metal ions form labile complexes. For labile complexes, it is impossible to physically separate isomers.

The two common Period 4 transition metal ions that form kinetically *inert* complexes are chromium(III) and cobalt(III). The former has a d^3 electron configuration; and the latter, a low-spin d^6 configuration (Figure 19.30). It is the stability of the half-filled and the filled lower set of d orbitals that precludes any low-energy pathway of reaction. For example, addition of concentrated acid to the hexaamminecobalt(III) ion should cause a ligand replacement according to free energy calculations. However, such a reaction occurs so slowly that several days must elapse before any color change is noticeable:

$$[Co(NH_3)_6]^{3+}(aq) + 6\ H_3O^+(aq) \rightleftharpoons [Co(OH_2)_6]^{3+}(aq) + 6\ NH_4^+(aq)$$

For this reason, to synthesize specific cobalt(III) or chromium(III) complexes, we usually find a pathway that involves synthesizing the complex of the respective labile 2+ ion and then oxidizing it to the inert 3+ ion. Isomers of inert complexes, including optical isomers, can be crystallized separately.

19.11 *Synthesis of Coordination Compounds*

There are many different routes of synthesizing coordination compounds, the two most common being ligand replacement and oxidation–reduction.

Ligand Replacement Reactions

Many transition metal complexes are synthesized in aqueous solution by displacement of the water ligands. For example, we can produce the hexaamminenickel(II) ion by adding an excess of aqueous ammonia to a solution of the hexaaquanickel(II) ion:

$$[Ni(OH_2)_6]^{2+}(aq) + 6\ NH_3(aq) \rightarrow [Ni(NH_3)_6]^{2+}(aq) + 6\ H_2O(l)$$

Ammonia, which is higher than water in the spectrochemical series, readily replaces the water ligands. In other words, the nickel–ammonia bond is stronger than the nickel–water bond, and the process is exothermic. Hence, the reaction is enthalpy driven.

If we are dealing with a chelating ligand, we find that the equilibrium is driven strongly to the right by entropy factors as well. This situation can be illustrated by the formation of the tris(1,2-diaminoethane)nickel(II) ion from the hexaaquanickel(II) ion:

$$[Ni(OH_2)_6]^{2+}(aq) + 3\ en(aq) \rightarrow [Ni(en)_3]^{2+}(aq) + 6\ H_2O(l)$$

In this case, we have a similar enthalpy increase, but there is also a major entropy increase because the total number of ions and molecules has increased from four to seven. It is the entropy factor that results in such strong complex formation by chelating ligands, behavior that is known as the *chelate effect*.

Oxidation–Reduction Reactions

Redox reactions are particularly important as a means of synthesizing compounds in "abnormal" oxidation states. High oxidation state fluorides can be synthesized by simple combination reactions, for example:

$$Os(s) + 3\ F_2(g) \rightarrow OsF_6(s)$$

Other high-oxidation-state compounds and polyatomic ions can be produced using oxidizing agents. For example, the red-purple ferrate(VI) ion, FeO_4^{2-}, can be prepared using the hypochlorite ion in basic solution:

$$Fe^{3+}(aq) + 8\ OH^-(aq) \rightarrow FeO_4^{2-}(aq) + 4\ H_2O(l) + 3\ e^-$$
$$ClO^-(aq) + 2\ H_2O(l) + 2\ e^- \rightarrow Cl^-(aq) + 2\ OH^-(aq)$$

The ion can then be precipitated as the barium salt:

$$FeO_4^{2-}(aq) + Ba^{2+}(aq) \rightarrow BaFeO_4(s)$$

As we mentioned in the previous section, because cobalt(III) complexes are kinetically inert, substitution reactions for cobalt(III) complexes are usually impractical. Cobalt(II) is the dominant oxidation state, and the hexaaqua ion can be used as a reagent to produce a range of cobalt(III) complexes. For example, we can synthesize the hexaamminecobalt(III) ion by the air oxidation of the hexaamminecobalt(II) complex. (As we shall see in Chapter 20, Section 20.7, substitution of ammonia for water as ligands changes the necessary redox potential substantially.)

$$[Co(OH_2)_6]^{2+}(aq) + 6\ NH_3(aq) \rightarrow [Co(NH_3)_6]^{2+}(aq) + 6\ H_2O(l)$$
$$[Co(NH_3)_6]^{2+}(aq) \rightarrow [Co(NH_3)_6]^{3+}(aq) + e^-$$
$$O_2(g) + 4\ H^+(aq) + 4\ e^- \rightarrow 2\ H_2O(l)$$

One method of preparing chromium(III) complexes is by reduction of the dichromate ion. In the following example, an oxalic acid/oxalate ion mixture is used as both the reducing agent and the ligand:

$$Cr_2O_7^{2-}(aq) + 14\ H^+(aq) + 6\ e^- + 5\ H_2O(l) \rightarrow 2\ [Cr(OH_2)_6]^{3+}(aq)$$
$$C_2O_4^{2-}(aq) \rightarrow 2\ CO_2(g) + 2\ e^-$$
$$[Cr(OH_2)_6]^{3+} + 3\ C_2O_4^{2-}(aq) \rightarrow [Cr(C_2O_4)_3]^{3-}(aq) + 6\ H_2O(l)$$

The large anion can then be crystallized as the potassium salt, $K_3[Cr(C_2O_4)_3]\cdot 3\ H_2O$.

Partial Decomposition Reactions

In a few cases, the identity of the coordinating ligands can be changed by gentle heating and vaporization of the more volatile ligand. An example in cobalt(III) chemistry is:

$$[Co(NH_3)_5(OH_2)]Cl_3(s) \xrightarrow{\Delta} [Co(NH_3)_5Cl]Cl_2(s) + H_2O(g)$$

and an example from chromium(III) chemistry is:

$$[Cr(en)_3]Cl_3(s) \xrightarrow{\Delta} cis\text{-}[Cr(en)_2Cl_2]Cl(s) + en(g)$$

19.12   Coordination Complexes and the HSAB Concept

Just as the HSAB concept can help us predict main group reactions (as we saw in Chapter 7, Section 7.9), so it can help us understand the ligand preferences of transition metals in different oxidation states. Table 19.4 shows the qualitative divisions into hard/borderline/soft acids of some of the first row transition elements while Figure 19.31 shows the similar division of ligand atoms into hard/borderline/soft. These ligand–atom categories generally apply, irrespective of the other atoms of the ligand. For example, all nitrogen–donor ligands of the form NR_3 are hard bases, where R is an alkyl group such as methyl or hydrogen. Conversely, all carbon–donor ligands, such as carbon monoxide and cyanide, are soft bases. Chloride is regarded as a hard base, but not as hard as fluoride or an oxygen–donor ligand. Hence, it is depicted in Figure 19.31 as half-white and half-shaded.

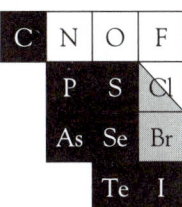

Figure 19.31   Classification of the HSAB ligand atoms into hard (white), borderline (shaded), and soft (black).

The HSAB concept accounts for the fact that high oxidation states of metal ions (hard acids) are found with fluoride or oxide ligands (hard bases). It is the low oxidation states (soft acids) that are stabilized by such ligands as the carbon-bonded carbonyl (a soft base). To provide a specific example, copper(II) fluoride is known, whereas copper(I) fluoride is not; conversely, copper(I) iodide is known whereas copper(II) iodide is not. We can use this principle to help us synthesize transition metal compounds in which the metal ions have abnormal oxidation states. For example, iron is usually found in +2 and +3 oxidation states. We can prepare a compound of the hard-acid iron(VI) using the hard-base oxide ion to give the ferrate ion, $[FeO_4]^{2-}$ (as described in the previous section). Similarly, we can prepare a compound of soft-acid iron(0) using the soft-base carbonyl molecule to give pentacarbonyliron(0), $Fe(CO)_5$.

The HSAB concept is also relevant to reactions of transition metal complexes. In particular, there is a general tendency for a complex ion to prefer ligand atoms of the same type. Thus, a complex with some hard-base ligands will prefer to add another hard-base ligand. Similarly, a complex with soft-base ligands will prefer to add another soft-base ligand. This preference for ligands of the same HSAB type is known as *symbiosis*.

Table 19.4   Some common first row transition metal ions categorized according to the HSAB concept

Hard	Borderline	Soft
Ti^{4+}, V^{4+}, Cr^{3+}, Cr^{6+}	Fe^{2+}, Co^{2+}, Ni^{2+}, Cu^{2+}	Cu^+
Mn^{2+}, Mn^{4+}, Mn^{7+}		All metals with 0 or
Fe^{3+}, Co^{3+}		negative oxidation states

Cobalt(III) chemistry is particularly useful for the illustration of symbiosis. For example, the complex $[Co(NH_3)_5F]^{2+}$ is much more stable in aqueous solution than $[Co(NH_3)_5I]^{2+}$. This can be explained in HSAB terms by considering that Co(III) is "hardened" by the presence of five hard-base ammonia ligands. Thus, the soft-base iodide ion is comparatively easily replaced by water (a hard-base ligand) to form $[Co(NH_3)_5(OH_2)]^{3+}$. On the other hand, $[Co(CN)_5I]^{3-}$ is more stable in water than $[Co(CN)_5F]^{3-}$. It is argued that the five soft-base cyanide ions "soften" the cobalt(III) complex, resulting in a preference for the soft-base iodide at the sixth coordination site rather than the hard-base water molecule.

A particularly interesting example of the application of the HSAB concept to transition metal complexes involves linkage isomerism. The thiocyanate ion, NCS^-, can bond either through the nitrogen atom, in which case it acts as a borderline base, or through the sulfur atom, in which case it behaves as a soft base. We find that in the pentamminethiocyanatocobalt(III) ion, $[Co(NH_3)_5(NCS)]^{2+}$, it is bonded through nitrogen, as we would expect, for the other ligands are hard bases, while in the pentacyanothiocyanatecobalt(III) ion, $[Co(CN)_5(SCN)]^{3-}$, coordination is through the sulfur as is consistent with the other ligands, cyanides, being soft bases.

19.13 Biological Aspects

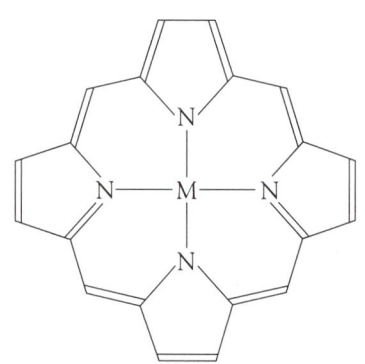

Figure 19.32 The core of metalloporphyrin complexes.

The chelate effect is important in biological complexes. There is a tetradentate ligand of particular importance to biological systems—the porphyrin ring. The basic structure of this complex is shown in Figure 19.32. It is an organic molecule in which alternating double bonds hold the structure in a rigid plane with four nitrogen atoms oriented toward the center. The space in the center is the right size for many metal ions.

In biological systems, the porphyrin ring carries different substituents on the edge of the molecule, but the core—a metal ion surrounded by four nitrogen atoms—is consistent throughout the biological world. Plant life depends on the chlorophylls, which contain magnesium–porphyrin units, for the process of photosynthesis. Animal life depends upon several metal porphyrin systems such as the hemoglobin molecule, used for oxygen transport, which contains four iron–porphyrin units. Thus, the porphyrin metal complexes are among the most important in the living world.

EXERCISES

19.1 Define the following terms: (a) transition metal; (b) ligand; (c) crystal field splitting.

19.2 Define the following terms: (a) coordination number; (b) chelate; (c) chelate effect.

19.3 Nickel forms two tetracyano complexes, $[Ni(CN)_4]^{2-}$ and $[Ni(CN)_4]^{4-}$. Explain why you might expect complexes with nickel in both normal ($+2$) and in low (0) oxidation states.

19.4 Suggest why the fluoride in which chromium has its highest oxidation state is CrF_6, whereas the highest oxidation state chromium assumes in a chloride is $CrCl_4$.

19.5 In addition to the two geometric isomers of $[Pt(NH_3)_2Cl_2]$, there is a third isomer. It has the same empirical formula and square planar geometry, but it is electrically conducting in solution. Write the structure of the compound.

19.6 Identify the probable type of isomerism for (a) $Co(en)_3Cl_3$; (b) $Cr(NH_3)_3Cl_3$.

19.7 Draw the geometric and optical isomers for the $[Co(en)_2Cl_2]^+$ ion.

19.8 Provide systematic names for each of the following compounds: (a) $Fe(CO)_5$; (b) $K_3[CoF_6]$; (c) $[Fe(OH_2)_6]Cl_2$; (d) $[CoCl(NH_3)_5]SO_4$.

19.9 Provide systematic names for each of the following compounds: (a) $(NH_4)_2[CuCl_4]$; (b) $[Co(NH_3)_5(OH_2)]Br_3$; (c) $K_3[Cr(CO)_4]$; (d) $K_2[NiF_6]$; (e) $[Cu(NH_3)_4](ClO_4)_2$.

19.10 Deduce the formula of each of the following transition metal complexes:
(a) hexamminechromium(III) bromide
(b) aquabis(1,2-diaminoethane)thiocyanato-cobalt(III) nitrate
(c) potassium tetracyanonickelate(II)
(d) tris(1,2-diaminoethane)cobalt(III) iodide

19.11 Deduce the formula of each of the following transition metal complexes:
(a) hexaaquamanganese(II) nitrate
(b) palladium(II) hexafluoropalladate(IV)
(c) tetraaquadichlorochromium(III) chloride dihydrate
(d) potassium octacyanomolybdenate(V)

19.12 Write the names of the following geometric isomers:

(a) $\qquad$ (b)

19.13 Construct energy level diagrams for both high and low spin situations for the d^6 electron configuration in (a) an octahedral field; (b) a tetrahedral field.

19.14 Which one of the iron(III) complexes, hexacyanoferrate(III) ion or tetrachloroferrate(III), is likely to be high spin, and which, low spin? Give your reasons (two in each case).

19.15 The crystal field splittings, Δ, are listed in the following for three ammine complexes of cobalt. Explain the differences in values.

Complex	$\Delta(cm^{-1})$
$[Co(NH_3)_6]^{3+}$	22 900
$[Co(NH_3)_6]^{2+}$	10 200
$[Co(NH_3)_4]^{2+}$	5900

19.16 The crystal field splittings, Δ, are listed in the following for four complexes of chromium. Explain the differences in values.

Complex	$\Delta(cm^{-1})$
$[CrF_6]^{3-}$	15 000
$[Cr(OH_2)_6]^{3+}$	17 400
$[CrF_6]^{2-}$	22 000
$[Cr(CN)_6]^{3-}$	26 600

19.17 For which member of the following pairs of complex ions would Δ_{oct} be greater? Explain your reasoning.
(a) $[MnF_6]^{2-}$ and $[ReF_6]^{2-}$
(b) $[Fe(CN)_6]^{4-}$ and $[Fe(CN)_6]^{3-}$

19.18 Construct a table of d electron configurations and corresponding number of unpaired electrons for a tetrahedral stereochemistry (similar to Table 19.2).

19.19 Construct a table of crystal field stabilization energies for the dipositive high-spin ions of the Period 4 transition metals in a tetrahedral field in terms of Δ_{tet} (similar to Table 19.3).

19.20 Would you predict that $NiFe_2O_4$ would adopt a normal spinel or an inverse spinel structure? Explain your reasoning.

19.21 Would you predict that $NiCr_2O_4$ would adopt a normal spinel or an inverse spinel structure? Explain your reasoning.

19.22 Ortho-phenanthroline, $C_8H_6N_2$, is a bidentate ligand, commonly abbreviated as phen. Explain why $[Fe(phen)_3]^{2+}$ is diamagnetic, while $[Fe(phen)_2(OH_2)_2]^{2+}$ is paramagnetic.

19.23 The ligand, $H_2NCH_2CH_2NHCH_2CH_2NH_2$, commonly known as det, is a tridentate ligand, coordinating to metal ions through all three nitrogen atoms. Write a balanced chemical equation for the reaction of this ligand with the hexaaquanickel(II) ion and suggest why the formation of this complex would be strongly favored.

BEYOND THE BASICS

19.24 Copper(II) normally forms a complex chloro-anion of formula $[CuCl_4]^{2-}$ with cations such as cesium. However, the cation $[Co(NH_3)_6]^{3+}$ can be used to precipitate a chloro-anion of formula $[CuCl_5]^{3-}$. Suggest two reasons why this cation stabilizes the formation of the $[CuCl_5]^{3-}$ anion. Also, give the correct name for the complete compound formed.

19.25 Iron(III) chloride reacts with triphenylphosphine, PPh_3, to form the complex $FeCl_3(PPh_3)_2$. However, with the ligand tricyclohexylphosphine, PCh_3, the compound $FeCl_3(PCh_3)$ is formed. Suggest a reason for the difference.

19.26 A complex of nickel, $Ni(PPh_3)_2Cl_2$ is paramagnetic, whereas the palladium analog, $Pd(PPh_3)_2Cl_2$, is diamagnetic. How many isomers would you predict for each compound? Give your reasoning.

19.27 With the availability of many oxidation states, Latimer, Frost, and Pourbaix diagrams are of considerable importance in the study of transition metal chemistry, as we will see in Chapter 20. For the imaginary transition metal, M, the Latimer diagram is as follows:

$$MO_3(aq) \xrightarrow{+0.50\text{ V}} MO_2(aq) \xrightarrow{+0.40\text{ V}} M^{3+}(aq) \xrightarrow{-0.20\text{ V}}$$
$$M^{2+}(aq) \xrightarrow{+0.05\text{ V}} M(s)$$

(a) Identify any species prone to disproportionation, and write the corresponding chemical equation.

(b) Calculate the pH below which the reaction:

$$2\ M^{2+}(aq) + 2\ H^+(aq) \rightarrow 2\ M^{3+}(aq) + H_2(g)$$

becomes spontaneous. Assume standard conditions except for $[H^+(aq)]$.

19.28 Which of the following ions will have the greater stability: $[AuF_2]^-$ or $[AuI_2]^-$. Give your reasoning.

19.29 Nickel forms an anion $[NiSe_4]^{2-}$ that is square planar, while the analogous zinc anion, $[ZnSe_4]^{2-}$, is tetrahedral. Suggest an explanation for the different structures.

19.30 Many tinned or bottled food products, such as commercial mayonnaise, salad dressings, and kidney beans, contain the hexadentate ligand ethylenediaminetetraacetate, $^{-2}(OOC)_2NCH_2CH_2N(COO)_2^{2-}$, abbreviated $edta^{4-}$, or a related compound. Suggest a reason for this.

19.31 Three hydrates of chromium(III) chloride are known: form A is a hexahydrate; form B, a pentahydrate; and form C, a tetrahydrate. Addition of excess silver ion solution to a solution of 1 mole of each form results in the precipitation of the following number of moles of silver chloride: form A—3; form B—2; form C—1. From this information, deduce the actual structure of each hydrate and write the corresponding name.

19.32 Should silver be designated a transition metal? Discuss.

19.33 The complex $[Co(NH_3)_6]Cl_3$ is yellow-orange, while $[Co(OH_2)_3F_3]$ is blue. Suggest an explanation for the difference in color.

19.34 The +3 to +2 reduction potentials for some first row transition metals are as follows:

$$Cr^{3+}(aq) + e^- \rightarrow Cr^{2+}(aq) \qquad E^\circ = -0.42\text{ V}$$
$$Mn^{3+}(aq) + e^- \rightarrow Mn^{2+}(aq) \qquad E^\circ = +1.56\text{ V}$$
$$Fe^{3+}(aq) + e^- \rightarrow Fe^{2+}(aq) \qquad E^\circ = +0.77\text{ V}$$
$$Co^{3+}(aq) + e^- \rightarrow Co^{2+}(aq) \qquad E^\circ = +1.92\text{ V}$$

Suggest why the manganese reduction potential is higher than might be expected, while the value for the iron reduction is lower than might be expected.

Properties of Transition Metals

Chapter **20**

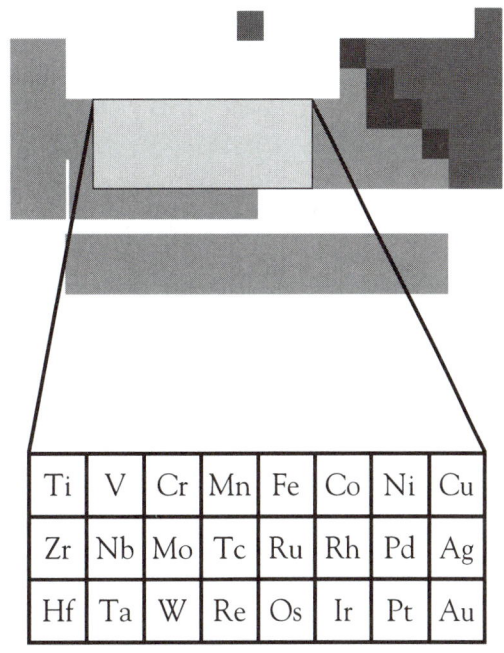

Ti	V	Cr	Mn	Fe	Co	Ni	Cu
Zr	Nb	Mo	Tc	Ru	Rh	Pd	Ag
Hf	Ta	W	Re	Os	Ir	Pt	Au

The most striking features of transition metal chemistry are the plethora of compounds and the variety of colors of the compounds. The number of compounds is a result of two factors: the many oxidation states that the metals exhibit and the ability to form complexes with a wide range of ligands. The colors usually result from electron transitions within the partially filled d orbitals in these species. The Period 4 members of the transition metal series are the most important; hence, these elements will be the focus of this chapter.

With its origins linked to the study of minerals, inorganic chemistry was the first branch of chemistry to be pursued rigorously. However, by the first part of the 20th century, it had fossilized into the memorization of long lists of compounds, their properties, and the methods of synthesis. The polymer and pharmaceutical industries became fields of rapid growth, and organic chemistry became the major focus of chemistry.

Ronald Nyholm, an Australian chemist, brought inorganic chemistry to life again. Nyholm was born in 1917 in Broken Hill, Australia, a mining

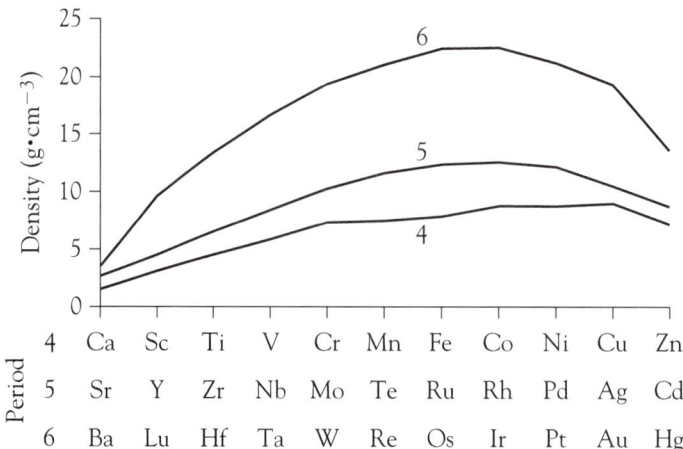

Figure 20.1 Densities of the transition metals in Periods 4, 5, and 6.

town where the streets have names such as Chloride, Sulphide, Oxide, and Silica. With this childhood environment—and an enthusiastic chemistry teacher in high school—it was natural for him to choose a career in chemistry. Nyholm moved to England to study with some of the great chemists of the time, where his research opened whole new directions of study by showing that it was the nature of the ligand that determined much of metal ion behavior. For example, using specific ligands, he was able to produce unusual oxidation numbers and coordination numbers in metal complexes. Nyholm, together with Ronald Gillespie, a British chemist, devised the VSEPR method of predicting molecular shape in 1957, and he was the first to argue that inorganic chemistry involved the understanding of molecular structure, not just memorization of formulas. Sadly, Nyholm was killed in an automobile accident in 1971, at the peak of his career.

20.1 Overview of the Transition Metals

A football made of iridium or osmium would have a mass of 320 kg.

All the transition metals are hard (except those of Group 11) and have very high melting points. In fact, 10 of the metals have melting points above 2000°C and 3 above 3000°C (tantalum, tungsten, and rhenium). The transition metals all have high densities, and the trends in this property are shown in Figure 20.1. The densities increase from the Period 4 elements to the Period 6 elements, with the highest values being those of osmium and iridium (23 g·cm^{-3}). Chemically, the metals themselves are comparatively unreactive. Only a few of the metals, such as iron, are electropositive enough to react with acids.

Group Trends

For the main group elements, there are clear trends down each group. For the transition metals, the elements of Periods 5 and 6 show very strong similarities in their chemistry within a group. This similarity is to a large extent a result of the filling of the $4f$ orbitals in the elements that lie between these two rows. Electrons in these orbitals are poor shielders of electrons in the outer $6s$ and $5d$ orbitals. With the greater effective nuclear charge, the atomic, covalent, and ionic radii of the Period 6 transition elements are reduced to almost the same as those in Period 5. This is known as the *lanthanoid contraction*. This effect is illustrated in Table 20.1, where the ionic radii of the Groups 2 and 5 metals are compared. The radii of the Group 2 metals

Table 20.1 Ion sizes of elements in Groups 2 and 5

Group 2 radius (pm) ion	Ion radius (pm)	Group 5 ion	Ion
Ca^{2+}	114	V^{3+}	78
Sr^{2+}	132	Nb^{3+}	86
Ba^{2+}	149	Ta^{3+}	86

increase down the group, whereas the niobium and tantalum ions have identical radii. It is the similarity in radii (and hence charge densities) that results in the strong resemblence in properties between the Period 5 and the Period 6 members of a group.

There are some superficial similarities in the chemistries of the Periods 5 and 6 elements to those of the Period 4 elements. For example, chromium, molybdenum, and tungsten all form oxides with an oxidation number of +6. However, chromium(VI) oxide, CrO_3, is highly oxidizing, whereas molybdenum(VI) oxide, MoO_3, and tungsten(VI) oxide, WO_3, are the "normal" oxides of these metals.

The limitations of such comparisons are also illustrated by the lower chlorides of chromium and tungsten. Chromium forms a compound, $CrCl_2$ (among others), whereas tungsten forms an apparently analogous compound, WCl_2. The former does contain the chromium(II) ion, but the latter is known to be $[W_6Cl_8]^{4+}\cdot 4\ Cl^-$, with the polyatomic cation containing a cluster of tungsten ions at the corners of an octahedron and chloride ions in the middle of the faces. The enthalpy of formation of the theoretical $W^{2+}\cdot 2\ Cl^-$ can be calculated as $+430\ kJ\cdot mol^{-1}$ (very different from the value of $-397\ kJ\cdot mol^{-1}$ for chromium(II) chloride), providing a thermodynamic reason for its nonexistence. The enthalpy difference is mostly due to the very much higher atomization energy of tungsten ($837\ kJ\cdot mol^{-1}$) relative to that of chromium ($397\ kJ\cdot mol^{-1}$). This high atomization energy reflects the strong metal–metal bonding in the Periods 5 and 6 transition metals. As a result, like WCl_2, many compounds of these elements contain groups of metal ions, and these are called *metal cluster compounds*.

The oxidation numbers of the transition metals are higher for the first half of each row than for the later members. The Period 5 and 6 elements usually have higher common oxidation numbers than does the analogous member of Period 4, as is shown in Table 20.2. Like the main group elements, the highest oxidation number of a transition metal is found in an oxide. Thus the +8 oxidation number of osmium occurs in osmium(VIII) oxide, OsO_4. Unlike the main group metals, transition metals exhibit

Table 20.2 The most common oxidation numbers of the transition metals

Ti	V	Cr	Mn	Fe	Co	Ni	Cu
+4	+3, +4	+3, +6	+2, +3, +7	+2, +3	+2, +3	+2	+1, +2
Zr	Nb	Mo	Tc	Ru	Rh	Pd	Ag
+4	+5	+6	—	+3	+3	+2	+1
Hf	Ta	W	Re	Os	Ir	Pt	Au
+4	+5	+6	+4, +7	+4, +8	+3, +4	+2, +4	+3

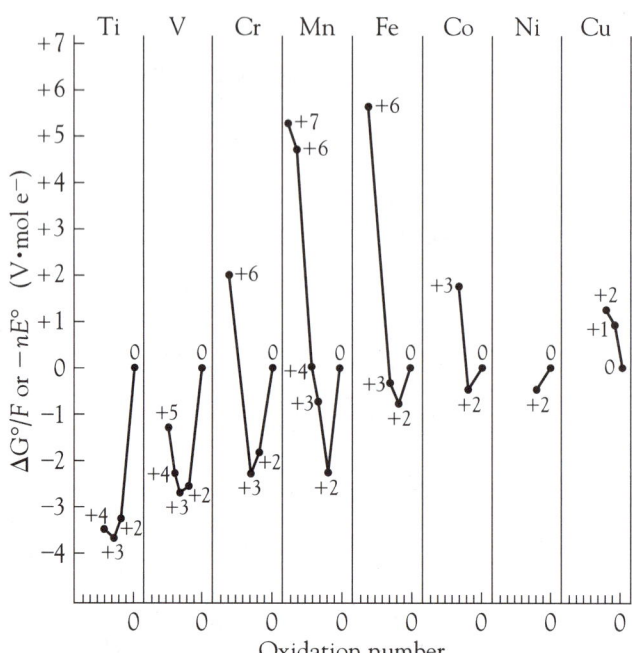

Figure 20.2 Frost diagrams for the Period 4 transition metals under acid conditions.

almost every possible oxidation number; for example, there are various compounds of manganese in which manganese has every oxidation number from $+7$ to -1.

One consistent factor found in each transition metal group is the increase in the crystal field splitting, Δ, from Period 4 to Period 6 elements. For example, in the series $[Co(NH_3)_6]^{3+}$, $[Rh(NH_3)_6]^{3+}$, and $[Ir(NH_3)_6]^{3+}$, the Δ_{oct} values are 23 000, 34 000, and 41 000 cm^{-1}, respectively. Because of the larger crystal field splittings for the Period 5 and 6 transition metals, almost all compounds of these elements are low spin.

Comparative Stability of Oxidation States of the Period 4 Transition Metals

The transition metals in Period 4 are most common and of greatest industrial importance. Furthermore, the patterns in their properties are the easiest to comprehend. Figure 20.2 summarizes the Frost diagrams for these elements. Titanium metal (oxidation state 0) is strongly reducing, but the elements become less so as we progress along the row. When we reach copper, the metal itself is the most thermodynamically stable oxidation state. As we move across the row, the highest oxidation state becomes less favored, and by chromium, it has become highly oxidizing. The most thermodynamically stable oxidation number is $+3$ for titanium, vanadium, and chromium, whereas $+2$ is favored by the other elements. For iron, the stabilities of the $+3$ and $+2$ oxidation states are very similar. Copper is unique in having a stable $+1$ oxidation number, but as is apparent from Figure 20.2, it is prone to disproportionation—to the $+2$ and 0 oxidation states.

20.2   Group 4: Titanium, Zirconium, and Hafnium

The only widely used element in this group is titanium. Titanium is the ninth most abundant element in Earth's crust, whereas zirconium and hafnium, like most of the Periods 5 and 6 transition metals, are rare.

Titanium

Titanium, a hard, silvery white metal, is the least dense (4.5 g·cm^{-3}) of the transition metals. This combination of high strength and low density makes it a preferred metal for military aircraft and nuclear submarines, where cost is less important than performance. The metal has more mundane applications in high-performance bicycle frames and golf clubs.

The pure metal is difficult to obtain from the most common titanium compound. Reduction of titanium(IV) oxide, TiO_2, with carbon produces the metal carbide rather than the metal. The only practical route (the *Kroll process*) involves the initial conversion of the titanium(IV) oxide to titanium(IV) chloride by heating the oxide with carbon and dichlorine:

$$TiO_2(s) + 2\ C(s) + 2\ Cl_2(g) \xrightarrow{\Delta} TiCl_4(g) + 2\ CO(g)$$

The titanium(IV) chloride gas is condensed at 137°C.

We can use an Ellingham diagram for chlorides to examine the possible reduction routes for this compound (Figure 20.3). Carbon is totally unsuitable for reducing the titanium(IV) chloride, because the slope of the line in the Ellingham diagram is the opposite of that required—in other words, the free energy line for the formation of carbon tetrachloride does not cross any metal chloride line. Hydrogen is also unsatisfactory, because it will only reduce titanium(IV) chloride above about 1700°C. The alternative is to find a metal whose metal–metal chloride line lies below that of the titanium–titanium(IV) chloride line. The choice of reactive metal is partially based on cost and partially on the ease of separating the titanium metal from the other metal chloride and from excess metal reactant. Magnesium is usually preferred, and at about 850°C, it will displace the titanium:

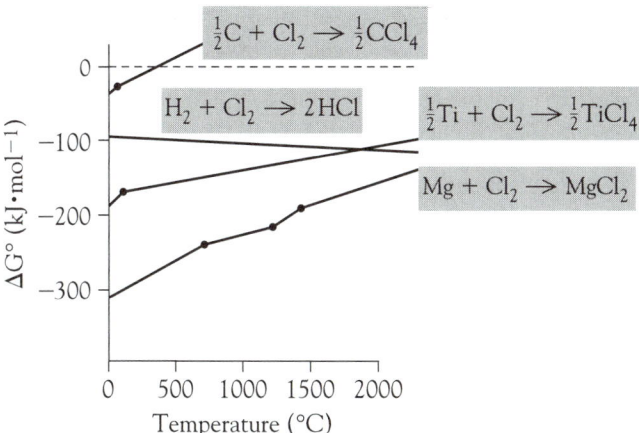

Figure 20.3 Ellingham diagram for various chloride species.

$$TiCl_4(g) + 2\ Mg(l) \xrightarrow{\Delta} Ti(s) + 2\ MgCl_2(l)$$

The spongy mass of titanium metal is porous, and the magnesium chloride and excess magnesium metal can be dissolved out by using dilute acid. The titanium metal granules are then fabricated into whatever shape is required.

Titanium(IV) Oxide

Although the production of titanium metal is vital for the defense industry, the enormous quantities of titanium ores mined each year are destined for a more innocuous purpose—as paint pigment. Of the 5 million tonnes of titanium ore produced each year, Canada provides about one-third and Australia provides about one-fourth. Although the element is often found as the dioxide (mineral name, rutile), it is too impure to be used directly.

The purification process involves the conversion of rutile to the chloride, as in the metal synthesis:

$$TiO_2(s) + 2\ C(s) + 2\ Cl_2(g) \xrightarrow{\Delta} TiCl_4(g) + 2\ CO(g)$$

The chloride is then reacted with dioxygen at about 1200°C to give pure white titanium(IV) oxide:

$$TiCl_4(g) + O_2(g) \xrightarrow{\Delta} TiO_2(s) + 2\ Cl_2(g)$$

The chlorine gas is recycled.

Prior to the use of titanium(IV) oxide in paints, the common white pigment was "white lead," $Pb_3(CO_3)_2(OH)_2$. Apart from its toxicity, it discolored in industrial city atmospheres to give black lead(II) sulfide. Titanium(IV) oxide, which is stable to discoloration in polluted air, has now replaced white lead completely. Not only is titanium(IV) oxide of very low toxicity, it has the highest refractive index of any white or colorless inorganic substance—even higher than diamond. As a result of this high light-scattering ability, it covers and hides previous paint layers more effectively. In addition to its use in white paint, it is also added to colored paints to make the colors paler and mask previous colors better.

Zirconium

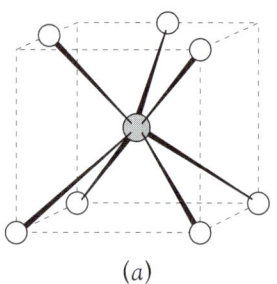

(a)

Zr^{4+} O^{2-}

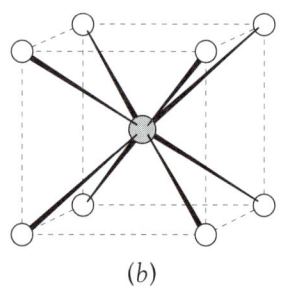

(b)

Figure 20.4 The zirconium(IV) oxide ion arrangements in (a) baddeleyite and (b) cubic zirconia.

Although a very rare metal, zirconium is used to make the containers for nuclear fuel because it has a low capture cross section for neutrons—that is, it does not absorb the neutrons that propagate the fission process. Unfortunately, hafnium has a high capture cross-section; thus, it is crucial to remove hafnium impurities from the chemically similar zirconium. To produce metallic zirconium, the ore baddeleyite (zirconium(IV) oxide), ZrO_2, is processed by a method similar to that for titanium:

$$ZrO_2(s) + 2\ C(s) + 2\ Cl_2(g) \xrightarrow{\Delta} ZrCl_4(g) + 2\ CO(g)$$

At this stage, the 2 percent impurity of hafnium(IV) chloride, $HfCl_4$, can be separated from the zirconium(IV) chloride, $ZrCl_4$, by fractional sublimation. The hafnium compound sublimes at 319°C; and the zirconium compound, at 331°C. (The closeness in sublimation temperatures shows the great similarity between the two elements.) Then the pure zirconium(IV) chloride is reduced with magnesium metal:

$$ZrCl_4(g) + 2\ Mg(l) \xrightarrow{\Delta} Zr(s) + 2\ MgCl_2(l)$$

In the baddeleyite crystalline form of zirconium(IV) oxide, each zirconium(IV) ion is surrounded by seven oxide ions (Figure 20.4a). Above 2300°C, the compound rearranges to an eight-coordinate fluorite structure (Figure 20.4b), cubic zirconia, which is an excellent diamond substitute in jewelry. Although the refractive index (and hence "sparkle") and hardness of cubic zirconia are less than those of diamond, its melting point of 2700°C makes it more thermally stable than diamond. By a patented process, zirconium(IV) oxide can be produced in a fibrous form. These silky fibers have nearly uniform dimensions: 3 mm in diameter and 2 to 5 cm long. They can be woven into a material that is stable up to 1600°C, making zirconia cloth very useful for high-temperature purposes.

20.3   *Group 5: Vanadium, Niobium, and Tantalum*

None of the Group 5 metals has any great usefulness, although vanadium is used for vanadium steels, a particularly hard alloy employed for knife blades and various workshop tools.

Oxidation States of Vanadium

The simple redox chemistry of vanadium is particularly interesting to inorganic chemists because vanadium readily exists in four different oxidation states: +5, +4, +3, and +2, corresponding to the d^0, d^1, d^2, and d^3 electron configurations. With an oxidation number of $+5$ for vanadium, the colorless vanadate ion, $[VO_4]^{3-}$, exists in very basic solution; under neutral conditions, conjugate acids such as the pale yellow dihydrogen vanadate ion, $[H_2VO_4]^-$, are formed.

A reducing agent, such as zinc metal in acid solution, can be used to reduce the vanadium(V) to give the characteristically colored ions of vanadium in lower oxidation states:

$$Zn(s) \rightarrow Zn^{2+}(aq) + 2\ e^-$$

Initial reduction of the dihydrogen vanadate ion by zinc metal in acid solution (or by a weak reducing agent, such as sulfur dioxide) gives the deep blue vanadyl ion, VO^{2+} (with $+4$ oxidation number):

$$[H_2VO_4]^-(aq) + 4\ H^+(aq) + e^- \rightarrow VO^{2+}(aq) + 3\ H_2O(l)$$

This ion is written more precisely as the $[VO(OH_2)_5]^{2+}$, because five water molecules occupy the other coordination sites.

As reduction continues, the bright blue color of the vanadyl ion is replaced by that of the green hexaaquavanadium(III) ion, $[V(OH_2)_6]^{3+}$ (or $V^{3+}(aq)$, for simplicity):

$$VO^{2+}(aq) + 2\ H^+(aq) + e^- \rightarrow V^{3+}(aq) + H_2O(l)$$

Provided air is excluded, further reduction results in the formation of the lavender hexaaquavanadium(II) ion, $[V(OH_2)_6]^{2+}$:

$$[V(OH_2)_6]^{3+}(aq) + e^- \rightarrow [V(OH_2)_6]^{2+}(aq)$$

As soon as this solution is exposed to air, it reoxidizes to the vanadium(III) ion and eventually to the vanadyl ion.

Biological Aspects

Vanadium is not widely used in nature, yet it does appear to be vital to one of the simplest groups of marine organisms: the tunicates, or sea squirts. These organisms belong somewhere between invertebrates and vertebrates. One family of tunicates utilizes very high levels of vanadium in its blood plasma for oxygen transport. Why the tunicates should have picked such a unique element for a biochemical pathway is still unclear. The element also appears to be used by a very different organism, the poisonous mushroom *Amanita muscaria*. Here, too, the reason for this element being utilized is not well understood.

20.4 *Group 6: Chromium, Molybdenum, and Tungsten*

All the stable Group 6 metals are utilized in the manufacture of metal alloys for specialized uses. In addition, chromium provides a shiny protective coating for iron and steel surfaces. Chromium metal is not inert in itself; instead, it has a very thin, tough, transparent oxide coating that confers the protection.

For molybdenum and tungsten, the oxidation number of $+6$ is thermodynamically preferred. However, for chromium, the $+6$ state is highly oxidizing; the oxidation number of $+3$ is most stable.

Chromates and Dichromates

Despite their thermodynamic instability, kinetic factors enable several chromium(VI) compounds to exist. The most important of these are the chromates and dichromates. The yellow chromate ion, $[CrO_4]^{2-}$, can only exist in solution under neutral or alkaline conditions, and the orange dichromate ion, $[Cr_2O_7]^{2-}$, only under acid conditions, because of the equilibrium

$$2[CrO_4]^{2-}(aq) + 2\ H^+(aq) \rightleftharpoons [Cr_2O_7]^{2-}(aq) + H_2O(l)$$

The chromate ion is the conjugate base of the hydrogen chromate ion, $[HCrO_4]^-$; thus, a chromate ion solution is always basic because of the equilibrium

$$[CrO_4]^{2-}(aq) + H_2O(l) \rightleftharpoons [HCrO_4]^-(aq) + OH^-(aq)$$

The equilibrium between the three species is shown in Figure 20.5.

Many chromates are insoluble, and they are often yellow if the cation is colorless, such as lead(II) chromate, $PbCrO_4$. The high insolubility of lead(II) chromate and its high refractive index (hence high opacity) have resulted in its use for yellow highway markings.

In both the chromate and the dichromate ions, chromium has an oxidation state of $+6$; hence, the metal has a d^0 electron configuration. Without d electrons, we might expect these, and all d^0 configurations, to be colorless. This is obviously not the case. The color comes from an electron transition from the ligand to the metal, a process known as *charge transfer*. That is, an electron is excited from a filled ligand p orbital through a π interaction into the empty metal ion d orbitals. We can depict the process simply as

$$Cr^{6+}\!\!-\!\!O^{2-} \rightarrow Cr^{5+}\!\!-\!\!O^-$$

Such transitions require considerable energy; hence, the absorption is usually centered in the ultraviolet part of the spectrum, with just the edge of the absorption in the visible region. Charge transfer is particularly evident when the metal is in high oxidation states, such as chromates and dichromates.

Silver(I) chromate, Ag_2CrO_4, has a unique brick-red color, making it a useful compound in the analysis of silver ion. One route is a precipitation titration (the *Mohr method*), in which silver ion is added to chloride ion to

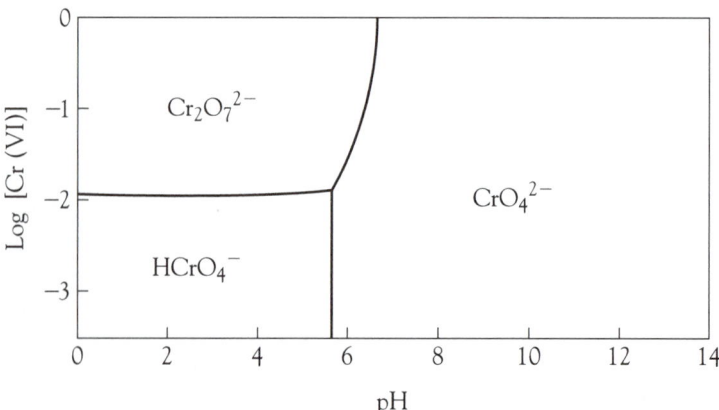

Figure 20.5 Predominance diagram for chromium(VI) species showing their dependence on pH and chromium concentration.

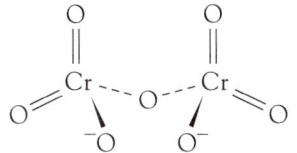

Figure 20.6 The dichromate ion, $[Cr_2O_7]^{2-}$.

The toxicity of chromium(VI) is the central theme of the movie Erin Brockovitch.

give a white precipitate of silver chloride:

$$Ag^+(aq) + Cl^-(aq) \rightarrow AgCl(s)$$

In the presence of chromate ion (usually about $0.01\ mol \cdot L^{-1}$), the brick-red, slightly more soluble silver chromate will form as soon as the chloride ion is completely consumed, the color change indicating that the equivalence point has been reached (actually, slightly exceeded):

$$2\ Ag^+(aq) + [CrO_4]^{2-}(aq) \rightarrow Ag_2CrO_4(s)$$

The dichromate ion has a structure involving a bridging oxygen atom (Figure 20.6). This ion is a strong oxidizing agent, although the carcinogenic nature of the chromium(VI) ion means that it should be treated with respect, particularly the powdered solid, which can be absorbed through the lungs. The orange dichromate ion is a good oxidizing agent and is reduced to the green hexaaquachromium(III) ion, $[Cr(OH_2)_6]^{3+}$, in the redox reaction

$$[Cr_2O_7]^{2-}(aq) + 14\ H^+(aq) + 6\ e^- \rightarrow 2\ Cr^{3+}(aq) + 7\ H_2O(l) \quad E^\circ = +1.33\ V$$

This reaction is used in breath analyzers for the detection of excessive alcohol intake. The ethanol in the breath is bubbled through an acidic solution of dichromate, the color change being detected quantitatively. In the reaction, the ethanol is oxidized to ethanoic (acetic) acid:

$$CH_3CH_2OH(aq) + H_2O(l) \rightarrow CH_3CO_2H(aq) + 4\ H^+(aq) + 4\ e^-$$

The oxidation of organic compounds with dichromate ion is a common reaction in organic chemistry. Sodium dichromate is preferred because it has a higher solubility than potassium dichromate.

For quantitative analysis, sodium dichromate cannot be used as a primary standard because of its deliquescence. However, potassium dichromate is an ideal primary standard because it does not hydrate and because it can be obtained in high purity by recrystallization—its solubility in water increases rapidly with increasing temperature. One application is the determination of iron(II) ion concentrations in an acidic solution. In this titrimetric procedure, the dichromate is reduced to chromium(III) ion and the iron(II) ion is oxidized to iron(III) ion:

$$Fe^{2+}(aq) \rightarrow Fe^{3+}(aq) + e^-$$

The characteristic color change of orange to green as the dichromate is reduced to the chromium(III) ion is not sensitive enough; thus, an indicator, barium diphenylamine sulfonate, has to be used. This indicator is less readily oxidized than iron(II) ions, but it is oxidized to give a blue color once all the iron(II) ions have been converted to the iron(III) state. Because free iron(III) ions affect the indicator and thus give rise to an inaccurate endpoint, some phosphoric acid is added before starting the titration. This reagent gives a stable iron(III) phosphate complex.

Ammonium dichromate, $(NH_4)_2Cr_2O_7$, is often used in "volcano" demonstrations. If a red-hot wire or a lit match is touched to a pile of ammonium dichromate, exothermic decomposition is initiated, emitting sparks and water vapor in a spectacular way. However, this is not a safe demonstration because a dust containing carcinogenic chromium(VI) compounds is usually released.

The reaction is nonstoichiometric, producing chromium(III) oxide, water vapor, nitrogen gas, and some ammonia gas. It is commonly represented as

$$(NH_4)_2Cr_2O_7(s) \rightarrow Cr_2O_3(s) + N_2(g) + 4\ H_2O(g)$$

The industrial production of dichromate provides some interesting chemistry. The starting material is a mixed oxide, iron(II) chromium(III) oxide, $FeCr_2O_4$ (commonly called iron chromite), an ore found in enormous quantities in South Africa. The powdered ore is heated to about 1000°C with sodium carbonate in air, thereby causing the chromium(III) to be oxidized to chromium(VI):

$$4\ FeCr_2O_4(s) + 8\ Na_2CO_3(s) + 7\ O_2(g) \xrightarrow{\Delta}$$
$$8\ Na_2CrO_4(s) + 2\ Fe_2O_3(s) + 8\ CO_2(g)$$

Addition of water dissolves the sodium chromate, a process called leaching; it leaves the insoluble iron(III) oxide. To obtain sodium dichromate, the Le Châtelier principle is applied. The following equilibrium lies to the left under normal conditions, but under high pressures of carbon dioxide (obtained from the previous reaction), the yield of sodium dichromate is high:

$$2\ Na_2CrO_4(aq) + 2\ CO_2(aq) + H_2O(l) \rightleftharpoons Na_2Cr_2O_7(aq) + 2\ NaHCO_3(s)$$

In fact, the aqueous carbon dioxide is really employed as a low-cost way of decreasing the pH to favor the dichromate ion in the chromate–dichromate equilibrium. It can be seen that the mole ratio of carbon dioxide to chromate produced in the previous step is exactly the same as that employed in this step. The slightly soluble sodium hydrogen carbonate has to be filtered off under pressure to prevent the equilibrium from shifting to the left. The sodium hydrogen carbonate is then reacted with an equimolar proportion of sodium hydroxide to obtain the sodium carbonate that can be reused in the first step. Thus, the ore and sodium hydroxide are the only bulk chemicals used in the process.

Chromium(VI) Oxide

The oxide of chromium in which chromium assumes its highest oxidation state is chromium(VI) oxide, CrO_3. It is a red crystalline solid that is prepared by adding concentrated sulfuric acid to a cold concentrated solution of potassium dichromate. The synthesis can be viewed as an initial formation of chromic acid followed by a decomposition to the acidic oxide and water:

$$K_2Cr_2O_7(aq) + H_2SO_4(aq) + H_2O(l) \rightarrow K_2SO_4(aq) + 2\ ``H_2CrO_4(aq)"$$

$$``H_2CrO_4(aq)" \rightarrow CrO_3(s) + H_2O(l)$$

Chromium(VI) oxide is an acidic oxide, as are most metal oxides in which the metal has a very high oxidation number. It is very soluble in water, forming "chromic acid," which is in fact a mixture of species. The strongly oxidizing (and acidic) nature of the solution results in its occasional use as a final resort for cleaning laboratory glassware. However, the hazard of the solution itself (it is carcinogenic and very acidic) and the potential danger from exothermic redox reactions with glassware contaminants make it a very unwise choice.

The oxide is also strongly oxidizing. For example, ethanol ignites on contact; it is oxidized to a mixture of ethanal (acetaldehyde), CH_3CHO, and ethanoic (acetic) acid, CH_3CO_2H, and the chromium(VI) oxide is reduced to chromium(III) oxide.

Chromyl Chloride

Chromyl chloride, a red, oily liquid of formula CrO_2Cl_2, is of interest only as a definitive means of identifying chloride ion if a halide ion is known to be present. When concentrated sulfuric acid is added to a mixture of solid potassium dichromate and an ionic chloride, such as sodium chloride, a dark red liquid is formed:

$$6 \text{ H}_2\text{SO}_4(l) + \text{K}_2\text{Cr}_2\text{O}_7(s) + 4 \text{ NaCl}(s) \rightarrow$$
$$2 \text{ CrO}_2\text{Cl}_2(l) + 2 \text{ KHSO}_4(s) + 4 \text{ NaHSO}_4(s) + 3 \text{ H}_2\text{O}(l)$$

When heated gently and very cautiously, a deep red, toxic vapor is produced. This gas can be collected and condensed to a dark red, covalent liquid, chromyl chloride. If this liquid is added to a basic solution, it immediately hydrolyzes to the yellow chromate ion:

$$\text{CrO}_2\text{Cl}_2(l) + 4 \text{ OH}^-(aq) \rightarrow \text{CrO}_4{}^{2-}(aq) + 2 \text{ Cl}^-(aq) + 2 \text{ H}_2\text{O}(l)$$

Because bromides and iodides do not form analogous chromyl compounds, this test is specific for chloride ions.

The molecule itself has a tetrahedral arrangement around the central chromium atom, and there is appreciable double bond character in the Cr—O bonds (Figure 20.7).

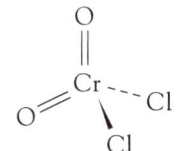

Figure 20.7 The chromyl chloride molecule, CrO_2Cl_2.

Chromium(III) Oxide

The green, powdery compound, chromium(III) oxide, Cr_2O_3, is an amphoteric oxide, as expected from the lower oxidation number of the metal. Just as lead(II) chromate (chrome yellow) is an important yellow pigment, so chromium(III) oxide is a common green pigment. It is chromium(III) oxide that has, since 1862, provided the pigment for the green U.S. currency ("greenbacks"). Because the pigment is a mineral rather than an organic dye, the green will not fade, nor is it affected by acids, bases, or either oxidizing or reducing agents. To prepare a pure pigment, sodium dichromate is reduced; the common reducing agent is sulfur at high temperatures:

$$\text{Na}_2\text{Cr}_2\text{O}_7(s) + \text{S}(l) \xrightarrow{\Delta} \text{Cr}_2\text{O}_3(s) + \text{Na}_2\text{SO}_4(s)$$

The sodium sulfate is washed away, leaving pure chromium(III) oxide.

Chromium(III) Chloride

Anhydrous chromium(III) chloride, $CrCl_3$, is a reddish violet solid obtained when chlorine is passed over strongly heated chromium metal:

$$2 \text{ Cr}(s) + 3 \text{ Cl}_2(g) \xrightarrow{\Delta} 2 \text{ CrCl}_3(s)$$

When crystallized from aqueous solution, a dark green hexahydrate is obtained. If a solution of this hydrated chromium(III) chloride is treated with a solution of silver nitrate, only one-third of the chloride precipitates out as silver chloride; that is, only one of the chlorides is present as the free ion. This result indicates that the formula of this compound is $[\text{Cr}(\text{OH}_2)_4\text{Cl}_2]^+\text{Cl}^-\cdot 2\text{H}_2\text{O}$. As mentioned in Chapter 19, Section 19.4, there are actually three hydration isomers of this compound: violet, $[\text{Cr}(\text{OH}_2)_6]^{3+}$ 3Cl^-; light green, $[\text{CrCl}(\text{OH}_2)_5]^{2+}$ $2\text{Cl}^-\cdot\text{H}_2\text{O}$; and dark green, $[\text{Cr}(\text{OH}_2)_4\text{Cl}_2]^+$ $\text{Cl}^-\cdot 2\text{H}_2\text{O}$.

Chromium(II) Acetate

Though chromium(II) is a low oxidation state for this element, there is one compound that is easy to synthesize—chromium(II) acetate. This insoluble, red compound is prepared by reducing chromium(III) ion with zinc metal:

$$2 \text{ Cr}^{3+}(aq) + \text{Zn}(s) \rightarrow 2 \text{ Cr}^{2+}(aq) + \text{Zn}^{2+}(aq)$$

then adding acetate ion. Acetates are generally soluble; in fact, they form the most soluble compounds except for nitrate. Thus, the insolubility of the compound is an indication that the product is not a simple compound. And, indeed, it is not. The complex is a dimer with the four acetate ligands acting as —O—C—O— bridges between the two chromium(II) ions which are directly linked by a chromium–chromium bond. The two water molecules occupy the sixth coordination sites on the ends of the molecule.

$$2\,Cr^{2+}(aq) + 4\,CH_3COO^-(aq) + 2\,H_2O(l) \rightarrow Cr_2(CH_3COO)_4(OH_2)_2(s)$$

Molybdenum (IV) Sulfide

Molybdenum(IV) sulfide is the only industrially important compound of molybdenum. It is the common ore of the metal, and nearly half of the world's supply is in the United States. The purified black molybdenum(IV) sulfide, MoS_2, has a layer structure that resembles graphite. This property has led to its use as a lubricant, both alone and as a slurry mixed with hydrocarbon oils.

Tungsten

Tungsten has the highest melting point of all metals ($3422°C$), and it is this property, together with its ductility, that results in tungsten's use in the incandescent light bulb. Passing an electric current through a wire generates heat—in the case of the light bulb, a high enough energy density to cause the wire to glow white-hot. Nevertheless, even with high-melting tungsten, atoms sublime off the hot metal surface, giving a dark deposit on the cool walls of the glass envelope, thinning the remaining wire until it finally breaks. To maximize filament lifetime, the current through the wire is the minimum necessary to provide acceptable incandescence. With the demand for high-intensity illumination, such as aircraft landing lights, a way of operating the filament at high temperature was required that provided greater light output. Fortunately, tungsten chemistry provided the answer: the thermal decomposition of tungsten(II) iodide. In these tungsten–halogen bulbs, the glass envelope is filled with iodine vapor. As the tungsten sublimes from the wire and migrates to the cooler parts of the bulb, the metal atoms combine with diiodine to form gaseous tungsten(II) iodide. When these molecules came close to the hot wire, the compound decomposes, re-depositing the tungsten metal back on the wire. In this way, the filament can be used at closer to the melting point of the metal, providing a much more intense light while still having an acceptable lifetime.

Tungsten Bronzes

In Chapter 16, Section 16.16, we described the perovskite unit cell which is common for oxides of formula $MM'O_4$, where M is a large, lower-charge cation and M′ is a small, higher charge cation. Sodium tungstate, $NaWO_3$, which adopts the perovskite structure, has very unusual properties. This compound can be prepared with less than stoichiometric proportions of sodium ion; that is, Na_xWO_3, where $x < 1$. The stoichiometric tungstate is white, but as the mole proportion of sodium drops to 0.9, it becomes a metallic golden yellow. As the proportion drops from 0.9 to 0.3, colors from metallic orange to red to bluish black are obtained. This material and its relatives, called the *tungsten bronzes,* are often used for metallic paints. Not only do the compounds look metallic but also their electrical conductivity approaches that of a metal. In the crystal, increasing proportions of the cell centers, where the large alkali metal would be found, are vacant. As a result, the conduction band, which in the stoichiometric compound would be full, is now partially empty. In these circumstances, electrons can move through the π system along the cell edges by using tungsten ion d orbitals and oxide ion p orbitals. It is this electron mobility that produces the color and the electrical conductivity.

Heteropoly-molybdates and Heteropoly-tungstates

Just as silicates can form clusters of SiO_4 tetrahedra to give polysilicates (Chapter 14, Section 14.19), so molybdates, MoO_4^{2-} and tungstates, WO_4^{2-}, can form clusters of octahedral MO_6 units. What is particularly interesting about this class of clusters is that they can incorporate within them a hetero-ion. For example, mixing molybdate and phosphate ions and acidifying give the phosphomolybdate ion, $[PMo_{12}O_{40}]^{3-}$:

$$PO_4^{3-}(aq) + 12\ MoO_4^{2-}(aq) + 24\ H^+(aq) \rightarrow [PMo_{12}O_{40}]^{3-} + 12\ H_2O(l)$$

The phosphorus(V) sits in a tetrahedral hole, surrounded by four oxygens, in the center of the cluster. The heteropoly ions are sometimes called *Keggin clusters* after J.F. Keggin who first determined their structures. The 1:12 clusters can be formed by both molybdenum and tungsten, while the hetero ion can be any small ion that can fit into the tetrahedral hole. These are ions whose respective oxoanion is tetrahedral, such as phosphorus(V), PO_4^{3-}, and silicon(IV), SiO_4^{4-}.

The formation of the phosphomolybdate is a useful test for the qualitative and quantitative analysis of phosphate ion. The intensity of the yellow color indicates the concentration of the complex ion. Alternatively, addition of a reducing agent reduces some of the molybdenum(VI) ions to molybdenum(V). With the resulting charge defects, the ion becomes intensely blue—the *heteropoly blues*—a very sensitive test for phosphate. Phosphomolybdate salts are good flame retardants on fabrics.

The salts of small cations and the parent acids are very water-soluble, while salts of the larger cations, such as cesium and barium, are insoluble. With the 3− or 4− ion charge delocalized over such an enormous cluster, the low charge density results in the acid being very strong and the conjugate base essentially neutral. The heteropolyacid itself can be synthesized at very low pH. The compound 12-tungstosilicic acid is one of the simplest to prepare:

$$SiO_4^{4-}(aq) + 12\ WO_4^{2-}(aq) + 28\ H^+(aq) \rightarrow H_4SiW_{12}O_{40}\cdot7H_2O(s) + 5\ H_2O(l)$$

The acid can then be titrated against sodium hydroxide like any typical strong acid:

$$H_4SiW_{12}O_{40}(aq) + 4\ OH^-(aq) \rightarrow [SiW_{12}O_{40}]^{4-}(aq) + 4\ H_2O(l)$$

 Biological Aspects

Although chromium(VI) is carcinogenic when ingested or absorbed through the skin, we require small quantities of chromium(III) in our diet. Insulin and the chromium(III) ion regulate blood glucose levels. A deficiency of chromium(III) or an inability to utilize the chromium ion can lead to one form of diabetes.

However, molybdenum is the most biologically important member of the group. It is the heaviest (highest atomic number) element to have a wide range of functions in living organisms. At the present time, dozens of known enzymes rely on molybdenum, which is usually absorbed as the molybdate ion, $[MoO_4]^{2-}$. The most crucial molybdenum enzyme (which contains iron as well) is nitrogenase. This family of enzymes occurs in bacteria that reduce the "inert" dinitrogen of the atmosphere to the ammonium ion, which is used in protein synthesis by plants. Some of these bacteria have a symbiotic relationship with the leguminous plants (pea and bean family), forming

Table 20.3   Some common pterin-containing molybdenum enzymes

Name	Metal atoms per molecule	Source
Sulfite reductase	2 Mo	Mammalian livers
Nitrate reductase	2 Mo, 2 Fe	Plants, fungi, algae, bacteria
Trimethylamine N-oxide reductase	2 Mo, 1 Fe, 1½ Zn	*Escherichia coli*
Xanthine oxidase	2 Mo, 4 Fe_2S_2	Cow's milk, mammalian liver, kidney
Formate dehydrogenase	Mo, Se, Fe_nS_n	Fungi, yeast, bacteria, plants
Carbon monoxide dehydrogenase	2 Mo, 4 Fe_2S_2, 2 Se	Bacteria

nodules on the roots. These bacteria process about 2×10^8 tonnes of nitrogen per year in the soils of this planet!

Outside of the nitrogenases, there is a family of molybdenum-containing enzymes that has a common core of molybdenum and an organic ring structure known as a pterin system. These enzymes often contain another metal, particularly an iron–sulfur system, and they perform the vital role of oxidants or reductants of toxic species in organisms. For example, sulfite oxidase oxidizes sulfite ion to sulfate ion and carbon monoxide dehydrogenase oxidizes carbon monoxide to carbon dioxide, while nitrate reductase reduces nitrate ion to nitirite ion. Some common pterin-containing molybdenum enzymes are listed in Table 20.3.

Why is a metal as rare as molybdenum so biologically important? There are a number of possible reasons: The molybdate ion has a high aqueous solubility at near-neutral pH values, making it easily transportable by biological fluids. The ion has a negative charge, making it more suitable for different environments than are the cations of the Period 4 transition metals. In fact, it is argued that the molybdate ion is transported by the same mechanism as the sulfate ion, SO_4^{2-}, another example of the similarities of ions of Group 6 and Group 16 (see Chapter 9, Section 9.5). The element has a wide range of oxidation states ($+4$, $+5$, and $+6$) whose redox potentials overlap with those of biological systems. Finally, molybdenum is about 18th in the order of abundances of metals in seawater, and much of the choice of elements for biochemical processes was probably determined when the only life on this planet was in the sea.

Tungsten enzymes are also known, and these are found in certain bacteria. In most cases, the bacteria also possess molybdenum-containing enzymes. However there are some bacteria, the hyperthermal archaea, that depend specifically upon tungsto-enzymes for their functioning. The tungsten center acts as an electron sink and source, oscillating between tungsten $+4$, $+5$, and $+6$ oxidation states. As these bacteria exist at very high temperatures, up to 110°C in some cases, it is argued that tungsten rather than molybdenum is utilized by the enzyme as tungsten has the stronger metal–ligand bond, enabling the enzyme to function at high temperatures without disintegrating.

20.5 Group 7: Manganese, Technetium, and Rhenium

Manganese is important as an additive in specific types of steel. Rhenium has little practical use, but technetium, all of whose isotopes are radioactive, has medical uses in radiotherapy and radioimaging.

Oxidation States of Manganese

Manganese readily forms compounds over a range of oxidation states that is wider than that of any other common metal. Figure 20.8 shows the relative stabilities for the oxidation states of manganese in acid solution. From this diagram, we can see that the permanganate ion, $[MnO_4]^-$, or tetraoxomanganate(VII), is very strongly oxidizing in acid solution. The deep green manganate ion, $[MnO_4]^{2-}$, or tetraoxomanganate(VI), is also strongly oxidizing, but it disproportionates readily to the permanganate ion and manganese(IV) oxide; thus, it is of little importance. Manganese(IV) oxide is oxidizing with respect to the most stable manganese species, the manganese(II) ion. In acid solution, the manganese(III) ion disproportionates, and it also is of little interest. Finally, the metal itself is reducing.

In basic solution, we find a different situation, as can be seen in Figure 20.9. The differences can be summarized as follows:

1. For a particular oxidation state, many of the compounds are unique. Manganese, like most metals, at high pH forms insoluble hydroxides (and oxide hydroxides) in which the metal has low oxidation states.

2. The higher oxidation states are not strongly oxidizing, as they are in acid solution. This difference can be explained simply in terms of reductions that involve hydrogen ion concentration, for these reactions will be strongly pH dependent. In the Frost diagram for acid solution, the concentration of hydrogen ion is $1 \ mol \cdot L^{-1}$; in the diagram for basic solution, the hydrogen ion concentration is $10^{-14} \ mol \cdot L^{-1}$ (that is, $1 \ mol \cdot L^{-1}$ concentration of hydroxide ion). Using the Nernst equation, we can show that this change in ion concentration will have a major effect on the standard reduction potential.

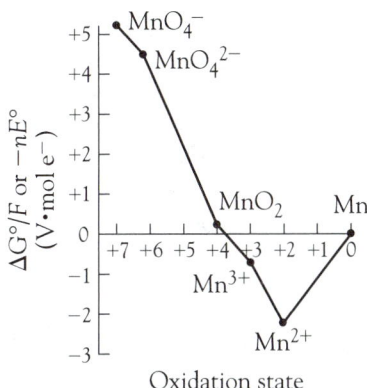

Figure 20.8 Frost (oxidation state) diagram for manganese in acidic solution.

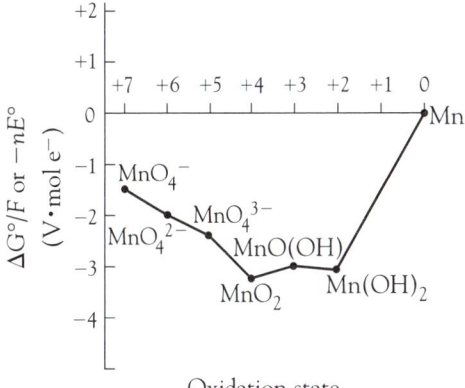

Figure 20.9 Frost (oxidation state) diagram for manganese in basic solution.

3. Oxidation states that are very unstable in acid can exist in basic solution (and vice versa). Thus, the bright blue manganite ion, $[MnO_4]^{3-}$, tetraoxomanganate(V), can be formed in basic solution.

4. In basic solution, the most thermodynamically stable species is manganese(IV) oxide, although the manganese(III) oxide hydroxide, $MnO(OH)$, and manganese(II) hydroxide also are both moderately stable. In fact, above pH 14, manganese(III) oxide hydroxide is thermodynamically more stable than manganese(II) hydroxide.

Potassium Permanganate

Potassium permanganate, $KMnO_4$, a violet-black solid, is the best known manganese compound with an oxidation number of +7. Like chromium(VI) compounds, the color in this d^0 ion is derived from charge transfer electron transitions. It dissolves in water to give a deep purple solution. The permanganate ion is an extremely powerful oxidizing agent, and under acid conditions, it is reduced to the colorless manganese(II) ion:

$$[MnO_4]^-(aq) + 8\,H^+(aq) + 5\,e^- \rightarrow Mn^{2+}(aq) + 4\,H_2O(l) \quad E° = +1.51\,V$$

It will oxidize concentrated hydrochloric acid to chlorine, and this is one way of producing chlorine gas in the laboratory:

$$2\,HCl(aq) \rightarrow Cl_2(g) + 2\,H^+(aq) + 2\,e^-$$

Potassium permanganate is an important reagent in redox titrations. Unlike potassium dichromate, it is not a suitable primary standard, because its purity cannot be guaranteed. Samples of the substance contain some manganese(IV) oxide, and aqueous solutions slowly deposit brown manganese(IV) oxide on standing. Its precise concentration is determined by titration against a standard solution of oxalic acid. Potassium permanganate solution is run into the oxalic acid solution from a buret, and the purple color disappears as the (almost) colorless manganese(II) ions and carbon dioxide are formed. The permanganate acts as its own indicator, for the slightest excess of permanganate gives a pink tint to the solution:

$$[MnO_4]^-(aq) + 8\,H^+(aq) + 5\,e^- \rightarrow Mn^{2+}(aq) + 4\,H_2O(l)$$

$$[C_2O_4]^{2-}(aq) \rightarrow 2\,CO_2(g) + 2\,e^-$$

This particular reaction has a high activation energy. To provide a reasonable reaction rate, the oxalate solution is initially warmed. However, once some manganese(II) ion is produced, it acts as its own catalyst, and the reaction occurs more quickly as the titration proceeds.

A standardized solution of potassium permanganate can be used for the quantitative determination of iron in samples such as mineral ores or foodstuffs. The iron is converted to iron(II) ion, which is then titrated with standardized permanganate ion solution, again using the permanganate ion as reagent and indicator:

$$[MnO_4]^-(aq) + 8\,H^+(aq) + 5\,e^- \rightarrow Mn^{2+}(aq) + 4\,H_2O(l)$$

$$Fe^{2+}(aq) \rightarrow Fe^{3+}(aq) + e^-$$

One of the few oxidizing agents even more powerful than permanganate is the bismuthate ion, $[BiO_3]^-$. A test for manganese(II) ion is the addition of sodium bismuthate to a sample under cold, acidic conditions. The purple permanganate ion is produced, thereby indicating the presence of manganese:

$$Mn^{2+}(aq) + 4\ H_2O(l) \rightarrow [MnO_4]^-(aq) + 8\ H^+(aq) + 5\ e^-$$

$$[BiO_3]^-(aq) + 6\ H^+(aq) + 2\ e^- \rightarrow Bi^{3+}(aq) + 3\ H_2O(l)$$

In organic chemistry, potassium permanganate is usually used under basic conditions. In base, purple permanganate ion is first reduced to green manganate ion and then to solid brown-black manganese(IV) oxide:

$$[MnO_4]^-(aq) + e^- \rightarrow [MnO_4]^{2-}(aq)$$

$$[MnO_4]^{2-}(aq) + 2\ H_2O(l) + 2\ e^- \rightarrow MnO_2(s) + 4\ OH^-(aq)$$

As an example, the permanganate ion can be used for oxidizing alkenes to diols:

$$-\overset{|}{C}=\overset{|}{C}- + 2\ OH^-(aq) \rightarrow -\overset{\overset{\displaystyle HO}{|}}{\underset{|}{C}}-\overset{\overset{\displaystyle OH}{|}}{\underset{|}{C}}- + 2\ e^-$$

Manganese (VII) Oxide

Manganese(VII) oxide, a reddish-brown liquid, is a strongly oxidizing covalent compound. It decomposes explosively to the more stable manganese(IV) oxide:

$$2\ Mn_2O_7(l) \rightarrow 4\ MnO_2(s) + 3\ O_2(g)$$

Potassium Manganate

Potassium manganate, a green solid, is about the only common compound of manganese(VI). It is only stable in the solid phase or in extremely basic conditions. When it is dissolved in water, it disproportionates, as predicted by the Frost diagram:

$$3\ [MnO_4]^{2-}(aq) + 2\ H_2O(l) \rightarrow 2[MnO_4]^-(aq) + MnO_2(s) + 4\ OH^-(aq)$$

Manganese (IV) Oxide

The only compound of manganese(IV) of any importance is the dioxide, MnO_2, which occurs naturally as the ore pyrolusite. Manganese(IV) oxide is a black insoluble solid and is considered to have an essentially ionic structure. The compound is a strong oxidizing agent; it releases chlorine from concentrated hydrochloric acid and is, at the same time, reduced to manganese(II) chloride:

$$MnO_2(s) + 4\ HCl(aq) \rightarrow MnCl_2(aq) + Cl_2(g) + 2\ H_2O(l)$$

Manganese (II) Compounds

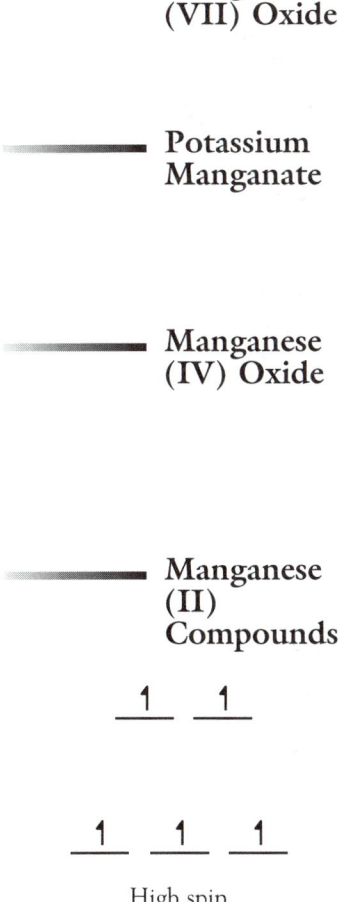

High spin

Figure 20.10 The *d* orbital occupancy for the d^5 electron configuration.

Under acid conditions, the most thermodynamically stable oxidation number of manganese is +2. Manganese in this oxidation state exists as a very pale pink ion, $[Mn(OH_2)_6]^{2+}$, a species present in all the common salts of manganese, such as nitrate, chloride, and sulfate. The very pale color of this ion contrasts with the strong colors of most other transition metal ions. The reason for the virtual absence of color can be deduced from our earlier remarks on the cause of color in transition metal compounds. The wavelengths absorbed correspond to the energy needed to raise a *d* electron from its ground state to an excited state. However, in the high-spin manganese(II) ion, each orbital already contains one electron. The only way an electron can absorb energy in the visible spectrum is by inverting its spin and pairing with another electron during the excitation (Figure 20.10). This process (a spin-forbidden transition) has a very low probability; hence, little visible light is absorbed by the manganese(II) ion.

Mining the Seafloor

We usually think of minerals as coming from mines bored into Earth's crust, yet there is increasing interest in mining ore deposits from the seafloor. In 1873, the Challenger expedition to the Pacific Ocean first dredged up mineral nodules from the bottom of the sea. We now know that nodules are widespread over the ocean floors. Generally, manganese and iron each make up between 15 and 20 percent of the content of these nodules; smaller concentrations of titanium, nickel, copper, and cobalt are also present. However, the composition varies from site to site, with some nodule beds containing up to 35 percent manganese.

The question of how such nodules formed puzzled chemists for a long time. It was the Swedish chemist I.G. Sillén who proposed that the oceans be considered as a giant chemical reaction vessel. As the metal ions accumulate in the seas from land runoff and undersea volcanic vents, the products of their reactions with anions in the seawater exceed the solubility product. The compounds then start to crystallize out very slowly over thousands and perhaps millions of years in the form of these nodules.

Because the nodules are such concentrated ores, there is much interest in using them for their metal content, particularly by the United States, which has to import much of the manganese, cobalt, and nickel it uses. A number of mining techniques are being developed to remove up to 200 tonnes of nodules per hour. However, there are two concerns, the first of which relates to the life on the seafloor; such large-scale excavations could have a major effect on bottom ecosystems. Furthermore, there is the question of ownership. Should the nodules be the property of whichever company and/or country that can mine them first, or, being in international waters, should they be the collective property of the world? Both of these issues need to be discussed and solved in the very near future.

When base is added to a solution containing manganese(II) ion, the white manganese(II) hydroxide is formed:

$$Mn^{2+}(aq) + 2\ OH^-(aq) \rightarrow Mn(OH)_2(s)$$

However, the manganese(III) state is favored under basic conditions, and the manganese(II) hydroxide oxidizes in the air to a brown hydrated manganese(III) oxide, $MnO(OH)$:

$$Mn(OH)_2(s) + OH^-(aq) \rightarrow MnO(OH)(s) + H_2O(l) + e^-$$

$$\tfrac{1}{2}\ O_2(g) + H_2O(l) + 2\ e^- \rightarrow 2\ OH^-(aq)$$

Biological Aspects

Manganese is a crucial element in a number of plant and animal enzymes. In mammals, it is used in the liver enzyme arginase, which converts nitrogen-containing wastes to the excretable compound urea. There is a group of enzymes in plants, the phosphotransferases, that incorporate manganese. Like most transition metals, the biological role of manganese seems to be as a redox reagent, cycling between the +2 and +4 oxidation states.

20.6 *Group 8: Iron, Ruthenium, and Osmium*

The Iron–Cobalt–Nickel Triad

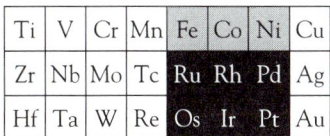

Figure 20.11 Iron, cobalt, and nickel have a horizontal relationship while the Periods 5 and 6 members of Groups 8, 9, and 10 have many similarities among themselves.

Under the old nomenclature system, Groups 8, 9, and 10 were collectively called Group VIII. In some respects this made chemical sense, for the three groups have some interesting relationships. In particular, the first row elements, iron, cobalt, and nickel have similarities in their chemistry while ruthenium, osmium, rhodium, iridium, palladium, and platinum have enough common chemistry that they have the collective name of the *platinum metals* (Figure 20.11).

There is a significant difference in the elements themselves between the triad and the platinum group. The metals of the *FeCoNi triad* are oxidized by dilute acids while the platinum metals are extremely inert to all but the most extreme reagents such as *aqua regia*, a mixture of concentrated hydrochloric and nitric acids. The triad are all ferromagnetic and they have almost identical melting points (between 1455 and 1535°C).

The most noticeable difference between the triad and the earlier first row transition metals is that the highest oxidation state is no longer that of the d^0 configuration (+4 for titanium, +5 for vanadium, +6 for chromium, and +7 for manganese). Each of the three FeCoNi triad metals has a common oxidation state of +2, and they all form the hexaaqua ions, $[M(OH_2)_6]^{2+}$, and the tetrahedral tetrachloro ions, $[MCl_4]^{2-}$. The +3 oxidation state is common for iron, less common for cobalt, and very rare for nickel.

It is important not to claim some exclusive link for the triad. In fact, there are similarities among many of the first row transition metals. For example, the $[M(OH_2)_6]^{2+}(aq)$ is air stable for all the metals from manganese through to copper. Also, the mixed metal oxide, M_3O_4, actually $(M^{2+})(M^{3+})_2(O^{2-})_4$, is known for manganese, iron, and cobalt.

Iron

Iron is believed to be the major component of Earth's core. This metal is also the most important material in our civilization. It does not hold this place because it is the "best" metal; after all, it corrodes much more easily than many other metals. Its overwhelming dominance in our society comes from a variety of factors.

1. Iron is the second most abundant metal in the Earth's crust, and concentrated deposits of iron ore are found in many localities, thus making it easy to mine.

2. The common ore can be easily and cheaply processed thermochemically to obtain the metal.

3. The metal is malleable and ductile, whereas many other metals are relatively brittle.

4. The melting point (1535°C) is low enough that the liquid phase can be handled without great difficulty.

5. By the addition of small quantities of other elements, alloys that have exactly the required combinations of strength, hardness, or ductility for very specific uses can be formed.

The one debatable factor is iron's chemical reactivity. This is considerably less than that of the alkali and alkaline earth elements but is not as low as that of many other transition metals. The relatively easy oxidation of iron is a major disadvantage—consider all the rusting automobiles, bridges, and other iron and steel structures, appliances, tools, and toys. At the same time, it does mean that our discarded metal objects will crumble to rust rather than remain an environmental blight forever.

Production of Iron

The most common sources of iron are the two oxides: iron(III) oxide, Fe_2O_3, and iron(II) iron(III) oxide, Fe_3O_4, together with a hydrated iron(II) oxide hydroxide, best represented as $Fe_2O_3 \cdot 1\frac{1}{2} H_2O$. These have the mineral names of hematite, magnetite, and limonite, respectively. The conventional extraction of iron is carried out in a *blast furnace* (Figure 20.12), which can be between 25 and 60 m in height and up to 14 m in diameter. The furnace itself is constructed of steel and has a lining of a heat- and corrosion-resistant material. The lining used to be brick, but it is now highly specialized ceramic materials. In fact, half of the high-temperature ceramics used today are produced for iron and steel smelter linings. The main ceramic material used for the lining is aluminum oxide (commonly called corundum), although the lining of the lower parts of the furnace consist of ceramic oxides of formula $Al_xCr_{2-x}O_3$. In these oxides, the chromium(III) ion has replaced some of the aluminum ions. These mixed metal oxide ceramics are more chemical- and temperature-resistant than the pure oxide ceramics.

A mixture of iron ore, limestone, and coke in the correct proportions is fed into the top of the blast furnace through a cone and hopper arrangement to prevent escape of the gases. Air, preheated to 600°C by combustion of the exhaust gases, is injected into the lower part of the furnace. The gases move up the furnace, while the solids descend as the products are drawn off from the bottom. The heat is generated by the reaction of the dioxygen in the air with the carbon (coke):

$$2\ C(s) + O_2(g) \rightarrow 2\ CO(g)$$

It is the hot carbon monoxide (initially at about 2000°C) that is the reducing agent for the iron ore.

At the top of the furnace, the temperature ranges from 200 to 700°C, a temperature sufficient to reduce iron(III) oxide to iron(II) iron(III) oxide, Fe_3O_4:

$$3\ Fe_2O_3(s) + CO(g) \rightarrow 2\ Fe_3O_4(s) + CO_2(g)$$

Lower in the furnace, at about 850°C, the iron(II) iron(III) oxide is reduced to iron(II) oxide,

$$Fe_3O_4(s) + CO(g) \rightarrow 3\ FeO(s) + CO_2(g)$$

and the temperature is high enough to also decompose the calcium carbonate (limestone) to calcium oxide and carbon dioxide:

$$CaCO_3(s) \rightarrow CaO(s) + CO_2(g)$$

As the mixture descends into the hotter regions (between 850 and 1200°C), the iron(II) oxide is reduced to iron metal and the carbon dioxide formed is re-reduced to carbon monoxide by the coke:

$$FeO(s) + CO(g) \rightarrow Fe(s) + CO_2(g)$$

$$C(s) + CO_2(g) \rightarrow 2\ CO(g)$$

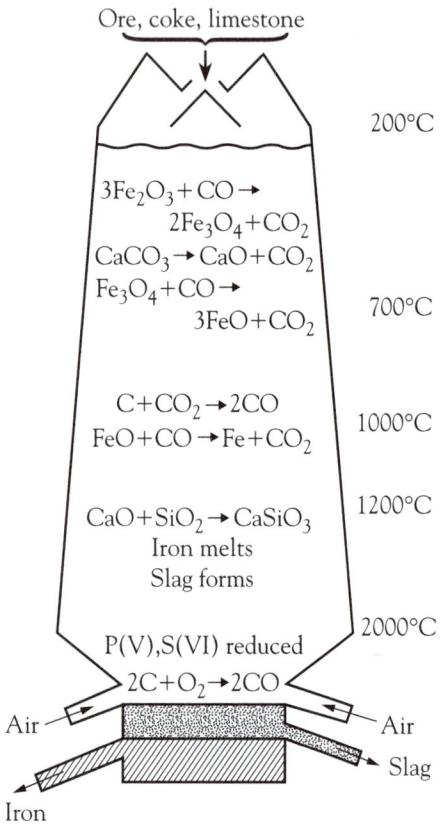

Ore, coke, limestone

200°C

$3Fe_2O_3 + CO \rightarrow$
$\quad 2Fe_3O_4 + CO_2$
$CaCO_3 \rightarrow CaO + CO_2$
$Fe_3O_4 + CO \rightarrow$
$\quad 3FeO + CO_2$

700°C

$C + CO_2 \rightarrow 2CO$
$FeO + CO \rightarrow Fe + CO_2$

1000°C

$CaO + SiO_2 \rightarrow CaSiO_3$
Iron melts
Slag forms

1200°C

2000°C

P(V),S(VI) reduced
$2C + O_2 \rightarrow 2CO$

Air

Air

Slag

Iron

Figure 20.12   A blast furnace.
[D. Shriver and P. Atkins, Inorganic Chemistry, 3rd. ed., Freeman, New York, 1998, p. 182.]

Lower still, at temperatures between 1200 and 1500°C, the iron melts and sinks to the bottom of the furnace, and the calcium oxide reacts with the silicon dioxide (and other impurities, such as phosphorus compounds) in the iron ore to give calcium silicate, commonly called *slag*. This is a high-temperature acid–base reaction between the basic metallic oxide, CaO, and the acidic nonmetal oxide, SiO_2.

$$CaO(s) + SiO_2(g) \rightarrow CaSiO_3(l)$$

The blast furnace is provided with two tapholes that are plugged with clay, the lower one for the denser iron metal and the upper one for the less dense slag. These plugs are periodically removed, releasing a stream of molten iron through the lower taphole and liquid slag through the upper. Blast furnaces are run 24 hours a day, and depending on its size, a furnace can produce from 1000 to 10 000 tonnes of iron every 24 hours.

The molten metal is usually conveyed directly in the liquid form to steel-making plants. The slag can be either cooled to the solid phase, ground, and used in concrete manufacture, or, while liquid, mixed with air and cooled into a "woolly" material that can be used for thermal insulation. The hot gases emerging from the top of the furnace contain appreciable amounts of carbon monoxide, and these are burnt to preheat the air for the furnace:

$$2\,CO(g) + O_2(g) \rightarrow 2\,CO_2(g)$$

The iron produced contains a wide range of impurities, such as silicon, sulfur, phosphorus, carbon, and oxygen. The carbon, which can be present in as great a proportion as 4.5 percent, is a particular contributor to the brittleness of the material. Iron is rarely used in pure form; more commonly, we require carefully controlled levels of impurities to provide exactly the properties required. One method for controlling content is the *basic oxygen process*. A schematic diagram of a typical furnace is shown in Figure 20.13.

Unlike the blast furnace, this process is not continuous. The converter is filled with about 60 tonnes of molten iron. A blast of oxygen diluted with carbon dioxide is blown through the converter. Oxygen is used instead of air because the nitrogen in air would react with the iron at these temperatures to form a brittle metal nitride. Oxygen reacts with the impurities and raises the temperature in the furnace to about 1900°C, and the diluent carbon dioxide prevents the temperature from increasing excessively. In addition, cold scrap metal is usually added to keep the temperature down.

In the basic oxygen process, carbon is oxidized to carbon monoxide, which burns at the top of the converter to carbon dioxide. The silicon, an impurity, is oxidized to silicon dioxide, which then reacts with the oxides of other elements to form a slag. The furnace also is lined with limestone (calcium carbonate), which reacts with acidic phosphorus-containing impurities. After several minutes, the flame at the top of the converter sinks, indicating that all the carbon has been removed. The slag is poured off, and any required trace elements are added to the molten iron. For normal steel, between 0.1 and 1.5 percent carbon is required. The carbon reacts with the iron to form iron carbide, Fe_3C, commonly called cementite. This compound forms separate small crystals among the crystals of iron. The ductility of the iron is reduced and the hardness increased by the presence of this impurity. To remove trapped oxygen from the iron, argon is blown through the liquid metal. About 3 m^3 of argon are used per tonne of iron.

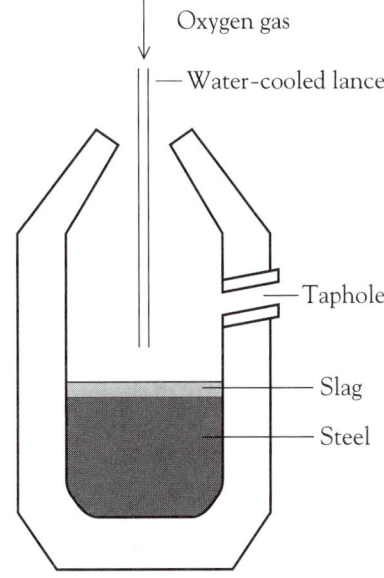

Oxygen gas

Water-cooled lance

Taphole

Slag

Steel

Figure 20.13 A furnace used for the basic oxygen process.

Table 20.4   Important alloys of iron

Name	Approximate composition	Properties and uses
Stainless steel	73% Fe, 18% Cr, 8% Ni	Corrosion resistant (tableware, cookware)
Tungsten steel	94% Fe, 5% W	Hard (high-speed cutting tools)
Manganese steel	86% Fe, 13% Mn	Tough (rock drill bits)
Permalloy	78% Ni, 21% Fe	Magnetic (electromagnets)

The properties of iron can be altered to suit our needs by adding controlled proportions of other elements. Examples of various iron alloys are given in Table 20.4.

Direct Reduction Iron (DRI)

An increasing proportion of iron is produced by direct reduction from the ore in the solid phase. For this method, high-purity iron ore must be used. Carbon monoxide and hydrogen, the reductants, are passed over heated iron ore. The iron(II) iron(III) oxide undergoes stepwise reduction:

$$Fe_3O_4(s) + CO(g) \rightarrow 3\ FeO(s) + CO_2(g)$$

$$Fe_3O_4(s) + H_2(g) \rightarrow 3\ FeO(s) + H_2O(g)$$

$$FeO(s) + CO(s) \rightarrow Fe(s) + CO_2(g)$$

$$FeO(s) + H_2(g) \rightarrow Fe(s) + H_2O(g)$$

The reducing gases are obtained by reforming methane with carbon dioxide and water. Thus, the economics of the process depend upon the price of cheap natural gas. In fact, this process can use poor quality methane, that is, methane deposits that contain high proportions of nonflammable gases. These deposits are unsuitable for most other commercial uses.

$$CH_4(g) + CO_2(g) \rightarrow 2\ CO(g) + 2\ H_2(g)$$

$$CH_4(g) + H_2O(g) \rightarrow CO(g) + 3\ H_2(g)$$

The advantage of DRI is that iron can be produced without the need for a massive expensive smelting operation while its primary disadvantage is that the product contains most element impurities from the natural ore. In developed countries, the major use of DRI is as "sweetener" or diluent in steel recycling. Scrap steel often contains unacceptable levels of metals such as copper, nickel, chromium, and molybdenum. DRI is low in these metals, thus mixing in a proportion of DRI results in an acceptable composition for electric arc steel manufacture. In developing countries, there is a shortage of scrap steel and shipping costs are high enough that importation of scrap is usually uneconomical. DRI provides an economical source of iron.

Iron(VI) Compounds

Beyond manganese, the Period 4 transition metals do not form compounds in which they have a d^0 electron configuration. In fact, a compound with a metal in an oxidation state higher than +3 is very difficult to prepare, and such compounds are only stable in the solid phase. The ferrate ion, $[FeO_4]^{2-}$, in which iron has an oxidation number of +6, is one of these rare compounds. This purple, tetrahedral ion can be stabilized by forming an insoluble ionic compound such as the red-purple solid barium ferrate, $BaFeO_4$. (The method of synthesis was described in Chapter 19, Section 19.11.) This mixed metal compound is a very powerful oxidizing agent.

The iron(III) ion itself is small and sufficiently polarizing that its anhydrous compounds exhibit covalent character. For example, iron(III) chloride is a red-black, covalent solid with a network covalent structure. When heated to the gas phase, it exists as the dimeric species Fe_2Cl_6 shown in Figure 20.14. Iron(III) chloride can be made by heating iron in the presence of dichlorine:

$$2\ Fe(s) + 3\ Cl_2(g) \rightarrow 2\ FeCl_3(s)$$

Iron(III) Compounds

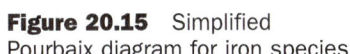

Figure 20.14   The diiron hexa-chloride molecule, Fe_2Cl_6.

The bromide is similar to the chloride, but the iodide cannot be isolated because the iodide ion reduces iron(III) to iron(II):

$$2\ Fe^{3+}(aq) + 2\ I^-(aq) \rightarrow 2\ Fe^{2+}(aq) + I_2(aq)$$

Anhydrous iron(III) chloride reacts exothermically with water, producing hydrogen chloride gas:

$$FeCl_3(s) + 3\ H_2O(l) \rightarrow Fe(OH)_3(s) + 3\ HCl(g)$$

This reaction contrasts with that of the golden yellow, ionic, hydrated salt, $FeCl_3 \cdot 6H_2O$, which simply dissolves in water to give the hexahydrate ion in solution. As we discussed in Chapter 9, Section 9.5, there are many parallels in the chloride chemistry of aluminum and iron(III).

The hexaaquairon(III) ion, $[Fe(OH_2)_6]^{3+}$, is very pale purple, a color that can be seen in the solid iron(III) nitrate nonahydrate. Like the manganese(II) ion, the iron(III) ion is a high spin d^5 species. Lacking any spin-allowed electron transitions, its color is very pale relative to that of other transition metal ions. The yellow color of the chloride compound is due to the presence of ions such as $[Fe(OH_2)_5Cl]^{2+}$ in which a charge transfer transition can occur:

$$Fe^{3+}\!\!-\!\!Cl^- \rightarrow Fe^{2+}\!\!-\!\!Cl^0$$

All the iron(III) salts dissolve in water to give an acidic solution, a characteristic of high charge density, hydrated cations. In such circumstances, the coordinated water molecules become sufficiently polarized that other water molecules can function as bases and abstract protons. The iron(III) ion behaves as follows:

$$[Fe(OH_2)_6]^{3+}(aq) + H_2O(l) \rightleftharpoons H_3O^+(aq) + [Fe(OH_2)_5(OH)]^{2+}(aq)$$

$$[Fe(OH_2)_5(OH)]^{2+}(aq) + H_2O(l) \rightleftharpoons H_3O^+(aq) + [Fe(OH_2)_4(OH)_2]^+(aq)$$

Figure 20.15   Simplified Pourbaix diagram for iron species.

and so on. The equilibria are pH dependent; thus, addition of hydronium ion will give the almost colorless hexaaquairon(III) ion. Conversely, addition of hydroxide ion gives an increasingly yellow solution, followed by precipitation of a rust-colored gelatinous (jellylike) precipitate of iron(III) oxide hydroxide, $FeO(OH)$:

$$Fe^{3+}(aq) + 3\ OH^-(aq) \rightarrow FeO(OH)(s) + H_2O(l)$$

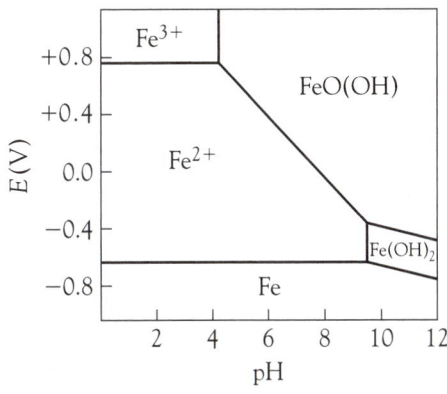

The pH and E dependence of the iron species are shown in Figure 20.15. For simplicity, the aqueous cations are simply shown as $Fe^{3+}(aq)$ and $Fe^{2+}(aq)$, respectively, although, as we have just seen, there is a whole range of different hydrated iron(III) ions that depend on the pH. The iron(III) ion is only thermodynamically preferred under oxidizing conditions (very positive E) and low pH. The iron(III) oxide hydroxide, however, predominates over much of the basic range. It is the iron(II) ion that is preferred over most of the E range and under acid conditions, whereas the iron(II) hydroxide, $Fe(OH)_2$, is only stable at high pH and strongly reducing conditions (very negative E).

The actual oxidation potential of iron(II) to iron(III) is very dependent on the ligands. For example, the hexacyanoferrate(II) ion, $[Fe(CN)_6]^{4-}$, is much more easily oxidized than the hexaaquairon(II) ion, $[Fe(OH_2)_6]^{2+}$:

$$[Fe(OH_2)_6]^{2+}(aq) \rightarrow [Fe(OH_2)_6]^{3+}(aq) + e^- \qquad E° = -0.77 \text{ V}$$

$$[Fe(CN)_6]^{4-}(aq) \rightarrow [Fe(CN)_6]^{3-}(aq) + e^- \qquad E° = -0.36 \text{ V}$$

This might seem surprising, considering that cyanide ion generally stabilizes low oxidation states, not high ones, and that, in fact, the iron–carbon bond is stronger in the iron(II) ion than in the iron(III) ion. But there is a thermodynamic aspect to the cyanide equilibrium: The aqueous $[Fe(CN)_6]^{4-}$ is of such high charge density that there is a strongly organized sphere of water molecules around it. Such an ion has a very negative entropy of hydration, but oxidation decreases the charge density, thereby reducing the organization of the hydration sphere and increasing the entropy. The oxidation, then, is entropy driven.

Although iron(III) species usually adopt an octahedral stereochemistry, the yellow tetrachloroferrate(III) ion, $[FeCl_4]^-$, is tetrahedral. This ion is easily formed by adding concentrated hydrochloric acid to a solution of the hexaaquairon(II) ion:

$$[Fe(OH_2)_6]^{3+}(aq) + 4\ Cl^-(aq) \rightleftharpoons [FeCl_4]^-(aq) + 6\ H_2O(l)$$

A specific test for the presence of the iron(III) ion is the addition of a solution of hexacyanoferrate(II) ion, $[Fe(CN)_6]^{4-}$, to give a dark blue precipitate of iron(III) hexacyanoferrate(II), $Fe_4[Fe(CN)_6]_3$:

$$4\ Fe^{3+}(aq) + 3[Fe(CN)_6]^{4-}(aq) \rightarrow Fe_4[Fe(CN)_6]_3(s)$$

In this compound, commonly called Prussian blue, the crystal lattice contains alternating iron(III) and iron(II) ions. The intense blue of this compound led to its use in the 19th century as a dye for the uniforms of Prussian soldiers, hence the origin of the name. The compound is used in blue inks and paints, and it is the blue pigment in traditional architectural and engineering blueprints.

The most sensitive test for the iron(III) ion is the addition of potassium thiocyanate solution. The appearance of the intense red color of the pentaaquathiocyanoiron(III) ion, $[Fe(SCN)(OH_2)_5]^{2+}$, indicates the presence of iron(III):

$$[Fe(OH_2)_6]^{3+}(aq) + SCN^-(aq) \rightarrow [Fe(SCN)(OH_2)_5]^{2+}(aq) + H_2O(l)$$

This test for iron(III) ion has to be used cautiously, for a solution of iron(II) ion usually contains enough iron(III) impurity to give some color.

A unique reaction of iron(III) ion is that with an ice-cold solution of thiosulfate. Mixing these two nearly colorless solutions gives the dark violet bis(thiosulfato)ferrate(III) ion:

$$Fe^{3+}(aq) + 2[S_2O_3]^{2-}(aq) \rightarrow [Fe(S_2O_3)_2]^-(aq)$$

When the solution is warmed to room temperature, the iron(III) is reduced to iron(II) and the thiosulfate is oxidized to the tetrathionate ion, $[S_4O_6]^{2-}$:

$$[Fe(S_2O_3)_2]^-(aq) + Fe^{3+}(aq) \rightarrow 2\ Fe^{2+}(aq) + [S_4O_6]^{2-}(aq)$$

Iron(II) Compounds

Anhydrous iron(II) chloride, $FeCl_2$, can be made by passing a stream of dry hydrogen chloride over the heated metal; the hydrogen produced acts as a reducing agent, preventing iron(III) chloride from being formed:

$$Fe(s) + 2\ HCl(g) \rightarrow FeCl_2(s) + H_2(g)$$

The pale green hexaaquairon(II) chloride, $Fe(OH_2)_6Cl_2$, can be prepared by reacting hydrochloric acid with iron metal. Both anhydrous and hydrated forms of iron(II) chloride are ionic.

All the common hydrated iron(II) salts contain the pale green $[Fe(OH_2)_6]^{2+}$ ions, although partial oxidation to yellow or brown iron(III) compounds is quite common. In addition, crystals of the simple salts, such as iron(II) sulfate heptahydrate, $FeSO_4·7H_2O$, tend to lose some of the water molecules (efflorescence). In the solid phase, the double salt, ammonium iron(II) sulfate hexahydrate, $(NH_4)_2Fe(SO_4)_2·6H_2O$ (or more correctly, ammonium hexaaquairon(II) sulfate), shows the greatest lattice stability. Commonly known as Mohr's salt, it neither effloresces nor oxidizes when exposed to the atmosphere. For this reason, it is used as a standard for redox titrations, especially for determining the concentration of potassium permanganate solutions. The tris(1,2-diaminoethane)iron(II) sulfate, $Fe(en)_3SO_4$, is also used as a redox standard.

In the presence of nitrogen monoxide, one molecule of water is displaced from the hexaaquairon(II) ion and replaced by the nitrogen monoxide to give the pentaaquanitrosyliron(II) ion, $[Fe(NO)(OH_2)_5]^{2+}$:

$$NO(aq) + [Fe(OH_2)_6]^{2+}(aq) \rightarrow [Fe(NO)(OH_2)_5]^{2+}(aq) + H_2O(l)$$

This complex is dark brown, and the previous reaction is the basis of the "brown-ring" test for ionic nitrates (the nitrate having been reduced to nitrogen monoxide by a reducing agent).

Addition of hydroxide ion to iron(II) gives an initial precipitate of green, gelatinous iron(II) hydroxide:

$$Fe^{2+}(aq) + 2\ OH^-(aq) \rightarrow Fe(OH)_2(s)$$

Table salt sometimes contains 0.01 percent sodium hexacyanoferrate(II), $Na_4[Fe(CN)_6]$, to stop it "caking" and enabling it to flow freely even in damp climates.

However, as Figure 20.14 shows, except for strongly reducing conditions (or the absence of air), it is the hydrated iron(III) oxide that is thermodynamically stable in basic solution over most of the potential range. Thus, the green color is replaced by the yellow-brown of the hydrated iron(III) oxide as the oxidation proceeds.

Just as iron(III) ion can be detected with the hexacyanoferrate(II) ion, $[Fe(CN)_6]^{4-}$, so can iron(II) ion be detected with the hexacyanoferrate(III) ion, $[Fe(CN)_6]^{3-}$, to give the same product of Prussian blue (formerly called Turnbull's blue when it was thought to be a different product):

$$3\ Fe^{2+}(aq) + 4[Fe(CN)_6]^{3-}(aq) \rightarrow Fe_4[Fe(CN)_6]_3(s) + 6\ CN^-(aq)$$

The Rusting Process

It is a common experiment in junior high science to show that the oxidation of iron (commonly called rusting) requires the presence of both dioxygen and water. By use of an indicator, it can also be shown that around parts of an iron surface, the pH rises. From these observations, the electrochemistry of the rusting process can be determined. This process is really a reflection of the Nernst equation, which states that potential is dependent on concentration—in this case, the concentration of dissolved dioxygen. At the point on the iron surface that has a higher concentration of dioxygen, the element is reduced to hydroxide ion:

$$O_2(aq) + 2\ H_2O(l) + 4\ e^- \rightarrow 4\ OH^-(aq)$$

The bulk iron acts like a wire connected to a battery, conveying electrons from another point on the surface that has a lower oxygen concentration, a point at which the iron metal is oxidized to iron(II) ions:

$$Fe(s) \rightarrow Fe^{2+}(aq) + 2\ e^-$$

These two ions diffuse through the solution, and where they meet, insoluble iron(II) hydroxide is formed:

$$Fe^{2+}(aq) + 2\ OH^-(aq) \rightarrow Fe(OH)_2(s)$$

Like hydrated manganese(III) oxide, the iron(III) oxide hydroxide (rust) is thermodynamically preferred to iron(II) hydroxide in basic solution:

$$Fe(OH)_2(s) + OH^-(aq) \rightarrow FeO(OH)(s) + H_2O(l) + 2\ e^-$$

$$\tfrac{1}{2}\ O_2(aq) + H_2O(l) + 2\ e^- \rightarrow 2\ OH^-(aq)$$

Iron Oxides

There are three common oxides of iron: iron(II) oxide, FeO; iron(III) oxide, Fe_2O_3; and iron(II) iron(III) oxide, Fe_3O_4. Black iron(II) oxide is actually a nonstoichiometric compound, always being slightly deficient in iron(II) ions. The most accurate formulation is $Fe_{0.95}O$. The oxide is basic, dissolving in acids to give the aqueous iron(II) ion:

$$FeO(s) + 2\ H^+(aq) \rightarrow Fe^{2+}(aq) + H_2O(l)$$

Iron(III) oxide, or hematite, is found in large underground deposits. The oldest beds of iron(III) oxide have been dated at about 2 billion years old. Because iron(III) oxide can only form in an oxidizing atmosphere, our current dioxygen-rich atmosphere must have appeared at that time. The appearance of dioxygen, in turn, indicates that photosynthesis (and life itself) became widespread about 2 billion years ago. The oxide can be made in the laboratory by heating the iron(III) oxide hydroxide, which was generated by adding hydroxide ion to iron(III) ion. The product formed by this route, α-Fe_2O_3, consists of a hexagonal close-packed array of oxide ions with the iron(III) ions in two-thirds of the octahedral holes. A different structural form, γ-Fe_2O_3, is produced by oxidizing iron(II) iron(III) oxide. In this form, the iron(III) oxide has a cubic close-packed array of oxide ions with the iron(III) ions distributed randomly among the tetrahedral and octahedral holes.

The third common oxide of iron contains iron in both the +2 and +3 oxidation states, and we have mentioned this compound, $(Fe^{2+})(Fe^{3+})_2(O^{2-})_4$, previously in the context of normal and inverse spinels (see Chapter 19, Section 19.8). This compound is found naturally as magnetite or lodestone, and a piece of this magnetic compound, suspended by a thread, was used as a primitive compass. Nature, as is often the case, beat us to that discovery. To enable them to navigate using the Earth's magnetic field, *magnetolactic* bacteria contain crystals of magnetite or, in sulfur-rich environments the sulfur equivalent, greigite, Fe_3S_4.

This oxide is now important as the usual magnetic component of *ferrofluids*—magnetic liquids. These liquids are attracted to magnetic fields, though in fact it is the colloidal solid iron oxide particles within the bulk liquid that are attracted into the magnetic field. A ferrofluid can be simply prepared by the reaction of iron(III) and iron(II) chlorides with aqueous ammonia in the presence of a surfactant that prevents the oxide from coagulating and settling out:

$$2\,FeCl_3(aq) + FeCl_2(aq) + 8\,NH_3(aq) + 4\,H_2O(l) \rightarrow Fe_3O_4(s) + 8\,NH_4Cl(aq)$$

The iron oxides are in great demand as paint pigments. Historically, colors such as yellow ochre, Persian red, and umber (brown) were obtained from iron ore deposits containing certain particle sizes of iron oxides, often with consistent levels of specific impurities. Most yellow, red, and black paints are still made from iron oxides, but they are industrially synthesized to give precise compositions and particle sizes to ensure the production of consistent colors. Of special note is the organic process for the manufacture of aniline, $C_6H_5NH_2$, an important organic chemical. This process used to produce waste iron oxides, but a modification of the synthesis has enabled the composition and particle size of the iron oxide to be adjusted. As a result, the oxide by-product can be marketed as a pigment rather than dumped:

$$4\,C_6H_5NO_2(l) + 9\,Fe(s) + 4\,H_2O(l) \xrightarrow{FeCl_2} 4\,C_6H_5NH_2(l) + 3\,Fe_3O_4(s)$$

If one were to consider which single chemical compound has most revolutionized our modern lives, it would be γ-iron(III) oxide. It is this form of iron(III) oxide that has precisely the magnetic characteristics (ferromagnetism—see Chapter 1) needed for audio- and videotapes and for the surfaces of the hard and floppy disks in computers. Thus, if any compound could be said to be indispensable, this would be it. It is somewhat ironic that iron(III) oxide, which is present on Earth in such vast amounts, should be so valuable in an ultrapure state of precise particle size range, the essential form for magnetic recording purposes.

Ferrites

It is not just iron oxides that are important magnetic materials. There are several mixed metal oxides, one metal being iron, that have valuable properties. These magnetoceramic materials are called *ferrites*. There are two classes of ferrites, the "soft" ferrites and the "hard" ferrites. These terms do not refer to their physical hardness but to their magnetic properties.

The soft ferrites can be magnetized rapidly and efficiently by an electromagnet, but they lose their magnetism as soon as the current is discontinued. Such properties are essential for the record—erase heads in videotape and audiotape systems and computer drive heads. The compounds have the formula MFe_2O_4, where M is a dipositive metal ion such as Mn^{2+}, Ni^{2+}, Co^{2+}, or Mg^{2+}, and the iron is in the form of Fe^{3+}. These compounds also have spinel structures.

The hard ferrites retain their magnetic properties constantly; that is, they are permanent magnets. These materials are used in DC motors, alternators, and other electrical devices. The general formula of these compounds is $MFe_{12}O_{19}$, where again the iron is Fe^{3+}; the two preferred dipositive metal ions are Ba^{2+} and Sr^{2+}. The hard ferrites adopt a more complex structure than the soft ferrites. The use of both ferrites is not particularly large in terms of mass, but in terms of value, annual world sales are several billion dollars.

Table 20.5 Major iron-containing proteins in an adult human

Name	Fe atoms per molecule	Function
Hemoglobin	4	O_2 transport in blood
Myoglobin	1	O_2 storage in muscle
Transferrin	2	Iron transport
Ferritin	Up to 4500	Iron storage in cells
Hemosiderin	10^3–10^4	Iron storage
Catalase	4	Metabolism of H_2O_2
Cytochrome c	1	Electron transfer
Iron–sulfur proteins	2–8	Electron transfer

Ruthenium and Osmium

These are the first two of the platinum metals. As a result of their lack of chemical reactivity, the platinum metals are sometimes called the *noble metals,* by analogy with the noble gases. These elements are all extremely rare, unreactive, silvery, lustrous metals that are found together in nature, the annual total production being about 300 tonnes. Some of that quantity is used to produce jewelry and bullion coinage, but most is used in the chemical industry as catalysts for various reactions, as we discuss in Chapter 22, Section 22.17. The densities of these metals show a strong horizontal relationship: The Period 5 metals have densities of about 12 $g \cdot cm^{-3}$, whereas those of the Period 6 metals are about 21 $g \cdot cm^{-3}$. The melting points of the platinum metals are also high, with values ranging from 1500 to 3000°C.

Though the elements themselves have many similarities, in their chemical compounds ruthenium and osmium behave more like a continuation of the earlier transition metals with their highest oxidation state being that of +8, the d^0 electron configuration. In fact, the only important compound of ruthenium or osmium is osmium(VIII) oxide, OsO_4, a useful oxidizing agent in organic chemistry. As we discussed in Chapter 9, Section 9.5, there are correspondences between osmium(VIII) and xenon(VIII) chemistry, and the formation of a tetraoxide (OsO_4 and XeO_4) is one of the similarities.

Biological Aspects

The biological roles of iron are so numerous that whole books have been written on the subject. Table 20.5 summarizes some major iron-containing proteins in an adult human.

Here we focus on three specific types of iron-containing macromolecules: hemoglobin, ferritin, and the ferredoxins. In hemoglobin, iron has an oxidation state of +2. (We mentioned this compound in the context of oxygen uptake in Chapter 16, Section 16.28.) There are four iron ions in each hemoglobin molecule, each iron ion being surrounded by a porphyrin unit (Figure 20.16). Each hemoglobin molecule reacts with four molecules of dioxygen to form oxyhemoglobin. The bonding to the dioxygen molecules is weak enough that, upon reaching the site of oxygen utilization such as the muscles, the oxygen can be released. Carbon monoxide is extremely toxic to mammals because the carbonyl ligand bonds very strongly to the iron of the hemoglobin, thus preventing it from carrying dioxygen molecules.

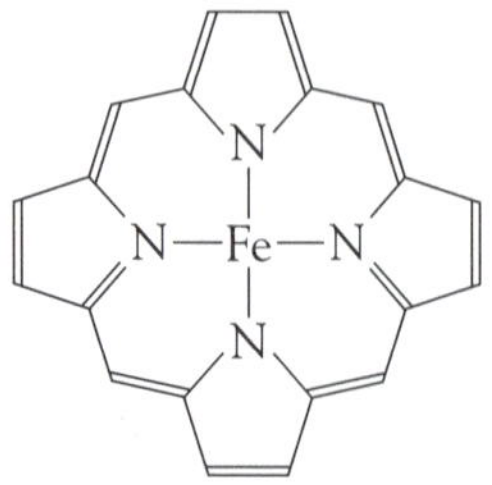

In oxyhemoglobin, the iron(II) is in the diamagnetic, low-spin state. It is just the right radius (75 pm) to fit in the plane of the porphyrin ring.

Figure 20.16 The simplified structure of an iron–porphyrin complex.

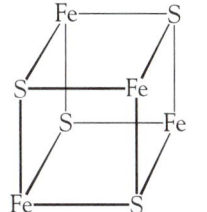

Figure 20.17 The Fe_4S_4 core of a four-iron ferredoxin protein.

Once the dioxygen is lost, iron in the deoxyhemoglobin molecule shifts below the plane of the porphyrin ring and away from the vacant coordination site, because it has become a larger (radius 92 pm), paramagnetic, high-spin iron(II) ion. Throughout the cycle, the iron stays in the iron(II) state, merely alternating between high- and low-spin forms. It is only when exposed to air that the red iron(II)-containing hemoglobin is oxidized to the brown iron(III) species, an irreversible reaction.

Both plants and animals need to store iron for future use. To accomplish this, members of an amazing protein family, the ferritins, are utilized. They consist of a shell of linked amino acids (peptides) surrounding a core of an iron(III) oxohydroxophosphate. This core is a cluster of iron(III) ions, oxide ions, hydroxide ions, and phosphate ions of average empirical formula $[FeO(OH)]_8[FeO(OPO_3H_2)]$. The molecule is very large, containing up to 4500 iron ions. With its hydrophilic coating, this large aggregate is water soluble, concentrating in the spleen, liver, and bone marrow.

Plants and bacteria use a family of iron(III)–sulfur structures as the core of their redox proteins, the ferredoxins. These proteins contain covalently bonded iron and sulfur, and they act as excellent electron transfer agents. Most interesting are the Fe_4S_4 cores, where the iron and sulfur atoms occupy the alternating corners of a cubic arrangement (Figure 20.17).

20.7 Group 9: Cobalt, Rhodium, and Iridium

Cobalt

Cobalt is a bluish white, hard metal, and, like iron, cobalt is a magnetic (ferromagnetic) material. The element is quite unreactive chemically. The most common oxidation numbers of cobalt are $+2$ and $+3$, the former being the "normal" state for simple cobalt compounds. For cobalt, the $+3$ state is more oxidizing than the $+3$ state of iron.

Cobalt(III) Compounds

All the cobalt(III) complexes are octahedral in shape, and, like chromium(III), the low-spin complexes are very kinetically inert, meaning that we can separate different optical isomers, where they are feasible. Typical examples of cobalt(III) complexes are the hexaamminecobalt(III) ion, $[Co(NH_3)_6]^{3+}$, and the hexacyanocobaltate(III) ion, $[Co(CN)_6]^{3-}$.

An unusual complex ion is the hexanitrocobaltate(III) ion, $[Co(NO_2)_6]^{3-}$, which is usually synthesized as the sodium salt, $Na_3[Co(NO_2)_6]$. As would be expected for an alkali metal salt, the compound is water soluble. However, the potassium salt is quite insoluble (as are the rubidium, cesium, and ammonium salts), the reason relating to relative ion sizes. The potassium ion is much closer in size to the polyatomic anion; hence, the balance of lattice energy and hydration energies favor a lower solubility for the compound. This is one of the few precipitation reactions that can be used for the potassium ion. We described this reaction in the context of similarities of the ammonium ion with the heavier alkali metals ions (Chapter 9, Section 9.12).

$$3\ K^+(aq) + [Co(NO_2)_6]^{3-}(aq) \rightarrow K_3[Co(NO_2)_6](s)$$

As noted for the iron ions, altering the ligands produces dramatic changes in $E°$ values, which in turn affects the stabilities of the various oxidation states. For example,

$$[Co(OH_2)_6]^{3+}(aq) + e^- \rightarrow [Co(OH_2)_6]^{2+}(aq) \qquad E° = +1.92\ V$$

$$[Co(NH_3)_6]^{3+}(aq) + e^- \rightarrow [Co(NH_3)_6]^{2+}(aq) \qquad E° = +0.10\ V$$

The value for the oxidation of the hexaamminecobalt(II) ion of $+0.10$ V is much less positive than the value $+1.23$ V for the reduction of oxygen:

$$\tfrac{1}{2}\,O_2(g) + 2\,H^+(aq) + 2\,e^- \rightarrow H_2O(l) \qquad\qquad E^\circ = +1.23\ V$$

Hence, oxygen should be potentially capable of oxidizing $[Co(NH_3)_6]^{2+}$ to $[Co(NH_3)_6]^{3+}$ in solution; and this is, in fact, found to be the case in the presence of charcoal as a catalyst:

$$[Co(OH_2)_6]^{2+}(aq) + 6\,NH_3(aq) \rightarrow [Co(NH_3)_6]^{2+}(aq) + 6\,H_2O(l)$$

$$[Co(NH_3)_6]^{2+}(aq) \xrightarrow{\ C\ } [Co(NH_3)_6]^{3+}(aq) + e^-$$

$$O_2(g) + 2\,H_2O(l) + 4\,e^- \rightarrow 4\,OH^-(aq)$$

This reaction must proceed in a series of steps, for it is possible in the absence of a catalyst to isolate an intermediate from the reaction, a brown compound containing the $[(H_3N)_5Co{-}O{-}O{-}Co(NH_3)_5]^{4+}$ ion. It is believed to form in a two-step process. Thus, it would seem that the dioxygen molecule attaches in a bidentate fashion to one cobalt(II) ion and then links two cobalt(II) complexes, oxidizing each cobalt(II) to cobalt(III), in the process being reduced to the peroxo unit, O_2^{2-}.

$$[Co(NH_3)_6]^{2+}(aq) + O_2(g) \rightarrow [Co(NH_3)_4O_2]^{2+}(aq) + 2\,NH_3(aq)$$

$$[Co(NH_3)_4O_2]^{2+}(aq) + [Co(NH_3)_6]^{2+}(aq) \rightarrow$$
$$[(H_3N)_5Co{-}O{-}O{-}Co(NH_3)_5]^{4+}(aq)$$

In a further series of steps, the peroxo unit is reduced to 2 moles of hydroxide ion and, for each molecule of dioxygen used, a total of 4 moles of the cobalt(II) complex is oxidized to the hexaamminecobalt(III) ion.

The product of oxidation is very sensitive to the precise conditions. For example, with hydrogen peroxide as the oxidizing agent, it is possible to form the pentaammineaquacobalt(III) ion and then to use concentrated hydrochloric acid to perform a ligand replacement to give the pentaamminechlorocobalt(III) ion:

$$[Co(OH_2)_6]^{2+}(aq) + 6\,NH_3(aq) \rightarrow [Co(NH_3)_6]^{2+}(aq) + 6\,H_2O(l)$$

$$[Co(NH_3)_6]^{2+}(aq) + H_2O(l) \rightarrow [Co(NH_3)_5(OH_2)]^{3+}(aq) + NH_3(aq) + e^-$$

$$H_2O_2(aq) + 2\,e^- \rightarrow 2\,OH^-(aq)$$

$$[Co(NH_3)_5(OH_2)]^{3+}(aq) + Cl^-(aq) \xrightarrow{\ H^+\ } [Co(NH_3)_5Cl]^{2+}(aq) + H_2O(l)$$

The monochloro- species is a useful reagent for the synthesis of other monosubstituted cobalt(III) complexes. In particular, we can synthesize the nitrito- and nitro-linkage isomers described in Chapter 19, Section 19.4:

$$[Co(NH_3)_5Cl]^{2+}(aq) + H_2O(l) \xrightarrow{\ OH^-\ } [Co(NH_3)_5(OH_2)]^{3+}(aq) + Cl^-(aq)$$

$$[Co(NH_3)_5(OH_2)]^{3+}(aq) + NO_2^-(aq) \rightarrow [Co(NH_3)_5(ONO)]^{2+}(aq) + H_2O(l)$$

$$[Co(NH_3)_5(ONO)]^{2+}(aq) \xrightarrow{\ \Delta\ } [Co(NH_3)_5(NO_2)]^{2+}(aq)$$

Using appropriate synthetic conditions, we can synthesize several permutations of cobalt(III) with ammonia and chloride ligands (Table 20.6). It is very easy to determine the formulation of these compounds. The number of ions can be identified from conductivity measurements, and the free chloride ion can be quantitatively precipitated with silver ion.

Table 20.6 Complexes derived from cobalt(III) chloride and ammonia

Formula	Color
$[Co(NH_3)_6]^{3+}$ $3Cl^-$	Orange-yellow
$[Co(NH_3)_5Cl]^{2+}$ $2Cl^-$	Purple
$[Co(NH_3)_4Cl_2]^+$ Cl^- (*cis*)	Violet
$[Co(NH_3)_4Cl_2]^+$ Cl^- (*trans*)	Green

High spin Co^{2+}

d^7

Low spin Co^{3+}

d^6

Oxidation

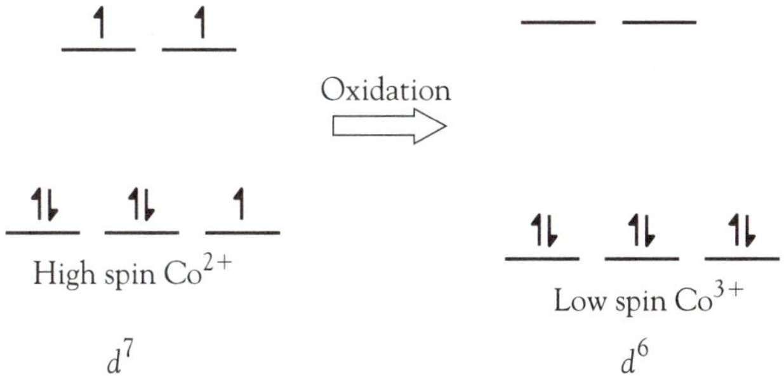

Figure 20.18 Comparison of cobalt(II) and cobalt(III) crystal field stabilization energies.

If 1,2-diaminoethane is employed as the ligand, cobalt(II) can be oxidized to the tris(1,2-diaminoethane)cobalt(III) ion and crystallized as the chloride, $[Co(en)_3]Cl_3$. This compound, as we mentioned in Chapter 19, Section 19.4, can be separated into the two optical isomers. By decreasing the ligand to metal ion ratio, it is possible to synthesize *trans*-dichlorobis(1,2-diaminoethane)cobalt(III) chloride, $[Co(en)_2Cl_2]Cl$. Dissolving in water and evaporating the solution causes the compound to isomerise to the *cis*-form.

By comparing the electron configurations in an octahedral field, we can see why ligands causing a larger crystal field splitting would enable cobalt(II) to be readily oxidized. For cobalt(II), nearly all the complexes are high spin, whereas cobalt(III), with its higher charge, is almost always low spin. Thus, the oxidation results in a much greater crystal field stabilization energy (Figure 20.18). The higher the ligands in the spectrochemical series, the greater the Δ_{oct} value and the greater the CFSE increase obtained by oxidation.

Cobalt(II) Compounds

In solution, cobalt(II) salts are pink, the color being due to the presence of the hexaaquacobalt(II) ion, $[Co(OH_2)_6]^{2+}$. When a solution of a cobalt(II) salt is treated with concentrated hydrochloric acid, the color changes to deep blue. This color is the result of the formation of the tetrahedral tetrachlorocobaltate(II) ion, $[CoCl_4]^{2-}$:

$$[Co(OH_2)_6]^{2+}(aq) + 4\,Cl^-(aq) \rightleftharpoons [CoCl_4]^{2-}(aq) + 6\,H_2O(l)$$
$$\text{pink} \qquad\qquad\qquad\qquad\qquad \text{blue}$$

This color change is characteristic for the cobalt(II) ion.

There is also a pink-to-blue transition when solid pink hexaaquacobalt(II) chloride is dehydrated. Paper impregnated with the blue form will turn pink on addition of water—a good test for water. Silica gel and calcium sulfate drying agents are often colored with cobalt(II) chloride. As long as the granules retain the blue color, they are an effective drying agent, with the appearance of pink indicating that the drying agent is water-saturated and must be heated to expel the absorbed moisture.

The addition of hydroxide ion to the aqueous cobalt(II) ion results in the formation of cobalt(II) hydroxide, which precipitates first in a blue form and then changes to a pink form on standing:

$$Co^{2+}(aq) + 2\,OH^-(aq) \rightarrow Co(OH)_2(s)$$

The cobalt(II) hydroxide is slowly oxidized by the dioxygen in the air to a cobalt(III) oxide hydroxide, $CoO(OH)$:

$$Co(OH)_2(s) + OH^-(aq) \rightarrow CoO(OH)(s) + H_2O(l) + e^-$$

Cobalt(II) hydroxide is amphoteric. When concentrated hydroxide ion is added to cobalt(II) hydroxide, a deep blue solution of the tetrahydroxocobaltate(II) ion, $[Co(OH)_4]^{2-}$, is formed:

$$Co(OH)_2(s) + 2\ OH^-(aq) \rightarrow [Co(OH)_4]^{2-}(aq)$$

Rhodium and Iridium

Both rhodium and iridium readily form complexes with the +3 oxidation state, which, like the cobalt analogs, are kinetically inert. The second most prevalent oxidation state is +1. Of the platinum metals, the chemistry of rhodium, iridium, palladium, and platinum are much closer to each other than to that of ruthenium and osmium. For example, the highest oxidation state for both Group 9 and Group 10 platinum metals is +6.

Biological Aspects

Cobalt is yet another essential element. Of particular importance, vitamin B_{12} has cobalt(III) at the core of the molecule, surrounded by a ring structure similar to the porphyrin ring. Injections of this vitamin are used in the treatment of pernicious anemia. Certain anaerobic bacteria use a related molecule, methylcobalamin, in a cycle to produce methane. Unfortunately, this same biochemical cycle converts elemental mercury and insoluble inorganic mercury compounds in mercury-polluted waters to soluble, highly toxic methyl mercury(II), $[HgCH_3]^+$, and dimethyl mercury(II), $Hg(CH_3)_2$.

Cobalt is also involved in some enzyme functioning. A deficiency disease among sheep in Florida, Australia, Britain, and New Zealand was traced to a lack of cobalt in the soil. To remedy this, sheep are given a pellet of cobalt metal in their food, some of which remains in their digestive system for life.

20.8 Group 10: Nickel, Palladium, and Platinum

Nickel

Nickel is a silvery white metal that is quite unreactive. In fact, nickel plating is sometimes used to protect iron. The only common oxidation number for nickel is +2. Most nickel complexes have an octahedral geometry, but some tetrahedral and square-planar complexes are known. Square-planar geometry is otherwise exceedingly rare for the compounds of Period 4 transition metals.

Extraction of Nickel

Although the extraction of nickel from its compounds is complex, the last two steps—the conversion of nickel(II) oxide to pure nickel metal—are of particular interest. As we saw from the Ellingham diagram for oxides (see Chapter 8, Figure 8.9), the line of the oxidation of carbon has a steep negative slope, crossing most of the metal oxides. The line for nickel(II) oxide is crossed at an obtainable temperature. Thus, the reduction can be performed by the inexpensive thermal smelting route rather than having to use a more expensive electrochemical process:

$$NiO(s) + CO(g) \xrightarrow{\Delta} Ni(s) + CO_2(g)$$

However, this nickel is impure. To extract the nickel from the other metals, such as cobalt and iron, there are two alternatives. One is an electrolytic process whereby impure nickel is cast into anodes and, by using solutions of nickel sulfate and chloride as electrolytes, 99.9 percent pure nickel is deposited at the cathode. The other process utilizes a reversible chemical reaction known as the *Mond process.* In this reaction, nickel metal reacts at about 60°C with carbon monoxide gas to form a colorless gas, tetracarbonylnickel(0), $Ni(CO)_4$ (bp 43°C):

$$Ni(s) + 4\ CO(g) \rightarrow Ni(CO)_4(g)$$

The highly toxic compound can be piped off, for nickel is the only metal to form a volatile carbonyl compound so easily. Heating the gas to 200°C shifts the equilibrium in the opposite direction, depositing 99.95 percent pure nickel metal:

$$Ni(CO)_4(g) \xrightarrow{\Delta} Ni(s) + 4\ CO(g)$$

The carbon monoxide can then be reused.

Nickel(II) Compounds

The hexaaquanickel(II) ion is a pale green color. Addition of ammonia gives the blue hexaamminenickel(II) ion:

$$[Ni(OH_2)_6]^{2+}(aq) + 6\ NH_3(aq) \rightarrow [Ni(NH_3)_6]^{2+}(aq) + 6\ H_2O(l)$$

Nickel(II) hydroxide can be precipitated as a green gelatinous solid by adding sodium hydroxide solution to a solution of a nickel(II) salt:

$$Ni^{2+}(aq) + 2\ OH^-(aq) \rightarrow Ni(OH)_2(s)$$

Like cobalt(II), the only common complexes having tetrahedral geometry are the halides, such as the blue tetrachloronickelate(II) ion. This complex is formed by adding concentrated hydrochloric acid to aqueous nickel(II) ion:

$$[Ni(OH_2)_6]^{2+}(aq) + 4\ Cl^-(aq) \rightleftharpoons [NiCl_4]^{2-}(aq) + 6\ H_2O(l)$$

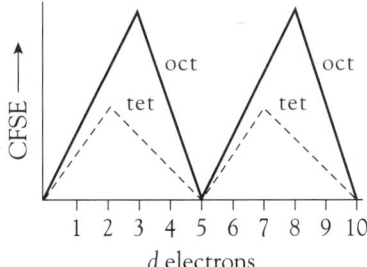

In addition to octahedral and tetrahedral complexes, nickel forms a few square-planar complexes. One such complex is the yellow tetracyanonickelate(II) ion, $[Ni(CN)_4]^{2-}$, and another is bis(dimethylglyoximato)nickel(II), $[Ni(C_4N_2O_2H_7)_2]$, which precipitates as a red solid when dimethylglyoxime is added to a solution of a nickel salt made just alkaline by the addition of ammonia. The formation of this characteristic red complex is used as a test for nickel(II) ions. Abbreviating dimethylglyoxime ($C_4N_2O_2H_8$), a bidentate ligand, as DMGH, we write the equation for its formation as

$$Ni^{2+}(aq) + 2\ DMGH(aq) + 2\ OH^-(aq) \rightarrow Ni(DMG)_2(s) + 2\ H_2O(l)$$

Figure 20.19 Variation of CFSE for octahedral and tetrahedral environments for M^{2+} ions with high-spin electron configuration.

Octahedral versus Tetrahedral Stereochemistry

Cobalt(II) readily forms tetrahedral complexes, but as we have just seen, nickel(II) complexes are usually octahedral; however, a few are square planar, and very few are tetrahedral. Several factors determine the choice of stereochemistry, but one in particular is the CFSE. We can calculate this for each configuration and plot the values as a function of the number of *d* orbital electrons. Because Δ_{tet} is four-ninths that of Δ_{oct}, the CFSE for a tetrahedral environment will always be less than that of the isoelectronic octahedral environment. Figure 20.19 shows the variation of CFSE with high-spin electron configurations. The differences between octahedral and tetrahedral CFSE energies are greatest for the high-spin d^3 and d^8 cases, and these are the electron configurations for which we find the fewest tetrahedral complexes.

Nevertheless, there are a few tetrahedral nickel(II) complexes. Such complexes are formed with large, negatively charged, weak field ligands (that is, ligands low in the spectrochemical series). In these cases, there will be considerable electrostatic repulsion between neighboring ligands, so four ligands would be preferred over six. Thus, the tetrahalonickelate(II) ions, $[NiX_4]^{2-}$, where X is chloride, bromide, or iodide, are the most common examples. Even then, to crystallize the tetrahedral ion, it is necessary to use a large cation; otherwise, the nickel ion will acquire other ligands (such as water molecules) to attain an octahedral environment.

Palladium and Platinum

The most common oxidation states for palladium and platinum are +2 and +4 (isoelectronic with the +1 and +3 states of rhodium and iridium). In the +2 state, the complexes are square planar.

Biological Aspects

The biochemistry of nickel is the most poorly understood of all those of the Period 4 transition elements. Nickel ions are present in some enzyme systems in the form of porphyrin-type complexes. Certain bacteria, such as those that reduce carbon dioxide to methane, need nickel. The requirement for nickel was explained when it was found that the most types of the enzyme *hydrogenase* contain nickel together with iron–sulfur clusters. Though in normal chemistry, the +3 oxidation state of nickel is very rare, nickel(III) is involved in the enzyme redox cycle. Nickel is also found in some plants that accumulate metals. In fact, certain tropical trees (nickel hyperaccumulators) concentrate nickel to such an extent that it makes up about 15 percent of their dry mass.

20.9   Group 11: Copper, Silver, and Gold

Copper, silver, and gold are sometimes called the *coinage metals*, because historically they were the three metals used for this purpose. The reasons for this were fourfold: They are readily obtainable in the metallic state; they are malleable, so disks of the metal can be stamped with a design; they are quite unreactive chemically, and, in the cases of silver and gold, the rarity of the metals meant that the coins had the intrinsic value of the metal itself. (These days, our coins are merely tokens with little actual value.)

Copper and gold are the two common yellow metals, although a thin coating of copper(I) oxide, Cu_2O, often makes copper look reddish. The color of copper is caused by the filled d-band in the metal being only about 220 kJ·mol^{-1} lower in energy than the s–p band. As a result, electrons can be excited to the higher band by photons of the corresponding energy range—the blue and green regions of the spectrum. Hence, copper reflects yellow and red. The band separation in silver is greater, and the absorption is in the ultraviolet part of the spectrum. Relativistic effects (see Chapter 2, Section 2.5) lower the s–p band energy in the case of gold, again bringing the absorption into the blue part of the visible range, resulting in the characteristic yellow color.

All three metals exhibit an oxidation number of +1. This group, together with the alkali metals and thallium, are the only metals that commonly do so. For copper, the oxidation number of +2 is more common than that of +1, whereas, for gold, the +3 state is thermodynamically preferred. The metals are not easily oxidized, as can be seen from the following (positive) reduction potentials:

$$Cu^{2+}(aq) + 2\ e^- \rightarrow Cu(s) \qquad\qquad E° = +0.34\ V$$

$$Ag^+(aq) + e^- \rightarrow Ag(s) \qquad\qquad E° = +0.80\ V$$

$$Au^{3+}(aq) + 3\ e^- \rightarrow Au(s) \qquad\qquad E° = +1.68\ V$$

Extraction of Copper

Although copper does not occur abundantly in nature, many copper-containing ores are known. The most common ore is copper(I) iron(III) sulfide, $CuFeS_2$, a metallic-looking solid that has the two mineralogical names of chalcopyrites and copper pyrites. A rarer mineral, $CuAl_6(PO_4)_4(OH)_8 \cdot 4H_2O$, is the valued blue gemstone, turquoise.

The extraction of copper from the sulfide can be accomplished by using either a thermal process (pyrometallurgy) or an aqueous process (hydrometallurgy). For the pyrometallurgical process, the concentrated ore is heated (a process called roasting) in a limited supply of air. This reaction decomposes the mixed sulfide, to give iron(III) oxide and copper(I) sulfide:

$$4\ CuFeS_2(s) + 9\ O_2(g) \xrightarrow{\Delta} 2\ Cu_2S(l) + 6\ SO_2(g) + 2\ Fe_2O_3(s)$$

Sand is added to the molten mixture, converting the iron(III) oxide into a slag of iron(III) silicate:

$$Fe_2O_3(s) + 3\ SiO_2(s) \rightarrow Fe_2(SiO_3)_3(l)$$

This liquid floats on the surface and can be poured off. Air is again added, causing sulfide to be oxidized to sulfur dioxide, and simultaneously copper(I) sulfide is converted to copper(I) oxide:

$$2\ Cu_2S(l) + 3\ O_2(g) \rightarrow 2\ Cu_2O(s) + 2\ SO_2(g)$$

The air supply is discontinued after about two-thirds of the copper(I) sulfide has been oxidized. The mixture of copper(I) oxide and copper(I) sulfide then undergoes an unusual redox reaction to give impure copper metal:

$$Cu_2S(l) + 2\ Cu_2O(s) \rightarrow 6\ Cu(l) + SO_2(g)$$

The pyrometallurgical process has a number of advantages: Its chemistry and technology are well known; there are many existing copper smelters; it is a fast process. The process also has disadvantages: The ore must be fairly concentrated; there is a large energy requirement for the smelting process; there are large emissions of sulfur dioxide.

Most metals are extracted by pyrometallurgical processes, using high temperatures and a reducing agent such as carbon monoxide. However, as mentioned, *pyrometallurgy* requires high-energy input, and the wastes are often major air and land pollutants. *Hydrometallurgy*—the extraction of metals by using solution processes—had been known for centuries but did not become widely used until the 20th century and then only for specific metals, such as silver and gold. This method has many advantages: Its by-products are usually less of an environmental problem than the flue gases and slag of a smelter; plants can be built on a small scale and then expanded, whereas a smelter needs to be built on a large scale to be economical; high temperatures are not required, so less energy is consumed than by smelting; hydrometallurgy can process lower grade ores (less metal content) than can pyrometallurgy.

In general, hydrometallurgical processes consist of three general steps: leaching, concentration, and recovery. The leaching is often accomplished by crushing and heaping the ore, then spraying it with some reagent, such as dilute acid (for copper extraction) or cyanide ion (for silver and gold extraction). Sometimes, instead of chemicals, solutions of the bacterium *Thiobacillus ferrooxidans* are used (this process is actually *biohydrometallurgical*). This bacterium oxidizes the sulfide in insoluble metal sulfides to a soluble sulfate. The dilute metal ion solution is then removed and concentrated by a variety of means. Finally, the metal itself is produced either by chemical precipitation using a single replacement reaction or by an electrochemical process. In the specific hydrometallurgical process for copper, copper pyrites is air oxidized in acid suspension to give a solution of copper(II) sulfate:

$$2 \text{ CuFeS}_2(s) + \text{H}_2\text{SO}_4(aq) + 4 \text{ O}_2(g) \rightarrow$$
$$2 \text{ CuSO}_4(aq) + 3 \text{ S}(s) + \text{Fe}_2\text{O}_3(s) + \text{H}_2\text{O}(l)$$

Thus, in this method, sulfur is released in the forms of sulfate ion solution and solid elemental sulfur, rather than as the sulfur dioxide produced in the pyrometallurgical method. The copper metal is then obtained by electrolysis, and the oxygen gas formed can be utilized in the first step of the process:

$$2 \text{ H}_2\text{O}(l) \rightarrow \text{O}_2(g) + 4 \text{ H}^+(aq) + 4 \text{ e}^-$$

$$\text{Cu}^{2+}(aq) + 2 \text{ e}^- \rightarrow \text{Cu}(s)$$

Copper is refined electrolytically to give a product that is about 99.95 percent pure. This impure copper (formerly the cathode) is now made the anode of an electrolytic cell that contains pure strips of copper as the cathode and an electrolyte of copper(II) sulfate solution. During electrolysis, copper is transferred from the anode to the cathode; an anode sludge containing silver and gold is produced during this process, thus helping to make the process economically feasible:

$$\text{Cu}(s) \rightarrow \text{Cu}^{2+}(aq) + 2 \text{ e}^-$$

$$\text{Cu}^{2+}(aq) + 2 \text{ e}^- \rightarrow \text{Cu}(s)$$

Because there is no net electrochemical reaction in this purification step, the voltage required is minimal (about 0.2 V), and the power consumption is very small. Of course, the environmentally preferred route for our copper needs is the recycling of previously used copper.

Pure copper has the highest thermal conductivity of all metals. For this reason, copper is used in premium cookware so that the heat is distributed rapidly throughout the walls of the container. An alternative approach is to apply a copper coating to the base of cookware made from other materials. Copper is second only to silver as an electrical conductor; hence, electrical wiring represents a major use of this metal. Copper is comparatively expensive for a common metal. To make penny coins of copper would now cost more than 1¢, so more recent coins have an outer layer of copper over a core of the lower priced zinc.

Though copper is normally considered an unreactive metal, over time it is slowly oxidized in moist air to give a coating of green *verdigris*, a copper(II) carbonate hydroxide, also called basic copper(II) carbonate, $\text{Cu}_2(\text{CO}_3)(\text{OH})_2$. This characteristic green color can be seen on the copper-covered roofs of

Table 20.7 Important alloys of copper

Alloy	Approximate composition	Properties
Brass	77% Cu, 23% Zn	Harder than copper
Bronze	80% Cu, 10% Sn, 10% Zn	Harder than brass
Nickel coins	75% Ni, 25% Cu	Corrosion resistant
Sterling silver	92.5% Ag, 7.5% Cu	More durable than pure silver

major buildings in Canada, such as the Parlaiment buildings in Ottawa, and parts of northern Europe.

Copper is a soft metal and it is often used in alloys—brass (for plumbing fixtures) and bronze (for statues). It is often a minor component in nickel and silver alloys. Some of the common alloy compositions are shown in Table 20.7.

Copper(II) Compounds

Although copper forms compounds in both the $+1$ and $+2$ oxidation states, it is the $+2$ state that dominates the aqueous chemistry of copper. In aqueous solution, almost all copper(II) salts are blue, the color being due to the presence of the hexaaquacopper(II) ion, $[Cu(OH_2)_6]^{2+}$. The common exception is copper(II) chloride. A concentrated aqueous solution of this compound is green, the color caused by the presence of complex ions such as the nearly planar tetrachlorocuprate(II) ion, $[CuCl_4]^{2-}$. When diluted, the color of the solution changes to blue. These color transformations are due to the successive replacement of chloride ions in the complexes by water molecules, the final color being that of the hexaaquacopper(II) ion. The overall process can be summarized as

$$[CuCl_4]^{2-}(aq) + 6\ H_2O(l) \rightleftharpoons [Cu(OH_2)_6]^{2+}(aq) + 4\ Cl^-(aq)$$

If a solution of ammonia is added to a copper(II) ion solution, the blue color of the hexaaquacopper(II) ion is replaced by the deep blue of the square-planar tetraamminecopper(II) ion, $[Cu(NH_3)_4]^{2+}$:

$$[Cu(OH_2)_6]^{2+}(aq) + 4\ NH_3(aq) \rightarrow [Cu(NH_3)_4]^{2+}(aq) + 6\ H_2O(l)$$

Although we commonly depict the process as a single reaction, the substitution of water ligands by ammonia is a stepwise process. Figure 20.20 shows the distribution of species with ammonia concentration. Because it is a log

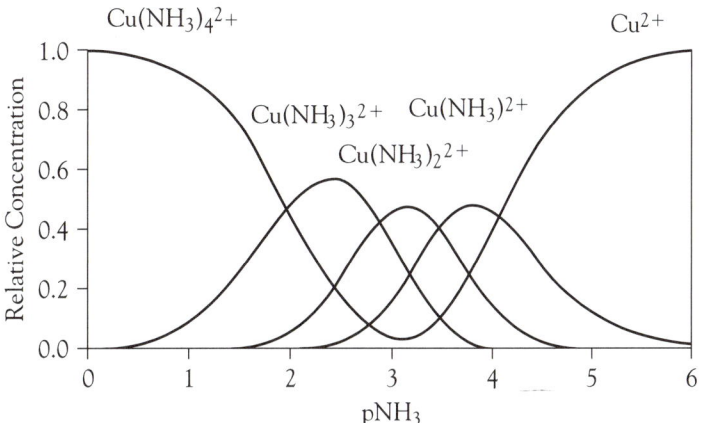

Figure 20.20 The relative concentrations of copper(II) species with increasing pNH$_3$ (to left). *[Adapted from A. Rojas-Hernández, et al., J. Chem. Educ., 72, 1100 (1995).]*

scale, pNH_3 (like pH), increasing ammonia concentration is to the left. For simplicity, the coordinated water molecules are not included. We see that, as the concentration of ammonia is increased, so the $[Cu(NH_3)_n]^{2+}$ complexes are formed in sequence.

Addition of hydroxide ion to a copper(II) ion solution causes the precipitation of the copper(II) hydroxide, a blue-green gelatinous solid:

$$Cu^{2+}(aq) + 2\ OH^-(aq) \rightarrow Cu(OH)_2(s)$$

However, warming the suspension causes the hydroxide to decompose to the black copper(II) oxide and water:

$$Cu(OH)_2(s) \rightarrow CuO(s) + H_2O(l)$$

Copper(II) hydroxide is insoluble in dilute base, but it dissolves in concentrated hydroxide solution to give the deep blue tetrahydroxocuprate(II) ion, $[Cu(OH)_4]^{2-}$:

$$Cu(OH)_2(s) + 2\ OH^-(aq) \rightarrow [Cu(OH)_4]^{2-}(aq)$$

Copper(II) hydroxide also dissolves in an aqueous solution of ammonia to give the tetraamminecopper(II) ion:

$$Cu(OH)_2(s) + 4\ NH_3(aq) \rightleftharpoons [Cu(NH_3)_4]^{2+}(aq) + 2\ OH^-(aq)$$

For most ligands, the copper(II) oxidation state is the more thermodynamically stable, although reducing ligands, such as iodide, will reduce copper(II) ions to the copper(I) state:

$$2\ Cu^{2+}(aq) + 4\ I^-(aq) \rightarrow 2\ CuI(s) + I_2(aq)$$

Jahn–Teller Effect

Copper(II) often forms complexes that are square planar. When copper(II) compounds with six occupied ligand sites are synthesized, it is usually found that the two axial ligands are more distant from the metal than those in the equatorial plane (though in a few cases the axial ligands are closer than those in the equatorial plane).

This preference for a distortion from a true octahedron can be explained simply in terms of the d orbital splittings. In an octahedral d^9 electron configuration, a slight energy benefit can be obtained by a splitting of the $d_{x^2-y^2}$ and the d_{z^2} energies, one increasing in energy and the other decreasing in energy by the same amount. The electron pair will occupy the lower orbital, and the single electron will occupy the higher energy orbital. Thus, two electrons will have lower energies and only one will have higher energy. This splitting is usually accomplished by lengthening the axial bond and weakening the electron–electron repulsion along the z-axis (Figure 20.21). The phenomenon of octahedral distortion is known as the *Jahn–Teller effect*. This effect can occur with other electron configurations, but it has been studied most in copper(II) compounds. A continuation of the distortion leads to the square-planar situation that is found for some d^8 ions (see Figure 19.19).

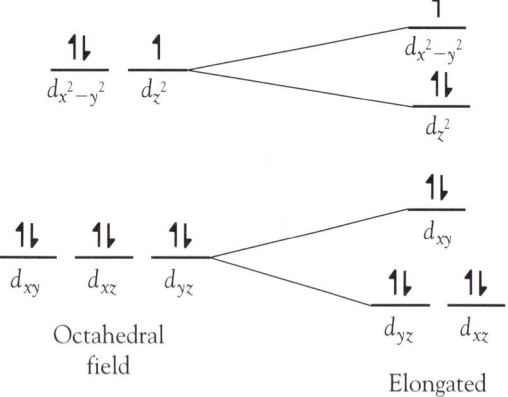

Figure 20.21 The splitting of the d orbital energies as a result of the Jahn–Teller effect.

Copper(I) Compounds

Even though copper is a relatively unreactive metal, it is attacked by concentrated acids. In particular, copper reacts with boiling hydrochloric acid to give a colorless solution and hydrogen gas. This reaction is particularly surprising because hydrochloric acid is not a strong oxidizing acid. The copper(I) ion formed in the oxidation is rapidly complexed by the chloride ion to produce the colorless dichlorocuprate(I) ion, $[CuCl_2]^-$. It is this second equilibrium step that lies far to the right and "drives" the first step:

$$2\ Cu(s) + 2\ H^+(aq) \rightleftharpoons 2\ Cu^+(aq) + H_2(g)$$

$$Cu^+(aq) + 2\ Cl^-(aq) \rightleftharpoons [CuCl_2]^-(aq)$$

When the solution is poured into air-free distilled water, copper(I) chloride precipitates as a white solid:

$$[CuCl_2]^-(aq) \rightarrow CuCl(s) + Cl^-(aq)$$

It must be rapidly washed, dried, and sealed in the absence of air, because a combination of air and moisture oxidizes it to copper(II) compounds.

In organic chemistry, the dichlorocuprate ion is used for converting benzene diazonium chloride into chlorobenzene (the *Sandmeyer reaction*):

$$[C_6H_5N_2]^+Cl^-(aq) \xrightarrow{[CuCl_2]^-} C_6H_5Cl(l) + N_2(g)$$

Generally, copper(I) compounds are colorless or white, because the ion has a d^{10} electron configuration. That is, with a filled set of d orbitals, there can be no d electron transitions to cause visible light absorption.

In aqueous solution, the hydrated copper(I) ion is unstable and disproportionates into the copper(II) ion and copper, as the Frost diagram in Figure 20.2 predicts:

$$2\ Cu^+(aq) \rightleftharpoons Cu^{2+}(aq) + Cu(s)$$

The Pourbaix diagram in Figure 20.22a shows the fields of stability for the various copper species. Again, we see that only the copper(II) ion is stable in aqueous solution. However, addition of cyanide ion changes the stability fields dramatically. In particular, Figure 20.22b shows that cyano species dominate over the entire range of accessible pH and E. And, the copper(I) complex, the tetracyanocuprate(I) ion, is preferred over all but the higher

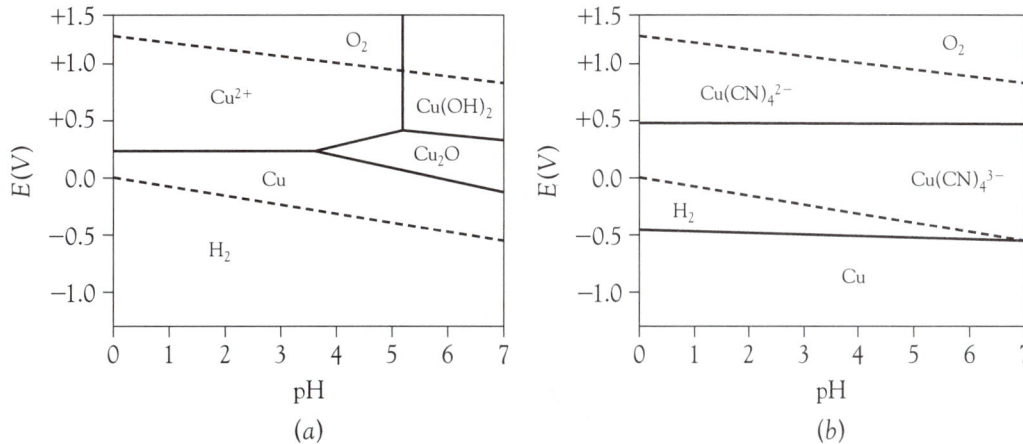

Figure 20.22 A comparison of (a) the "normal" Pourbaix diagram for copper with (b) the Pourbaix diagram for copper in the presence of cyanide ion. *[Modified from A. Napoli and L. Pogliani, Educ. Chem.* **34**, 51 (1997).]

E values. This is not surprising, for as we discussed earlier, cyanide stabilizes low oxidation states.

Silver

Silver is found mostly as the free element and as silver(I) sulfide, Ag_2S. Significant amounts of silver are also obtained during the extraction of lead from its ores and from the electrolytic refining of copper. One method of extraction of the metal involves the treatment of pulverized silver(I) sulfide with an aerated solution of sodium cyanide, a process that extracts the silver as the dicyanoargentate(I) ion, $[Ag(CN)_2]^-$:

$$2\ Ag_2S(s) + 8\ CN^-(aq) + O_2(g) + 2\ H_2O(l) \rightarrow$$
$$4[Ag(CN)_2]^-(aq) + 2\ S(s) + 4\ OH^-(aq)$$

Addition of metallic zinc causes a single replacement reaction in which the very stable tetracyanozincate ion, $[Zn(CN)_4]^{2-}$, is formed:

$$2\ [Ag(CN)_2]^-(aq) + Zn(s) \rightarrow 2\ Ag(s) + [Zn(CN)_4]^{2-}(aq)$$

There is one case of a country being named after a chemical element: Argentina named from the Latin for silver, argentum.

The pure metal is obtained by electrolysis, using an electrolyte of acidified silver nitrate solution with the impure silver as the anode and a pure strip of silver as the cathode:

$$Ag(s) \rightarrow Ag^+(aq) + e^- \qquad \text{(anode)}$$

$$Ag^+(aq) + e^- \rightarrow Ag(s) \qquad \text{(cathode)}$$

Silver Compounds

In almost all the simple compounds of silver, the metal has a +1 oxidation number, and the Ag^+ ion is the only water-stable ion of the element. Hence, it is common to substitute "silver" for "silver(I)."

The most important compound of silver is white silver nitrate. One of only two highly water-soluble salts of silver (the other being silver fluoride), it gives the colorless, hydrated silver ion when dissolved in water. Silver nitrate is used industrially for the preparation of other silver compounds, particularly the silver halides that are used in photography. In the laboratory, a standard solution of silver nitrate is used to test for the presence of chloride, bromide, and iodide ion. In qualitative analysis, the halide can be identified by color:

$$Ag^+(aq) + Cl^-(aq) \rightarrow AgCl(s) \text{ (white)} \qquad K_{sp} = 2 \times 10^{-10}$$

$$Ag^+(aq) + Br^-(aq) \rightarrow AgBr(s) \text{ (cream)} \qquad K_{sp} = 5 \times 10^{-13}$$

$$Ag^+(aq) + I^-(aq) \rightarrow AgI(s) \text{ (yellow)} \qquad K_{sp} = 8 \times 10^{-17}$$

Because the intensity of the color depends on particle size, it can be difficult to differentiate chloride from bromide or bromide from iodide. Hence, there is a secondary confirmatory test. This test involves addition of dilute ammonia solution. Silver chloride reacts with dilute ammonia solution to give the diamminesilver(I) ion:

$$AgCl(s) + 2\ NH_3(aq) \rightarrow [Ag(NH_3)_2]^+(aq) + Cl^-(aq)$$

Silver bromide is only slightly soluble, and silver iodide is insoluble in dilute ammonia. However, silver bromide will react with concentrated ammonia:

$$AgBr(s) + 2\ NH_3(aq) \rightarrow [Ag(NH_3)_2]^+(aq) + Br^-(aq)$$

To understand this difference in behavior, we must compare the equation for the precipitation reaction (where X is any of the halides) with the equation for the complexation reaction:

$$Ag^+(aq) + X^-(aq) \rightleftharpoons AgX(s)$$

$$Ag^+(aq) + 2\,NH_3(aq) \rightleftharpoons [Ag(NH_3)_2]^+(aq) \qquad K_{stab} = 2 \times 10^7$$

There are two competing equilibria for the silver ion. In qualitative terms, it is the one with the larger equilibrium constant that will predominate. Hence, in the case of very insoluble silver iodide, it is the precipitation equilibrium that will dominate. Conversely, the more soluble silver chloride will result in a silver ion concentration high enough to drive the complexation reaction:

$$Ag^+(aq) + Cl^-(aq) \rightleftharpoons AgCl(s)$$
$$\downarrow$$
$$Ag^+(aq) + 2\,NH_3(aq) \rightarrow [Ag(NH_3)_2]^+(aq)$$

The quantitative estimation of chloride, bromide, and iodide ions can be accomplished either gravimetrically, by weighing the silver halide produced or titrimetrically, by using an indicator such as potassium chromate (the Mohr method) that we discussed earlier in the context of the chromate ion.

The insolubility of the silver chloride, bromide, and iodide was explained in terms of covalent character in Chapter 5, Section 5.3. The silver fluoride, AgF, is a white, water-soluble solid and is considered to be ionic in both solid and aqueous solution.

Silver chloride, bromide, and iodide are sensitive to light, and the ready reduction of the silver ion results in a darkening of the solid. (This is why silver compounds and their solutions are stored in dark bottles.)

$$Ag^+(s) + e^- \rightarrow Ag(s)$$

This reaction is the key to the traditional photographic process. In black-and-white photography, it is simply the impact of light on sensitized silver halide microcrystals that initiates the production of the negative image. For color photography, the film is composed of layers, with organic dyes acting as color filters for the silver bromide. Thus, only light from a particular spectral region will activate the silver bromide in a specific layer. The photographic process was briefly described in the Feature "Chemistry and Photography" in Chapter 16.

Although nearly all simple silver compounds exhibit the +1 oxidation state, there are exceptions. For example, silver metal can be oxidized to black AgO, which is actually silver(I) silver(III) oxide, $(Ag^+)(Ag^{3+})(O^{2-})_2$. This compound reacts with perchloric acid to give the paramagnetic tetraaquasilver(II) ion, $[Ag(OH_2)_4]^{2+}$. The reaction is the reverse of a disproportionation (called *conproportionation*), and the highly oxidizing perchlorate ion stabilizes the +2 oxidation state of the silver:

$$AgO(s) + 2\,H^+(aq) \rightarrow Ag^{2+}(aq) + H_2O(l)$$

Gold

With its very high reduction potential, this element is usually found in nature as the free metal. As gold is a very soft acid, the gold minerals that are known, such as calaverite, $AuTe_2$, and sylvanite, $AuAgTe_4$, involve the very soft base, tellurium. For extraction of metallic gold from rock, the same cyanide process as that for silver metal is used. Gold does form a variety of complexes but few simple inorganic compounds. Gold(I) oxide, Au_2O, is one of the few stable gold compounds in which the metal has an oxidation number of $+1$. Like copper, this oxidation state is only stable in solid compounds, because solutions of all gold(I) salts disproportionate to gold metal and gold(III) ions:

$$3\ Au^+(aq) \rightarrow 2\ Au(s) + Au^{3+}(aq)$$

One of the most common compounds of gold is gold(III) chloride, $AuCl_3$; this can be prepared simply by reacting the two elements together:

$$2\ Au(s) + 3\ Cl_2(g) \rightarrow 2\ AuCl_3(s)$$

Dissolving gold(III) chloride in concentrated hydrochloric acid gives the tetrachloroaurate(III) ion, $[AuCl_4]^-$, an ion that is one of the components in "liquid gold," a mixture of gold species in solution that will deposit a film of gold metal when heated.

To diminish the need for summer air conditioning, the windows in some office high-rise buildings are coated with a reflective layer of 10^{-11} m film of gold.

Biological Aspects

Copper is the third most biologically important transition metal after iron and zinc. About 5 mg are required in the daily human diet. A deficiency of this element renders the body unable to use iron stored in the liver. There are numerous copper proteins throughout the living world, the most intriguing being the hemocyanins. These molecules are common oxygen carriers in the invertebrate world: Crabs, lobsters, octopi, scorpions, and snails all have bright blue blood. In fact, there are parallel iron and copper compounds (with very different structures) for many biological functions (Table 20.8). As it is the invertebrates that contain the copper systems, it can be argued that early life developed with copper as the functional metal and that only later did iron systems develop.

At the same time, an excess of copper is highly poisonous, particularly to fish. This is why copper coins should never be thrown into fish pools for "good luck." Humans usually excrete any excess, but a biochemical (genetic) defect can result in copper accumulation in the liver, kidneys, and brain. This illness, Wilson's disease, can be treated by administering chelating agents, which complex the metal ion and allow it to be excreted harmlessly.

Both silver and gold have specific medical applications. Silver ion is a bactericide, and dilute solutions of silver nitrate are placed in the eyes of newborn babies to prevent infection. Gold compounds, such as the drug auranofin, are used in the treatment of rheumatoid arthritis.

Table 20.8 Some parallels between iron and copper proteins

Function	Iron protein	Copper protein
Oxygen transport	Hemoglobin	Hemocyanin
Oxygenation	Cytochrome P-450	Tyrosinase
Electron transfer	Cytochromes	Blue copper proteins
Antioxidative function	Peroxidases	Superoxide dismutase
Nitrite reduction	Heme-containing Nitrite reductase	Copper-containing Nitrite reductase

20.10 *Element Reaction Flowchart*

Flowcharts are shown for titanium, vanadium, chromium, manganese, iron, nickel, cobalt, and copper.

$$TiO_2 \underset{O_2}{\overset{\Delta/C/Cl_2}{\rightleftarrows}} TiCl_4 \xrightarrow{Mg} Ti$$

$$[H_2VO_4]^- \xrightarrow{Zn} VO^{2+} \xrightarrow{Zn} V^{3+} \xrightarrow{Zn} V^{2+}$$

$$MnO_4^- \xrightarrow{+e^-} Mn(OH_2)_6^{2+} \xrightarrow{OH^-} Mn(OH)_2 \xrightarrow{-e^-} MnO(OH)$$

$$NiCl_4^{2-}$$

$$\uparrow Cl^-$$

$$Ni(CO)_4 \underset{\Delta}{\overset{CO}{\rightleftharpoons}} Ni \xleftarrow{+e^-} Ni(OH_2)_6^{2+} \underset{OH^-}{\overset{H^+}{\rightleftharpoons}} Ni(OH)_2$$

$$\downarrow NH_3$$

$$Ni(NH_3)_6^{2+}$$

$$CuCl_2^- \qquad CuCl_4^{2-}$$

$$\uparrow HCl \qquad \uparrow Cl^-$$

$$Cu \underset{-e^-}{\overset{Zn}{\rightleftharpoons}} Cu(OH_2)_6^{2+} \underset{OH^-}{\overset{H^+}{\rightleftharpoons}} Cu(OH)_2 \underset{OH^-}{\overset{H^+}{\rightleftharpoons}} Cu(OH)_4^{2-}$$

$$\downarrow NH_3 \qquad \overset{H^+}{\searrow} \quad \swarrow \Delta$$

$$Cu(NH_3)_4^{2+} \qquad CuO$$

EXERCISES

20.1 Write balanced equations for
(a) the reaction between titanium(IV) chloride and oxygen gas
(b) sodium dichromate with sulfur at high temperature
(c) the warming of copper(II) hydroxide
(d) the reaction between the dicyanoargentate ion and zinc metal
(e) the reaction between gold and dichlorine

20.2 Write balanced equations for
(a) the vanadyl ion with zinc metal in acidic solution (two equations)
(b) the oxidation of chromium(III) ion to dichromate ion by ferrate ion, which itself is reduced to iron(III) ion in acid solution (write initially as two half-equations)
(c) the warming of copper(II) hydroxide
(d) the reaction between copper(II) ion and iodide ion
(e) the reaction between gold and dichlorine

20.3 Discuss briefly how the stability of the oxidation states of the Period 4 transition metals changes along the row.

20.4 Identify uses for (a) titanium(IV) oxide; (b) chromium(III) oxide; (c) molybdenum(IV) sulfide; (d) silver nitrate.

20.5 What evidence do you have that titanium(IV) chloride is a covalent compound? Suggest why this is to be expected.

20.6 The equation for the first step in the industrial extraction of titanium is

$$TiO_2(s) + 2\ C(s) + 2\ Cl_2(g) \rightarrow TiCl_4(g) + 2\ CO(g)$$

Which element is being oxidized and which is being reduced in this process?

20.7 Write half-equations for the reduction of permanganate ion in (a) acidic solution; (b) basic solution.

20.8 Aluminum is the most abundant metal in the Earth's crust. Discuss the reasons why iron, not aluminum, is the more important metal in the world's economy.

20.10 Contrast how iron(II) chloride and iron(III) chloride are synthesized.

20.11 In the purification of nickel metal; tetracarbonylnickel(0) is formed from nickel at a lower temperature, while the compound decomposes at a higher temperature. Qualitatively discuss this equilibrium in terms of the thermodynamic factors, enthalpy and entropy.

20.12 Identify the elements that are called (a) the coinage metals; (b) the noble metals.

20.13 Identify each metal from the following tests and write balanced equations for each reaction:
(a) Addition of chloride ion to a pink aqueous solution of this cation gives a deep blue solution.

(b) Concentrated hydrochloric acid reacts with this metal to give a colorless solution. On dilution, a white precipitate is formed.

(c) Addition of acid to this yellow anion results in an orange-colored solution.

20.14 Identify each metal from the following tests and write balanced equations for each reaction:

(a) Acidifying a solution of this cation gives a pale violet solution and, on addition of chloride ion, a yellow solution is formed.

(b) Addition of ammonia to this pale blue cation gives a deep blue solution.

(c) Addition of thiocyanate solution to this almost colorless cation gives a deep red color.

20.15 Addition of a solution of this halide ion to a silver ion solution gives a whitish precipitate that is insoluble in dilute ammonia solution but soluble in a concentrated ammonia solution. Identify the halide ion and write a balanced chemical equation for each step.

20.16 A solution containing a colorless anion is added to a cold solution containing a pale yellow solution of a cation. A violet solution is formed that becomes colorless upon warming to room temperature. Identify the ions and write a balanced chemical equation for each step.

20.17 You wish to prepare a tetrahedral complex of vanadium(II). Suggest the best choice of a ligand (two reasons).

20.18 The highest oxidation state for nickel in a simple compound is found in the hexafluoronickelate(IV) ion, $[NiF_6]^{2-}$.

(a) Why would you expect fluoride to be the ligand?

(b) Would you expect the complex to be high spin or low spin? Give your reasoning.

20.19 Suggest why aluminum oxide is amphoteric and iron(III) oxide is basic.

20.20 The ferrate(VI) ion, FeO_4^{2-}, is such a strong oxidizing agent that it will oxidize aqueous ammonia to nitrogen gas, itself being reduced to iron(III) ion. Write a balanced equation for the reaction.

20.21 When iron(III) forms a complex with dimethylsulfoxide, $(CH_3)_2SO$, is the ligating atom likely to be the oxygen or the sulfur? Explain your reasoning.

20.22 Of the two common oxides of chromium, chromium(VI) oxide, and chromium(III) oxide, which should have the lower melting point? Explain your reasoning.

20.23 Of the two common oxides of chromium, chromium(VI) oxide and chromium(III) oxide, which should be acidic? Explain your reasoning.

20.24 Suggest why chromium(III) nitrate dissolves in water to form an acidic solution.

20.25 There is only one simple anion of cobalt(III) that is high spin. Identify the likely ligand and write the formula of this octahedral ion.

20.26 Suggest why copper(I) chloride is insoluble in water.

20.27 Suggest why the silver bromide and iodide are colored, even though both silver and halide ions are colorless.

20.28 Taking the Jahn–Teller effect into account, how many absorptions would you expect from d electron transitions for the octahedral copper(II) ion?

20.29 In an early version of the Periodic Table (Chapter 2, Figure 2.1), the alkali metals and the coinage metals were placed in the same group and, until recently, Groups 1 and 11 were known as Groups IA and IB to indicate the similarities between the groups. Discuss the chemical reasons for linking the two groups and explain the significant differences between them.

20.30 Use the Pourbaix diagram in Figures 8.11 and 20.15 to suggest what is the most likely form of iron in (a) a well-aerated lake; (b) a lake suffering from the effects of acid rain; (c) bog water.

20.31 Identify each of the following ions and write net ionic equations for each reaction:

(a) A colorless cation that gives a white precipitate with chloride ion and a red-brown precipitate with chromate ion.

(b) A pale pink cation that gives a deep blue color with chloride ion. The cation gives a blue solid with hydroxide ion.

(c) A colorless anion that gives a yellowish precipitate with silver ion. Addition of aqueous chlorine to the anion produces a deep brown color that extracts into organic solvents as a purple color.

(d) A yellow anion that gives a yellow precipitate with barium ion. Addition of acid to the anion causes a color change to orange. The orange anion is reduced by sulfur dioxide to give a green cation; the other product is a colorless anion that gives a white precipitate with barium ion.

(e) A pale blue cation that reacts with zinc metal to give a red-brown solid. Addition of the pale blue cation to excess ammonia gives a deep blue color.

20.32 Identify which transition metal(s) is(are) involved in each of the following biochemical molecules: (a) hemocyanin; (b) ferrodoxin; (c) nitrogenase; (d) vitamin B_{12}.

20.33 Write balanced chemical equations corresponding to each transformation in each Element Reaction Flowchart.

■ BEYOND THE BASICS

20.34 On the basis of the following thermodynamic data
(a) Calculate the equilibrium constant, K, for the formation of tetracarbonylnickel(0) from nickel and carbon monoxide under standard conditions.
(b) Calculate the temperature above which the complex is favored, that is, when K becomes 1.00 or greater.

Explain how these calculations relate to the purification of nickel metal.

	$\Delta H_f^\circ (\text{kJ·mol}^{-1})$	$S^\circ (\text{J·mol}^{-1}\cdot\text{K}^{-1})$
$Ni(s)$	0.0	29.9
$CO(g)$	−110.5	197.7
$Ni(CO)_4(g)$	−602.9	410.6

Assume ΔH and S are temperature independent.

20.35 Calculate the equilibrium constant (stability constant) for the complexation of gold(I) ion with cyanide ion given:

$$Au^+(aq) + e^- \rightarrow Au(s) \qquad E^\circ = +1.68 \text{ V}$$

$$Ag(CN)_2{}^-(aq) + e^- \rightarrow Au(s) + 2\ CN^-(aq)$$
$$E^\circ = -0.60 \text{ V}$$

20.36 Cobalt(II) undergoes the equilibrium:

$$\underset{\text{pink}}{Co(OH_2)_6{}^{2+}(aq)} + 4\ Cl^-(aq) \rightarrow \underset{\text{blue}}{CoCl_4{}^{2-}(aq)} + 6\ H_2O(l)$$

Suggest an explanation why addition of an anhydrous calcium compound to the mixture drives the equilibrium to the right, while addition of an anhydrous zinc compound drives the equilibrium to the left.

20.37 Copper(I) fluoride crystallizes in a sphalerite structure while copper(II) fluoride adopts a rutile structure. Calculate the enthalpy of formation of each of these compounds. What are the crucial terms in the respective Born–Haber cycles that contribute to the stability of each one. Qualitatively, what influence does the entropy of formation play in the feasibility of the reactions?

20.38 When lead(II) ion is added to a solution of dichromate ion, lead(II) chromate precipitates. Using chemical equations, explain why this happens.

20.39 Cobalt(II) perchlorate reacts with dimethylsulfoxide, $(CH_3)_2SO$, abbreviated to DMSO, to form a pink compound that is a 1:2 (one cation to two anions) electrolyte in DMSO solution. Cobalt(II) chloride reacts with DMSO to give a 1:1 electrolyte. The cations are the same in both cases, but the anion in the latter compound also contains a complex ion of cobalt. Deduce the identity of the two compounds.

20.40 (a) When potassium cyanide is added to aqueous nickel(II) ions, a green precipitate initially forms. What is the identity of the product.
(b) Further addition of cyanide ion results in the dissolution of the precipitate to give a yellow solution. A salt can be isolated from the yellow solution. What is the identity of this product? What is the geometry of the complex ion?
(c) A large excess of potassium cyanide gives a red solution. Isolation of the product gives a compound that is a 3:1 electrolyte. What is the identity of the compound?

20.41 For the (unbalanced) reaction

$$VO^{2+}(aq) + Cr^{2+}(aq) \rightarrow V^{2+}(aq) + Cr^{3+}(aq)$$

write a balanced redox equation then use appropriate data tables to calculate the equilibrium constant and deduce a probable two-step mechanism for the process.

20.42 Explain why:
(a) Iron(III) perchlorate is soluble in water, whereas iron(III) phosphate is insoluble in water.
(b) Complexes with the ligands NH_3 and H_2O are very common while those of PH_3 and H_2S are quite unusual.
(c) Iron(III) bromide is more intensely colored than iron(III) chloride.

20.43 Tungsten forms iodides of empirical formula WI_2 and WI_3. What do you suggest is the likely product from the reaction of tungsten metal with fluorine gas? Give your reasoning.

20.44 Rhenium forms the unusual compound, $K_2[ReH_9]$. What is the oxidation state of rhenium in this compound? Is this an expected oxidation state of rhenium? Is the oxidation state expected with hydrogen as a ligand?

20.45 Nickel forms a compound of formula NiS_2. What are the probable oxidation states of nickel and sulfur? Give your reasoning.

20.46 The precipitation of iron(III) hydroxide is used to clarify waste waters because the gelatinous compound is very efficient at the entrapment of contaminants. Ignoring the many hydroxoiron(III) species, we can write a simplified equilibrium as:

$$Fe^{3+}(aq) + 3\ H_2O(l) \rightleftharpoons Fe(OH)_3(s) + 3\ H^+(aq)$$

(a) Using the ion product constant at 25°C of 1.0×10^{-14}, and given the solubility product for iron(III) hydroxide as 2.0×10^{-39}, calculate the mathematical relationship between $[Fe^{3+}]$ and $[H^+]$.

(b) If iron(III) hydroxide is used to clarify a water supply, what concentration of free iron(III) ion will enter the water system if the water supply has a pH of 6.00?

(c) What mass of iron(III) hydroxide will be dissolved during the passage of 1×10^6 L of water?

20.47 Dilute hydrochloric acid was added to a metallic-looking compound (A). A colorless gas (B) with a characteristic odor was formed together with a pale green solution of the cation (C).

The gas (B) was burned in air to give another colorless gas (D) that turned yellow dichromate paper green. Mixing (B) and (D) gave a yellow solid element (E). Depending on the mole ratios, (E) reacted with chlorine gas to give two possible chlorides, (F) and (G).

Addition of ammonia to a sample of the green cation solution (C) gave a pale blue complex ion (H). Addition of hydroxide ion to another sample of the green solution gave a green gelatinous precipitate (I). Addition of zinc metal to a third sample of the green solution gave a metal (J) that on drying could be reacted with carbon monoxide to give a compound (K) with a low boiling point.

Identify each of the substances and write balanced chemical equations for each reaction.

20.48 When a very pale pink salt (A) is heated strongly, a brown-black solid (B) is produced; a deep brown gas (C) is the only other product. Addition of concentrated hydrochloric acid to (B) gives a colorless solution of salt (D), a pale green gas (E), and water. When the pale green gas is bubbled into a solution of sodium bromide, the

solution turns brown. The yellow-brown substance (F) can be extracted into dichloromethane and other low-polarity solvents.

The brown solid (B) can also be produced when a deep purple solution of the anion (G) is reacted in basic solution with a reducing agent, such as hydrogen peroxide. The other product is a gas (H), which will relight a glowing splint. The anion of compound (A) does not form any insoluble salts, whereas the gas (C) is in equilibrium with colorless gas (I), the latter being favored at low temperatures.

Identify (A) through (I), writing balanced equations for each step.

20.49 A transition metal, M, reacts with dilute hydrochloric acid in the absence of air to give $M^{3+}(aq)$. When the solution is exposed to air, an ion $MO^{2+}(aq)$ is formed. Suggest the identity of metal M.

20.50 When chromium(II) oxide, CrO, is synthesized from pure chromium and oxygen, the actual stoichiometry is found to be $Cr_{0.92}O_{1.00}$, yet the crystals are electrically neutral. Suggest an explanation.

20.51 Suggest why 84 percent of all chromium atoms are the chromium-52 isotope.

20.52 Iron(III) iodide has been synthesized in the total absence of air and water. Suggest a mechanism for the decomposition of this compound involving these two reagents.

20.53 Contrast the solubilities of the silver halides with those of calcium. Suggest an explanation for the difference.

20.54 On the basis of its chemistry, should silver be regarded as a transition metal? Discuss. Which main group element does silver most resemble?

20.55 Palladium metal can absorb up to 935 times its own volume of dihydrogen. Assuming conditions of SATP, calculate the approximate formula to which this corresponds. The α form of the hydrogen-saturated palladium has approximately the same density as palladium itself (12.0 g·cm^{-3}). Hence, also calculate the density of dihydrogen within the palladium. Compare the value to that of pure liquid dihydrogen (0.070 g·cm^{-3}).

20.56 Chromium forms a variety of dimeric species such as the blue $[(H_3N)_5Cr—O—Cr(NH_3)_5]^{4+}$ ion. What is the formal oxidation state of each chromium atom? The ion has a linear Cr—O—Cr arrangement. Suggest an explanation for this. If it existed, would the equivalent ion with cobalt instead of chromium probably be linear or bent? Explain.

20.57 To stoichiometrically oxidize a particular solution of iron(II) ion to iron(III) ion, 20 mL of an acidified permanganate ion solution was required. However, in the presence of a large excess of fluoride ion, 25 mL of the permanganate ion solution was required. Suggest an explanation.

20.58 Sulfur dioxide will reduce iron(III) ion to iron(II) ion in a solution acidified with a few drops of dilute hydrochloric acid. However, in a solution of concentrated hydrochloric acid, very little reduction occurs. Suggest an explanation.

20.59 Copper(I) cyanide, though insoluble in water, dissolves in aqueous potassium cyanide. Write a balanced net ionic equation to explain why the solution process occurs.

20.60 The colors of the silver halides are the result of charge–transfer processes. Explain why the color becomes more intense in the order I > Br > Cl.

20.61 Manganese forms five common oxides: MnO, Mn_3O_4, Mn_2O_3, MnO_2, and Mn_2O_7.
(a) Calculate the oxidiation number of manganese in each oxide.
(b) Suggest an explanation for your answer for the oxidation number of manganese in Mn_3O_4. Which other transition metal forms an oxide of this stoichiometry?

(c) Which oxide should be basic? Which oxide should be strongly acidic? Give your reasoning.
(d) Which common oxide of manganese has the same element to oxygen ratio as a common oxide of its main group analog, chlorine?

20.62 Calculate the enthalpy change for the two methane reforming reactions used for the production of direct-reduced iron:

$$CH_4(g) + CO_2(g) \rightarrow 2\ CO(g) + 2\ H_2(g)$$

$$CH_4(g) + H_2O(g) \rightarrow CO(g) + 3\ H_2(g)$$

20.63 The iron produced by the DRI process is a low density porous lump material. The high surface area makes it more reactive than bulk iron. DRI is usually stored in heaps open to the air.
(a) Write a chemical equation for the reaction that will occur over the surface air-exposed layer.
(b) A sodium silicate solution can be sprayed over the surface to minimize oxidation. Explain why this treatment is effective.
(c) In 1996, there was a fire and explosion on a freighter carrying DRI. The DRI had begun to oxidize rapidly and exothermically. The crew had sprayed sea water over the DRI to cool it down. How would the water have reacted with the hot iron? What would have caused the explosion?

Chapter 21

The Group 12 Elements

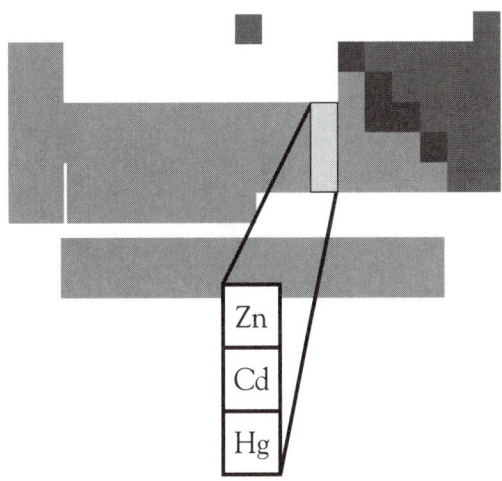

Zn
Cd
Hg

The Group 12 elements, although at the end of the transition metal series, behave like main group metals. Zinc is the most commonly encountered member of the group, both chemically and biochemically.

Mercury, the metal that flows like water, has been a source of fascination for millennia. Accounts of the metal are found in ancient Chinese and Indian writings, and Egyptian specimens date to about 1500 B.C. From 200 B.C., a mine in Spain supplied mercury (as mercury(II) sulfide) to the Roman Empire. One of the most feared punishments for Roman convicts was a sentence to the mercury mine, for working in this mercury vapor-rich atmosphere almost guaranteed an agonizing death within months. The same mine continues in production to the present day. Not until 1665 was it made illegal to work in the mine more than 8 days a month and more than 6 hours a day—although this concern for workers' health was more related to production efficiency than to the welfare of the workers themselves.

Medieval alchemists used mercury in their attempts to turn common metals into gold. However, the surge in demand for mercury came when, in about 1570, it was realized that mercury could be used to extract silver from silver-containing ores. The solution of silver in mercury was separated from the solid residue and then heated strongly—the mercury vaporized and dissipated into the atmosphere. This process was obviously dangerous for the workers, and we are still living with the resulting mercury pollution 300 years later. It has been estimated that in the Americas alone about 250 000 tonnes of mercury have been released into the environment by precious metal extraction, and the fate of most of this mercury is unknown. Even today, this primitive and environmentally dangerous method is being used, this time to extract gold from gold deposits in the Amazon basin.

21.1   *Group Trends*

This group of silvery metals superficially appears to belong to the transition metals, but in fact the chemistry of these elements is distinctly different. For example, the melting points of zinc and cadmium are 419 and 321°C, respectively, much lower than the typical values of the transition metals, which are close to 1000°C. The liquid phase of mercury at room temperature can be best explained in terms of relativistic electron effects—namely, that the contraction of the outer orbitals makes the element behave more like a "noble liquid."

The Group 12 elements have filled d orbitals in all their compounds, so they are better considered as main group metals. Consistent with this assignment, most of the compounds of the Group 12 metals are white, except when the anion is colored. Zinc and cadmium are very similar in their chemical behavior, having an oxidation number of +2 in all their simple compounds. Mercury exhibits oxidation numbers of +1 and +2, although the Hg^+ ion itself does not exist. Instead, a Hg_2^{2+} ion is formed. The only real similarity between the Group 12 elements and the transition metals is complex formation, particularly with ligands such as ammonia, cyanide ions, and halide ions. All of the metals, but especially mercury, tend to form covalent rather than ionic compounds.

As we discussed in Chapter 9, Section 9.5, there are strong similarities in behavior between the chemistry of magnesium and that of zinc (the n and $n + 10$ relationship). Also, we mentioned in Chapter 9, Section 9.8, that there is a "Knight's Move" link between Zn(II) and Sn(II) and between Cd(II) and Pb(II).

21.2   *Zinc and Cadmium*

These two soft metals are chemically reactive. For example, zinc reacts with dilute acids to give the zinc ion:

$$Zn(s) + 2\ H^+(aq) \rightarrow Zn^{2+}(aq) + H_2(g)$$

The metal also burns when heated gently in chlorine gas:

$$Zn(s) + Cl_2(g) \rightarrow ZnCl_2(s)$$

Extraction of Zinc

The principal source of zinc is zinc sulfide, ZnS, the mineral sphalerite (or zinc blende) the prototypical structure for tetrahedral ionic lattice structures (Chapter 5, Section 5.5). Sphalerite occurs in Australia, Canada, and the United States. Extraction of the metal is not simple, as the Ellingham diagram shows (Figure 21.1).

For oxides, the carbon–oxygen line has a low free energy that becomes even more negative with increasing temperature. For the carbon–sulfur line, the free energy change and the slope are both close to zero. The difference between the lines is, in part, accounted for by the very different enthalpy of formation values, which reflect the fact that the C$=$S bond is much weaker

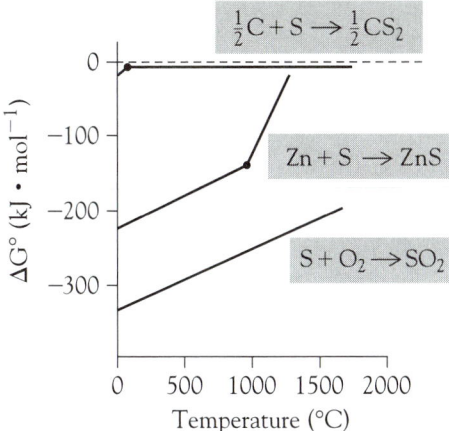

Figure 21.1   Ellingham diagram for selected sulfur compounds.

than the C=O bond. As a result, carbon reduction of sulfides is impractical.

$$C(s) + ¼\ S_8(s) \rightarrow CS_2(l) \qquad \Delta H° = +117\ \text{kJ·mol}^{-1}$$

$$C(s) + O_2(g) \rightarrow CO_2(g) \qquad \Delta H° = -394\ \text{kJ·mol}^{-1}$$

However, the diagram also shows that the sulfur–oxygen line is below that of the zinc–zinc sulfide line, a relation indicating that we can use the oxidation of the sulfide ion by oxygen gas as a source of free energy to convert zinc sulfide to zinc oxide. Thus, the first step in the extraction of zinc is "roasting" the zinc sulfide in air at about 800°C, converting it to the oxide:

$$2\ ZnS(s) + 3\ O_2(g) \xrightarrow{\Delta} 2\ ZnO(s) + 2\ SO_2(g)$$

It is then possible to use coke to reduce the metal oxide to the metal, as is apparent from the oxide Ellingham diagram (Figure 21.2):

$$ZnO(s) + C(g) \xrightarrow{\Delta} Zn(g) + CO(g)$$

Unlike the smelting of other metals, the two lines do not cross until zinc is in the gas phase. In fact, a temperature of about 1400°C is used. At these temperatures, the zinc is readily reoxidized, for example, by any carbon dioxide formed:

$$Zn(g) + CO_2(g) \xrightarrow{\Delta} ZnO(s) + CO(g)$$

To prevent this reaction, an excess of carbon is used, so that any carbon dioxide is reduced to carbon monoxide:

$$CO_2(g) + C(s) \xrightarrow{\Delta} 2\ CO(g)$$

In addition, the hot gas produced is rapidly cooled by spraying it with molten lead. The two metals are then easily separated, because the liquid metals are immiscible. The zinc (density 7 g·cm^{-3}) floats on the lead (density 11 g·cm^{-3}), and the lead is recycled.

Zinc is mainly used as an anticorrosion coating for iron. This process is called galvanizing, a term that recognizes the electrochemical nature of the process. Actually, the metal is not quite as reactive as one would expect. This results from the formation of a protective layer in damp air. Initially this is the oxide, but over a period of time the basic carbonate, $Zn_2(OH)_2CO_3$, is

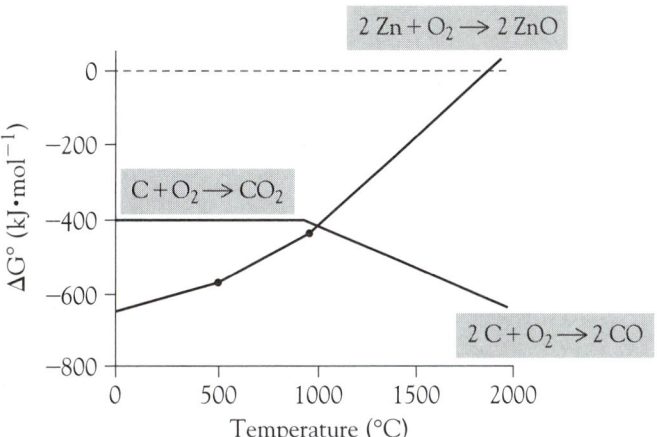

Figure 21.2   Ellingham diagram for the oxides of zinc and carbon.

formed. The advantage of zinc plating is that the zinc will be oxidized in preference to the iron, even when some of the iron is exposed. This is a result of the more negative reduction potential of the zinc than of the iron, the zinc acting as a sacrificial anode:

$$Zn(s) \rightarrow Zn^{2+}(aq) + 2\ e^- \qquad\qquad E° = +0.76 \text{ V}$$

$$Fe^{2+}(aq) + 2\ e^- \rightarrow Fe(s) \qquad\qquad E° = -0.44 \text{ V}$$

Zinc Salts

Most zinc salts are soluble in water, and these solutions contain the colorless hexaaquazinc(II) ion, $[Zn(OH_2)_6]^{2+}$. The solid salts are often hydrated; for example, the nitrate is a hexahydrate; and the sulfate, a heptahydrate, just like those of magnesium and cobalt(II). The structure of the sulfate heptahydrates is $[Zn(OH_2)_6]^{2+}\ [SO_4·H_2O]^{2-}$.

The zinc ion has a d^{10} electron configuration, so there is no crystal field stabilization energy. Hence, it is often the anion size and charge that determine whether the zinc ion adopts octahedral or tetrahedral stereochemistry. Solutions of zinc salts are acidic as the result of a multistep hydrolysis similar to that of aluminum or iron(III):

$$[Zn(OH_2)_6]^{2+}(aq) \rightleftharpoons H_3O^+(aq) + [Zn(OH)(OH_2)_3]^+(aq) + H_2O(l)$$

Addition of hydroxide ion causes precipitation of white, gelatinous zinc hydroxide, $Zn(OH)_2$:

$$Zn^{2+}(aq) + 2\ OH^-(aq) \rightarrow Zn(OH)_2(s)$$

With excess hydroxide ion, the soluble tetrahydroxozincate(II) ion, $[Zn(OH)_4]^{2-}$, is formed:

$$Zn(OH)_2(s) + 2\ OH^-(aq) \rightarrow [Zn(OH)_4]^{2-}(aq)$$

The precipitate will also react with ammonia to give a solution of the tetraamminezinc(II) ion, $[Zn(NH_3)_4]^{2+}$:

$$Zn(OH)_2(s) + 4\ NH_3(aq) \rightarrow [Zn(NH_3)_4]^{2+}(aq) + 2\ OH^-(aq)$$

Zinc chloride is one of the most commonly used zinc compounds. It is obtainable as the dihydrate, $Zn(OH_2)_2Cl_2$, and as sticks of the anhydrous zinc chloride. The latter is very deliquescent and extremely soluble in water. It is also soluble in organic solvents such as ethanol and acetone, and this property indicates the covalent nature of its bonding. Zinc chloride is used as a flux in soldering and as a timber preservative. Both uses depend on the ability of the compound to function as a Lewis acid. In soldering, the oxide film on the surfaces to be joined must be removed, otherwise the solder will not bond to these surfaces. Above 275°C, the zinc chloride melts and removes the oxide film by forming covalently bonded complexes with the oxide ions. The solder can then adhere to the molecular-clean metal surface. When it is applied to timber, zinc chloride forms covalent bonds with the oxygen atoms in the cellulose molecules. As a result, the timber is coated with a layer of zinc chloride, a substance toxic to living organisms.

Zinc Oxide

Zinc oxide can be obtained by burning the metal in air or by the thermal decomposition of the carbonate:

$$2\ Zn(s) + O_2(g) \xrightarrow{\Delta} 2\ ZnO(s)$$

$$ZnCO_3(s) \xrightarrow{\Delta} ZnO(s) + CO_2(g)$$

In the crystal, each zinc ion is surrounded tetrahedrally by four oxygen ions, and each oxygen ion is likewise surrounded by four zinc ions. Unlike other white metal oxides, zinc oxide develops a yellow color when heated. The reversible change in color that depends on temperature is known as *thermochromism*. In this case, the color change results from the loss of some oxygen from the lattice, thus leaving it with an excess negative charge. The excess negative charge (electrons) can be moved through the lattice by applying a potential difference; thus, this oxide is a semiconductor. Zinc oxide returns to its former color when cooled, because the oxygen that was lost during heating returns to the crystal lattice.

Zinc oxide is the most important compound of zinc. It is used as a white pigment, as a filler in rubber, and as a component in various glazes, enamels, and antiseptic ointments. In combination with chromium(III) oxide, it is used as a catalyst in the manufacture of methanol from synthesis gas.

Cadmium Sulfide

The only commercially important compound of cadmium is cadmium sulfide, CdS. Whereas zinc sulfide has the typical white color of Group 12 compounds, cadmium sulfide is an intense yellow. As a result, the compound is used as a pigment. Cadmium sulfide is prepared in the laboratory and industry by the same route, the addition of sulfide ion to cadmium ion:

$$Cd^{2+}(aq) + S^{2-}(aq) \rightarrow CdS(s)$$

Even though cadmium compounds are highly toxic, cadmium sulfide is so insoluble that it presents little hazard.

21.3 Mercury

With the weakest metallic bonding of all, mercury is the only liquid metal at 20°C. Mercury's weak bond also results in a high vapor pressure at room temperature. Because the toxic metal vapor can be absorbed through the lungs, spilled mercury globules from broken mercury thermometers are a major hazard in the traditional chemistry laboratory. Mercury is a very dense liquid (13.5 g·cm^{-3}). It freezes at $-39°C$ and boils at 357°C.

Extraction of Mercury

The only mercury ore is mercury(II) sulfide, HgS, the mineral cinnabar, although mercury is occasionally found as the free liquid metal. The deposits of mercury(II) sulfide in Spain and Italy account for about three-quarters of the world's supply of the metal. Many mercury ores contain considerably less than 1 percent of the sulfide, which accounts for the high price of the metal. Mercury is readily extracted from the sulfide ore by heating it in air. Mercury vapor is evolved and is then condensed to the liquid metal:

$$HgS(s) + O_2(g) \xrightarrow{\Delta} Hg(l) + SO_2(g)$$

Mercury is used in thermometers, barometers, electrical switches, and mercury arc lights. Solutions of other metals in mercury are called amalgams. Sodium amalgam and zinc amalgam are used as laboratory reducing agents, and the most common amalgam of all, dental amalgam (which contains mercury mixed with one or more of the metals silver, tin, and copper), is used for filling cavities in back teeth. It is suitable for this purpose for several reasons: It expands slightly as the amalgam forms, thereby anchoring the filling to the surrounding material. It does not fracture easily under the extreme localized pressures exerted by our grinding teeth. And it has a low coefficient of thermal expansion; thus, contact with hot substances will not cause it to expand and crack the surrounding tooth. In terms of total consumption, the major uses of mercury compounds are in agriculture and in horticulture; for example, organomercury compounds are used as fungicides and as timber preservatives.

Mercury(II) Compounds

Virtually all mercury(II) compounds utilize covalent bonding. Mercury(II) nitrate is one of the few compounds believed to contain the Hg^{2+} ion. It is also one of the few water-soluble mercury compounds.

Mercury(II) chloride can be formed by mixing the two elements:

$$Hg(l) + Cl_2(g) \rightarrow HgCl_2(s)$$

This compound dissolves in warm water, but the nonelectrically conducting behavior of the solution shows that it is present as $HgCl_2$ molecules, not as ions. Mercury(II) chloride solution is readily reduced to white insoluble mercury(I) chloride and then to black mercury metal by the addition of tin(II) chloride solution. This is a convenient test for the mercury(II) ion:

$$2\,HgCl_2(aq) + SnCl_2(aq) \rightarrow SnCl_4(aq) + Hg_2Cl_2(s)$$

$$Hg_2Cl_2(s) + SnCl_2(aq) \rightarrow SnCl_4(aq) + 2\,Hg(l)$$

Red mercury(II) oxide forms when mercury is heated for a long time in air at about 350°C:

$$2\,Hg(l) + O_2(g) \xrightarrow{\Delta} 2\,HgO(s)$$

Mercury(II) oxide is thermally unstable and decomposes into mercury and dioxygen when heated more strongly. This decomposition is a visually interesting demonstration because mercury(II) oxide, a red powder, "disappears" as silvery globules of metallic mercury form on the cooler parts of the container. However, the reaction is very hazardous because a significant portion of the mercury metal escapes as vapor into the laboratory. The experiment is of historical interest, for it was the method used by Joseph Priestley to obtain the first sample of pure oxygen gas:

$$2\,HgO(s) \xrightarrow{\Delta} 2\,Hg(l) + O_2(g)$$

Mercury(I) Compounds

An interesting feature of mercury is its ability to form the $[Hg-Hg]^{2+}$ ion in which the two mercury ions are united by a single covalent bond. In fact, there are no known compounds containing the simple mercury(I) ion.

Mercury(I) chloride, Hg_2Cl_2, and mercury(I) nitrate, $Hg_2(NO_3)_2$, are known; but compounds with other common anions, such as sulfide, have never been synthesized. To understand the reason for this, we must look at the disproportionation equilibrium,

$$Hg_2{}^{2+}(aq) \rightleftharpoons Hg(l) + Hg^{2+}(aq)$$

which has an equilibrium constant, K_{dis}, of about 6×10^{-3} at 25°C. The low value for the equilibrium constant implies that, under normal conditions, there is little tendency for the mercury(I) ion to disproportionate into the mercury(II) ion and mercury. However, anions such as sulfide form highly insoluble compounds with mercury(II) ion:

$$Hg^{2+}(aq) + S^{2-}(aq) \rightarrow HgS(s)$$

This precipitation "drives" the disproportionation equilibrium to the right. As a result, the overall reaction of mercury(I) ion with sulfide ion becomes

$$Hg_2^{2+}(aq) + S^{2-}(aq) \rightarrow Hg(l) + HgS(s)$$

21.4 Batteries Containing Group 12 Elements

One common thread among the Group 12 elements is their use in batteries, each metal being important in a different type of cell. It is unfortunate that some of the best battery materials—mercury, cadmium, and lead (discussed in Chapter 14, Section 14.27)—are among the most toxic elements, hence presenting major disposal problems.

The Alkaline Battery

This cell has become the most popular household battery. The cell consists of a zinc case (the anode) with a central rod as cathode. This rod consists of a compressed mixture of graphite (a good electrical conductor) and manganese(IV) oxide. The electrolyte is a potassium hydroxide solution. In the cell reaction, the zinc is oxidized to zinc hydroxide while the manganese(IV) oxide is reduced to the manganese(III) oxide hydroxide, $MnO(OH)$:

$$Zn(s) + 2\ OH^-(aq) \rightarrow Zn(OH)_2(s) + 2\ e^-$$

$$2\ MnO_2(s) + 2\ H_2O(l) + 2\ e^- \rightarrow 2\ MnO(OH)(s) + 2\ OH^-(aq)$$

In the overall process, 1 mole of hydroxide ion is consumed at the anode and 1 mole of hydroxide ion is produced at the cathode. As a result of the constancy of hydroxide ion concentration, the cell potential remains constant, a useful advantage over the old "dry cell" in which the deliverable voltage drops over the lifetime of the battery (Figure 21.3).

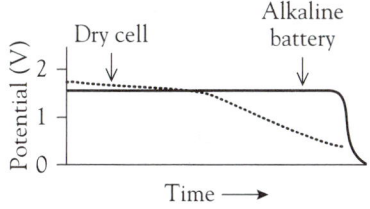

Figure 21.3 Decrease in voltage with time for an alkaline battery and a "dry cell."

The Mercury Cell

For very compact power needs, such as hearing aids, the mercury cell is often used. In this cell, again zinc is the anode, but mercury(II) oxide (mixed with conducting graphite) is the cathode. The zinc is oxidized to zinc hydroxide while the mercury(II) oxide is reduced to mercury metal:

$$Zn(s) + 2\ OH^-(aq) \rightarrow Zn(OH)_2(s) + 2\ e^-$$

$$HgO(s) + H_2O(l) + 2\ e^- \rightarrow Hg(l) + 2\ OH^-(aq)$$

Again the electrolyte concentration remains constant, thus providing a steady cell potential.

The NiCad Battery

Unlike the previous two cells, the NiCad battery can be recharged. In the discharge cycle, cadmium is oxidized to cadmium hydroxide, while nickel is reduced from the unusual oxidation state of $+3$ in nickel(III) oxide hydroxide, $NiO(OH)$, to the more normal $+2$ state, as nickel(II) hydroxide. Once more the electrolyte is hydroxide ion:

$$Cd(s) + 2\ OH^-(aq) \rightarrow Cd(OH)_2(s) + 2\ e^-$$

$$2\ NiO(OH)(s) + 2\ H_2O(l) + 2\ e^- \rightarrow 2\ Ni(OH)_2(s) + 2\ OH^-(aq)$$

In the charging process, the reverse reactions occur. There are two major reasons for using a basic reaction medium: The nickel(III) state is only stable in base, and the insolubility of the hydroxides means that the metal ions will not migrate far from the metal surface, thus allowing the reverse reactions to happen readily at the same site. This battery, which is commonly used in rechargeable flashlights, portable home appliances, and inexpensive portable computers has one disadvantage—a charging "memory." This unusual phenomenon means that if the NiCad battery is only partially discharged, then recharged, it "remembers" the previous level of discharge and will only discharge to this level again. Thus, it is important to discharge the cell completely each time before recharging.

21.5 *Biological Aspects*

This group contains one essential element (zinc) and two very toxic elements.

The Essentiality of Zinc

Among trace essential elements, zinc is second only to iron in importance. Around the world, a lack of zinc is the most common soil deficiency. Beans, citrus fruit, coffee, and rice are the crops most susceptible to zinc deficiency.

Zinc is an essential element for animals. Over 200 zinc enzymes have been identified in living organisms and their roles determined. Zinc enzymes that perform almost every possible type of enzyme function are known, but the most common function is hydrolysis; for example, the zinc-containing hydrolases are enzymes that catalyze the hydrolysis of P—O—P, P—O—C and C—O—C bonds. With such a dependence upon zinc enzymes, it is understandable that zinc is one of the most crucial elements in our diet. Yet it has been estimated that up to one-third of people in the western world suffer from zinc deficiency. Such deficiencies are not life-threatening, but they do contribute to fatigue, lethargy, and related symptoms (and possibly to diminished disease resistance).

The question arises as to what makes zinc such a useful ion, considering that it cannot serve a redox function. There are several reasons:

1. Zinc is widely available in the environment.

2. The zinc ion is a strong Lewis acid and zinc functions as a Lewis acid in enzymes.

3. Zinc, unlike many other metals, prefers tetrahedral geometries, a key feature of the metal site in most zinc enzymes. Five- and six-coordinate geometries are also available, making transition states involving these coordination numbers possible.

4. The zinc ion has a d^{10} electron configuration, so there is no crystal field stabilization energy associated with exact geometries as there are with the transition metals. Hence, the environment around the zinc can be distorted from the exact tetrahedral to allow for the precise bond angles needed for its function without an energy penalty.

5. The zinc ion is completely resistant to redox changes at biological potentials; thus, its role cannot be affected by changing redox potentials in the organism.

6. The zinc ion undergoes extremely rapid ligand exchange, facilitating its role in enzymes.

The Toxicity of Cadmium

Cadmium is a toxic element that is present in our foodstuffs and is normally ingested at levels that are close to the maximum safe level. The kidney is the organ most susceptible to cadmium; about 200 ppm cause severe damage. Cigarette smokers absorb significant levels of cadmium from tobacco smoke.

Exposure to cadmium from industrial sources is a major concern. In particular, the nickel–cadmium battery is becoming a major waste-disposal problem. Many battery companies now accept return of defunct NiCad batteries so that the cadmium metal can be safely recycled. Cadmium poisonings in Japan have resulted from cadmium-contaminated water produced by mining operations. The ensuing painful bone degenerative disease has been called *itai-itai*.

The Many Hazards of Mercury

As mentioned earlier, mercury is hazardous because of its relatively high vapor pressure. The mercury vapor is absorbed in the lungs, dissolves in the blood, and is then carried to the brain, where irreversible damage to the central nervous system results. The metal is also slightly water soluble, again a result of its very weak metallic bonding. The escape of mercury metal from leaking chlor-alkali electrolysis plants into nearby rivers has led to major pollution problems in North America.

Inorganic compounds of mercury are usually less of a problem because they are not very soluble. A note of historical interest: At one time, mercury ion solutions were used in the treatment of animal furs for hat manufacture. Workers in the industry were prone to mercury poisoning, and the symptoms of the disease were the model for the Mad Hatter in the book *Alice in Wonderland*.

The organomercury compounds pose the greatest danger. These compounds, such as the methylmercury cation, $HgCH_3^+$, are readily absorbed and are retained by the body much more strongly than the simple mercury compounds. The symptoms of methylmercury poisoning were first established in Japan, where a chemical plant had been pumping mercury wastes into Minamata Bay, a rich fishing area. Inorganic mercury compounds were converted by bacteria in the marine environment to organomercury compounds. These compounds, particularly CH_3HgSCH_3, were absorbed by the fatty tissues of fish, and the mercury-laden fish were consumed by the unsuspecting local inhabitants. The unique symptoms of this horrible poisoning have been named Minamata disease. Another major hazard are the organomercury fungicides. In one particularly tragic case, farm families in Iraq were sent mercury fungicide–treated grain, some of which they used for bread making instead of planting, being unaware of the toxicity. As a result, 450 people died, and over 6500 became ill.

21.6   Element Reaction Flowchart

The only flowchart shown is that for zinc. Notice the similarities with the flowchart for copper (Chapter 20).

$$Zn \underset{H^+}{\overset{+e^-}{\rightleftharpoons}} Zn(OH_2)_6^{2+} \underset{OH^-}{\overset{H^+}{\rightleftharpoons}} Zn(OH)_2 \underset{OH^-}{\overset{H^+}{\rightleftharpoons}} Zn(OH)_4^{2-}$$

$$\downarrow NH_3 \qquad \nwarrow H^+ \quad \Delta \nearrow$$

$$ZnO$$

$$Zn(NH_3)_4^{2+} \qquad \uparrow \Delta$$

$$ZnCO_3$$

EXERCISES

21.1 Write balanced chemical equations for the following chemical reactions:
(a) zinc metal with liquid bromine
(b) the effect of heat on solid zinc carbonate

21.2 Write balanced chemical equations for the following chemical reactions:
(a) aqueous zinc ion with ammonia solution
(b) heating mercury(II) sulfide in air

21.3 Suggest a two-step reaction sequence to prepare zinc carbonate from zinc metal.

21.4 Explain briefly the reasons for considering the Group 12 elements separately from the transition metals.

21.5 Compare and contrast the properties of (a) zinc and magnesium; (b) zinc and aluminum.

21.6 Normally, metals in the same group have fairly similar chemical properties. Contrast and compare the chemistry of zinc and mercury by this criterion.

21.7 Write the two half-equations for the charging process of the NiCad battery.

21.8 Cadmium sulfide adopts the zinc sulfide (wurtzite and zinc blende) structure but cadmium oxide adopts the sodium chloride structure. Suggest an explanation for the difference.

21.9 Cadmium-coated paper clips were once common. Suggest why they were used and why their use was discontinued.

21.10 Compare and contrast the chemistry of calcium (Group 2) and cadmium (Group 12).

21.11 Write balanced chemical equations corresponding to each transformation in the Element Reaction Flowchart.

BEYOND THE BASICS

21.12 Both cadmium ion, Cd^{2+}, and sulfide ion, S^{2-}, are colorless. Suggest an explanation for the color of cadmium sulfide.

21.13 Mercury(I) selenide is unknown. Suggest an explanation.

21.14 Mercury(II) iodide is insoluble in water. However, it will dissolve in a solution of potassium iodide to give a dinegative anionic species. Suggest a formula for this ion.

21.15 In the industrial extraction of zinc, molten lead is used to cool the zinc vapor until it liquefies. The molten zinc and molten lead do not mix; thus, they can be easily separated. Suggest why the two metals do not mix to any significant extent.

21.16 You are an artist and you wish to make your "cadmium yellow" paint paler. Why is it not a good idea to mix in some "white lead," $Pb_3(CO_3)_2(OH)_2$, to accomplish this?

21.17 When mercury(II) forms a complex with dimethylsulfoxide, $(CH_3)_2SO$, is the ligating atom likely to be the oxygen or the sulfur? Explain your reasoning.

21.18 The only common ore of mercury is mercury(II) sulfide, whereas zinc is found as a sulfide and a carbonate. Suggest an explanation.

21.19 The acid–base chemistry of liquid ammonia often parallels that of aqueous solutions. On this basis, write a balanced equation for:
(a) the reaction of zinc amide, $Zn(NH_2)_2$, with ammonium ion in liquid ammonia
(b) the reaction of zinc amide with amide ion, NH_2^-, in liquid ammonia

21.20 A compound (A) of a dipositive metal ion is dissolved in water to give a colorless solution. Hydroxide ion is added to the solution. A gelatinous white precipitate (B) initially forms, but in excess hydroxide ion, the precipitate redissolves to give a colorless solution of complex ion (C). Addition of concentrated ammonia solution to the precipitate (B) gives a colorless solution of complex ion (D). Addition of sulfide ion to a solution of compound (A) gives a highly insoluble white precipitate (E). Addition of silver ion to a solution of compound (A) results in a yellow precipitate (F). Addition of aqueous bromine to a solution of (A) gives a black solid (G), which can be extracted into an organic solvent and gives a purple solution. The solid (G) reacts with thiosulfate ion to give a colorless solution containing ions (H) and (I), the latter being an oxyanion.
Identify (A) to (I) and write balanced equations for each reaction.

21.21 Which would you expect to have a higher melting point, zinc oxide or zinc chloride? Explain your reasoning.

21.22 For the reaction:

$$Hg(l) + \tfrac{1}{2}O_2(g) \rightarrow HgO(s)$$

the value of the free energy change, ΔG, reverses its sign from negative to positive above about 400°C. Explain why this happens.

21.23 When hydrogen sulfide is bubbled into a neutral solution of zinc ion, zinc sulfide precipitates. However, if the solution is first acidified, no precipitate forms. Suggest an explanation.

21.24 The following are four different forms of mercury, each of which poses a different level of health hazard: $Hg(l)$, $Hg(CH_3)_2(l)$, $HgCl_2(aq)$, $HgS(s)$. Which of these forms upon ingestion:

(a) will pass unchanged through the digestive tract (for digestion, substances must be water or fat soluble)?

(b) will be most easily eliminated through the kidneys?

(c) will be the greatest hazard for absorption through the skin?

(d) will most readily pass from blood into the (nonpolar) brain tissue?

(e) will be absorbed by inhalation through the lungs?

Organometallic Chemistry

Chapter **22**

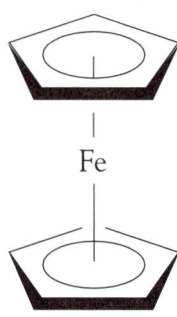

Approximately half the world's research publications in chemistry are currently about organometallic compounds. This field, bridging inorganic and organic chemistry, will be of continuing importance throughout the 21st century. Organometallics play a vital role in the economy with about 10 of the world's top 30 chemicals being produced using organometallic catalysts.

Organometallic compounds straddle both inorganic and organic chemistry. To be a member of this group, the compound must contain at least one direct metal-to-carbon covalent bond. The metal can be a transition, main group, or *f*-group metal, and the term "metal" is often stretched to include boron, silicon, germanium, arsenic, antimony, selenium, and tellurium. The carbon-containing group(s) may be a carbonyl, alkyl, alkene, alkyne, aromatic, cyclic, or heterocyclic compound.

The chemistry of organometallic compounds is interesting in that the compounds occur in a wide variety of molecular shapes and coordination numbers. Organometallic compounds are often *pyrophoric* (spontaneously flammable) and thermodynamically unstable. The central metal atom in an organometallic compound is often in a very low oxidation state.

The first organometallic compound was synthesized over 200 years ago when, in 1760, the French chemist L.C. Cadet was trying to make invisible inks. He produced a foul-smelling liquid from an arsenate salt, and it was later identified as dicacodyl (from the Greek for "stink"), As_2Me_4. This is a typical main group organometallic compound, in which the bonding utilizes *s* and *p* electrons.

However, it is the transition metal organometallic compounds that provide us with richness and variety in their structures and bonding types. This richness arises from the fact that transition metals can utilize *s*, *p*, and *d* orbitals in bonding. The orbitals can both donate and accept electron density, and the *d* orbitals are particularly well suited to interact with orbitals on the organic species in similar ways to the bonding in the coordination compounds studied in Chapter 19. As a result of the flexibility in bonding and the ability to transfer electron density, transition metal organometallic compounds are of industrial importance as catalysts.

22.1 Naming Organometallic Compounds

In addition to the general rules that we use to name simple inorganic compounds or transition metal complexes, there are a few supplementary rules that are used to give additional information about the nature of the bonding within the organometallic molecule. First, we need to introduce a selection of the many organic species which act as ligands in organometallic chemistry (Table 22.1).

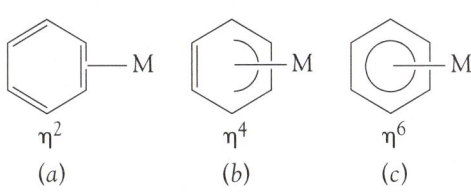

η^2 η^4 η^6
(a)   (b)   (c)

Figure 22.1 The benzene molecule bonded to a metal via (a) two carbon atoms (b) four carbon atoms, and (c) all six carbon atoms.

The number of carbon atoms within an organic species which are directly interacting with the metal is specified by the prefix η (the Greek letter *eta*). This is called *hapticity*, and most ligands bond through one atom only, so they are described as monohapto. Some ligands, especially those with multiple π bonds, may bond in more than one way. For example, benzene may bond to a metal center through 1, 2, or 3 π bonds. Consequently, we can describe benzene as mono-, bi-, or trihapto, and use the notation η^2, η^4, and η^6, respectively, as we consider the bonding to be across two-, four-, and six-ligand atoms, respectively (Figure 22.1).

Species which bridge two metal centers are indicated by the prefix μ (the Greek letter *mu*). This can apply to carbonyl ligands, halide, or carbene species (Figure 22.2).

In main group chemistry, *s*-block compounds are named according to the substituent names used in organic chemistry, as in methyllithium, $Li_4(CH_3)_4$. Ionic compounds are named as salts, such as sodium naphthalide, $Na^+[C_{10}H_8]^-$. *p*-Block compounds are named as simple organic species, such as trimethylboron, $B(CH_3)_3$. Alternatively, they may be named as derivatives of the hydride, for example, trimethylborane.

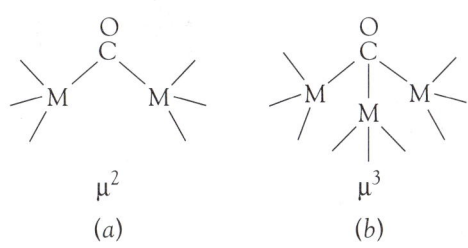

μ^2 μ^3
(a)   (b)

Figure 22.2 Carbonyl ligands bridging (a) two metal centers and (b) three metal centers.

Table 22.1 Some common ligands in organometallic chemistry

Formula	Name	Abbreviation
CO	Carbonyl	
CH_3	Methyl	Me
$CH_3CH_2CH_2CH_2$	*n*-Butyl	^nBu
$(CH_3)_3C$	Tertiary-butyl	^tBu
$[C_5H_5]^-$	Cyclopentadienyl	Cp
C_6H_5	Phenyl	Ph
C_6H_6	Benzene	
$(C_6H_5)_3P$	Triphenylphosphine	PPh_3

Table 22.2 Details of some common organometallic ligands

Ligand	Formal charge	Electrons donated
H	-1	2
F, Cl, Br, I	-1	2
CN	-1	2
μ-F, Cl etc	-1	4
CO	0	2
μ-CO	0	2
PR_3, PX_3	0	2
CH_3, C_2H_5 etc	-1	2
μ-CH_3	-1	2
NO	0	3

In d- and f-block species, the usual rules for naming coordination compounds are followed, with the additional use of η and μ. For example, $(\eta^5\text{-}C_5H_5)Mn(CO)_3$ is named pentahaptocyclopentadienyltricarbonylmanganese(I).

22.2 Counting Electrons

Formal oxidation numbers (see Chapter 8) of metallic species can be useful in keeping track of electrons in both the structures and the reactions of organometallic compounds. For organometallics, the electron count helps in predicting stabilities of compounds (Table 22.2). The oxidation number assigned does not necessarily relate to the actual charge on the metal. The convention used is that the organic moiety is usually assigned a charge of -1. For example, lithium in $Li_4(CH_3)_4$ has a formal oxidation state of $+1$; iron in $(\eta^5\text{-}C_5H_5)_2Fe$ has an oxidation state of $+2$. Other ligands are always neutral, for example, CO and C_6H_6.

There are two conventions when it comes to counting electrons and charges on some ligands. Some ligands, such as the cyclopentadienyl ligand, can be considered as an anionic species, for example, $Na^+C_5H_5^-$, formed on removal of a proton from C_5H_6. This is known as the *ionic convention*. Alternatively, the cyclopentadienyl ligand can be regarded as a neutral radical, $C_5H_5\cdot$. This is known as the *covalent* or *radical convention*. The choice has an effect on the formal charge that is assigned to the metal center, an important point when considering the relative stabilities of low and high oxidation states. The two conventions are shown here for $(\eta^5\text{-}C_5H_5)_2Fe$, commonly called ferrocene (Figure 22.3):

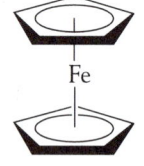

Figure 22.3 $(\eta^5\text{-}C_5H_5)_2Fe$, ferrocene.

Ionic convention		Covalent or radical convention	
$C_5H_5^-$	6 e$^-$	$C_5H_5\cdot$	5 e$^-$
$C_5H_5^-$	6 e$^-$	$C_5H_5\cdot$	5 e$^-$
Fe^{2+}	d^6	Fe	d^8
Total electrons	18 e$^-$	Total electrons	18 e$^-$

You can see that the convention chosen makes no difference to the final number of electrons. In fact, both systems are in widespread use, and,

Table 22.3 Some typical solvents for organometallic chemistry

Name	Formula	Common name/abbreviation
Dichloromethane	CH_2Cl_2	DCM
2-Propanol	$CH_3CH(OH)CH_3$	Isopropanol
2-Propanone	CH_3COCH_3	Acetone
Diethylether	$CH_3CH_2OCH_2CH_3$	Ether
Oxacyclopentane		Tetrahydrofuran, THF

provided there is consistency, it really does not matter which one is adopted. The approach used throughout this chapter will be the ionic convention.

22.3 Solvents for Organometallic Chemistry

Many reactions in inorganic chemistry are carried out in aqueous solution. Synthesis of transition metal coordination compounds and their reactions often take place either in aqueous solution or in moderately polar solvents such as ethanol and acetone. One of the striking features of organometallic chemistry is that reactions are almost never carried out in an aqueous solution. Most reactions take place in organic solvents, many of which are of low polarity. This is because most organometallic compounds are unstable in water—another illustration of their "organic" nature. (In addition, the synthesis of organometallic compounds is carried out under an inert atmosphere, such as nitrogen or argon, as many organometallic compounds are also unstable in air.) Some common solvents for organometallic chemistry are given in Table 22.3.

Unlike the other chapters in this text, the chemical equations in this chapter do not include symbols indicating solid, gas, liquid phase, or solution. This is the normal convention in organometallic chemistry, and it should be assumed that all reactions are carried out in an organic solvent.

22.4 Main Group Organometallic Compounds

Organometallic compounds of the main group elements have many structural and chemical similarities to the analogous hydrogen compounds. This is because the electronegativities of carbon and hydrogen are similar; hence, M—C and M—H bonds have similar polarities.

Organometallic Compounds of the Alkali Metals

All organometallic compounds of Group 1 elements are unstable and pyrophoric. Organic species which readily lose protons form ionic compounds with the Group 1 metals. For example, cyclopentadiene reacts with sodium metal.

$$Na + C_5H_6 \rightarrow Na^+[C_5H_5]^- + \frac{1}{2}H_2$$

Sodium and potassium form intensely colored compounds with aromatic species. The oxidation of the metal results in transfer of an electron to the aromatic system which produces a radical anion. A radical anion is an anion which possesses an unpaired electron, for example, the naphthalide anion in the deep blue sodium naphthalide.

$$Na + \quad \text{[naphthalene]} \quad \longrightarrow \quad Na^+[C_{10}H_8]^-$$

Colorless sodium and potassium alkyls are solids that are insoluble in organic solvents and, when stable enough, have fairly high melting temperatures. They are produced by *transmetallation reactions*. Transmetallation is a common method for the synthesis of main group organometallic compounds. It involves the breaking of the metal–carbon bond and the forming of a metal–carbon bond to a different metal. Alkyl-mercury compounds are often the starting materials in these reactions. For example, we can synthesize methylsodium by reacting sodium metal with dimethylmercury:

$$Hg(CH_3)_2 + 2\ Na \rightarrow 2\ NaCH_3 + Hg$$

Lithium alkyls and aryls are by far the most important Group 1 organometallics. They are liquids or low melting point solids, are more thermally stable than other Group 1 organometallic compounds, and are soluble in organic and non-polar solvents. Synthesis can be accomplished from an alkyl halide and lithium metal or by reacting the organic species with *n*-butyl lithium, $Li(C_4H_9)$, commonly abbreviated to ^nBuLi.

$$^nBuCl + 2\ Li \rightarrow {}^nBuLi + LiCl$$

$$^nBuLi + C_6H_6 \rightarrow Li(C_6H_5) + C_4H_{10}$$

A feature of many main group organometallic compounds is the presence of bridging alkyl groups. When ethers are the solvent, methyl lithium exists as $Li_4(CH_3)_4$, with a tetrahedron of lithium atoms and bridging methyl groups, with each carbon atom essentially six coordinate (three hydrogens of the methyl and three lithiums; see Figure 22.4). In hydrocarbon solvents, $Li_6(CH_3)_6$ is present and its structure is based upon an octahedral arrangement of lithium atoms. Other lithium alkyls adopt similar structures except when the alkyl groups become very bulky, as in the case of *t*-butyl, $—C(CH_3)_3$, when tetramers are the largest species formed.

    Organolithium compounds are very important in organic synthesis. They act in a similar way to Grignard reagents but are much more reactive. Among their many uses, they are used to convert *p*-block halides to organocompounds, as we will see later in this chapter. For example, boron trichloride reacts with *n*-butyl lithium to give the organo-boron compound:

$$BCl_3 + 3\ {}^nBuLi \rightarrow ({}^nBu)_3B + 3\ LiCl$$

The driving force for this and many other reactions of organometallic compounds is the formation of the halide compound of the more electropositive metal. This is a feature we will encounter many times in organometallic chemistry.

    Lithium alkyls are important industrially in the stereospecific polymerization of alkenes to form synthetic rubber. *n*-Butyl lithium is used as an initiator in solution polymerization to produce a wide range of elastomers and polymers. The composition and molecular weight of the polymer can be carefully controlled to produce very varied products suitable for wide range of different uses, such as footware, hoses and pipes, adhesives, sealants, and resins. Organolithium compounds are also used in the synthesis of a range of pharmaceuticals including vitamins A and D, analgesics, antihistamines, antidepressants, and anticoagulants.

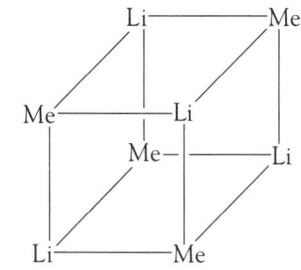

Figure 22.4 The structure of methyl lithium, $Li_4(CH_3)_4$.

Organometallic Compounds of the Alkaline Earth Metals

The compounds of calcium, strontium, and barium are generally ionic and very unstable. The organometallic compounds of beryllium and magnesium are the most important and will be discussed in more detail.

Organometallic compounds of beryllium are pyrophoric and readily hydrolyzed. They can be prepared by transmetallation from methylmercury. The synthesis of dimethylberyllium by transmetallation is shown here:

$$Hg(CH_3)_2 + Be \rightarrow Be(CH_3)_2 + Hg$$

Another synthetic route is by *halogen exchange* or *metathesis* reactions in which a metal halide reacts with an organometallic compound of a different metal. The products are the halide of the second metal and the organo-derivative of the first metal. In this way the halide and organo groups are effectively "transferred" between the two metals. Once again, the halide of the more electropositive metal is formed. Here are two examples:

$$2\ ^nBuLi + BeCl_2 \rightarrow Be(^nBu)_2 + 2\ LiCl$$

$$2\ Na(C_5H_5) + BeCl_2 \rightarrow Be(C_5H_5)_2 + 2\ NaCl$$

Methylberyllium, $Be(CH_3)_2$, is a monomer in the vapor phase and polymerizes in the solid (Figure 22.5). There are insufficient electrons, so the polymer is held together by 3-center 2-electron bridging bonds. This bond type was described in Chapter 10, Section 10.5, in the context of diborane. Bulkier alkyl groups lead to a lower degree of polymerization, and *t*-butyl beryllium is a monomer.

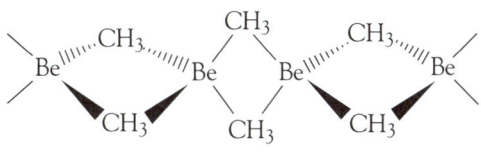

Figure 22.5 The structure of methyl beryllium in the solid phase.

Alkyl and aryl magnesium halides are very well known as Grignard reagents and are widely used in synthetic organic chemistry. They are prepared from magnesium metal and an organohalide. The reaction is carried out in ether and is initiated by a trace of iodine. A general reaction is shown in the following, where R is used as a generic symbol for any alkyl group.

$$Mg + RBr \rightarrow RMgBr$$

The compounds produced in this way are not pure, often containing other species such as R_2Mg. To synthesize the pure compounds, we employ transmetallation using a mercury compound:

$$Mg + RHgBr \rightarrow RMgBr + Hg$$

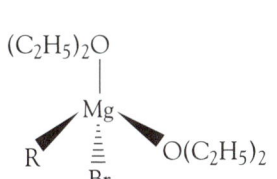

Figure 22.6 A solvated molecule of a Grignard reagent, $RMgBr[O(C_2H_5)_2]_2$.

The structure of these compounds is not simple. They have a coordination number of 2 only in solution and when the alkyl group is bulky. Otherwise, they are solvated with a tetrahedral magnesium atom (Figure 22.6).

Organometallic Compounds of the Group 13 Elements

In Group 13, the organometallic compounds of beryllium and aluminum are the most important, and we will focus mainly on these.

Organoboranes of the type BR_3 can be prepared by the reaction of an alkene with diborane. This is an example of *hydroboration* and involves insertion of the alkenyl group between a boron–hydrogen bond.

$$B_2H_6 + 6\ CH_2{=}CH_2 \rightarrow 2\ B(CH_2CH_3)_3$$

Alternatively, they can be produced from a Grignard reagent (where X is a halogen):

$$(C_2H_5)_2O:BF_3 + 3\ RMgX \rightarrow BR_3 + 3\ MgXF + (C_2H_5)_2O$$

Grignard Reagents

Victor Grignard, born in Cherbourg, France, in 1871, started his academic career in mathematics at the University of Lyons (pronounced *Lee-on*) but later switched to chemistry. The reagent was actually discovered by Phillipe Barbier in 1899. Barbier had been searching for a metal to replace zinc in reactions used to insert a methyl group into organic compounds. The disadvantage with zinc was that the zinc compounds caught fire on contact with air. Barbier found that magnesium was a superior substitute. (Of course, we would expect that to be a first choice on the basis of the (n) and ($n + 10$) relationship—see Chapter 9, Section 9.5.)

Barbier asked his junior colleague, Grignard, to study the reaction in more detail. The comprehensive study of the reaction and of its applicability to a wide range of syntheses formed the basis of Grignard's doctoral thesis in 1901. Though originally called the Barbier–Grignard reaction, Barbier generously insisted that credit should go to Grignard, even though it had been Barbier who first synthesized the key species, methyl magnesium iodide. In 1912, Grignard was awarded the Nobel Prize for chemistry. Though it was certainly true that Grignard's demonstration of the many applications of Grignard reactions revolutionized synthetic organic chemistry, it is unfortunate that Barbier's pioneering contribution and kind gesture toward his young colleague has been forgotten. Grignard later succeeded Barbier as senior professor at the University of Lyons.

Grignard reagents must be produced in an ether solvent, and the reagents must be absolutely dry. The reagent is never isolated but used *in situ*. Virtually all alkyl halides will form Grignards. Reactions of Grignards generally fall into two types: either attack on a hydrogen attached to an O, N, or S, or addition to a compound containing multiple bonds such as C=O, C=S, N=O. It was soon found that Grignard reagents would react with many classes of compound and within 8 years of his original paper there were 500 other research papers published on the topic of organomagnesium compounds.

Alkylboranes are all stable to water but are pyrophoric. The aryl species are more stable. They are all monomeric and planar. Like other boron compounds, the organoboron species are electron deficient and consequently act as Lewis acids and form adducts easily (Figure 22.7).

An important anion is the tetraphenylborate ion, $[B(C_6H_5)_4]^-$, or more commonly written as $[BPh_4]^-$, analogous to the tetrahydridoborate ion, $[BH_4]^-$ (see Chapter 13, Section 13.5). The sodium salt can be obtained by a simple addition reaction:

$$BPh_3 + NaPh \rightarrow Na^+[BPh_4]^-$$

The sodium salt is water soluble, but the salts of most massive, monopositive ions are insoluble. Consequently, the anion is useful as a precipitating agent and can be used in gravimetric analysis.

Alkyl aluminum compounds can be prepared on a laboratory scale by transmetallation of a mercury compound:

$$4\,Al + 3\,Hg(CH_3)_2 \rightarrow Al_2(CH_3)_6 + 3\,Hg$$

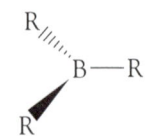

Figure 22.7   A molecule of an alkyl borane, BR$_3$.

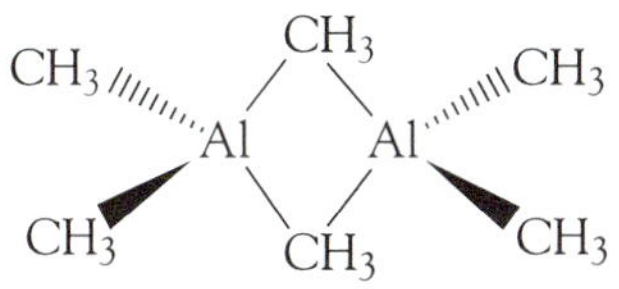

Figure 22.8 Representation of trimethyl aluminum, Al_2Me_6.

Trimethyl aluminum (Figure 22.8) is prepared commercially by the reaction of aluminum metal with chloromethane to give $Al_2Cl_2(CH_3)_4$. This is then reduced with sodium to give $Al_2(CH_3)_6$. These dimers are similar in structure to the dimeric halides (Chapter 13, Section 13.9), but the bonding is different. In the halides, the bridging Al—Cl—Al bonds are 2-center 2-electron bonds; that is, each Al—Cl bond involves an electron pair. In the alkyls, the Al—C—Al bonds are longer than the terminal Al—C bonds suggesting that they are 3-center 2-electron bonds; with one bonding pair shared across the Al—C—Al, somewhat analogous to the bonding in diborane, B_2H_6 (Chapter 10, Section 10.5).

Triethylaluminum and higher alkyl compounds are prepared from the metal, an appropriate alkene and hydrogen gas at elevated temperatures and pressures.

$$2\ Al + 3\ H_2 + 6\ CH_2{=}CH_2 \xrightarrow{\text{60–110°C, 10–20 MPa}} Al_2(CH_2CH_3)_6$$

This route is relatively cost-effective, and this means that aluminum alkyls have found many commercial applications. Triethylaluminum, often written as the monomer, $Al(C_2H_5)_3$, is an organometallic complex of aluminum of major industrial importance as we shall see in Section 22.17.

Steric factors have a powerful effect on the structures of aluminum alkyls. While dimers are favored, the long, weak, bridging bonds are easily broken, and this tendency increases with the bulkiness of the ligand. So for example, triphenyl aluminum is a dimer, but the mesityl, $2,4,6\text{-}(CH_3)_3C_6H_2{-}$, compound is a monomer.

Organometallic Compounds of the Group 14 Elements

Some Group 14 organometallic compounds are of great commercial importance. Silicon forms the very important silicones, which yield oils, gels, and rubbers (see Chapter 14, Section 14.22). Organotin compounds are used to stabilize PVC and as antifouling agents on ships, as wood preservatives and pesticides. Tetraethyllead has been widely used as an antiknock agent in leaded fuels (see Chapter 14, Section 14.26). Generally, organometallic compounds of this group are tetravalent and have low polarity bonds. Their stability decreases from silicon to lead.

All silicon tetraalkyls and tetraaryls are monomeric with a tetrahedral silicon center and resemble the carbon analogs. The carbon–silicon bond is strong, so all the compounds are fairly stable. They can be prepared in a variety of ways, examples of which are shown in the following:

$$SiCl_4 + 4\ RLi \rightarrow SiR_4 + 4\ LiCl$$

$$SiCl_4 + RLi \rightarrow RSiCl_3 + LiCl$$

The *Rochow process* provides industry with a cost-effective route to methylchlorosilane, an important starting material:

$$n\ MeCl + Si/Cu \rightarrow Me_nSiCl_{(4-n)}$$

These methylchlorosilanes, $Me_nSiCl_{(4-n)}$, where $n = 1$–3, can be hydrolyzed to form silicones and siloxanes.

$$(CH_3)_3SiCl + H_2O \rightarrow (CH_3)_3SiOH + HCl$$

$$2\ (CH_3)_3SiOH \rightarrow (CH_3)_3SiOSi(CH_3)_3 + H_2O$$

The reaction yields oligomers which contain the tetrahedral silicon group and oxygen atoms which form Si—O—Si bridges. Dimers can condense to form

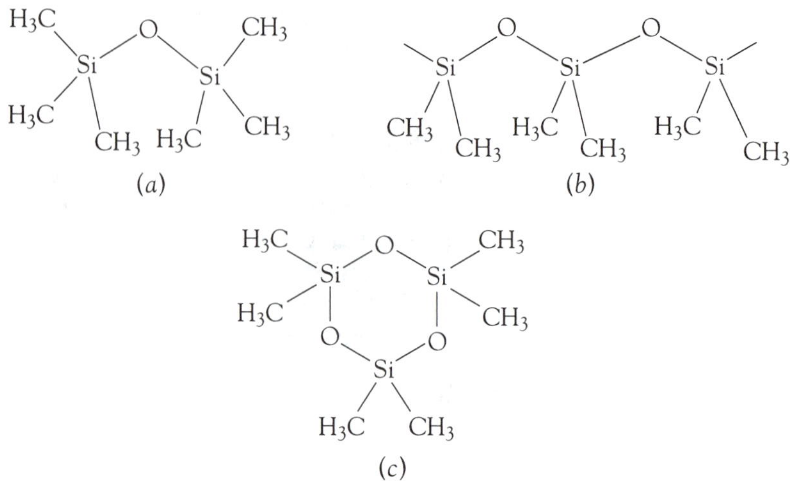

Figure 22.9 Structures of some silicones: (*a*) a dimer, (*b*) a chain, and (*c*) a ring structure.

(*a*)

(*b*)

(*c*)

chains or rings, (Figure 22.9). The hydrolysis of $MeSiCl_3$ produces a cross-linked polymer.

Silicone polymers have a range of structures and uses. Their properties depend on the degree of polymerization and cross-linking. These are influenced by both the choice and the mix of reactants and by the cross-linking and the use of dehydrating agents such as sulfuric acid and elevated temperatures.

Silicone products have many commercial uses. At one end of the spectrum they are essential in the personal-care products such as shampoos, conditioners, shaving foams, hair gels, and toothpastes. It is the silicone additives which are responsible for the "silky" feel of such products. At the other end of the spectrum, silicone greases, oils, and resins are used as sealants, lubricants, varnishes, waterproofing, synthetic rubbers, and hydraulic fluids.

Organotin compounds differ from silicon and germanium compounds in several ways. There is a greater occurrence of $+2$ oxidation state and a greater range of coordination numbers, and halide bridges are often present.

Most organotin compounds tend to be colorless liquids or solids which are stable to air and water. The structures of R_4Sn compounds are all similar with a tetrahedral tin atom (Figure 22.10).

The halide derivatives, R_3SnX, often contain Sn—X—Sn bridges and form chain structures. The presence of bulky R groups may effect the shape. For example, in $(CH_3)_3SnF$, the Sn—F—Sn backbone is in a zigzag arrangement (Figure 22.11); in Ph_3SnF the chain has straightened, and $(Me_3SiC)Ph_2SnF$ is a monomer. The haloalkyls are more reactive than the tetraalkyls and are useful in the synthesis of tetraalkyl derivatives.

Alkyl tin compounds may be prepared in a variety of ways, including via a Grignard reagent and by metathesis.

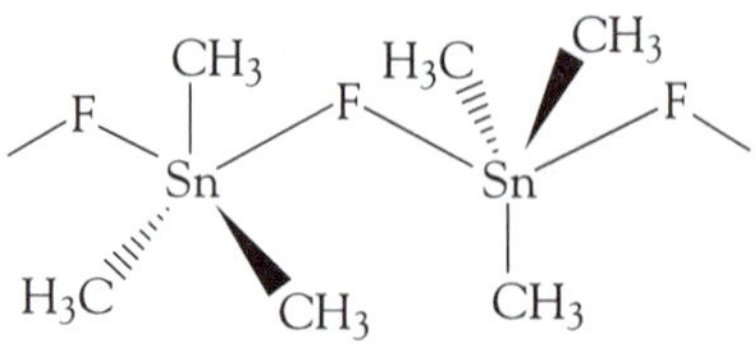

$$SnCl_4 + 4\ RMgBr \rightarrow SnR_4 + 4\ MgBrCl$$

$$3\ SnCl_4 + 2\ Al_2R_6 \rightarrow 3\ SnR_4 + 2\ Al_2Cl_6$$

Organo-tin compounds have the widest range of uses of all main group organometallic compounds, and the worldwide industrial production of organotin complexes has probably now passed the 50 000-tonne level. The major use of tin organometallic compounds is in the stabilization of PVC (polyvinyl chloride) plastics. Without the additive, halogenated polymers

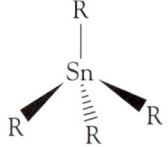

Figure 22.10 A tetralkyl tin molecule.

Figure 22.11 The zigzag backbone in $(CH_3)_3SnF$.

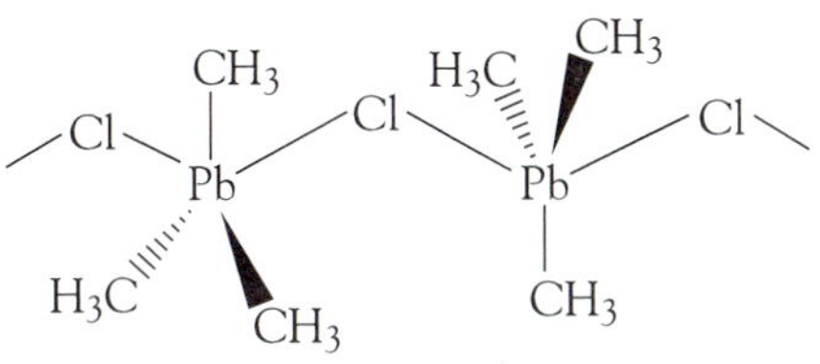

Figure 22.12 The chain structure of $Pb(CH_3)_3Cl$.

are rapidly degraded by heat, light, or atmospheric dioxygen to give discolored, brittle products.

Organotin (IV) compounds have a wide range of applications relating to their biocidal effects. They are used as fungicides, algaecides, wood preservatives, and antifouling agents. Their widespread use has caused environmental concerns. High levels of organotin compounds have been found in harbor regions where boats are treated with organotin compounds to prevent fouling, barnacles etc. There is evidence that high concentrations of organotin compounds may kill some species of marine life and affect the growth and reproduction of others. Many nations now restrict the use of organotin compounds to vessels over 25 m long.

R_4Pb compounds can be made in the laboratory via either a Grignard reagent or an organolithium compound.

$$2\ PbCl_2 + 4\ RLi \rightarrow R_4Pb + 4\ LiCl + Pb$$

$$2\ PbCl_2 + 4\ RMgBr \rightarrow 2\ R_4Pb + Pb + 4\ MgBrCl$$

They are all monomeric molecules with a tetrahedral lead center. The halide derivatives may contain bridging halide atoms to form chains. Monomers are favored with more bulky organic substituents. For example, $Pb(CH_3)_3Cl$ (Figure 22.12) exists as a chain structure with bridging chloride atoms, whereas the mesityl derivative $Pb(Me_3C_6H_2)_3Cl$ is a monomer.

Organometallic Compounds of the Group 15 Elements

Organometallic compounds of arsenic(III), antimony(III), and bismuth(III) can be prepared by the use of a Grignard or organolithium compound or from the element and an organohalide. The three alternatives are shown in the following:

$$AsCl_3 + 3\ RMgCl \rightarrow AsR_3 + 3\ MgCl_2$$

$$2\ As + 3\ RBr \xrightarrow{Cu/\Delta} AsRBr_2 + AsR_2Br$$

$$AsR_2Br + R'Li \rightarrow AsR_2R' + LiBr$$

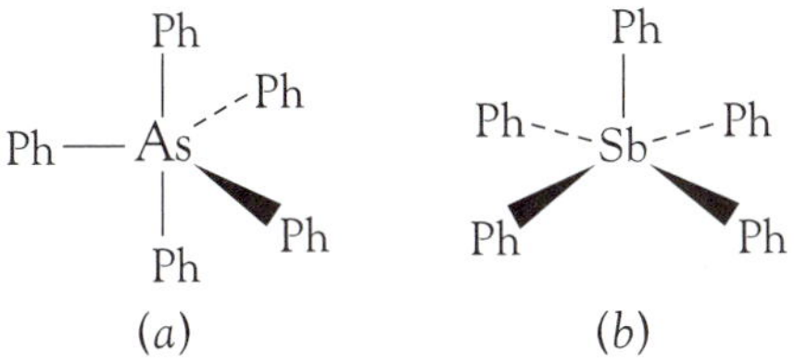

The compounds are all readily oxidized but stable to water. The aryl compounds are more stable than the alkyls. They are all trigonal pyramidal, and the C—M—C bond strength decreases for a given R group in the order As > Sb > Bi.

MR_5 compounds are usually trigonal bipyramidal, for example, $BiMe_5$ and $AsPh_5$. However, $SbPh_5$ is a square pyramid. They are all thermally unstable, and the stability decreases down the group (Figure 22.13).

Figure 22.13 The structures of (a) $AsPh_5$ and (b) $SbPh_5$.

Organometallic Elements of the Group 12 elements

The chemistry of the organometallic compounds of zinc, cadmium, and mercury is much more similar to that of the Group 2 elements than to that of the transition metals.

The alkyl compounds are linear, monomeric species with 2-center 2-electron bonds. Unlike the Group 2 analogs, they do not form polymeric chains through alkyl bridges. They can be prepared by metathesis with an aluminium alkyl:

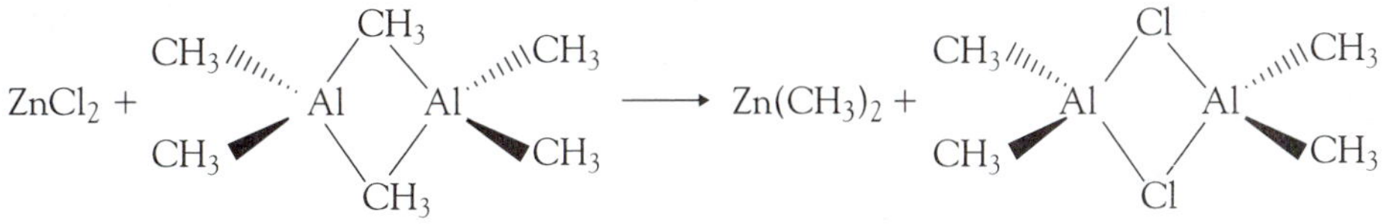

The Death of Karen Wetterhahn

August 14, 1996, would seem like a normal day for Karen Wetterhahn, professor of chemistry at Dartmouth College, New Hampshire. Wetterhahn was one of the foremost experts on metal toxicology, the study of the effects of heavy metals in biological systems. She was probably the most knowledgeable person in the world on chromium toxicology, having devised the uptake-reduction model of chromium(VI) toxicity. On that day, she was using dimethylmercury, $(CH_3)_2Hg$, as a standard for NMR. She took all the required precautions: lab coat, goggles, and disposable latex gloves. In view of its high vapor pressure, she handled the compound in a fume hood at all times. As she transferred the compound to the NMR tube, one or two drops of the liquid dripped from the pipette onto her left glove. Completing the task, she removed the gloves and thoroughly washed her hands.

Five months later, Wetterhahn found herself walking into walls and slurring her speech. She was admitted to hospital where the symptoms were matched to severe mercury poisoning. The only mercury incident that she could recall was the drips of dimethylmercury—but she had used gloves as required. She urged her colleagues to disseminate the news of the extreme toxicity of this compound. The chelation therapy administered had no effect. Three weeks later, she went into a coma and died on the June 8, 1997.

All of the material safety data sheets (MSDS) stated that gloves were needed when handling dimethylmercury, but they did not stipulate any type. No research had been done on the permeability of materials to dimethylmercury. Her colleagues found a testing laboratory. The results were horrifying. It took less than 15 seconds—and perhaps much less—for the dimethylmercury to pass through the glove. Other glove types were no better. Only a special laminated glove, SilverShield®, delayed the passage of the liquid for any significant length of time. Her colleagues rushed to warn scientists throughout the world of the hazard.

Dimethylmercury was known to be very toxic. In fact, the two British chemists who first synthesized the compound both died from mercury poisoning. However, no-one realized the extreme toxicity of the compound. As one chemist commented, if compounds were rated on a safety scale from 1 to 10, with 10 being the most toxic, we now realize dimethylmercury would rate a "15."

Laboratory chemistry is always accompanied by some degree of risk, as is crossing a highway. For chemists, the MSDS provide an awareness of the hazards of particular compounds and of classes of compounds. It is essential for any practicing chemist to read such safety information. Unfortunately for Wetterhahn, the extreme risk of this particular volatile compound had not previously been realized.

Alkyl zinc compounds are pyrophoric and are readily hydrolyzed in air. The alkyl cadmium compounds are less reactive. Alkyl mercury compounds are prepared by metathesis reactions between mercury(II) halides and a Grignard or organolithium compound.

$$2\ CH_3MgCl + HgCl_2 \rightarrow Hg(CH_3)_2 + MgCl_2$$

Dimethylmercury is stable to air oxidation. As already discussed, alkyl mercury compounds are versatile starting materials for the synthesis of the organometallic compounds of more electropositive metals.

22.5 Organometallic Compounds of the Transition Metals

Looking at the formulas of many transition metal organometallic compounds, one might expect their physical and chemical properties to be very similar to many of the coordination compounds. In fact, the properties of organometallics are much more "organic" in nature, as the contrast in the following shows.

Properties of typical coordination compounds	Properties of typical organometallic compounds
Water-soluble	Hydrocarbon soluble
Air-stable	Air sensitive
High melting solids ($>250°C$)	Low melting solids, or liquids

This difference in properties between transition metal complexes and transition metal compounds involving metal–carbon bonds can be explained in terms of bonding as will be discussed in the following section.

The 18-Electron Rule

As we have seen, main group organometallic compounds generally obey the "octet rule" and share their valence electrons to form σ bonds with the "organo" group. For example, tin forms the stable tetramethyltin, $SnMe_4$.

The *18-electron rule* for transition metal organometallics is based on a similar concept—the central transition metal ion can accommodate electrons in the d, s, and p orbitals, giving a maximum of 18. Thus, to the number of electrons in the outer electron set, a metal can add electron pairs from Lewis bases to bring the total up to 18. This rule is not satisfactory for "classical" transition metal complexes for, as we saw in Chapter 19, coordination chemistry is dominated by the presence of incompletely filled d orbitals. However, organometallic complexes of transition metals do, to a significant extent, obey the 18-electron rule.

The classic examples are the complexes in which carbon monoxide is the ligand. In Chapter 20, Section 20.8, for example, we mentioned tetracarbonylnickel(0), $Ni(CO)_4$, a compound used in the purification of nickel metal. In this compound, nickel has an oxidation state of zero. It is important to realize that, although in gaseous atoms the $4s$ level is filled before the $3d$, for example, $Ni = [Ar]4s^2 3d^8$, in a chemical environment, the $4s$ is always higher in energy; that is, $Ni^0 = [Ar]3d^{10}$.

As we discussed in Section 22.2, each carbon monoxide molecule is taken as a two-electron donor. Thus, the bonding of four carbon monoxide molecules would provide eight additional electrons, resulting in a total of 18.

$$
\begin{array}{ll}
\text{nickel(0) electrons } (3d^{10}) & = 10 \\
\text{carbon monoxide electrons} = 4 \times 2 = & \underline{8} \\
\text{total} = & 18
\end{array}
$$

Similarly, pentacarbonyliron(0), $Fe(CO)_5$, has iron in a zero oxidation state, giving eight d electrons, and a total of 10 electrons from the five CO groups. The total is once again 18, and this compound too is stable.

16-Electron Species

Most stable organometallic compounds obey the 18-electron rule. However, stable complexes do exist with electron counts other than 18, as factors such as crystal field stabilization energy and the nature of the bonding between the metal and the ligand affect the stability of the compound.

The most widely encountered exceptions to the rule are 16-electron complexes of the transition metals on the right-hand side of the d block, particularly Groups 9 and 10. These 16-electron, square-planar complexes commonly have d^8 electron configurations, for example Rh(I), Ir(I), Ni(II), and Pd(II). Examples of such complexes include the anion of *Zeise's salt*, $K^+[PtCl_3C_2H_4]^-$, and the iridium complex $IrCl(CO)(PPh_3)_2$, *Vaska's compound*. Crystal field stabilization energy favors low-spin square-planar d^8 configurations for large values of Δ. Values of Δ are larger in Periods 5 and 6 Consequently, there are many square-planar complexes of rhodium, iridium, palladium, and platinum. These square-planar complexes are all low spin (Chapter 19, Section 19.7). The d_{xy}, d_{xz}, d_{yz}, and d_{z^2} orbitals all contain two electrons, while the high-energy $d_{x^2-y^2}$ orbital remains empty. The greater the crystal field splitting, the more stable the complexes will be.

Odd Electron Species

Odd electron complexes may achieve stability by accepting an electron. For example, $V(CO)_6$ is a 17-electron species. It readily completes the 18-electron configuration by accepting an electron from a reducing agent.

$$
\begin{array}{lll}
V(CO)_6 & + \ Na \rightarrow Na^+ \ + & [V(CO)_6]^- \\
17 \ e^- & & 18 \ e^-
\end{array}
$$

Other odd electron species may acquire an additional electron by dimerizing with another molecule. For example, $Mn(CO)_5$ has 17 electrons. Two molecules "share" their odd electron in order to form a Mn—Mn bond. Consequently, each Mn becomes an 18-electron species.

$$
\begin{array}{ll}
2 \ Mn(CO)_5 \rightarrow & Mn_2(CO)_{10}. \\
17 \ e^- & 18 \ e^-
\end{array}
$$

Metal–Metal Bonding and the 18-Electron Rule

The 18-electron rule can be useful in predicting the number of metal–metal bonds in an organometallic compound which contains multiple metal atoms. Such a molecule will be most stable if the number of electrons around each metal atom is 18. As we have seen from the example discussed previously, the metal may gain additional electrons by forming covalent bonds to another metal atom. For example, if we examine the compound $[(\eta^5\text{-}C_5H_5)_2Mo(CO)_2]_2$, we see that for each molybdenum center there are 5 $4d$ electrons from the molybdenum(I), 6 electrons from the cyclopentadienyl ligand, and 4 electrons from the carbonyl ligands. This gives a total of 15 electrons per molybdenum, which is 3 short of 18. This deficit is made up

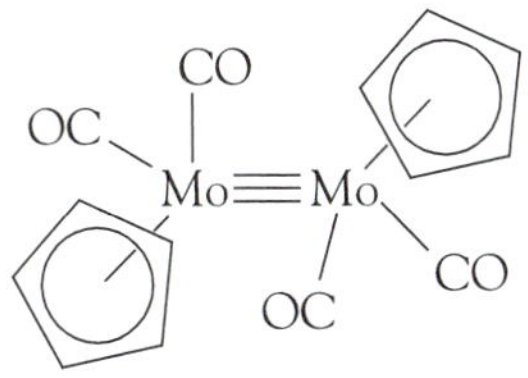

Figure 22.14 Structure of $[(\eta^5\text{-}C_5H_5)_2Mo(CO)_2]_2$.

by forming 3 bonds to the other molybdenum, and they result in the structure shown in Figure 22.14.

$$
\begin{aligned}
\text{molybdenum(1) electrons } (4d^5) &= 5 \\
\text{cyclopentadienyl electrons} &= 6 \\
\text{carbon monoxide electrons} = 2 \times 2 &= 4 \\
\text{Mo--Mo shared electrons} &= \underline{3} \\
\text{total} &= 18
\end{aligned}
$$

22.6 Transition Metal Carbonyls

Transition metal carbonyls are the most important class of transition metal organometallic compounds. The classic σ-bonding ligands form complexes with both main group and transition metals. This is not true of carbon monoxide as a ligand. With the exceptions of borane carbonyl, H_3BCO, and the potassium carbonyl, $K_6(CO)_6$, the only known carbonyls are those of transition metals. In transition metal carbonyls the σ bonding is reinforced by additional π bonding that stabilizes the complexes and also stabilizes very low oxidation states of the metal. Many carbonyl compounds exist quite happily with the metal in a zero oxidation state, for example, hexacarbonylchromium(0), $Cr(CO)_6$. These very low oxidation states are not found with σ-bonding-only ligands such as water and ammonia.

Carbonyl compounds are volatile and toxic. Their toxicity arises from their interaction with hemoglobin in red blood cells. In hemoglobin, iron has an oxidation state of $+2$. There are four iron ions in each hemoglobin molecule, each iron ion being surrounded by a porphyrin unit (see Chapter 20, Section 20.6). Each hemoglobin molecule reacts with four molecules of dioxygen to form oxyhemoglobin. The bonding to the dioxygen molecules is weak and the oxygen can be released quite readily. However, because of the nature of the bonding between iron and carbon monoxide, which comprises both σ and π contributions, the carbonyl ligand bonds almost irreversibly to the iron of the hemoglobin, thus preventing it from carrying dioxygen molecules.

Bonding in Carbonyl Compounds

As mentioned previously, it is the nature of the bonding between carbon monoxide and a transition metal that makes carbon monoxide so lethal. This bonding is also the reason why so many transition metal carbonyl compounds exist, why they are so stable, why they can exist in low oxidation states, and why main group carbonyls are very rare. Thus, we need to look at the bonding between transition metals and carbon monoxide in some detail.

In Chapter 3, Section 3.5, we saw that we could represent the bonding in heteronuclear diatomic molecules by a molecular orbital diagram. As the effective nuclear charge differs between the constituent atoms, the orbital energies are lower for the atom with the higher effective nuclear charge. Figure 22.15 shows a simplified molecular orbital diagram for the carbon monoxide molecule (a more sophisticated representation would mix in some $2s$ component to the bonding).

For the carbon monoxide molecule, the highest energy occupied molecular orbital (HOMO) is a σ_{2p} orbital essentially derived from the high energy $2p$ carbon and oxygen atomic orbitals. We assume this orbital resembles a lone pair on the carbon atom. The lowest energy unoccupied molecular orbitals (LUMOs) are the π^*_{2p} antibonding orbitals. Again, the predominant

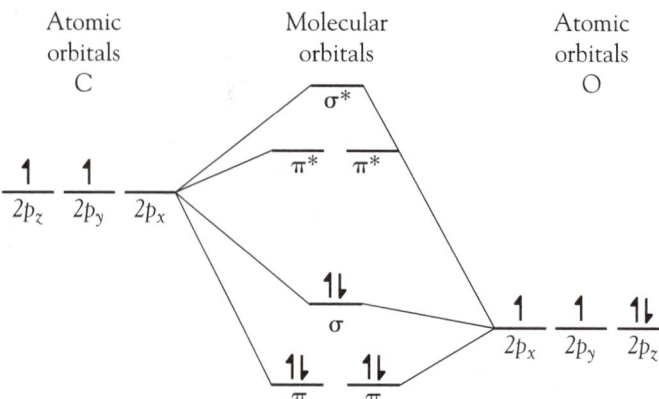

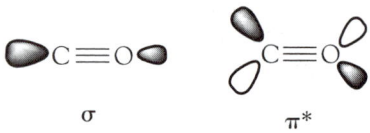

Figure 22.15 Partial simplified molecular orbital energy level diagram for carbon monoxide.

Figure 22.16 The highest energy occupied molecular orbital (σ) and the lowest energy unoccupied molecular orbital (π^*) for carbon monoxide.

contribution comes from the $2p$ atomic orbitals of carbon, so they, too, are focused around the carbon rather than around the oxygen atom. Approximate shapes of these orbitals are shown in Figure 22.16.

We can picture an overlap of the end of the σ HOMO of the carbon monoxide with an empty d orbital of the metal (Figure 22.17); that is, the carbon monoxide is acting as a Lewis base and donating a pair of electrons to the metal that acts as a Lewis acid.

This leads to high electron density on the metal. Imagine six ligands simultaneously donating electrons to a metal center, which, if it is in a low oxidation state, will already be electron rich. Simultaneously, there is an overlap of a full d orbital on the metal with the π^* LUMO of the carbon monoxide (Figure 22.18). These two orbitals have the correct symmetry to allow this interaction and thus the electron density is removed from the metal center back onto the carbonyl ligand to some extent.

This additional bond is a π bond. So, the carbon monoxide is said to be a σ-donor and a π-acceptor, and the metal is a σ-acceptor and a π-donor. Thus, there would be a flow of electrons from the carbon monoxide to the metal through the σ system and a flow through the π system in the reverse direction. This interaction is known as *back-bonding* or synergistic bonding. This *synergistic effect* leads to a strong, short, almost double, covalent bond between the metal and carbon atoms.

Figure 22.17 Sigma donation from the carbonyl ligand to the metal.

It is the removal of electron density from the metal by the carbonyl ligand that enables us to account for the stabilization of low oxidation states of transition metals. In a low or zero oxidation state, the metal would have a full, or nearly full, compliment of electrons even before the bonding of the carbonyl ligands. The even higher electron density which results from coordination is then effectively removed to the ligands by the synergistic effect.

According to this representation of the bonding, electrons would be "pumped" into the π antibonding orbital of the carbon monoxide. An increased occupancy of antibonding orbitals would lead to a reduction of the bond order below its value of 3 in the free carbon monoxide molecule. Experimental measurements have shown that, indeed, the carbon–oxygen bond in these carbonyl compounds is longer and weaker than that in carbon monoxide itself. This is good evidence for the validity of our molecular orbital bonding model. As added evidence, nearly all stable neutral metal carbonyl compounds are found in the middle transition metal groups (Groups 6 to 9), where the metal has some d electrons available for donation

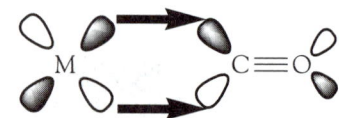

Figure 22.18 Pi "back-bonding" from the metal to the carbonyl ligand.

into the carbonyl π system but few enough that other d orbitals are empty and capable of accepting electron pairs from carbonyl ligands.

Other ligands can bond to transition metals in this way, for example, alkenes and phosphines, but they are not as good π acceptors as carbon monoxide.

Evidence for Synergistic Bonding

Infrared spectroscopy is a useful tool for investigating the structures of carbonyl compounds. The $C\equiv O$ bond vibrates around 2143 cm^{-1} in the gaseous state. The vibration frequency in carbonyl compounds lies within the range 2150–1850 cm^{-1}, depending on the structure and nature of other ligands present.

If the synergistic bonding model is valid, then we would expect the length and strength of the C—O bond to be affected as electrons are pushed into the π^* orbital. As the bond becomes weaker and longer it should vibrate at a lower frequency. (Think about "twanging" a tight elastic band. This is like a strong bond and vibrates at a high frequency. Now compare this to what happens when you twang a less taught elastic band. This represents the "weaker" bond and will vibrate at a lower frequency.) The fact that $C\equiv O$ bonds in carbonyl compounds do vibrate at lower frequencies than gaseous carbon monoxide does, indeed, support this model.

The C—O stretching frequency is very sensitive to the electronic environment around the metal. For example, a higher electron density on the metal will cause an increase in the extent of the "back-bonding" as more electron density is removed. This leads to an increase in the electron density in the π^* orbital, a lengthening of the bond, and the decrease is the C—O stretching frequency. This effect can be seen in the isoelectronic series:

$$Fe(CO)_4^{2-} \qquad 1790 \text{ cm}^{-1}$$

$$Co(CO)_4^{-} \qquad 1890 \text{ cm}^{-1}$$

$$Ni(CO)_4 \qquad 2060 \text{ cm}^{-1}$$

All of these species have eight d electrons. However, the effective nuclear charge increases from Fe to Ni (see Chapter 2, Section 2.5). This means that the *residual* negative charge of the metal is greater for iron than for nickel. As the negative charge on the complex increases, the metal has a higher electron density to be dispersed and so the back-bonding is increased. Consequently, the electron density in the π^* will be increased, and the bond between the carbon and oxygen will become weaker and vibrate at a lower frequency.

In substituted carbonyls, the extent of back bonding can be observed by examining CO frequencies.

$$Ni(CO)_3PMe_3 \qquad 2064 \text{ cm}^{-1}$$

$$Ni(CO)_3PPh_3 \qquad 2069 \text{ cm}^{-1}$$

$$Ni(CO)_3PF_3 \qquad 2111 \text{ cm}^{-1}$$

The methyl groups on PMe$_3$ show an inductive effect. An *inductive effect* is the shifting of electrons in a bond in response to the electronegativity of nearby atoms. In other words, a highly electronegative atom will tend to pull electron density toward itself, while an electropositive group will act as an electron source. In this case, electron density will be pushed toward the metal. This leads to an increase in the extent of back bonding, stronger M—C bonds, and a weaker C—O interaction, shown by the lower infrared stretching frequency. By contrast, fluorine in PF_3 is electron withdrawing and will remove electron density from the metal. This leads to a reduction in the extent of back-bonding required, and so the C—O bond remains stronger and shorter, indicated by the higher stretching frequency.

Types of Carbonyl Ligand

The carbonyl ligand may bond to transition metal centers in several ways. It may bond as a terminal ligand, as in simple complexes such as $Mo(CO)_6$, in which it acts as a two-electron donor. The carbonyl infrared stretching frequency for these groups is usually in the range $2010–1850$ cm^{-1}.

The carbonyl ligand may also bridge two or more metal centers (Figure 22.19). In this case, the carbonyl infrared stretching frequency is in the range $1850–1750$ cm^{-1}. A carbonyl which bridges two metals atoms is identified by the prefix μ or more precisely, μ^2, and donates one electron to each metal species. An example of a complex with bridging carbonyl groups is $Co_2(CO)_8$ (see Section 22.7).

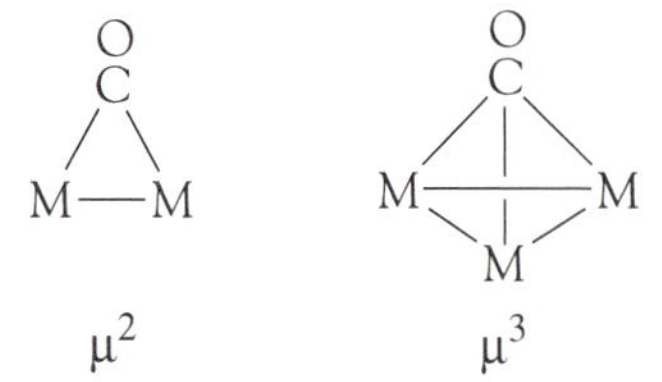

Some complexes have carbonyl ligands which bridge three metal centers and give carbonyl infrared stretching frequencies in the range $1675–1600$ cm^{-1}. These most often occur in "cluster" compounds and are given the prefix μ^3, for example, $Rh_6(CO)_6$.

Figure 22.19   Doubly and triply bridging carbonyl groups.

22.7   Synthesis and Properties of Simple Metal Carbonyls

Carbonyls of the Group 4 Elements

Titanium has a d^4 electron configuration. The 18-electron rule would suggest that the stable carbonyl would be $Ti(CO)_7$, but titanium has insufficient electron density for this to be formed. However, substituted carbonyls are known. For example, $(\eta^5\text{-}C_5H_5)_2Ti(CO)_2$ is a red, 18-electron compound which is prepared from titanium(IV) chloride.

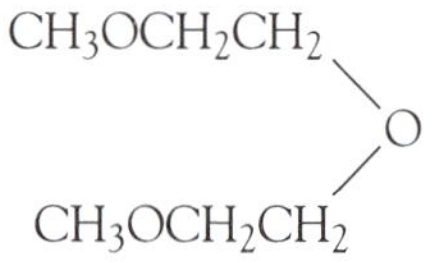

Figure 22.20   Structure of diglyme.

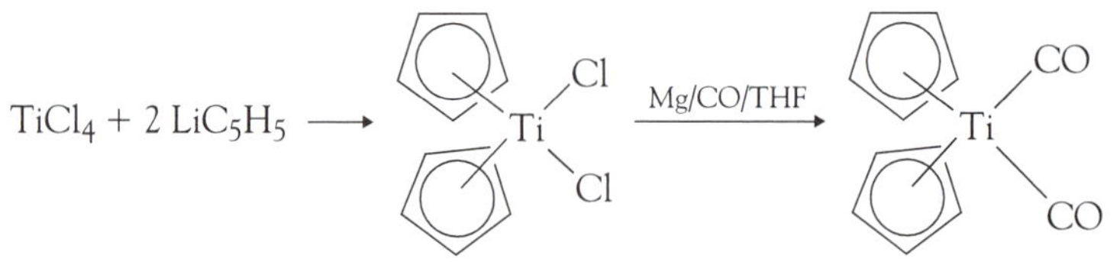

$$TiCl_4 + 2\,LiC_5H_5 \longrightarrow \quad \xrightarrow{Mg/CO/THF} \quad$$

Carbonyls of the Group 5 Elements

Hexacarbonylvanadium(0), $V(CO)_6$, is a green-black, paramagnetic solid. It is a 17-electron species which decomposes at $70°C$. It can be prepared from vanadium(III) chloride via an 18-electron anionic intermediate using 2-methoxyethyl ether, abbreviated to diglyme (from the common name diethyleneglycol dimethyl ether) as solvent (Figure 22.20).

$$VCl_3 + 4\,Na + CO \xrightarrow{\text{diglyme}} [Na(diglyme)_2]^+[V(CO)_6]^- \xrightarrow{H^+} V(CO)_6$$

Carbonyls of the Group 6 Elements

Hexacarbonylchromium(0), $Cr(CO)_6$, is a stable, 18-electron, octahedral molecule. It is synthesized in the same way as $V(CO)_6$. All three Group 6 hexacarbonyls, $Cr(CO)_6$, $Mo(CO)_6$ and $W(CO)_6$, are white crystalline solids which will sublime under vacuum. They are the most stable carbonyl compounds and do not react until heated.

Carbonyls of the Group 7 Elements

Manganese forms the 17-electron pentacarbonylmanganese(0), $Mn(CO)_5$. It readily dimerizes to give the 18-electron, yellow, crystalline $Mn_2(CO)_{10}$, as we mentioned in Section 22.5 (Figure 22.21).

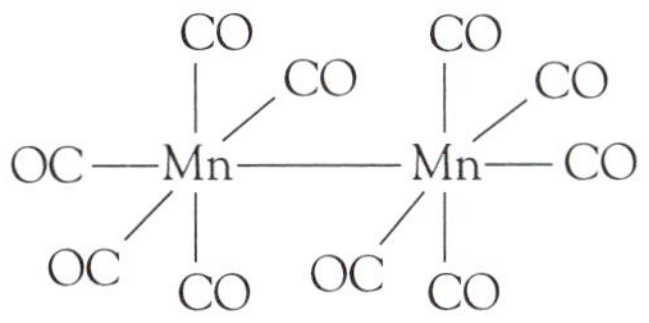

The Mn—Mn bond is long and weak and is easily broken. For example, the reaction with sodium amalgam leads to cleavage of the Mn—Mn bond and reduction of the manganese to manganese(−I).

$$Mn_2(CO)_{10} + 2\ Na \rightarrow 2\ Na[Mn(CO)_5]$$

The reaction with a halogen leads to cleavage of the bond and oxidation of the manganese to manganese(I).

$$Mn_2(CO)_{10} + Br_2 \rightarrow 2\ Mn(CO)_5Br$$

The corresponding carbonyls of the lower members of the group, $Tc_2(CO)_{10}$ and $Re_2(CO)_{10}$, are both white, crystalline solids.

Figure 22.21 Structure of $Mn_2(CO)_{10}$.

Carbonyls of the Group 8 Elements

Pentacarbonyliron(0), $Fe(CO)_5$, is a yellow, toxic liquid, which is used for making magnets and iron films. It can be prepared by heating finely divided iron under carbon monoxide. $Fe(CO)_5$ reacts photochemically to give the yellow dimer $Fe_2(CO)_9$. When heated, it forms the black solid, $Fe_3(CO)_{12}$.

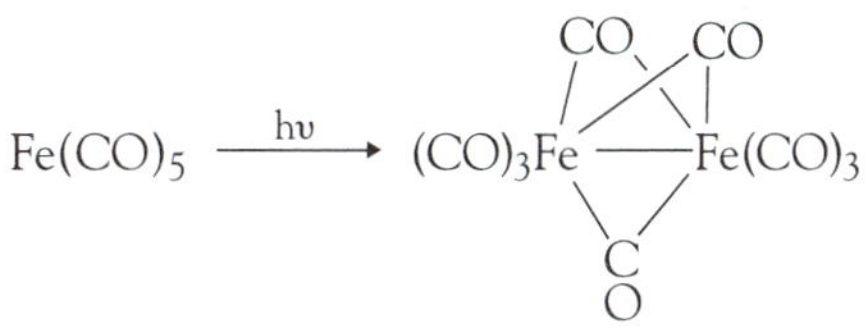

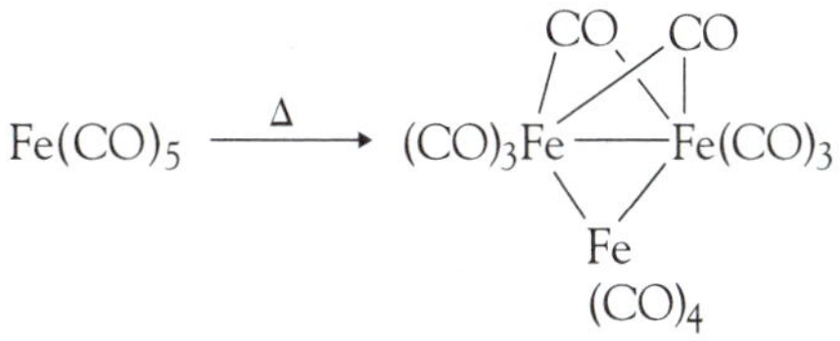

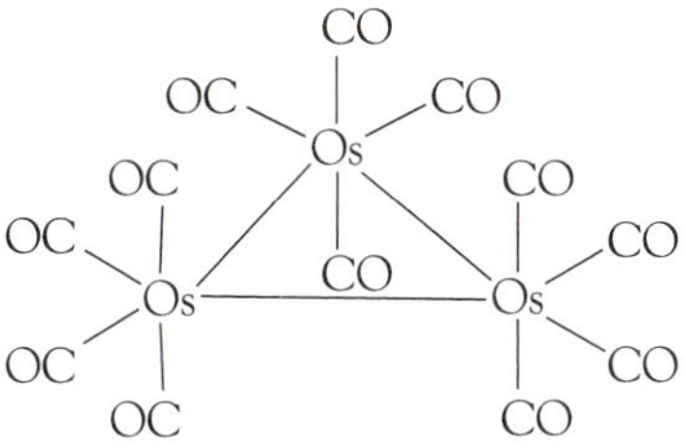

Ruthenium and osmium form $Ru(CO)_5$ and $Os(CO)_5$ which are both colorless liquids. They also form corresponding cluster compounds, $Ru_3(CO)_{12}$ and $Os_3(CO)_{12}$, although their structures are different from that of $Fe_3(CO)_{12}$. The structure of $Os_3(CO)_{12}$ is shown in Figure 22.22.

Figure 22.22 Representation of $Os_3(CO)_{12}$ cleavage of the metal–metal bond.

Carbonyls of the Group 9 Elements

Cobalt has an odd number of electrons, so the carbonyl compound, $Co(CO)_4$, has 17 electrons and dimerizes to give $Co_2(CO)_8$, which is an orange, low-melting solid. This compound is interesting because it exists in two isomeric forms. The solid contains a metal–metal bond and bridging carbonyl groups. When the solid is dissolved in hexane, the bridging carbonyl bands disappear from the infrared spectrum and the staggered structure is formed. The energy difference between these two forms is only approximately 5 kJ·mol^{-1}, so the interconversion occurs easily. Such interconversions and intramolecular rearrangements are common in organometallic chemistry.

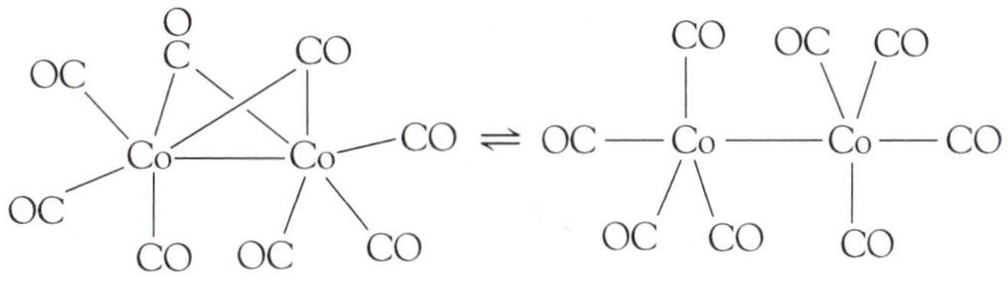

Cobalt also forms the cluster compounds $Co_4(CO)_{12}$ and $Co_6(CO)_{16}$ which are both black solids. Rhodium and iridium also form cluster compounds of the same formula. The structures are similar to those of the cobalt compounds. $Ir_6(CO)_{16}$ exists as a red isomer and a black isomer.

Carbonyls of the Group 10 Elements

Tetracarbonylnickel(0), $Ni(CO)_4$, is a toxic colorless liquid synthesized by direct interaction of carbon monoxide with the finely divided metal. The reaction takes place a little above room temperature and at atmospheric pressure. The reaction is the basis of the Mond process for extraction and purification of nickel (Chapter 20, Section 20.8).

$$Ni + 4\ CO \rightleftharpoons Ni(CO)_4$$

Carbonyls of the Group 11 Elements

Copper has very high electron density, and no vacant d orbitals. There are a few substituted carbonyl compounds known, but they are all very unstable.

22.8   Reactions of Transition Metal Carbonyls

The most important reactions of metal carbonyls are substitution reactions. Carbonyl ligands can be displaced by other ligands such as phosphines and unsaturated hydrocarbons. The substitution may be activated by either heat or light, and the products usually still obey the 18-electron rule. For example, for an octahedral complex such as $Co(CO)_6$, the reaction with another ligand may lead to a trisubstituted carbonyl.

Subsequent substitutions by the incoming ligand always happen *cis* to the initial ligand. This is because the substitution of a ligand which is a better σ donor but poorer π acceptor than carbonyl will lead to an increase in the back-bonding between the metal and the carbonyl ligand. The substitution reaction never proceeds further than $M(CO)_3L_3$ because the electron density on the metal would be too great.

These substitution reactions of 18-electron complexes follow a *dissociative mechanism*, which means that a species which is coordinatively unsaturated is produced as an intermediate. An associative mechanism would produce a 7-coordinate intermediate, a complex with more than 18 electrons.

$$\begin{array}{ccc} M(CO)_6 \rightarrow & M(CO)_5 \rightarrow & M(CO)_5L \\ 18\ e^- & 16\ e^- & 18\ e^- \end{array}$$

Complexes of Period 5 metals react much more quickly than those of Period 6. This leads to the fact that Period 5 elements, such as ruthenium, rhodium, and palladium, are much more widely used in catalysis than those of Period 6, as we will see later in the chapter. Some typical substitution reactions are given in the following:

$$Cr(CO)_6 + 3\ MeCN \rightarrow Cr(CO)_3(NCMe)_3 + 3\ CO$$

$$Ni(CO)_4 + 2\ PF_3 \rightarrow Ni(CO)_2(PF_3)_2 + 2\ CO$$

$$Mo(CO)_6 + C_6H_6 \rightarrow Mo(CO)_3(\eta^6\text{-}C_6H_6) + 3\ CO$$

$$Mo(CO)_6 + CH_2{=}CHCH{=}CH_2 \rightarrow (CO)_4Mo \,]\!\!= + 2\ CO$$

Another important type of reaction is the formation of carbonylate anions by reaction with reducing agents or alkali. The species which form anions most readily are odd electron species, particularly 17-electron species, and dimers. The species formed are 18-electron complexes. For example,

$$Fe(CO)_5 + 3\ NaOH \xrightarrow{H_2O} Na[HFe(CO)_4] + Na_2CO_3 + H_2O$$

$$Co_2(CO)_8 + 2\ Na/Hg \xrightarrow{THF} 2\ Na[Co(CO)_4]$$

A related reaction of dimers is the formation of halides by reaction with a halogen and cleavage of the metal–metal bond.

$$Mn_2(CO)_{10} + Br_2 \longrightarrow 2\ Mn(CO)_5Br$$

22.9 *Other Carbonyl Compounds*

Metal Carbonyl Anions

As we have already mentioned, the reduction of metal carbonyls gives rise to anionic species which are very reactive. These are most often prepared by reacting a carbonyl compound with an alkali metal or sodium borohydride.

$$Cr(CO)_6 \xrightarrow{NaBH_4} Na_2[Cr_2(CO)_{10}]$$

$$Mn_2(CO)_{10} \xrightarrow{Li} Li[Mn(CO)_5]$$

Some of these compounds form hydrides, such as $HMn(CO)_5$, when acidified. Manganese, iron, and cobalt carbonyl hydrides are colorless or yellow liquids. Studies have shown that the hydrogen is directly attached to the metal. The $HMn(CO)_5$ and $H_2Fe(CO)_4$ compounds are quite acidic in aqueous solution. The cobalt compound, $HCo(CO)_4$, is insoluble in water but is a strong acid in methanol.

Metal Carbonyl Hydrides

Metal carbonyl hydrides are highly reactive, and most reactions involve the insertion of another species into the M—H bond. Some typical reactions of metal hydrides are given in the following:

$$(CO)_5MnH + CH_2{=}CH_2 \longrightarrow (CO)_5MnCH_2CH_3$$

$$(CO)_5MnH + CO_2 \longrightarrow (CO)_5MnCO_2H$$

$$(CO)_5MnH + CH_2N_2 \longrightarrow (CO)_5MnCH_3 + N_2$$

Metal Carbonyl Halides

Most metals that form stable carbonyls also form carbonyl halides. Their structures are analogous to the carbonyl for monomeric species. Dimers are always bridged through the halide rather than through the carbonyl. They are usually white or yellow solids and are made by the reaction of a halogen with a metal carbonyl at high temperatures and pressures. For example,

$$Fe(CO)_5 + I_2 \longrightarrow Fe(CO)_4I_2 + CO$$

They are soluble in organic solvents but decompose in water. Most of the carbonyl halides obey the 18-electron rule. The most notable exception is Vaska's compound, *trans*-$(Ph_3P)_2Ir(CO)Cl$. This compound undergoes a wide range of addition reactions during which the metal atom achieves 18 electrons in the valence shell. During these reactions, the square-planar compound is converted to an octahedral one and the iridium is oxidized by 2 charge units. This type of reaction is known as *oxidative addition*. The scheme for the oxidative addition of Vaska's complex with hydrogen is shown in the following. Note the 3-centered intermediate formed between the metal and the H_2.

Oxidative addition is a common reaction of organometallic compounds and is key to many catalytic reactions. For an oxidative addition reaction to take place, there must be two spare sites for coordination of incoming ligands and the metal must be able to exist in stable oxidation states separated by 2 units. During the reaction, two ligands are associated to the metal which is simultaneously oxidized by 2 units. The reverse of the reaction is, not surprisingly, called *reductive elimination*.

22.10   *Complexes with Phosphine Ligands*

We have already seen that metal carbonyl complexes may react with ligands such as triphenyl phosphine and phosphorus trichloride. Phosphine ligands are so important that they are worth a whole section for discussion.

Phosphines are able to accept some electron density through back-bonding via the P—C (antibonding) σ^* orbital. The extent of this back-bonding depends upon the nature of the ligand. For example, alkyl phosphines, such as $P(CH_3)_3$, are strong electron donors but fairly weak electron acceptors, due to the inductive effect of the alkyl group. Conversely, the phosphine halides are weak donors but strong electron acceptors, due to the electron-withdrawing properties of the halide atoms. π-Acidity is the term used to describe the ability of a species to accept electron density through a π-type overlap of orbitals. The order of increasing π acidity is given by

$$P(^tBu)_3 < P(Me)_3 < P(OMe)_3 < PCl_3 < CO$$

So, the stability of a phosphine-containing species is affected by the electronic characteristics of the phosphine. A further important factor in determining the stability and structure of phosphine-containing ligands is the shape and size of the ligand, the steric bulk. The "bulkiness" of a ligand is defined by the *Tolman cone angle* of the ligand. This is shown in Figure 22.23.

It can be seen that a small, compact, substituted ligand gives a small Tolman cone angle, whereas a large, bulky, substituted phosphine gives a large cone angle. Some examples are given in Table 22.4.

With the crowding resulting from bulky ligands, it is not surprising that a ligand can sometimes be expelled. For example, tetra(triphenylphosphine)platinum(0), $Pt(PPh_3)_4$, readily loses a ligand to form tris(triphenylphosphine)platinum(0), $Pt(PPh_3)_3$. These two factors, the amount of electron density on the metal and the bulkiness of the ligand, determine the reactivity and the coordination number of the complex.

Table 22.4   Tolman cone angles for phosphine ligands

Ligand	θ/deg
PH_3	87
PF_3	104
PMe_3	118
PMe_2Ph	122
$PMePh_2$	136
PPh_3	145
P^tBu_3	183

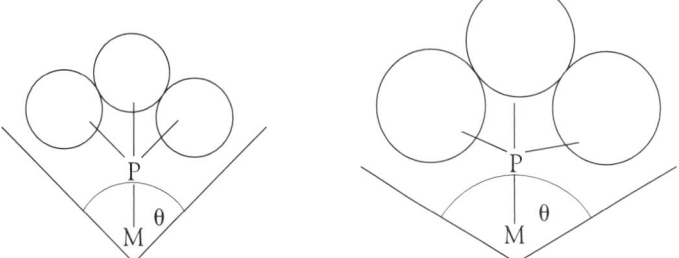

Figure 22.23   The Tolman cone angle for a small and a bulky phosphine ligand.

22.11   Complexes with Alkyl, Alkene, and Alkyne Ligands

The making and breaking of M—C bonds play an important role in organometallic chemistry and are central to its application in catalysis. Whenever alkanes, alkenes, or alkynes are generated, polymerized, or functionalized, metal alkyl intermediates are involved. About 75 percent of all products produced by the chemical industry pass through a catalytic cycle involving an organometallic catalyst at some stage.

Transition metals form simple alkyls, but, with the exception of zinc and mercury, they are unstable. Most stable organometallic compounds are formed with alkenes, alkynes, and unsaturated ring systems. The first truly organometallic compound was the yellow, crystalline Zeise salt, $K[PtCl_3(C_2H_4)]$, which was discovered in 1830.

Transition metal organometallic compounds with metal–alkyl bonds are known with both σ and π interactions. In general, most compounds obey the 18-electron rule with the exception of alkyl groups which are only σ bonded to the metal. Some of these, such as Zeise's salt mentioned previously, have only 16 electrons. They are square planar and are said to be *coordinatively unsaturated,* which means that they can accept other ligands to give a maximum of six-coordination. Such coordinative unsaturation is an extremely important feature of transition metal homogeneous catalysis. Simple σ-bonding alkyls can be understood by analogy with metal halides and hydrides. So a methyl ligand will be considered to have a formal charge of -1 and donate two electrons to the metal.

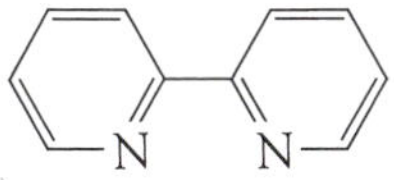

Alkyls that contain no other groups are very unstable. For example, $Ti(CH_3)_4$ decomposes at $-50°C$, whereas $Ti(bipy)(CH_3)_4$ (*bipy* is the abbreviation for bipyridine, Figure 22.24) can be warmed to $30°C$. Alkyls are stabilized by the presence of π-bonding ligands such as bipyridine, carbonyl, and triphenylphosphine, which results in an increase in stability as the electron density can be removed from the metal.

Figure 22.24   Bipyridine.

Synthesis of Transition Metal Alkyls

The most widely used method for preparing transition metal alkyls is alkylation, often using a Grignard reagent or a lithium alkyl:

$$(C_5H_5)_2MoCl_2 + 2\ CH_3Li \longrightarrow (C_5H_5)_2Mo(CH_3)_2 + 2\ LiCl$$

$$(R_3P)_2PtCl_2 + LiCH_2CH_2CH_2CH_2Li \longrightarrow \quad + 2\ LiCl$$

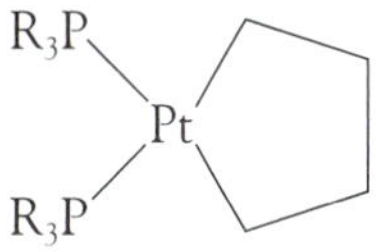

Low-valent complexes, especially Ir^0, Ni^0, Pd^0 and Pt^0 stabilized by phosphines, may be synthesised by oxidative addition with an alkyl halide. In this case, we have a coordinatively unsaturated, square-planar complex containing labile ligands which is simultaneously oxidized and has the coordination number expanded to 6. For example, for Vaska's compound:

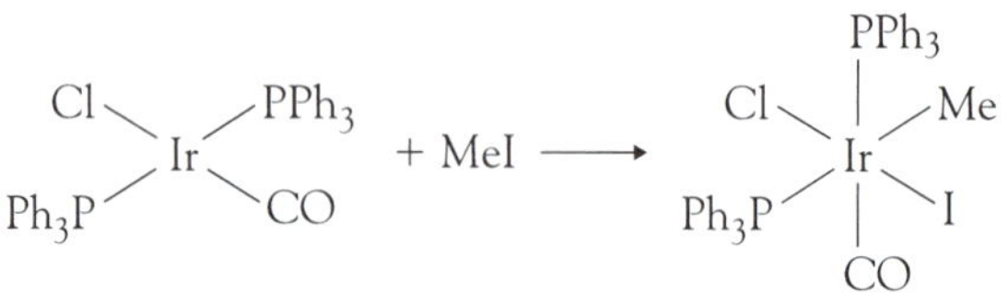

Complexes of alkenes, alkynes, and polyconjugated systems are prepared when other ligands, often carbonyl ligand, are displaced. The 18-electron

The Preservation of Books

Most inexpensive paper, such as newsprint, discolors and rots from reactions that produce acid within the fibers of the paper. For archivists, the rotting and decay of rare books, manuscripts, and old newspapers is a worrisome problem. In recent years, there have been many attempts to find a means of preserving large archives by a low-cost route that would not damage the paper or the ink.

The most promising solution utilizes diethylzinc, $Zn(C_2H_5)_2$, first synthesized by Edward Frankland in 1849. In the preservation process used by the Library of Congress, up to 9000 books are placed in a chamber. The air is pumped out and the chamber refilled with pure nitrogen gas under low pressure. It is essential to remove all dioxygen from the chamber because diethylzinc is highly flammable:

$$Zn(C_2H_5)_2(g) + 7\ O_2(g) \rightarrow ZnO(s) + 4\ CO_2(g) + 5\ H_2O(l)$$

Next, diethylzinc vapor is pumped into the chamber, permeating the pages of the books. There it reacts with any hydrogen ions to give zinc ions and ethane gas:

$$Zn(C_2H_5)_2(g) + 2\ H^+(aq) \rightarrow Zn^{2+}(aq) + 2\ C_2H_6(g)$$

The compound also reacts with any moisture in the paper to form zinc oxide:

$$Zn(C_2H_5)_2(g) + H_2O(l) \rightarrow ZnO(s) + 2\ C_2H_6(g)$$

The zinc oxide, being a basic oxide, serves as a reserve of alkalinity in case any more acid is produced by continued rotting of the paper.

The excess diethylzinc and the ethane formed during the reaction are pumped out, and the chamber is flushed with dinitrogen and then air, after which the books can be removed. This procedure takes from 3 to 5 days for each batch of books, a slow process, but one that results in the survival of many precious documents.

rule can be used to predict the formulas of the products as each π bond can replace a lone pair donated by another ligand.

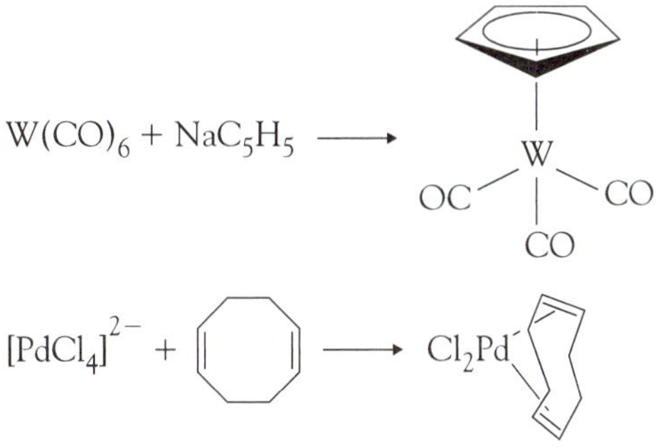

$$W(CO)_6 + NaC_5H_5 \longrightarrow$$

$$[PdCl_4]^{2-} + \quad \longrightarrow \quad Cl_2Pd$$

Another common method of synthesis is to insert a molecule into an M—H bond. For example,

$$PtCl(PPh_3)_2H + C_2H_4 \longrightarrow PtCl(PPh_3)_2C_2H_5$$

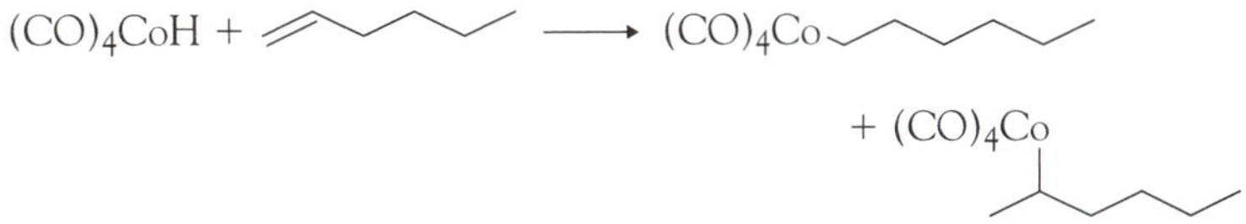

$$(CO)_4CoH + \quad \longrightarrow \quad (CO)_4Co$$

$$+ (CO)_4Co$$

Polar, unsaturated ligands (for example, carbonyl) are susceptible to nucleophilic attack. A *nucleophile* is a reagent that is "nucleus-loving," having electron-rich sites that can form a bond by donating an electron pair to an electron-poor site. Nucleophiles are often, though not always, negatively charged. The converse is an *electrophile,* an electron-poor species that will accept an electron pair, sometimes a positively charged species. This pair of definitions might sound familiar (see Chapter 7, Section 7.8, Lewis Theory). Lewis bases are electron donors and usually behave as nucleophiles, whereas Lewis acids are electron acceptors and usually behave as electrophiles. The main difference is that the terms nucleophile and electrophile are normally used in the context of bonds to carbon. In the following reaction, the ethyl anion, $C_2H_5^-$, is the nucleophile, and the carbon of a carbonyl is the electrophile.

$$Fe(CO)_5 + LiC_2H_5 \longrightarrow Li^+[(CO)_4Fe\overset{\displaystyle O}{\overset{\|}{C}}C_2H_5]^-$$

As with the carbonyl complexes, most transition metal alkyls obey the 18-electron rule (Figure 22.25), or have 16 electrons for the square-planar d^8 system.

On the other hand, there are numerous thermally stable complexes for which the stability cannot be explained on the basis of the 18-electron rule. These species often have very low numbers of electrons and owe their stability to the kinetic stabilization provided by sterically demanding, bulky ligands. These ligands are often too bulky to allow the metal to bind to enough donor ligands in order to achieve the higher electron count but provide an effective "umbrella," shielding the metal from incoming ligands. Examples are found mainly among early transition metals (Figure 22.26).

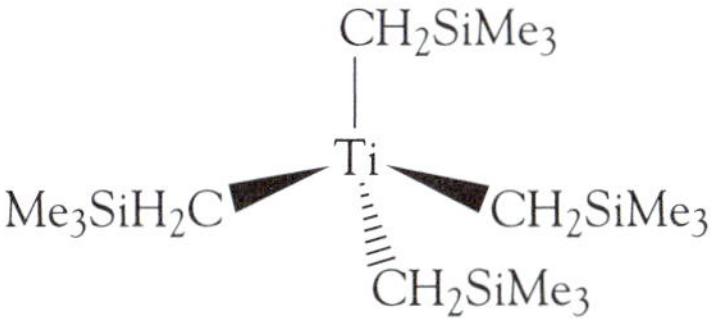

Metal alkyls, alkenes, and alkynes are highly reactive. The reactions generally involve M—C bond cleavage and insertion reactions. Simple diatomic molecules such as halogens and hydrogen promote bond cleavage reactions

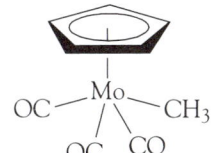

Figure 22.25 Structure of the 18-electron compound, $(\eta^5\text{-}C_5H_5)Mo(CO)_3(CH_3)$.

$$\begin{array}{c} CH_2SiMe_3 \\ | \\ Me_3SiH_2C \quad \text{——} \quad Ti \quad \text{——} \quad CH_2SiMe_3 \\ | \\ CH_2SiMe_3 \end{array}$$

Figure 22.26 Structure of the 8-electron compound, $Ti(CH_2SiMe_3)_4$.

and lead to the formation of halides and hydride species.

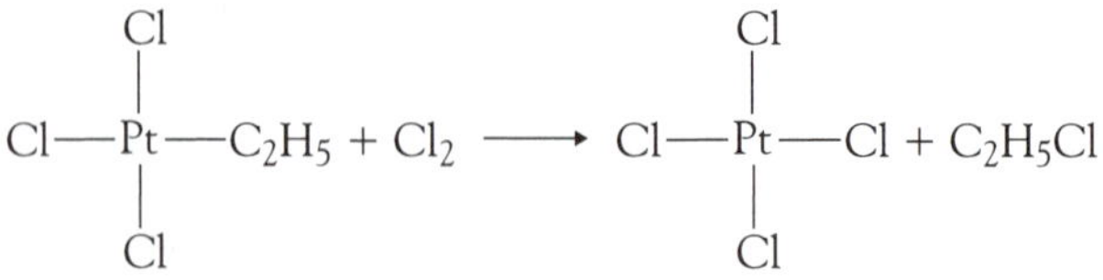

Another common reaction is the insertion of carbon monoxide into a metal–carbon bond to form a metal acyl derivative. This reaction is very important industrially.

$$C_2H_5Mn(CO)_5 + CO \rightarrow C_2H_5COMn(CO)_5$$

The mechanism of this reaction is not as straightforward as it might seem at first glance. The inserted carbonyl group is actually one of those originally coordinated to the metal center. The reaction takes place by intramolecular nucleophilic attack followed by alkyl migration.

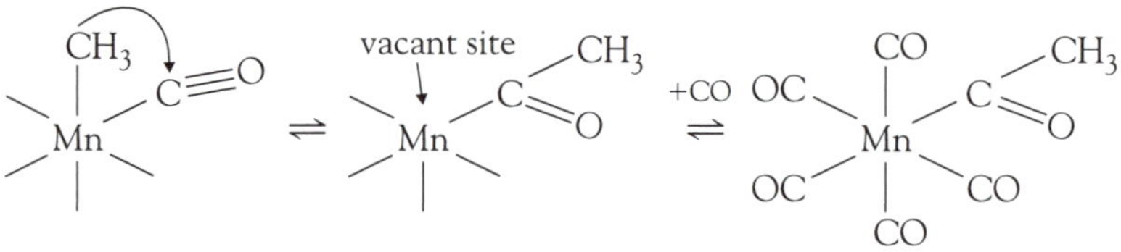

Alkenes and alkynes may also insert between the metal-carbon bond.

$$(C_5H_5)(CH_2CH_2)NiR \xrightarrow[\text{5 MPa}]{CH_2CH_2} (C_5H_5)(CH_2CH_2)NiCH_2CH_2R$$

An example of this which has enormous industrial importance is the Ziegler–Natta catalyst used for alkene polymerisation (see Section 22.15).

22.12 Complexes with Allyl and 1,3-Butadiene Ligands

Propenyl species may bond to transition metals via the terminal carbon atom or, more usually, through the delocalized allyl system, forming an η^3 complex (Figure 22.27).

η^3-Allyl-containing molecules may be formed via a σ-bonded η^1-intermediate followed by expulsion of another ligand, often carbon monoxide. For example,

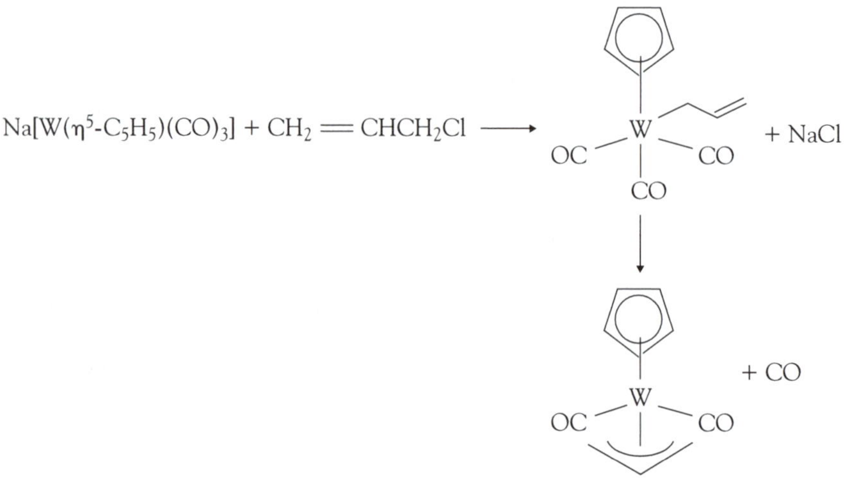

$$M\!-\!CH_2CH\!=\!CH_3$$

2 electron donor

(a)

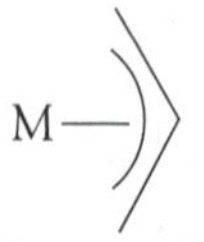

$$M\!-\!\text{(allyl)}$$

4 electron donor

(b)

Figure 22.27 A propenyl species bonding via (a) the terminal carbon atom and (b) the delocalized allyl system.

$$Na[W(\eta^5\text{-}C_5H_5)(CO)_3] + CH_2\!=\!CHCH_2Cl \longrightarrow$$

+ NaCl

+ CO

The η^3-complexes may be prepared in a variety of ways, including by the deprotonation of a coordinated propene ligand and by the protonation of coordinated 1,3-butadiene ligands as shown in the following:

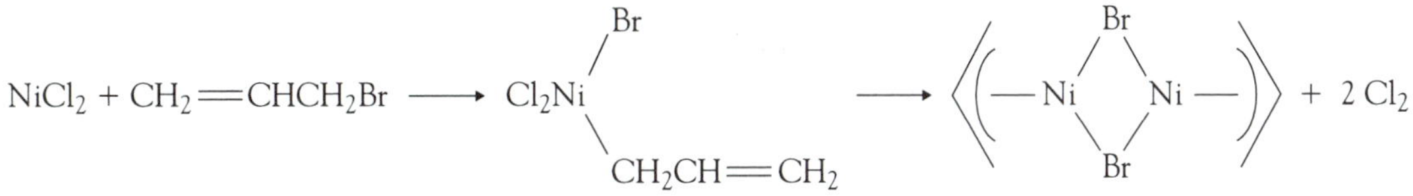

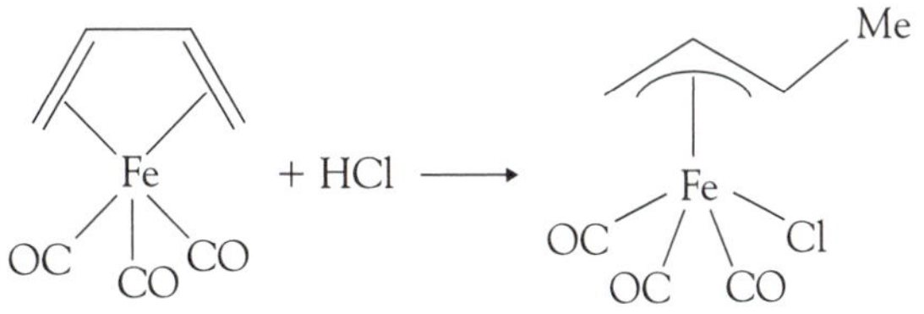

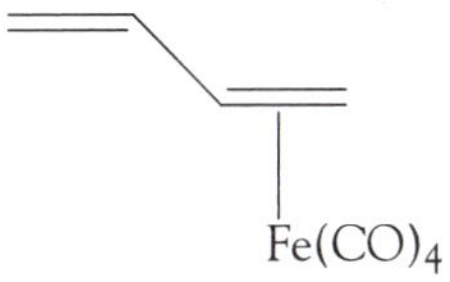

$$NiCl_2 + CH_2=CHCH_2Br \longrightarrow Cl_2Ni\underset{CH_2CH=CH_2}{\overset{Br}{\diagup}} \longrightarrow \left(\!\!\left(-Ni\underset{Br}{\overset{Br}{\diagdown\diagup}}Ni-\right)\!\!\right) + 2\,Cl_2$$

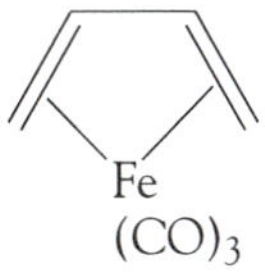

1,3-Butadiene ligands may bond to transition metals via either one or both π bonds, thereby donating either two or four electrons. The number of other coordinating ligands reduces accordingly. The most important of these compounds are the iron carbonyl derivatives (Figure 22.28).

The iron tricarbonyl-1,3-diene derivatives are important in organic synthesis. The coordinated diene is difficult to hydrogenate and does not undergo the classical organic *Diels–Alder reactions*, typical of 1,3-dienes. The $Fe(CO)_3$ group acts as a protecting group for the diene, preventing additions to the double bonds and allowing reactions to be carried out on other parts of the molecules. In the example shown, the iron tricarbonyl is used to protect two $C=C$ bonds against hydrogenation while it takes place readily at the third $C=C$ bond.

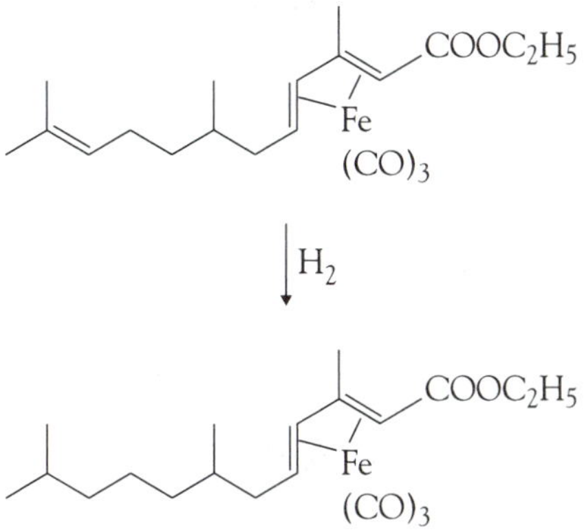

Figure 22.28 Structure of (a) $Fe(CO)_4(\eta^2\text{-}CH_2CHCHCH_2)$ and (b) $Fe(CO)_3(\eta^4\text{-}CH_2CHCHCH_2)$.

22.13 Metallocenes

Metallocenes are sandwich compounds in which the metal center lies between two π-bonded η^5-cyclopentadienyl rings. By far, the most important of these is ferrocene, $(\eta^5\text{-}C_5H_5)_2Fe$, the structure of which was shown in Figure 22.2. The synthesis of ferrocene in 1951 by Kealy and Pauson was one of the greatest chemical discoveries of the 20th century, and it stimulated a great increase in interest in organometallic chemistry.

Ferrocene is a paramagnetic, orange solid with a melting point of 174°C. It is a very stable compound and can be heated to 400°C without decomposing. In the gas phase, the two cyclopentadienyl rings are eclipsed but, in the solid phase, several structures exist with different orientations of the rings. However, rotation of the rings occurs, even in the solid phase at 25°C, and consequently all the hydrogen atoms appear equivalent. Ferrocene is available commercially, and a large number of derivatives can be formed. The cyclopentadiene rings are aromatic and can be derivatized in many ways (that is, substitution reactions can be performed on the rings themselves). A selection of reactions is shown in the following:

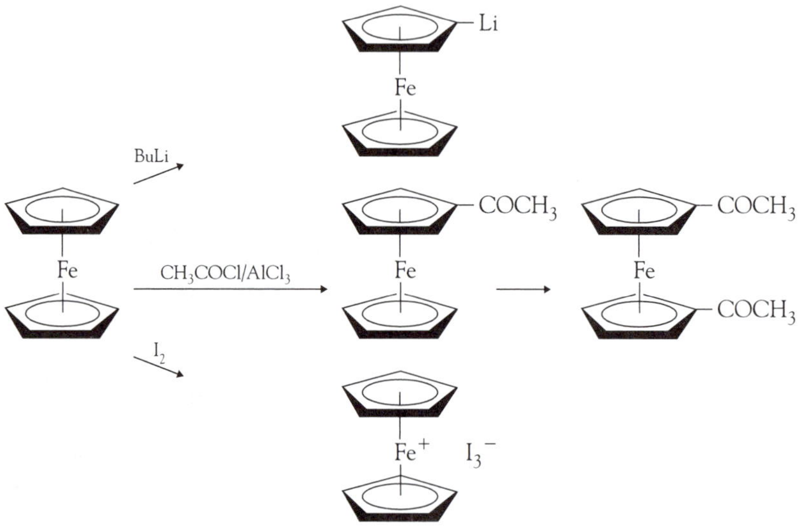

Uses of Ferrocene

Ferrocene and ferrocene-related compounds find many different applications because of the interesting redox properties of the iron, the fact that ferrocene acts as an aromatic compound, and the fact that the cyclopentadienyl rings can be derivatized.

The compound has found uses as a fuel additive. Its addition to various fuels, such as diesel, leads to smoke reduction and increased fuel economy. Ferrocene is also used in the formulation of high-grade lead-free fuels, being one of the compounds that has replaced tetraethyllead as an antiknock agent. The additive improves the performance of these fuels, because the combustion of ferrocene produces iron ions which react with oxygen to give iron oxides, promoters of the hydrocarbon combustion reaction.

Ferrocene compounds have been developed as electron transfer catalysts in the formation of compounds with specific magnetic and conducting properties. It is the redox properties of the ferrocene that make these applications possible. Another application arising from the redox properties is the use of ferrocene derivatives as a molecular switch.

Ferrocene derivatives are also used in biosensors. One such example uses vinyl ferrocene cross-linked with an acrylamide monomer to form a conducting polymer gel. Enzymes may be trapped in the gel, and then this is used in sensors for determining amounts of the enzyme in solution.

Other Metallocenes

Metallocenes are known for first row transition metals: vanadium(II), chromium(II), manganese(II), cobalt(II), and nickel(II). With the exception of vanadium where the starting compound is vanadium(III) chloride, they can all be prepared by the following reaction:

$$MCl_2 + 2\,Na[C_5H_5] \rightarrow (\eta^5\text{-}C_5H_5)_2M + 2\,NaCl$$

Unlike ferrocene, most of the other metallocenes are air sensitive or pyrophoric as they are not 18-electron systems:

$(\eta^5\text{-}C_5H_5)_2V$ is an air-sensitive, violet solid.

$(\eta^5\text{-}C_5H_5)_2Cr$ is an air-sensitive red solid.

$(\eta^5\text{-}C_5H_5)_2Mn$ is a pyrophoric brown solid. At room temperature manganocene is polymeric, while at higher temperatures its structure is related to that of ferrocene.

$(\eta^5\text{-}C_5H_5)_2Co$ is an air-sensitive, black solid which has 19 electrons and is easily oxidized to $[(\eta^5\text{-}C_5H_5)_2Co]^+$.

$(\eta^5\text{-}C_5H_5)_2Ni$ is a green solid which has 20 electrons. The reactions of nickelocene result in an 18-electron species.

22.14 *Complexes with η^6-Arene Ligands*

Species such as benzene or toluene can act as 6-electron donors. Complexes containing these ligands can be prepared from a carbonyl or substituted carbonyl. For example,

$$Cr(CO)_6 + C_6H_6 \rightarrow (\eta^6\text{-}C_6H_6)Cr(CO)_3 + 3\,CO$$

Compounds of this type, with just one ring system, are often referred to as *half-sandwich compounds*. The arene ring in these compounds can be subjected to *lithiation* (that is, the substitution of a lithium atom for a hydrogen atom) and may then take part in many reactions. The electron-withdrawing nature of the carbonyl groups means that the arene ring is much more reactive than it would be normally.

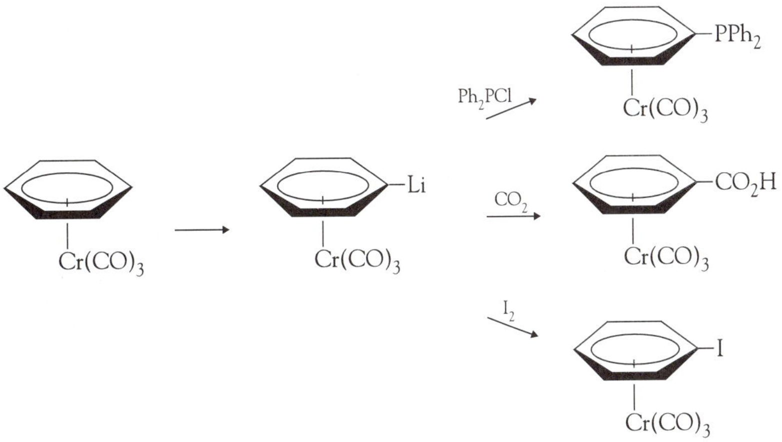

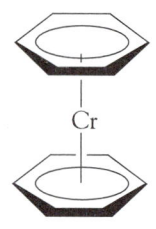

Figure 22.29 The sandwich compound, dibenzenechromium(0), $(\eta^6\text{-}C_6H_6)_2Cr$.

The carbonyl ligands may also be susceptible to substitution by other ligands, for example, phosphines:

$$(\eta^6\text{-}C_6H_6)Cr(CO)_3 + PPh_3 \rightarrow (\eta^6\text{-}C_6H_6)Cr(CO)_2PPh_3 + CO$$

Sandwich compounds also exist and can be prepared by co-condensing the metal and arene vapors.

$$Cr + 2\ C_6H_6 \rightarrow (\eta^6\text{-}C_6H_6)_2Cr$$

Chromium, molybdenum, and tungsten form air-sensitive 18-electron complexes. In the solid state the two benzene rings are eclipsed and the C—C bond lengths are slightly longer than in benzene (Figure 22.29).

22.15   Complexes with Cycloheptatriene and Cyclooctatetraene Ligands

If cycloheptatriene is reacted with hexacarbonylchromium(0), $Cr(CO)_6$, it will replace three carbonyl ligands and bond to the metal as a 6-electron donor via the three π bonds. It bonds as a classic η^6 triene with the molecule being folded with the CH_2 group directed away from the metal. Under some conditions the cycloheptatriene gives the tropylium or cycloheptatrienylium cation, $C_7H_7^+$, which is aromatic and bonds through all the seven carbon atoms. In this case, the ligand is planar, a 6-electron donor like cycloheptatriene but bonds as a η^7-species. In this situation all of the C—C bond lengths are equal, unlike cycloheptatriene. Under other conditions, a proton may be removed from the cycloheptatriene species giving $C_7H_7^-$, the cycloheptatrienyl anion, which may then act as an 8-electron donor (Figure 22.30).

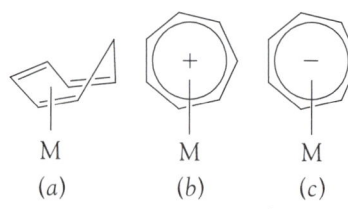

Figure 22.30 The (a) η^6-cycloheptatriene, C_7H_7, (b) η^7-cycloheptatrienylium, $C_7H_7^+$, and (c) η^7-cycloheptatrienyl, $C_7H_7^-$, ligands.

Cyclooctatetraene (Figure 22.31) is a large ligand which may bond as an η^2, η^4, η^6, or an η^8 species. In the case of η^2, η^4, and η^6, the ring is puckered. In the η^8 species, the ring is planar and is best considered to exist as the $C_8H_8^{2-}$ group. Cyclooctatetraene complexes are best prepared from the photochemical reactions of carbonyl compounds.

$$Fe(CO)_5 + C_8H_8 \rightarrow (\eta^4\text{-}C_8H_8)Fe(CO)_3 + 2\ CO$$

22.16   Fluxionality

Figure 22.31 Cyclooctatetraene, C_8H_8.

One of the remarkable features of many complexes with cyclic polyene ligands is their structural nonrigidity. For example the two rings in ferrocene rotate rapidly relative to each other. This form of *fluxionality* is called internal rotation.

More interesting is the fluxionality that is observed when a poly-ene ligand is attached to the metal by some, but not all, of the carbon atoms. In these cases the metal–carbon interaction may hop around the ring. This is known as *ring whizzing*. ^1H NMR spectroscopy provides evidence of this process. For example, at room temperature the NMR spectrum of $(\eta^4\text{-}C_8H_8)Ru(CO)_3$ consists of a single sharp peak, which might suggest that the ligand is attached by all eight carbons. However, when the temperature is reduced, the signal broadens to give four peaks, which would be expected from the four different hydrogen environments in the η^4-bonded ligand. At room temperature, the "ring whizzing" occurs too quickly for the NMR experiment to observe the individual environments so an averaged signal is recorded. However, at low temperatures the ring motion is slowed down, and so the NMR experiment can "see" each hydrogen atom environment.

22.17 *Organometallic Compounds in Industrial Catalysis*

A catalyst increases the rate, and sometimes the selectivity, of a reaction without itself being consumed during the reaction. Catalysts are widely employed in nature and in industry, and many of the world's bulk chemicals are produced via catalysis. Catalysts play an important role in the production of organic chemicals and petrochemical products and often provide routes to cleaner technology. In addition to playing an important economic and environmental role, catalyzed reactions are interesting because the exact reaction pathways are by no means certain, and there is still scope for much further research. Organometallic compounds are crucial to many of the more important catalytic processes. Some of these are described here.

Acetic Acid Synthesis: The Monsanto Process

The age-old method for producing acetic acid is by fermentation of ethanol, which produces vinegar. However, this is an inefficient way to produce concentrated acetic acid for industrial applications. The Monsanto Company developed a catalytic method for the production of acetic acid by the carbonylation of methane. The method employs a rhodium complex and is so successful that it is used throughout the world. Over 1 million tonnes of acetic acid is produced every year using the *Monsanto process*. The reaction is highly selective, goes in high yield, and is extremely fast.

$$CH_3OH + CO \xrightarrow{\text{Rh catalyst}} CH_3COOH$$

The catalytic cycle for the Monsanto process is shown in Figure 22.32.

The catalyst is the 4-coordinate $[RhI_2(CO)_2]^-$. This is a 16-electron species and is coordinatively unsaturated. The first step in the reaction is the oxidative addition of iodomethane to give the 18-electron, 6-coordinate species. This oxidative addition to coordinatively unsaturated species is very common in catalytic cycles.

This step is followed by migration of the methyl group to the carbonyl group. This results in another 16-electron species which can gain a carbonyl ligand to form an 18-electron species. This can then lose acetyl iodide,

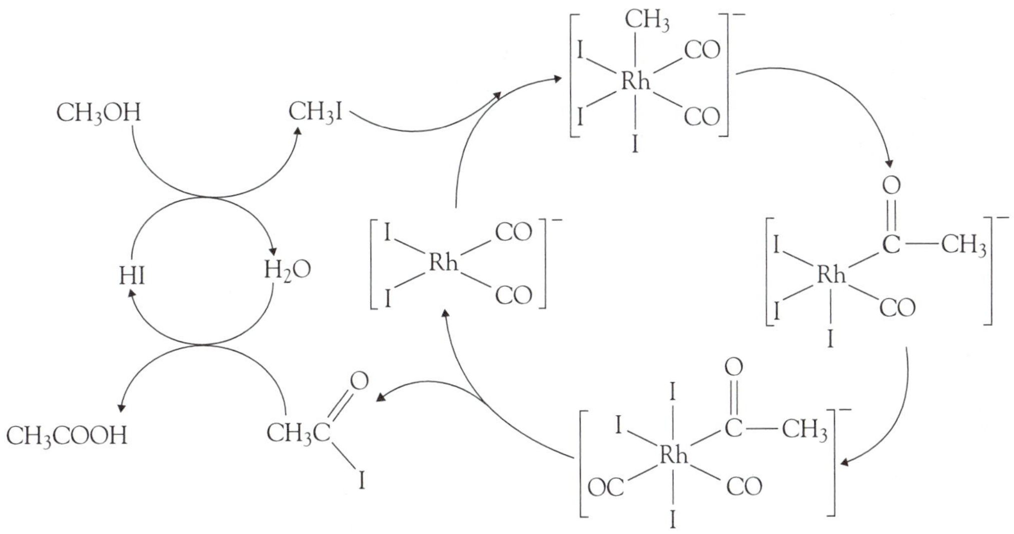

Figure 22.32 The catalytic cycle for the Monsanto Process.

CH_3COI, by reductive elimination, which is another common step in catalytic cycles. This final step regenerates the catalyst, and acetic acid is formed by hydrolysis of the acetyliodide.

$$CH_3COI + H_2O \rightarrow CH_3COOH + HI$$

There are two major concerns with the Monsanto process. First, rhodium is an expensive catalyst. Second, iodine is cheap but is extremely corrosive. Other halogens or halogen substitutes do not work as well.

Alkene Polymerization: The Ziegler–Natta Catalyst

The polymerization of alkenes to form polythenes is extremely important commercially. The most useful polythenes are the stiff, high-density ones which are produced via stereospecific polymerization. These polymers are described as *isotactic*, as all the branched groups lay on the same side of the polymer chain. This leads to efficient packing and a highly ordered, crystalline polymer.

It was the German chemist K. Ziegler, who found that mixing triethylaluminum with titanium(IV) chloride in a hydrocarbon solvent gave a brown suspension that caused ethene (old name, ethylene) to polymerize to polythene (polyethylene) at room temperature and pressure. The high-density polymer produced had different uses compared to the low-density form produced by the high temperature and very high pressures that were traditionally used to synthesize polyethylene. Ziegler and the Italian chemist G. Natta, who utilized this catalyst for the stereospecific polymerization of propene, were awarded a Nobel Prize in 1963 for their development of the organoaluminum catalyst (the *Ziegler–Natta catalyst*). This catalyst is now used for the production of about 5×10^7 tonnes of polyalkenes per year.

The catalyst forms a solid mass so the catalysis is heterogeneous. The reaction takes place at the coordinatively unsaturated titanium centers. The exact mechanism is still not entirely known, but the *Cosee–Arlmann mechanism* is widely accepted as a plausible one and is shown in Figure 22.33.

The triethylaluminum, $Al(CH_2CH_3)_3$, alkylates the titanium species before the incoming alkene molecule coordinates to a neighboring vacant site on the titanium. The alkene then undergoes insertion into the Al—C bond,

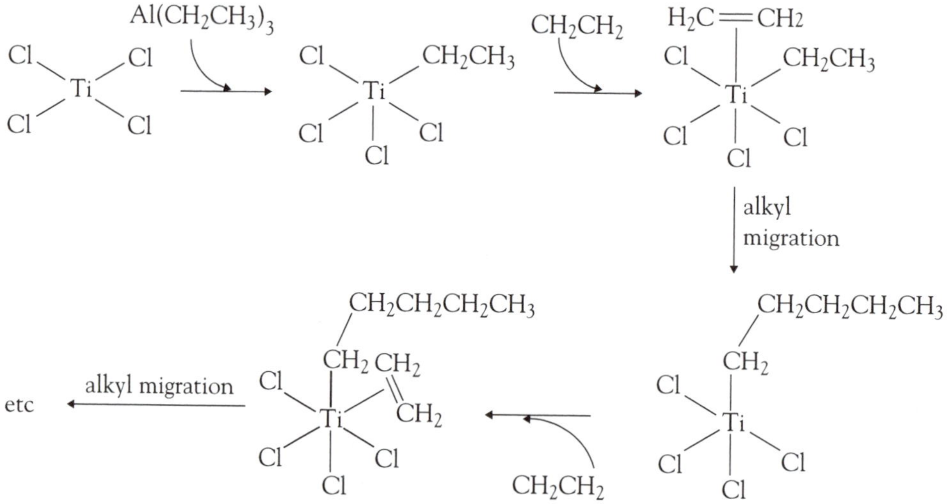

Figure 22.33   Catalytic cycle for Ziegler–Natta polymerization of ethane.

leaving another site vacant for coordination of another alkene molecule. The insertion process is repeated, thus building the polymer chain. The polymer can ultimately be cleaved from the catalyst by β-hydrogen elimination, but in practice some catalyst remains in the polymer. The amount of this is negligible as the polymerization reaction is so effective.

Hydrogenation of Alkenes: Wilkinson's Catalyst

Hydrogenation of alkenes by catalytic processes is of enormous industrial importance. It is used in the manufacture of such diverse products as margerine, pharmaceuticals, and petrochemicals.

The most studied catalytic system for these hydrogenation reactions is [RhCl(PPh$_3$)$_3$], which is known as *Wilkinson's catalyst* after its discoverer, Sir Geoffrey Wilkinson, who received the Nobel Prize for Chemistry in 1973. You may not think that this molecule belongs in a chapter on organometallic chemistry as there are no metal–carbon bonds. However, the phosphine ligands are very similar to carbonyl ligands in many ways, and this is such an important industrial process that it is appropriate to include it here.

The catalyst hydrogenates a wide range of alkenes at atmospheric or reduced pressure. The catalytic cycle involves the oxidative addition of hydrogen to the 16-electron rhodium(I) species to give an 18-electron rhodium(III) species. A phosphine ligand is then lost, giving a coordinatively unsaturated molecule which interacts with the alkene. Hydrogen transfer from the rhodium to the alkene is followed by reductive elimination of the alkane (Figure 22.34).

These catalysts can be used in *enantioselective reactions*. Enantioselective reactions produce products which are chiral. In organic chemistry, a chiral compound is one which has four different groups attached to a carbon atom within the molecule. Chiral compounds are *optically active*, which means that

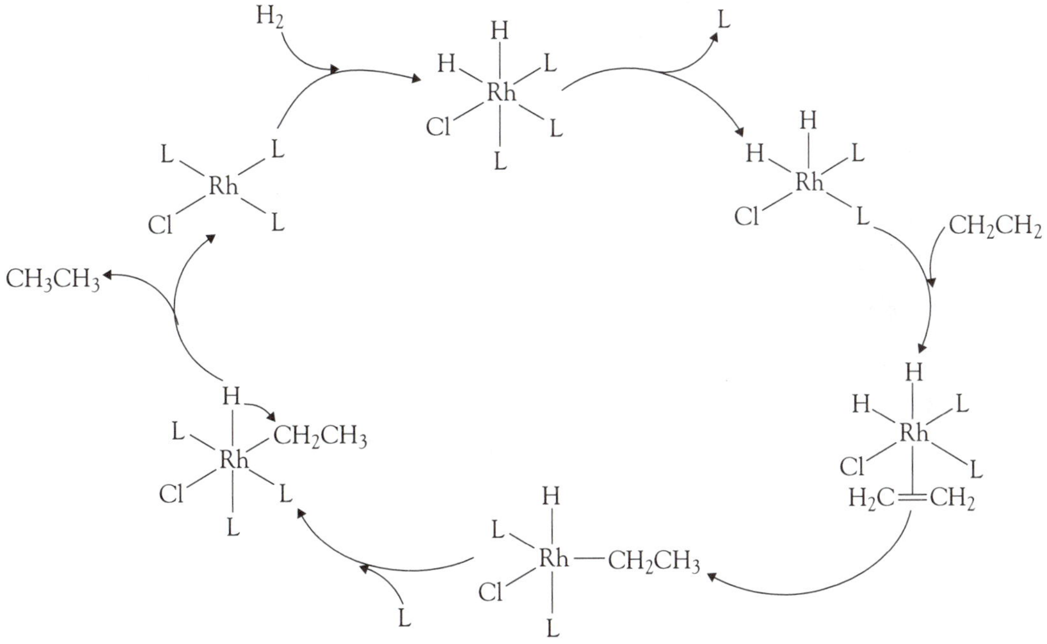

Figure 22.34 The catalytic cycle for the hydrogenation of ethene using Wilkinson's catalyst where L = PPh$_3$.

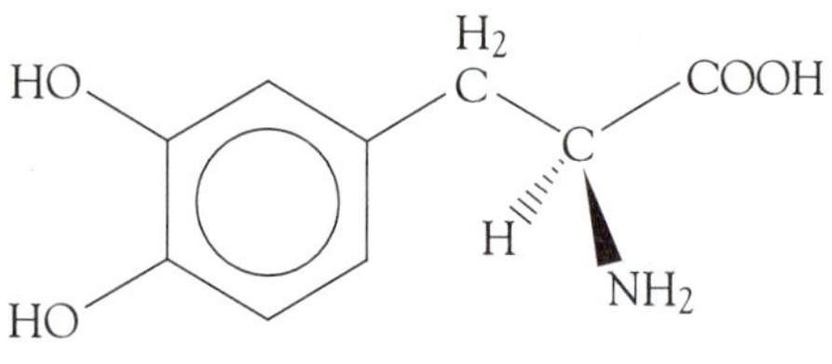

Figure 22.35 L-dopa.

they rotate the plane of polarization of light. Chiral compounds exist in two isomers called enantiomers. (We discussed related coordination isomers in Chapter 19, Section 19.4.) In biological chemistry it is very important to know which enantiomer is being used, as one isomer may be beneficial, while the other may be inactive or harmful. An important example of this is the enantioselective synthesis of L-dopa by the Monsanto Company. L-dopa (Figure 22.35) is used in the treatment of Parkinson's disease.

Hydroformylation

In a *hydroformylation reaction,* an alkene reacts with carbon monoxide and hydrogen over a rhodium or cobalt catalyst to form an aldehyde containing one more carbon atom than the original alkene.

$$RCH{=}CH_2 + CO + H_2 \rightarrow RCH_2CH_2CHO$$

The aldehydes that are produced are usually converted to alcohols that go on to be utilized in a wide range of products including solvents, plasticizers, and detergents. The reaction produces millions of tonnes of product per year.

The catalytic cycle for the hydroformylation reaction was first proposed by Heck and Breslow in 1961. Their cycle, shown in Figure 22.36, is still used today but, as with many other catalytic processes, it has proven difficult to verify experimentally.

The catalyst use is of the type $[Co_2(CO)_8]$, and this initially reacts with hydrogen to break the Co—Co bond and to form the hydride complex, as shown in the following:

$$[Co_2(CO)_8] + H_2 \rightarrow 2[HCo(CO)_4]$$

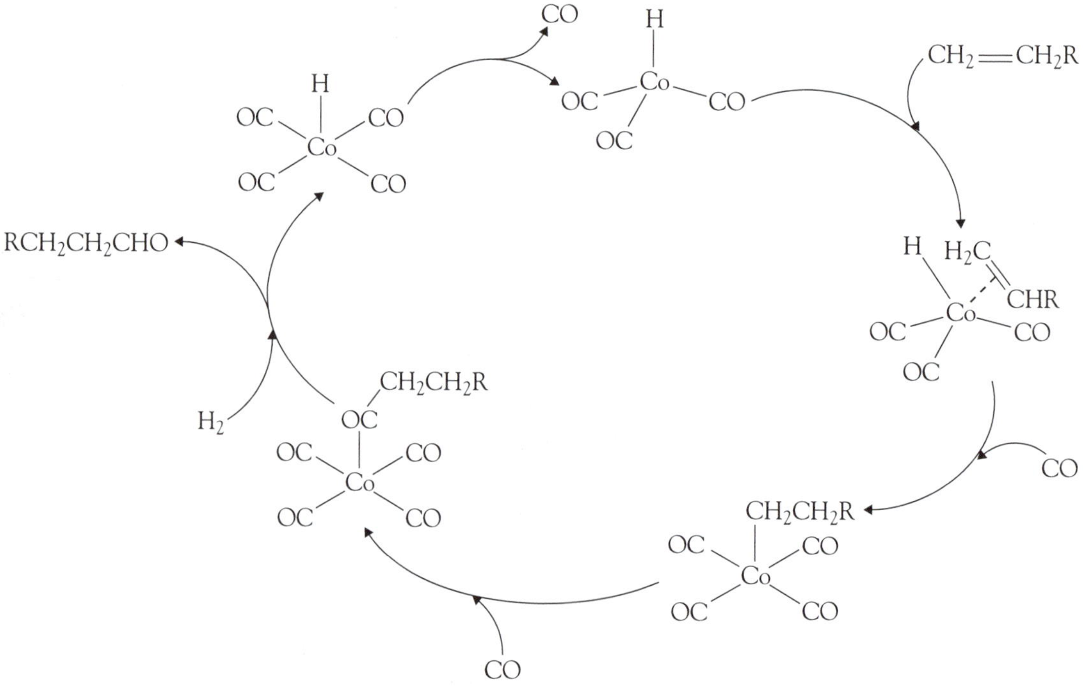

Figure 22.36 The catalytic cycle for a hydroformylation reaction.

This product then loses carbon monoxide to form $[HCo(CO)_3]$ which can then coordinate to the alkene. The hydrogen attached to the cobalt inserts into the alkene, giving a coordinated alkane. Under high pressures of carbon monoxide, the carbonyl inserts into the metal–alkane bond. The aldehyde is finally formed by attack of hydrogen, which also regenerates the catalyst.

EXERCISES

22.1 Decide whether each of the following compounds should be described as organometallic:
(a) $B(CH_3)_3$
(b) $B(OCH_3)_3$
(c) $Na_4(CH_3)_4$
(d) $N(CH_3)_3$
(e) CH_3COONa
(f) $Si(CH_3)_4$
(g) $SiH(C_2H_5)_3$

22.2 Name each of the compounds in Exercise 22.1.

22.3 Write the formula for each of the following species. Where appropriate, give the alternative name based on the hydrogen compound:
(a) methylbismuth
(b) tetraphenylsilicon
(c) potassium tetraphenylboron
(d) methyllithium
(e) ethylmagnesium chloride

22.4 Sketch the structures of the following compounds
(a) $Li_4(CH_3)_4$
(b) $Be(CH_3)_2$
(c) $B(C_2H_5)_3$
(d) $(CH_3)_3SnF$
(e) $(CH_3)_3PbCl$

22.5 Discuss the difference that you might expect between the structures of the two Grignard compounds, C_2H_5MgBr and $[2,4,6-(CH_3)_3C_6H_2]MgBr$.

22.6 Discuss how the steric features of the alkyl group affect the structures of main group organometallic compounds.

22.7 Give an example of a transmetallation reaction.

22.8 Give an example of a halogen exchange reaction.

22.9 Predict the products of the following reactions:
(a) $CH_3Br + 2\,Li \rightarrow$
(b) $MgCl_2 + LiC_2H_5 \rightarrow$
(c) $Mg + (C_2H_5)_2Hg \rightarrow$
(d) $C_2H_5Li + C_6H_6 \rightarrow$
(e) $Mg + C_2H_5HgCl \rightarrow$
(f) $B_2H_6 + CH_3CH{=}CH_2 \rightarrow$
(g) $SnCl_4 + C_2H_5MgCl \rightarrow$

22.10 Compare the nature of the bonding in the aluminum compounds Al_2Cl_6, $Al_2(CH_3)_6$, and $Al_2(CH_3)_4(\mu\text{-}Cl)_2$.

22.11 Name each of the following species:
(a) $Cr(CO)_6$
(b) $(\eta^5\text{-}C_5H_5)_2Fe$
(c) $(\eta^6\text{-}C_6H_6)Mo(CO)_3$
(d) $(\eta^5\text{-}C_5H_5)W(CO)_3$
(e) $Mn(CO)_5Br$

22.12 For each of the following compounds, determine the formal oxidation state of the transition metal and the corresponding number of d electrons. State whether or not each one is likely to be stable enough to be characterized.
(a) $Re(CO)_5$
(b) $[HFe(CO)_4]^-$
(c) $(\eta^5\text{-}C_5H_5)_2Fe$
(d) $(\eta^6\text{-}C_6H_6)_2Cr$
(e) $(\eta^5\text{-}C_5H_5)ZrCl(OCH_3)$
(f) $[IrCl(PPh_3)_3]$
(g) $Mo(CO)_3(PPh_3)_3$
(h) $Fe(CO)_4(C_2H_4)$
(i) $[W(CO)_5Cl]^-$
(j) $Ni(CO)_4$

22.13 Deduce the probable formula of the simplest carbonyl compounds of chromium, iron and nickel. Show your calculations.

22.14 Chromium forms two common anionic carbonyls: $Cr(CO)_5^{n-}$ and $Cr(CO)_4^{m-}$. Deduce the probable charges, n and m, on these ions.

22.15 Suggest why $V(CO)_6$ is easily reduced to $V(CO)_6^-$.

22.16 Use the 18-electron rule to predict the number of carbonyl ligands, n, in each of the following complexes:
(a) $Cr(CO)_n$
(b) $Fe(CO)_n(PPh_3)_2$
(c) $Mo(CO)_n(PMe_3)_3$
(d) $W(CO)_n(\eta^6\text{-}C_6H_6)$

22.17 Assuming that each of the following obeys the 18-electron rule, determine the number of metal–metal bonds in each complex. Sketch a possible structure in each case:
(a) $Mn_2(CO)_{10}$
(b) $[(\eta^5\text{-}C_5H_5)Mn(CO)_2]_2$
(c) $\mu\text{-}CO\text{-}[(\eta^4\text{-}C_4H_4)Fe(CO)_2]_2$
(d) $(\mu\text{-}Br)_2\text{-}[Mn(CO)_4]_2$

22.18 Consider each pair of carbonyl complexes. In each case decide which one would have the lower infrared CO stretching frequency. Explain your choice.
(a) $Fe(CO)_5$ and $Fe(CO)_4Cl$
(b) $Mo(CO)_6$ and $Mo(CO)_4(PPh_3)_2$
(c) $Mo(CO)_4(PPh_3)_2$ and $Mo(CO)_4(PMe_3)_2$

22.19 Predict the products from each of the following reactions:

(a) $Cr(CO)_6 + CH_3CN \rightarrow$

(b) $Mn_2(CO)_{10} + H_2 \rightarrow$

(c) $Mo(CO)_6 +$
$(CH_3)_2PCH_2CH_2P(Ph)CH_2 \ CH_2P(CH_3)_2 \rightarrow$

(d) $Fe(CO)_5 + $ 1,3-cyclohexadiene $\rightarrow$

(e) $NaMn(CO)_5 + CH_2{=}CHCH_2Cl \rightarrow$

(f) $Cr(CO)_6 + C_6H_6 \rightarrow$

(g) $PtCl_2(PMe_3)_2 + LiCH_2CH_2CH_2CH_2Li \rightarrow$

(h) $Ni(CO)_4 + PF_3 \rightarrow$

(i) $Mn_2(CO)_{10} + Br_2 \rightarrow$

(j) $HMn(CO)_5 + CO_2 \rightarrow$

22.20 The compound, $IrCl(CO)(PPh_3)_2$, Vaska's compound, is used for the study of oxidative addition processes. What is the formal oxidation number of iridium in this compound?

22.21 Iridium forms a compound $[Ir(C_5H_5)(H_3)(PPh_3)]^+$. Two possible structures were proposed, one containing three separate hydride ions and the other containing the unusual trihydrogen ligand, H_3.

(a) What would be the oxidation state of the iridium if the compound contained three hydride ions?

(b) If the compound contained a single H_3 ligand, and the iridium had the same oxidation state as that in Vaska's compound (Exercise 22.20), what would be the charge on the H_3 unit? By comparison with diborane, suggest why this trihydrogen ion might indeed exist.

22.22 Predict the products from the oxidative addition reactions of

(a) dihydrogen

(b) dinitrogen

(c) hydrogen chloride

(d) dioxygen
    to Vaska's compound, *trans*-$IrCl(CO)(PPh_3)_2$.

BEYOND THE BASICS

22.23 When nickelocene, $(\eta^5\text{-}C_5H_5)_2Ni$, and tetracarbonylnickel, $Ni(CO)_4$, are refluxed together in benzene in a 1:1 molar ratio, the product is a red-purple crystalline compound. This compound has the empirical formula C_6H_5ONi and has a relative molecular mass of 302. Suggest a possible structure for the compound.

22.24 When molybdenum hexacarbonyl is reacted with an excess of acetonitrile, CH_3CN, a pale yellow product, A, is formed.

When compound A is refluxed with benzene the pale yellow product, B, is obtained which has the molecular formula $C_9H_6O_3Mo$ and shows a sharp singlet at 5.5 ppm in the ^1H NMR spectrum.

When compound A is refluxed with 1,3,5,7-cyclooctatetraene in hexane, compound C is produced which has the molecular formula $C_{11}H_8O_3Mo$.

Use this information to identify compounds A, B, and C and suggest a name for each of them.

22.25 Sodium tricarbonyl(pentahaptocyclopentadienyl) tungsten(0) reacts with 3-chloropropene to give a solid, A, which has the molecular formula $(C_3H_5)(C_5H_5)(CO)_3W$. This compound loses carbon monoxide on exposure to light and forms the compound B, which has the formula $(C_3H_5)(C_5H_5)(CO)_2W$. Treating compound A with hydrogen chloride and then potassium hexafluorophosphate, $K^+PF_6^-$, results in the formation of a salt, C. Compound C has the molecular formula $[(C_3H_6)(C_5H_5)(CO)_3W]PF_6$. When this is left to stand for some time, a hydrocarbon is produced.

Use this information and the 18-electron rule to identify the compounds A, B, and C. Show how the hydrocarbon interacts with the metal in each case. Name each of the compounds and identify the hydrocarbon formed.

22.26 Draw the catalytic cycle for the Ziegler–Natta polymerization of propene. What do you notice about the polymer which is formed?

22.27 Predict what the product would be if carbon disulfide were interacted with the Ziegler–Natta titanium species. (Hint: Consider the structure of carbon disulphide, $S{=}C{=}S$.)

22.28 The trend in the rate of hydrogenation of some alkenes by Wilkinson's catalyst is given in the following:

Cyclohexene > hexane > *cis*-4-methyl-2-pentene > 1-methylcyclohexene

Explain this trend and identify the step in the catalytic cycle which is most affected.

The Rare Earth and Actinoid Elements

Chapter **23**

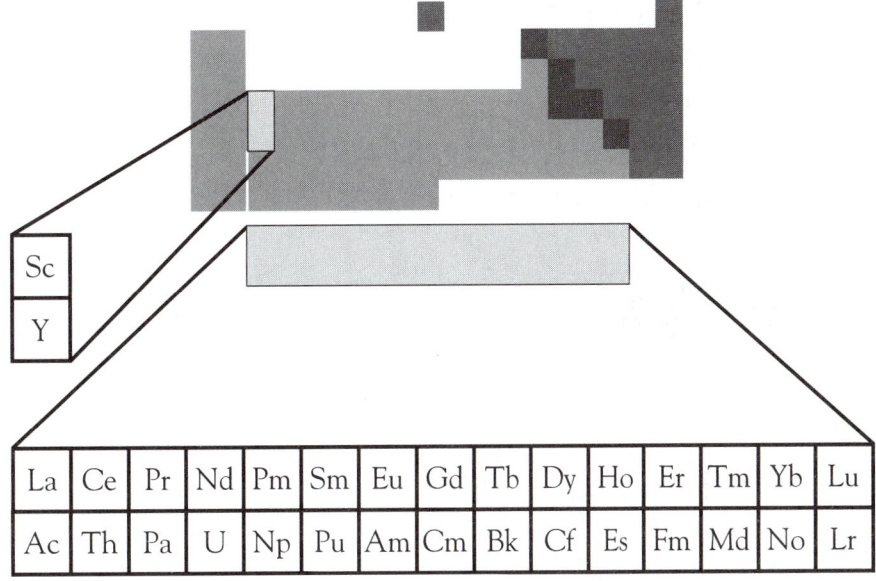

Sc
Y

La	Ce	Pr	Nd	Pm	Sm	Eu	Gd	Tb	Dy	Ho	Er	Tm	Yb	Lu
Ac	Th	Pa	U	Np	Pu	Am	Cm	Bk	Cf	Es	Fm	Md	No	Lr

For Chapter 23, see http://www.whfreeman.com/rayner

Thermodynamic Properties of Some Selected Inorganic Compounds

Appendix 1

A s thermodynamic data are experimental, their values differ from one source to another. A consistent set of values have been used here, summarized from G. Aylward and T. Findlay, *SI Chemical Data*, 3rd ed., New York: Wiley, 1994.

Compound Name	Formula	$\Delta H°$ $(kJ \cdot mol^{-1})$	$S°$ $(J \cdot mol^{-1} \cdot K^{-1})$	$\Delta G°$ $(kJ \cdot mol^{-1})$
Aluminum	$Al(s)$	0	+28	0
	$Al(g)$	+330	+165	+290
	$Al^{3+}(aq)$	−538	−325	−492
aluminate ion	$Al(OH)_4^-(aq)$	−1502	+103	−1305
bromide	$AlBr_3(s)$	−511	+180	−489
carbide	$Al_4C_3(s)$	−209	+89	−196
chloride	$AlCl_3(s)$	−704	+111	−629
chloride hexahydrate	$AlCl_3 \cdot 6H_2O(s)$	−2692	+318	−2261
fluoride	$AlF_3(s)$	−1510	+66	−1431
iodide	$AlI_3(s)$	−314	+159	−301
nitride	$AlN(s)$	−318	+20	−287
oxide	$Al_2O_3(s)$	−1676	+51	−1582
phosphate	$AlPO_4(s)$	−1734	+91	−1618
sulfate	$Al_2(SO_4)_3(s)$	−3441	+239	−3100
Ammonium	$NH_4^+(aq)$	−133	+111	−79
bromide	$NH_4Br(s)$	−271	+113	−175
chloride	$NH_4Cl(s)$	−314	+95	−203
fluoride	$NH_4F(s)$	−464	+72	−349
iodide	$NH_4I(s)$	−201	+117	−113
nitrate	$NH_4NO_3(s)$	−366	+151	−184
sulfate	$(NH_4)_2SO_4(s)$	−1181	+220	−902
vanadate	$NH_4VO_3(s)$	−1053	+141	−888
Antimony	$Sb(s)$	0	+46	0
	$Sb(g)$	+262	+180	+222
pentachloride	$SbCl_5(l)$	−440	+301	−350
pentaoxide	$Sb_2O_5(s)$	−972	+125	−829
tribromide	$SbBr_3(s)$	−259	+207	−239
trichloride	$SbCl_3(s)$	−382	+184	−324
trihydride	$SbH_3(g)$	+145	+233	+148

Compound Name	Formula	$\Delta H°$ (kJ·mol^{-1})	$S°$ (J·mol^{-1}·K^{-1})	$\Delta G°$ (kJ·mol^{-1})
triodide	$SbI_3(s)$	-100	$+215$	-99
trioxide	$Sb_2O_3(s)$	-720	$+110$	-634
trisulfide	$Sb_2S_3(s)$	-175	$+182$	-174
Arsenic	$As(s)$ (grey)	0	$+35$	0
	$As(g)$	$+302$	$+174$	$+261$
pentafluoride	$AsF_5(g)$	-1237	$+317$	-1170
pentaoxide	$As_2O_5(s)$	-925	$+105$	-782
tribromide	$AsBr_3(s)$	-130	$+364$	-159
trichloride	$AsCl_3(l)$	-305	$+216$	-259
trifluoride	$AsF_3(l)$	-786	$+289$	-771
trihydride (arsine)	$AsH_3(g)$	$+66$	$+223$	$+69$
triiodide	$AsI_3(s)$	-58	$+213$	-59
trioxide	$As_2O_3(s)$	-657	$+107$	-576
trisulfide	$As_2S_3(s)$	-169	$+164$	-169
Barium	$Ba(s)$	0	$+63$	0
	$Ba(g)$	$+180$	$+170$	$+146$
	$Ba^{2+}(aq)$	-538	$+10$	-561
bromide	$BaBr_2(s)$	-757	$+146$	-737
carbonate	$BaCO_3(s)$	-1216	$+112$	-1138
chloride	$BaCl_2(s)$	-859	$+124$	-810
chloride dihydrate	$BaCl_2·2H_2O(s)$	-1460	$+203$	-1296
fluoride	$BaF_2(s)$	-1207	$+96$	-1157
hydroxide	$Ba(OH)_2(s)$	-945	$+101$	-856
hydroxide octahydrate	$Ba(OH)_2·8H_2O(s)$	-3342	$+427$	-2793
iodide	$BaI_2(s)$	-605	$+165$	-601
nitrate	$Ba(NO_3)_2(s)$	-992	$+214$	-797
nitride	$Ba_3N_2(s)$	-363	$+152$	-292
oxide	$BaO(s)$	-554	$+70$	-525
peroxide	$BaO_2(s)$	-634		
sulfate	$BaSO_4(s)$	-1473	$+132$	-1362
sulfide	$BaS(s)$	-460	$+78$	-456
Beryllium	$Be(s)$	0	$+9$	0
	$Be(g)$	$+324$	$+136$	$+287$
	$Be^{2+}(aq)$	-383	-130	-380
bromide	$BeBr_2(s)$	-356	$+100$	-337
chloride	$BeCl_2(s)$	-490	$+83$	-445
fluoride	$BeF_2(s)$	-1027	$+53$	-979
hydroxide	$Be(OH)_2(s)$	-903	$+52$	-815
iodide	$BeI_2(s)$	-189	$+120$	-187
oxide	$BeO(s)$	-609	$+14$	-580
Bismuth	$Bi(s)$	0	$+57$	0
chloride	$BiCl_3(s)$	-379	$+177$	-315
oxide	$Bi_2O_3(s)$	-574	$+151$	-494
oxide chloride	$BiOCl(s)$	-367	$+120$	-322
sulfide	$Bi_2S_3(s)$	-143	$+200$	-141
Boron	$B(s)$	0	0	6
	$B(g)$	$+565$	$+153$	$+521$
boric acid	$H_3BO_3(s)$	-1095	$+90$	-970

Compound Name	Formula	$\Delta H°$ (kJ·mol^{-1})	$S°$ (J·mol^{-1}·K^{-1})	$\Delta G°$ (kJ·mol^{-1})
carbide	$B_4C(s)$	−71	+27	−71
decaborane(14)	$B_{10}H_{14}(g)$	+32	+353	+216
diborane	$B_2H_6(g)$	+36	+232	+87
nitride	$BN(s)$	−254	+15	−228
pentaborane(9)	$B_5H_9(l)$	+43	+184	+172
tribromide	$BBr_3(l)$	−240	+230	−238
trichloride	$BCl_3(g)$	−404	+290	−389
trifluoride	$BF_3(g)$	−1136	+254	−1119
trioxide (di−)	$B_2O_3(s)$	−1273	+54	−1194
trisulfide (di−)	$B_2S_3(s)$	−252	+92	−248
Bromine	$Br_2(l)$	0	+152	0
	$Br_2(g)$	+31	+245	+3
	$Br(g)$	+112	+175	+82
	$Br^-(aq)$	−121	+83	−104
bromate ion	$BrO_3^-(aq)$	−67	+162	+19
hypobromite ion	$BrO^-(aq)$	−94	+42	−33
monochloride	$BrCl(g)$	+15	+240	−1
monofluoride	$BrF(g)$	−94	+229	−109
pentafluoride	$BrF_5(g)$	−429	+320	−351
trifluoride	$BrF_3(g)$	−256	+293	−229
Cadmium	$Cd(s)$	0	+52	0
	$Cd(g)$	+112	+168	+77
	$Cd^{2+}(aq)$	−76	−73	−78
bromide	$CdBr_2(s)$	−316	+137	−296
carbonate	$CdCO_3(s)$	−751	+92	−669
chloride	$CdCl_2(s)$	−391	+115	−344
fluoride	$CdF_2(s)$	−700	+77	−648
hydroxide	$Cd(OH)_2(s)$	−561	+96	−474
iodide	$CdI_2(s)$	−203	+161	−201
nitrate	$Cd(NO_3)_2(s)$	−456		
oxide	$CdO(s)$	−258	+55	−228
sulfate	$CdSO_4(s)$	−933	+123	−823
sulfide	$CdS(s)$	−162	+65	−156
Calcium	$Ca(s)$	0	+42	0
	$Ca(g)$	+178	+155	+144
	$Ca^{2+}(aq)$	−543	−56	−553
bromide	$CaBr_2(s)$	−683	+130	−664
carbide	$CaC_2(s)$	−60	+70	−65
carbonate	$CaCO_3(s)$ (calcite)	−1207	+93	−1129
chloride	$CaCl_2(s)$	−796	+105	−748
fluoride	$CaF_2(s)$	−1220	+69	−1167
hydride	$CaH_2(s)$	−186	+42	−147
hydroxide	$Ca(OH)_2(s)$	−986	+83	−898
iodide	$CaI_2(s)$	−533	+142	−529
nitrate	$Ca(NO_3)_2(s)$	−938	+193	−743
oxide	$CaO(s)$	−635	+38	−603
phosphate	$Ca_3(PO_4)_2(s)$	−4121	+236	−3885
silicate	$CaSiO_3(s)$	−1567	+82	−1499
sulfate	$CaSO_4(s)$	−1434	+107	−1332
sulfate hemihydrate	$CaSO_4·\frac{1}{2}H_2O(s)$	−1577	+131	−1437

Compound Name	Formula	$\Delta H°$ (kJ·mol^{-1})	$S°$ (J·mol^{-1}·K^{-1})	$\Delta G°$ (kJ·mol^{-1})
sulfate dihydrate	$CaSO_4 \cdot 2H_2O(s)$	-2023	$+194$	-1797
sulfide	$CaS(s)$	-482	$+56$	-477
Carbon	$C(s)$ (graphite)	0	$+6$	0
	$C(s)$ (diamond)	$+2$	$+2$	$+3$
	$C(g)$	$+717$	$+158$	$+671$
carbonate ion	$CO_3^{2-}(aq)$	-675	-50	-528
chloride (–yl) (phosgene)	$COCl_2(g)$	-219	$+284$	-205
cyanide ion	$CN^-(aq)$	$+151$	$+94$	$+172$
dioxide	$CO_2(g)$	-394	$+214$	-394
dioxide	$CO_2(aq)$	-413	$+119$	-386
disulfide	$CS_2(l)$	$+90$	$+151$	$+65$
ethane	$C_2H_6(g)$	-85	$+230$	-33
hydrogen carbonate ion	$HCO_3^-(aq)$	-690	$+98$	-587
methane	$CH_4(g)$	-75	$+186$	-51
monoxide	$CO(g)$	-111	$+198$	-137
tetrabromide	$CBr_4(s)$	$+19$	$+213$	$+48$
tetrachloride	$CCl_4(l)$	-135	$+216$	-65
tetrafluoride	$CF_4(g)$	-933	$+262$	-888
thiocyanate ion	$NCS^-(aq)$	$+76$	$+144$	$+93$
Cesium	$Cs(s)$	0	$+85$	0
	$Cs(g)$	$+76$	$+176$	$+49$
	$Cs^+(aq)$	-258	$+132$	-291
bromide	$CsBr(s)$	-406	$+113$	-391
carbonate	$Cs_2CO_3(s)$	-1140	$+204$	-1054
chloride	$CsCl(s)$	-443	$+101$	-415
fluoride	$CsF(s)$	-554	$+93$	-526
iodide	$CsI(s)$	-347	$+123$	-341
nitrate	$CsNO_3(s)$	-506	$+155$	-407
sulfate	$Cs_2SO_4(s)$	-1443	$+212$	-1324
Chlorine	$Cl_2(g)$	0	$+223$	0
	$Cl_2(aq)$	-23	$+121$	$+7$
	$Cl(g)$	$+121$	$+165$	$+105$
	$Cl^-(aq)$	-167	$+57$	-131
chlorate ion	$ClO_3^-(aq)$	-104	$+162$	-8
dioxide	$ClO_2(g)$	$+102$	$+257$	$+120$
hypochlorite ion	$ClO^-(aq)$	-107	$+42$	-37
monofluoride	$ClF(g)$	-54	$+218$	-56
oxide (di-)	$Cl_2O(g)$	$+80$	$+266$	$+98$
perchlorate ion	$ClO_4^-(aq)$	-128	$+184$	-8
trifluoride	$ClF_3(g)$	-163	$+282$	-123
Chromium	$Cr(s)$	0	$+24$	0
	$Cr(g)$	$+397$	$+175$	$+352$
	$Cr^{2+}(aq)$	-139		-165
	$Cr^{3+}(aq)$	-256		-205
(II) chloride	$CrCl_2(s)$	-395	$+115$	-356
(III) chloride	$CrCl_3(s)$	-556	$+123$	-486
chromate ion	$CrO_4^-(aq)$	-881	$+50$	-728

Compound Name	Formula	$\Delta H°$ (kJ·mol^{-1})	$S°$ (J·mol^{-1}·K^{-1})	$\Delta G°$ (kJ·mol^{-1})
dichromate ion	$Cr_2O_7^{2-}(aq)$	-1490	$+262$	-1301
(III) oxide	$Cr_2O_3(s)$	-1140	$+81$	-1058
(VI) oxide	$CrO_3(s)$	-580	$+72$	-513
(III) sulfate	$Cr_2(SO_4)_3(s)$	-2911	$+259$	-2578
Cobalt	$Co(s)$	0	$+30$	0
	$Co(g)$	$+425$	$+180$	$+380$
	$Co^{2+}(aq)$	-58	-113	-54
	$Co^{3+}(aq)$	$+92$	-305	$+134$
(II) carbonate	$CoCO_3(s)$	-713	$+89$	-637
(II) chloride	$CoCl_2(s)$	-313	$+109$	-270
(II) chloride hexahydrate	$CoCl_2·6H_2O(s)$	-2115	$+343$	-1725
(II) hydroxide	$Co(OH)_2(s)$ (pink)	-540	$+79$	-454
(II) oxide	$CoO(s)$	-238	$+53$	-214
(II) sulfate	$CoSO_4(s)$	-888	$+118$	-782
(II) sulfate heptahydrate	$CoSO_4·7H_2O(s)$	-2980	$+406$	-2474
Copper	$Cu(s)$	0	$+33$	0
	$Cu(g)$	$+337$	$+166$	$+298$
	$Cu^+(aq)$	$+72$	$+41$	$+50$
	$Cu^{2+}(aq)$	$+65$	-98	$+65$
(I) chloride	$CuCl(s)$	-137	$+86$	-120
(II) chloride	$CuCl_2(s)$	-220	$+108$	-176
(II) chloride dihydrate	$CuCl_2·2H_2O(s)$	-821	$+167$	-656
(II) hydroxide	$Cu(OH)_2(s)$	-450	$+108$	-373
(I) oxide	$Cu_2O(s)$	-169	$+93$	-146
(II) oxide	$CuO(s)$	-157	$+43$	-130
(II) sulfate	$CuSO_4(s)$	-771	$+109$	-662
(II) sulfate pentahydrate	$CuSO_4·5H_2O(s)$	-2280	$+300$	-1880
(I) sulfide	$Cu_2S(s)$	-80	$+121$	-86
(II) sulfide	$CuS(s)$	-53	$+67$	-54
Fluorine	$F_2(g)$	0	$+203$	0
	$F(g)$	$+79$	$+159$	$+62$
	$F^-(aq)$	-335	-14	-281
Gallium	$Ga(s)$	0	$+41$	0
	$Ga(g)$	$+277$	$+169$	$+239$
	$Ga^{3+}(aq)$	-212	-331	-159
bromide	$GaBr_3(s)$	-387	$+180$	-360
chloride	$GaCl_3(s)$	-525	$+142$	-455
fluoride	$GaF_3(s)$	-1163	$+84$	-1085
iodide	$GaI_3(s)$	-239	$+204$	-236
oxide	$Ga_2O_3(s)$	-1089	$+85$	-998
Germanium	$Ge(s)$	0	$+31$	0
	$Ge(g)$	$+372$	$+168$	$+331$
dioxide	$GeO(s)$	-262	$+50$	-237
tetrachloride	$GeCl_4(g)$	-496	$+348$	-457
tetraoxide	$GeO_2(s)$	-580	$+40$	-521

Compound Name	Formula	$\Delta H°$ (kJ·mol^{-1})	$S°$ (J·mol^{-1}·K^{-1})	$\Delta G°$ (kJ·mol^{-1})
Hydrogen	$H_2(g)$	0	+131	0
	$H(g)$	+218	+115	+203
	$H^+(aq)$	0	0	0
bromide	$HBr(g)$	−36	+199	−53
chloride	$HCl(g)$	−92	+187	−95
fluoride	$HF(g)$	−273	+174	−275
hydrobromic acid	$HBr(aq)$	−122	+82	−104
hydrochloric acid	$HCl(aq)$	−167	+56	−131
hydrofluoric acid	$HF(aq)$	−333	−14	−279
hydroiodic acid	$HI(aq)$	−55	+111	−52
iodide	$HI(g)$	+26	+207	+2
oxide (water)	$H_2O(l)$	−286	+70	−237
	$H_2O(g)$	−242	+189	−229
hydroxide ion	$OH^-(aq)$	−230	−11	−157
peroxide	$H_2O_2(l)$	−188	+110	−120
selenide	$H_2Se(g)$	+30	+219	+16
sulfide	$H_2S(g)$	−21	+206	−34
telluride	$H_2Te(g)$	+100	+229	+85
Indium	$In(s)$	0	+58	0
	$In(g)$	+243	+174	+209
	$In^{3+}(aq)$	−105	−151	−98
(I) chloride	$InCl(s)$	−186	+95	−164
(III) chloride	$InCl_3(s)$	−537	+141	−462
oxide	$In_2O_3(s)$	−926	+104	−831
Iodine	$I_2(s)$	0	+116	0
	$I_2(g)$	+62	+261	+19
	$I(g)$	+107	+181	+70
	$I^-(aq)$	−55	+106	−52
iodate ion	$IO_3^-(aq)$	−221	+118	−128
heptafluoride	$IF_7(g)$	−944	+346	−818
monochloride	$ICl(g)$	+18	+248	−5
triodide ion	$I_3^-(aq)$	−51	+239	−51
Iron	$Fe(s)$	0	+27	0
	$Fe(g)$	+416	+180	+371
	$Fe^{2+}(aq)$	−89	−138	−79
	$Fe^{3+}(aq)$	−49	−316	−5
(II) carbonate	$FeCO_3(s)$	−741	+93	−667
(II) chloride	$FeCl_2(s)$	−342	+118	−302
(III) chloride	$FeCl_3(s)$	−399	+142	−334
(II) disulfide	$FeS_2(s)$ (pyrite)	−178	+53	−167
(II) hydroxide	$Fe(OH)_2(s)$	−569	+88	−487
(III) hydroxide	$Fe(OH)_3(s)$	−823	+107	−697
(II) oxide	$FeO(s)$	−272	+61	−251
(II)(III) oxide	$Fe_3O_4(s)$	−1118	+146	−1015
(III) oxide	$Fe_2O_3(s)$	−824	+87	−742
(II) sulfate	$FeSO_4(s)$	−928	+108	−821
(II) sulfate heptahydrate	$FeSO_4 \cdot 7H_2O(s)$	−3015	+409	−2510
(III) sulfate	$Fe_2(SO_4)_3(s)$	−2582	+308	−2262
(II) sulfide	$FeS(s)$	−100	+60	−100

Compound Name	Formula	$\Delta H°$ (kJ·mol^{-1})	$S°$ (J·mol^{-1}·K^{-1})	$\Delta G°$ (kJ·mol^{-1})
Lead	Pb(s)	0	+65	0
	Pb(g)	+196	—	—
	Pb^{2+}(aq)	+1	+18	−24
(II) carbonate	PbCO$_3$(s)	−699	+131	−626
(II) chloride	PbCl$_2$(s)	−359	+136	−314
(IV) chloride	PbCl$_4$(g)	−552	+382	−492
(II) oxide	PbO(s)	−217	+69	−188
(IV) oxide	PbO$_2$(s)	−277	+69	+217
(II) sulfate	PbSO$_4$(s)	−920	+149	−813
(II) sulfide	PbS(s)	−100	+91	−99
Lithium	Li(s)	0	+29	0
	Li(g)	+159	+139	+127
	Li$^+$(aq)	−278	+12	−293
bromide	LiBr(s)	−351	+74	−342
carbonate	Li$_2$CO$_3$(s)	−1216	+90	−1132
chloride	LiCl(s)	−409	+59	−384
fluoride	LiF(s)	−616	+36	−588
hydride	LiH(s)	−91	+20	−68
hydroxide	LiOH(s)	−479	+43	−439
iodide	LiI(s)	−270	+87	−270
nitrate	LiNO$_3$(s)	−483	+90	−381
nitride	Li$_3$N(s)	−164	+63	−128
oxide	Li$_2$O(s)	−598	+38	−561
sulfate	Li$_2$SO$_4$(s)	−1436	+115	−1322
sulfide	Li$_2$S(s)	−441	+61	−433
tetrahydridoaluminate	LiAlH$_4$(s)	−116	+79	−45
Magnesium	Mg(s)	0	+33	0
	Mg(g)	+147	+149	+112
	Mg^{2+}(aq)	−467	−137	−455
bromide	MgBr$_2$(s)	−524	+117	−504
carbonate	MgCO$_3$(s)	−1096	+66	−1012
chloride	MgCl$_2$(s)	−641	+90	−592
chloride hexahydrate	MgCl$_2$·6H$_2$O(s)	−2499	+366	−2115
fluoride	MgF$_2$(s)	−1124	+57	−1071
hydride	MgH$_2$(s)	−75	+31	−36
hydroxide	Mg(OH)$_2$(s)	−925	+63	−834
iodide	MgI$_2$(s)	−364	+130	−358
nitrate	Mg(NO$_3$)$_2$(s)	−791	+164	−589
nitrate hexahydrate	Mg(NO$_3$)$_2$·6H$_2$O(s)	−2613	+452	−2080
nitride	Mg$_3$N$_2$(s)	−461	+88	−401
oxide	MgO(s)	−602	+27	−569
sulfate	MgSO$_4$(s)	−1285	+92	−1171
sulfate heptahydrate	MgSO$_4$·7H$_2$O(s)	−3389	+372	−2872
sulfide	MgS(s)	−346	+50	−342
Manganese	Mn(s)	0	+32	0
	Mn(g)	+281	+174	+238
	Mn^{2+}(aq)	−221	−74	−228
(II) carbonate	MnCO$_3$(s)	−894	+86	−817
(II) chloride	MnCl$_2$(s)	−481	+118	−441

Compound Name	Formula	$\Delta H°$ (kJ·mol^{-1})	$S°$ (J·mol^{-1}·K^{-1})	$\Delta G°$ (kJ·mol^{-1})
(II) fluoride	$MnF_2(s)$	-803	$+92$	-761
(III) fluoride	$MnF_3(s)$	-1004	$+105$	-935
(II) hydroxide	$Mn(OH)_2(s)$	-695	$+99$	-615
(II) oxide	$MnO(s)$	-385	$+60$	-363
(III) oxide	$Mn_2O_3(s)$	-959	$+110$	-881
(IV) oxide	$MnO_2(s)$	-520	$+53$	-465
permanganate ion	$MnO_4{}^-(aq)$	-541	$+191$	-447
(II) sulfate	$MnSO_4(s)$	-1065	$+112$	-957
(II) sulfide	$MnS(s)$	-214	$+78$	-218
Mercury	$Hg(l)$	0	$+76$	0
	$Hg(g)$	$+61$	$+175$	$+32$
	$Hg_2{}^{2+}(aq)$	$+167$	$+66$	$+154$
	$Hg^{2+}(aq)$	$+170$	-36	$+165$
(I) chloride	$Hg_2Cl_2(s)$	-265	$+192$	-211
(II) chloride	$HgCl_2(s)$	-224	$+146$	-179
(II) oxide	$HgO(s)$	-91	$+70$	-59
(I) sulfate	$Hg_2SO_4(s)$	-743	$+201$	-626
(II) sulfate	$HgSO_4(s)$	-708	$+140$	-595
Nickel	$Ni(s)$	0	$+30$	0
	$Ni(g)$	$+430$	$+182$	$+385$
	$Ni^{2+}(aq)$	-54	-129	-46
(II) bromide	$NiBr_2(s)$	-212	$+136$	-198
(II) carbonate	$NiCO_3(s)$	-681	$+118$	-613
(II) chloride	$NiCl_2(s)$	-305	$+98$	-259
(II) chloride hexahydrate	$NiCl_2·6H_2O(s)$	-2103	$+344$	-1714
(II) fluoride	$NiF_2(s)$	-651	$+74$	-604
(II) hydroxide	$Ni(OH)_2(s)$	-530	$+88$	-447
(II) iodide	$NiI_2(s)$	-78	$+154$	-81
(II) oxide	$NiO(s)$	-240	$+38$	-212
(II) sulfate	$NiSO_4(s)$	-873	$+92$	-760
(II) sulfate heptahydrate	$NiSO_4·7H_2O(s)$	-2976	$+379$	-2462
(II) sulfide	$NiS(s)$	-82	$+53$	-80
tetracarbonyl (0)	$Ni(CO)_4(l)$	-633	$+313$	-588
Nitrogen	$N_2(g)$	0	$+192$	0
	$N(g)$	$+473$	$+153$	$+456$
ammonia	$NH_3(g)$	-46	$+193$	-16
azide ion	$N_3{}^-(aq)$	$+275$	$+108$	$+348$
dinitrogen oxide	$N_2O(g)$	$+82$	$+220$	$+104$
dinitrogen pentaoxide	$N_2O_5(g)$	$+11$	$+356$	$+115$
dinitrogen tetraoxide	$N_2O_4(g)$	$+9$	$+304$	$+98$
dinitrogen trioxide	$N_2O_3(g)$	$+84$	$+312$	$+139$
hydrazine	$N_2H_4(l)$	$+51$	$+121$	$+149$
hydrogen azide	$HN_3(l)$	$+264$	$+141$	$+327$
hydrogen nitrate	$HNO_3(l)$	-174	$+156$	-81
nitrate ion	$NO_3{}^-(aq)$	-207	$+147$	-111
nitrite ion	$NO_2{}^-(aq)$	-105	$+123$	-32
nitrogen dioxide	$NO_2(g)$	$+33$	$+240$	$+51$
nitrogen monoxide	$NO(g)$	$+90$	$+211$	$+87$

Compound Name	Formula	$\Delta H°$ (kJ·mol^{-1})	$S°$ (J·mol^{-1}·K^{-1})	$\Delta G°$ (kJ·mol^{-1})
Oxygen	$O_2(g)$	0	+205	0
	$O_3(g)$	+143	+239	+163
	$O(g)$	+249	+161	+232
	$O^-(g)$	+102	+158	+92
difluoride	$OF_2(g)$	+25	+247	+42
Phosphorus	$P_4(s)$ (white)	0	+41	0
	$P(s)$ (red)	−18	+23	−12
	$P_4(g)$	+59	+280	+24
	$P(g)$	+317	+163	+278
hydrogen phosphate	$H_3PO_4(s)$	−1279	+110	−1119
pentachloride	$PCl_5(g)$	−375	+365	−305
pentafluoride	$PF_5(g)$	−1594	+301	−1521
phosphate ion	$PO_4^{3-}(aq)$	−1277	−220	−1019
phosphoryl chloride	$POCl_3(l)$	−597	+222	−521
tetraphosphorus decaoxide	$P_4O_{10}(s)$	−2984	+229	−2700
trichloride	$PCl_3(l)$	−320	+217	−272
trifluoride	$PF_3(g)$	−919	+273	−898
trihydride (phosphine)	$PH_3(g)$	+5	+210	+13
Potassium	$K(s)$	0	+65	0
	$K(g)$	+89	+160	+61
	$K^+(aq)$	−252	+101	−284
bromide	$KBr(s)$	−394	+96	−381
carbonate	$K_2CO_3(s)$	−1151	+156	−1064
chlorate	$KClO_3(s)$	−398	+143	−296
chloride	$KCl(s)$	−437	+83	−409
chromate	$K_2CrO_4(s)$	−1404	+200	−1296
cyanide	$KCN(s)$	−113	+128	−102
dichromate	$K_2Cr_2O_7(s)$	−2062	+291	−1882
dioxide(2−) (peroxide)	$K_2O_2(s)$	−494	+102	−425
dioxide(1−) (superoxide)	$KO_2(s)$	−285	+117	−239
fluoride	$KF(s)$	−567	+67	−538
hydride	$KH(s)$	−58	+50	−53
hydrogen carbonate	$KHCO_3(s)$	−963	+116	−864
hydrogen sulfate	$KHSO_4(s)$	−1161	+138	−1031
hydroxide	$KOH(s)$	−425	+79	−379
iodide	$KI(s)$	−328	+106	−325
nitrate	$KNO_3(s)$	−495	+133	−395
nitrite	$KNO_2(s)$	−370	+152	−307
oxide	$K_2O(s)$	−363	+94	−322
perchlorate	$KClO_4(s)$	−433	+151	−303
permanganate	$KMnO_4(s)$	−837	+172	−738
peroxodisulfate	$K_2S_2O_8(s)$	−1916	+279	−1697
pyrosulfate	$K_2S_2O_7(s)$	−1987	+225	−1792
sulfate	$K_2SO_4(s)$	−1438	+176	−1321
sulfide	$K_2S(s)$	−376	+115	−363
tetrafluoroborate	$KBF_4(s)$	−1882	+152	−1786
Rubidium	$Rb(s)$	0	+77	0
	$Rb(g)$	+81	+170	+53
	$Rb^+(aq)$	−251	+122	−284

Compound Name	Formula	$\Delta H°$ (kJ·mol^{-1})	$S°$ (J·mol^{-1}·K^{-1})	$\Delta G°$ (kJ·mol^{-1})
bromide	$RbBr(s)$	−395	+110	−382
carbonate	$Rb_2CO_3(s)$	−1179	+186	−1096
chloride	$RbCl(s)$	−435	+96	−408
fluoride	$RbF(s)$	−558	+75	−521
iodide	$RbI(s)$	−334	+118	−329
nitrate	$RbNO_3(s)$	−495	+147	−396
sulfate	$Rb_2SO_4(s)$	−1436	+197	−1317
Selenium	$Se(s)$ (grey)	0	+42	0
	$Se(g)$	+227	+177	+187
hexafluoride	$SeF_6(g)$	−1117	+314	−1017
selenate ion	$SeO_4^{2-}(aq)$	−599	+54	−441
tetrachloride	$SeCl_4(s)$	−183	+195	−95
Silicon	$Si(s)$	0	+19	0
	$Si(g)$	+450	+168	+406
carbide	$SiC(s)$	−65	+17	−63
dioxide (quartz)	$SiO_2(s)$	−911	+41	−856
tetrachloride	$SiCl_4(l)$	−687	+240	−620
tetrafluoride	$SiF_4(g)$	−1615	+283	−1573
tetrahydride (silane)	$SiH_4(g)$	+34	+205	+57
Silver	$Ag(s)$	0	+43	0
	$Ag(g)$	+285	+173	+246
	$Ag^+(aq)$	+106	+73	+77
bromide	$AgBr(s)$	−100	+107	−97
carbonate	$Ag_2CO_3(s)$	−506	+167	−437
chloride	$AgCl(s)$	−127	+96	−110
chromate	$Ag_2CrO_4(s)$	−732	+218	−642
cyanide	$AgCN(s)$	+146	+107	+157
fluoride	$AgF(s)$	−205	+84	−187
iodide	$AgI(s)$	−62	+115	−66
nitrate	$AgNO_3(s)$	−124	+141	−33
oxide	$Ag_2O(s)$	−31	+121	−11
sulfate	$Ag_2SO_4(s)$	−716	+200	−618
sulfide	$Ag_2S(s)$	−33	+144	−41
Sodium	$Na(s)$	0	+51	0
	$Na(g)$	+107	+154	+77
	$Na^+(aq)$	−240	+58	−262
azide	$NaN_3(s)$	+22	+97	+94
bromide	$NaBr(s)$	−361	+87	−349
carbonate	$Na_2CO_3(s)$	−1131	+135	−1044
carbonate monohydrate	$Na_2CO_3 \cdot H_2O(s)$	−1431	+168	−1285
carbonate decahydrate	$Na_2CO_3 \cdot 10H_2O(s)$	−4081	+563	−3428
chlorate	$NaClO_3(s)$	−366	+123	−262
chloride	$NaCl(s)$	−411	+72	−384
cyanide	$NaCN(s)$	−87	+116	−76
dihydrogen phosphate	$NaH_2PO_4(s)$	−1537	+127	−1386
dioxide(2−) (peroxide)	$Na_2O_2(s)$	−511	+95	−448
fluoride	$NaF(s)$	−574	+51	−544
hydride	$NaH(s)$	−56	+40	−33

Compound Name	Formula	$\Delta H°$ (kJ·mol^{-1})	$S°$ (J·mol^{-1}·K^{-1})	$\Delta G°$ (kJ·mol^{-1})
hydrogen carbonate	NaHCO$_3$(s)	-951	$+102$	-851
hydrogen phosphate	Na$_2$HPO$_4$(s)	-1748	$+150$	-1608
hydrogen sulfate	NaHSO$_4$(s)	-1126	$+113$	-993
hydroxide	NaOH(s)	-425	$+64$	-379
iodide	NaI(s)	-288	$+99$	-286
nitrate	NaNO$_3$(s)	-468	$+117$	-367
nitrite	NaNO$_2$(s)	-359	$+104$	-285
oxide	Na$_2$O(s)	-414	$+75$	-375
perchlorate	NaClO$_4$(s)	-383	$+142$	-255
phosphate	Na$_3$PO$_4$(s)	-1917	$+174$	-1789
silicate	Na$_2$SiO$_3$(s)	-1555	$+114$	-1463
sulfate	Na$_2$SO$_4$(s)	-1387	$+150$	-1270
sulfide	Na$_2$S(s)	-365	$+84$	-350
sulfite	Na$_2$SO$_3$(s)	-1101	$+146$	-1012
tetrahydroborate	NaBH$_4$(s)	-189	$+101$	-124
thiosulfate	Na$_2$S$_2$O$_3$(s)	-1123	$+155$	-1028
thiosulfate pentahydrate	Na$_2$S$_2$O$_3$·5H$_2$O(s)	-2608	$+372$	-2230
Strontium	Sr(s)	0	$+52$	0
	Sr(g)	$+164$	$+165$	$+131$
	Sr^{2+}(aq)	-546	-33	-559
carbonate	SrCO$_3$(s)	-1220	$+97$	-1140
chloride	SrCl$_2$(s)	-829	$+115$	-781
oxide	SrO(s)	-592	$+54$	-562
sulfate	SrSO$_4$(s)	-1453	$+117$	-1341
Sulfur	S$_8$(s) (rhombic)	0	$+32$	0
	S$_8$(s) (monoclinic)	$+0.3$	$+33$	$+0.1$
	S$_8$(g)	$+102$	$+431$	$+50$
	S(g)	$+227$	$+168$	$+236$
dichloride	SCl$_2$(l)	-50	$+184$	-28
dichloride (disulfur)	S$_2$Cl$_2$(l)	-58	$+224$	-39
dioxide	SO$_2$(g)	-297	$+248$	-300
hexafluoride	SF$_6$(g)	-1209	$+292$	-1105
hydrogen sulfate	H$_2$SO$_4$(l)	-814	$+157$	-690
hydrogen sulfide ion	HS$^-$(aq)	-16	$+67$	$+12$
peroxodisulfate ion	S$_2$O$_8^{2-}$(aq)	-1345	$+244$	-1115
sulfate ion	SO$_4^{2-}$(aq)	-909	$+19$	-744
sulfide ion	S^{2-}(aq)	$+33$	-15	$+86$
sulfite ion	SO$_3^{2-}$(aq)	-635	-29	-487
thiosulfate ion	S$_2$O$_3^{2-}$(aq)	-652	$+67$	-522
trioxide	SO$_3$(g)	-396	$+257$	-371
Thallium	Tl(s)	0	$+64$	0
	Tl(g)	$+182$	$+181$	$+147$
	Tl$^+$(aq)	$+5$	$+125$	-32
	Tl^{3+}(aq)	$+197$	-192	$+215$
(I) chloride	TlCl(s)	-204	$+111$	-185
(III) chloride	TlCl$_3$(s)	-315	$+152$	-242
Tin	Sn(s) (white)	0	$+51$	0
	Sn(s) (grey)	-2	$+44$	$+0.1$

Compound Name	Formula	$\Delta H°$ (kJ·mol^{-1})	$S°$ (J·mol^{-1}·K^{-1})	$\Delta G°$ (kJ·mol^{-1})
	$Sn(g)$	+301	+168	+266
(II) chloride	$SnCl_2(s)$	−331	+132	−289
(IV) chloride	$SnCl_4(l)$	−551	+259	−440
hydride	$SnH_4(g)$	+163	+228	+188
(II) hydroxide	$Sn(OH)_2(s)$	−561	+155	−492
(II) oxide	$SnO(s)$	−281	+57	−252
(IV) oxide	$SnO_2(s)$	−578	+49	−516
(II) sulfide	$SnS(s)$	−100	+77	−98
(IV) sulfide	$SnS_2(s)$	−154	+87	−145
Titanium	$Ti(s)$	0	+31	0
	$Ti(g)$	+473	+180	+428
(II) chloride	$TiCl_2(s)$	−514	+87	−464
(III) chloride	$TiCl_3(s)$	−721	+140	−654
(IV) chloride	$TiCl_4(l)$	−804	+252	−737
(IV) oxide	$TiO_2(s)$ (rutile)	−944	+51	−890
Vanadium	$V(s)$	0	+29	0
	$V(g)$	+514	+182	+469
(II) chloride	$VCl_2(s)$	−452	+97	−406
(III) chloride	$VCl_3(s)$	−581	+131	−511
(IV) chloride	$VCl_4(l)$	−569	+255	−504
(II) oxide	$VO(s)$	−432	+39	−404
(III) oxide	$V_2O_3(s)$	−1219	+98	−1139
(IV) oxide	$VO_2(s)$	−713	+51	−659
(V) oxide	$V_2O_5(s)$	−1551	+131	−1420
Xenon	$Xe(g)$	0	+170	0
difluoride	$XeF_2(g)$	−130	+260	−96
tetrafluoride	$XeF_4(g)$	−215	+316	−138
trioxide	$XeO_3(g)$	+502	+287	+561
Zinc	$Zn(s)$	0	+42	0
	$Zn(g)$	+130	+161	+94
	$Zn^{2+}(aq)$	−153	−110	−147
carbonate	$ZnCO_3(s)$	−813	+82	−732
chloride	$ZnCl_2(s)$	−415	+111	−369
hydroxide	$Zn(OH)_2(s)$	−642	+81	−554
nitride	$Zn_3N_2(s)$	−23	+140	+30
oxide	$ZnO(s)$	−350	+44	−320
sulfate	$ZnSO_4(s)$	−983	+110	−872
sulfate heptahydrate	$ZnSO_4 \cdot 7H_2O(s)$	−3078	+389	−2563
sulfide	$ZnS(s)$ (wurtzite)	−193	+68	−191
sulfide	$ZnS(s)$ (sphalerite)	−206	+58	−201

Appendix 2

Charge Densities of Selected Ions

Charge densities ($C \cdot mm^{-3}$) are calculated according to the formula

$$\frac{ne}{(4/3)\pi r^3}$$

where the ionic radii r are the Shannon–Prewitt values in millimeters (*Acta Cryst.*, 1976, A32, 751), e is the electron charge (1.60×10^{-19} C), and n represents the ion charge. The radii used are the values for six-coordinate ions except where noted by (T) for four-coordinate tetrahedral ions; (HS) and (LS) designate the high spin and low spin radii for the transition metal ions.

Cation	Charge density	Cation	Charge density	Cation	Charge density
Ac^{3+}	57	B^{3+}	1663	Cl^{7+}	3880
Ag^{+}	15	Ba^{2+}	23	Cm^{3+}	84
Ag^{2+}	60	Be^{2+}	1108 (T)	Co^{2+}	155 (LS)
Ag^{3+}	163	Bi^{3+}	72	Co^{2+}	108 (HS)
Al^{3+}	770 (T)	Bi^{5+}	262	Co^{3+}	349 (LS)
Al^{3+}	364	Bk^{3+}	86	Co^{3+}	272 (HS)
Am^{3+}	82	Br^{7+}	1796	Co^{4+}	508 (HS)
As^{3+}	307	C^{4+}	6265 (T)	Cr^{2+}	116 (LS)
As^{5+}	884	Ca^{2+}	52	Cr^{2+}	92 (HS)
At^{7+}	609	Cd^{2+}	59	Cr^{3+}	261
Au^{+}	11	Ce^{3+}	75	Cr^{4+}	465
Au^{3+}	118	Ce^{4+}	148	Cr^{5+}	764
B^{3+}	7334 (T)	Cf^{3+}	88	Cr^{6+}	1175

Cation	Charge density	Cation	Charge density	Cation	Charge density
Cs^+	6	Mn^{4+}	508	Sb^{3+}	157
Cu^+	51	Mn^{7+}	1238	Sb^{5+}	471
Cu^{2+}	116	Mo^{3+}	200	Sc^{3+}	163
Dy^{2+}	43	Mo^{6+}	589	Se^{4+}	583
Dy^{3+}	99	NH_4^+	11	Se^{6+}	1305
Er^{3+}	105	Na^+	24	Si^{4+}	970
Eu^{2+}	34	Nb^{3+}	180	Sm^{3+}	86
Eu^{3+}	88	Nb^{5+}	402	Sn^{2+}	54
F^{7+}	25 110	Nd^{3+}	82	Sn^{4+}	267
Fe^{2+}	181 (LS)	Ni^{2+}	134	Sr^{2+}	33
Fe^{2+}	98 (HS)	No^{2+}	40	Ta^{3+}	180
Fe^{3+}	349 (LS)	Np^{5+}	271	Ta^{5+}	402
Fe^{3+}	232 (HS)	Os^{4+}	335	Tb^{3+}	96
Fe^{6+}	3864	Os^{6+}	698	Tc^{4+}	310
Fr^+	5	Os^{8+}	2053	Tc^{7+}	780
Ga^{3+}	261	P^{3+}	587	Te^{4+}	112
Gd^{3+}	91	P^{5+}	1358	Te^{6+}	668
Ge^{2+}	116	Pa^{5+}	245	Th^{4+}	121
Ge^{4+}	508	Pb^{2+}	32	Ti^{2+}	76
Hf^{4+}	409	Pb^{4+}	196	Ti^{3+}	216
Hg^+	16	Pd^{2+}	76	Ti^{4+}	362
Hg^{2+}	49	Pd^{4+}	348	Ti^+	9
Ho^{3+}	102	Pm^{3+}	84	Tl^{3+}	105
I^{7+}	889	Po^{4+}	121	Tm^{2+}	48
In^{3+}	138	Po^{6+}	431	Tm^{3+}	108
Ir^{3+}	208	Pr^{3+}	79	U^{4+}	140
Ir^{5+}	534	Pr^{4+}	157	U^{6+}	348
K^+	11	Pt^{2+}	92	V^{2+}	95
La^{3+}	72	Pt^{4+}	335	V^{3+}	241
Li^+	98 (T)	Pu^{4+}	153	V^{4+}	409
Li^+	52	Ra^{2+}	18	V^{5+}	607
Lu^{3+}	115	Rb^+	8	W^{4+}	298
Mg^{2+}	120	Re^{7+}	889	W^{6+}	566
Mn^{2+}	144 (LS)	Rh^{3+}	224	Y^{3+}	102
Mn^{2+}	84 (HS)	Ru^{3+}	208	Yb^{3+}	111
Mn^{3+}	307 (LS)	S^{4+}	1152	Zn^{2+}	112
Mn^{3+}	232 (HS)	S^{6+}	2883	Zr^{4+}	240

HS, high spin; LS, low spin; T, four-coordinate tetrahedral ions.

Anion	Charge density	Anion	Charge density	Anion	Charge density
As^{3-}	12	I^-	4	O_2^{2-}	19
Br^-	6	MnO_4^-	4	OH^-	23
CN^-	7	N^{3-}	50	P^{3-}	14
CO_3^{2-}	17	N_3^-	6	S^{2-}	16
Cl^-	8	NO_3^-	9	SO_4^{2-}	5
ClO_4^-	3	O^{2-}	40	Se^{2-}	12
F^-	24	O_2^-	13	Te^{2-}	9

Appendix 3

Selected Bond Energies

For homonuclear diatomic molecules, such as dihydrogen, precise measured values of bond energies are listed. For most of the heteronuclear bonds, only average values are given and these tend to differ among literature sources. All values here are in units of $kJ \cdot mol^{-1}$.

Hydrogen

H—H	432	H—S	363
H—B	389	H—F	565
H—C	411	H—Cl	428
H—N	386	H—Br	362
H—O	459	H—I	295

Group 13

B—C	372	B—F	613
B—O	536	B—Cl	456
		B—I	377

Group 14

C—C	346	C—O	358
C=C	602	C=O	799
C≡C	835	C≡O	1072
C—N	305	C—F	485
C=N	615	C—Cl	327
C≡N	887	C—Br	285
C—P	264	C—I	213
Si—Si	222	Si—Cl	381
Si—O	452	Si—Br	310
Si—F	565	Si—I	234

Group 15

N—N	247	N—O	201
N=N	418	N=O	607
N≡N	942	N—F	278
		N—Cl	192
P—F	490	P—Br	264
P—Cl	326	P—I	184

Group 16

O—O	207	S—F	327
O=O	494	S—Cl	271
O—F	190	S—S	266
O—Cl	218		
O—Br	201		
O—I	201		

Group 17

F—F	155	F—I	278
F—Cl	249	F—Xe	130
F—Br	249		
Cl—Cl	240	Cl—I	208
Cl—Br	216		
Br—Br	190	Br—I	175
I—I	149		

Ionization Energies of Selected Metals

Appendix 4

These ionization energies are in units of $MJ \cdot mol^{-1}$ and they have been summarized from G. Aylward and T. Findlay, *SI Chemical Data*, 3rd ed., New York: Wiley, 1994. Only selected ionization energies for outer (valence) electrons are listed.

The 1st ionization energy represents the energy required for the process:

$$M_{(g)} \Rightarrow M^{+}_{(g)} + e^{-}$$

while that of the 2nd ionization process represents that for:

$$M^{+}_{(g)} \Rightarrow M^{2+}_{(g)} + e^{-}$$

and successive ionization energies are defined similarly as one-electron processes.

Element	Ionization energy				
	1st	2nd	3rd	4th	5th
Lithium	0.526				
Beryllium	0.906	1.763			
Sodium	0.502				
Magnesium	0.744				
Beryllium	0.906	1.763			
Sodium	0.502				
Magnesium	0.744	1.457			
Aluminum	0.584	1.823	2.751		
Potassium	0.425				
Calcium	0.596	1.152			
Scandium	0.637	1.241	2.395		

Element	Ionization energy				
	1st	2nd	3rd	4th	5th
Titanium	0.664	1.316	2.659	4.181	
Vanadium	0.656	1.420	2.834	4.513	6.300
Chromium	0.659	1.598	2.993		
Manganese	0.724	1.515	3.255		
Iron	0.766	1.567	2.964		
Cobalt	0.765	1.652	3.238		
Nickel	0.743	1.759			
Copper	0.752	1.964			
Zinc	0.913	1.740			
Lead	0.722	1.457			

Electron Affinities of Selected Nonmetals

Appendix 5

These ionization energies are in units of $kJ \cdot mol^{-1}$ and they have been summarized from J.E. Huuhey et al., *Inorganic Chemistry*, 4th ed., New York: HarperCollins, 1993.

The 1st electron affinity represents the energy required for the process:

$$X(g) + e^- \Rightarrow X^-(g)$$

while that of the 2nd electron affinity represents that for:

$$X^-(g) + e^- \Rightarrow X^{2-}(g)$$

and that of the 3rd electron affinity represents that for:

$$X^{2-}(g) + e^- \Rightarrow X^{3-}(g)$$

Element	Electron Affinity		
	1st	2nd	3rd
Nitrogen	−7	+673	+1070
Oxygen	−141	+744	
Fluorine	−328		
Phosphorus	−72	+468	+886
Sulfur	−200	+456	
Chlorine	−349		
Hydrogen	−79		
Bromine	−331		
Iodine	−301		

Appendix 6

Selected Lattice Energies

These lattice energies are in units of $kJ \cdot mol^{-1}$ and they have been calculated from Born–Haber cycles. The values are summarized from G. Aylward and T. Findlay, *SI Chemical Data*, 3rd ed., New York: Wiley, 1994.

Ion	Fluoride	Chloride	Bromide	Iodide	Oxide	Sulfide
Lithium	1047	862	818	759	2806	2471
Sodium	928	788	751	700	2488	2199
Potassium	826	718	689	645	2245	1986
Rubidium	793	693	666	627	2170	1936
Cesium	756	668	645	608	—	1899
Magnesium	2961	2523	2434	2318	3800	3323
Calcium	2634	2255	2170	2065	3419	3043
Strontium	2496	2153	2070	1955	3222	2879
Barium	2357	2053	1980	1869	3034	2716

Appendix 7

Selected Hydration Enthalpies

These hydration enthalpies are in units of $kJ \cdot mol^{-1}$ and the values were obtained from J.G. Stark and H.G. Wallace, *Chemistry Data Book*, London: John Murray, 1990.

Element	ΔH_f° ($kJ \cdot mol^{-1}$)
Lithium	−519
Sodium	−406
Potassium	−322
Rubidium	−301
Cesium	−276
Magnesium	−1920
Calcium	−1650
Strontium	−1480
Barium	−1360
Aluminum	−4690
Silver	−464
Fluorine	−506
Chlorine	−364
Bromine	−335
Iodine	−293

Appendix 8 Selected Ionic Radii

These values of ionic radii are the Shannon–Prewitt values in pm (*Acta Cryst.*, 1976, A32, 751) for six-coordinate ions except where noted by (T) for four-coordinate tetrahedral ions; (HS) and (LS) designate the high-spin and low-spin radii for the transition metal ions. The values for polyatomic ions are adapted from Jenkins and Thakur (*J. Chem. Educ.*, 1979, 56, 576).

Ion	Ionic radii
Li^+ (T)	73
Na^+	116
K^+	152
Rb^+	166
Cs^+	181
Mg^{2+}	86
Ca^{2+}	114
Sr^{2+}	132
Ba^{2+}	149
Al^{3+}	68
Fe^{2+}	92
Fe^{3+} (HS)	78
Fe^{3+} (LS)	69
Co^{2+} (HS)	88
Co^{3+} (LS)	68
Ni^{2+}	83

Ion	Ionic radii
Cu^+	91
Cu^{2+}	87
Zn^{2+}	88
F^- (T)	117
Cl^-	167
Br^-	182
I^-	206
NH_4^+	151
CO_3^{2-}	164
NO_3^-	165
OH^-	119
SO_4^{2-}	244
O^{2-}	126
S^{2-}	170

Index